W9-AYC-350

# THE FLIGHT FROM SCIENCE AND REASON

# THE FLIGHT FROM SCIENCE AND REASON

*Edited by Paul R. Gross, Norman Levitt, and Martin W. Lewis*

*The New York Academy of Sciences*
*New York, New York*

*Distributed by The Johns Hopkins University Press*
*Baltimore and London*

© 1996 by the New York Academy of Sciences
All rights reserved. Published 1997
Printed in the United States of America on acid-free paper
06 05 04 03 02 01 00 99 98 97 5 4 3 2 1

This volume represents the proceedings of a conference entitled "The Flight from Science and Reason," which was sponsored by the New York Academy of Sciences and held in New York, New York, on May 31-June 2, 1995.

The New York Academy of Sciences believes it has a responsibility to provide an open forum for discussion of scientific questions. The positions taken by the participants in the reported conferences are their own and not necessarily those of the Academy. The Academy has no intent to influence legislation by providing such forums.

Distributed by
The Johns Hopkins University Press
2715 North Charles Street
Baltimore, Maryland 21218-4319
The Johns Hopkins Press Ltd., London

**Library of Congress Cataloging-in-Publication Data**

The flight from science and reason / edited by Paul R. Gross, Norman Levitt, and Martin W. Lewis.
p. cm.
"This volume represents the proceedings of a conference entitled: 'The Flight From Science and Reason,' which was sponsored by the New York Academy of Sciences and held in New York, New York, on May 31-June 2, 1995"—T.p. verso.
Originally published: New York : New York Academy of Sciences, 1996.
Includes bibliographical references and index.
ISBN 0-8018-5676-0 (pbk. : alk. paper)
I. Science—Philosophy—Congresses. 2. Science—Social aspects—Congresses. 3. Women in science—Congresses. 4. Religion and science—Congresses. I. Gross, Paul R. II. Levitt, N. (Norman), 1943- . III. Lewis, Martin W.
[Q175.F8 1997]
500—dc21 96-53491
CIP

A catalog record for this book is available from the British Library.

# CONTENTS

Introductory Remarks: Medicine Took an Earlier Flight.
*By* HENRY GREENBERG . . . . . . . . . . . . . . . . . . . . . . . . . . . . . . . . . . . ix

Introduction. *By* PAUL R. GROSS . . . . . . . . . . . . . . . . . . . . . . . . . . . . . . . 1

**The Public Image of Science** . . . . . . . . . . . . . . . . . . . . . . . . . . . . . . 9

Imaginary Gardens with Real Toads. *By* DUDLEY R. HERSCHBACH . . . . . . . . 11

Conduct and Misconduct in Science. *By* DAVID GOODSTEIN. . . . . . . . . . . . 31

Mathematics as the Stepchild of Contemporary Culture.
*By* NORMAN LEVITT. . . . . . . . . . . . . . . . . . . . . . . . . . . . . . . . . . . . . . 39

**Reasonable Foundations** . . . . . . . . . . . . . . . . . . . . . . . . . . . . . . . 55

Concern for Truth: What It Means, Why It Matters. *By* SUSAN HAACK . . . . 57

The Propensity to Believe. *By* JAMES E. ALCOCK . . . . . . . . . . . . . . . . . . . 64

Flights of Fancy: Science, Reason, and Common Sense.
*By* BARRY R. GROSS. . . . . . . . . . . . . . . . . . . . . . . . . . . . . . . . . . . . . . 79

Feelings and Beliefs. *By* LOREN FISHMAN . . . . . . . . . . . . . . . . . . . . . . . 87

In Praise of Intolerance to Charlatanism in Academia. *By* MARIO BUNGE . . 96

**The Foundations of Physics** . . . . . . . . . . . . . . . . . . . . . . . . . . . . . 117

Quantum Philosophy: The Flight from Reason in Science.
*By* SHELDON GOLDSTEIN. . . . . . . . . . . . . . . . . . . . . . . . . . . . . . . . . . . 119

Physics and Common Nonsense. *By* DANIEL KLEPPNER . . . . . . . . . . . . . . . 126

Science of Chaos or Chaos in Science? *By* JEAN BRICMONT . . . . . . . . . . . . 131

**Health** . . . . . . . . . . . . . . . . . . . . . . . . . . . . . . . . . . . . . . . . . . . . 177

"Sucking with Vampires": The Medicine of Unreason.
*By* GERALD WEISSMANN . . . . . . . . . . . . . . . . . . . . . . . . . . . . . . . . . . 179

Antiscience Trends in the Rise of the "Alternative Medicine" Movement. *By* Wallace Sampson . . . . . . 188

Constructivism in Psychotherapy: Truth And Consequences. *By* Barbara S. Held . . . . . . 198

**Environment** . . . . . . 207

Radical Environmental Philosophy and the Assault on Reason. *By* Martin W. Lewis . . . . . . 209

Is Environmental Cancer a Political Disease? *By* Stanley Rothman and S. Robert Lichter . . . . . . 231

Old Messages: Ecofeminism and the Alienation of Young People from Environmental Activism. *By* Rene Denfeld . . . . . . 246

**Social Theories of Science** . . . . . . 257

Towards a Sober Sociology of Science. *By* Susan Haack . . . . . . 259

Wrestling with the Social Constructor. *By* Noretta Koertge . . . . . . 266

Voodoo Sociology: Recent Developments in the Sociology of Science. *By* Stephen Cole . . . . . . 274

The Allure of the Hybrid: Bruno Latour and the Search for a New Grand Theory. *By* Oscar Kenshur . . . . . . 288

**History, Society, Politics** . . . . . . 299

Whatever Happened to Historical Evidence? *By* Mary Lefkowitz . . . . . . 301

Building Bridges to Afrocentrism: A Letter to My Egyptological Colleagues. *By* Ann Macy Roth . . . . . . 313

State of the Art/Science in Anthropology. *By* Robin Fox . . . . . . 327

Liberalism, Public Opinion, and Their Critics: Some Lessons for Defending Science. *By* Simon Jackman . . . . . . 346

Pathological Social Science: Carol Gilligan and the Incredible Shrinking Girl. *By* Christina Hoff Sommers . . . . . . 369

**Feminisms** . . . . . . 383

Why Feminist Epistemology Isn't. *By* Janet Radcliffe Richards . . . . . . 385

Feminist Epistemology: Stalking an Un-Dead Horse. *By* Noretta Koertge . . . . . . 413

The Science Question in Postcolonial Feminism. *By* Meera Nanda . . . . . . 420

Are "Feminist Perspectives" in Mathematics and Science Feminist? *By* Mary Beth Ruskai . . . . . . 437

**Humanities** . . . . . . . . . . 443

On Sitting Down to Read *King Lears* Once Again: The Textual Deconstruction of Shakespeare. *By* PAUL A. CANTOR . . . . . . . . . . 445

Constructing Literature: Empiricism, Romanticism, and Textual Theory. *By* GEORGE BORNSTEIN . . . . . . . . . . 459

Freudian Suspicion versus Suspicion of Freud. *By* FREDERICK CREWS . . . . . 470

Ecosentimentalism: The Summer Dream beneath the Tamarind Tree. *By* GERALD WEISSMANN . . . . . . . . . . 483

**Religion** . . . . . . . . . . 491

Two Sources of Unreason in Democratic Society: The Paranormal and Religion. *By* PAUL KURTZ . . . . . . . . . . 493

Creationism, Ideology, and Science. *By* EUGENIE C. SCOTT . . . . . . . . . . 505

The Flight from Reason: The Religious Right. *By* LANGDON GILKEY . . . . . . 523

Doubt, Certainty, Faith, and Ideology. *By* OSCAR KENSHUR . . . . . . . . . . 526

**Education** . . . . . . . . . . 537

Science, Reason, and Education. *By* HENRY ROSOVSKY . . . . . . . . . . 539

Scientific Literacy. *By* JAMES TREFIL . . . . . . . . . . 543

Science Education and the Sense of Self. *By* GERALD HOLTON . . . . . . . . . . 551

Afrocentric Pseudoscience: The Miseducation of African Americans. *By* BERNARD R. ORTIZ DE MONTELLANO . . . . . . . . . . 561

Notes on Contributors . . . . . . . . . . 573

Index of Contributors . . . . . . . . . . 577

Subject Index . . . . . . . . . . 579

# INTRODUCTORY REMARKS
## Medicine Took an Earlier Flight

HENRY GREENBERG

THIS VOLUME EMERGED out of the primary concerns of the hard physical and biological sciences. The proponents and organizing committee of the conference on which this volume is based reflect this. And the individual chapters concentrate on the threats to these disciplines. Although a majority of the authors come from a background in the social sciences, the issues under scrutiny relate to scientific reasoning, logical deduction, and professional expertise.

Let me take a brief look at a soft science that engages all levels of society—namely, medicine. Last year I organized a conference, sponsored by this Academy, that sought to enlarge the view of health care reform so that potential long-term results could be included in the debate. Its title was ***Beyond the Crisis: Preserving the Capacity for Excellence in Health Care and Medical Science.***[1] In my talk I focused on threats to excellence in medicine.[2] Toward that end, I explored the genesis of the term ***health care provider***, a description with which we are all familiar and one which infuriates physicians. I tried to show that the apparent democratizing of the team, the cultural leveling of the peak usually inhabited by the physician, was a genuine threat to the professional uniqueness of the physician. The social construction of reality has come to medicine. Borrowing from Larry Churchill,[3] I showed that without his or her own professional ethic, a physician who is dependent only upon the usual guiding ethics—law, custom, and common sense—would not be able to defend the best interests of his patient when law, common practice, and conventional wisdom defined a patient's interests in terms that best serve society and not the individual. The loss of the professional ethic will curtail the physician's ability to defend his patient's best interests.

Since then—and those of us in the east are behind—a new threat has emerged. The for-profit health maintenance organization (HMO) has arrived in Gotham. Everyone—hospitals, doctors, medical schools—is fighting to join. Packaging themselves in new organizations, bundling their services in financially pleasing ways, and then devaluing these services so as to be the lowest bidder, they are part of the great game to survive; at least we all think so. And the HMOs, with their ten-million-dollar CEOs, paying out sixty-eight cents on the dollar for care, understand and like what they see. With little extra effort they will reduce the physician to an employee. Again, the unique professional ethic will be stripped away, and the doctor will measure his life in degrees of compliance with cost-driven algorithms and will hone his skills to reduce an office visit to seven minutes.

One HMO circulated a memorandum stating that it wanted all its specialists to have at least twenty office hours a week. Since a physician could arrange to see a patient within a day should the need arise, why are his ten hours a week inadequate? I think I know the answer. He must have a financial base outside the office. If he does, he may not be as subservient to the HMO requirements as he should be. But if he is dependent upon his office practice, the HMO will gain the control it wants. The physician can be kept in place.

Medicine, then, is being put at grave risk by many strong forces. The social construction of its reality is only one of these forces. This on-going assault has, however, weakened medicine. The profession's inability or unwillingness to confront, let alone recognize, this attack has sapped it of much of its vigor. The other criticisms have a core validity. Because of accentuated attention to these—arrogance, greed, and excessive paternalism—medicine feels constrained to defend itself and is paralyzed when it comes to speaking clearly about its strengths. The rare but highlighted focus on fraud and the less rare examples of marginally scrupulous physicians feeding at the "pass through" trough of Medicare and other insurance plans are commonplace headlines. However, the near-comprehensive inability to distinguish the profession from the practitioner has inhibited the defense of professionalism. And yet when the profession is dead, it will be missed, warts and all.

There is a parallel development in the research science environment. If the incursion of attempted invalidation of scientific reasoning gains a foothold and then saps significant energy for its refutation, the body politic of science will be weakened. If the defense is incomplete or intellectually ineffective, the situation will be frighteningly similar to medicine's. Such a failure to defend science from its irrational critics can set the stage for a lethal blow when the real budgetary attacks arrive, and they are nearly upon us. Some would say they are here, but I am not so optimistic. If science cannot claim a preeminence for its intellectual virtues or an excellence for its methodologies and sense of design, then it will have great difficulty laying claim to a rational share of the nation's resources for its perpetuation. This volume has an important role to play.

## NOTES

1 H. M. Greenberg & S. U. Raymond, eds., *Beyond the Crisis: Preserving the Capacity for Excellence in Health Care and Medical Science.*
2 H. M. Greenberg, "Three Threats to the Capacity for Excellence in Medicine."
3 L. R. Churchill, "Reviving a Distinct Medical Ethic."

## REFERENCES

CHURCHILL, L. R. "Reviving a Distinct Medical Ethic." Hastings Center Report 19 (May–June, 1989): 28–34.

GREENBERG, H. M. & S. U. RAYMOND, eds. *Beyond the Crisis: Preserving the Capacity for Excellence in Health Care and Medical Science. Annals of the New York Academy of Sciences* 729 (1994).

GREENBERG, H. M. "Three Threats to the Capacity for Excellence in Medicine." In *Beyond the Crisis: Preserving the Capacity for Excellence in Health Care and Medical Science*, edited by H. M. Greenberg & S. U. Raymond. *Annals of the New York Academy of Sciences* 729 (1994): 8–18.

# THE FLIGHT FROM SCIENCE AND REASON

# INTRODUCTION

PAUL R. GROSS

*Mephistopheles* (disguised, in Faust's academic gown, ready to receive a fawning student whom the sage refuses to see; sotto voce, after the withdrawing protagonist):

*Scoff at all knowledge and despise*
*reason and science, those flowers of mankind.*
*Let the father of all lies*
*with dazzling necromancy make you blind,*
*then I'll have you unconditionally—*

—GOETHE, *Faust*[1]

EVIDENCE OF A FLIGHT from reason is as old as human record-keeping; the *fact* of it certainly goes back an even longer way. Flight from science specifically, among the forms of rational inquiry, goes back as far as science itself. Whenever and wherever what we generally mean by "science" *really* began, it is true that rejection of it began in earnest with the Enlightenment itself. Nor has it been a rejection solely by the ignorant, by professional irrationalists, or by minor figures. William Blake rejected all the forms of inquiry upon which modern science was built; Goethe, himself a keen scientific investigator, author of the diabolical whisper used here as an epigraph, nevertheless opposed, finally, the very reasoning that allowed science to escape the straitjacket of *naturphilosophie.* Examples like these could be multiplied. So flight from reason, and from science, is nothing new. We owe the reader some explanation for a new volume on the subject.

There are common, and obvious, forms of antireason and antiscience. Religious obscurantism has always been with us. The mountebank healer and dispenser of potions is an archaic figure of fun on the stage and in print; and there are plenty of them on the bench today, and the sales of snake oil will expand, for reasons given in this book. The cults of UFO-watchers, victims of abduction by extraterrestrials, spoon benders, adjusters of human energy fields, reincarnates recovering their past lives, have had equivalents and defenders in every period since empirical science arose and posed them a terrible threat. At no time since the late nineteenth century, when the concordances of evidence for an evolutionary history of Earth and its biosphere became evident, has wrathful opposition been lacking, opposition not on the basis of evidence but by denial of the efficacy of rational inquiry—or insistence upon the equal epistemic merit of alternatives.[2]

Humbug and credulousness, however, are one thing. Their widespread promotion by intellectuals and in the respected media of communications is another. We believe that there is today in the West, among professors and others who are paid, in principle, to think and teach, a new and more systemic flight from science and reason.[3] It is given endless and contradictory justifications; but its imperialism—for example under the banner of "science studies"[4]—and the high esteem in which it holds the trendiest irrationalisms, are undeniable. This has brought with it, from that unexpected academic quarter, a truculent defense in the name of "democracy" of New Age and traditional forms of sophistry and charlatanism.[5] Younger programs of antilogic and antiscience are both diffuse (oppositionist movements being by their nature fractious) and angrier than the siege, already a few decades old, of "objectivity" in the social sciences and humanities.

But rejection of reason is now a pattern to be found in most branches of scholarship and in all the learned professions. Antirealism has taken hold even in such surprising venues as psychotherapy.[6] For readers of this volume I need not, presumably, argue that such a movement, if it is as pervasive as we think, has consequences, short- and long-term, for the culture as a whole. Those go far beyond the nonfiction book business, the haruspicies of media pundits, the sedations of the seminar room. On the record of history, they are evil consequences,[7] and they affect, more and more as students become alumni, hierarchs, and opinion makers, the way everybody thinks and lives.

If, then, there *is* a new onslaught against science and reason, sufficiently pervasive and well enough supported to deserve examination, and—if appropriate—rebuttal of its ritual pejoratives, then scholars of competence and integrity, representing a broad range of disciplines, need a forum in which to measure the phenomenon. Such was the purpose of the conference of which this volume is the product; and it is the purpose of the volume itself. Not only are all the contributors to this volume such scholars and writers, but included among them are honored figures in disciplines matching the range of a modern university: philosophy, history, religion, literature, economics, sociology, political science, psychology, biology, geography, chemistry, physics, mathematics, medicine. It is not their honors, however, that we hope will be convincing; it is what they said at the conference and have now written here, in their unique ways and from unique points of view. Those points of view reflect not only a broad range of disciplines, but a similar range of political interests and affiliations. At the conference, liberals and conservatives, socialism and capitalism, shared the panel table; feminism, multiculturalism, anticolonialism, and environmental activism were as central to the positions of some contributors as they were remote from the interests of others. ***The issues are of reason and one of its applications—science—and their status in our time; not of politics.***

## THEORY

That acidulous balloon pricker, the Underground Grammarian, inveighed, some time ago, against the following product of educational Theory: the confident assertion that young children can and must be equipped with correct views on complex questions (the cold war was his case in point), despite absent knowledge (often matching that of the teacher) of the facts of the case. Thus:

> This is the fact that lies at the heart of all our troubles in "education," the fact that must ultimately defeat all attempts at reform. The children in the schools are just children, who might someday, if left unmolested, put away childish things. But the *other* people in the schools, the teachers and teacher-trainers, the educrats and theory-mongers, are confirmed children. They are, indeed and alas, exactly what they claim to be—"role models." And they represent the end of that process to which schooling is the means: the subversion of knowledge and reason . . .[8]

Not much has happened, in the twelve years since these words were published in book form, to reduce their acid *or* their sting, but a great deal has happened to extend their applicability. It would then have been blatant exaggeration to substitute "universities" for "schools," "professors" for "teachers." It would be, while still an exaggeration, rather less blatant today. Not only have the definition of, and the need for "role models" been inflated, but—and this is much more important—the conviction is stronger than ever that inculcating correct *attitudes* toward big questions is far more important than conveying actual knowledge of their substance. Indeed, admired Theory has it that there is no such thing as actual knowledge, all "knowledge" (the scare quotes are obligatory; the well-spoken postmodernist refers to *knowledges*) being social, a social construction, whose content flows from, and is designed to support, the elites of power. Thus all teaching must be indoctrination.

I do *not*, in this, refer to "political correctness." That term refers to certain unpleasant institutional styles, activities, and policies. To be sure, derogation of science and reason, the oxymoronic multiplication of "knowledges," are among its characteristics. But the flight from—the "demythologizing" of—science and reason is not so limited: those flying from reason are far more numerous. There are well-known instances of it at both ends of the political spectrum. It just happens that derogation of science and reason is an old story on the far reaches of the right. It is something relatively new coming from the left, where until a decade or two ago naming oneself a "progressive" meant aligning oneself with logic, science, and the truth; associating oneself, as it were, with the future.

Over a wide area of academic life today, these self-characterizations have been reversed. One need not, as a "progressive," align oneself with science; one "critiques" it from a suitable "standpoint." To show respect for science, or for universalism, to allude to any sort of "foundations" of knowledge in a truly with-it college (or in some parts of the media world) today, is to show oneself incapable of "the vision-thing," a stick-in-the-mud, a dinosaur—or worse, insensitive. Among philosophers, beleaguered realists and other nonrelativists are well aware of all this; but academic courage being what it is, they do not say much about it directly (although there are some splendid exceptions,[9] including contributors to this volume). Among scientists, whose work is the most recent target of pop epistemology and "standpoint" (that is, *political*) criticism from commentators who may know little or nothing about the work itself, awareness of what is going on is spotty at best.

To explore the reasons would require an essay much too long for this collection. Suffice it to say that awareness among scientists has begun lately to increase a bit, not least because of such discreditable, high-profile performances of the critics as the Smithsonian exhibit "Science in American Life" and the earliest versions of the National Science Education Standards.[10] It is a sad comment on our era of expertise that such performances leave most scientists *surprised,* and hurt—but unsure of why they are hurt and whether an effort to respond would be in good taste. So urgent is the demand for Theory, and so suspicious the idea that a case can have *facts*, that it is now acceptable to assert, and to get away with, not only the claim that authors of literary works do not really know or mean what they say, but the same for the recorders of laws, of all manner of social and political facts, and of scientific observations.

There is Theory and there are theories; some are serious and some are not. Judgments of seriousness, furthermore, will differ among honest commentators on theory that *is* theory. Thus it is consistent with the catholicity of tastes, backgrounds, political views, and technical commitments among our contributors that the responses of some to particular theories differ from those of others to the same, even when both recognize and hope to eliminate an irrational element upon whose existence they agree. Cases in point are the views of Bornstein and Cantor on the new Oxford Shakespeare, of Lefkowitz and Roth on Afrocentric history, and of Kurtz and Scott (and to some extent Gilkey) on the proper response of reason and science to certain forms of religion. Conversely, speakers approaching a vexed issue of theory from quite different but equally expert points of view offer conclusions that converge in a satisfactory way, as for example in the judgments offered by Goldstein and Kleppner on the quantum formalisms, their meaning, and their utility.

But a collection of impressions and assertions about impressions is in any case not a theory, by any reasonable definition, let alone a tenable one. The flight from reason of which our contributors write is in large measure, among contemporary scholars (as opposed to innocent believers in the unbelievable), based upon a body of modish Theory that is not theory and upon theories that are untenable. A number of these essays concern themselves, therefore, quite directly with theory, grand and otherwise.

## CREDITS

The New York Academy of Sciences provided appropriate and effective sponsorship, for which we are grateful. One or another of the humanities or social sciences has lately been defended against standpoint attacks by official or unofficial voices, however thinly and sporadically; but science has not done so to even that level of seriousness. In fact, the politics of organized science, and of many distinguished scientists, seem to be limited to avoiding offense to all critics whatsoever, perhaps so as to ward off empty accusations of biases, crimes, and *systematic* misconduct.[11] The burgeoning interest of the New York Academy of Sciences in the social and broader intellectual implications of natural science is thus a welcome development; it will, we hope, extend a regular, civilized welcome to honest social analysis and provide a platform for cogent response from scientists willing to take the time to study and respond thereto; and vice versa! Perhaps such a movement will receive some stimulus from the conference and this volume.

Norman Levitt (who contributes an essay of his own to the collection) and I first spoke of this effort in 1994, and were encouraged in it by many proposals from colleagues; but it was early evident to us that our competences fell far short of those needed to plan the kind of conference and volume we believed to be needed. Accordingly, we took our courage (and all the spare time left from teaching, research, and family) in hand and inveigled into joining us as an organizing committee the following equally busy scholars: Gerald Holton (Physics, History of Science), Susan Haack (Philosophy), Martin Lewis (Geography, Environmental Science), and Loren Fishman (Medicine). That group designed the conference. Included here are the written versions of conference presentations (but not *ad hoc* remarks by the session chairs and others), with a few late additions from contributors who could not be present. In aid of continuity of argument, the grouping of essays here is somewhat altered from that followed in the sessions.

Professor Levitt and I agreed to edit the published proceedings; but of course much of the real editorial labor was accomplished by the staff of the Academy's *Annals*, especially Mr. Richard Stiefel. Among the external sponsors whose contributions supported conference operations and participant travel were the Lucille P. Markey Charitable Trust, the John M. Olin Foundation, the Lynde and Harry Bradley Foundation, the Russell Sage Foundation, and the Center for Advanced Studies, University of Virginia. We are deeply indebted to them all for their confidence in the timeliness and value of this undertaking. As appropriate to its purpose as general sponsorship by the New York Academy of Sciences was conduct of the sessions in the historic home of the New York Academy of Medicine. This is so, not only because a disproportionate volume of trendy science and cultural criticism is directed at medicine and the biomedical sciences, but also because "alternativism,"—in philosophy, history, and literature as well as in natural science—whose benign exterior can cover dangerous and exploitative interior machinery, is at the moment a more serious problem for the sciences of health than for any others. Because of the Human Genome Project, because of AIDS, because of the eagerness of the media, some of the new biotechnology companies, and some investigators to report (often inaccurately and prematurely) and to solemnize heavily on every new finding in medical or behavioral genetics, biomedicine has joined high-energy physics under the pejorative rubric, "Big Science"—the instrument of oligarchs. By contrast, "alternative," or "nontraditional," or "natural" healing is portrayed, even in the nominally most serious outlets,[12] as small, beleaguered, innocent, and above all, *democratic,* the consolation of the people. Some of our contributors offer here the fruits of long and practical experience with these forms of consolation.

The essays to follow are grouped by broad categories of concern: Literary Studies, Education, Reason and the Social Sciences, Medical Science, Sociology of Science, Religion, Feminisms, Physical Science, Inquiry, Belief and Credulity, Environmentalism. Inevitably, there is considerable overlap of categories, and the categories do not correspond to any university's—or political party's—catalog of departments and offerings. They are, explicitly, *categories of concern;* there is no reasonable way to eliminate the core epistemic issues, to which a separate group of papers is devoted, from, say, an examination of current styles and practices in sociology of science, or in literary studies. Nevertheless these groups, and the essays they contain, cover a large part of the academic acreage and a good deal of the terrain over which there is, or should be, dispute among intellectuals. We hope that readers of these proceedings will find, whether or not they agree with particular arguments, that the contributions are in general not only competent but convincing.

## NOTES

1 The stage-direction is mine; from Part One, II, 1851–1855, Goethe's lines are:

> Verachte nur Vernunft und Wissenschaft,
> Des Menschen allerhöchste Kraft,
> Lass nur in Blend- und Zauberwerken
> Dich von dem Lügengeist bestärken,
> So hab' ich dich schon unbedingt—

here given in the translation of Carlyle F. MacIntyre. I am grateful to my Virginia colleague, Professor Paul A. Cantor, for reminding me of this episode.

2 A recent and very influential example of the promotion of "alternative" modes of inquiry, even about the physical world, has been the antievolutionist tract by Phillip E. Johnson. *Darwin on Trial.* For a sober review of this book, see David L. Hull, "The God of the Galápagos."

3 A bestiary, reasonably current, is given in Paul R. Gross and Norman Levitt, *Higher Superstition.* Its fourteen-page, small-type bibliography can hardly scratch the surface, however. To see what goes on in particular disciplines, *à la mode*, one needs concentration, as in the study, by a committed and serious environmentalist thinker, of ecoradical mythology: Martin W. Lewis, *Green Delusions.*

4 The celibate (as regards "STS"—Sci-Tech-Studies) reader can get a sense of its agreements, which are few but unquestioned (science is a social process, not a mode of inquiry, it has no special claim to efficiency in getting at "truth"), and its internecine wars, which are deadly and amusing, from Andrew Pickering, *Science as Practice and Culture.* Its imperialistic hopes, however, the size of its literature, and its wildly variable standards are even more evident in a massive document that recently surfaced on the World Wide Web. This is, apparently, a manifesto from a Summer Institute sponsored by the National Endowment for the Humanities in the Summer of 1991: "Science as Cultural Practice," edited by Steve Fuller and Sujatha Raman. It would overburden this note to list even its many authors; but here is a characteristic passage:

> . . . So far, the public has been able to view the divergence of expert opinion in public commissions and courts, and still retain confidence in the institution of science. This confidence seems to hinge on the notion that what is being debated is the "truth of the matter" which can be determined by the scientific method. Discoveries of flaws are treated as human error.
>
> Given that this image does not stand up to scrutiny, and this becomes public knowledge, the system could be in a state of uncertainty. In this context, STS is a two-edged sword capable of precipitating a backlash triggered by the already existing disillusionment over the consequences of science. Alternatively, STS can steer the response to an examination of why and how is science successful, followed by a discussion of what we can expect from science if it is to be worthy of public support. Science can still be worthy of respect, although not for the reasons that are

commonly projected. Hence the significance of having people trained in STS move into policy negotiation roles and science teaching is that they are part of the vanguard that has already faced up to the new conception of science and begun rethinking the role of science in society.

5 Nearby on a bookshelf, as I write this, is an example whose title is enough, given as a gift (that is, approved and strongly recommended) to a member of my family by someone who should know better: Deepak Chopra, M.D., *Ageless Body, Timeless Mind: The Quantum Alternative to Growing Old.*

6 For which see the recent book by a psychologist who is also a contributor to this volume: Barbara S. Held, *Back to Reality: A Critique of Postmodern Theory in Psychotherapy.*

7 See, for example, Gerald Holton, *Science and Anti-Science.*

8 Richard Mitchell, *The Leaning Tower of Babel,* p. 266.

9 Examples: John R. Searle, *The Construction of Social Reality;* David Stove, *The Plato Cult and Other Philosophical Follies.*

10 A response to the complaint that the scientific oligarchy has overreacted to the Smithsonian exhibit can be found in Paul R. Gross, "Response to Tom Gieryn."

11 An encouraging and very recent sign of change, on the eve of putting this book to bed, is the devastating review written by Nobel Laureate biophysicist M. F. Perutz, of a recent, debunking biography of Pasteur (by Gerald L. Geison). Perutz's essay, "The Pioneer Defended," is explicitly an effort "to deconstruct his [Geison's] deconstruction."

12 Try PBS or the *New York Times,* or any day on one of the numerous internet discussion lists devoted to science studies.

## REFERENCES

CHOPRA, DEEPAK. *Ageless Body, Timeless Mind: The Quantum Alternative to Growing Old.* New York, NY: Harmony Books, 1993.

FULLER, STEVE and SUJATHA RAMAN. "Science as Cultural Practice." Summer Institute sponsored by the National Endowment for the Humanities, 1991.

GROSS, PAUL R. "Response to Tom Gieryn." *Science, Technology, and Human Values,* 21,1 (Winter 1994): 116–120.

GROSS, PAUL R. & NORMAN LEVITT. *Higher Superstition: The Academic Left and Its Quarrels with Science.* Baltimore, MD: Johns Hopkins University Press, 1994.

GOETHE. *Faust.* Translated by Carlyle F. MacIntyre. Norfolk, CT: New Directions, 1941.

HELD, BARBARA S. *Back to Reality: A Critique of Postmodern Theory in Psychotherapy.* New York, NY: W. W. Norton, 1995.

HOLTON, GERALD. *Science and Anti-Science.* Cambridge, MA: Harvard University Press, 1993.

HULL, DAVID L. "The God of Galápagos." *Nature* 352(1991): 485–486.

JOHNSON, PHILLIP E. *Darwin on Trial.* Washington, DC: Regnery Gateway, 1991.

LEWIS, MARTIN W. *Green Delusions.* Durham, NC: Duke University Press, 1992.

MITCHELL, RICHARD. *The Leaning Tower of Babel.* New York, NY: Simon and Schuster, 1984.

PERUTZ, M. F. "The Pioneer Defended." *The New York Review of Books,* December 25, 1995, pp. 54–58.

PICKERING, ANDREW. *Science as Practice and Culture.* Chicago, IL: University of Chicago Press, 1992.

SEARLE, JOHN R. *The Construction of Social Reality.* New York, NY: The Free Press, 1995.

STOVE, DAVID. *The Plato Cult and Other Philosophical Follies.* Oxford: Basil Blackwell, 1991.

# THE PUBLIC IMAGE OF SCIENCE

THIS OPENING SECTION addresses the public perception of science. How well does the general population comprehend what science is, how scientists do it, what tools, intellectual and moral, are necessary? How well do nonscientists understand, and how far do they misperceive? What can be done to remedy misperception?

Nobel Laureate Herschbach ponders the misunderstanding—indeed, the hostility—towards science and scientists that is frequently found within the community of humanist intellectuals. Much of this derives from the sense that doing science is an inflexible, mechanistic procedure, the prime example of "linear thinking," to use a pejorative now much in vogue. By analysis and example, Herschbach shows that on the contrary, science is better compared to a tapestry of variable design, constantly under construction, with seemingly unrelated threads combining in unexpected and beautiful ways. This makes clear that imagination, insight, and even poetry are requisite for scientific success. In short, science is as "holistic" as any critic of linearity could wish.

David Goodstein, administrator, physicist, and gifted scientific pedagogue, examines a question that has greatly agitated the public in recent years: how pervasive is fraud in science, and how far does it bring into question the reliability of scientific knowledge? He shows that actual fraud in the scientific world is quite rare and its effect in misdirecting scientific work minimal. But he also acknowledges that there are other kinds of misconduct, sometimes confused with fraud, that cause public unease. In Goodstein's view, the public must acquire more sophistication in sorting out such matters. For their part, scientists should be more candid in admitting to the human elements in scientific work. The image of scientists as a neutral and aloof priesthood, whatever its origins, invites disenchantment that, in today's climate, is damaging to science.

Norman Levitt, a Rutgers mathematician, finds that the public's estrangement from science is in large measure traceable to a distaste for the mathematical language in which so much of science is embodied. He notes that the exile of mathematics from the humanistic worldview is a relatively new phenomenon, reversing a long tradition that incorporated mathematics into the background of most serious thinkers. He analyzes the effects of this blind spot on the worldviews of prominent intellectuals, and speculates on how a reunion of mathematics with contemporary humanism might improve discourse among the disciplines.

# IMAGINARY GARDENS WITH REAL TOADS

DUDLEY R. HERSCHBACH

MARIANNE MOORE SAID: "Poetry is about imaginary gardens with real toads in them." That applies just as well to many aspects of human cultures and especially to science. Scholars and scientists of all kinds are gardeners of ideas, trying to cultivate lovely flowers and fruits of understanding. Such intellectual gardeners usually pay little attention to toads or other uninvited creatures residing among the flowers—unless those creatures begin to munch or trample the plants. Visitors who seem greatly concerned about the toads but unappreciative of the fruits and flowers naturally dismay the gardeners, especially if such visitors even mistake the gardeners for toads.

This horticultural whimsy explains my title and offers rustic metaphors for some of the phenomena addressed in this volume. In my view, the blossoming of "science studies" by historians, sociologists, and philosophers is to be welcomed. It is seeding fallow fields that have long separated scientific and humanistic gardens. In the process, some dust and mud is being stirred up, coating alike flowers and toads. That is inevitable. Science has always drawn lively critics, often motivated by ideology or mysticism. However, I am startled to find a philosopher suggesting that Newton's *Principia* could be regarded as a "rape manual," or radical postmodernist scholars asserting that scientific discoveries are "socially constructed fictions." Such claims, and the disregard for rational analysis that often indulges them, deprecate not only science but all objective scholarship and public discourse. As in horticulture, blights readily spread to neighboring gardens.

In this paper, my chief aim is to welcome any visitors or critics, friendly or not, to tour three patches of my personal scientific garden. The first deals with what for me was a rehearsal for this conference. It was a public television program, titled "The Nobel Legacy" and aired last May, in which I had to counter doleful complaints about science delivered by Anne Carson, a poet and professor of classics at McGill University. The second part comments on aspects of scientific work that are often misunderstood. These pertain to science education as well as to the pursuit of new knowledge. The third part sketches a saga of scientific discovery, submitted to exemplify favorite themes and to confound postmodern nihilists. I conclude by emphasizing the kinship of natural science with other liberal arts.

A few caveats. For brevity, I say just "science" when usually I mean to include all or most of natural science, mathematics, and some engineering and technology. I do not mean to include social science; that would stretch too far my unruly metaphors and commentary.

## SCENES FROM *THE NOBEL LEGACY*

The television program was part of a trilogy comprising Medicine, hosted by J. Michael Bishop; Physics, hosted by Leon Lederman; and Chemistry, hosted by me. The series was directed by Adrian Malone, known for his many previous science productions, including "The Ascent of Man" with Jacob Bronowski and "Cosmos" with Carl Sagan. As described in a synopsis, *The Nobel Legacy* "seeks to reveal the beauty and humanity of science to audiences not generally drawn to science programming . . . and [to] encourage viewers to become more aware of the scientific strides transforming their lives." Each segment featured one or two scientists whose work ushered in major revolutions: Jim Watson and Francis Crick, Werner Heisenberg, Antoine Lavoisier, and Robert Woodward, my late colleague at Harvard who transformed synthetic organic chemistry. The filming was done at sites of great historical or esthetic interest, ranging from Venice, Florence, and Paris to Hawaii and Carlsbad Caverns.

Anne Carson appears in two or three scenes in each of the segments. Her role was "to provide a provocative philosophical counterpoint . . . the spark of dissenting opinion." Typical of the sparks she struck were her scenes in the Chemistry program. It opens with Carson in Paris, on a bridge over the Seine, and she says:

> The Nobel Prize idealizes the notion of progress. My problem is that I don't believe in progress, and I am skeptical of how chemistry is contributing to my humanity. . . . Now that we've filled the world with Styrofoam cups, carbon monoxide and holes in the ozone, maybe it's time . . . to stand still and pay attention to the real relation between our humanity and our progress.

From this beautiful scene, there's a quick flash, without explanation, to the guillotine (in anticipation of Carson's later description of Lavoisier's execution). Next Carson appears at the Institute de France, the famous "home of the immortals." She explains that it was previously a school attended by Lavoisier, and continues:

> Before the Revolution this place had a reputation for reason . . . Lavoisier's teachers believed that using the light of reason, old ideas and superstitions could be burned away . . . [and] a truly rational society lay around the corner.

As she stands by the bust of Lavoisier, with ominous flashbacks to the guillotine punctuating her sentences, she delivers a scornful verdict:

> What an irrational idea. . . . This happy delusion that there are such things as facts, and they do not deceive us, underlies the whole progress of science and chemistry down to the present day.

A later scene, my favorite among hers, has Carson in the lovely garden within the inner courtyard of the Gardner Museum in Boston, where she says:

> Like every science, chemistry promises to use technology to bring us to paradise. . . . This garden is a little like paradise. It's a small enclosed area of symbolic perfection, an illusion created by the idea that you can have perfect control of nature and all external conditions. It would be a mistake to confuse that illusion with reality, or to think that tinkering with the chemistry of the human condition can ever bring us to paradise. I don't want scientists messing around in the garden of my soul.

Some of my scientific colleagues were quite annoyed with Anne Carson; they regarded her remarks as nonsensical, a silly distraction.[1] Others were annoyed with me, because in interviews I defended her role in the program.[2] I regarded it as a forthright way to acknowledge grim concerns that many people have, misplaced as some of them are. At the outset of the filming, I thought it odd that Adrian Malone did not allow the scientists serving as hosts to reply directly to Carson's criticisms. He wanted to maintain dramatic tension. Later, I came to appreciate that, at least for TV, his indirect approach was more effective. It left the hosts free to extol the wonder and beauty of science and to give our own perspective on toadish issues like pollution. Likewise, it invited the viewers to think for themselves about the issues rather than passively watching a battle of sound bites.

## NOTES FOR A CONVERSATION

Someday I would like to meet Carson and discuss her concerns, with the hope of understanding them better. As Malone liked to emphasize, TV is an emotional medium, not suited to comprehensive explanations. Having experienced that, I do not suppose Carson is so dogmatic as some of the caustic lines she delivered on the program. In any case, again I do not feel obliged to refute her comments directly. Instead, I offer here notes on six points I would want to bring up if I ever do have the opportunity for a conversation with Anne Carson.

1. However questionable the moral or spiritual progress of humanity may be, science has vastly enhanced our capacities and perspectives. As just one example, plucked from a cornucopia, consider the germ theory of disease.[3] For millennia, humankind was almost helpless to combat frequent epidemics of infectious diseases. Two of Pasteur's five children died of typhoid fever, not an uncommon toll in his time. His work establishing the germ theory transformed medicine. Beyond the enormous practical impact of the vaccines thereby developed, the conceptual advance opened the way to the incredible modern era of biomolecular science. Ultimately, this will profoundly impact cultural perspectives too. Just as the Copernican revolution drastically recast mankind's place in the outer world, the "DNA revolution" has transformed our conception of life's inner cosmos.

Exhilarating as they are, the achievements of science cannot create an earthly paradise or even a humane civilization. Much beyond science is required for that. Today, about two million children die each year of diseases for which vaccines are already available.

2. Most scientists would agree with Carson's concerns about toads nurtured in the gardens of science. Our modern technological society has fostered reckless proliferation of people, weapons, pollution, and environmental damage. These are not inevitable consequences, but cannot be overcome without unprecedented economic and political initiatives. Such problems have often plagued preindustrial societies too.

The specialization and sophistication of modern technology have induced another pervasive anxiety. People understand less and less about the many machines, techniques, and underlying concepts on which they depend. This has increasingly serious political consequences; it prevents objective assessment of important questions of public policy involving science, and fosters both distrust and credulity. Here science education, which I discuss later, has a crucial task.

3. There is another kind of pollution, not spawned by science, and far more terrible and intractable than any chemical garbage. It is a pollution by words, toads in humanistic gardens. Perceptions shaped by the power of words have given rise to countless wars and all sorts of political and social folly. Most often that power is illegitimate, not drawn from objective referents. Science can seldom solve sociopolitical problems. Yet sometimes it can change conditions enough to enable resolution, or provide evidence and analysis to help contend with wordly pollution.

4. The notion, advocated by Francis Bacon, that the aim of science is to attain "power over nature" has long drawn attacks as a foolish and arrogant quest. In my view, the aim is not control but comprehension. When science enables us to cope better with microbes and lightning bolts, we should be grateful. When it lets us glimpse more deeply into mysteries of Nature, we should be awed.

At a time when lightning was considered a divine punishment, Ben Franklin gave a sparkling response to a critic of his lightning rod:

> He speaks as if he thought it Presumption in man, to propose guarding himself against the Thunders of Heaven! Surely the Thunder of Heaven is no more supernatural than the Rain, Hail or Sunshine of Heaven, against the inconveniences of which we guard by Roofs & Shades without Scruple.[4]

The alleged arrogance of science also does not accord with what I regard as the most important aspect of the modern scientific "world view." As stated by Richard Feynman: "Science is not about what we know; it's about what we don't know." Science lives at its frontiers, looking to the future, aware of the enormous scope of our ignorance.

This is painfully evident in medicine; Judah Folkman, a distinguished physician, points out that the medical encyclopedias describe some fifteen to twenty thousand recognized human diseases, but fewer than a thousand can be "cured" or completely reversed.[5] Many more cannot be affected at all by modern medicine, even to relieve symptoms. That is what allows "alternative medicines" to flourish, some of it outright fraudulent.

5. Before our conversation, I would recommend to Anne Carson, as I do to my science students, a wise and delightfully engaging book: ***The Limits of***

*Science* by Peter Medawar. In particular, Medawar points out that, while Bacon is identified with seeking power over Nature, his writings have "many more typical passages in which he advocated a much more humbly meliorist position," and quotes from his preface to *The Great Instauration*:

> I would advise all in general that they take into serious consideration the true and genuine ends of knowledge; that they seek it neither for pleasure, or contention, or contempt of others or for profit or fame, or for honor and promotion; or suchlike adulterate or inferior ends: but for the merit and emolument of life, and that they regulate and perfect the same in charity: for the desire of power was the fall of angels.[6]

Like most philosophers and all politicians, Bacon said different things at different places. But I do believe that this statement, rather than any urge to control nature, and despite familiar human frailties, represents the chief motivation of most scientists. Recently, Henry Rosovsky proposed, in discussing university governance and contentions, that there should be a creed that professors subscribe to just as do physicians.[7] Perhaps Bacon's preface might serve for any field of scholarship (although I could not forswear seeking knowledge for pleasure!). It would be interesting to find out how Carson and other critics of science might respond.

6. Another book I especially recommend, both to scientists and critics, is *Science and Human Values* by Jacob Bronowski. It comprises three eloquent essays, devoted to showing that science imposes "inescapable conditions for its practice" which compel an ethic. Far from being neutral with respect to human values, science like art requires freedom, honesty, and tolerance in order to foster originality and creativity. Bronowski was both a scientist and author of a major study of William Blake, a Romantic poet vehemently hostile to science. In the revised edition of *Science and Human Values*, Bronowski added a dramatic dialogue, "The Abacus and the Rose." This presents a vigorous debate between a literary scholar, "bitter because he feels helpless in a changing time," and a molecular biologist, "slow to see that there really are other points of view." In his preface, Bronowski says he tried "to put the arguments on each side fairly . . . in words which do not caricature its case." As the conversation spirals between the two cultures, their perspectives gradually intertwine and become complementary. Bronowski concludes with a sonnet; his first lines are:

> I, having built a house, reject
> The feud of eye and intellect,
> And find in my experience proof
> One pleasure runs from root to roof

This humane unity in spirit, more often praised by scientists[8] than others, should be more widely acknowledged.

## SOME METAPHORS FOR SCIENCE

At its frontiers, science-in-the-making is inevitably a messy and uncertain business, easily misunderstood by policy makers, funding agencies, reporters, students, and sometimes even the researchers. Those who decry science often confuse its rude frontiers with its civilized domains or foundations. Many examples of such confusion, inadvertent or deliberate, are noted and analyzed in *Higher Superstition* by Paul Gross and Norman Levitt and in *Science and Antiscience* by Gerald Holton. Here I simply outline aspects of frontier science that I think should be considered by anyone approaching it, pro or con. My sermonette begins with three favorite metaphors: language, pathfinding, and puzzle solving.

1. Nature speaks to us in many tongues. They are all alien. In frontier research, the scientist is trying to discover something of the grammar and vocabulary of at least one of these dialects. To the extent the scientist succeeds, we gain the ability to decipher many messages that Nature has left for us, blithely or coyly. No matter how much human effort and money we might devote to solve a practical problem in science or technology, failure is inevitable unless we can read the answers that Nature is willing to give us. That is why basic research is an essential and practical investment, and why its most important yield is ideas and understanding.

The language metaphor occurred to me when I received a letter in Braille. We are all born blind to Nature's language, and it takes much persistent groping and guessing to learn something of it. Ironically, our academic science courses inhibit the willingness of students to guess. In my teaching, I try to counteract that with questions that cannot be approached otherwise. Also, I tell my students:

> Not so many years from now, most of you will be considered expert in something. Then you will find that clients often come to ask your opinion, not because of what you know, but because they think as an expert you can guess better than they can.

Of course, the expert must also devise means to test the guesses. That is a crucial part of science; like children playing with new words, we want to try to converse with Nature. In some fields, few of Nature's words are yet known; in others, many volumes.

2. Science is often described as finding a way up an unexplored mountain. This emphasizes a tremendous advantage enjoyed by science: the goal, call it truth or understanding, waits patiently to be discovered at the top of the mountain. Thus marvelous advances can be achieved by ordinary human talent, given sustained effort and freedom in the pursuit. Far more formidable are enterprises such as business or politics; there the objectives may shift kaleidoscopically, so a brilliant move often proves a fiasco rather than a triumph because it comes a little too soon or too late.

The patience of scientific truth has another important consequence, often puzzling to the public. Frequently, what might appear as the most promising path up the mountain does not pan out; there are unanticipated roadblocks. Then it is vital to have some scientists willing to explore unorthodox paths, perhaps straying far from the route favored by consensus. By going off in what is deemed the wrong way, such a maverick may discern the right path. Hence in science, it is not even desirable, much less necessary or possible, to be right at each step. Since the truth waits patiently, and we must grope for it, wrong steps are intrinsic to the search. Adventurous scientists are heading in wrong directions much of the time, optimistically looking for new perspectives that may show the way up the mountain.

Again, this pathfinding contrasts markedly with introductory science courses. Many students have told me about a disheartening syndrome: the questions and problems seem to have only one right answer, to be found by some canonical procedure. The student who does not quickly grasp the "right" way, or finds it uncongenial, is soon likely to become alienated from science. There seems to be very little scope for a personal, innovative experience. Nothing could be further from what actual frontier science is like. At the outset, nobody knows the "right" answer, so the focus is on asking an interesting question or casting the familiar in a new light. In my freshman chemistry course, I explain this to the students and ask them to write poems about major concepts, because that is much more like doing real science than the usual textbook exercises. I also show them quite a few poems that pertain to science, often without intending to. For instance, here is a quatrain by Jan Skacel, a Czech poet:

> Poets don't invent poems
> The poem is somewhere behind
> It's been there a long long time.
> The poet merely discovers it.[9]

This is the crux of our pathfinding metaphor. I hope the poems help students realize that there is much in science that transcends its particulars.

3. Science is also often likened to assembling a giant jigsaw puzzle. This metaphor was invoked by Michael Polanyi in his classic essay on the organization of scientific research, "The Republic of Science." He envisions competing teams with exactly the same talent, some structured in the hierarchical ways customary in practical affairs but one enjoying the chaotic freedom of science. In the hierarchical organizations, the team would be divided into units directed by a chain of officers; one unit might focus on yellow pieces of the puzzle, another on blue pieces, etc. Each unit would report up the chain of officers to a central authority that assigns tasks. Science proceeds very differently and much more efficiently. Each unit is on its own, free to look at whatever parts of the puzzle interest it. Nonetheless, the independent units are coordinated by "an invisible hand," because each has the opportunity to observe and apply the results found by the others. This creates a community of scientists that fosters and amplifies individual initiatives. Such a community, as Polanyi says, is "a society of explorers, poised to examine any new understanding of Nature."

Again, there is an ironic contrast between such intrinsic cooperation and the artificial competition among students that is imposed in typical courses. In my general chemistry course, we use an absolute grading scale, so students compete against my standard, not each other. Also, some work is designed to be done by teaming up with other students. Of course, in research there is some competition, but it is superficial in the context of the essential collaboration with predecessors and colleagues. Since results are reported openly, the work of a competitor often proves especially helpful, as it is likely to provide a complementary perspective.

## IMAGINARY TOADS IN REAL GARDENS

As suggested by these metaphors, creativity in science has much in common with the arts. New insights often seem strange and idiosyncratic; blunders and breakthroughs at first look very much alike. The typically fitful evolution of germinal work into a widely accepted paradigm or movement is also in some respects similar for science and the arts. These aspects were brought out clearly by Thomas Kuhn[10] as well as Michael Polanyi.[11] Both wanted to counter simplistic depictions of the scientific method as a rather mechanical collecting of facts and calmly logical testing of hypotheses. In my laboratory, visitors are reminded of this theme by an epigram of Vladimir Nabokov: "There's no science without fancy, and no art without facts."

Such cheerful bows to the artistic qualities of science may be considered damning by critics like radical cultural constructionists and postmodernists. Among them some think that the political, class, and gender bias of scientists renders them incapable of finding objective knowledge about Nature. Those critics cannot imagine that toadish scientists could discover real gardens.

When asked to refute Bishop Berkeley, Ben Johnson simply kicked a rock. To a cultural constructionist, we might likewise point out an airplane, say. Rather than being limited to "socially constructed fictions," the natural sciences attain objective truths by functioning as a constructive society. A major part of Michael Polanyi's *Personal Knowledge* is devoted to showing in detail just how this works, despite human frailties and cultural contingencies.

For brevity, again I invoke my horticultural whimsy. In essence, the validity and range of applicability of new knowledge is established by examining many generations of its progeny, when it is mated with other plants already well established in the garden. In this propagation, errors are revealed as unviable mutations and die out. The stock that emerges is hardy, but not an eternal, universal tree of knowledge. Eventually, it will be hybridized anew to produce further progeny. Any part of the garden (as well as the toads) can be transplanted to another culture, so long as it gets suitable care, sun and soil there.

Now I turn from genial horticulture to consider the rape motif favored by Sandra Harding.[12] Like many scientists, male and female, I am sympathetic to feminist issues. But not to Harding's notion that Newton's *Principia* could be regarded as a "rape manual," later generalized by comparing science to "marital rape, the husband as scientist forcing nature to his wishes." As noted already, like Medawar and Bacon in his meliorist mood, I do not regard "forcing nature" as an aim of science. Certainly, it is absurd to tar Newton with that; he was an extremely pious fellow who regarded his work as revealing a divine order. For me, the chief lesson of all natural science, typified by Newton's mechanics, exalts Nature: she is the boss; we try to discover her rules; she lets us know the extent to which we have done so. Harding should consider Richard Feynman's remark:

> For a successful technology, reality must take precedence over public relations, for Nature cannot be fooled.[13]

This was the last sentence of his report on the tragic explosion of the Challenger space shuttle.

Science, like so much else, has often been marred and misled by sexual bias.[14] It is thus gratifying that embryology now indicates the Garden of Eden story needs to be recast, as Eve came before Adam.[15] For its first thirty-five days in the womb, every human fetus is female, whether or not it has XX or XY chromosomes. On about the thirty-sixth day, a genetically programmed switch kicks in. For a normal XY fetus, some already developing female structures then fade away and male structures begin to form. In rare cases, however, the switch is defective; the XY fetus then continues on to become a woman, with external anatomy correct in all respects, although infertile. (Olympic athletes have since 1968 been subjected to a chromosome test, and allegedly a number of XY women athletes have been barred from competing.)[16] The most common defect in the genetic switch has recently been shown to occur on a single amino acid residue, at a critical location on the Y chromosome.[17] In fact, that defect is a missing methyl group, just a carbon atom with three attached hydrogen atoms. Thus, a single methyl group can determine the sex of an XY fetus. If it were missing much more often, human males might be as outnumbered as male bees or ants.

This evidence, indicating Eve's primacy and shrinking Adam's rib to a methyl group, was found by male scientists. Ultimately, Nature forces scientists, willing or not, to her truths.

## AN EXEMPLARY SAGA OF SCIENCE

Several garden paths crisscross my field of research; here I retrace one such path.[18] It begins in Frankfurt with the work of Otto Stern, seventy-five years ago. The invention of the high speed vacuum pump had made it possible for him to undertake experiments using "molecular rays," now called "beams." These are tenuous, ribbonlike streams of molecules traveling in a vacuum at sufficiently low pressures to prevent disruption of the beams by collisions with background gas. The most celebrated of Stern's experiments using these beams was devised to test a curious prediction of the early form of quantum mechanics proposed by Niels Bohr. His atomic model postulated that an electron circulated around the nucleus only in certain definite orbits, whereas the classical mechanics of Newton imposed no such restriction. Stern's experiment focused on that key difference.

His simple apparatus was set up inside an evacuated glass tube, about the size of a quart jar. Silver atoms evaporated from a heated wire passed through a slit which collimated them into a narrow beam (only 0.1 mm wide, about the breadth of a human hair). The beam then passed between the poles of a small magnet and deposited its silver atoms on a glass plate. Stern's magnet was special: one pole came to a sharp edge, like the roof of a house; the opposite pole had a wide groove, like a trench. The magnetic field therefore was much stronger near the sharp pole, weaker near the grooved pole.

According to the Bohr model, the orbital motion of its outer electron makes a silver atom act like a tiny bar magnet. Furthermore, the model predicted that the atomic magnets could point only in either of two directions, "up or down," say, when the atoms interact with an external field, like that of Stern's magnet. Consequently, as indicated in FIGURE 1, atoms with one of these two allowed orientations would move towards the strong field of the

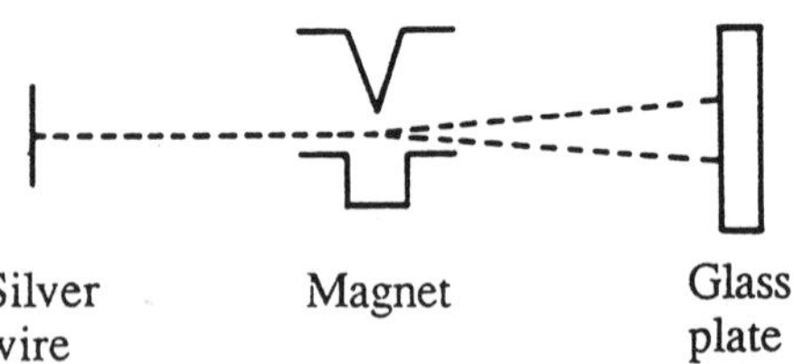


FIGURE 1 Stern's experiment: a beam of silver atoms splits into two on traversing a magnetic field, revealing "space quantization."

sharp pole, the other towards the weak field of the grooved pole: the beam would split into two components. In contrast, if Newton's mechanics were applicable to atoms as it is to planets, the atomic magnets could have any orientation in space: in passing through Stern's magnet, the beam would not split but merely broaden.

The experiment showed unequivocally that the beam split in two. This discovery, revealing the existence of what is now called "space quantization," was shocking to physicists of the day. It provided one of the most compelling items of evidence that a new mechanics was required to describe the atomic world. Two aspects are especially pertinent here. First, Stern himself did not want to accept the quantum picture; he hoped and expected that his experiment would support a Newtonian description of atoms. Second, the atomic magnetism demonstrated by Stern's experiment actually does not come from orbital electronic motion but from a different intrinsic property, called "spin," not discovered until a few years later. The Bohr model was soon discarded; it was seen to have merely simulated, in a limited way, some features of a far more comprehensive quantum theory. I note these aspects because they so clearly show how myopic is the view of cultural constructivists. In its historical context, Stern's experiment gave an "antisocial" result. Other experiments revealed the Bohr model was not correct; yet it was not "fiction" but rather scaffolding for the emergent quantum theory.

Descendants of Stern's beam technique and his concept of sorting quantum states are legion. The prototypes for nuclear magnetic resonance (NMR), radioastronomy, and the laser all derived from space quantization. NMR spectroscopy, first developed at Columbia University by Isidor Rabi in the late 1930s, deals with the orientation of nuclear magnets or spins in an external field. FIGURE 2 indicates the key aspects. For the different orientations, the energy of interaction of the nuclear spin with the field differs slightly. The higher energy state corresponds to an unfavorable orientation, the lower energy to favorable orientation. By means of radio waves of an appropriate "resonant" frequency, the nuclear spin can be induced to change its orientation, or "flip." However, it was not obvious that the minute nucleus, with dimensions about 100,000-fold smaller than the atomic electron distribution, would interact appreciably with the electrons. If not, the different nuclear spin orientations (favorable: "up" or unfavorable: "down") would remain equally probable even when subjected to an external magnetic field. There would then be no net absorption of resonant radio waves, since spin flips induced by the radiation are equally likely up or down. Net absorption can only occur if interactions permit the populations of the spin orientation states to be unequal.

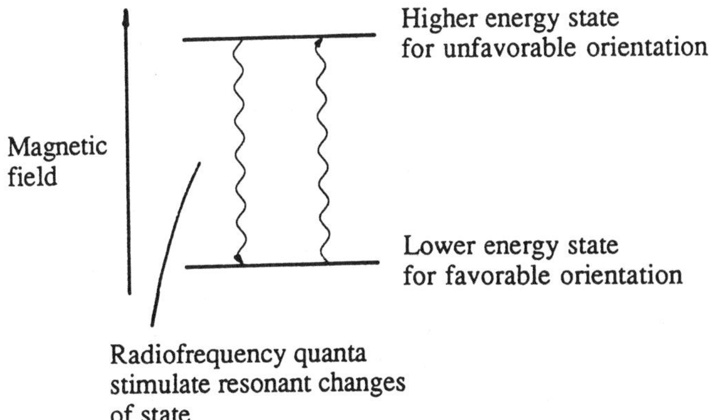


FIGURE 2 Rabi's experiment: transitions between energy levels, corresponding to different spin orientations, are induced by absorption or emission of quanta.

Rabi devised an elegant way to use atomic beams to escape this constraint. In an evacuated chamber, he introduced two magnets like Stern's, but with their fields in opposite directions. A beam of atoms traversing the first field (denoted A) is split into its nuclear spin components, but on passing through the second field (denoted B) these are recombined. Between these magnets, which act like diverging and converging lenses, Rabi introduced a third magnet (the C-field). This had flat pole pieces, so had no lens action, but served to define the "up" and "down" directions. Since atoms in the beam experience virtually no collisions, the populations of the component spin orientation states remain equal. Also, radio waves in the C-field region will at resonance flip equal numbers of spins up or down. But now any spin changing its orientation after the A-field will not be refocused in the B-field. That enabled Rabi to detect which radio frequencies produced resonances. He thus created a versatile new spectroscopic method with extremely high resolving power. His work provided a wealth of information about nuclear structure and led to many other fruitful developments.

The scope of NMR has expanded vastly since 1945, when resonances were first detected in liquids and solids, at Harvard by Edward Purcell and at Stanford by Felix Bloch and their coworkers. In these prototype experiments, the key discovery was that the populations of the "up" and "down" nuclear spin orientations in a magnetic field (C-field) become unequal by virtue of intra- and intermolecular interactions (so Rabi's A- and B-fields are not needed). Over the intervening decades, many sophisticated variants have been devised, with myriad applications. Now NMR spectra are routinely used to determine detailed structural features of proteins containing thousands of atoms. For instance, that was how the missing methyl group in the defective switch on the Y chromosome was detected.[19] Another marvelous offspring is magnetic resonance imaging, which provides far higher resolution than x-rays. It is applicable also to soft tissues such as the brain; we can now literally catch sight of glimmering thoughts.

Radioastronomy vastly extended what we can see of the heavens. The prototype experiment, conducted at Harvard by Purcell in 1951, detected radio frequency emission from the spin flip of the nucleus of a hydrogen atom, subject to the magnetic field generated by orbital motion of the atom's lone electron. Hydrogen is by far the most abundant element in the universe, and the ability to map its distribution soon revealed unsuspected aspects of the structure of galaxies. Kindred techniques employing radio frequency and microwave spectroscopy also revealed a remarkable variety of molecules in interstellar clouds, products of previously unsuspected galaxies of chemical reactions.

Beyond leading to ways to probe nuclei, proteins, and galaxies, the path stemming from Otto Stern even brought forth a new form of light itself: laser light. The name is an acronym for **l**ight **a**mplified by **s**timulated **e**mission of **r**adiation. The prototype experiment, done at Columbia University in 1955 by Charles Townes and his students, employed a beam of ammonia molecules subjected to an electric field which acted as a state-selection device quite analogous to Stern's magnet. The device selected an energetically unfavorable molecular state. When illuminated with microwaves, the molecule reverted to the energetically favorable state, emitting the excess energy as radiation. Unlike ordinary light, this emission has a special property, called "coherence"; the difference is analogous to that between a meandering crowd of people and a marching band. Thus was born molecular amplifiers and oscillators and other wonders of quantum electronics.[20] Laser light now performs eye surgery, reads music or data from compact disks, and scans bar codes on grocery packages or DNA base pairs in the human genome.

As Stern was a physical chemist, he would be pleased that his beam techniques, augmented both by magnetic and electric resonance spectroscopy and by lasers, have also evolved powerful tools for study of molecular structure and reactivity. In my own work, the basic method simply involves crossing two molecular beams in a vacuum and detecting the products in free flight, before subsequent collisions degrade the information they carry about the intimate dynamics of the reactive encounter. This method, applied and refined in many laboratories over more than thirty years, has enabled the forces involved in making and breaking chemical bonds to be resolved and related to the electronic structure of the reactant molecules.

From this point, I want to trace a trail that backtracks from Stern's path. Usually, an atomic or molecular beam comes from a small source chamber mounted within a vacuum apparatus. Stern had stressed that the pressure in the source chamber should be kept low enough so that the atoms or molecules, as they emerge from the exit orifice, do not collide with each other. The emergent beam then provides a true random sample of the gas within the source, undistorted by collisions in the exit orifice. Of course, this canonical ideal was blatantly violated by chemists who wanted to study reactions in crossed beams. Such experiments desperately needed intensity, so much higher source pressures were used. Collisions within the exit orifice then produce supersonic flow. This has advantageous properties. In addition to high intensity, supersonic beams have narrow distributions in both direction and molecular speeds. Under suitable conditions, the rotational and vibrational motions of the beam molecules can also be markedly reduced, in effect producing a very low internal temperature.

These properties of supersonic beams, all resulting from collisions as molecules crowd out the exit aperture, are readily understood by anyone who has attended a Saturday morning sale at a department store such as Filene's in Boston. Typically, a dense crowd gathers (like the high-pressure gas within the beam source). When the doors are thrown open and the crowd rushes in, collisions induce everybody to flow in the same direction with the same speed, whether they want to or not. Moreover, if some customers are excited at the prospect of a bargain and therefore leap about or turn handstands (like vibrating or rotating molecules), they suffer more frequent and harder collisions—even black eyes or bloody noses. Thereby such lively customers are calmed down (to low effective temperatures).

The drastic cooling that occurs in supersonic beams has proved extremely useful. In particular, it led to the discovery a few years ago of a new form of elemental carbon.[21] The molecules have sixty carbon atoms, arrayed in linked hexagons and pentagons in the same highly symmetrical pattern as a soccer ball, and whimsically named "Buckyball" to celebrate the architectural resemblance to Buckminster Fuller's geodesic domes. It is striking that the study that led to Buckyball was motivated by interstellar spectra. The question had been raised whether a series of absorption lines, long unidentified, might arise from clusters of carbon atoms. That seems not to be the case. Ironically, however, it was eventually shown that Buckyball and kindred molecules can be extracted in quantity from soot. For synthetic chemists, this has opened up a huge realm of potential molecular structures, built with a form of carbon that has sixty valences rather than just four. By looking at the heavens, scientists came to find, in ashes that had lain under the feet of ancestral cavemen, a Cinderella-like molecule.

Our path next pauses at a computer screen.[22] Three years ago a graduate student at the University of California at San Francisco was trying to see if he could find or design a molecule that would inactivate an enzyme that is crucial for replication of the AIDS virus. The inactivating molecule needs to have the right size and shape to fit into a cleft in the enzyme. The molecule also needs to present a water-repellent torso, since the cleft is hydrophobic. A friend of the student, listening to him lament about what he tried that did not fit, jokingly suggested he try Buckyball. He went back to the computer and found it was just right. Soon a chemist at the University of California at Santa Barbara synthesized a derivative of carbon-sixty which has a pair of wings that make the molecule water soluble despite its hydrophobic torso. Physiologists at Emory University in Georgia then found that, at least in a test tube, this Buckyball derivative does indeed totally inactivate the enzyme that governs replication of the AIDS virus.

This is still a long way from having an actual AIDS drug that can function in living cells. However, the path from Stern's atomic magnets to the AIDS virus, like many others, serves to emphasize that research on fundamental questions inevitably creates a host of unforeseeable opportunities. No funding agency would find plausible a research proposal requesting support of work on supersonic beams or interstellar spectra as an approach to AIDS. But many such historical paths can be traced that celebrate hybridizing discoveries from seemingly unrelated patches of scientific gardens.

## SCIENCE AMONG THE LIBERAL ARTS

In *As You Like It*, set in an imaginary garden, Shakespeare said: "Sweet are the uses of adversity, which, like the toad, ugly and venomous, wears yet a precious jewel in his head." The fruits and flowers of science and rationalism are, I believe, hardy enough to withstand even venomous critiques. Provided we gardeners respond well, the critiques can help us learn how to nurture better our precious flora. Like our gardens, the newly luxuriant growth of "science criticism" in academic fields will be subject to the scrutiny of a coming generation of scholars. With that in mind, I conclude with three comments emphasizing why and how we need to cultivate common ground, shared by science and the liberal arts.

1. The masterworks of natural science and human arts alike testify to the marvelous creative capacities of mankind. Wherever in the world these emerged, they now are a rich legacy cherished by all cultures. This should be a powerful force for amity. Forty years ago, as a beginning graduate student, I heard this theme sounded by Rabi in a fervent lecture to the Harvard physics department:

> How can we hope to obtain wisdom, the wisdom which is meaningful in our own time? [The wisdom of] balanced judgment based on . . . a well-stored mind and feeling heart as expressed in word and action. . . . We certainly cannot attain it as long as the two great branches of human knowledge, the sciences and the humanities, remain separate and even warring disciplines. . . .
>
> To my mind, the value of science or [the humanities] lies not in the subject matter alone, or even in greater part. It lies chiefly in the spirit and living tradition in which these [different] disciplines are pursued. . . . Our problem is to blend these two traditions. . . . The greatest difficulty which stands in the way . . . [is] communication. The nonscientist cannot listen to the scientist with pleasure and understanding.
>
> Only by the fusion of science and the humanities can we hope to reach the wisdom appropriate to our day and generation. The scientists must learn to teach science in the spirit of wisdom, and in the light of the history of human thought and human effort, rather than as the geography of a universe uninhabited by mankind. Our colleagues in the nonscientific faculties must understand that if their teachings ignore the great scientific tradition and its accomplishments, however eloquent and elegant in their words, they will lose meaning for this generation and be barren of fruit. Only with a unified effort . . . can we hope to succeed in discovering a community of thought, which can lead us out of the darkness, and the confusion, which oppress all mankind.[23]

Rabi later discussed these concerns with C. P. Snow, who developed them further in his famous *Two Cultures.*

Today, the cultural gulf seems wider and more ominous. The recently proposed new national standards for teaching American history are blatant evidence.[24] From a process extending over three years and involving many historians, there emerged a 250-page document describing the new curriculum for grades 5–12. It makes no mention of any scientific discovery or any scientist, not even Benjamin Franklin.

2. Yet there is also much to hearten both scientists and humanists responsive to Rabi's appeal. Many efforts are now under way to foster major reforms in science education at all levels. The past forty years have in fact produced an outpouring of articles, books, films, and TV programs that make science accessible to a wide public. Most large cities now have thriving science museums; for instance, last year the Boston Museum of Science drew nearly 1.7 million visitors (compared with 2.2 million for a typical Red Sox season). The Internet soon will greatly enhance such resources.

Excellent science writing in a humanistic mode is now abundant. Much of this stems from the history of science, approaching full bloom as an academic discipline (even if often ignored by other historians). Much also comes directly from scientists writing of "human thoughts and effort," exemplified by Lewis Thomas, Victor Weisskopf, Philip and Phylis Morrison, Gerald Holton, Freeman Dyson, Jeremy Bernstein, Carl Sagan, Jared Diamond, James Trefil, Roald Hoffmann, and many others.[25] Another genre effective in conveying both specific and cultural aspects is fiction incorporating genuine science (in contrast to science fiction);[26] a fine example is Carl Djerrasi's recent novel, *The Bourbaki Gambit.*

A liberal arts education worthy of the tradition surely must aim to integrate science into our general culture. But we cannot expect to accomplish much if we continue to confine science to separate courses. Even a "physics for poets" course reinforces the prevalent view that science belongs solely to its professionals. There are abundant opportunities to include some science in many other subjects—and vice versa! Often such items, used to broaden perspectives, take the form of parables. Here are a few examples, appropriate for various courses.

### *Economics*

In 1880, aluminum metal was more precious than gold. Although aluminum ores are plentiful, the metal is tightly bound to oxygen, and no cheap means had yet been found to free it. Consequently, the Washington Monument is capped with a small pyramid of aluminum; The Danish King Christian X wore a crown of aluminum; and at dinner parties of Napoleon III, plates of gold were placed before ordinary guests, an aluminum plate before the guest of honor. In a chemistry class at Oberlin College, Charles Hall heard his professor assert that "whoever frees aluminum from its oxide, will make his fortune." Hall undertook a long series of experiments, working in a woodshed with a frying pan and jars borrowed from his mother's kitchen. By 1886, he found he could dissolve aluminum oxide in a molten mineral, install electrodes, and run electricity to free the metal.[27] (This is still the way the metal is obtained.) Within a few years a new industry had emerged and the price of aluminum metal had dropped to about fifty cents a pound. The price has remained fairly constant for the past century, but economists seem not to have proposed using aluminum as a monetary standard.

*History*

Polymer synthesis had a crucial role in enabling the United States to enter World War II.[28] The Japanese attack on Pearl Harbor on December 7, 1941, was much less disastrous than the fall of Singapore three months later. That deprived the United States and Britain of virtually their sole supply of rubber. As stated in a report by Bernard Baruch:

> Of all the critical and strategic materials, rubber presents the greatest threat to the success of the Allied cause. . . . If we fail to secure quickly a large new rubber supply, our war effort and our domestic economy both will collapse.[29]

This report launched a crash program to produce synthetic rubber, using a method developed and implemented in Germany. In effect, our enemy had provided the saving key. Some 50 plants were quickly built, enabling military action to start in late 1942. The Allied victory could not have been achieved without this enormous rubber project. However, for it to succeed on such an urgent time scale, we had to have a sufficient corps of polymer chemists and engineers. It was actually the pursuit of artificial silk started 15 years before and culminating in nylon that largely created those vital human resources.

*Political Science*

The eminent historian of science, I. Bernard Cohen, has just published a remarkable book, *Science and the Founding Fathers.* This traces the role of science in the political thought of Franklin, Jefferson, Adams, and Madison, an aspect "conspicuously absent from the usual textbooks." Here I will mention just a bit about Franklin. As an American icon, he is aptly portrayed as wise, witty, and pragmatic in business and politics, but his scientific work is greatly undervalued. He is represented as a chubby, comic fellow flying a kite or a clever tinkerer coming up with useful devices, such as the lightning rod, bifocals, and an efficient stove. As Cohen emphasizes, in the early eighteenth century, electricity was a greater mystery than gravity had been a century earlier. Franklin, almost entirely self-educated and far from any center of learning, solved that mystery. His book, *Experiments and Observations on Electricity*, was a sensation in Europe: it went through five editions in English and was translated into French, German, and Italian. It was read not only by scholars but by the literate public, including the clergy and aristocracy. Many high honors were bestowed on Franklin, including election as a foreign associate of the French Academy; he was the first American elected and the only one for another century.

His scientific stature, comparable to Newton in his day or Einstein in ours, had a significant role in the success of the American Revolution. Franklin's arrival in Paris in 1776 coincided with the signing of a nonaggression pact, which specified that France not aid any rebellion in the British Colonies. His fame as the tamer of lightning helped him gain influence with the French court and immense popularity with the public. That accelerated and enlarged the vital flow of arms and funds supplied by the French to the American rebels. Indeed, Franklin's celebrity was like that of a rock star today; although T-shirts were not yet fashionable, his image appeared everywhere in Paris on medallions and banners, often with the motto *Eripuit celeo fulmen sceptrumque tyrannis* ("he snatched lightning from the sky and the scepter from tyrants"). Louis XVI became so annoyed by this veneration that he gave his favorite mistress a chamber pot with a Franklin medallion at the bottom of the bowl.

*Mathematics*

The account of Thomas Jefferson's science in Cohen's book offers several items that could provide an appealing and instructive context for standard topics. My favorite concerns Jefferson's mathematical skill. Among the extraordinary range of his intellectual interests, Jefferson ranked science as his "supreme delight." He especially esteemed Isaac Newton, studied avidly the *Principia*, and mastered geometry and calculus. When Jefferson undertook to design an improved plow, a key question was the optimum shape of the moldboard, the part that peels back and turns over the sod as the blade cuts through the soil. At first, he pursued this by trial and error, using only simple geometrical calculations. On consulting some mathematical friends, however, he realized that finding the optimum shape required a solution of a calculus problem. He was thus able to treat the plowing of earthly furrows by means of a mathematical tool that Newton had developed to analyze the orbits of heavenly bodies. Jefferson's moldboard design remained the standard for a century; it might open vistas for calculus students today.

3. A liberal arts education aims above all to instill the habit of self-generated questioning and adventurous thinking. In pursuit of that aim, as urged by Bronowski and Rabi, the humanities and sciences should be complementary. Each academic subculture deals with entirely different kinds of questions, develops very different criteria for evaluating answers, and evolves its own language. That handicaps interdisciplinary communication, but need not be allowed to engender disrespect. Science or any subculture can only treat questions amenable to its methods. We should strive for an inclusive intellectual society in which the whole can greatly exceed the sum of its parts.

In conclusion, I return to Peter Medawar, because he stated so well what I believe:

> I am a rationalist, but I'm usually reluctant to declare myself to be so because of the widespread misunderstanding or neglect of the distinction there must always be drawn in philosophic discussion between the sufficient and the necessary. I do not believe, indeed I deem it a comic blunder to believe, that the exercise of reason is *sufficient* to explain our condition and where necessary to remedy it. But I do believe that the exercise of reason is at all times unconditionally *necessary* and that we disregard it at our peril. I and my kind believe that the world can be made a better place to live in . . . believe, indeed, that it has already been made so by an endeavor in which, in spite of shortcomings which I do not conceal, natural science has played an important part. . . . To people of sanguine temperament, the thought that this is so is a source of strength and the energizing force of a just and honorable ambition. . . . The dismay that may be aroused by the inability of science to answer the ultimate first and last questions is really something for which ordinary people have long since worked out for themselves Voltaire's remedy: "We must cultivate our garden."[30]

## NOTES

1 R. Seltzer, "New Television Series May Draw both Kudos and Arrows."
2 M. C. Seidel, "The Forgotten Science"; D. R. Herschbach, "On the Nobel Legacy."
3 F. Ashall, ***Remarkable Discoveries***, pp. 137–151, 160–170.
4 Quoted in R. W. Clark, ***Benjamin Franklin***, p. 89.
5 Personal communication from J. Folkman, derived from ***The International Classification of Diseases***, 9th Revision, 1996.
6 Quoted in P. Medawar, ***The Limits of Science***, p. 40.
7 H. Rosovsky, "Appearing for the Defense Once Again."
8 See, for example, D. W. Curtin, ed., ***The Aesthetic Dimension of Science***; S. Chandrasekhar, ***Truth and Beauty***; R. Hoffmann, ***The Same and Not the Same.***
9 Quoted in M. Kundera, ***The Art of the Novel***, p. 99.
10 T. S. Kuhn, ***The Structure of Scientific Revolutions.***
11 M. Polanyi, ***Personal Knowledge***; ***The Tacit Dimension.***
12 S. Harding, ***The Science Question in Feminism***; "Value-Free Research Is a Delusion."
13 Quoted in J. Mehra, ***The Beat of a Different Drum: The Life and Science of Richard Feynman***, p. 599.
14 E. F. Keller, ***Reflections on Gender and Science***; E. Martin, "The Egg and the Sperm: How Science Has Constructed a Romance Based on Stereotypical Male-Female Roles."
15 T. W. Sadler, ***Langman's Medical Embryology.***
16 J. D. Wilson, "Sex Testing in International Athletics."
17 C. M. Haqq, C.-Y. King, E. Ukiyama, S. Falsafi, T. N. Haqq, P. K. Donahoe & M. A. Weiss, "Molecular Basis of Mammalian Sexual Determination."
18 D. R. Herschbach, "Molecular Dynamics of Elementary Chemical Reactions."
19 Haqq ***et al.***, "Molecular Basis."
20 C. H. Townes, "Quantum Electronics, and Surprise in the Development of Technology."
21 R. F. Curl & R. E. Smalley, "Fullerenes"; H. W. Kroto, "$C_{60}$: The Celestial Sphere That Fell to Earth."
22 R. Baum, "Fullerene Bioactivity."
23 Quoted in J. S. Rigden, ***Rabi, Scientist and Citizen***, pp. 256–257.
24 R. L. Park, "The Danger of Voodoo Science."
25 A sampling of recent vintage: Gerald Holton, ***The Advancement of Science and Its Burdens***; Philip Morrison & Phylis Morrison, ***The Ring of Truth***; James S. Trefil, ***Meditations at Sunset***; C. Tanfold, ***Ben Franklin Stilled the Waves***; R. Levi-Montalcini, ***In Praise of Imperfection***; Victor Weisskopf, ***The Joy of Insight***; S. J. Gould, ***Wonderful Life***; B. Hölldobler and E. O. Wilson, ***Journey to the Ants***; P. Ball, ***Designing the Molecular World.***
26 L. Herschbach, " 'True Clinical Fictions': Medical and Literary Narratives from the Civil War Hospital."
27 N. C. Craig, "Charles Martin Hall—The Young Man, His Mentor, and His Metal."
28 H. F. Mark, ***Giant Molecules.***
29 Quoted in H.F. Mark, ***Giant Molecules***, p. 130.
30 Peter Medawar, ***The Limits of Science.***, pp. 98–99.

## REFERENCES

ASHALL, F. ***Remarkable Discoveries.*** Cambridge: Cambridge University Press, 1994.

BALL, P. ***Designing the Molecular World.*** Princeton, NJ: Princeton University Press, 1994.

BAUM, R. "Fullerene Bioactivity." ***Chemical and Engineering News*** (August 2, 1993): 3.

BRONOWSKI, JACOB. ***Science and Human Values.*** Rev. edit. New York, NY: Harper and Row, 1965.

CHJANDRASEKHAR, S. ***Truth and Beauty.*** Chicago, IL: University of Chicago Press, 1987.

CLARK, R. W. ***Benjamin Franklin.*** New York, NY: Random House, 1983.

COHEN, I. BERNARD. *Science and the Founding Fathers.* New York, NY: W. W. Norton, 1995.
CRAIG, N. C. "Charles Martin Hall—The Young Man, His Mentor, and His Metal." *Journal of Chemical Education* 63 (1986): 557.
CURL, R. F. & R. E. SMALLEY. "Fullerenes." *Scientific American*, October 1991, p. 54.
CURTIN, D. W., ed. *The Aesthetic Dimension of Science.* New York, NY: Philosophical Library, 1983.
DJERRASI, CARL. *The Bourbaki Gambit.* Athens, GA: University of Georgia Press, 1994.
GOULD, S. J. *Wonderful Life.* New York, NY: W. W. Norton, 1993.
GROSS, PAUL R. & NORMAN LEVITT. *Higher Superstition: The Academic Left and Its Quarrels with Science.* Baltimore, MD: Johns Hopkins University Press, 1994.
HAQQ, C. M., C.-Y. KING, E. UKIYAMA, S. FALSAFI, T. N. HAQQ, P. K. DONAHOE & M. A. WEISS. "Molecular Basis of Mammalian Sexual Determination." *Science* 266 (1994): 1494.
HARDING, SANDRA. *The Science Question in Feminism.* Ithaca, NY: Cornell University Press, 1986.
———. "Value-Free Research Is a Delusion." *New York Times*, October 22, 1989, p. E24.
HERSCHBACH, D. R. "Molecular Dynamics of Elementary Chemical Reactions." *Angewandte Chemie International Edition in English* 26 (1987): 1221.
———. "On the Nobel Legacy." *Chemistry and Engineering News* 73 (October 2, 1995): 4.
HERSCHBACH, LISA. "'True Clinical Fictions': Medical and Literary Narratives from the Civil War Hospital." *Culture, Medicine and Psychiatry* 19 (1995): 183.
HOFFMANN, R. *The Same and Not the Same.* New York, NY: Columbia University Press, 1995.
HÖLLDOBLER, B. & E. O. WILSON. *Journey to the Ants.* Cambridge, MA: Harvard University Press, 1994.
HOLTON, GERALD. *The Advancement of Science and Its Burdens.* Cambridge: Cambridge University Press, 1986.
———. *Science and Anti-Science.* Cambridge, MA: Harvard University Press, 1993.
KELLER, E. F. *Reflections on Gender and Science.* New Haven, CT: Yale University Press, 1986.
KROTO, H. W. "$C_{60}$: The Celestial Sphere That Fell to Earth." *Angewandte Chemie International Edition in English* 31 (1992): 111.
KUHN, THOMAS S. *The Structure of Scientific Revolutions.* 2nd ed. Chicago, IL: University of Chicago Press, 1970.
KUNDERA, M. *The Art of the Novel.* New York, NY: Grove Press, 1986.
LEVI-MONTALCINI, R. *In Praise of Imperfection.* New York, NY: Basic Books, 1988.
MARK, H. F. *Giant Molecules.* Life Science Library. New York, NY: Time, 1966.
MARTIN, E. "The Egg and the Sperm: How Science Has Constructed a Romance Based on Stereotypical Male-Female Roles." *Signs: Journal of Women in Culture and Society* (Spring, 1991): 37.
MEDAWAR, P. *The Limits of Science.* Oxford: Oxford University Press, 1984.
MEHRA, J. *The Beat of a Different Drum: The Life and Science of Richard Feynman.* Oxford: Clarendon Press, 1994.
MORRISON, PHILIP & PHYLIS MORRISON. *The Ring of Truth.* New York, NY: Random House, 1987.
PARK, R. L. "The Danger of Voodoo Science." *New York Times*, July 9, 1995, p. E15.
POLANYI, MICHAEL. *Personal Knowledge.* Chicago, IL: University of Chicago Press, 1962.
———. "The Republic of Science: Its Political and Economic Theory." *Minerva* 1 (1962): 54.
———. *The Tacit Dimension.* New York, NY: Doubleday, 1966.
RIGDEN, J. S. *Rabi, Scientist and Citizen.* New York, NY: Basic Books, 1987.
ROSOVSKY, H. "Appearing for the Defense Once Again." *Bulletin of the American Academy of Arts and Sciences* 49, 2 (November 1995): 25–39.
SADLER, T. W. *Langman's Medical Embryology.* 6th ed. London: Williams and Wilkins, 1990.

SEIDEL, M. C. "The Forgotten Science." *Chemistry and Engineering News* 73 (June 26, 1995): 4.

SELTZER, R. "New Television Series May Draw both Kudos and Arrows." *Chemistry and Engineering News* 73 (April 24, 1995): 52.

SNOW, C. P. *The Two Cultures and the Scientific Revolution.* New York, NY: Cambridge University Press, 1962.

TANFOLD, C. *Ben Franklin Stilled the Waves.* Durham, NC: Duke University Press, 1989.

TOWNES, C. H. "Quantum Electronics, and Surprise in the Development of Technology." *Science* 159 (1968): 699.

TREFIL, JAMES S. *Meditations at Sunset.* New York, NY: Collier Books, 1987.

WEISSKOPF, VICTOR. *The Joy of Insight.* New York, NY: Basic Books, 1991.

WILSON, J. D. "Sex Testing in International Athletics." *Journal of the American Medical Association* 267 (1992): 853.

# CONDUCT AND MISCONDUCT IN SCIENCE

DAVID GOODSTEIN

My CAREER IN SCIENTIFIC FRAUD began some years ago when, as Caltech's vice provost, I became aware that federal regulations would soon make it necessary for us, for the first time ever, to have in place formal rules about what to do if the unthinkable were to happen on our own campus, the very inner sanctum of pure science. Since then it has virtually become an academic subspecialty for me. I have given lectures, written articles,[1] and taught courses about it. I have written the regulations, seen them adopted by Caltech (and copied by other universities), and, much to my dismay, seen them put into action in a high-profile case at Caltech. During that case I had the remarkable experience of seeing a skilled lawyer, with a copy of my regulations highlighted and underlined in four colors, guide us in following every word I had written, whether I meant it or not. Through all of that, I have learned a few things about conduct and misconduct in science that I would like to share with you.

Let me begin by stating right up front a few of the things I have come to believe. Outright fraud in science is a special kind of transgression, different from civil fraud. When it does occur, it is almost always found in the biomedical sciences, never in fields like physics or astronomy or geology, although other kinds of misconduct do occur in these other fields. Science is self-correcting, in the sense that a falsehood injected into the body of scientific knowledge will eventually be discovered and rejected; but that does not protect us against fraud, because injecting falsehoods into the body of science is never the purpose of those who perpetrate fraud. That is why science needs active measures to protect itself. Unfortunately, the government has so far made a mess of trying to do that job. Part of the reason government agencies have performed so poorly in this arena is that they have mistakenly tried to obscure the important distinction between real fraud and lesser forms of scientific misconduct. I also believe that fraud and other forms of serious misconduct have been and still are quite rare in science, but there are reasons to fear that they may become less rare in the future. Finally, I believe we scientists are responsible for complicity in presenting to the public a false image of how science works—one that can sometimes make normal behavior by scientists appear to be guilty. Let me try to explain what I mean by all of this.

Fraud means serious misconduct with intent to deceive. Intent to deceive is the very antithesis of ethical behavior in science. When you read a scientific paper, you are free to agree or disagree with its conclusions, but you must always be confident that you can trust its account of the procedures that were used and the results produced by those procedures. There are, to be sure, minor deceptions in virtually all scientific papers, as there are in all other aspects of human life. For example, scientific papers typically describe investigations as they logically should have been done rather than as they actually were done. False steps, blind alleys, and outright mistakes are usually omitted once the results are in and the whole experiment can be seen in proper perspective. Also, the list of authors may not reveal who deserves most of the credit (or blame) for the work. Behavior of these kinds may or may not be correct or laudable, but they do not amount to fraud. Real fraud occurs only if the procedures needed to replicate the results of the work or the results themselves are in some way knowingly misrepresented.

This view of scientific fraud differs from civil fraud as described in tort law in that in civil fraud there must be a plaintiff, who brings the case to court and who must be able to prove that the misrepresentation was believed and led to actual damages. By contrast, if a scientist makes a serious misrepresentation in a scientific presentation, knowing that it is false or recklessly disregarding whether it is false, then, virtually all scientists would agree, scientific fraud has been committed. It is unnecessary to prove that anyone in particular believed the statement or was damaged by believing it. In fact, it makes no difference at all whether the conclusions reached by the scientist are correct or not. In this stern view of scientific fraud, all that matters is whether procedures and results are reported honestly or not.

This kind of misbehavior seems to be restricted largely to the biomedical and closely related sciences. A study by sociologist Pat Woolf[2] of some twenty-six cases that surfaced one way or another between 1980 and 1986 revealed that twenty-one came from biomedical science, two from chemistry and biochemistry, one from physiology, and the other two from psychology. I do not know of any more recent studies of this kind, but one cannot help noticing that the Office of Research Integrity of the Public Health Service, which investigates misconduct in research supported by the National Institutes of Health—in other words, biomedical research—seems constantly to be embroiled in controversy; while the Inspector General of the National Science Foundation, which supports all of the sciences, including biology, conducts its business in relative anonymity, unmolested by serious attention from the press.[3]

There are undoubtably many reasons for this curious state of affairs. For example, many of these cases involve M.D.'s, rather than Ph.D.'s, who are trained to do research. To an M.D. the welfare of the patient may be more important than scientific truth. In a recent case, a physician in Montreal was found to have falsified the records of participants in a large-scale breast cancer study. Asked why he did it, he said it was in order to get better medical care for his patients.[4] However, the larger number of cases arise from more self-interested motives.

In the cases of scientific fraud that I have looked at, three motives, or risk factors, have always been present. In all cases, the perpetrators

1. were under career pressure;
2. knew, or thought they knew, what the answer would turn out to be if they went to all the trouble of doing the work properly; and
3. were working in a field where individual experiments are not expected to be precisely reproducible.

It is by no means true that fraud always occurs when these three factors are present; quite the opposite; they are often present, and fraud is quite rare. But they do seem to be present whenever fraud occurs. Let us consider them one at a time.

*Career Pressure.* This is included because it is clearly a motivating factor, but it does not provide any distinctions. All scientists, at all levels from fame to obscurity, are pretty much always under career pressure. On the other hand, simple monetary gain is seldom, if ever, a factor in scientific fraud.

*Knowing the Answer.* If we defined scientific fraud to mean knowingly inserting an untruth into the body of scientific knowledge, it would be essentially nonexistent, and of little concern in any case because science is self-correcting. Scientific fraud is always a transgression against the *methods* of science, never purposely against the body of knowledge. Perpetrators always think they know how the experiment would come out if it were done properly, and decide it is not necessary to go to all the trouble of doing it properly. The most obvious seeming counterexample to this assertion is Piltdown Man, a human skull and ape jaw planted in a gravel pit in England around 1908. If ever a fraudulent physical artifact was planted in the scientific record, this was it. Yet it is quite possible that the perpetrator was only trying to help along what was known or thought to be the truth. Prehistoric remains had been discovered in France and Germany, and there were even rumors of findings in Africa. Surely human life could not have started in those uncivilized places. And, as it turned out, the artifact was rejected by the body of scientific knowledge. Long before modern dating methods showed it to be a hoax in 1954, growing evidence that our ancestors had ape skulls and human jaws made Piltdown Man an embarrassment at the fringes of anthropology.

*Reproducibility.* In reality, scientific experiments are seldom repeated by others. When a wrong result is found out, it is almost always because new work based on the wrong result does not proceed as expected. Nevertheless, the belief that someone else can repeat an experiment and get the same result can be a powerful deterrent to cheating. This appears to be the chief difference between biology and the other sciences. Biological variability—the fact that the same procedure performed on two organisms as nearly identical as possible is not expected to give exactly the same result—may provide some apparent cover for a biologist who is tempted to cheat. This last point, I think, explains why scientific fraud is found mainly in the biomedical area.

Federal agencies, particularly the Public Health Service (PHS, parent of the National Institutes of Health, NIH) and the National Science Foundation (NSF), take quite a different view from that outlined above of what constitutes serious misconduct in science. The word *fraud* is never used at all, because fraud means intent to deceive, and the agencies do not want investigations hung up on issues of motive. Instead they (both PHS and NSF) define serious misconduct to be

> . . . fabrication, falsification, plagiarism, or *other practices that seriously deviate from those that are commonly accepted within the scientific community* for proposing, conducting and reporting research.[5]

Controversy has swirled around this statement ever since it was first proposed in 1988 and issued as a "final rule" in 1990. No one takes issue with "fabrication, falsification (and) plagiarism," (ffp, mnemonic, "frequent flier plan" for the cognoscenti). The controversial part is the catch-all phrase "practices that seriously deviate from those commonly accepted. . . ." To the agencies, that phrase is a mandate needed to enable them to carry out their mission as stewards of public funds. To many scientists and other observers it raises the horrifying specter of the government forcing scientists into some preconceived mold of orthodox thought. To me it seems poor public policy to create a government bureaucracy mandated to root out transgressions that are not specified in advance. The NSF Inspector General's Office has argued, for example, that the catch-all phrase was needed for it to be able to deal with a principal investigator (government phrase for one of their grantees) who committed sexual harassment against female students during an NSF-sponsored archeological dig.[6] Surely that was serious misconduct, it deviated seriously from practices commonly accepted within the scientific community, and the NSF had a duty to do something about it. Nevertheless, there are other laws governing that kind of misbehavior. It was not scientific misconduct.

Even the uncontroversial "ffp" part of the final rule is different from the view of scientific fraud outlined above, and also different from the definition of research fraud given in the Caltech rules. The federal rule includes plagiarism, and the Caltech rules, not content with plagiarism alone, specifies ". . . faking data, plagiarism and misappropriation of ideas." The idea here is that "faking data" includes both "fabrication and falsification," and that one can misappropriate ideas in science without committing plagiarism, a rather technical term that means copying a certain number of consecutive words or consecutive musical notes or the like without proper attribution.

Early on, when I had just begun to think and write about this subject, one of my very distinguished Caltech colleagues came to see me, and said, in effect, "Look, son, let me explain to you how all this works. Scientific truth is the coin of the realm. Faking data means counterfeiting the coin. That's a serious crime. All this other stuff you talk about, plagiarism, authorship problems, and so on, that's just a matter of who's been handling the coins. That's much less important." In light of this "coin of the realm" view, one can easily summarize everything that has been said above. Science is a special currency that needs to be protected. I have outlined a strict view, in which only counterfeiting the coin is real scientific fraud. The federal agencies see themselves instead as central banks, charged with overseeing all transactions and catching all wrongdoers, not merely counterfeiters. Thus, they wish to protect not only the integrity of the "monetary" system, but all other forms of behavior by scientists as well. Plagiarism and misappropriation of ideas are intermediate cases because, although they do not threaten the integrity of the coin itself, they do deal with the orderly assigning of credit for scientific discovery, an important motivating factor in scientific progress. One would think that actual plagiarism would be rare in science, where the precise words used are less important than the substantive content; but it turns out that outright plagiarism does show up, especially in the frantic cut-and-paste procedures that are used in preparing grant proposals for federal funds (for currency of the real kind, not the metaphorical kind).

Federal regulations (both PHS and NSF) call for each university to investigate any case that turns up before turning it over to the sponsoring agencies (that is why universities are required to have regulations of their own). The Caltech regulations call for a scientific investigation rather than a judicial proceeding. There is no confrontation of the accuser, cross-examination of witnesses, and so on. In fact, private attorneys are strongly discouraged. Nevertheless, sufficient safeguards are built in to protect both accused and accuser so that the courts have many times ruled that this sort of proceeding does have the requisite degree of fairness or, to use the inappropriate constitutional term, of "due process." In limited experience up to now, these proceedings have been more successful than those of the federal agencies, where the Office of Research Integrity of PHS has seen virtually all of its major decisions overturned on appeal, and where the Inspector General of NSF has had little to do in this area.

Although I have said I believe fraud in science to be quite rare, the popular press, certain congressmen, and some others inside and outside science sometimes seem to believe otherwise. These suspicions were given a boost recently by an article published in the Sigma Xi journal *American Scientist*[7] and widely noted in the press. Based on questionnaires sent to 2000 graduate students and 2000 faculty in chemistry, civil engineering, microbiology, and sociology, it concluded that ". . . such problems are more pervasive than many insiders believe." A more thoughtful analysis, however, might lead to a different conclusion. To indicate what I mean, let us leave aside for the moment the perceived behavior of students (who, after all, may not yet have completed their ethical as well as their scientific educations) and lesser misdeeds ("keeping poor research records," "violation of government regulations") and get to the core of the matter: how often were faculty members perceived by fellow faculty members or by students to have committed "coin of the realm"–type misbehavior (defined in the article as "falsifying or 'cooking' research data")? The answer is that six percent of the faculty and eight percent of the students responding to the survey reported having seen direct evidence of such misbehavior once or twice (none more than once or twice, and ninety-four percent and ninety-one percent, respectively, none at all). What are we to make of such figures?

Needless to say, the response to the questionnaire was less than universal, and those who responded may be imperfectly representative of the whole group. Furthermore, of the six to eight percent who think they saw instances of real scientific fraud, only a fraction will be sufficiently confident or sufficiently disturbed to bring charges before an appropriate authority. As I can attest on the basis of first-hand experience at Caltech, of the charges that do come forth to an appropriate authority only a fraction (one-third or one-fourth) survive the initial discreet inquiry phase and proceed to full investigation (the inquiry usually finds that the perceived misconduct was a misunderstanding that can be resolved to everyone's satisfaction, including the "whistleblower's"). Finally, of those that go to investigation, not all are found guilty. Thus, the correct conclusion to be drawn from the "data" (if that is what they are) is that real fraud in science is exceedingly rare. However, I think a more sensible conclusion is that this kind of study is not very valuable, and that we do not know much about the incidence of fraud in science.

Whatever the situation is now and has been in the past, it seems likely to change for the worse in the future. Throughout most of its recent history, science was constrained only by the limits of imagination and creativity of its participants. In the past couple of decades that state of affairs has changed dramatically. Science is now constrained primarily by the number of research posts, and the amount of research funds available. What had previously always been a purely intellectual competition has now become an intense competition for scarce resources. This change, which is permanent and irreversible, is likely to have an undesirable effect in the long run on ethical behavior among scientists. Instances of scientific fraud are almost sure to become more common, but so are other forms of scientific misbehavior. For example, the institution of peer review is now in critical danger.

Peer review is used by scientific journals to decide what to publish and by granting agencies to decide what research to support. Obviously, sound decisions on what to publish and what research to support are crucially important to the proper functioning of science. Journal editors usually send manuscripts submitted to them to referees, who remain anonymous to the authors of the manuscript. Funding agencies sometimes do the same, especially for small projects, and sometimes instead assemble panels of referees to judge proposals for large projects.

Peer review is quite a good way to identify valid science. It was wonderfully well suited to an earlier era, when progress in science was limited only by the number of good ideas available. Peer review is not at all well suited, however, to adjudicate an intense competition for scarce resources such as research funds or pages in prestigious journals. The reason is obvious enough. The referee, who is always among the few genuine experts in the field, has an obvious conflict of interest. It would take impossibly high ethical standards for referees to fail to use their privileged anonymity to their own advantage, but as time goes on more and more referees have their ethical standards eroded by receiving unfair reviews when they are authors. Thus the whole system of peer review is in peril.

Editors of scientific journals and program officers at the funding agencies have the most to gain from peer review, and they steadfastly refuse to believe that anything might be wrong with the system. Their jobs are made easier because they have never had to take responsibility for decisions. They are also never called to account for their choice of referees, who in any case always have the proper credentials. Since the referees perform a professional service, almost always without pay, the primary responsibility of the editor or program officer is to protect the referee. Thus, referees are never called to account for what they write in their reviews. As a result, referees are able, with relative impunity, to delay or deny funding or publication to their rivals. When misconduct of this kind occurs, it is the referee who is guilty; but it is the editors and program officers who are responsible for perpetuating a corrupt system that makes misconduct almost inevitable.

This is the kind of misconduct that is, I fear, rampant in all fields of science, not only biomedical science. Recently, as part of a talk to a large audience of mostly young researchers at an extremely prestigious university, I outlined this analysis of the crisis of peer review. The moderator, a famous senior scientist, was incredulous. He asked the audience how many disagreed with my heresy. No one responded. Then he asked how many agreed. Every hand in the house went up. Many of us in my generation wish to believe that nothing important has changed in the way we conduct the business of doing science. We are wrong. Business as usual is no longer a real option for how we conduct the enterprise of science.

Finally, I think we scientists are guilty of promoting, or at least tolerating, a false popular image of ourselves that may be flattering but that, in the long run, leads to real difficulties when the public finds out that our behavior does not match that image. I like to call it the Myth of the Noble Scientist. It arises, I think out of the long-discredited Baconian view of the scientist as disinterested seeker of the truth, gathering facts with a mind cleansed of prejudices and preconceptions. Thus, the ideal scientist would be more honest than ordinary mortals, certainly immune to such common human failings as pride or personal ambition. When it turns out, as invariably it does, that scientists are not at all like that, the public that we have misled may react with understandable anger or disappointment.

The fact is that scientists are usually rigorously honest about the things that really matter to them, such as accurate reporting of procedures and data. In other arenas, such as disputes over priority or credit, they tend to behave like the ordinary mortals they are. Furthermore, scientists are not disinterested truth seekers; they are more like players in an intense, winner-take-all competition for scientific prestige, or perhaps merchants in a no-holds-barred marketplace of ideas. The sooner we learn to admit to those facts, and to distinguish carefully between serious scientific misconduct and common human conduct by scientists, the better off we all will be.

## NOTES

1 David L. Goodstein, "Scientific Fraud"; "Science Fraud."

2 Patricia K. Woolf, *Project on Science Fraud and Misconduct, Report on Workshop Number One*, p. 37.

3 The Office of the Inspector General of the National Science Foundation recounts its activities twice each year in its Semiannual Report to the Congress.

4 Reported, e.g., *New York Times*, April 4 and 12, 1994; *Time*, March 28, 1994.
5 See, for example, the Office of Inspector General of the National Science Foundation, Semiannual Report to the Congress, number 9, April 1, 1993–September 30, 1993, p. 22, for a discussion of this definition.
6 D. E. Buzzelli, "The Definition of Misconduct in Science: A View from the NSF."
7 J. P. Swazey, M. S. Anderson & K. S. Lewis, "Ethical Problems in Academic Research."

## REFERENCES

ALTMAN, LAWRENCE K. "Fall of a Man Pivotal in Breast Cancer Research." *New York Times*, April 4, 1994.

———. "Scientist Ousted from Cancer Study Declines to Testify to House Panel." *New York Times*, April 12, 1994.

BUZZELLI, D. E. "The Definition of Misconduct in Science: A View from the NSF." *Science* (January 29, 1993): 584, 647–648.

GOODSTEIN, DAVID L. "Scientific Fraud."*Engineering and Science* (Winter 1991): 10–19.

———. "Science Fraud." *American Scholar* 60 (Autumn 1991): 505–515.

GORMAN, CHRISTINE. "Breast Cancer: A Diagnosis of Deceit." *Time Magazine*, March 28, 1994.

OFFICE OF INSPECTOR GENERAL OF THE NATIONAL SCIENCE FOUNDATION. Semiannual Report to the Congress, number 9, April 1, 1993–September 30, 1993.

SWAZEY, J. P., M. S. ANDERSON & K. S. LEWIS. "Ethical Problems in Academic Research." *American Scientist* 81 (1993): 542–553.

WOOLF, PATRICIA K. *Project on Science Fraud and Misconduct, Report on Workshop Number One.* Washington, DC: American Association for the Advancement of Science, 1988.

# MATHEMATICS AS THE STEPCHILD OF CONTEMPORARY CULTURE

NORMAN LEVITT

MY TITLE WILL DOUBTLESS strike some as impertinent, others as paranoid. Mathematicians in particular will hardly be likely to view themselves as pariahs skulking at the margins of academic and intellectual life. Mathematical research, partly because it is incredibly cheap by the standards of experimental physics, chemistry, or biology, tends to be resistant to the ups and downs of funding policy. It flourishes with exponential vigor. Every year, hundreds of thousands of papers appear in thousands of journals, and journals themselves proliferate at the rate of dozens per year. Math libraries groan with the burden of incoming publications. Math department bulletin boards are scarcely able to hold all the conference announcements that arrive with "please post" requests. In that sense, mathematics and related fields can hardly be thought of as the starveling outcasts of intellectual life.

And yet, I will argue, mathematics itself occupies an ambiguous place in the larger world of ideas. It is as much resented as admired. Even worse, there is a widespread strategy, among humanist intellectuals, of cheerfully conceding that one is mystified and paralyzed by mathematics, with the clear corollary that this defect, like lack of perfect pitch, is to be regretted mildly or not at all. Implicitly, one categorizes mathematics, and those deeply involved with its ideas and methods, as an arcane cult, mostly harmless, sometimes useful, but, all in all, inexplicable and extraneous to one's deepest interests, whatever those might be. Occasionally, the affectation of disinterest falls away, and comments on mathematics and mathematicians become positively envenomed. Yet even discounting this occasional spitefulness, and concentrating solely on the rather more widespread combination of ignorance and indifference, we find a deep fault line in the culture of those who take ideas seriously, or pretend to.

This is the theme I want to explore. At the same time, I propose to understand the gulf between *mathematics*—and, as will be seen, I am content to use that term rather broadly—and *humanistic culture* as something rather novel in Western intellectual life, if one takes the long view. This exclusion of mathematics from the general traffic in ideas is not only recent, but goes against the grain of our deepest traditions. A recent piece by the cosmologist David N. Shramm[2] puts it well:

> The beauty of the mathematics and physics that describe nature is repeatedly noted by scientists and mathematicians. One's impression is that in the past it was not uncommon for intellectual leaders in the humanities to appreciate it as well. Mr. Osserman quotes Voltaire as saying a great mathematician has at least as much imagination as a great poet. He could find similar testimony from any number of other generally cultured people in the 18th and 19th centuries.

Indeed, he could. "Euclid alone has looked on beauty bare," says Edna St. Vincent Millay, a sentiment hardly less enthusiastic than Bertrand Russell's own dictum:[3]

> Mathematics possesses not only truth, but supreme beauty—a beauty cold and austere, like that of sculpture, without appeal to any part of our weaker nature, sublimely pure and capable of a stern perfection such as only the greatest art can show.

Yet, much nearer to our own day, we find the distinguished humanist Jacques Barzun dismissing Russell's very words as "sheer affectation" and "nonsense."[4] Millay may be seen as speaking for a tradition nearing its end, that end having been heralded by William Blake:

> May God keep us
> From single vision and Newton's sleep.

Here, I cannot help pointing out that it is Blake who seems to be retreating into "single vision." Nonetheless, it is Blake to whom Barzun recurs, citing the quoted line with evident enthusiasm.[5]

The historical links, within the Western tradition, between mathematical competence and the general ability to think deeply and effectively go back at least as far as the ancient Athenians, for whom mathematical training, specifically in the rigorous technique of axiomatic geometry, was the necessary precursor to any significant philosophical speculation. Nearly two millennia on, the "liberal arts" of the Schoolmen were: grammar, rhetoric, logic, geometry, arithmetic, music, and astronomy. (Notwithstanding, my everyday dictionary has it that "liberal arts" are to be defined as "the course of study, including literature, philosophy, history, etc., distinguished from professional and technical subjects." In other words, to be a liberal arts major nowadays is to declare oneself hell-bent on avoiding anything mathematical in the curriculum.) The great English composer of the 15th century, John Dunstable, was famed throughout Europe as a mathematician simply because, in the high culture of his day, musical composition, *ars combinatoria*, *was* a branch of mathematics. A modern student of his music reminds us that "The composer's task could therefore be described as reflecting the ordered perfection of the universe in musical structures that obeyed the same numerical principles."[6] From Walter of Chatillion, the 12th-century poet and scholar, we have:[7]

Creatori serviunt omnia subjecta,
sub mensure, numero, pondere perfecta.
Ad invisibilia, per haec intellecta
sursum trahit hominem ratio directa

(The Creator orders all matters
by measure, number, just weight
Through this intellect, under right reason
He raises Man to the Invisible Realm)

We might reflect as well that the finest of Middle English poets, Chaucer, was an astronomer of some skill (as was Omar Khayyam, at the other end of the great arc of Euro-Islamic civilization). As for medieval architecture, "With few exceptions the Gothic builders have been tight-lipped about the symbolic significance of their projects, but they are unanimous in paying tribute to *geometry* as the basis of their art.[8] (Contrast Barzun:[9]

> A general recognition that science is at the other extreme from art and cosmology [sic![10]]; that at least a double view of nature is possible; that much of life is to be dealt with anthropomorphically, because *anthropos* is the customer to be served—this might begin to clear the ground of errors and stumbling blocks.)

It might be noted that, much as the medieval scholar doted on mathematics, he viewed it as rather a static and settled matter, a body of law that had been set down by the ancient pagan philosophers, much as Scripture had been set down, by divinely inspired prophets, Evangelists, and Apostles. In that sense, mathematics was something to be displayed or exhibited, rather than put to use in the modern fashion. Inevitably, though, an architect—a composer as well—would have found himself reckoning and computing by means of arithmetic and geometry, in addition to contemplating their perfection. Nonetheless, the view of "mathematics" as an open-ended, creative process, capable of indefinite expansion and of being applied to detailed explication of a vast range of natural phenomena, was not that of the European Middle Ages.

The mathematics of the 17th century is, to indulge in gross understatement, something else again. Partly through the inspiration of Arab and Persian mathematicians and astronomers, partly through what must be called (though to say so is highly unfashionable these days) the singular genius of the culture of Western Europe in that age, mathematics burgeoned with a vigor that is, even today, astounding to contemplate. What the cultural roots of this phenomenon were, how it might have been generated by the *Zeitgeist* or the deep structure of socioeconomic imperatives, I shall not presume to say, for the simple reason that the whole matter seems deeply obscure to me. Such analyses along these lines as I have seen strike me as shallow and unconvincing, at best *post-hoc* fables to be admired, if at all, more for ingenuity than persuasiveness. We do not, however, need any solid theory of origins, we don't even need to believe such a theory is accessible, to note just how deeply and firmly the ideal of mathematical virtuosity was embedded in the tissue of 17th-century intellectual life. It was an era in which there was a near-universal supposition that serious thinking of any kind, participation in the debates and contentions that roiled the landscape, necessitated a panoply of mental equipment within which mathematical ability took a prominent place, possibly the leading role. Of course, there were other requirements as well. One had to be a decent Latinist, at the least, and to be well acquainted with a range of thinkers both ancient and contemporary. But beyond these, there was the further assumption that one was as much at home in Euclid as in the Latin poets and the Gospel. Even more, since mathematics was now a feverishly active science, and not just a received text, there was a presumption that one could follow arcane disputes in geometry, algebra, and physics.

We find ample documentary confirmation of this widespread mathematical competence. Pepys's diary reveals a man who read Boyle's physics simply for his pleasure. The lively, wildly disjointed memoranda of John Aubrey comprise another highly amusing source. Though of modest mathematical ability in his own right, Aubrey took the subject seriously and was personally acquainted with most of the major English mathematical thinkers of his time, Hooke, Wallis, and Newton among them. In the anthology of his *Brief Lives*[11] that I consulted, I found about twenty entries devoted to eminent mathematicians, with about as many references to others scattered through the text. On the basis of his antiquarian interests, Aubrey was a founding member of the Royal Society, as were his fellow-antiquarian Ashmole and the poet John Dryden. It was expected of all these gentlemen that they could follow with unfeigned interest discussions and disputes in which intense mathematical reasoning played a central role. There were, it is true, some wryly comical counterpoints to this phenomenon. The political philosopher Thomas Hobbes suffered greatly from "delusions of adequacy" in the mathematical realm, and his attempts at circle squaring (quite accurately denounced by Huygens as "absurd, childish nonsense") led to a long series of fractious disputes wherein Hobbes, thoroughly outclassed in matters mathematical by opponents and onlookers alike, made a rather pathetic spectacle of himself.[12]

Equally interesting testimony comes from the journal *Acta Eruditorum*, published in Leipzig, which first appeared in 1683. The closest contemporary analog (in English) to this serial is the *New York Review of Books*, or possibly, the *Times Literary Supplement. Acta Eruditorum*, a sort of pan-European intellectual-journal-of-record, was principally composed of careful, lengthy accounts of significant new books—accounts that were extensive summaries rather than mere reviews. Occasional original publications were included as well. Topics ranged from theology to jurisprudence to botany to narratives of exploration. But mathematics and mathematical physics were also well represented. We find tables of new astronomical observations, Hooke's analysis of a proposed perpetual motion machine, and papers by Leibniz in pure and applied mathematics. There is also a long, detailed summary of Newton's *Principia*. It's possible to wonder how many subscribers (aside from Leibniz) read the *Acta* from cover to cover, but the presumption that there were such readers hardly requires an extraordinary leap of faith.

Mathematical ability was a vital component of the typical scholar's intellectual arsenal well into the 18th century. The celebrated naturalist Buffon, for instance, made the first translation of Newton's *Method of Fluxions* (i.e., calculus) into French. Voltaire was conversant with geometry and physics, as was Diderot. Another Encyclopedist, d'Alembert, was a mathematician of the first rank. Hume and Kant were at home in mathematical discourse, and in some respects initiated the inquiry into the nature of mathematical knowledge that was to engage so many 19th- and 20th-century mathematicians.

Notwithstanding this long history, the 19th century witnessed a growing disjunction between what was regarded as general scholarly competence and the particular talent for doing and understanding mathematics. This development was, perhaps, foreshadowed by certain strains within 18th-century thought itself. Even Diderot anticipated a "physics" of the future wherein mathematical thinking would become marginal or irrelevant. And Berkeley challenged the growing influence of a mechanistic, essentially atheistic worldview on the basis of a conceptually shrewd but ideologically biased examination of the foundations of the new mathematics, as they were then widely understood.[13] But, whatever the underlying causal pattern, the arrival of the Romantic style in art and thought seems to coincide with the point at which humanistic and scientific thought began seriously to split apart in Western culture. It almost goes without saying that one must use a diffuse term like *Romantic* only with the greatest caution and with full knowledge of its fragility as an analytic category. Nonetheless, it almost takes willful blindness not to see in some primal Romantic figures the very personification of the reaction against rational, analytically reductive thought, against science, certainly, but particularly against the formalization of science that mathematics embodies. Both Blake and Goethe, polar opposites in so many respects, loudly despised a cosmology the deepest insights of which were embedded in what to them was an alien, disagreeable tissue of computation and mathematical abstraction. Both men constructed private demonologies centered on the figure of Newton; and, it is interesting to note, in both cases Sir Isaac was assigned his ungrateful role after the poet had made an intense but unsuccessful attempt to master Newtonian physics and the mathematical methods—those of Newton's calculus—central to it.

Whatever one thinks of Romantic poets, it is undeniable that the 19th century witnessed the growth of increasingly impermeable barriers between humanistic culture and the world of scientific ideas, especially highly mathematical ones. I do not believe that there is any easy, reductive explanation for this alienation of philosophers, historians, and men of letters from mathematical thought. It strikes me, however, that there are certain obvious factors that no thorough account can omit. First of all, the growing difficulty, both technical and conceptual, of mathematics itself played an important role. Calculus and mathematical physics, at any reasonable level of sophistication, are harder than basic geometry and elementary algebra.[14] The savant of the 17th century needed to know his Euclid and might perhaps be expected to follow with a modicum of understanding some of the newer ideas in algebra and the evaluation of infinite series. By the early 18th century, however, a sophisticated student of mathematics and physics had to know not only Newton, but the further elaborations that had emerged with bewildering swiftness in

Newton's wake from the fertile minds of the Bernoullis, Hamilton, Taylor, Euler, Lagrange, Laplace, and, of course, Gauss. Those readers who know mathematics will appreciate the vast intensification and deepening of mathematical thought that this involved. The chief consequence was that whereas until Newton's era the intelligent and well-educated layman could follow the progress of mathematical work and even comment upon it astutely, a hundred years later such capabilities were reserved to those with substantial mathematical talent and extensive mathematical training.

Another factor lay in the emergence of a professional class of technocrats, and an educational system designed to nurture it. Engineering became a learned profession with its own schools and institutes, usually distinct from those of traditional humanistic learning. Mathematics was increasingly adjoined to these studies, where its methods were indispensable. There are, it is true, some curious exceptions to this. In England, at Oxford and Cambridge, mathematics remained the special province of aspiring clergy of the Church of England (C. L. Dodgson—Lewis Carrol—being the best-known example). But, generally speaking, mathematics increasingly came to be thought of as the special province of engineers and builders. To a corresponding extent, it lost its longstanding associations with the worlds of classical learning and systematic philosophy. Naturally, the associated exact sciences—physics, astronomy, and, eventually, chemistry—came to have these new affiliations as well. Correspondingly, humane studies generally lost touch with the great corpus of mathematical ideas, even those ancient ones that had once constituted part of humanism's root system.

Matthew Arnold's essay *Literature and Science*[15] provides clear evidence of this. Indignantly refuting Huxley's insistence that science ought to be central to education, Arnold allows that an educated man should have some knowledge of the most significant scientific facts and theories. But he is clearly uneasy about the demand that *details* of science should be mastered. He seems to dismiss the possibility that education might be designed to provide the nonspecialist with some insight into the method and internal structure of the scientific disciplines. On mathematics itself, he has this to say:

> My friend, Professor Sylvester, who is one of the first mathematicians in the world, holds transcendental doctrines as to the virtue of mathematics, but those doctrines are not for common men. In the very Senate House and heart of our English Cambridge I once ventured, though not without an apology for my profaneness, to hazard the opinion that for the majority of mankind, a little mathematics, even, goes a long way. Of course, this is quite consistent with their being of immense importance as an instrument to something else; but it is the few who have the aptitude for thus using them, not the bulk of mankind.

Now Arnold, as we know, famously and admirably insists that esthetics and ethics are "for the bulk of mankind," and that the finest discernment in these matters is the goal of education, not just for the elite but for the mass of humanity, in whom the highest ideals are to be evoked. But it is pretty clear that in regard to mathematics, he much prefers to be classed with those "common men" who are free to live in blessed ignorance of deep mathematical notions. I venture to say that Barzun's book, cited above, is very much in the Arnoldian tradition, although possibly with an added element of resentment at the central, ineluctable role of science in 20th-century thought and society. With variations in emphasis and tone, this has been pretty much the standard line of humanistic scholars for decades. Occasionally, it is inflected with outright venom, as displayed by the jeremiad of the fictitious but representative Professor Nightingale in Tom Stoppard's recent play *Arcadia.*[16]

What, however, ought we to make of this great divide in intellectual life? Even if we accept the proposition, declared a generation ago by C. P. Snow, that humanists ought to shape up and learn at least the rudiments of contemporary science, does it then follow that in the process they must acclimate themselves to the intricacies of technical mathematics? Does their failure to do so have any grave consequences, or is it merely the corollary of a diverse, complicated, rather messy world of ideas where division of labor is inevitable and attempts to remove the supposed blinders from specialists' eyes will do more harm than good, provoking intense resentment in the process? In other words should we mathematicians simply leave well enough alone?

My position is that the situation is not benign, although its ill effects are long-term and subtle, rarely manifesting themselves acutely. Just as ideas have consequences, so does a hole in the realm of ideas. The consequences may build slowly and circuitously, and by the time they are discernible, their tracks may be difficult to read; but I think they are nonetheless real and unwelcome.

I am not trying to mandate the intense study of modern mathematics by present and future humanities scholars. There might be some mathematicians whose enthusiasm would be ignited by such a project, but I am not among them; and, frankly, I have never met anyone who is. I do not believe in Leibniz's dream of a universal discursive calculus that could resolve all questions in every subject of interest to humanity. It makes an amusing metaphor in the form, say, of Hesse's *Glass-Bead Game*; and it survives in attenuated form as the thesis, appealing to most computer scientists, that intelligence itself is computational in the technical sense. But even within mathematics, the death of the Hilbert program at the hands of Gödel's incompleteness theorem annihilated the Leibnizian conceit that knowledge could be at once exhaustive and reducible to an algorithmic procedure.

At a more quotidian level, it is clear that, while there may be some clever specialists in history or archaeology or even literary study who will be able to exploit mathematical technique of some sort, they will be rare exceptions. Most humanists will never need to know very much about 18th-century, let alone 20th-century, mathematics in order to get on with their work. Nor will mathematicians have any particular head start should they wish to do some serious study of the humanities.

I want to argue, therefore, on a more general level than that of academic specialization and its singular product, the publishable paper. In that regard, let me say that I want to use the term "intellectual" in a more general sense than "professional academic." I want to define it, risking vagueness, to mean, simply, someone who takes abstract ideas seriously and is willing to put in hard (and, to some degree, disinterested) work in order to clarify and evaluate them. By the same token, and in the face of four decades of my own rather snobbish habits, I want to use the term "mathematician" in a somewhat wider sense than the current academic taxonomy allows. Hereafter, "mathematician" is understood to mean anyone who is familiar with a body of nontrivial mathematics and who uses it in his work in a way that demands genuine understanding of what is involved. Thus, in my usage, "mathematician" naturally embraces physicists and computer scientists but also most other physical scientists and engineers, many biologists, psychologists, and linguists, and, to the bargain, a host of economists, sociologists, demographers, and professional gamblers. My question, then, in bluntest form, is: What does an intellectual lose by not being a mathematician?

Frankly, I do not think the loss is all that trivial. Let me start with a well-known example that is as vulgar as it is obvious. The central point in the O. J. Simpson trial, which, as of this writing, colonizes an absurd amount of television airtime, is a probabilistic one. The notion that the chances are "one in six billion" (or whatever) that a drop of blood came from anyone but the defendant sounds almost mystical in its astronomical resonance. In fact, it is the result of a straightforward computation that crucially depends for its relevance on the notion of "independence" of random variables. I doubt this is at all well understood even by most of the well (outside of mathematics) educated who follow the trial.[17] I would go so far as to say that most of the lawyers don't understand it. I think most educated people believe the experts, on the whole.[18] Yet they do so somewhat grudgingly, with that faint animus reserved for oracles who are known to be trustworthy but whose methods are maddeningly inscrutable. They are probably made somewhat uneasy by questions like: "If you've taken samples from only a few thousand people, how can you talk about odds like several billion to one?" Of course, a little knowledge of elementary statistics and probability would dispel that uneasiness. But such a background is precisely what most of the well educated, most "intellectuals," lack.

This is just one instance of a broad phenomenon. Uninstructed human intuition is simply not very good at estimating probabilities, judging the significance of apparent coincidences, discerning actual randomness in patterns. Evolution seems not to have done a particularly good job in that respect. By way of experiment, I have tried simple probability questions—simpler, even, than those that come up in the Simpson trial—on very bright people whose familiarity with the relevant mathematics has lapsed, if it ever existed. A surprising number not only guess wrong, but stick adamantly to their guns for hours, despite hearing what I judge to be clear, cogent explanations of the correct answer. Systematic analysis and a certain amount of work are needed to rectify the defects of most people's instinctive habits of thought in this area. In a word, one must study some mathematics. Not to do so excludes one from a decent comprehension of vast realms of experience.

But again, this is not the whole story. The stunting of one's mathematical intelligence simply leaves one at a loss to understand what goes on in the world. It cuts one off from any sense of how our most accurate analyses of physical and natural reality are structured. It reduces them to disjointed collections of odd, unaccountable "facts," naked propositions without context, logic, or elegance. This is precisely what happened to poor Matthew Arnold, thus rendering his view of science so "philistine." The thinker whose mathematics lags behind his general intelligence finds that, for him, our deepest, best-founded theories about how the world is put together and how phenomena are linked are severely occluded. Their skeletons, muscles, and sinews are forever invisible. One need not take mathematical theory building in the exact sciences as the prototype for all meaningful knowledge to consider that a deficient understanding of how this kind of theory-building proceeds—which may, to be old-fashioned about it, constitute a failure to understand how reality itself is structured—leaves one, at the very least, with a deficient appreciation of human capability and human aspiration. It represents a vast hole in one's grasp of psychology and a lack of suppleness in one's capacity to accommodate the insights and to appreciate the striving of one's fellows.

I assert as well that the smug homilies of generations of high school geometry teachers were, in fact, correct; that is to say, the wisdom of the Hellenic philosophers, from Pythagoras to Aristotle, was, in fact, wisdom. Training in systematic, rigorously deductive thought, experience of the process and results of that kind of thinking, is an aspect of intellectual development whose neglect invites, if it does not quite insure, a measure of intellectual slovenliness. Only a woefully small number of intellectuals have actually engaged in the kind of mental calisthenics where principles of logical inference are rigidly applied, ambiguities of language sifted out, unstated assumptions kept from the premises, words prohibited from sliding unannounced from one meaning to another, and appeals to emotion or cultural prejudice or moral indignation despised. This is an exercise well worth going through, just to see that it can, in fact, be done. Furthermore, there are extra rewards to doing the kind of mathematical thinking that genuinely stretches one's intuition. At the very least, it greatly enriches one's library of phenomenological models.

It would be wonderful (mathematical snobbery alert!), even for a Milton scholar, to learn the machinery of algebraic topology or differential geometry! Granted, that may be a bit utopian. Yet is it too much to ask that the well-rounded scholar know what the fundamental theorem of algebra is or even, if that is too demanding, why two angle bisectors of an isosceles triangle are equal?[19] Just this last modest project is incontestably worthwhile, if done honestly and thoroughly.

The real value, as I see it, of even a modest mathematical education is that it breeds a certain salutary impatience, a distaste for intellectual flatulence, for otiose pseudotheorizing, for argument by browbeating. It breeds a certain shrewdness, as well, in all sorts of odd corners of modern life. It helps purge the staleness, the laziness, the careless propensity to accept unexamined clichés, from ones thinking, simply because it provides a rich array of mental patterns and the habit of looking for instances where they are applicable. The recent book by John Alan Paulos[20] provides numerous examples of this. Many of his illustrations are homely, and some are whimsical; but modest as they are, they throw into relief the limitations with which even the well educated are afflicted for want of cultivating mathematical habits of mind. At one point, Paulos throws out the following catechism:

> What's the difference between an empirical statement and an a priori one, between scientific induction and deduction? What about the difference between scientific induction and mathematical induction? Does some implication go both ways or is its converse false? Is such and such a claim falsifiable?

I am willing to bet that most intellectuals will someday come to grief somewhere in this thicket. Paulos is on the mark in claiming that the proper inoculation against such gaffes lies in getting on reasonably friendly terms with some serious, formal mathematics. The great mathematician Poincaré once remarked that mathematics may be defined as "the exact part of our thinking." He was making a point in connection with the debates about the epistemological status of mathematics that were beginning to quicken in his day, but I think his assertion can be glossed in another sense. Not to know mathematics at a mature level is to open oneself, however unwittingly, to the definite possibility of inexactitude in one's thinking.

Let us consider whether the institutionalization of mathematical illiteracy—for that is what we are talking about, after all—has any dire consequences for culture and politics on a global, transpersonal level. It is one thing to talk about the debilities, even the follies, of one's fellow intellectuals as individual quirks, and, simply because to do so as a mathematician constitutes, *ipso facto*, a kind of sneering, a trace of nastiness is inevitably involved. But do these individual shortcomings have, cumulatively, any larger impact upon the quality and content of discourse? Do they steer us, in the collective sense, into blind alleys or weigh us down with nonsense or place us in danger of ignoring things, as a culture, that we cannot afford to ignore? In keeping with the theme of this volume, I think that the answer is "yes." This, it must be said, is tricky ground. I don't want to hedge; my convictions on this question are not at all tentative. I think that over the years, the world of ideas has come to pursue its concerns with an increasingly impaired organ. Yet, as I have indicated above, the particulars can be devilishly hard to trace, and it would be futile to try to describe a simple mechanism by which all the putative damage is being done. Yet something has gone wrong with the machinery of our higher-order cogitation, outside the realm of the sciences themselves and more broadly, those areas where the "mathematicians" function. In the latter precincts, things are going along nicely, as I see it, at least on the purely intellectual front. But much of humanistic thought and much of what purports to be political thought is in rather poor shape. Windbags, bluffers, and moral one-uppers are having a field day. The daft and the silly are raised on high. A palpable impatience reigns—not the healthy impatience of the logician, the impatience that cauterizes sloppy thinking, but rather an impatience with the attitude that thinks it at all important to challenge sloppy thinking. It no longer comes as news, for instance, that logical thought itself is in bad odor in certain scholarly circles; indeed, that realization is what occasions this volume in the first place. The denial that rational, systematic, largely disinterested inquiry is even possible is a virtual rite-of-passage in some corners of the academic world, most notably postmodern literary criticism, post-Geertzian anthropology, and, tragically, women's studies.[21] The social sciences are importuned, with increasing success, to regard the exact sciences as delusory.[22] The most prominent spokesmen for this odd view are gleefully touted as the elect of contemporary scholarship.[23]

The one comforting fact in all of this is that those most eager to challenge the efficacy and reliability of rational thought have not, as a general rule, shown themselves to be particularly good at it. There are some exceptions, but these seem to be cases of conversion through embitterment. The giddy enthusiasms that have roiled the academy, reduced hundreds of administrators to speaking in psychobabble, commandeered the coffers of dozens of foundations, and bored the living hell out of most sensible undergraduates reflect precisely those deficiencies of thought against which the study of mathematics is supposed to be an effective prophylactic. And, indeed, these fads hardly provide a counterexample to this supposition. Mathematical literacy is only rarely to be found among the cheerleaders of the current academic foofaraw. This is particularly noteworthy in "science studies," a new hybrid ostensibly comprising sociological, historical, political and cultural analysis of science, with some philosophical aspects thrown in as well. To date, this has been a rather uneven enterprise, but I think it is safe to say that flashy and politically emphatic work has tended to overshadow sober competence. A recurrent theme among the Young Turks of the field is that a close familiarity with the methodology and conceptual content of science is not only unnecessary but positively detrimental. Science, in this view, is an anthropological phenomenon; it may be studied in ignorance of the truths it discovers, or even with disdain for the notion that it discovers truth. Mathematics is especially suspect. It is perceived as an arcane, exclusionary code. One even hears mutterings to the effect that the mathematization of science is "historically contingent" and that equally valid "alternative science"—including physics—might well have come into being without drawing upon the machinery of abstract mathematics.[24] Naturally, the people who make these claims seem unimpaired by any deep understanding of what mathematics is. More tellingly, they are locked into the delusion, which arises repeatedly in the scholarly world of the postmodernists, that mere assertion is equivalent to argument. This quirky logic underlies the high reputation of many of the most admired books in the field. Quite clearly, if any fantasy is likely to be cured by a comfortable familiarity with mathematics, it is this one.

"Science Studies," in this vein, clearly seems to be an enterprise driven by resentment. Since most of its practitioners and fans clearly identify themselves, in some fashion, with the political left, it is exquisitely ironic that they seem to have found the politicians of their dreams in the new Republican Congress. The scientific philistinism of the ax-wielding freshmen legislators seems to echo perfectly, in the realm of practical politics, that of the science studies academics. This is not so much an instance of "strange bedfellows" as of the metaphorical truth (for which I am indebted to the superb mathematician V. Arnold) that the best geometrical model for politics is projective space![25]

All this having been said, one may take quizzical note of the fact that resentment of mathematics has an envious aspect to it, and that such envy, as it exists among faddish humanists, emerges in a minor fad for studying mathematics under the rubric of cultural studies, poststructuralist discourse theory, and the like. There are entire tomes in this mode, and celebrities like Lyotard, Baudrillard, Irigaray, Lacan, and Derrida have frequently been known to dress up their pages with mathematical, or quasimathematical, or pseudomathematical phraseology. On the cultural studies end of things, the received doctrine is that the methodology of this brand-new field is such a powerful intellectual solvent that anything may be brought under its purview. The dictum that everything that people do is "cultural" in its essence licenses the idea that the cultural critic can meaningfully analyze even the most intricate accomplishments of art and science—and of mathematics, when it comes to that—solely by recourse to the tool-kit of "cultural theory." This need not—in plain fact, it does not—involve analyzing, nor even comprehending, the particular details of what one is supposedly studying. All the methodological problems are dispelled at a stroke by invoking the doctrine of "interdisciplinarity," which means simply (to the eyes of this cynical observer) that if one has bluffed one's way to success in the fields of postmodern literary analysis and "critical theory," it is open season on everything else. For a mathematician, encountering this stuff can produce some dizzying effects. It is distinctly weird to listen to pronouncements on the nature of mathematics from the lips of someone who cannot tell you what a complex number is! Much as one might wish to see the people who propound this sort of cant get their comeuppance, it is probably an idle fantasy. Folks like this have had the intellectual shame wrung out of them by a perverse educational system, and it will take considerable ingenuity, perhaps on the part of a practical joker of genius, to shake them out of their complacency.

It is worth noting that much of the recent fad for cultural criticism of what the critics take to be mathematics centers on the relatively new discipline of "chaos theory." I conjecture that the name itself has something to do with this. So, too, the opportunity to see it as the embodiment of a seismic "paradigm shift," à la Thomas Kuhn. Chaos theory has produced important new insights, as well as a host of delightful multicolored pictures; at its best, it is fascinating and highly useful mathematics. But the mathematical issues it raises are subtle and delicate. One has to have swum fairly far upstream, mathematically speaking, to be able to talk about it sensibly and coherently. The chief virtue, then, of its popularity among cultural theorists is that it gives those prone to doing so every opportunity to make asses of themselves. For that, at least, we should be grateful.[26]

In the exclusion of the habits of mind associated with careful mathematical thinking from so much of contemporary intellectual work, we find both symptom and cause of the weakness, the outright frivolity, of much academic discourse. Mathematics, one surmises, has always made a lot of people nervous. It is hard! It is exacting and unforgiving. It does not let you squirm out of things. The "coldness" of it, which Bertrand Russell spoke of as "without appeal to any part of our weaker nature," is understandably off-putting to many. We are not a species that has evolved to be able to separate thought and feeling with good grace, but mathematics demands this of us. Confronted with an obligation to deal seriously with mathematics, individuals often look for a way out. So, apparently, do cultures.

This, I submit, has cost us something over the long haul. It has deadened the ears of the culture to exactitude in argument and discourse. It has disarmed suspicions that ought to have remained acute, while nurturing suspicions that are poorly founded, when they are not outright silly. It has allowed the general culture's ties with reality to fray, simply because science, I insist, conveys many important things about reality to those equipped to listen, and because a minimum of mathematical skill is a vital component of the listening equipment.

I am not trying to claim that, on the individual level, mathematics is the guarantor of either wisdom or virtue. There are plenty of skillful mathematicians who are bullies, bigots, selfish neurotics, fanatics, and damned fools. At best, the cultivation of mathematical skill provides but a fraction of what one needs to deal successfully with the endless complexities of life and thought, of morality and politics. It only marginally improves the capacity of the scholar or thinker who works at some remove from the mathematical sciences. For someone in that situation, it is just one auxiliary device among many others; it may do little more than hone certain intellectual edges and provide a testing ground for certain capacities.

Yet on the collective level the effects of being at home in mathematics would be stronger and deeper. Let us suppose, contrary to current fact, a broad intellectual community in which a repertoire of nontrivial mathematical skills and insights is accessible to pretty much anyone. There, I think, we would find some pleasant contrast to our present situation. The cement of common rationality and mutual intelligibility among thinkers of all kinds would be appreciably stronger. The edginess one now meets when forcing people to deal with scientific matters—or economic or demographic matters—would be substantially softened. Bluffs would be called when they ought to be. Analysis and exposition would reacquire a sureness of touch that has, I submit, decayed as general familiarity with mathematics has leached out of the culture. Above all, effective canons of rationality, of coherence, of consistency would be restored to their appropriate place.

What can we do to realize this pipe dream? Nothing about it seems particularly easy to me. It will not be a matter of simply making some adjustments, even far-reaching ones, in the system of formal education. This is not a task for educational bureaucrats, even competent, high-minded, and well-intentioned ones. It is not a matter of suddenly forcing a horde of humanities and social science majors to enroll in the math courses that heretofore have been forced upon only economics, biology, and engineering majors. What is involved is, in the deepest and most mystifying sense, cultural or even moral. What puts a culture at ease with mathematics, and with the world of golden ideas and quicksilver intellectual deftness that it opens up, is very difficult to specify. But we know that such a cultural possibility exists because history shows us cultures where it has been realized. There have been times and places where mathematics was a vital element in the play of ideas that edified and enriched the entire society, where it lent a quality of easefulness and repose to thought, where it obliged thinkers of all kinds to be more elegant and accurate. There is no royal road that will swiftly transport our contemporary culture to a similar state of grace. But we ought now to be scrutinizing our cultural road maps for whatever guidance they can offer, in hopes of finding an accessible pathway.

## NOTES

1 I wish to thank the following for helpful and instructive comments and conversations: Renate Blumenfeld-Kosinski, Felix Browder, Jeff Cheeger, Robin Fox, Sheldon Goldstein, Paul R. Gross, Antoni Kosinski, and Hector Sussmann.

2 David N. Schramm, "Nature Throws a Curve."

3 Bertrand Russell, "The Study of Mathematics."

4 Jacques Barzun, *Science: The Glorious Entertainment*, p. 115.

5 Ibid., p. 295.

6 Paul Hillier, Notes to *Dunstable: Motets.*

7 Quoted in Otto von Simson, *The Gothic Cathedral*, p. 13.

8 Otto von Simson, *The Gothic Cathedral*, p. 13.

9 Barzun, *Science: The Glorious Entertainment*, p. 116.

10 There is an unintended irony in Barzun's complaint that physics has abandoned cosmological questions. His book appeared just as the discovery of the cosmological "background radiation" put cosmological questions at the center of the physics agenda, while making the "Big Bang" theory the central theme in cosmological thinking.

11 *Aubrey's Brief Lives.*

12 This point is thoroughly ignored, or perhaps misunderstood, in Shapin and Schaffer's well-known *Leviathan and the Air-Pump.* Paul R. Gross and I have analyzed this misreading at length in *Higher Superstition: The Academic Left and Its Quarrels with Science.*

13 Berkeley's challenge was to the ontological reality of the "infinitesimal quantities" favored by Leibniz and his followers. This, as the astute freshman calculus student knows, is something of a red herring. Newton certainly understood the essential role of the notion of limit (although his published work on "fluxions" avoids a detailed discussion), and D'Alembert proffered definitions that were essentially equivalent to the modern ones.

14 Take this with a grain of salt; it is very much an "other things being equal" kind of statement. A rigorous and exacting course in Euclidean geometry will be very much *harder* than a conceptually vacuous powder-puff calculus course such as is usually offered by high schools (and colleges).

15 Matthew Arnold, *Literature and Science.*

16 One of the ironies of *Arcadia*—fully intended by Stoppard, of course—is that it is nearly impossible for a playgoer to follow the piece fully without a pretty thorough grounding in contemporary mathematics, or at least a companion who is well equipped in this respect.

17 Let us be honest; damn few of us are so abstemious as not to follow it at all!

18 My apologies to Judge Ito for commenting on a case that is still *sub judice* at this writing, but I doubt this essay will reach the snugly sequestered jurors in time to pervert the course of justice.

19 If you are ambitious, you might try the converse: show that a triangle with two equal angle bisectors must be isosceles. It is trickier than you might think.

20 John Alan Paulos, *A Mathematician Reads the Newspaper.*

21 See, *inter alia*, Christina Hoff Sommers, *Who Stole Feminism* and Daphne Patai and Noretta Koertge, *Professing Feminism.* See as well the articles by Martha Nussbaum ("Feminists and Philosophy") and Susan Haack ("Knowledge and Propaganda: Reflections of an Old Feminist").

22 See Mario Bunge, "A Critical Examination of the New Sociology of Science" (Parts 1 and 2).
23 See David Berreby, "That Damn'd Elusive Bruno Latour."
24 See, in particular, the work of the up-and-coming Steve Fuller. Fuller seems to think that "set theory" alone might have been adequate to the formulation of physics, which is very odd because (a) set theory is quintessentially mathematical and (b) once set theory is in the door, the rest of mathematics tends to come tumbling in close behind it. Fuller ("Can Science Studies Be Spoken in a Civil Tongue?" p. 164, note 8) attributes to W. V. O. Quine the view that for physics, "set theory" is adequate and "number" a superfluous ontological luxury. The present author, who personally watched Quine construct the classical continuum from the elements of set theory in a couple of courses, has no choice but to regard this as bizarre.
25 "Projective space" is derived from the geometry of the sphere by thinking of antipodal points as identical. It is an important model for axiomatic non-Euclidean geometry.
26 Gross & Levitt, *Higher Superstition: The Academic Left and Its Quarrels with Science*, contains a brief analysis of the popularity of chaos theory among cultural critics.

## REFERENCES

AUBREY, JOHN. *Aubrey's Brief Lives.* Edited by Oliver Lawton Dick. Harmondsworth: Penguin Books, 1972.

ARNOLD, MATTHEW. "Literature and Science." In *The Portable-Matthew Arnold*, edited by Lionel Trilling. New York, NY: The Viking Press, 1949.

BARZUN, JACQUES. *Science: The Glorious Entertainment.* New York, NY: Harper and Row, 1964.

BERREBY, DAVID. "That Damn'd Elusive Bruno Latour." *Lingua Franca* (October 1994): 1 ff.

BUNGE, MARIO. "A Critical Examination of the New Sociology of Science, Part I." *Philosophy of the Social Sciences* 21 (1991): 524–560.

———. "A Critical Examination of the New Sociology of Science, Part 2." *Philosophy of the Social Sciences* 22 (1992): 46–76.

FULLER, STEVE. "Can Science Studies be Spoken in a Civil Tongue?" *Social Studies of Science* 24 (1994): 143–168.

HAACK, SUSAN. "Knowledge and Propaganda: Reflections of an Old Feminist." *Partisan Review*, Fall 1993, pp. 556–564.

GROSS, PAUL R. & NORMAN LEVITT. *Higher Superstition: The Academic Left and Its Quarrels with Science.* Baltimore, MD: Johns Hopkins University Press, 1994.

HILLIER, PAUL. Notes to *Dunstable: Motets.* EMI Records Compact Disc CDC 7 49002 2 (1987).

NUSSBAUM, MARTHA. "Feminists and Philosophy" *New York Review of Books*, October 20, 1994, pp. 59–63.

PATAI, DAPHNE & NORETTA KOERTGE. *Professing Feminism.* New York, NY: Basic Books, 1994.

PAULOS, JOHN ALAN. *A Mathematician Reads the Newspaper.* New York, NY: Basic Books, 1995.

RUSSELL, BERTRAND. "The Study of Mathematics." In *Mysticism and Logic and Other Essays.* New York, NY: Longmans, Green and Co., 1918.

SCHRAMM, DAVID N. "Nature Throws a Curve." *New York Times Book Review*, May 21, 1995, p. 38.

SHAPIN, STEVEN & SIMON SCHAFFER. *Leviathan and the Air Pump.* Princeton, NJ: Princeton University Press, 1985.

SOMMERS, CHRISTINA HOFF. *Who Stole Feminism?* New York, NY: Simon and Schuster, 1994.

VON SIMSON, OTTO. *The Gothic Cathedral: Origins of Architecture and the Medieval Conception of Order.* New York, NY: Harper and Row, 1964.

# REASONABLE FOUNDATIONS

IN RECENT YEARS, fashionable thinkers, particularly those associated with postmodernism and some brands of feminism, have challenged the traditional veneration of reason, claiming, variously, that reason is chimerical, impossible, or simply oppressive to groups outside the magic circle of white, male privilege. This section evaluates the standing of reason from a number of viewpoints.

Philosopher Susan Haack decries attempts by some contemporary philosophers to join the antireason bandwagon. She notes that the abandonment of a genuine commitment to truth as the purpose of inquiry leads inevitably to sophistry, frivolity, speciousness, or outright bad faith. For Haack, truth and reason (which is the chief instrument for the pursuit of truth) deserve their traditional place of honor.

Psychologist James Alcock has a different concern. He shows how the "reasonableness" of human intuition is unreliable. In particular, we are, as a species, rather poorly equipped to make good instinctive judgments of probability, randomness, coincidence, and so forth; and countless errors of judgment result. Exact thinking, indispensable as it is, thus goes in many respects against the evolutionary grain. Critical analytic thought, vital to an accurate picture of the world, faces a continuing struggle to maintain itself against a deeply embedded predisposition toward belief in magic.

Philosopher Barry Gross (who died, tragically, shortly after writing this piece) traces much of the confusion in contemporary academic and intellectual life to a reverence for inappropriate abstraction at the expense of common sense and practical wisdom. He finds that theory builders are often willfully blind to prosaic evidence that inconveniently contradicts them, and that their stubbornness, these days, is often reinforced by a persistent utopianism.

Physician Loren Fishman evaluates some recent cynicisms as to the efficacy of reason. The first derives from the argument that social convention is not only a factor in determining belief, but a force powerful enough to overwhelm mere logic even within the thinking that makes up scientific theory. The second challenges scientific modes of thought on the ground that, since scientists often do rely on intuition and flashes of insight, science cannot claim the rationality usually associated with it. Fishman finds deep flaws in both arguments.

Finally, the distinguished philosopher Mario Bunge, noting that disparagement and outright dismissal of reasoned inquiry have become commonplace in the academic world, delivers a stinging rebuke. He calls for a renewal of the academy's traditional devotion to canons of reason, to be enforced by an implicit code of intellectually responsible conduct that would turn its back on postmodern extravagances.

# CONCERN FOR TRUTH: WHAT IT MEANS, WHY IT MATTERS[a]

SUSAN HAACK

A CENTURY OR SO AGO, C. S. Peirce wrote, "in order to reason well . . ., it is absolutely necessary to possess . . . such virtues as intellectual honesty and sincerity and a real love of truth,"[1] and that genuine reasoning consists "in actually drawing the bow upon truth with intentness in the eye, with energy in the arm."[2] Forty years or so ago, C. I. Lewis observed that "we presume, on the part of those who follow any scientific [he means, "intellectual"] vocation, . . . a sort of tacit oath never to subordinate the motive of objective truth-seeking to any subjective preference or inclination or any expediency or opportunistic consideration."[3] These philosophers had some insight into what the life of the mind demands.

Now, however, it is fashionable to suggest that such insights are really illusions. Stephen Stich professes a sophisticated disillusionment, writing that "once we have a clear view of the matter, most of us will not find any value . . . in having true beliefs."[4] Richard Rorty refers to those of us who are willing to describe ourselves as seeking the truth as "lovably old-fashioned prigs"[5] boasting that *he* "do[es]n't have much use for notions like 'objective truth' "[6] since, after all, to call a statement true "is just to give it a rhetorical pat on the back."[7] Jane Heal concludes with evident satisfaction that "there is no goddess, Truth, of whom academics and researchers can regard themselves as priests or devotees."[8] These philosophers reveal a startling failure, or perhaps a refusal, to grasp what intellectual integrity is, or why it is important.

Still, as the saying goes, those who know only their own side of a case know very little of that; so perhaps it is healthy to be obliged to articulate, as I shall try to do, what concern for truth means, why it matters, and what has gone wrong in the thinking of those who denigrate it.

[a] This paper was prepared for publication with the help of N.E.H. Grant #FT-40534-95. I wish to thank Paul Gross for helpful comments on a draft, and Mark Migotti for supplying the quotation from Nietzsche in Note 9. I draw in what follows on my *Evidence and Inquiry*, especially Chapter 9; "The First Rule of Reason"; "The Ethics of Belief"; and "Preposterism and Its Consequences."

The first step is to point out that the concept of truth is internally related to the concepts of belief, evidence, and inquiry. To believe that *p* is to accept *p* as true. Evidence that *p* is evidence that *p* is true, an indication of the truth of *p*. And to inquire into whether *p* is to inquire into whether *p* is true; if you aren't trying to get the truth, you aren't really inquiring.

Of course, both pseudobelief and pseudoinquiry are commonplace. Pseudobelief includes those familiar psychological states of obstinate loyalty to a proposition that one half suspects is false, and of sentimental attachment to a proposition to which one has given no thought at all. Samuel Butler puts it better that I could when, after describing Ernest Pontifex's sudden realization that "few care two straws about the truth, or have any confidence that it is righter or better to believe what is true than what is untrue," he muses, "yet it is only these few who can be said to believe anything at all; the rest are simply unbelievers in disguise."[9]

And pseudoinquiry is so far from unusual that, when the government or our university institutes an Official Inquiry into this or that, some of us reach for our scare quotes. Peirce identifies one kind of pseudoinquiry when he writes of "sham reasoning": attempts, not to get to the truth of some question, but to make a case for the truth of some proposition one's commitment to which is already evidence- and argument-proof. He has in mind theologians who devise elaborate metaphysical underpinnings for theological propositions that no evidence or argument would induce them to give up; but his concept applies equally to the advocacy "research" and the politically motivated "scholarship" of our times. And then there is what I have come to think of as fake reasoning: attempts not to get to the truth of some question, but to make a case for the truth of some proposition to which one's only commitment is a conviction that advocating it will advance oneself—also a familiar phenomenon when, as in some areas of contemporary academic life, a clever defense of a startlingly false or impressively obscure idea is a good route to reputation and money.

But we need to go beyond the tautology that sham inquirers and fake inquirers aren't really inquiring to see what, substantively, is wrong with sham and fake reasoning. Sham and fake inquiries aim, not to find the truth, but to make a case for some proposition identified in advance of inquiry. So they are motivated to avoid careful examination of any evidence that might impugn the proposition for which they are seeking to make a case, to play down or obfuscate the importance or relevance of such evidence, to contort themselves explaining it away. The genuine inquirer, by contrast, wants to get to the truth of the matter that concerns him, whether or not that truth comports with what he believed at the outset of the investigation, and whether or not his acknowledgment of that truth is likely to get him tenure, or make him rich, famous, or popular. He is motivated, therefore, to seek out and assess the worth of evidence and arguments thoroughly and impartially; to acknowledge, to himself as well as others, where his evidence and arguments seem shakiest and his articulation of problem or solution vaguest; to go with the evidence even to unpopular conclusions or conclusions that undermine his formerly deeply held convictions; and to welcome someone else's having found the truth he was seeking.

This is not to deny that sham and fake reasoners may hit upon the truth, and, when they do, may come up with good evidence and arguments; nor that genuine inquirers may come to false conclusions or be led astray by misleading evidence. Commitment to a cause and desire for reputation can prompt energetic intellectual effort. But the intelligence that will help a genuine inquirer figure things out will help a sham or fake reasoner suppress unfavorable evidence more effectively, or devise more impressively obscure formulations. A genuine inquirer, by contrast, will not suppress unfavorable evidence, nor disguise his failure with affected obscurity; so, even when he fails, he will not impede others' efforts.

The genuine inquirer's love of truth, as this reveals, is not like the love of a collector for the antique furniture or exotic stamps he collects, nor is it like a religious person's love of God. He is not a collector of true propositions, nor is he a worshipper of an intellectual ideal. He is a person of intellectual integrity. He is not, like the fake reasoner, indifferent to the truth of the propositions for which he argues. He is not, like the sham reasoner, unbudgeably loyal to some proposition, committed however the evidence turns out. Whatever question he investigates, he tries to find the truth of that question, whatever the color of that truth may be.

The argument thus far has taken us beyond the tautology that genuine inquiry aims at the truth, to the substantive claim that lack of intellectual integrity is apt, in the long run and on the whole, to impede inquiry. But why, it will be asked, should we care about that? After all, in some circumstances one may be better off not inquiring, or better off having an unjustified belief than one well grounded by evidence, or better off having a false belief than a true one; and some truths are boring, trivial, unimportant, some questions not worth the effort of investigating.

Intellectual integrity is instrumentally valuable, because, in the long run and on the whole, it advances inquiry; and successful inquiry is instrumentally valuable. Compared with other animals, we are not especially fleet or strong; our forte is a capacity to figure things out, hence to anticipate and avoid danger. Granted, this is by no means an unmixed blessing; the capacity that, as Hobbes puts it, enables men, unlike brutes, to engage in ratiocination, also enables men, unlike brutes, "to multiply one untruth by another."[10] But who could doubt that our capacity to reason is of instrumental value to us humans?

And intellectual integrity is morally valuable. This is suggested already by the way our vocabulary for the epistemic appraisal of character overlaps with our vocabulary for the moral appraisal of character: e.g., "responsible," "negligent," "reckless," "courageous," and, of course, "honest." And "He is a good man but intellectually dishonest" has, to my ear, the authentic ring of oxymoron.

As courage is the soldier's virtue par excellence, one might say, oversimplifying a little, so intellectual integrity is the academic's. (The oversimplification is that intellectual integrity itself requires a kind of courage, the hardihood needed to give up long-standing convictions in the face of contrary evidence, or to resist fashionable shibboleths.) I would say, more bluntly than Lewis, that it is downright indecent for one who denigrates the importance or denies the possibility of honest inquiry to make his living as an academic.

This explains why intellectual integrity is morally required of those of us who have a special obligation to undertake inquiry; but the explanation of why it is morally important for all of us has to be more oblique. Over-belief (believing beyond what one's evidence warrants) is not always consequential, nor is it always something for which the believer is responsible. But sometimes it is both; and then it is morally culpable. Think of W. K. Clifford's striking case of the ship owner who knows his ship is elderly and decaying, but doesn't check and, managing to deceive himself into believing that the vessel is seaworthy, allows it to depart; he is, as Clifford rightly says, "verily guilty" of the deaths of passengers and crew when the ship goes down.[11] The same argument applies, mutatis mutandis, for under-belief (not believing when one's evidence warrants belief). Intellectual dishonesty, a habit of reckless or feckless self-deceptive belief formation, puts one at chronic risk or morally culpable over- and underbelief.

So, what has gone wrong in the thinking of those who denigrate concern for truth? Unfortunately, even with just the three writers I quoted at the beginning of this paper, not quite the same thing in each case.

Stich begins by ignoring the internal connection of the concepts of belief and truth, and misconstruing belief as nothing more than "a brain-state mapped by an interpretation-function into a proposition," or, as he likes to say to make the idea vivid, a sentence inscribed in a box in one's head marked "Beliefs." This encourages him in the mistaken idea that truth would be a desirable property for a belief to have only if truth is either intrinsically or instrumentally valuable. He then compounds the confusion with two manifest non sequiturs: that since truth is only one of a whole range of semantic properties a sentence in one's head could have, truth is not intrinsically valuable; and that since one may sometimes be better off with a false belief than a true one, truth is not instrumentally valuable either.

With Heal one encounters a different kind of misdirection. She points out, correctly, that not every true proposition is worth knowing; again correctly, that, like courage, intellectual integrity can be useful in the service of morally bad projects as well as good ones; correctly once more, that what an inquirer wants to know is the answer to the question into which he inquires. Even her conclusion—that there is no goddess, Truth, of whom academics can regard themselves as devotees—is true enough. What is wrong with it is not that it is false, but that it suggests that if one takes concern for truth to matter, one must deny it. The instrumental value of intellectual integrity does not require that all truth be worth knowing; its moral value does not require that it be a character trait capable only of serving good uses; and valuing intellectual integrity is not, as Heal's conclusion hints, a kind of superstition.

And as Rorty more than hints when he tells us that he sees the intellectual history of the West as an attempt "to substitute a love of truth for a love of God,"[12] Rorty is of the party that urges that there is no one truth, but many truths. If this means that different but compatible descriptions of the world may be both true, it is trivial; if it means that different and incompatible descriptions of the world may be both true, it is tautologically false. More likely, Rorty has confused it with the claim that there are many incompatible truth *claims.*

This reveals a connection with a ubiquitous fallacy. What passes for known truth is often no such thing, and incompatible truth claims are often pressed by competing interests. But it obviously doesn't follow, and it isn't true, that incompatible truth claims can be both true; or that to call a claim true is just to make a kind of rhetorical gesture or power grab on its behalf. This latter mistaken inference, like the inference from the true premise that what passes for objective evidence is often no such thing to the false conclusion that the idea of objective evidence is just ideological humbug, is an instance of what I have come to dub the "passes for" fallacy.[13] Rorty transmutes this fallacy into a shallow misconception that identifies "true" and " 'true'," the true with what passes for true. "True" is a word that *we apply* to statements about which we agree, simply because, if we agree that *p*, we agree that *p* is true. But we may agree that *p* when *p* is *not* true. So "true" is not a word that *truly applies* to all or only statements about which we agree; and neither, of course, does calling a statement "true" *mean that* it is a statement we agree about.

Here is Peirce again, describing what happens if pseudoinquiry becomes commonplace: "man loses his conceptions of truth and of reason . . . [and comes] to look upon reasoning as mainly decorative. The result . . . is, of course, a rapid deterioration of intellectual vigor."[14] This is the very debacle taking place before our eyes. Sham reasoning, in the form of "research" bought and paid for by bodies with an interest in its turning out this way rather than that, or motivated by political conviction, and fake reasoning, in the form of "scholarship" better characterized as a kind of self-promotion, are all too common. When people are aware of this, their confidence in what passes for true declines, and with it their willingness to use the words "truth," "evidence," "objectivity," "inquiry," without the precaution of scare quotes. And as those scare quotes become ubiquitous, people's confidence in the concepts of truth, evidence, inquiry, falters; and one begins to hear, from Rorty, Stich, Heal and company, that concern for truth is just a kind of superstition—which, I should add, in turn encourages the idea that there is, after all, nothing wrong with sham or fake reasoning . . . and so on.

One thinks of Primo Levi on the subject of Fascism and chemistry: "the chemistry and physics on which we fed, besides being nourishments vital in themselves, were the antidote to Fascism . . ., because they were clear and distinct and verifiable at every step, and not a tissue of lies and emptiness, like the radio and newspapers."[15] I would put it more prosaically, but perhaps a little more precisely: the antidote to pseudoinquiry, and to the loss of confidence in the importance of intellectual integrity it engenders, is real inquiry and the respect for the demands of evidence and argument it engenders. Real inquiry of any kind, I should say: scientific, historical, textual, forensic, . . ., even philosophical. But there is a reason for putting "scientific" first on this list, the same reason that led Lewis to write "scientific vocation," meaning "intellectual vocation," and that led Peirce sometimes to describe the genuine inquirer's concern for truth as "the scientific attitude":[16] not that all or only scientists have the scientific attitude, but that this is the attitude that made science possible. It is not concern for truth, but the idea that such concern is superstition, that is superstitious.

## NOTES

1 C. S. Peirce, *Collected Papers*, 2.82.
2 Ibid., 1.235.
3 C. I. Lewis, *The Ground and Nature of the Right*, p. 34.
4 Stephen P. Stich, *The Fragmentation of Reason*, p. 101.
5 Richard Rorty, *Essays on Heidegger and Others*, p. 86.
6 Richard Rorty, "Trotsky and the Wild Orchids," p. 141.
7 Richard Rorty, *Consequences of Pragmatism*, p. xvii.
8 Jane Heal, "The Disinterested Search for Truth," p. 108.
9 Samuel Butler, *The Way of All Flesh*, p. 259. See also Friedrich Nietzsche, *The Gay Science*, p. 76; "*[t]he great majority of people* does not consider it contemptible to believe this or that and to live accordingly, without first having given themselves an account of the final and most certain reasons pro and con, and without even troubling about such reasons afterward."
10 Thomas Hobbes, *Human Nature*, p. 23.
11 W. K. Clifford, "The Ethics of Belief," p. 70.
12 Richard Rorty, *Contingency, Irony and Solidarity*, p. 22.
13 A term I introduced in "Knowledge and Propaganda: Reflections of an Old Feminist."
14 Peirce, *Collected Papers*, 1.57–59.
15 Primo Levi, *The Periodic Table*, p. 42. I owe this reference to Cora Diamond, "Truth: Defenders, Debunkers, Despisers," to which I refer readers for illuminating discussion of Rorty and Heal.
16 And another reason, too: that, in scientific inquiry the circumpressure of facts, of evidence, is relatively direct (though not, I think, as direct as the quotation from Levi suggests). It may be worth recalling in this context that Peirce, a working scientist as well as the greatest of American philosophers, was trained as a chemist.

## REFERENCES

BUTLER, SAMUEL. *The Way of All Flesh* (1903). New York, NY: Signet Books, The New American Library of World Classics, 1960.

CLIFFORD, W. K. "The Ethics of Belief" (1877). In *The Ethics of Belief and Other Essays*. London: Watts and Co., 1947.

DIAMOND, CORA. "Truth: Defenders, Debunkers, Despisers." In *Commitment in Reflection*, edited by L. Toker. New York, NY: Garland, 1994.

HAACK, SUSAN. "The First Rule of Reason." In *The Rule of Reason: The Philosophy of C. S. Peirce*, edited by J. Brunning & P. Forster. Toronto: Toronto University Press. In press.

———. *Evidence and Inquiry: Towards Reconstruction in Epistemology*. Oxford: Blackwell, 1993.

———. "Knowledge and Propaganda: Reflections of an Old Feminist." *Partisan Review* (Fall 1993): 556–564. (Reprinted in *Our Country, Our Culture*, edited by E. Kurzweil & W. Phillips. Boston, MA: Partisan Review Press, 1994.)

———. " 'The Ethics of Belief' Reconsidered." In *The Philosophy of R. M. Chisholm*. La Salle, IL: Open Court. In press.

———. "Preposterism and Its Consequences." *Social Philosophy and Policy* 13 no. 2 (1996): 296–315. (Also in *Scientific Innovation, Philosophy and Public Policy*, edited by E. Frankel Paul. Cambridge: Cambridge University Press. In press.)

HEAL, JANE. "The Disinterested Search for Truth." *Proceedings of the Aristotelian Society* 88 (1987–1988): 108.

HOBBES, THOMAS. *Human Nature* (1650). In *Hobbes Selections*, edited by J. E. Woodbridge. New York, NY: Charles Scribners Sons, 1930.

LEVI, PRIMO. *The Periodic Table*. Translated by R. Rosenthal. New York, NY: Schocken Books, 1984.

LEWIS, C. I. *The Ground and Nature of the Right*. New York, NY: Columbia University Press, 1955.

NIETZSCHE, FRIEDRICH. *The Gay Science*. Translated by W. Kaufman. New York, NY: Vintage Books, 1974.

PEIRCE, C. S. *Collected Papers*. Edited by C. Hartshorne, P. Weiss, and A. Burks. Cambridge MA: Harvard University Press, 1931–58. References by volume and paragraph number.

RORTY, RICHARD. *Consequences of Pragmatism*. Hassocks, Sussex, England: Harvester Press, 1982.

———. *Contingency, Irony and Solidarity*. Cambridge: Cambridge University Press, 1989.

———. *Essays on Heidegger and Others*. Cambridge: Cambridge University Press, 1991.

———. "Trotsky and the Wild Orchids." *Common Knowledge* 1.3, 1992, 141.

STICH, STEPHEN P. *The Fragmentation of Reason*. Cambridge, MA: Bradford Books, MIT Press, 1990.

# THE PROPENSITY TO BELIEVE

JAMES E. ALCOCK

WE ARE MAGICAL BEINGS in a scientific age. Notwithstanding all the remarkable achievements of our species in terms of understanding and harnessing nature, we are born to magical thought and not to reason.

The nervous system of the newborn is primed to learn. Entering this world with little to guide it apart from some basic reflexes and some rudimentary perceptual ability, the infant immediately sets out on the long journey of trying to make sense of the vast swirl of sensory stimulation with which it is bombarded. Gradually, the child develops an internal representational model of the outside world, a model so powerful that the resultant perceptions of the world seem to occur outside the brain. We logically know that perception takes place in the brain, but when we look at objects or listen to sounds or sniff the air, it is as though our perceptions of reality actually *are* reality. We do not see an image of an apple; we see the apple.

It is very difficult for most people to accept that what they perceive is not always isomorphic with what is really "out there." Our perceptual apparatus constantly and automatically seeks meaningful patterns among the myriad stimuli that impinge upon our senses. Many of these patterns that we detect are truly meaningful, while some are not; and it is difficult, if not impossible, for us to tell the difference on the basis of perceptual processing alone. We do such pattern finding unconsciously and automatically, but we are limited in two ways. First, there are limits imposed by the incoming data: it may be distorted or fragmentary. For example, walking to our campsite late at night, we see a shadow in the woods, ill-defined, fleeting. What is it? Our mental set and anxiety level and past experience will influence what we perceive. Second, there are limits imposed by the perceptual process itself. Our attention is selective. We cannot focus on everything at once, and so we collect only some of the information, particularly in situations of emotional impact.

However, our perception goes beyond these limits; it goes beyond the available information. We fill in the gaps—again usually automatically and without awareness—and perceive and recognize whole objects or events. Much of the time this works well. Sometimes it leads to egregious errors. Our perception of a bear in the darkness may be based on nothing more than a bush swaying in the breeze.

Under some conditions, perceptions have little or nothing at all to do with the outside world. The brain in some circumstances cannot tell the difference between external perceptual information and information originating within other parts of the brain itself. Dreams and nightmares are examples. However, we systematically teach children that dreams and nightmares are the product of fantasy, and we all learn to recognize them as such. Once we realize that we were asleep at the time of their occurrence, we categorize the experience as one of fantasy. Even then, the emotional impact may be so great that an individual is tempted to find some link—possibly precognitive—between the experience and external reality. There are other situations, which are not so common, where internally generated perceptions are taken to be actual perceptions of external reality. Hallucinations are, of course, one example. The so-called near-death experience—in which a person sees himself or herself as outside the body, being drawn away down a long tunnel, meeting deceased loved ones and so on—is almost certainly another. The experience is an emotionally powerful one, the perceptions are vivid and seem to be "real," and thus there is no reason for most people not to treat the experience as one of literal reality. If such experiences happened frequently in everyone, then no doubt we would be prepared for their occurrence and would come to see them as fantasy based.

## EXPERIENTIAL LEARNING

The brain and nervous system are "hardwired" to learn about the world, to learn about associations among objects and actions and events. This learning occurs on the basis of two major factors. The first of these is stimulus similarity, or *resemblance*: what we learn about one object or event will automatically be attributed to similar objects or events, unless we have enough experience to permit differentiation between them. If we taste guava for the first time and do not like it, we are unlikely to try other guavas in future. If we are thrown from a horse the first time we try to ride one, we will be cautious about all the horses, at least until we gain enough experience to see differences among them.

The second factor upon which learning is based is *temporal contiguity*: our nervous systems are set up so that salient events that occur closely together in time automatically come to be associated in the brain, as though they in some way "go together." Thus, Pavlov's dogs came to salivate at the sound of the bell that had been paired several times with the presentation of food. This is referred to as "classical conditioning"—a previously neutral stimulus becomes capable of eliciting a physiological response on the basis of being paired with another stimulus that naturally produces that response. There is another form of basic learning. Skinner's rat, having happened to press a bar that led to the presentation of a food pellet, quickly learned to press it over and over to get more food. This is referred to as "operant conditioning"—a behavior that is followed by some desirable consequence is likely to be repeated in order to gain the same consequence again. The desired consequence or "reinforcer" might be food for a hungry organism, water for a thirsty one, or relief from anxiety for a frightened one.

This learning, or "conditioning," occurs in fish, fowl, animals and humans, infants and adults. It provides the basis for experiential learning. Humans of course "think," and such automatic learning will feed into our mentation. However, such conditioning generally occurs without awareness, or in some cases despite awareness, and thus it happens without benefit of logical scrutiny. This sometimes condemns us to learning things that we do not want to learn. If we are attacked by a vicious dog, we may well develop a phobic fear of all canines that has nothing to do with any theory or logic. Emotional reactions can be readily conditioned, and emotion is often involved in experiential learning.

Indeed, an emotional response may be elicited, on the basis of past learning, by stimuli from the outside world even before we are consciously aware of the stimuli. For example, if while one is walking through the woods at night, one's visual system picks up some vague movement of a shadowy form among the trees, this information, once it reaches the limbic system, may lead to autonomic arousal before the information has arrived at the cortex—before we have any conscious perception of it.[1] When the cortex becomes aware of this information, it is also aware of the concomitant autonomic arousal, the emotion of fear, that has developed; and this emotion may influence the cognitive appraisal to the extent that the cognitive representation develops in line with the fear—"it's a bear!"

## INTELLECTUAL LEARNING

Experiential learning, based on resemblance and temporal contiguity, continues to play a powerful role throughout an individual's life. However, as the child grows a second form of knowledge acquisition gradually develops. This form is based on the mental manipulations of ideas and concepts, rather than being tied directly to personal experience. It is usually deliberate instead of automatic, and it can be carried on quite independently of emotion. As the child grows, this form of information processing becomes more and more dominated by language. It is in this *intellectual* system that nonexperiential information is absorbed—through books, teachers, stories from other people. And gradually, even without any instruction, a native logic takes shape: "Let me see, I did not have my hat on when I went into the barn, so there is no point in looking for it there." Through all the years spent in school, the child is given the benefit of thousands of years of human experience and deliberation, rather than having to experience everything first hand.

## INTERPLAY OF EXPERIENTIAL AND INTELLECTUAL LEARNING

Thus, human beings relate to and learn about the world in two distinctly different ways.[2,3] One of these ways is tied very much to the automatic associations that the nervous system constructs on the basis of temporal contiguity and resemblance and is more likely to involve emotion, while the other is a logical, analytical, and largely verbal approach. Epstein has written at length about the ways in which these two modes of information processing interact with each other.[2] He points out that people's thinking is transformed in the presence of emotional arousal; and the more aroused they become, the more their thinking is nonanalytical, concrete, and action oriented—all of which are attributes of the experiential system. Moreover, thinking that occurs in a highly emotional state often seems to be more obviously valid to the thinker, and thus high confidence in the products of one's thinking is not always reflective of accuracy or cogency of thought.

These two learning systems, being more or less independent, sometimes work in opposition. A person may have a dread for high places and yet "know" logically that a particular trip up an elevator in a high building presents no real danger. Yet an attempt to overcome the phobic learning can lead to such distress that the individual fails and actually strengthens the phobia, all the while recognizing logically that there is nothing to fear. We may disavow a belief in ghosts, and yet—having experienced chills while listening to ghost stories in our childhoods—may feel uncomfortable or even frightened while walking through a graveyard late at night, all the while telling ourselves that ghosts do not exist. Such anxieties are not easily staved off by reason. People who have given up a childhood belief in God may be surprised to find themselves silently praying for divine assistance in a moment of great fear where intellectual problem solving proves insufficient.

## SOCIAL TRANSMISSION OF BELIEF

While language provides a powerful means for the symbolic analysis of events, it also provides a basis for the transfer of massive amounts of information from others. In childhood, of course, such information is usually provided by older people, whose attractiveness or authority is such that the child rarely questions the accuracy of what he or she is taught. Whether instructed on the benefits of regular dental hygiene or on the spooky effects of the full moon on people's behavior, the child has no reason to reject what is being taught. A society may teach that all perception is illusion, or that the earth is flat, or that we are born and reborn millions of times, bound to the wheel of life. The child will have no reason and no means to reject such teaching.

We generally come to think that our beliefs are justified and sensible. In maintaining a healthy self-image, it is difficult to view our own beliefs as silly, irrational, chauvinistic, prejudiced, or unjustified; and we invest considerable effort in maintaining a belief system that is credible to ourselves. Rarely do we question most of our beliefs. We believe the earth to be a globe, but when asked to justify that belief, people cannot generally point to personal experience. They have always believed that "the world is round," since that is what they have been taught, even though the flat earth notion often fits better with everyday experience. And most people believe in God for similar reasons. While most North Americans may react with disbelief to Hindu pantheism, and they may find bizarre the Zoroastrian belief in the sanctity of earth, air, fire, and water that leads to the practice of disposing of dead bodies by leaving them atop tall towers to be consumed by birds, little such reaction attends thoughts of a Christian God who sent His son to earth to die for our sins against Him.

It is not only information that is transmitted from one generation to the next. Families and societies also strive to inculcate certain forms of thinking. While the individual may and does develop some form of logical analysis through simple personal experience, the larger society makes considerable efforts to teach logic of some sort. The logic may be basically hypothetico-deductive, or it may be based in magical thought. Usually, both logical and magical forms of analysis are taught, to be applied to different situations. In Western society, we systematically immerse children in two belief systems, which Frank has referred to as the *scientific-humanist* belief system and the *transcendental* belief system.[4] The former is allied with intellectual learning processes, while the second is more closely, although not exclusively, tied to automatic, experiential, intuitive learning. In the first belief system, children are taught to value logic and critical thinking to some degree. They are encouraged to question, to analyze. However, in the second belief system, they are taught to turn off analytical skills and to accept ideas on the basis of faith or experience. With regard to transcendental belief, religion generally teaches children to accept on faith phenomena that would make little sense if viewed from an analytical perspective. Often, guilt feelings are associated with any attempts to examine the religious belief from an intellectual point of view. Even children who have little exposure to organized religion are led by popular culture to believe in various aspects of reality—God and Heaven, for example—that are not easy to reconcile with the prevailing scientific viewpoint. Even in these days of declining religious participation in Western countries, religious and quasi-religious belief is abundant. United States currency affirms a belief in trust in God. The Canadian national anthem asks God to maintain freedom, while the British national anthem prays for the well-being of the monarch. Courts try to ensure a high standard of truth by means of an oath before God. Weddings and funerals usually involve prayers to God. It is interesting that during the recent Gulf War, the most highly technological war in history, major leaders including the president of the United States, the British prime minister, and the Iraqi president all publicly prayed to God for victory. Thus do modern technology and ancient theology work together towards a common purpose.[5]

Little wonder, then, that a Gallup poll in 1990 reported that 90% of Americans express a belief in God. Moreover, more than seventy percent indicated a belief in life after death. Most people reported belief in some sort of paranormal phenomenon: only seven percent indicated rejected all of a list of eighteen such beliefs presented in the poll.[6] Forty-nine percent reported a belief in extrasensory perception, and half of these claimed to have had experienced the phenomenon personally. One person in four reported a belief in the reality of ghosts, and one in ten claimed to have experienced the presence of a ghost. Thus, the large majority of individuals believe in some aspect of reality that supposedly lies beyond the realm of science, that lies beyond the application of quantum mechanics, and that has nothing to do with quarks or photons or blood corpuscles or neurons.

Virtually every society is a society of contradiction, valuing and promoting reason and logic to some degree, but at the same time valuing and promoting deeply held transcendental beliefs that defy that logic. In consequence, individuals learn to question, to analyze, to apply logic in one belief system, but not to do so in the other. From then on, whether or not critical thinking is brought to bear will depend a great deal on what sort of phenomenon is involved. The child is taught, essentially, that certain classes of problems and events require logic and others do not. People will be doubtful about claims that a new car can run without fuel, but will be much less ready to analyze faith healing from a skeptical perspective.

## MAGICAL THINKING

As discussed above, automatic, experiential learning is tied to resemblance and temporal contiguity. *Magical thinking* is characterized by taking resemblance and/or co-occurrence as indicators of some sort of link or causal relationship between two objects or actions or events without any concern about the actual reality of, or the nature, of that link or relationship. Magical thinking was a topic of considerable interest to anthropologists earlier in this century, as they were confronted by the superstitious beliefs and practices that they found rampant in the so-called primitive societies that were being examined in various corners of the world. However, contemporary anthropologists recognize that the propensity for magical thinking does not distinguish "primitive" societies from technologically advanced ones. It is a feature of the intellectual activity observed in all societies.[7] Just as has been observed in pretechnical societies,[8] it has been observed in modern industrial societies that during stressful times superstitiousness, magical thinking, and belief in the paranormal increase.[9,10]

The automatic power of temporal contiguity to produce a perceived relationship between two events promotes survival, as well as understanding of the world around us. It also leaves us open to egregious error from time to time. Crossing one's fingers may *seem* to produce desired outcomes, just as taking large doses of Vitamin C may *seem* to protect against colds, or supplication to Jehovah or Krishna or Zeus may *seem* to ward off danger. And of course, *seeming* is reinforcing; one might even say that *seeming* is believing. Although the ritual appears to bring the desired outcomes only occasionally, that is all that is necessary to establish a powerful belief in it.

Causality is important to humans, and most make considerable effort to explain the reason behind events that affect them. As early as eight months of age, the infant begins to demonstrate some attention to cause and effect. The infant's first concept of causality develops through observation of co-occurring events (events occurring in temporal contiguity). The great Swiss psychologist Jean Piaget studied how infants come to see cause and effect in situations where two unrelated events occur one after the other.[11] A baby laughs, and by chance the mobile above her head is moved by a gust of air. The baby laughs again and again, while staring at the mobile, in the apparent belief that her laughing will produce more movement. He referred to this as *magico-phenomenalistic causality*, "magico" because there is no comprehension or concern about the link between the cause and the sought-after effect, and "phenomenalistic" because the personal perception of the coincidental occurrence of the two events, one shortly after the other, leads the infant to react as though they are causally related. Such magical thought reflects the fundamental nature of the nervous system, as discussed earlier, which links together two stimuli that occur closely together in time. Only with repeated experience and explicit teaching does the child gradually acquire a more accurate understanding of causal relationships in the physical world. Yet, the brain by its very nature remains forever vulnerable to magical thinking.

Magical learning in its simplest form is referred to *superstitious conditioning*, which occurs without any necessary cognitive component, and is observed in animals as well as humans. The pigeon that finds itself in a cage where grains of food fall from a chute on a random basis can be observed to develop superstitious behaviors based on what it was doing at the time of a food delivery. It if happened to have lifted one foot just as the food appeared, it is likely to continue in future to hop on one foot from time to time, as if it is trying to produce the food. Occasionally, the food will appear again while it is hopping on one foot, reinforcing the behavior that in reality has nothing to do with the appearance of food. It is operantly conditioned by the happenstance pairing of a particular behavior with a particular event. Human superstitious behaviors are little different, except that they often become part of the belief system of a society, so that individuals learn them without the need for direct original experience. If one has been told often enough that a black cat crossing one's path leads to bad luck, then when one sees a black cat cross one's path, one may become slightly anxious. Such superstitious beliefs are readily maintained by anxiety reduction—see a black cat ahead, avoid it, feel better—and by salient events that occur at random, producing partial reinforcement. And it is important to recognize that partial reinforcement—that is, reinforcement that occurs sporadically and not on every occasion—leads to patterns of behavior and belief that are very resistant to extinction. For example, after having had a black cat cross our path, we may interpret subsequent events as though they reflect the portended bad luck. Again, we may well experience conflict with our intellectual system—we *know* that the belief in the jinx brought by a black cat is not something we should take seriously, and yet, if we have been reared with that belief, it may be very hard to shake off.

Transcendental beliefs can provide a sense of understanding of the world and an avenue for controlling undesirable elements in the world, and this is especially so in areas where science can offer little comfort: Why was my friend killed in the accident while everyone else survived? I am getting old, life is almost over; is this all there is? Why does this drought go on so long; how will we survive it, and who can help us? Transcendental beliefs that provide explanation or offer hope or intervention result in a lessening of anxiety, which is itself a powerful reinforcement of the belief. If woodland sprites or omnipotent deities are the cause of our problems or control our fates, then if we can find out how to please them, we should be able to rest easier. If our God is a loving and omnipotent being, then He will save us if we pray. Such is the basis of what Kurtz has called the "transcendental temptation;"[12] it is tempting to believe in transcendental reality because such a belief offers us respite, comfort, and salvation.

It is also important to understand that such beliefs may help calm an individual who is in a crisis; and he or she may, in turn, react in a calmer, more intellectual manner, solving the problem and then attributing the solution to the magical belief. Thus, the frightened individual who believes that praying to Zeus or waving an amulet will aid escape from a rapidly pursuing wolf may be in a better position to take defensive action once he or she has prayed or rubbed the amulet. By calming down a little, and perhaps deciding to climb a tree to escape the wolf, the individual comes to believe even more powerfully in the efficacy of the prayer or charm.

While experiential learning is the wellspring of magical thinking, intellectual activity is often harnessed in service of magical/transcendental beliefs. If we believe that dowsers can locate water or gold by means of a forked stick, we may generate quasi-scientific explanations to make our belief more palatable to ourselves. Quantum mechanics has been bent and twisted in all sorts of ways by those anxious to accommodate belief in precognition or psychokinesis (the putative movement of physical objects directly by the power of the mind).

## PROBABILISTIC REASONING

We rely upon both logical thought and intuitive thought in our day-to-day lives. Intuitive thought—"experientially derived knowledge"—comes to us without effort and is often very compelling and may influence behavior to a greater extent than does abstract intellectual knowledge.[13] Reed suggested that intuitive thought may be so very compelling because its products spring full-blown into consciousness.[14] Unlike the products of our logic, we have no access to how the thought developed.

A well-known example of how intuitive thought can overwhelm logic is the Monte Carlo, or "gambler's," fallacy. Although we take for granted that playing cards and dice and roulette wheels have no memory and that each outcome is independent of the preceding outcomes, this is not the way our intuition works. If the person playing the roulette wheel has observed that the outcome has been black twenty times in a row, it may seem compellingly obvious that red is "overdue," and that one should bet on red. Even people who are aware of the illogic of this conclusion can be caught up in it. Again, we have the conflict between experiential and intellectual systems.

Indeed, intuition is often very wrong when we deal with subject matter removed from the realm of everyday life. While an apocryphal scientist may have declared that a honey bee cannot possibly fly because its wing span is not great enough, for some lay people it is in some sense hard to believe that a jumbo jet can ever get off the runway, even though they have in the past witnessed this happen. They *know* that the airplane can fly, and they may even understand Bernoulli's principle and know enough about the physics of flight not to be surprised that it can indeed get into the air. Yet, it does not really fit in with common sense. It is much easier at an intuitive level to accept that small model airplanes can fly. They are light, and we all "understand" intuitively how birds and newspapers and other light things can be lifted on currents of air. The further we are removed from everyday commonplace experience, the more likely our intuition is to be wrong.

Psychologist Ray Hyman provides an interesting example of how intuition fails when we stray from the realm of ordinary experience.[15] Imagine taking a sheet of paper and folding it exactly into two. Repeat this operation a total of fifty times, ignoring—since this is a thought experiment—the difficulty of making all those tiny folds. The question is this: how high from the floor would the folded paper extend? When I pose this question to my students, some venture that it would go from the floor to the top of the desk. None ever believe that it would reach the ceiling. The correct answer is that a piece of paper of ordinary thickness would extend out beyond the sun, over 93,000,000 miles, after fifty folds. This is very counterintuitive, and students are always quick to demand proof—which is, of course, readily given.

There are other areas where intuition fails us. Our brains are very poor at keeping track of relative frequencies and thinking in terms of correlations between co-occurring events.[2] It is interesting to observe that despite thousands of years of exploration and discovery in mathematics, the study of probability did not begin until the 17th century. The Babylonians were skilled in several areas of mathematics—for example, they developed enough sophistication to be able to solve for the positive roots of any quadratic equation and even some cubic equations, and they developed compound interest tables. The ancient Greeks are well-known for their prowess in geometry—Pythagoras, Euclid, Archimedes, for example. In the Middle Ages, Muslim scholars made significant discoveries relating to number theory and algebra. Yet, through all those years, and even though games of chance had been a part of every age, notions of probability were ignored. Only in the 17th century, when a gambling devotee who happened also to be an amateur mathematician approached Pascal seeking assistance in his efforts to understand the likelihood of winning or losing in a particular game of chance, did humanity for the first time bring intellect to bear on the subject of probability. The lack of interest in this domain across the centuries possibly mirrors the inherent inability we have with regard to processing relative frequencies and thinking in terms of correlations.

We are inherently poor at correlational thinking because of an asymmetry that exists in the nervous system that makes it difficult for us to intuit co-occurrence rates. If a child touches a hot stove and feels pain, and then later accidently touches it again but this time it is cool and there is no pain, the association between stove and pain is not simply nullified. Many pairings of stove/no pain would be needed to overcome the association that was made on the basis of one experience. The co-occurrence of two significant events, even if purely coincidental, can have an overwhelming effect on our judgement of co-occurrence likelihood and our interpretation of cause and effect. A motorcycle gang rides into town, and that night the town hall burns down under suspicious circumstances. Just get into the shower, and the telephone rings, as it so often seems to do on such occasions. Stare at someone's back and the person turns around, as if the stare had caused this. Dream about an accident, and the next day someone runs into your car. The brain does not automatically keep track of nonpairings of the two events that have on this particular occasion occurred closely together in time. If the motorcycle gang rides into town and nothing bad happens, this does not get stored away to cancel out a subsequent pairing with some negative event. If one has a dream of death and the next day one learns that a friend has died, logically we would want to know how many times the individual has had dreams of death and nothing happened. In other words, was this simply a coincidence? However, situations where people dream of death and no one dies, or where people die and there was no prior dream, do not make an impact, again because of the architecture of the nervous system. Yet, logically we cannot evaluate the correlation between dreaming and subsequent events without a tally of those situations as well. Thus, we store relevant information not in terms of probabilities but rather in terms of frequencies of salient co-occurrences. Given that a predictor event has occurred, people have great difficulty in using nonconfirmatory information coming from a nonoccurrence of the second event.

Many research studies have shown that individuals have little ability to judge correlations between a series of events. For example, subjects in one study were shown a whole series of trials relating to cloud seeding. On each trial they were informed that cloud seeding had or had not been carried out, and that it had or had not rained. Although there was no actual statistical association between seeding and rain, the subjects judged that such an association indeed existed. This judgment was based on those times when seeding and rain were paired. In other words, the subjects were much more influenced by the pairing of seeding and rain than they were by other pairings—seeding, no rain; no seeding, rain; no seeding, no rain. They incorrectly intuited that seeding did indeed make rain more likely.[16]

People have considerable difficulty organizing information into a format that is amenable to correlational inference. We do not normally need to worry about our correlational deficiencies because in general in everyday life events that are highly correlated statistically also tend to follow each other closely in time.[16] Thus temporal contiguity is normally enough to inform us about which events are related to each other. Tversky and Kahneman have demonstrated that it is not only the general public that is poor at probabilistic reasoning; so are professionals, even professional statisticians in situations outside their work.[17]

Even if paranormal phenomena do not exist, the workings of the brain are such that it is likely that most people will sooner or later experience something that seems to be paranormal—be it in the form of telepathy, psychokinesis, precognition, or whatever. The experience of the paranormal most often involves an emotionally striking co-occurrence of two events. One is about to call Uncle Harry, to whom one has not spoken for months; as one reaches for the telephone, it rings, and it is Harry calling. A person has a dream about a fire, and the next day there is a major explosion in the city. In each of these and many other situations, the nervous system of the individual says to him or her that there is a significant correspondence between these events. If we are not careful to subject these events to a probabilistic analysis, if we do not keep records about how often we dream with no correspondence to events of the next day, then we are unable to evaluate properly the likelihood that the correspondence occurred by chance. And given the emotional zing that such correspondence often produce, we are more likely to categorize the occurrence as something meaningful and be loath to think of it in terms of coincidence.

## SCIENCE AND CRITICAL THINKING

Science, religion, and even natural magic all attempt to make sense out of nature. Co-occurrences are important in all three. However, unlike magico-religious thought, a basic goal of science is to discriminate between patterns that are meaningful and those that are not. It is only science that attempts to evaluate whether or not coincidence accounts for observed co-occurrence. Scientists have to learn to bend their own thinking—which is subject to intuitive and magical influences just as is everyone else's—to the rules of science, through the use of socially developed and socially agreed-upon tools to help avoid error.[18] Science is in part a social enterprise, in which the observations and conclusions of a single individual are fully accepted only if they can be verified independently by other researchers. Hence, collective, rather than individual, validation of findings is paramount.

However, when science is taught, it is almost always taught without any reference to existential or transcendental questions, thus allowing students' transcendental beliefs to continue to escape logical analysis and reinforcing the notion that science need not concern itself with transcendent beliefs. This in turn helps nourish the notion that science is limited and that putatively transcendent phenomena lie in a realm of reality beyond science's ken. Great care is generally taken to avoid challenge to religious beliefs. Hence, why should we expect scientists to be any different from the general public when it comes to transcendental belief?

Many scientists are religious; many believe in psychic phenomena based on their own experiences and emotional needs. We are left, then, in a situation of conflict between experiential learning and logic, between faith and science. As William James, himself deeply interested both in science and the paranormal, wrote, "At one hour scientists, at another they are Christians or common men, with the will to live burning hot in their breasts; and holding thus the two ends of the chain, they are careless of the intermediate connection."[19]

## PARAPSYCHOLOGY: AT THE FRONTIER BETWEEN SCIENCE AND RELIGION

Science and organized religion have clashed mightily in the past, but today there is a generally peaceful coexistence between them. Parapsychology lies at the frontier between science and religion. It is the quest to find evidence that there is more to the human personality than materialistic science allows. Parapsychological phenomena are by definition those that involve human (or animal) agency, that are not explainable in terms of modern science, and that involve some radically different relationship between consciousness and the physical world than is held to be possible by contemporary science.[20]

When the Society for Psychical Research (SPR) was organized in England in 1882, with well-known Cambridge philosopher Henry Sidgwick as its president, a central part of its mission was to prove scientifically the existence of the soul, or at the very least to demonstrate that there is a nonmaterial aspect to human existence.[21] Sidgwick and other leaders of the SPR had suffered tremendous personal conflicts between their religious beliefs and the implications of developing science, especially the implications of Darwinian theory.[22] They felt a strong need to reconcile science and religion. By giving up the apparent mythology of religion and focusing instead on survival of the human personality, Sidgwick was hopeful that the postmortem survival of the human personality could be put on a solid scientific footing.

Some years later in the United States a young botanist, Joseph Banks Rhine, who was to become the preeminent parapsychologist in the world and transform parapsychology into a laboratory discipline, was also caught in a struggle between religious beliefs and scientific materialism. He had planned to become a minister until science got in the way. He came to believe that parapsychology could serve as a bridge between science and religion and believed that in order to demonstrate the existence of a soul or of at least mind-body dualism, it was first necessary to demonstrate that extrasensory perception and psychokinesis were genuine phenomena.

Such conflict between science and religion was experienced by many of the most influential figures in parapsychology. Indeed, it has been observed that most parapsychologists, after a fall into agnosticism, have been motivated by a search for evidence that would show that life has meaning.[23]

There is an important difference between the role of religion in the lives of conventional scientists and the role it plays in the lives of many if not most parapsychologists. In the former science is pursued in spite of religious needs, whereas with parapsychologists it is typical that a dissatisfaction with or renunciation of traditional religious teaching has promoted a search for transcendent phenomena *because* of such needs.[22] This reflects a dissatisfaction both with traditional religious "mythology" and with the materialism inherent in modern science. The inherent difficulty, if not impossibility, of reconciling science and religion underlies the difficulties that separate parapsychology from conventional science.

It is noteworthy that there is nothing emerging from conventional scientific research that would suggest the existence of psychic (psi) phenomena. Physicists do not report that their experiments turn out differently depending on the wishes and motivations and thoughts of the researcher. Yet for more than a century various researchers trained in the ways of science, some of them preeminent scientists in their own right before they turned to parapsychology, have tried without success to produce empirical evidence of a quality that would convince mainstream science of the existence of psi phenomena. How is the belief in such phenomena sustained despite such a history? In my view, the belief comes first: the quest is to find data to substantiate the belief, rather than to find an explanation for an observable phenomenon.

Parapsychology attempts to be a science. Many of its practitioners are devoted to the application of scientific research methods. Yet it has not been able to gain real acceptance within the halls of science, and this for a number of reasons. First of all, its subject matter is defined in exclusively negative terms. One can demonstrate that a psychic phenomenon has occurred only if one can rule out all normal explanations; and since one can never be certain that all such explanations can be ruled out, one can never be certain that a particular event lies outside the range of normal explanation. Second, even using this highly unsatisfactory negative definition, there has never been a demonstration of a psychic phenomenon that is, in the words of parapsychologist John Beloff "strongly replicable,"[24] meaning that any competent researcher can follow the specified methodology and produce the same effect. Unrepeatability is such a problem in parapsychology that some parapsychologists have argued that criterion should be abandoned, that psi is not only elusive but may also be inherently unlawful. The suggestion that the very criteria that scientists have developed to help protect them from error should be lowered in order to accommodate the elusiveness of psi is hardly likely to find attentive ears in the mainstream scientific community. Add to this the fact that parapsychological research continues to be plagued by methodological flaws (e.g., Refs. 25–27). Equally important, parapsychologists render their hypotheses virtually unfalsifiable by the employment of various "effects" to explain away failures in their data. One of the more notable of these is the "psi experimenter effect," an effect that is cited so often that it has been described by one parapsychologist as parapsychology's one and only finding.[28] It suggests that failures to replicate may be due to differences in the psychic abilities of the researchers. Of course, this would explain the inability of skeptical researches to obtain results. Indeed, if this hypothesis were correct, it would seem that science as a whole would be impossible, for the theoretical predilections of researchers would influence the results they obtain.

Finally, parapsychology cannot and will not specify when psi should *not* occur. It supposedly transcends space and time, and according to some studies works just as powerfully across the world as across the room, and forward and backward in time as well as contemporaneously. Humans, animals, insects, and even fertilized eggs have supposedly proved to be successful psi operators. Psi apparently knows no bounds.

Belief in parapsychological phenomena, including the belief that words or actions lead to physical effects not accountable for in terms of normal science, falls within the definition of magical thinking, since an effect is imputed to a cause without knowledge of, or particular concern about, the putative causal link.[29] Some parapsychologists admit this and view natural magic simply as a manifestation of psychic power; and, indeed, it has been stated in the parapsychological literature that if one takes away the rituals, natural magic is nothing more than psychokinesis.[30]

In conclusion, parapsychology reflects a duality in our thinking, a duality that can be traced back to the inherent conflict between the two ways in which we learn—between experiential, magical thought and logical, analytical thought. Reflecting these two different and often conflicting information processing systems, human beings have constructed two major belief systems, one scientific-humanist and the other transcendental. They are difficult to reconcile. Some people resolve the conflict by compartmentalization. Others reject science and logic and go with intuition; there is abundant worship at this temple these days. Others choose logic and science, and then overcome or suppress whatever existential anxieties may present themselves. Parapsychologists attempt to find a single system to accommodate the two.

In any case, it is an ongoing struggle. Every one of us is vulnerable to magical thinking whether we recognize it or not; and those who fail to recognize it are most likely to be vulnerable to a conversion experience. The critical thinking and data gathering methods that make up science have proved to be the most effective system we have for protecting ourselves from error and self-delusion. They are the product of centuries of intellectual activity, but they do not come naturally or easily to the human brain. We need to be as concerned for their survival as we are about the survival of rain forests and other endangered treasures.

## NOTES

1 J. E. LeDoux, "Emotion, Memory and the Brain."
2 S. Epstein, "Integration of the Cognitive and Psychodynamic Unconscious."
3 J. E. Alcock, *Parapsychology: Science or Magic?*
4 J. T. Frank, "Nature and Function of Belief Systems."
5 J. E. Alcock, "Religion and Rationality."
6 G. H. Gallup, Jr. & F. Newport, "Belief in Paranormal Phenomena among Adult Americans."
7 R. A. Shweder, "Likeness and Likelihood in Everyday Thought: Magical Thinking in Judgments about Personality."
8 B. Malinowski, *Magic, Science, Religion and Other Essays.*
9 V. R. Padgett & D. O. Jorgenson, "Superstition and Economic Threat: Germany, 1918–1940."
10 G. Keinan, "Effects of Stress and Tolerance of Ambiguity on Magical Thinking."
11 J. Piaget, *The Construction of Reality in the Child.*
12 P. Kurtz, *The Transcendental Temptation.*
13 R. M. Shiffrin & W. Schneider, "Controlled and Automatic Human Information Processing 2: Perceptual Learning, Automatic Attending and a General Theory."
14 G. Reed, "Superstitious Beliefs and Cognitive Processes."
15 R. Hyman, personal communication.
16 H. M. Jenkins & W. C. Ward, "Judgments of Contingency between Responses and Outcomes."
17 A. Tversky & D. Kahneman, "Judgment under Uncertainty: Heuristics and Biases."
18 D. W. Fiske, "Comments."
19 W. James, *The Will to Believe.*
20 J. E. Alcock, "Parapsychology: Science of the Anomalous or Search for the Soul?"
21 J. J. Cerullo, *The Secularization of the Soul.*
22 J. E. Alcock, "Parapsychology as a 'Spiritual Science.'"
23 R. L. Moore, *In Search of White Crows.*
24 J. Beloff, "Research Strategies for Dealing with Unstable Phenomena."
25 J. E. Alcock, *Science and Supernature.*
26 C. Akers, "Methodological Criticisms of Parapsychology."
27 R. Hyman, "The Ganzfeld/psi Experiment: A Critical Appraisal."
28 A. Parker, "A Holistic Methodology in psi Research."
29 L. Zusne & W. H. Jones, *Anomalistic Psychology.*
30 L. M. Beynam, "Quantum Physics and Paranormal Events."

## REFERENCES

ALCOCK, J. E. "Parapsychology as a 'Spiritual Science.'" In *A Skeptic's Handbook of Parapsychology*, edited by P. Kurtz. Buffalo, NY: Prometheus Books, 1985.

———. "Parapsychology: Science of the Anomalous or Search for the Soul?" *Behavioral and Brain Sciences* 10, no. 4 (1987): 553–565.

———. *Parapsychology: Science or Magic?* Oxford: Pergamon, 1981.

———. "Religion and Rationality." In *Religion and Mental Health*, edited by J. F. Schumaker. Oxford: Oxford University Press, 1992.

———. *Science and Supernature.* Buffalo, NY: Prometheus Books, 1990.

AKERS, C. "Methodological Criticisms of Parapsychology." *Advances in Parapsychological Research* 4 (1984): 112–164.

BELOFF, J. "Research Strategies for Dealing with Unstable Phenomena." *Parapsychological Review* 1 (1984): 1–7.

BEYNAM, L. M. "Quantum Physics and Paranormal Events." In *Future Science*, edited by J. White & S. Krippner. Garden City, NY: Doubleday, 1977.

CERULLO, J. J. *The Secularization of the Soul.* Philadelphia, PA: Institute for the Study of Human Issues, 1982.

EPSTEIN, S. "Integration of the Cognitive and Psychodynamic Unconscious." *American Psychologist* 49 (1994): 709–724.

FISK, D. W. "Comments." *Current Anthropology* 18 (1977): 649.

FRANK, J. T. "Nature and Function of Belief Systems." *American Psychologist* 32 (1977): 555–559.

GALLUP, G. H., JR. & F. NEWPORT. "Belief in Paranormal Phenomena among Adult Americans." *The Skeptical Inquirer* 15 (1991): 137–146.

HYMAN, R. "The Ganzfeld/psi Experiment: A Critical Appraisal." *Journal of Parapsychology* 49 (1985): 3–49.

———. Personal Communication, 1990.

JAMES, W. *The Will to Believe.* New York, NY: Dover Publications, 1956.

JENKINS, H. M. & W. C. WARD. "Judgments of Contingency between Responses and Outcomes." *Psychological Monographs* 79 no. 1 (1965).

KEINAN, G. "Effects of Stress and Tolerance of Ambiguity on Magical Thinking." *Journal of Personality and Social Psychology* 67 (1994): 48–55.

KURTZ, P. *The Transcendental Temptation.* Buffalo, NY: Prometheus Books, 1986.

LEDOUX, J. E. "Emotion, Memory and the Brain."*Scientific American* 270 (1994): 50–57.

MALINOWSKI, B. *Magic, Science, Religion and Other Essays.* New York, NY: Free Press, 1948.

MOORE, R. L. *In Search of White Crows.* New York, NY: Oxford University Press, 1977.

PADGETT, V. R. & D. O. JORGENSON. "Superstition and Economic Threat: Germany, 1918–1940." *Personality and Social Psychology Bulletin* 8 (1982): 736–741.

PARKER, A. "A Holistic Methodology in psi Research." *Parapsychology Review* 9 (1978): 1–6.

PAIGET, J. *The Construction of Reality in the Child.* New York, NY: Basic Books, 1954.

REED, G. "Superstitious Beliefs and Cognitive Processes." Paper presented at the Symposium on Anomalistic Psychology, Annual Meeting of the American Psychological Association, Toronto, 1984.

SHRIFFIN, R. M. & W. SCHNEIDER. "Controlled and Automatic Human Information Processing 2: Perceptual Learning, Automatic Attending and a General Theory." *Psychological Review* 84 (1977): 127–190.

SHWEDER, R. A. "Likeness and Likelihood in Everyday Thought: Magical Thinking in Judgments about Personality." *Current Anthropology* 18 (1977): 637–648.

TVERSKY, A. & D. KAHNEMAN. "Judgment under Uncertainty: Heuristics and Biases." *Science* 185, no. 4157 (1974): 1124–1131.

ZUSNE, L. & W. H. JONES. *Anomalistic Psychology.* Hillsdale, NJ: Erlbaum, 1989.

# FLIGHTS OF FANCY
## Science, Reason, and Common Sense

BARRY R. GROSS

A CONSIDERABLE PART though by no means the whole of philosophy is given to the task of constructing accounts of how one or another interesting aspect of human activity that works actually does work. To give such an account turns out to be much harder than one might have thought, and I know of no example of one that is agreed by all knowledgeable observers to be successful. For example, we know certain things. But how do we know that we know them? What is knowledge and how does it come about? In the *Theaetetus*, Plato attempts a definition of "knowledge" that, after a lengthy discussion he admits collapses into circularity.[1] And in his *Sophist* he attempts an account of how we can predicate, as he calls it, "nonbeing"—how we can say, for instance, "There are no unicorns" without finding ourselves speaking about nothing and, therefore, not having succeeded in speaking at all. Here he is less unsuccessful, if one may speak that way, but unsuccessful still.[2] More recent examples are Susan Haack's neat reconstruction of the theory of justification[3] and John Searle's attempt to show how and in what sense social reality, but only social reality, is constructed.[4]

Another part of philosophy is given over to the task of saying how arguments or ontological commitments that seem to go wrong do, in fact, go wrong. Here one thinks of Aristotle on sophistical refutation,[5] Aquinas on the ontological proof,[6] Russell on denoting,[7] Ryle on category mistakes,[8] Goodman on the logic of evidence,[9] and Wittgenstein on everything.[10]

I associate the quite modest remarks that follow with this second philosophical tradition. I hope to give an outline of what has gone wrong in the thinking of the critics of science in particular, and of many of the critics of the successes of the developed West in general. For we shall not be able to combat antiscientism and its epigones with any great hope of success unless we understand quite clearly just what the problem is. As I see it, the problem is intellectually not very deep. That is the good news. The bad news is that it appears to be psychologically quite deep in two ways. First, those attracted to such modes of thinking are tenacious, and no amount of argument or counterevidence is likely to move many of them. Second, the antiauthoritarianism and Utopian modes of thought are perennial in Western history.

The sole remedy at our disposal is to quarantine the antiscience brigades and inoculate the rest of the population against them. This requires that those who understand something about science—I mean scientists—and have some common sense, too, will have to devote some of their energy to systematic confrontation with the enemies of science and continually to make the case for science for those, including our political masters, who have not thought much about it. And that means that scientists will have to ally themselves with intelligent nonscientists and learn how to turn aside the rhetoric of the Luddites, uncongenial as that task may seem. To rework a phrase, eternal vigilance is the price of a reasonable society.

My late and sometime mother-in-law was no intellectual. She moved in romantic poetic circles in the Riga of the twenties, thirties, and forties. Not only did she not understand science, she was unreasonably, or perhaps reasonably, suspicious of it—of medicine in particular, whose principle and nefarious task she thought was the use of one's body for experimental purposes. Even so, she would avail herself of its services as old age crept up, provided always that she could summon a Latvian doctor in New York. However, she did know a thing or two and hewed to an old Baltic proverb as she watched human antics: smart people have smart problems—perhaps a colloquially expressed version of Hume's first inquiry.[11]

Peasants and other plain folk don't mess with epistemology. They are unlikely even to know the word. Only smart people do that. The organizers of this conference did not ask me to suggest a title for it. Had they sought my counsel, I should have suggested that it wasn't any flight from reason that fuels most of the problems that confront us. I should have suggested that, on the contrary, the difficulty we face stems from hewing to reason, but hewing to it unreasonably—to sophistic and sophomoric reason, in the manner of Buridan's ass,[12] but to reason nevertheless.

There is a flight, but it is a flight from common sense. What is common sense? I suppose a decent approximation is that it is reason robustly laced with experience, to which experience one pays serious attention. A rational man proportions his belief to the evidence, as Hume puts it.[13] *Reason* has many different meanings, one of which is, of course, just *common sense.* It will emerge shortly why I find it useful to set common sense off against the other meanings.

About reason it is probably helpful to keep in mind the distinctions among (1) the multiple and particular uses of reasoning that any of us is likely to employ in his daily and professional life, (2) the defenses we might mount if challenged on any one or few of these particular uses, and (3) a general defense of reason that philosophers (who else?) might make against a global attack not on this or that particular piece of reason but on reason *tout court.*

I would be very curious to read a global attack on reason itself. That term denotes so many different things that the attack would have to cover vast territory. And how could it be made without presupposing some canons of what was supposed to be under attack—reason? Of course, there always have been glancing attacks—some skeptical, like Epictetus; some religious, like

Tertullian, Paul, Luther, Calvin, or Kierkegaard; some literary, like Swift, Dostoevsky, or Sartre; and some conversational, like the late Feyerabend and the still quick Rorty. But to my knowledge no one has challenged, say, Aristotle's syllogistic, or the proofs of consistency and completeness for the propositional and first order predicate calculi. As Aristotle would no doubt point out were he still with us and a *Metaphysics*-ish (Book Gamma) mood, the Sandra Hardings and Latours of this world still do count their change at the checkout counter and negotiate their fringe benefits with care. Radical *aporia* raises its ugly head only in the lecture hall or the seminar room. Those who claim to deny reason's claims give pragmatic refutation of their denial.

But in the seminar room and in print there is a flight from science. Though I have not tried to apply my thesis to all the fauna catalogued in *Higher Superstition*,[14] I think it would be tedious but not difficult to show that the various attacks on science all accept implicitly what we loosely call reason. They employ reason to attack science, sometimes as merely misapplied—like radical environmentalists—sometimes as a complete enterprise—e.g., Harding-type feminists.[15] Harding herself is not good at making arguments. A shockingly high percentage of hers are either invalid or unsound. But even bad arguments are arguments. So she uses reason. Her problem is that she misconstrues her target. She focuses not on central issues but on ancillary ones, which she wrongly thinks are central. To put it neutrally in a formulation I borrow from Susan Haack,[16] Harding and her allies have different background beliefs from those of us who think current science is in general a salutary enterprise. Insofar as her background beliefs miss the central points, Harding lacks not reason but reasonableness or judgment or common sense.

Among the background beliefs shared by the antiscience brigades we may pick out one general belief and seven particular ones. The general belief is that what most of us take to be ancillary to the scientific enterprise is really central. The particular beliefs are (1) authority is always wrong and, consequently, (2) so is hierarchy; (3) populations should be represented in the various sciences proportionately by race, sex, and ethnicity; (4) one can give meaningful analyses and criticisms of a technical subject though completely ignorant of it; (5) abstract argument supersedes particular results; (6) there exist non-Western ways and feminine ways to do science that are, root and branch, different from the ways in which science is now done; and (7) Utopia cannot merely be agreed upon in principle but can actually be achieved.

Consider Harding's complaints,[17] among others, that (1) science is dominated largely by men, (2) it is hierarchical, and (3) it is authoritarian. Her strong point is that all three claims are true; her weakness is, of course, that they are all irrelevant, for they do not constitute an attack on anything central in what we value in science. What do we value in the various sciences? We value the production of knowledge and its practical application to ease our lives.

Now Harding says in several places that the fact that science works is not enough to justify it. *A fortiori*, she admits that it works. And if it works, then the facts that it is male dominated, hierarchical, and authoritarian are no reasons to change it unless one could be reasonably sure that it would work better were these things not the case. Let's leave male bashing aside for the moment. Could science be egalitarian? Only if all scientists had equal and equally important levels of knowledge and ideas. To state this condition is to see its absurdity. While it is logically possible that we should all be mental clones of each other, nothing that we know about the world or about evolution suggests there is the remotest possibility of such a state. Indeed, if this state came about, the world would be a most uninteresting and unfruitful place. If we all had the same knowledge and ideas, then there would be far less new knowledge. Science would advance like a slowly moving wave front, and I suppose we could make do with only one scientist.

But the way the world actually is, we find in athletics, entertainment, science, and mathematics that some of us will be better than others, a few will be outstanding, and a very few, towering. So some of us will know a great deal more than others and have a great many more new and successful ideas. This hierarchy is as necessary a concomitant of science as it is of tennis or plumbing.

In fields of knowledge as in other fields authority goes roughly along with achievement. Those who know and have achieved more will have more authority. Of course personality traits enter the picture, but this complication does not alter the main principle that hierarchy and authority are systematically correlated with high levels of knowledge and achievement. So if you want knowledge, you must accept the correlatives. Now, attacks on hierarchy and authority are not attacks on reason. They are attacks aimed to buttress a set of background beliefs. Among them are the beliefs that hierarchy and authority are, in themselves, evil and that one can have science without them. Given what we know about science, these beliefs are false. They are poor judgments and unreasonable beliefs, but they are not flights from reason itself.

Turning to the question of male "domination" we see that the background beliefs are two: that proportionality by race or sex in every occupation is achievable and that absence of proportionality is morally wrong. Of course, this indicts virtually all activities, since such proportionality is never, as a matter of sociological fact, found. Baseball, football, Grand Prix racing, government, the arts—virtually all occupations (shall we indict childbearing here, too?) will, on this view, turn out to be morally culpable. For reasonable people this would constitute a kind of *reductio.* But, *credo, quia ineptum.*[18] Some people are not reasonable. Them we could never convince. And, of course, proof in the strict sense is out of the question, since there never could be strict proof of such things. In any case, those who hold such absurd background beliefs are rather like my logic students. They do not understand how a formal proof works and so never see that any particular proof proves any result. What would be morally culpable, of course, is not if the population of immunologists or particle physicists exhibited one or another racial, ethnic, or sexual pattern, but rather if talented particle physicists or immunologists were prevented from a career because of race or sex. Here as elsewhere the disparate impact argument has much to answer for.[19]

Another absurd background belief held by most, though not quite all, of the antiscience brigade—the late Paul Feyerabend[20] was an exception, the still quick Dick Rorty[21] is not—is that one can make useful, even profound, criticisms about a technical subject of which one is entirely ignorant. Now, we might have to show some sympathy with this notion, for it is a very American one, seen especially in our penchant to moralize about politics and international relations in deep ignorance of history, diplomacy, and the current situation. The fact is, though, that people who have not been trained in one or another science and have not practiced at it are very unlikely to know how it works on the ground, or even in theory. They will not know how research is produced, how it is written up, how it is announced at meetings or in journals, or what counts as success. Lacking this technical knowledge they will tend to focus on those aspects of science they think they understand—the social aspects.

Consider the following example. The mythical, proverbial middle American who has never been abroad and is innocent of the rules of any card games of skill—suppose him transported, say, to a London bridge party about which he is to write a sociological analysis. Suppose, further, that the dominant player, the player who takes most hands and takes them brilliantly, and to whom the others defer, rejoices in a hereditary title with accent to match. What will our plain-speaking sociologist think? What will he write? Very likely that dominance at the bridge table is a function of ancient peerage, plumy voice, and arrogant behavior.

Now you may think this unkind. But I have a real-life example. Last fall one prominent proponent of this sort of nonsense submitted to the National Association of Scholars journal, *Academic Questions*, a paper that claimed—*en passant*, as so often—that the major reasons German physicists accepted Heisenberg's uncertainty principle were to be found in money and politics; that the reputation of physics in general had fallen in Weimar, Germany—why he did not say—and that control of funds fell largely into Heisenberg's hands—why he did not say. Wishing to get funds, physicists were obliged to climb aboard the Heisenberg express. We did not publish this article.

If you are ignorant of a subject, the best available way to talk about it is in the abstract. Newspaper and television editorials provide many amusing examples. So do the antiscience brigades. Strong social constructivism trades on this, and it would be tedious to enumerate every example. It is this reference to the abstract that allows such silly formulations as "reality is socially constructed." Is one meant to think that if "society" (another abstraction) changed its mind and came to a different consensus, then one might happily drink $H_2SO_4$ with the results one now expects from a glass of $H_2O$, or hurl himself from the top of the World Trade Center to float down like a feather?

Even in cases of those aspects of the world where it is not altogether wrong to say, very carefully, they are socially constructed—e.g., money and markets—there is a recalcitrant reality one does bump up against, as you easily may find out by attempting to evade payment for goods purchased. And as Aristotle pointed out some time ago, it is a non-socially constructed fact about the world that money and markets tend to make life much easier and more efficient than it would be in their absence.[22]

About the possibility of any science completely and utterly different from science as we know and practice it I have written at length elsewhere.[23] Here I merely point out that it is possible to think and utter such phrases only *in abstracto*, only if one gives no thought to any content they may have. For to try to think of a possible content for them is to see them as entirely empty.

Not so for Utopias. Everyone can imagine his own. For example, the recent death of Burl Ives recalled to me his rendition of "Big Rock Candy Mountain" and the buzzing of the bees in the politically incorrect cigarette trees. And, of course, there is Verlaine's naughty vision of a ballroom constructed entirely of bits of female anatomy.[24] What are absurd are the notions (1) that we should all agree in theory on what Utopia is, and (2) that any Utopias at all are on offer. With thoughts of Utopias all common sense has vanished, and we see plainly that no amount of argument or reason or evidence will carry weight with the antiscience brigades.

There are two more sources of antiscientism that should be kept in mind. First is the general antiintellectualism that pervades American life.[25] One of innumerable examples that I could give is the article that appeared in *The New York Times* on the O. J. Simpson trial and supposed juror reaction to the DNA evidence.[26] I quote briefly:

> For the squirming jurors held captive in Judge Lance A. Ito's Los Angeles courtroom, the last couple of weeks of DNA testimony must seem like the worst possible visitation of a well-known recurrent nightmare: It's the last week of the semester and suddenly you remember the science class you forgot and signed up for and never thought to attend. Too late to drop it now. In a vast silent hall, you stare at the final exam with its sickening thicket of meaningless symbols and equations. Your head swimming with the nausea of incomprehension, you run for the exit. . . . You're being asked to comprehend the incomprehensible. . . .

Does one have to point out the sophomorically fictitious character of this passage and what it says about the mind-sets of the author who wrote it and the editor who permitted it to be printed?

The second sort of antiscientism is that of the political paranoia[27] that runs deep in some souls. One need merely notice those AIDS activists who blame science and government both for the origins of AIDS and for the lack of a vaccine and a cure, or that small spectrum of black nationalists who claim that AIDS is black genocide practiced by white doctors and whose rantings keep many an AIDS victim from seeking help.[28]

While writing out this paper I had the nagging feeling that I ought to apologize for outlining such obvious points to this distinguished audience. But it would be wrong of me to do so. It is a *déformation professionelle* of scholars and scientists to prize depth and originality in every context. In the battle for the hearts and minds of the public, the government, and the media it is far better to make a few obvious and important points as accurately and clearly as possible and to hold to them tenaciously. If you wish large numbers of people to accept what you say, then the very last thing you want to be is original. Moreover, what is obvious to you as scholars and scientists will be quite mysterious to those who have not spent years thinking about it.

We must make no mistake about what it will take to show up the antiscience Luddites for what they are and to defeat them. The two tasks are by no means the same. It will take organized, continuous, strong, and clear opposition.

## NOTES

1 Plato, *The Theaetetus*, pp. 209d, ff.
2 Plato, *The Sophist*, pp. 263b, ff.
3 Susan Haack, *Evidence and Inquiry: Towards Reconstruction in Epistemology.*
4 John Searle, *The Construction of Social Reality.*
5 Aristotle, *Sophistical Refutation.*
6 Saint Thomas Aquinas, "Question 2: The Existence of God: Article 1, Whether the Existence of God Is Self-Evident."
7 Bertrand Russell, "On Denoting." See also P. F. Strawson, "On Referring."
8 Gilbert Ryle, *The Concept of Mind*, pp. 15–24; and "Systematically Misleading Expressions."
9 Nelson Goodman, *Fact, Fiction, and Forecast.*
10 Ludwig Wittgenstein. See, e.g., *Philosophical Investigations.*
11 David Hume, *Enquiries Concerning Human Understanding and Concerning the Principles of Morals.*
12 John Buridan (1295?–1365?), philosopher and logician. This epithet refers to an ass who, finding himself equidistant from two equally succulent bales of hay, starves to death, there being no reason to incline him towards one or the other. It is absent from Buridan's extant works. Some speculate that it arose as a counterexample to Buridan's theory of the will: that one must will that which presents itself to reason as the greater good, though choice may be delayed until reason has examined the alternatives thoroughly. For the origins of the problem see Aristotle, *De Caelo*, 295B:32.
13 David Hume, *An Enquiry Concerning Human Understanding*, sec. x, pt. 1. para. 88.
14 Paul R. Gross & Norman Levitt, *Higher Superstition: The Academic Left and Its Quarrels with Science.*
15 Sandra Harding, *Whose Science? Whose Knowledge?*
16 Susan Haack, "Response to Thomas Nagel, 'Relativism and Reason.'"
17 Sandra Harding, *The Science Question in Feminism.* See almost any page.
18 "I believe because it is absurd." Tertullian, *On the Flesh of Christ*, ch. 5. Quoted in Etienne Gilson, *A History of Christian Philosophy in the Middle Ages*, p. 45.
19 For the theory that all conduct that adversely affects minority groups in employment should be legally defined as discriminatory, see Alfred Blumrosen, *Blacks, Employment, and the Law.* The first use of this analysis of which I am aware is in the landmark case *Griggs v. Duke Power Company*, 401 U.S. 424 (1971). The term "disparate impact" is first used in Barbara Lindemann Schlei and Paul Grossman, eds., *Employment Discrimination Law.*
20 Paul Feyerabend, *Against Method.*
21 Richard Rorty. See, e.g., *The Consequences of Pragmaticism.*
22 Aristotle, *The Nichomachean Ethics*, bk. V 1133a:7, *et seq.*
23 Barry R. Gross, "What Could a Feminist Science Be?"
24 Paul Verlaine, "At the Dance."
25 Richard Hofstadter, *Anti-Intellectualism in American Life.*
26 *The New York Times*, "News of the Week in Review," May 21, 1995, p. 1.
27 Richard Hofstadter, *The Paranoid Style in American Politics.*
28 See, e.g., Hanna Rosin, "The Homecoming"; Ann Louise Bardach, "The White Cloud"; and Dante Ramos, "A Second Wave."

## REFERENCES

AQUINAS, SAINT THOMAS. "Question 2: The Existence of God: Article 1, Whether the Existence of God is Self-Evident." In *The Basic Works of Saint Thomas*, edited by Anton Pegis, vol. 1. New York, NY: Random House, 1944.

ARISTOTLE. *De Caelo.*

———. *The Nichomachean Ethics.*

———. *Sophistical Refutation.* Translated by W. A. Pickford-Cambridge. In *The Com-*

*plete Works of Aristotle: The Revised Oxford Translation*, edited by Jonathan Barnes, vol. 1. The Bollingen Series LXXI-2. Princeton, NJ: Princeton University Press, 1984.

BARDACH, ANN LOUISE. "The White Cloud." *The New Republic*, June 5, 1995.

BLUMROSEN, ALFRED. *Blacks, Employment, and the Law.* New Brunswick, NJ: Rutgers University Press, 1971.

FEYERABEND, PAUL. *Against Method.* London: New Left Books, 1975.

GILSON, ETIENNE. *A History of Christian Philosophy in the Middle Ages.* London: New Left Books, 1955.

GOODMAN, NELSON. *Fact, Fiction, and Forecast.* New York, NY: Bobbs Merrill, 1965.

*Griggs* v. *Duke Power Company*, 401, U.S. 424 (1971).

GROSS, BARRY R. "What Could a Feminist Scientist Be?" *The Monist* 77, no. 4 (1994): 434–444.

GROSS, PAUL R. & NORMAN LEVITT. *Higher Superstition: The Academic Left and Its Quarrels with Science.* Baltimore, MD: Johns Hopkins University Press, 1994.

HAACK, SUSAN. *Evidence and Inquiry: Towards Reconstruction in Epistemology.* Oxford: Basil Blackwell, 1993.

———. "Response to Thomas Nagel, 'Relativism and Reason.'" Trilling Seminar, Columbia University, April 6, 1995, p. 5.

HARDING, SANDRA. *The Science Question in Feminism.* Ithaca, NY: Cornell University Press, 1986.

———. *Whose Science? Whose Knowledge?* Ithaca, NY: Cornell University Press, 1991.

HOFSTADTER, RICHARD. *Anti-Intellectualism in American Life.* New York, NY: Alfred Knopf, 1963.

———. *The Paranoid Style in American Politics.* New York, NY: Alfred Knopf, 1965.

HUME, DAVID. *Enquiries Concerning Human Understanding and Concerning the Principles of Morals.* Oxford: Clarendon Press, 1955.

PLATO. *The Sophist.* Translated by F. M. Cornford. London: Routledge & Kegan Paul, 1935.

———. *The Theaetetus.* Translated by M. J. Levett, revised by Miles Burnyeat. Indianapolis, IN: The Hackett Publishing Company, 1990.

RAMOS, DANTE. "A Second Wave." *The New Republic*, June 5, 1995, p. 29.

RORTY, RICHARD. *The Consequences of Pragmaticism.* Minneapolis, MN: University of Minnesota Press, 1982.

ROSIN, HANNA. "On Denoting." In *Essays in Analysis*, edited by Douglas Lackey. New York, NY: George Braziller, 1973.

RYLE, GILBERT. *The Concept of Mind.* London: Hutchinson & Co., 1949.

———. "Systematically Misleading Expressions." In *Gilbert Ryle: Collected Papers*, vol. 2. London: Hutchinson & Co., 1971.

SCHLEI, BARBARA LINDEMANN & PAUL GROSSMAN, EDS. *Employment Discrimination* Law. Washington, DC: Bureau of National Affairs, 1976.

SEARLE, JOHN. *The Construction of Social Reality.* New York, NY: The Free Press, 1995.

STRAWSON, P. F. "On Referring." *Mind* 59 no. 235 (1950).

VERLAINE, PAUL. "At the Dance." In *Femmes et Hombres*, translated by Alistair Elliot. London: Anvil Press Poetry, 1979.

WITTGENSTEIN, LUDWIG. *Philosophical Investigations.* Translated by G. E. M. Anscombe. Oxford: Basil Blackwell, 1958.

# FEELINGS AND BELIEFS

LOREN FISHMAN

IT IS A COMMONPLACE of psychology that beliefs may be engendered by feelings. Unfortunately, in putting things so bluntly we run the danger of conflating different categories, both of "feeling" and "belief." I want to explore two contexts in which there is, or at least there is asserted to be, a link between feeling and belief. One is the postmodern sociology of knowledge, and of science in particular. Here it is asserted that belief is always, or almost always, a social construction, and therefore to ask whether it is rational or irrational is beside the point and possibly meaningless. I contend that the mechanism invoked here, though rarely explicitly, is one that supposedly generates a "belief" from a "feeling." In this case, the belief might, for the sake of argument, be one concerning a scientific theory or the relevance of an observation to the truth of the theory. The general tenor of the constructivist view is that one believes because the psychic pressures of the social context in which one is embedded force one to believe: bluntly, "A believes X because A is afraid not to."

In contrast, there is another kind of situation in which feelings are said to generate beliefs, one that assuredly does occur in the practice of science and of scientific medicine. This is the realm of intuition, where hunches, surmises, "gut feelings" are brought into play in a manner that undergirds beliefs, at least conditional, provisional beliefs. In a practical science like medicine, however, these beliefs may be the only ones available as a guide to immediate action.

My aim is twofold. I want to assert that the first kind of mechanism, the one implicitly invoked by the constructivist, is highly problematical and certainly cannot be so commonplace or invariant in its effects as to account for the construction of something as self-consistent and reliable as a body of scientific theory. In other words, I assert that there are canons of rationality, that scientific theories adhere to these, and that the constructivist view of how scientific convictions are generated badly fails to account for this. On the other hand, I want to rescue science from the accusation that, since "intuition" or "hunch" or "educated guess" plays such an important role, science is ipso facto irrational in the pejorative sense implied by the constructivist literature.[1]

Let us explore the difference between feelings and beliefs. The vast majority of what we call beliefs, everyday beliefs if you will, are propositions that we are willing to defend if challenged. You have certain premises, which you can make explicit, and a high degree of confidence that others who agree with the premises and who have a chance to scrutinize your reasoning will agree with you on the point at issue. This doesn't end all arguments, of course. By invoking "premises," we open the door to indefinite regress, and even to questions of "foundational" beliefs. But in practice this is what rational discussion is about. One cannot believe (or, indeed, disbelieve) just because one wants to. On the contrary, belief must, by variable contributions of observation and persuasion, be earned.

Thus, as I propose to use the term, a belief is not just a proposition that is assented to in some haphazard fashion. Rather, it is a deep, intricate, and finely coordinated mental structure that bears the traces of complex inference. Characteristically, the scientific beliefs of scientists, pure and applied, have these qualities, and this is an important part of what makes them scientific.

Feelings, in practice, are different animals. Like beliefs, they cannot just be willed. One cannot be grateful or happy by dint of a willful act, any more than one can intentionally forget. Unlike beliefs, however, feelings cannot be justified from premises. Either a feeling exists, or it does not exist. You can persuade someone else to believe as you do through an examination of premises and reasoning, but you cannot persuade someone else to feel as you do, absent some predilection independent of the tools of persuasion. You can verbally induce somebody to feel something, at least with a high degree of reliability—"Your wife and baby just died in a traffic accident!"—but this is not persuasion, as it applies to supporting or changing belief.

Note that such common expressions as: "I guess," "I think," "I feel"—as was pointed out long ago by Gilbert Ryle[2]—can be understood *not* to represent an inner state of mind. Rather, they denote a certain hedge, a weakening of an assertion. "I feel that it will rain" is not a report of an internal mental event, but rather an overtly tentative step down from "It will rain" or "I am certain that it will rain." We must beware of figurative language as well; "I like Road Warrior in the third" is not a report of affection.

Also, we have to distinguish utterances from either beliefs or feelings. This is merely to say that people may make utterances that do not comport with their beliefs or their feelings. Lies are the most common instances, but the category also includes ironic comments, rote recitations, and playacting. Clearly, feelings may generate utterances that are not mere expressions of the feelings in question. If someone puts a gun to my head and demands that I recite the Ruritanian pledge of allegiance, I probably shall do so (if I know the words) because "I am afraid not to." Obviously, this will not correspond to a belief in me that Ruritania is supreme among the nations of the earth, nor to any feelings of deep nostalgia for the fields and woodlands of Ruritania. If the gunman then demands of me that I justify my stated loyalty to Ruritania, presumably I shall try to give utterance to such a justification; but this will be spurious in the sense that the purported justification will either invoke reasoning that I consider unsound or recur to premises that I do not really believe (unless, of course, I really am a Ruritanian patriot).

Finally, we note that "intuitions," "hunches," and so forth deserve to be classed with feelings rather than beliefs for the primary reason that those who hold them cannot explicitly construct a justification for holding them from an adequate stock of premises and a compelling line of reasoning. This is precisely why there are special terms for them, distinguishing them from beliefs. You cannot persuade someone of a hunch, although you may possibly evoke a similar hunch in him. At the same time, hunches and the like are characterized by the same inscrutability usually acknowledged as such by the holder, that is associated with the realm of feeling. Note, however, that this is the way we treat such remarks. It is not, in itself, sufficient grounds for categorizing hunches, nor the use of hunches as "irrational" in any pejorative sense. How rational or irrational reliance upon hunches may be is a function of the origin of hunches and of the availability of sounder, timely guidance. About this we may make some reasonable guesses, as I hope to show.

## THE CONSTRUCTIVIST VIEW OF KNOWLEDGE

Let us consider a couple of points that come up in the work of Bruno Latour, since among postmodernist and "constructivist" theorizers about science he is as widely known and well regarded as any. In *Science in Action* he tells us "The fate of what we say and make is in later users' hands." Indeed, he states this as his "First Principle." This is a typical move—invoking an obvious tautology as though it were a profundity. But, to make a long story short, it comes as part of a strategy designed to make some highly contestable points. In Latour's view, as a general rule scientific statements:

1. Are often made from ulterior motives
2. Often have downright poor and incomplete justification
3. Are subject to inevitable vagaries of meaning and understanding because of changing historical and cultural contexts.

If this were so, we would have to concede that the Dark Ages continue and that we are fated to live narrow intellectual lives, necessarily confined to a body of observation and thought circumscribed by its time, locale, and language.

The contrasting and, I think, much stronger view is implicit in Galileo: "The secrets of nature are written in the language of mathematics."[3] He was referring to the symbolic representation of relationships expressed as mathematics usually expresses them, in the timeless present. Here we have no tenses; we are dealing, as in logic, with rules and their sequelae, which are timelessly true. Thus, whatever inheres in the logical structure of the representations and the deductions that result from them is transferable across periods and cultures, at least in principle. In practice, this is near enough to the truth. We have no more trouble in deciphering Galileo's mathematical arguments, or Archimedes's for that matter, than those of contemporary mathematical physics—rather less, in fact. This is despite the fact that, unless we are specialist historians, we know very little about Galileo's ambient culture, and even less about Archimedes's. Moreover, the skill involved passes effortlessly to anyone trained in mathematical science, even if he (or she) is Peruvian or Papuan. There is no further need of acculturation to "Western" ways.

But the arguments of the constructivists fail in other ways than these. For they depend, essentially, on the view that feelings—in particular, feelings generated by social demands—can generate beliefs of the kind that scientists typically traffic in. Again, to telescope the argument, they assert, in effect, that a scientist believes a given belief "because he is afraid not to." Simply put, they assert that the feeling of fear constructs the psychic condition of belief. Let us see if this is at all plausible.

First of all, the historical example of Galileo gives us a prime instance of the kind of evidence that will *not* prop up the constructivist argument. Consider Galileo's famous recantation, an utterance made from fear if ever there was one. On the constructivist view, one ought to hold that, given the intense persuasive force directed at him by his society and its most powerful institutions, Galileo not only renounced his former heliocentric views verbally, but actually readopted the Ptolemaic system as a matter of *belief.* After all, the pressures upon him were far more severe than anything experienced by the average working scientist; so, too, the feelings they evoked. Thus, the "constructive" power of his experience must have been exceptionally mighty!

Of course, very few of us infer that Galileo, whatever his utterances before the Inquisition, actually changed his beliefs. History is against this. A belief is, as we have said, something more than the propositional content of an utterance. It is a whole system of justifications, prior beliefs, and potential arguments and lines of inference from those prior beliefs. "I used to think X; now I think Y" conveys much more than the assertion that X is false, while Y is true. In particular, if made in earnest, it means that the speaker is prepared to give a cogent account of the truth of Y and of the invalidity of the previously credited arguments for X.

Now let us look at constructivism carefully. One's beliefs, remember, are not simply what follows the words "I believe." Consider the following sentences, quite similar as to grammatical structure, each purporting to explain something.

1. He believed it turned red because of oxidation.

This is an *explanation.* If it turned red for another reason, his belief is false. If we change "he" to "I" and contemplate the speaker, we merely have an instance of someone applying a theory of causation (right or wrong) to an observed fact.

2. He believed it turned red because he suddenly saw a red reflection in the glass.

This reports a belief and a piece of evidence for that belief. It is, of course, reasonable to believe that the color of an object changes if its reflection changes color. But the sentence itself is true precisely when it correctly reports the subject's thinking. There is no particular problem involved in grasping its truth conditions. If it turned red, but that was totally independent of any red reflection, the sentence is still true. Again, changing "he" to "I" merely produces a report by an observant individual.

Now consider another purported reason for believing:

3. He believed it turned red because he was afraid not to.

Surprisingly, many of M. Latour's comments about "explicit interests," etc., fit this model. It may superficially resemble sentence 2, but there is no empirical or logical connection with the world of color here. Instead, there is an imputation of motive. Sentence 3 is about *him*, the "he" referred to in it, and in no way explains or supports anything's redness, insofar as anyone else in the world is concerned. Whereas in the prior case we too could well infer "redness" if we believed the subject's report about reflected light, nothing about his "fear" persuades us of anything.

Again, change "he" to "I." Now the insincerity is out in the open: his tongue spoke but not his heart. The explanatory *form* of the words suggests a relationship between the beginnings and ends of all of these sentences, separated by the word *because.* But in this case the relationship is set up between his *saying* something and his fear. An individual emotion-and-action is described here, not a generalizable rule or an observation about why or when things turn red. Doing something—here the act of making a statement—has been connected to its reason. This is like a motive or feeling, a reason for *doing* something, not something others can *believe* for the same or different reasons, like a prediction, observation, or opinion. Yet this is exactly what authors confuse who cite "ulterior motives" as reasons for believing (as opposed to saying that one believes)!

Note that in 1 and 2 the belief (that something turned red) is connected with what follows (oxidation or red reflections). Yet in 3 the belief in redness is actually undermined if what follows is true—i.e., if it were "held because of fear."

This is a revered maneuver of sophists: attaching irrelevant truth conditions to a given statement, thereby making the latter appear spurious. But here, as often is the case, the meanings of the words resist the form into which they have been forced: being afraid is not a good, not a mediocre, not even a bad reason for believing something has turned red. It is an impossible reason. Recall that one cannot believe (or indeed disbelieve) just because one wants to, any more than one can forget, or be grateful or happy by willful act.

One notes an intentionality about what one feels and what one says, that does not apply to what one silently believes. Many sophistic arguments fasten what one says to what one believes, attempting to convert agreement based on different premises into the seeming disagreement seen in *oratio obliqua.* Yet: "Socrates believed in the Forms because he reasoned that they were necessary," and "Plato believed in the Forms because Socrates taught him" suggests that they agreed, not that they disagreed.

Thus, while the contention "A said that he believed X because he was afraid not to" is unproblematical—it is the story we tell schoolchildren about Galileo—the contention "A believed X because he was afraid not to" leaves one with many problems indeed. What, precisely, is the mechanism for the construction of this belief, remembering now that a belief is not merely assent to an isolated proposition? How is "society" supposed to construct the apparatus, which, remember, includes ready recourse to a line of argument—often multiple lines of argument, in fact? If it is merely asserted that whoever adapted the belief absorbed, in some fashion, all the lines of argument that point to the belief, then it hardly makes any sense to say that "he believes because he is afraid not to." The notion that he believes because he has been persuaded by a certain line of argument makes infinitely better sense in this context. On the other hand, to assert that he believes something (in the sense that a scientist believes) without having at hand the rather elaborate justificatory mechanisms that warrant calling a belief "scientific" is fatuous. At best, he has acquired the habit of rote recitation of a proposition. This may constitute "credence" at some level, but certainly not "belief" in accordance with the everyday habits of scientists and other rational inquirers.

This being said, the constructivist is either left with no argument, or with an argument for something much weaker than what he really wants to assert. Of course, social factors limit the evidence available to an inquirer and direct his attention to one class of questions rather than another; this kind of "construction" is more or less universally conceded. But that beliefs, in and of themselves, with their necessary mechanisms of justification standing at the ready, are adopted in toto independently of internal logical coherence and with indifference to observable evidence, out of the need or desire to conform with some kind of shadowy social ethos, defies plausibility. It is doubtful that Latour or anyone else has ever seen this phantom "in action." The best evidence on offer, when examined, seems to stem from versions of the Galileo fallacy—that a compelled utterance is evidence of a compelled belief.

## INTUITION AND BELIEF

As I noted above, an intuitive feeling—a hunch—cannot be a belief in the sense that a scientific belief is. To say it again, the very condition of being "intuitive" means that sufficient grounds of explicit persuasion are absent. Yet hunches and the like are often the chief guides to scientific research. It is not too much to say that no science could proceed, in a practical sense, without them. Even more, hunches, often dignified as educated guesses or disparaged as "gut feelings," frequently underlie the decisions that are made in medical practice as the basis for diagnostic tests and, in dire emergency, for vital decisions or "judgment calls." Hunches are, in short, grounds for action. This is often adduced as evidence for the "irrationality" of science and of the scientific community, as though this reliance on hunches was tantamount to the reliance of other cultures on belief systems that Western rationality decries as superstition. Is this comparison at all apt? In what sense is a hunch an "alternative way of knowing" like those so dear to the devotees of the postmodern?

Consider something that happens in the everyday practice of medicine, which I will call the "difficult case" scenario. A physician is having difficulty with a case. After the results of indicated tests have been analyzed and evaluated in light of the signs, symptoms, and history, the physician cannot diagnose the patient or even determine a means of proceeding further with the diagnosis. Typically, a specialist is called, and, since medicine is highly pragmatically oriented, he knows which specialist to seek. He does not pause to savor the problem as a problem. In all probability the problem has, indeed, been solved before, but that is not of the essence here, nor is it particularly helpful; time, not the sorting out of scholarly priority, is what is of the essence.

Why a specialist? What does that specialist have that the physician calling him in lacks? Almost inevitably, the answer given is "more experience." But of what advantage is that greater experience? Certainly, a bigger database has something to do with it; but most physicians would say that a significant factor, if not the most significant one, is intuition. Practically, this means making a decision, proceeding along a path that in others would be called "guesswork," and doing so, moreover, without being able to bring forth an explicit set of reasons for making that decision.

I believe that what the specialist physician does to "solve" a problem by intuition is to recognize, by some mechanism, the pattern that most likely fits the case, and to extend that pattern in time from what is currently known to what can be expected.

Two elements come into play. There is rational deduction, with facts being adduced, accepted theories brought consciously into focus, and the like. But this works in conjunction with intuitions and hunches that are not readily put into this framework. Why, then, is the specialist who works in this hybrid fashion trusted? Presumably, where "beliefs" are involved, the specialist could assure his colleagues of the rational grounds of his judgment, citing facts, studies, methodologies, and so forth, that fall within his particular ambit. But why trust his intuitions? Does this not carry us out of the realm of rationality? Not really, because part of the database of the physician and of his trusted colleagues is that so-and-so is a reliable specialist whose intuitions, whose "feeling" for cases of a certain sort, has proved reliable in the past.

Inevitably, there is some tension in such cases between the physician's real need to rely on the intuitive gifts of the specialist and his desire to have a plain line of fact and inference that can justify the guesswork. Yet nothing like "black magic" is involved here. We must note that these hintings and promptings and intuitions are not feelings like pain or giddiness. They have a cognitive content, and their import is directed to solving concrete problems, as they frequently manage to do. There are cerebral phenomena here, at times nonlinguistic, at times not self-conscious; but not, on that ground, necessarily *irrational.*

Recent advances in computer science, based on observations of actual neuronal function, as well as discoveries in mathematics may help us to understand what is involved. We certainly need not invoke emotion or irrational bias to explain it. Recall that artificial neural networks (ANNs), now a popular tool among computer researchers, perform functions that seem closely analogous to the intuitive judgments of human experts, including medical specialists. (Needless to say, they may do so without any knowledge, or bias, of race, sex, age, employment status, feelings, politics, or even their own future.)

ANNs can pick up, analyze, and indeed begin to describe patterns far more elaborate and subtle than humans have been able to discern. In fact, by searching for so-called grandmother neurons one can actually identify correlations that were not suspected by the people who put the data into a computer file in the first place. Basically, these networks function by strengthening the predictive connections between depictions of (sets of) events that are observed to occur together. While humans are good at identifying one, two, or possibly even five causes and effects operating simultaneously, the ANNs can do so for fifty variables with no trouble at all. These are not to be confused with "rule-following" expert systems. In such systems, the programmer provides a decision algorithm based on his own understanding of the factors involved and their interrelationships. ANNs, on the other hand, develop their own "rules" autonomously, and it may well be that the human programmer who sets the system going in the first place will be unable to decode the "rule" that emerges in comprehensible terms.

The beauty of ANNs for our purpose is that they refine and clarify our notions of "intuitions." At root, it is simply a question of perceiving the relative strengths of repeating patterns and "going with them" as much as the facts warrant. I emphatically suggest that this may be one of the things that we commonly identify as "operating by intuition." If this is so, then intuition, as we may plausibly argue, loses its connotations of subjectivism and prejudice. There may well be a psychological mechanism (or mechanisms) that provide us with cognitive rules and generalizations from experience of which we are consciously unaware, and which we find impossible to articulate or formalize. Nonetheless, the knowledge or surmise that intuition generates may be defended, in principle, as perfectly rational—no less so, really, than the beliefs whose roots in conscious inference we can trace. Thus, the fact that scientists exploit intuition in a variety of ways is no argument for asserting that science depends either on irrationality or the clairvoyant. Nor does it give any real comfort to constructivists, who might wish to argue that the "intuitive" is intrinsically more subject to socially induced biases and prejudices than is self-conscious reasoning. For one thing, the charge that the "intuitive" is inevitably contaminated by whim, emotion, or prejudice is seen to be dubious, at the least; for the cognitive sciences as well as computer models offer us paradigms of "intuitive" process from which these elements are clearly absent. Intuitive rationality, though its mechanisms still remain mysterious, is not an oxymoron.

## CONCLUSION

The influence of cultural factors on thought and feeling is undeniable. The power of culture to dictate which facts we focus on, which conclusions we are most interested in, is all too apparent. Yet all facts are equally accessible to logic. The consistency of the principles of science is what holds science together. The characteristics of a society may hold its science together—for example, through practical respect for the scientific method, widely applied standards of measurement, a cultural commitment to free speech and telling the truth, and so forth. But in general the idiosyncracies of factious, polymorphous society are not part of science.

Wittgenstein remarked that "Philosophy arises when language goes on holiday."[4] We might say that postmodernism, in its various extravagant formulations—deconstruction, relativism, extreme social determinism—would have us put language—and science—on indefinite furlough. Under the postmodernist regime, there can be no confidence in data-based conclusions, because, it is held, language is incapable of bearing facts or of being used logically. Still less are we to trust reports of investigators; when they come from even the not-too-distant past, they are held to be "incommensurable" with contemporary thought.

Postmodernist understanding of statements and their truth leaves us paralyzed as regards predictions, for prediction depends on the truth of premises and on the validity of inference. If, as postmodernists would have it, meaning and truth are inexorably bound to context and historical setting, then the whole point of scientific theorizing would vanish, and science itself would have to be abandoned.

As I hope I have shown, the situation is not all that dire. The main thrust of the postmodernist assault on science is that its rationality is illusory, because what purports to be scientific rigor is helpless against irrationality in various forms. I have argued above that two of the chief props of the postmodernist critique—that scientific beliefs are mysteriously constructed by (unspecifiable) social processes, and that the reliance of science, and of medicine, on intuitive judgments ipso facto introduces nonrational elements into scientific discourse—cannot stand close examination. The defenders of science have a robust and vigorous client. It is unlikely to melt away under any version of the postmodernist critique.

## NOTES

1 See, for instance, Bruno Latour, *Science in Action*, pp. 21–48, for one highly influential version of the constructivist position.
2 Gilbert Ryle, *The Concept of Mind.*
3 Galileo Galilei, *Discoveries and Opinions.*
4 Ludwig Wittgenstein, *Philosophical Investigations.*

## REFERENCES

GALILEI, GALILEO. *Discoveries and Opinions.* New York, NY: Bantam, 1989.

LATOUR, BRUNO. *Science in Action.* Cambridge, MA: Harvard University Press, 1988.

RYLE, GILBERT. *The Concept of Mind.* Chicago, IL: University of Chicago Press, 1984.

WITTGENSTEIN, LUDWIG. *Philosophical Investigations.* 3d edit. London: McMillan, 1987.

# IN PRAISE OF INTOLERANCE TO CHARLATANISM IN ACADEMIA[a]

MARIO BUNGE

UP UNTIL THE MID-1960s whoever wished to engage in mysticism or freewheeling, intellectual deceit or antiintellectualism had to do so outside the hallowed groves of academe. For nearly two centuries before that time the university had been an institution of higher learning, where people cultivated the intellect, engaged in rational discussion, searched for the truth, applied it, or taught it to the best of their abilities. To be sure once in a while a traitor to one of these values was discovered, but he was promptly ostracized. And here and there a professor, once tenured, refused to learn anything new and thus became quickly obsolete. But he seldom lagged more than a couple of decades, was still able to engage in rational argument as well as to distinguish genuine knowledge from bunk, and did not proclaim the superiority of guts over brains or of instinct over reason—unless, of course, he happened to be an irrationalist philosopher.

This is no longer the case. Over the past three decades or so very many universities have been infiltrated, though not yet seized, by the enemies of learning, rigor, and empirical evidence: those who proclaim that there is no objective truth, whence "anything goes," those who pass off political opinion as science and engage in bogus scholarship. These are not unorthodox original thinkers; they ignore or even scorn rigorous thinking and experimenting altogether. Nor are they misunderstood Galileos punished by the powers that be for proposing daring new truths or methods. On the contrary, nowadays many intellectual slobs and frauds have been given tenured jobs, are allowed to teach garbage in the name of academic freedom, and see their obnoxious writings published by scholarly journals and university presses. Moreover, many of them have acquired enough power to censor genuine scholarship. They have mounted a Trojan horse inside the academic citadel with the intention of destroying higher culture from within.

[a] The research leading to this paper was supported in part by the Humanities and Social Sciences Research Council of Canada. Some paragraphs of this paper have been taken from M. Bunge, *Finding Philosophy in Social Science* and *Social Science under Debate.*

The academic enemies of the very raison d'être of the university can be grouped into two bands: the antiscientists, who often call themselves "postmodernists," and the pseudoscientists. The former teach that there are no objective and universal truths, whereas the academic pseudoscientists smuggle fuzzy concepts, wild conjectures, or even ideology as scientific findings. Both gangs operate under the protection of academic freedom, and often at the taxpayer's expense, too. Should they continue to use these privileges, misleading countless students and misusing public funds in defaming the search for truth, or should they be expelled from the temple of higher learning? This is the main problem to be tackled in the present paper. But first let us sample the production of the academic antiscientists and pseudoscientists, restricting ourselves to the humanities and social studies.

## ACADEMIC ANTISCIENCE

Academic antiscience is part of the counterculture movement. It can be found in nearly all departments of any contemporary faculty of arts, particularly in the advanced countries. Let us take a look at a small sample of the antiscientific reaction inside the gates of Academia: existentialism, phenomenology, phenomenological sociology, ethnomethodology, and radical feminist theory.

### *Example 1: Existentialism*

Existentialism is a jumble of nonsense, falsity, and platitude. Let the reader judge by himself from the following sample of Heidegger's celebrated *Sein und Zeit*, dedicated to Edmund Husserl, his teacher and the founder of phenomenology. On human existence or being-there (*Dasein*): "Das Sein des Daseins besagt: Sich-vorweg-schon-sein-in-(der Welt-) als Sein-bei (innerweltlich begegnendem Seienden)."[1] On time: "Zeit ist ursprünglich als Zeitigung der Zeitlichkeit, als welche sie die Konstitution der Sorgestruktur ermöglicht."[2] I dare anyone to make sense of these wordplays, or even to translate them into standard German. Other famous formulas of Heidegger's, such as *Die Welt weltet* ("The world worlds"), *Das Nichts nichtet* ("Nothingness nothings"), *Die Sprache spricht* ("Language speaks"), and *Die Werte gelten* ("Values are valuable"), have the virtue of brevity but are just as nonsensical as the former.

Not content with writing nonsense and torturing the German language, Heidegger heaped scorn on "mere science" for being allegedly incapable of "awakening the spirit."[3] He also denigrated logic, "an invention of schoolteachers, not of philosophers."[4] Last, but not least, Heidegger was a Nazi ideologist and militant, and remained unrepentant until the end.[5] (No mere coincidence here: the training of obedient soldiers ready to die for an insane criminal cause starts by discouraging clear critical thinking.) In short, existentialism is no ordinary garbage: it is unrecyclable rubbish. Its study in academic courses is justified only as an illustration of, and warning against, irrationalism, academic imposture, gobbledygook, and subservience to reactionary ideology.

*Example 2: Phenomenology*

This school, the parent of existentialism, is characterized by opaqueness. Let the reader judge from this sample of its founder's celebrated attack upon the exact and natural sciences: "I as primaeval I [*Ur-Ich*] construct [*konstituire*] my horizon of transcendental others as cosubjects of the transcendental intersubjectivity that constructs the world."[6] Phenomenology is also a modern paragon of subjectivism. In fact, according to its founder the gist of phenomenology is that it is a "pure egology," a "science of the concrete transcendental subjectivity."[7] As such, it is "in utmost opposition to the sciences as they have been conceived up until now, i.e., as *objective* sciences."[8] The very first move of the phenomenologist is the "phenomenological reduction" or "bracketing out" (*époché*) of the external world. "One must lose the world through *époché* in order to regain it through universal self-examination."[9] He must do this because his "universal task" is the discovery of himself as transcendental (i.e., nonempirical) ego.[10]

Having feigned that real things such as chairs and colleagues do not exist, the phenomenologist proceeds to uncover their essences. To this end he makes use of a special intuition called "vision of essences" (*Wesensschau*), the nature of which is not explained, and for which no evidence at all is offered. The result is an a priori and intuitive science.[11] This "science" proves to be nothing but transcendental idealism.[12] This subjectivism is not only epistemological but also ontological: "the world itself is an infinite idea."[13]

How could anyone think that this wild fantasy could shed any light on anything except the decadence of German philosophy? This extravagance can only have at least one of two negative effects on social studies. One is to focus on individual behavior and deny the real existence of social systems and macrosocial facts; these would be the products of such intellectual procedures as aggregation and "interpretation" (guessing). The other possible negative effect is to alienate students from empirical research, thus turning the clock back to the times of armchair ("humanistic") social studies. The effect of the former move is that *social* science is impossible; that of the second is that social *science* is impossible. Either or both of these effects are apparent in the two schools to be examined next.

*Example 3: Phenomenological Sociology*[14]

This school is characterized by spiritualism and subjectivism, as well as by individualism (both ontological and methodological) and conservatism—ethical and political. The first two features are obvious: according to phenomenology social reality is a construction of the knower, not a given; for all social facts would be "meaningful" (have a purpose) and the subject of "interpretation" (guessing), whence everything social would be spiritual and subjective, or at most intersubjective, rather than material and observer independent. The ontological individualism of phenomenology derives from its subjectivism. Because individuals are said to "interpret" themselves and others, without ever facing any brute social facts, the task of the sociologist is to grasp "subjective meaning structures" rather than to construct or test models of social systems or processes. In particular, he must study the *Lebenswelt* or everyday life of individuals, skirting such macrosocial issues as gender and race discrimination, mass unemployment, social conflict, and war. The phenomenological sociologist claims to grasp directly the objects of his study, alleging that they are ordinary. Moreover, let us remember that he is graced with the "vision of essences," which gives him instant insight. Hence he can dispense with statistics, mathematical modeling, tedious argument, and empirical test. In short, phenomenological sociology is avowedly nonscientific and an invitation to sloth.

*Example 4: Ethnomethodology*[15]

This is the offspring of the union of phenomenology with symbolic interactionism. The members of this school practice what phenomenological sociologists preach: they observe at first hand and record trivial events in the *Lebenswelt* or everyday life, focus on symbols and communication, and skirt any important activities, processes, and issues, particularly large-scale social conflicts and changes. They engage in participant (short-range) observation but shun experimentation, which they disapprove of on philosophical grounds. Lacking theories of their own, the ethnomethodologists invoke the murky pronouncements of hermeneutics, phenomenology, and even existentialism—all of them declared enemies of science. Obviously an antiscientific philosophy that opposes the search for objective truth could hardly inspire scientific research. Mercifully the ethnomethodologists make no use of these doctrines in their empirical work. As a matter of fact, in field work they behave as positivists—even while vehemently denouncing positivism—inasmuch as they spend most of their time collecting data, which they are unable to interpret correctly for want of theory.

In fact, the ethnomethodologist audiotapes and videotapes "the detailed and observable practices which make the incarnate [?] production of ordinary social facts, for example, order of service in a queue, sequential order in a conversation, and the order of skillfully embodied [?] improvised conduct."[16] Possible English translation: "The ethnomethodologists record observable ordinary life events." The data thus collected are audible or visible traces left by people who presumably behave purposefully and intelligently. These traces are the only clues the ethnomethodologists can go by, for, lacking a theory, they cannot tell us what makes people tick—i.e., they cannot explain the behavior they observe and record. Their practice does not differ from that of the empiricist and, in particular, the behaviorist—as even Atkinson, a sympathizer of the school, has admitted.[17] In short, they behave like positivists even while engaging in positivism bashing—actually a devious way of attacking the scientific approach.

Only the ethnomethodologists' convoluted lingo suggests intimate contact with their philosophical mentors. For example, Garfinkel starts one of his books by stating that ethnomethodology "recommends" that "the activities whereby members [of a group?] produce and manage settings [?] of organized everyday affairs are identical with members' procedures for making those settings 'account-able'[?]. The 'reflexive'[?] or 'incarnate'[?] character of accounting [?] practices and accounts makes up the crux of that recommendation."[18] Or consider the same author's definition of ethnomethodology as "the investigation of the rational [intelligible?] properties of indexical [context-dependent] expressions and other practical actions as contingent [?] ongoing accomplishments [outcomes?] of organized artful [purposive?] practices of everyday life."[19] Why use extraordinary prose to describe ordinary accounts of ordinary life?

This is not to deny the value of observing everyday life occurrences, such as casual encounters and conversations—the favorite material of ethnomethodologists. Such observation, a common practice of anthropologists, yields raw material for the scientist to process in the light of hypotheses and with a view to coming up with new hypotheses. But that empirical material is of limited use unless it is accompanied by reliable information concerning the role that the observed subject enacts, e.g., boss or employee. The reason is that such roles—in other words, the system in which the protagonists are embedded—largely determine the "meaning" (purpose) of everyday actions and the content of conversations.[20] But ethnomethodologists overlook the macrosocial context and are not interested in any large social issues. This fact, combined with the absence of tests of the proposed "interpretations" (hypotheses) and the lack of theory, explains the paucity of findings of ethnomethodology.

A characteristic product of this school is Lynch's study "Sacrifice and the Transformation of the Animal Body into a Scientific Object: Laboratory Culture and Ritual Practice in the Neurosciences." Taking his cue from Durkheim's studies in the sociology of religion, Lynch claims that the killing of laboratory animals at the end of a run of experiments is part of a ritual practice whereby the body of the animal is transformed into "a bearer of transcendental significances." Characteristically, he presents no evidence for the extraordinary claim that the laboratory bench is just a sacrifice altar.

### *Example 5: Radical Feminist Theory*

The word "feminism" nowadays denotes three very different objects: the movement for women's emancipation from male domination; the scientific study of the feminine biological, psychological, and social condition; and radical feminist "theory." While the first two are legitimate and laudable endeavors, the third is an academic industry that makes no use of science. It is, moreover, hostile to science and is characterized by pseudoproblems and wild speculation. Some radical feminist theorists have promised a "successor science" that would eventually replace or at least complement what they call "male-dominated science." Others, more consistent, are dead against all science, because they believe that reason and experiment are weapons of male domination. They hold that the scientific method is part of the "male-stream." They denounce precision—in particular, quantitation, rational argument, the search for empirical data, and the empirical testing of hypotheses as so many tools of male domination. They are constructivist-relativists: they denounce what they call "the myth of objectivity." (More on this below under ACADEMIC PSEUDOSCIENCE.)

For example, the feminist theorists Belenky, Clinchy, Goldberger, and Tarule hold that truth is context dependent and that "the knower is an intimate part of the known"—just because some of the women they interviewed felt so.[21] Sandra Harding goes as far as to assert that it would be "illuminating and honest" to call Newton's laws of motion "Newton's rape manual."[22] (The rape victim would be Mother Nature, which of course is feminine.) Moreover, basic science would be indistinguishable from technology, and the search for scientific knowledge would be just a disguise for the struggle for power—as Herbert Marcuse[23] and Michel Foucault[24] had claimed earlier on the strength of the same empirical evidence, namely none. The radical feminist philosophers are interested in power, not in truth. They want to undermine science, not to advance it. In this way they do a double disservice to the cause of feminine emancipation: they discredit feminism by making it appear to be barbaric, and they deprive it of a strong lever—namely the scientific research of the spurious causes and the pernicious effects of gender discrimination. Moreover, their attack on science alienates women from scientific studies and thus reinforces their subordinate position in modern society.[25]

To sum up, our antiscience colleagues are characterized by their appalling ignorance of the very object of their attack, namely science.[26] Lacking intellectual discipline and rigor, they have been utterly barren. This has not prevented them from misleading countless students, encouraging them to choose the wide door, incapacitating them to think straight and get their facts right, and in many cases even write intelligibly.[27] Why should any serious and socially responsible scholar tolerate barbarians intent on discrediting genuine scholarly pursuits and even destroying modern culture?

## ACADEMIC PSEUDOSCIENCE

To paraphrase Groucho Marx: the trademark of modern culture is science; if you can fake this, you've got it made. Hence the drive to clothe groundless speculations, and even old superstitions, with the gown of science. The popular pseudosciences, such as astrology, pyramidology, graphology, UFOlogy, "scientific" creationism, parapsychology, and psychoanalysis, are easy to spot, for they are obviously at variance with what is being taught at the science faculties. (Psychoanalysis would seem to refute this assertion, but it does not. Indeed, nowadays psychoanalysis is taught in only some psychiatry departments, which are part of medical schools, not of science faculties.) On the other hand, the academic pseudosciences are harder to spot partly because they are taught at university departments the world over. A second reason is that these pseudosciences abide by reason, or at least seem at first sight to do so. Their main flaws are that their constructions are fuzzy and do not match reality. (Some of them, such as neo-Austrian economics, even claim that their theories are true a priori.) Let us take a small sample, restricting our discussion to two trends: the love of spurious precision (in particular, pseudoquantification) and the post-Mertonian sociology of science.

### *Example 1: Pseudomathematical Symbolism*

Vilfredo Pareto, an original, insightful, and erudite student of society who used mathematics in economics, passes for being one of the founders of mathematical sociology merely because in this field he used some symbols other than words. Thus, in his massive and famous *Trattato di sociologia generale* Pareto listed a number of "residues" or "forces," among them sentiments, abilities, dispositions, and myths.[28] He assumed tacitly that the "residues" are numerical variables. But, since he failed to define them, the symbols he used are mere abbreviations for intuitive notions. Unaware of this confusion between arbitrary symbols and symbols designating mathematical concepts, he wrote about the composition of such "forces."[29] Further down he introduced the formula "$q = A/B$," where $A$ stands for "the force of class I residues," and $B$ for "the force of class II residues" in a given social group or nation.[30] Roughly, $q$ would be the ratio of progressivism to conservatism. Since Pareto made no attempt to define any of these "magnitudes," he had no right to divide them or to assert that they increased or decreased quantitatively over time in any group or nation. Ironically, earlier in the same work (p. 509) he had warned that "Residues correspond to certain instincts in human beings, and for that reason they are usually wanting in definiteness, in exact delimitation."[31] And even earlier in the same work he had devoted an entire chapter to characterizing and criticizing pseudoscientific theories.[32] Likewise Pitirim Sorokin, one of the founders of American sociology and an early critic of what he called "quantophrenia," sometimes indulged in the latter.[33] For example, he defined the freedom of an individual as the quotient of the sum of his wishes by the sum of his means for gratifying them.[34] But since he did not bother to define wishes and means in a mathematically correct way, he "divided" words. In sum, the symbols he used in this case were mere shorthand for intuitive notions.

Professor Samuel Huntington, the famous Harvard political scientist, was far sloppier. In fact he proposed the following "equations" concerning the impact of modernization in developing nations:

Social mobilization/Economic development = Social frustration,
Social frustration/Mobility opportunities = Political participation,
Political participation/Political institutionalization = Political instability.[35]

Huntington did not define any of these "variables," he did not explain how numerical values could be assigned to them, and he did not even bother to tell us their dimensions and units. Obviously, he was unaware that he had "divided" words, not numerical values of honest functions. This was pointed out by the mathematician Neal Koblitz in a paper titled "Mathematics as Propaganda," which led Yale mathematician Serge Lang to campaign successfully against the induction of Professor Huntington into the United States Academy of Sciences. Regrettably, many political scientists and sociologists defended Huntington, thereby exhibiting their mathematical and methodological naiveté.[36]

Professor Gary Becker, a Nobel laureate at the University of Chicago, is famous for his economic approach to the study of human behavior. Unfortunately he leans heavily on undefined utility functions and tends to pepper his writings with symbols that do not always represent concepts. For example, a key formula of his theory of social interactions reads thus: "$R = D_i + h$."[37] Here $i$ labels an arbitrary individual, and $R$ is supposed to stand for "the opinion of $i$ held by other persons in the same occupation"; and "$h$ measures the effect of $i$'s efforts, and $D_i$ the level of $R$ when $i$ makes no effort; that is, $D_i$ measures $i$'s 'social environment.'" Becker christens these "functions" but does not specify them. Consequently he adds words, not functions. We are not even told what the dimensions and units of these pseudomagnitudes are. Therefore, we would not know how to measure the corresponding properties and so to test for the adequacy of the formula.

Of course, pseudoquantitation is sufficient but not necessary to engage in pseudoscience. An alternative is to relate precise magnitudes in imprecise ways, such as "$Y$ is some function of $X$," where $X$ and $Y$ are well defined but the function is left unspecified. Milton Friedman's "theoretical framework for monetary analysis" is a case in point.[38] Indeed, it revolves around three undefined function symbols ($f$, $g$, and $l$). Hence it may at most pass for a research proposal, an aim of which would be to find the precise form of the hopeful functions in question. But the project does not seem to have been carried out. And in any case, given the bankruptcy of monetarism, the project does not seem worthy of being carried out.

*Example 2: Subjective Probability*

When confronted with a random or seemingly random process, one attempts to build a probabilistic model that could be tested against empirical data; no randomness, no probability. Moreover, as Poincaré pointed out long ago, talk of probability involves some knowledge; it is no substitute for ignorance. This is not how the Bayesians or personalists view the matter: when confronted with ignorance or uncertainty, they use probability—or rather their own version of it. This allows them to assign prior probabilities to facts and propositions in an arbitrary manner—which is a way of passing off mere intuition, hunch, or guess for scientific hypothesis. In other words, in the Bayesian perspective there is no question of objective randomness, randomization, random sample, statistical test, or even testability; it is all a game of belief rather than knowledge.

This approach contrasts with science, where gut feelings and wild speculations may be confided over coffee breaks but are not included in scientific discourse, whereas (genuine) probabilities are measured (directly or indirectly), and probabilistic models are checked experimentally. (Think of models of radiative and radioactive decay, Brownian motion, gene mutation, or random mating.) This is not to write off the scientific study of belief. Such study is important; and, precisely for this reason, it belongs in experimental psychology and sociology, and it should be conducted scientifically. There is no reason to believe that probability theory, a chapter of pure mathematics, is the ready-made (a priori) empirical theory of belief. In fact, there is reason to believe that credences are not probabilities, if only because we seldom know all the branches of any given decision tree.[39]

In the field of jurisprudence the so-called new evidence scholarship, born in the mid-1960s, claims to use probability to measure credence and in particular the credibility of legal evidence. In this connection there is even talk of "trial by mathematics."[40] I submit that probability hardly belongs in legal argument because probability measures only the likelihood of random events, not the plausibility of a piece of evidence, the veracity of a witness, or the likelihood that a court of law will produce the just verdict. Consequently, talk of probability in law is pseudoscientific. Worse, the American and other criminal codes require the death penalty when "there is a probability that the defendant would commit criminal acts of violence"—as if such a "probability" (actually a mere plausibility) could be either measured or calculated. Thus sometimes not only property and freedom but even life hang on epistemologies that would not stand a chance in science or engineering, and whose only function is to justify an academic industry.

*Example 3: Subjective Utility*

Most of the utility "functions" occurring in neoclassical microeconomics and its applications to other social studies are not well defined—as Henri Poincaré pointed out to Léon Walras.[41] In fact, the only conditions required of them is that they be twice differentiable, the first derivative being positive and the second negative. Obviously, infinitely many functions satisfy these mild requirements. This often suffices in some branches of pure mathematics. (Likewise the general theory of metric spaces does not require the specification of the distance function.) But the factual (or empirical) sciences are more demanding: here one uses only functions that are defined explicitly (e.g., by infinite series or products) or implicitly (e.g., by differential equations together with initial or boundary conditions). Such specification makes for definite meaning, more exacting testability, and more rigorous measurement. Finally, experimental studies have shown that preferences and subjective estimates of utility and risk do not satisfy the assumptions of expected utility theory.[42]

In short, the use of utility functions is often mathematically sloppy and empirically unwarranted. Now, rational choice models make heavy use of both subjective utilities and subjective probabilities, as well as of the simplistic hypothesis that selfishness is the only motivation of human behavior. Not surprisingly, none of these models fits the fact. Hence, although at first sight they look scientific, as a matter of fact they are pseudoscientific.[43]

*Example 4: Loose Talk of Chaos Theory*

James N. Rosenau, a well-known politologist, has claimed that political instability and turbulence are similar to the instabilities and vortices of fluids, and, moreover, that they satisfy chaos theory.[44] However, he did not write, let alone solve, any nonlinear differential or finite difference equation for political processes; all he did was some hand-waving. Another politologist, Courtney Brown, does write some equations, but they happen to concern two key variables—level of public concern and environmental damage—that he fails to define, so that the formulas are only ornamental.[45]

All of the above-mentioned examples are exercises in either shorthand or mathematical name-dropping, not in genuine mathematical social science. What we have here is some of the accoutrements of science without its substance; i.e., we are in the presence of pseudoscience.

### *Example 5: Post-Mertonian Sociology of Science*

The modern sociology of science is a scientific discipline born in the 1930s around Robert K. Merton.[46] It attempts to investigate in a scientific way scientific communities and the interactions between scientific research and social structure; and it holds the former to be realist, disinterested, critical, and subject to a moral code. In the mid-1960s an irrationalist and idealist reaction against the Merton school was born.[47]

The pseudoscientific sociology of science, usually described as constructivist-relativist, claims to paint a far more realistic image of scientific research through jettisoning what are called the "myths" of disinterested research and objective truth. However, most of the new-style sociologists of science mistrust or even attack science. They regard it as an ideology, a power tool, an inscription-making device with no legitimate claim to universal truth, one more social construction on a par with myths, dress codes, and a variety of politicking. They regard scientists as skilled craftsmen but somewhat unscrupulous wheelers-dealers and unprincipled politicians. In short, they laugh at Merton's classical characterization of the scientific ethos.

The members of this school regard all facts, or at least what they call scientific facts, as constructions, none as given. (Thus, the book that earned Latour and Woolgar instant fame is titled *Laboratory Life: The Social Construction of Scientific Facts.*) But actually in matters of knowledge the only genuine social constructions are the exceedingly uncommon scientific forgeries committed by a team. A famous forgery of this kind was the Piltdown fossil man, "discovered" by two pranksters in 1912, certified as authentic by a number of experts (among them Father Teilhard de Chardin), and unmasked as a fake only in 1950. According to the existence criterion of constructivism-relativism we should admit that the Piltdown man did exist—at least between 1912 and 1950—just because the scientific community believed in it. Are we prepared to believe this, or rather to suspect that the self-styled post-Mertonians are incapable or even unwilling to tell hot air from cold fact?

Because the constructivist-relativists deny that there is any conceptual difference between science and other human endeavors, they feel entitled to pass judgment on the content of science, not only on its social context. Thus, after reading one of Einstein's popularizations of special relativity, Latour concludes that the poor man was wrong in believing that it deals with "the electrodynamics of moving bodies," the title of the founding paper—one that Latour could not possibly understand for lack of mathematical and physical competence.[48] The theory, he reveals to us, is about long distance travelers. Not only this: it renders everything physical relative to the knower (not to the reference frame), thus confirming subjectivism—the misinterpretation popular among idealist philosophers at the beginning of this century. There is no telling what further wonders these modern day "Darwins of science"—as Latour calls himself and his friends[49]—may bring.

Because the constructivist-relativists ignore science, they are incapable of distinguishing it from pseudoscience. Thus Michael Mulkay, a pioneer of the movement, waxed indignant over the way the scientific community treated Immanuel Velikovsky's allegedly revolutionary *Worlds in Collision* of 1950.[50]

He scolded scientists for their "abusive and uncritical rejection" of Velikovsky's fantasies and for clinging to their "theoretical and methodological paradigms"—among them the equations of celestial mechanics. He claimed that the astronomers had the duty to put Velikovsky's fantasies to the test. Obviously Mulkay ignores that the burden of proof rests on the would-be innovator, that nearly all of Velikovsky's claims have been proved wrong, and that scientists have more important tasks than to test fantasies that collide head-on with the bulk of scientific knowledge. However, a number of scientists, headed by Carl Sagan, did take their time to criticize in detail Velikovsky's fantasies, and the American Association for the Advancement of Science devoted an entire symposium to them.[51]

Other vocal constructivist-relativists have mounted spirited defenses of astrology and parapsychology.[52] They attack the critics of these pseudosciences for espousing what they call "the standard model of science," which they dub "ideology." Regrettably they do not propose an alternative "model" of science. They only call for a "reappraisal of scientific method" to make room for astrology, parapsychology, psychoanalysis, and other "extraordinary sciences." It would go against the grain of their school to propose its own clear-cut criteria of scientificity, since it holds science to be an ordinary "social construction." But how is it possible to discuss rationally the scientific status of an idea or practice otherwise than in the light of *some* definition of scientificity? As for the truth values of the alleged findings of astrologers, parapsychologists, and the like, how can we discuss them in the constructivist-relativist framework, where truth is said to be a social convention on a par with table manners?[53]

### *Example 6: "Scientific" Racism*

Racism is very old, but "scientific" racism is a 19th-century invention that culminated with the Nazi *Rassenkunde* and the accompanying extermination camps. The American version of this doctrine was introduced by some psychologists on the basis of flawed IQ measurements, and it was entrenched in the American legislation restricting immigration from Southern Europe and other regions.[54] It was muted for a while in the wake of the revelation of the Nazi horrors, but it was resuscitated in 1969 by the Harvard professor Arthur Jensen, who, on the basis of some IQ measurements, asserted the innate intellectual inferiority of Afroamericans. This "finding" was unanimously rejected by the scientific community. In particular the Genetics Society of America warned against "the pitfalls of naive hereditarian assumptions."[55]

Yaron Ezrahi, a member of the constructivist-relativist pseudosociology of science, claimed that this denial was due to ideological reasons.[56] He held that the geneticists were particularly vehement in their criticisms of Jensen's work for being concerned, at least in part, with their own "public image and support." Ezrahi did not bother to analyze the very IQ tests from which Jensen had derived his "conclusions." Had he done so he might have learned that (a) such tests were indeed culture bound and thus likely to favor whites over blacks, and (b) no IQ test will be fully reliable unless it is backed up by a well-confirmed theory of intelligence—a theory that is overdue.[57]

Undaunted by such methodological criticisms, Richard Herrnstein and Charles Murray repeated the racist claim in their best seller *The Bell Curve* without adding any new evidence.[58] Their book was promoted by the American Enterprise Institute and widely publicized by right-wing journalists, who saw in this book the "scientific" basis for their proposal to eliminate all the social programs aimed at giving a chance to Afroamerican children and youngsters. The idea is, of course, that no amount of money, particularly if public, can correct for an allegedly genetic deficiency. This time around geneticists and psychologists were slow to react: perhaps they took the book for what it is, namely a political tract. On the other hand some journalists and sociologists did point out the methodological flaws of the book, uncovered its ideological sources, and denounced its implications for public policy.[59]

*Example 7: Feminist Technology*

Since technology is the art and science of getting things done, maintained, and repaired, psychotherapy and jurisprudence should be regarded as technologies. Now, in recent years these technologies have acquired a sex: there is now talk of feminist psychotherapy and feminist jurisprudence. Let us take a quick look at the former. A forte of feminist psychotherapy is "recovered memory therapy," consisting in "enhancing" a woman's memory—if necessary, with the help of hypnosis and drugs—until she "remembers" having been sexually abused by her father during childhood. The patient is then encouraged to take her father to court, in order to punish him and extract from him the maximum possible monetary compensation—to be shared with the therapist. This racket flourished during the past decade in the United States until the American Medical Association and above all the False Memory Syndrome Foundation warned the courts of law that they were being taken in. Thanks to this reaction the number of lawsuits of that type has started to decline. This is not to deny that many children are sexually abused by their relatives. What is objectionable is planting by the therapist of false memories into her patient and the "theory" that underlies this practice: the former is unscrupulous, and the latter false. Indeed, the theory in question is psychoanalysis, a pseudoscience according to which we never forget anything unless it is repressed by the "superego." This hypothesis is false: psychologists know that memory is not photographic but selective, distorting, and constructive. They also know that many people are suggestible, so that unscrupulous psychotherapists can successfully plant false memories in their brains.

To sum up, academic pseudoscience is just as toxic as academic antiscience. Why should serious and socially responsible scholars tolerate it? Being a travesty of scientific research, it should be dissected and exposed, taught only to exemplify bogus science.[60]

## TWO KINDS OF IGNORANCE: NATURAL OR STRAIGHT, AND CONTRIVED OR WILLFUL

No chemistry department would hire an alchemist. A department of crystallography is no place for believers in the psychic power of crystals. No engineering school would keep someone intent on designing a perpetual motion machine. An astronomical observatory is no place for people who believe that the planets are pushed by angels. A biology department would close its doors to anyone who rejects genetics. No one who denies the existence of Nazi concentration camps or Communist labor camps would be able to teach history at a decent university. No mathematics department would tolerate anyone holding that logic is a tool of male domination and quantity is masculine. No Jungian psychology is taught in any self-respecting department of psychology. Whoever believes in homeopathy cannot make it into an accredited medical school. To generalize: neither proven falsities nor lies are tolerated in any scientific or technological institution. And for a good reason, too: namely, because such institutions are set up with the specific purpose of finding, refining, applying, or teaching truths, not just any old opinions.

Walk a few steps away from the faculties of science, engineering, medicine, or law, towards the faculty of arts. Here you will meet another world, one where falsities and lies are tolerated, nay manufactured and taught, in industrial quantities. Here the unwary student may take courses in all manner of nonsense and falsity. Here some professors are hired, promoted, or given power for teaching that reason is worthless, empirical evidence unnecessary, objective truth nonexistent, basic science a tool of either capitalist or male domination, and the like. Here we find people who reject all the knowledge painstakingly acquired over the past half-millennium. This is the place where students can earn credits for learning old and new superstitions of nearly all kinds, and where they can unlearn to write, so as to sound like phenomenologists, existentialists, deconstructionists, ethnomethodologists, or psychoanalysts. This is where taxpayers' moneys are squandered in the maintenance of the huge industry of cultural involution centered around the deliberate rejection of rational discussion and empirical testing. This fraud has got to be stopped in the name of intellectual honesty and social responsibility.

Let there be no mistake: I am not proposing that we teach only what can be ascertained as true. On the contrary, we must doubt our learning, and we must continue teaching that we are all ignorant in most respects and to some degree or other. But we must also teach that ignorance can be gradually overcome by rigorous research, that falsity can be detected, that partial truth can be attained and perfected—the way Archimedes illustrated with his method for computing successive approximations to the exact value of the area of the circle.

We must also realize and teach that there are two kinds of ignorance: natural and willful, traditional and postmodern. The former is unavoidable and its admission mandatory; it is part of being a curious learner and an honest teacher. By contrast, willful or postmodern ignorance is the deliberate refusal to learn items relevant to one's interests. Examples: the refusal of the psychotherapist and the philosopher of mind to learn some experimental psychology and neuropsychology; the refusal of the literary critic with sociological interests to learn some sociology; and the refusal of the philosopher of science to learn a bit of the science he pontificates about. All these are instances of willful ignorance. This is the only intolerable kind of ignorance, for it is a form of dishonesty. And yet this kind of ignorance is being peddled nowadays in many faculties of arts.

Willful ignorance comes in two guises: naked or naive, and disguised or contrived. Naked or *indocta ignorantia* is the clear rejection of science, or—what amounts to the same—the denial of any differences between science and nonscience, in particular pseudoscience. This is what the irrationalists and the relativist-constructivists preach: it is part of the radical feminist and environmentalist "theories," as well as of existentialism, poststructuralism, general semiotics, philosophical hermeneutics, deconstructionism, and similar obscurantist fads.

The first to deny the difference between science and nonscience was Paul K. Feyerabend, the philosophical godfather of the "new" philosophy and sociology of science. He has been listened to because he was wrongly believed to know some physics. But in fact his ignorance of this, the one science he tried to learn, was abysmal. Thus he misunderstood the only two formulas that occur in his *Against Method*, the book that earned him instant celebrity.[61] The first formula, which he calls "the equipartition principle," is actually the Maxwell-Boltzmann distribution function for a system of particles in thermal equilibrium. (Incidentally, the constant occurring in the correct formula is not $R$, the universal gas constant, but Boltzmann's far more universal $k$. This is no small mistake, because it renders Feyerabend's formula dimensionally wrong.) The second formula, Lorentz's, does not give "the *energy* of an *electron* moving in a *constant magnetic* field" (my emphases), as Feyerabend claims. Instead, the formula gives the *force* that an *arbitrary electromagnetic* field $<E, B>$ exerts on a particle with an *arbitrary* electric charge. (Incidentally, the constant $c$ is missing in Feyerabend's copy—which, again, makes his formula dimensionally incorrect.) To top it all, Feyerabend substitutes the second formula into the first; and, not surprisingly, he gets an odd result that, in a mysterious way, leads him to speculate on the (nonexistent) magnetic monopoles imagined by his teacher Felix Ehrenhaft. But the substitution cannot be made, because (a) the second formula does not give us an energy, which occurs in the first one; (b) the first formula refers to a system of particles, whereas the second concerns a single particle; and (c) unlike the energy, which is a scalar, the force is a vector, and therefore it cannot occur by itself in the argument of an exponential function, which is defined only for scalars.[62] None of Feyerabend's critics detected these elementary errors—a disturbing indicator of the present state of the philosophy of science. In sum, one of the gurus of the new philosophy of science was guilty of *indocta ignorantia*. He was also seen as a guru of the student leftist movement.

However, irrationalism, in particular the distrust of science, has no political color; it is found left, center, and right. Still, in most cases it is passive: Babbitt is not Torquemada but is just indifferent to and suspicious of intellectual pursuits. On the other hand militant philistinism is strong in the New Left, the Old Right, and the religious wing of the New Right. This is no coincidence: all of these groups are authoritarian. And, as Popper pointed out half a century ago, authoritarianism is incompatible with rationalism in the broad sense, i.e., "the readiness to listen to critical arguments and to learn from experience."[63] Indeed, the citizen of a democracy is supposed to form his own opinions on matters of public interest, to debate them in the agora, and to participate to some extent in the management of the commonwealth. Rationality is thus a necessary component of democratic life, just as irrationality is a necessary ingredient of the *dressage* of a faithful loyal subject of a totalitarian regime. Remember Mussolini's commandment: "Believe, obey, fight." So much for academic antiscience.

Academic pseudoscience is a different ball game: it is far more subtle and therefore harder to diagnose and uproot. Indeed, it wears some of the accoutrements of genuine science, in particular an esoteric jargon that fools the unwary, or even a symbolic apparatus that intimidates the innumerate. It looks like science, but is not scientific because it does not enrich knowledge; and, far from having a self-correcting mechanism, it is dogmatic. Because it misleads the innocent, academic pseudoscience is at least as damaging as outright antiscience.

## CONCLUSION

I submit that the academic charlatans have not earned the academic freedom they enjoy nowadays. They have not earned it because they produce or circulate cultural garbage, which is not just a nonacademic activity but an antiacademic one. Let them do that anywhere else they please, but not in schools; for these are supposed to be places of learning. We should expel the charlatans from the university before they deform it out of recognition and crowd out the serious searchers for truth. They should be criticized, nay denounced, with the same rigor and vigor that Julien Benda attacked the intellectual mercenaries of his time (1927) in his memorable *La trahison des clercs*—which, incidentally, earned him the hatred of the so-called organic intellectuals of all political hues. Spare the rod and spoil the charlatan. Spoil the charlatan and put modern culture at risk. Jeopardize modern culture and undermine modern civilization. Debilitate modern civilization and prepare for a new Dark Age.

In former times higher learning was only a refined form of entertainment and a tool of social control. Today it is all that and more: scientific knowledge, science-based technology, and the rationalist humanities are not only intrinsically valuable public goods but also means of production and welfare, as well as conditions of democratic debate and rational conflict resolution. The search for authentic knowledge should therefore be protected from attack and counterfeit both inside and outside Academia. To this end I propose the adoption of the following Charter of Intellectual Academic Rights and Duties:

1. Every academic has the duty to search for the truth and the right to teach it.
2. Every academic has the right and the duty to question anything that interests him, provided he does it in a rational manner.
3. Every academic has the right to make mistakes and the duty to correct them upon detecting them.
4. Every academic has the duty to expose bunk, whether popular or academic.
5. Every academic has the duty to express himself in the clearest possible way.

6. Every academic has the right to discuss any unorthodox views that interest him, provided those views are clear enough to be discussed rationally.
7. No academic has the right to present as true ideas that he cannot justify in terms of either reason or experience.
8. Nobody has the right to engage knowingly in any academic industry.
9. Every academic body has the duty to adopt and enforce the most rigorous known standards of scholarship and learning.
10. Every academic body has the duty to be intolerant to both counterculture and counterfeit culture.

To conclude. Let us tolerate, nay encourage, all search for truth, however eccentric it may look, as long as it abides by reason or experience. But let us fight all attempts to suppress, discredit, or fake this search. Let all genuine intellectuals join the Truth Squad and help dismantle the "postmodern" Trojan horse stabled in Academia before it destroys them.

## NOTES

1 M. Heidegger, *Sein und Zeit*, p. 192.
2 Ibid., p. 331.
3 Heidegger, *Einführung in die Metaphysik*, pp. 20, 37.
4 Ibid., p. 92.
5 Ibid., p. 152.
6 E. Husserl, *Die Krisis der europäischen Wissenschaftgen und die tranzendentale Phänomenologie*, p. 187.
7 Husserl, *Cartesianische Meditationen*, p. 68.
8 Ibid.
9 Ibid., p. 183.
10 Ibid., p. 76.
11 Ibid., section 34.
12 Ibid., p. 118.
13 Ibid., p. 97.
14 E.g., A. Schu[e]tz, *The Phenomenology of the Social World*; and P. Berger & T. Luckmann, *The Social Construction of Reality.*
15 E.g., H. Garfinkel, *Studies in Ethnomethodology*; and E. Goffman, *Behavior in Public Places.*
16 M. Lynch, E. Livingston & H. Garfinkel, "Temporal Order in Laboratory Work," p. 206.
17 P. Atkinson, "Ethnomethodology: A Critical Review."
18 H. Garfinkel, *Studies in Ethnomethodology*, p. 1.
19 Ibid., p. 11.
20 R. Collins, "Interaction Ritual Chains, Power and Property."
21 M. F. Belenky, B. McV. Clinchy, N. R. Goldberger & J. M. Tarule, *Women's Ways of Knowing. The Development of Self, Voice, and Mind.*
22 Sandra Harding, *The Science Question in Feminism*, p. 113.
23 Herbert Marcuse, *One-Dimensional Man: Studies in the Ideology of Industrial Society.*
24 Michel Foucault, *Discipline and Punish.*
25 D. Patai & N. Koertge, *Professing Feminism. Cautionary Tales from the Strange World of Women's Studies*, p. 157.
26 P. R. Gross & N. Levitt, *Higher Superstition: The Academic Left and Its Quarrels with Science.*
27 For more on antiscience, particularly in social studies, see M. Bunge, *Finding Philosophy in Social Science.*

28 Vilfredo Pareto, *A Treatise on General Sociology*, section 2087.
29 Ibid., e.g., section 2148.
30 Ibid., section 2466.
31 Ibid., p. 509.
32 Ibid., chapter 5.
33 Pitirim Sorokin, *Fads and Foibles in Modern Sociology and Related Sciences.*
34 Pitirim Sorokin, *Social and Cultural Dynamics*, vol. 3, p. 162.
35 Samuel Huntington, *Political Order in Changing Societies*, p. 55.
36 See S. Lang, *The File.*
37 Gary S. Becker, *The Economic Approach to Human Behavior*, p. 257.
38 Milton Friedman, "A Theoretical Framework for Monetary Analysis."
39 See, e.g., D. Kahnemann, P. Slovic & A. Tversky, eds., *Judgment under Uncertainty: Heuristics and Biases*; and M. Bunge, "Two Faces and Three Masks of Probability."
40 See P. Tillers, "Decision and Inference" and the subsequent papers.
41 H. Poincaré, *Correspondence of Léon Walras and Related Papers*, vol. 3, pp. 164–165.
42 M. Allais, "The Foundations of a Positive Theory of Choice Involving Risk and a Criticism of the Postulates and Axioms of the American School"; A. Tversky, "A Critique of Expected Utility Theory: Descriptive and Normative Considerations"; J. W. Hernstein, "Rational Choice Theory: Necessary but Not Sufficient."
43 M. Bunge, "Game Theory is Not a Useful Tool for the Political Scientist"; "The Poverty of Rational Choice Theory"; *Philosophy in Social Science*; D. P. Green & I. Shapiro, *Pathologies of Rational Choice Theory: A Critique of Applications in Political Science.*
44 J. N. Rosenau, *Turbulence in World Politics. A Theory of Change and Continuity.*
45 Courtney Brown, "Politics and the Environment: Nonlinear Instabilities Dominate."
46 See, e.g., R. K. Merton, *The Sociology of Science. Theoretical and Empirical Investigations.*
47 See, e.g., the journal *Social Studies of Science*; B. Barnes, ed., *Sociology of Science. Selected Readings*; D. Bloor, *Knowledge and Social Imagery*; K. D. Knorr-Cetina & M. Mulkay, eds., *Science Observed. Perspectives on the Social Study of Science*; and B. Latour & S. Woolgar, *Laboratory Life: The Social Construction of Scientific Facts.*
48 B. Latour, "A Relativistic Account of Einstein's Relativity."
49 B. Latour, "Who Speaks for Science?"
50 M. Mulkay, "Some Aspects of Cultural Growth in the Natural Sciences."
51 D. Goldsmith, ed., *Scientists Confront Velikovsky. Papers from an AAAS Symposium.*
52 See, e.g., T. J. Pinch & H. M. Collins, "Is Anti-science Not-science?" and "Private Science and Public Knowledge: The Committee for the Scientific Investigation of the Claims of the Paranormal and Its Use of the Literature."
53 More criticisms in M. Bunge, "A Critical Examination of the New Sociology of Science, Part 1"; "A Critical Examination of the New Sociology of Science, Part 2"; L. Wolpert, *The Unnatural Nature of Science*; R. Boudon & M. Clavelin, eds., *Le relativisme est-il irrésistible? Regards sur la sociologie des sciences*; and R. Boudon, *Le juste et le vrai.*
54 See, e.g., S. J. Gould, *The Mismeasure of Man.*
55 E. S. Russell, "Report of the Ad Hoc Committee."
56 Y. Ezrahi, "The Political Resources of American Science."
57 See, e.g., M. Bunge & R. Ardila, *Philosophy of Psychology.*
58 R. J. Herrnstein & C. Murray, *The Bell Curve. Intelligence and Class Structure in American Life.*
59 C. Lane, "The Tainted Sources of 'The Bell Curve' " and the March 1995 issue of *Contemporary Society.*
60 More on pseudoscience in social studies may be found in M. Bunge, *Finding Philosophy in Social Science* and *Social Science under Debate.*

61 P. K. Feyerabend, *Against Method. Outline of an Anarchistic Theory of Knowledge*, p. 62.

62 More on Feyerabend's scientific incompetence in M. Bunge, "What is Science? Does It Matter to Distinguish It from Pseudoscience? A Reply to My Commentators."

63 K. R. Popper, *The Open Society and Its Enemies*, chapter 24.

## REFERENCES


ALEXANDER, J. C., B. GIESEN, R. MÜNCH & N. J. SMELSER, eds. *The Micro-Macro Link*. Berkeley, CA: University of California Press, 1987.

ALLAIS, M. "The Foundations of a Positive Theory of Choice Involving Risk and a Criticism of the Postulates and Axioms of the American School." In *Expected Utility Hypotheses and the Allais Paradox*, edited by M. Allais & O. Hagen. Boston: Reidel, 1979.

ATKINSON, P. "Ethnomethodology: A Critical Review." *Annual Review of Sociology* 14 (1988): 441–465.

BARNES, B., ed. *Sociology of Science. Selected Readings.* London: Penguin, 1972.

BECKER, G. S. *The Economic Approach to Human Behavior.* Chicago, IL: University of Chicago Press, 1976.

BELENKY, M. F., B. MCV. CLINCHY, N. R. GOLDBERGER & J. M. TARULE. *Women's Ways of Knowing. The Development of Self, Voice, and Mind.* New York, NY: Basic Books, 1986.

BENDA, J. *La trahison des clercs.* 2nd edit. Paris: Grasset, 1946.

BERGER, P. & T. LUCKMANN. *The Social Construction of Reality.* Garden City, NJ: Doubleday, 1966.

BLOOR, D. *Knowledge and Social Imagery.* London: Routledge and Kegan Paul, 1976.

BOUDON, R. *Le juste et le vrai.* Paris: Fayard, 1995.

BOUDON, R. & M. CLAVELIN, eds. *Le relativisme est-il irrésistible? Regards sur la sociologie des sciences.* Paris: Presses Universitaires de France, 1994.

BROWN, C. "Politics and the Environment: Nonlinear Instabilities Dominate." *American Political Science Review* 88 (1994): 292–303.

BUNGE, M. "Two Faces and Three Masks of Probability." In *Probability in the Sciences*, edited by E. Agazzi. Dordrecht: Reidel, 1988.

———. "Game Theory is Not a Useful Tool for the Political Scientist." *Epistemologia* 21 (1989): 195–212.

———. "What is Science? Does It Matter to Distinguish It from Pseudoscience? A Reply to My Commentators." *New Ideas in Psychology* 9 (1991): 245–283.

———. "A Critical Examination of the New Sociology of Science, Part 1." *Philosophy of the Social Sciences* 21 (1991): 524–560.

———. "A Critical Examination of the New Sociology of Science, Part 2." *Philosophy of the Social Sciences* 22 (1992): 46–76.

———. "The Poverty of Rational Choice Theory." In *Critical Rationalism, Metaphysics and Sciences*, Vol. 1, edited by I. C. Jarvie & N. Laor. Dordrecht: Kluwer Academic Publishers, 1995.

———. *Finding Philosophy in Social Science.* New Haven, CT: Yale University Press. In press.

———. *Social Science under Debate.* Forthcoming.

BUNGE, M. & R. ARDILA. *Philosophy of Psychology.* New York, NY: Springer-Verlag, 1987.

COLLINS, R. "Interaction Ritual Chains, Power and Property." In *The Micro-Macro Link*, edited by J. C. Alexander, B. Giesen, R. Münch & N. J. Smelser. Berkeley, CA: University of California Press, 1987.

EZRAHI, Y. "The Political Resources of American Science." In *Sociology of Science. Selected Readings*, edited by B. Barnes. London: Penguin Books, 1972.

FEYERABEND, P. K. *Against Method. Outline of an Anarchistic Theory of Knowledge.* London: Verso, 1978.

FOUCAULT, M. *Discipline and Punish.* New York, NY: Vintage Books, 1977.

FRIEDMAN, M. "A Theoretical Framework for Monetary Analysis." In *Milton Friedman's Monetary Framework*, edited by R. J. Gordon. Chicago, IL: University of Chicago Press, 1970.

GARFINKEL, H. *Studies in Ethnomethodology.* Englewood Cliffs, NJ: Prentice-Hall, 1967.

GOFFMAN, E. *Behavior in Public Places.* New York, NY: Free Press, 1963.

GOLDSMITH, D., ed. *Scientists Confront Velikovsky. Papers from an AAAS Symposium.* Ithaca, NY: Cornell University Press, 1977.

GOULD, S. J. *The Mismeasure of Man.* New York, NY: W. W. Norton, 1981.

GREEN, D. P. & I. SHAPIRO. *Pathologies of Rational Choice Theory: A Critique of Applications in Political Science.* New Haven, CT: Yale University Press, 1994.

GROSS, PAUL R. & NORMAN LEVITT. *Higher Superstition: The Academic Left and Its Quarrels with Science.* Baltimore, MD: Johns Hopkins University Press, 1994.

HARDING, S. *The Science Question in Feminism.* Ithaca, NY: Cornell University Press, 1986.

HEIDEGGER, M. *Sein und Zeit.* 16th edit. Tübingen: Max Niemeyer, 1986.

———. *Einführung in die Metaphysik.* 5th edit. Tübingen: Max Niemeyer, 1987.

HERRNSTEIN, J. W. "Rational Choice Theory: Necessary but Not Sufficient." *American Psychologist* 45 (1990): 356–367.

HERRNSTEIN, R. J. & C. MURRAY. *The Bell Curve. Intelligence and Class Structure in American Life.* New York, NY: Free Press, 1994.

HUNTINGTON, S. P. *Political Order in Changing Societies.* New Haven, CT: Yale University Press, 1968.

HUSSERL, E. "Cartesianische Meditationen." In *Husserliana: Gesammelte Werke*, Vol. 1. The Hague: Martinus Nijhoff, 1950.

———. Die Krisis der europäischen Wissenschaften und die tranzendentale Phänomenologie. In *Husserliana: Gesammelte Werke*, Vol. 6. The Hague: Martinus Nijhoff, 1954.

KAHNEMAN, D., P. SLOVIC & A. TVERSKY, eds. *Judgment under Uncertainty: Heuristics and Biases.* Cambridge: Cambridge University Press, 1982.

KNORR-CENTINA, K. D. & M. MULKAY, eds. *Science Observed. Perspectives on the Social Study of Science.* London: Sage Publications, 1983.

KOBLITZ, N. "Mathematics as Propaganda: A Tale of Three Equations; or, the Emperors Have No Clothes." *Mathematical Intelligencer* 10 (1988): 4–10.

LANE, C. "The Tainted Sources of 'The Bell Curve.'" *New York Review of Books* 41, no. 20 (1994): 14–19.

LANG, S. *The File.* New York, NY: Springer-Verlag, 1981.

LATOUR, B. "Give Me a Laboratory and I Will Raise the World." In *Science Observed*, edited by K. D. Knorr-Cetina & M. Mulkay. London: Sage Publications, 1983.

———. "A Relativistic Account of Einstein's Relativity." *Social Studies of Science* 18 (1988): 3–44.

———. "Who Speaks for Science?" *The Sciences* 35 (1995): 6–7.

LATOUR, B. & S. WOOLGAR. *Laboratory Life: The Social Construction of Scientific Facts.* Beverly Hills, CA: Sage Publications, 1979.

LYNCH, M. E. "Sacrifice and the Transformation of the Animal Body into a Scientific Object: Laboratory Culture and Ritual Practice in the Neurosciences." *Social Studies of Science* 18 (1988): 265–289.

LYNCH, M. E., E. LIVINSTON & H. GARFINKEL. "Temporal Order in Laboratory Work." In *Science Observed. Perspectives on the Social Study of Science*, edited by K. D. Knorr-Cetina & M. Mulkay. London: Sage Publications, 1983.

MACKINNON, C. *Toward a Feminist Theory of the State.* Cambridge, MA: Harvard University Press, 1989.

MARCUSE, H. *One-Dimensional Man: Studies in the Ideology of Industrial Society.* Boston, MA: Beacon, 1964.

MERTON, R. K. *Social Theory and Social Structure.* Rev. edit. New York, NY: The Free Press, 1957.

———. *The Sociology of Science. Theoretical and Empirical Investigations.* Chicago, IL: University of Chicago Press, 1973.

MULKAY, M. "Some Aspects of Cultural Growth in the Natural Sciences." Reprinted in *Sociology of Science. Selected Readings*, edited by B. Barnes. London: Penguin, 1972.

PARETO, V. *A Treatise on General Sociology.* 4 vols. New York, NY: Harcourt, Brace and Co., 1935. Reprint. New York, NY: Dover Publications, 1963.

PATAI, D. & N. KOERTGE. *Professing Feminism. Cautionary Tales from the Strange World of Women's Studies.* A New Republic Book. New York, NY: Basic Books, 1994.

PINCH, T. J. & H. M. COLLINS. "Is Anti-science Non-science?" *Sociology of the Sciences Yearbook*, Vol. 3: 221–250.

———. "Private Science and Public Knowledge: The Committee for the Scientific Investigation of the Claims of the Paranormal and Its Use of the Literature." *Social Studies of Science* 14 (1984): 521–546.

POINCARÉ, H. "Letter to L. Walras." In *Correspondence of Léon Walras and Related Papers*, Vol. 3, edited by W. Jaffé. Amsterdam: North Holland, 1965.

POPPER, K. R. *The Open Society and Its Enemies.* 2 vols. 2nd edit. London: Routledge and Kegan Paul, 1962.

ROSENAU, J. N. *Turbulence in World Politics. A Theory of Change and Continuity.* Princeton, NJ: Princeton University Press, 1990.

RUSSELL, E. S. "Report of the Ad Hoc Committee." *Genetics* 83 (1976): s99–s101.

SCHU[E]TZ, A. *The Phenomenology of the Social World.* Evanston, IL: Northwestern University Press, 1967.

SOROKIN, P. A. *Social and Cultural Dynamics*, Vol. 3. London: Allen and Unwin, 1937.

———. *Fads and Foibles in Modern Sociology and Related Sciences.* Chicago, IL: Henry Regnery, 1956.

TILLERS, P. "Decision and Inference." *Cardozo Law Review* 13 (1991): 253–256.

TVERSKY, A. "A Critique of Expected Utility Theory: Descriptive and Normative Considerations." *Erkenntnis* 9 (1975): 163–173.

WOLPERT, L. *The Unnatural Nature of Science.* London: Faber and Faber, 1992.

# THE FOUNDATIONS OF PHYSICS

PHYSICS HAS OFTEN been regarded as the model of incontrovertible rigor in the empirical sciences, whence the prevalence of "physics envy" in other disciplines. Yet there is much about the foundations of the subject that is still incompletely understood; and when people outside the discipline (and some within) are brought face to face with this fact, the consequence can be an exaggerated despair over the possibility of reasoned inquiry, or, even worse, an outright celebration of unreason and incoherence.

This section deals with several aspects of the situation. It begins with an exchange between physicists Sheldon Goldstein of Rutgers and Daniel Kleppner of MIT. Goldstein points out that the undoubted perplexities of quantum mechanics, along with the philosophical predispositions of Niels Bohr and Werner Heisenberg, among the principal founders, led to the emergence of a mystifying "quantum philosophy" that not only left the apparent contradictions of quantum mechanics unresolved, but declared, in advance, that any attempt to resolve them is foredoomed to failure. Even more curiously, the preeminence of the "Copenhagen interpretation" masked the work of David Bohm and John Bell, which did, in fact, unravel many of the conundra. In Goldstein's view, physicists (if not physics itself) bear some responsibility for the current celebration of unreason.

Daniel Kleppner responds as an experimentalist. For him, the most striking thing about quantum mechanics is its uncanny accuracy in predicting phenomena that can be confirmed by observation and experiment. Such striking empirical success convinces Kleppner that even though foundational loose ends may remain to be tied up to the satisfaction of philosophers, quantum mechanics is simply additional evidence that physics embodies a true, or at least highly accurate, picture of physical reality.

In recent years, with the emergence of such notions as "chaos" and "self-organization," there has been renewed attention to the foundations of classical thermodynamics and statistical mechanics. Some thinkers, notably Ilya Prigogine, have offered a postmodern interpretation, in which indeterminacy and a kind of uncertainty principle make their appearance in what has traditionally been an offshoot of classical deterministic mechanics. These interpretations have been welcomed by literary theorists and poststructuralist social thinkers as validation of the postmodern viewpoint. In the current essay, however, Belgian physicist Jean Bricmont argues in detail that many of the claims of Prigogine, *et al.*, are misconceived, deriving from errors that are as much mathematical as philosophical. Bricmont vindicates the insights of Ludwig Boltzmann, the founder of statistical mechanics, and shows that contemporary challenges to his views are poorly supported.

# QUANTUM PHILOSOPHY
## The Flight from Reason in Science[a]

SHELDON GOLDSTEIN

I WANT TO DISCUSS a rather delicate matter concerning a notoriously difficult subject, the foundations of quantum mechanics, a subject that has inspired a great many peculiar proclamations. Some examples:

> The idea of an objective real world whose smallest parts exist objectively in the same sense as stones or trees exist, independently of whether or not we observe them . . . is impossible.[1]

and

> We can no longer speak of the behavior of the particle independently of the process of observation. As a final consequence, the natural laws formulated mathematically in quantum theory no longer deal with the elementary particles themselves but with our knowledge of them. Nor is it any longer possible to ask whether or not these particles exist in space and time objectively. . . .
>
> Science no longer confronts nature as an objective observer, but sees itself as an actor in this interplay between man and nature. The scientific method of analysing, explaining, and classifying has become conscious of its limitations . . . method and object can no longer be separated.[2]

and

> A complete elucidation of one and the same object may require diverse points of view which defy a unique description. Indeed, strictly speaking, the conscious analysis of any concept stands in a relation of exclusion to its immediate application.[3]

This last quotation is an expression of what has traditionally been called complementarity—but what might nowadays be called multiphysicalism.

For my purposes here, what is most relevant about these sentiments is that they were expressed, not by lay popularizers of modern science, or by its post-modern critics, but by Werner Heisenberg and Niels Bohr, the two physicists

[a] I am grateful to Rebecca Goldstein and Eugene Speer for their comments and suggestions. This work was supported in part by National Science Foundation Grants DMS-9305930 and DMS-9504556.

most responsible, with the possible exception of Erwin Schrödinger, for the creation of quantum theory. It does not require great imagination to suggest that there is little in these sentiments with which a postmodernist would be inclined to disagree and much that he or she would be happy to regard as compelling support for the postmodern enterprise (see, for example, Notes 4–6).

The "quantum philosophy" expressed by such statements is part of the Copenhagen interpretation of quantum theory, which, in addition to the vagueness and subjectivity suggested by the preceding quotes, also incorporated as a central ingredient the notion that in the microscopic quantum domain the laws of nature involve irreducible randomness. The Copenhagen interpretation was widely, I would say at one time almost universally, accepted within the physics community, though there were some notable exceptions, such as Einstein and Schrödinger. Here is Schrödinger in 1926:

> Bohr's . . . approach to atomic problems . . . is really remarkable. He is completely convinced that any understanding in the usual sense of the word is impossible. Therefore the conversation is almost immediately driven into philosophical questions, and soon you no longer know whether you really take the position he is attacking, or whether you really must attack the position he is defending.[7]

and Schrödinger in 1959:

> With very few exceptions (such as Einstein and Laue) all the rest of the theoretical physicists were unadulterated asses and I was the only sane person left. . . . The one great dilemma that ails us . . . day and night is the wave-particle dilemma. In the last decade I have written quite a lot about it and have almost tired of doing so: just in my case the effect is null . . . because most of my friendly (truly friendly) nearer colleagues (. . . theoretical physicists) . . . have formed the opinion that I am—naturally enough—in love with "my" great success in life (viz., wave mechanics) reaped at the time I still had all my wits at my command and therefore, so they say, I insist upon the view that "all is waves." Old-age dotage closes my eyes towards the marvelous discovery of "complementarity." So unable is the good average theoretical physicist to believe that any sound person could refuse to accept the Kopenhagen oracle. . . .[8]

Einstein in 1949 offered a somewhat more constructive response:

> I am, in fact, firmly convinced that the essentially statistical character of contemporary quantum theory is solely to be ascribed to the fact that this (theory) operates with an incomplete description of physical systems. . . .
>
> [In] a complete physical description, the statistical quantum theory would . . . take an approximately analogous position to the statistical mechanics within the framework of classical mechanics.[9]

Part of what Einstein is saying here is that (much of) the apparent peculiarity of quantum theory, and in particular its randomness, arises from mistaking an incomplete description for a complete one.

In view of the radical character of quantum philosophy, the arguments offered in support of it have been surprisingly weak. More remarkable still is the fact that it is not at all unusual, when it comes to quantum philosophy, to find the very best physicists and mathematicians making sharp emphatic claims, almost of a mathematical character, that are trivially false and profoundly ignorant. For example, John von Neumann, one of the greatest mathematicians of this century, claimed in 1932 to have mathematically proved that Einstein's dream, of a deterministic completion or reinterpretation of quantum theory, was mathematically impossible. He concluded:

> It is therefore not, as is often assumed, a question of a re-interpretation of quantum mechanics—the present system of quantum mechanics would have to be objectively false, in order that another description of the elementary processes than the statistical one be possible.[10]

This claim of von Neumann was, of course, just about universally accepted. For example, Max Born, who formulated the statistical interpretation of the wave function, assures us:

> No concealed parameters can be introduced with the help of which the indeterministic description could be transformed into a deterministic one. Hence if a future theory should be deterministic, it cannot be a modification of the present one but must be essentially different.[11]

However, in 1952 David Bohm, through a refinement of de Broglie's pilot wave model of 1927, found just such a reformulation of quantum theory.[12] Bohm's theory, Bohmian mechanics, was precise, objective, and deterministic—not at all congenial to quantum philosophy and a counterexample to the claims of von Neumann. Nonetheless, we still find, more than a quarter of a century after the discovery of Bohmian mechanics, statements such as these:

> The proof he [von Neumann] published . . ., though it was made much more convincing later on by Kochen and Specker, still uses assumptions which, in my opinion, can quite reasonably be questioned. . . . In my opinion, the most convincing argument against the theory of hidden variables was presented by J. S. Bell (1964).[13]

and

> This [hidden variables] is an interesting idea and even though few of us were ready to accept it, it must be admitted that the truly telling argument against it was produced as late as 1965, by J. S. Bell. . . . This appears to give a convincing argument against the hidden variables theory.[14]

Now there are many more statements of a similar character that I could have cited; I chose these partly because Wigner was not only one of the leading physicists of his generation, but, unlike most of his contemporaries, he was also profoundly concerned with the conceptual foundations of quantum mechanics and wrote on the subject with great clarity and insight.

There was, however, one physicist who wrote on this subject with even greater clarity and insight than Wigner himself, namely, the very J. S. Bell whom Wigner praises for demonstrating the impossibility of a deterministic completion of quantum theory such as Bohmian mechanics. So let us see how Bell himself reacted to Bohm's discovery:

> But in 1952 I saw the impossible done. It was in papers by David Bohm. Bohm showed explicitly how parameters could indeed be introduced, into nonrelativistic wave mechanics, with the help of which the indeterministic description could be transformed into a deterministic one. More importantly, in my opinion, the subjectivity of the orthodox version, the necessary reference to the "observer," could be eliminated.[15]

and Bell again:

> Bohm's 1952 papers on quantum mechanics were for me a revelation. The elimination of indeterminism was very striking. But more important, it seemed to me, was the elimination of any need for a vague division of the world into "system" on the one hand, and "apparatus" or "observer" on the other. I have always felt since that people who have not grasped the ideas of those papers . . . and unfortunately they remain the majority . . . are handicapped in any discussion of the meaning of quantum mechanics.[16]

Wigner to the contrary notwithstanding, Bell did not establish the impossibility of a deterministic reformulation of quantum theory, nor did he ever claim to have done so. On the contrary, over the course of the past several decades, until his untimely death several years ago, Bell was the prime proponent, for a good part of this period almost the sole proponent, of the very theory, Bohmian mechanics, that he is supposed to have demolished. What Bell did demonstrate is the remarkable conclusion that nature, if governed by the predictions of quantum theory, must be nonlocal, exhibiting surprising connections between distant events. And unlike the claims of quantum philosophy, this nonlocality *is* well founded, and, with the experiments of Aspect,[17] rather firmly established. Nonetheless, *it* is far from universally accepted by the physics community. Here is Bell, expressing his frustration at the obtuseness of his critics, and insisting that his argument for nonlocality involves no unwarranted assumptions:

> Despite my insistence that the determinism was inferred rather than assumed, you might still suspect somehow that it is a preoccupation with determinism that creates the problem. Note well then that the following argument makes no mention whatever of determinism. . . . Finally you might suspect that the very notion of particle, and particle orbit . . . has somehow led us astray. . . . So the following argument will not mention particles . . . nor any other picture of what goes on at the microscopic level. Nor will it involve any use of the words "quantum mechanical system," which can have an unfortunate effect on the discussion. The difficulty is not created by any such picture or any such terminology. It is created by the predictions about the correlations in the visible outputs of certain conceivable experimental set-ups.[18]

So what is the relevance of what I have described to the theme of this volume? Well, there is some bad news and some good news. The bad news, nothing you did not already know anyway, is that objectivity is difficult to maintain and that physicists, even in their capacity as scientists, are only human. Nothing new. I must say, however, that the complacency of the physics establishment with regard to the foundations of quantum mechanics has been, it seems to me, somewhat astonishing, though I must admit to lacking sufficient historical perspective to have genuine confidence that what has occurred is at all out of the ordinary. But let me once again quote Bell:

> But why then had Born not told me of this "pilot wave"? If only to point out what was wrong with it? Why did von Neumann not consider it? . . . Why is the pilot wave picture ignored in text books? Should it not be taught, not as the only way, but as an antidote to the prevailing complacency? To show us that vagueness, subjectivity, and indeterminism, are not forced on us by experimental facts, but by deliberate theoretical choice?[19]

The last quoted sentence refers, of course, to the good news: that when we consider, not the behavior of physicists, but the physics itself, we find, in the stark contrast between the claims of quantum philosophy and the actual facts of quantum physics, compelling support for the objectivity and rationality of nature herself.

Here is one more bit of information somewhat relevant in this regard. You may well be wondering how, in fact, Bohm managed to accomplish what was so widely regarded as impossible, and what his completion of quantum theory involves. But you probably imagine that what eluded so many great minds could not be conveyed in but a few minutes, even were this an audience of experts. However, the situation is quite otherwise. In order to arrive at Bohmian mechanics from standard quantum theory one need do almost nothing! One need only avoid quantum philosophy and complete the usual quantum description in what is really the most obvious way: by simply including the positions of the particles of a quantum system as part of the state description of that system, allowing these positions to evolve in the most natural way.[20] The entire quantum formalism, including the uncertainty principle and quantum randomness, emerges from an analysis of this evolution.[21] My long-time collaborator, Detlef Dürr, has expressed this succinctly—though in fact not succinctly enough—by declaring that the essential innovation of Bohmian mechanics is the insight that *particles move*![22] Bell, referring to the double-slit interference experiment, put the matter this way:

> Is it not clear from the smallness of the scintillation on the screen that we have to do with a particle? And is it not clear, from the diffraction and interference patterns, that the motion of the particle is directed by a wave? De Broglie showed in detail how the motion of a particle, passing through just one of two holes in screen, could be influenced by waves propagating through both holes. And so influenced that the particle does not go where the waves cancel out, but is attracted to where they cooperate. This idea seems to me so natural and simple, to resolve the wave-particle dilemma in such a clear and ordinary way, that it is a great mystery to me that it was so generally ignored.[23]

I think this should be a bit of a mystery for all of us!

## NOTES

1 W. Heisenberg, *Physics and Philosophy*, p. 129.
2 W. Heisenberg, *The Physicist's Conception of Nature*, pp. 15, 29.
3 N. Bohr (1934), quoted by M. Jammer, in *The Philosophy of Quantum Mechanics*, p. 102.

4 A. Plotnitsky, ***Complementarity: Anti-Epistemology after Bohr and Derrida.***
5 S. Aronowitz, ***Science as Power: Discourse and Ideology in Modern Society.***
6 E. Fox Keller, "Cognitive Repression in Contemporary Physics."
7 E. Schrödinger, Letter to Wien, quoted by W. Moore in ***Schrödinger***, p. 228.
8 E. Schrödinger, Letter to Synge, ibid., p. 472.
9 P. A. Schilpp, ed., ***Albert Einstein, Philosopher-Scientist***, pp. 666, 672.
10 J. von Neumann, ***Mathematical Foundations of Quantum Mechanics***, p. 325.
11 ***Natural Philosophy of Cause and Chance***, p. 109.
12 "A Suggested Interpretation of the Quantum Theory in Terms of "Hidden" Variables, Parts I and II.
13 E. P. Wigner, "Interpretation of Quantum Mechanics," p. 291.
14 E. P. Wigner, "Review of Quantum Mechanical Measurement Problem," p. 53.
15 J. S. Bell, ***Speakable and Unspeakable in Quantum Mechanics***, p. 160.
16 Ibid., p. 173.
17 A. Aspect, J. Dalibard & G. Roger, "Experimental Test of Bell's Inequalities Using Time-Varying Analyzers."
18 Bell, p. 150.
19 Ibid., p. 160.
20 D. Dürr, S. Goldstein & N. Zanghì. "Quantum Equilibrium and the Origin of Absolute Uncertainty"; "Quantum Mechanics, Randomness, and Deterministic Reality."
21 Ibid. and K. Berndl, M. Daumer, D. Dürr, S. Goldstein & N. Zanghì. "A Survey of Bohmian Mechanics."
22 Private communication.
23 Bell, p. 191.

## REFERENCES

ARONOWITZ, S. ***Science as Power: Discourse and Ideology in Modern Society.*** Minneapolis, MN: University of Minnesota Press, 1988.

ASPECT, A., J. DALIBARD & G. ROGER. "Experimental Test of Bell's Inequalities Using Time-Varying Analyzers." ***Physical Review Letters*** 49 (1982): 1804–1807.

BELL, J. S. ***Speakable and Unspeakable in Quantum Mechanics.*** Cambridge: Cambridge University Press, 1987.

BERNDL, K., M. DAUMER, D. DÜRR, S. GOLDSTEIN & N. ZANGHÌ. "A Survey of Bohmian Mechanics." ***Il Nuovo Cimento.*** 110B (1995): 737–750.

BOHM, DAVID. "A Suggested Interpretation of the Quantum Theory in Terms of "Hidden" Variables, Parts 1 and 2." ***Physical Review*** 85 (1952): 166–179, 180–193.

BORN, MAX. ***Natural Philosophy of Cause and Chance.*** Oxford: Oxford University Press, 1949.

DÜRR, D., S. GOLDSTEIN & N. ZANGHÌ. "Quantum Equilibrium and the Origin of Absolute Uncertainty." ***Journal of Statistical Physics*** 67 (1992): 843–907.

———. "Quantum Mechanics, Randomness, and Deterministic Reality." ***Physics Letters A*** 172 (1992): 6–12.

HEISENBERG, W. ***The Physicist's Conception of Nature.*** Translated by Arnold J. Pomerans. New York, NY: Harcourt Brace, 1958.

———. ***Physics and Philosophy.*** New York, NY: Harper and Row, 1958.

JAMMER, M. ***The Philosophy of Quantum Mechanics.*** New York, NY: Wiley, 1974.

KELLER, E. FOX. "Cognitive Repression in Contemporary Physics." ***American Journal of Physics*** 47 (1979): 718–721.

MOORE, W. ***Schrödinger.*** Cambridge: Cambridge University Press, 1989.

PLOTNITSKY, A. ***Complementarity: Anti-Epistemology after Bohr and Derrida.*** Durham, NC: Duke University Press, 1994.

SCHILPP, P. A., ed. ***Albert Einstein, Philosopher-Scientist.*** Evanston, IL: Library of Living Philosophers, 1949.

VON NEUMANN, J. ***Mathematical Foundations of Quantum Mechanics.*** Translated by R. T. Beyer. Princeton, NJ: Princeton University Press, 1955.

WIGNER, E. P. "Interpretation of Quantum Mechanics." In *Quantum Theory and Measurement*, edited by J. A. Wheeler & W. H. Zurek. Princeton, NJ: Princeton University Press, 1983.

———. "Review of Quantum Mechanical Measurement Problem." In *Quantum Optics, Experimental Gravity and Measurement Theory*, edited by P. Meystre & M. O. Scully. New York, NY: Plenum, 1983.

# PHYSICS AND COMMON NONSENSE

DANIEL KLEPPNER

SCIENTISTS are naturally dismayed when the concepts, goals, and achievements of science are misrepresented in newspapers and books, and particularly when they are misrepresented in the halls of Congress. The problem is more than a matter of professional pride. At a time when society is undergoing profound changes—sometimes for the better but often, it seems, for the worse—the rejection of scientific reasoning and science itself is potentially suicidal. Under such circumstances, scientists cannot afford to remain silent.

Reasoned criticism of science can be constructive, but unreasoned criticisms seem to attract the most popular attention. In particular, two cornerstones of physics, relativity and quantum mechanics, are irresistible to science bashers who happen to have a philosophical bent. I propose to comment on relativity and quantum mechanics from the point of view of a practicing physicist for whom these cornerstones are familiar everyday objects. The broader issues—the growing manifestations of antiscience and the historical and social forces that are contributing to a deteriorating climate for rational thought—are discussed elsewhere in these proceedings and have been documented in books by three of the conference organizers. *Higher Superstition* by Paul R. Gross and Norman Levitt provides a chart for voyagers who set out to navigate the shoals of antiscience. Gerald Holton's *Science and Anti-Science* and *Einstein, History and Other Passions* not only discuss these issues but reward the reader with fascinating essays on science and the history of science.

A recurrent complaint among science critics is that physics no longer makes sense, that it has parted ways with reality. An example is an op-ed essay in the *New York Times* by John Lukacs (June 17, 1993) which argued that the proposed superconducting supercollider, a major high-energy accelerator, was unwise. His objections, however, were not really to the accelerator, which in any case was subsequently canceled, but to science itself. He saw little difference between modern science and medieval attempts to weigh the soul. Because of Heisenberg's uncertainty principle, he argued, it is impossible to speak either of "nature" or of "matter." In essence, physics deserves no respect because it has parted company with reality.

Brian Appleyard, in "Understanding the Present," also bemoans the split between physics and reality. He views physics as an "affront to common sense."[1] Somehow Appleyard came to believe that knowledge fails at the quantum level. "If reality behaved differently in different circumstances, did this mean there was no reality?" The same confusion is echoed by Stanley Aronowitz in "Science as Power," as described in "Higher Superstition."[2] "The argument roughly but accurately paraphrased . . . is that since physics has discovered the uncertainty principle, it can no longer provide reliable information about the physical world, has lost its claim to objectivity."

Whether or not physics has parted ways with common sense can only be answered by turning to the physics itself. So, here are a few comments on the two theories invariably cited to illustrate the alleged breakdown of common sense: relativity and quantum mechanics.

Years ago, popular accounts of special relativity were filled with awe at its apparently radical view of space and time. However, the theory is intensely conservative as Einstein himself recognized. Without relativity, the great classical edifice of electromagnetic theory would crumble, the underlying unity of electricity and magnetism would vanish, and we would be left with innumerable paradoxes. With relativity, however, the structure of electromagnetic theory is united so transparently that it can be summarized in a single equation.

Happily, no sophisticated knowledge of physics or mathematics is needed to understand relativity—indeed, it is often taught in the freshman year. Furthermore, the experimental evidence for relativity is overwhelming. For particle physicists, relativistic mechanics is as immediate and intuitive as the law of the lever. And for all physicists, relativity's intimate linkage between space and time provides a harmonious picture of events, much as Newton's mechanics provides a harmonious picture of planetary motion. Like every fundamental theory of physics, relativity has been tested time and again. Some tests have been at a precision of a part per billion. So far, there have been no discrepancies. Far from being an esoteric theory that baffles common sense, relativity is a simple, reliable, and essential tool for physicists.

General relativity, Einstein's theory of gravity, is far more complicated than special relativity. The geometry of curved space with its bending light beams and distortion of time seems totally alien to our everyday experience. Such behavior, however, is not as remote as one might guess. According to general relativity, gravity effects time, though the effect at the earth is fantastically small. The flow of time at points that differ in height by one meter is about one second in three hundred million years. This is small but by no means negligible, for the effect of gravity on time must be taken into account in the recently developed global positioning system for navigation. This system allows observers to find their positions within a few hundred feet anywhere on earth. It works by comparing timing signals from various satellite-borne clocks. If general relativity were overlooked, a plane from San Francisco that was aiming for the middle of a runway at LaGuardia airport would find itself somewhere in New Jersey. That the flow of time depends on gravity may defy common sense, but rejecting that fact would be roughly tantamount to joining the Flat Earth Society.

One can gain a reasonable knowledge of relativity from any number of popular books, and it is often introduced in elementary physics courses. Quantum mechanics requires a little more work. A sophomore in physics can learn a good deal about it, and a junior can easily achieve a working mastery. To gain real intuition into the quantum world, however, requires some experience. For an experimenter, that could mean listening to the clicks or seeing the blips that indicate something has happened, or watching patiently as an interference pattern, resonance signal, or some other picture of quantum reality emerges from the data. It means designing an apparatus based on the laws of quantum mechanics and watching it come alive in harmony with those laws.

At a deeper level, understanding quantum mechanics means appreciating that it makes everything we measure measurable. At such a level of familiarity, there is nothing vague or tenuous about it. Quantum mechanics provides the language for describing what we see in the world of atoms and molecules. That world is no longer remote from our immediate experience. Today one can see with the naked eye the light radiated by a single atom, take photographs of atoms resting on a crystalline surface, and move them much as children move building blocks on a table. Quantum mechanics provides the language for describing this world. To those who understand it, the language is beautiful.

Unfortunately, quantum mechanics carries a mystique of mathematical complexity. C. P. Snow, in his autobiographical novel *The Search*, describes the enviable career of a young scientist at Cambridge University, a crystallographer who achieved eminence at a young age. The world was his oyster when he abruptly turned his back on science and started a new career in literature. The reason? The problem was not spiritual, religious, or romantic. It was mathematical. Quantum mechanics had just been invented, and though he was a brilliant scientist, he concluded that he lacked the mathematical talent needed to become one of the new scientists. I recall reading that story as I myself was struggling to learn quantum mechanics, and feeling rather annoyed that Snow had quit science so easily over material my teachers now expected me to take in stride. They were, of course, correct.

From time to time, Heisenberg's uncertainty principle, a central tenet of quantum mechanics, is cited by many critics, including Appleyard, as evidence that modern science has abandoned the search for objective knowledge. The impossibility of simultaneously measuring certain variables to arbitrary accuracy is unacceptable to these critics because it defies common sense. By the familiar rules of common sense, however, the very existence of matter is absurd. Here is the reason. There is abundant evidence that atoms are composed of negative electrons bound electrically to a positive nucleus. By the elementary law of electrical attraction, these particles should simply collapse into each other. If this happened, atoms and molecules—in fact, all of matter as we know it—could not exist. Fortunately, quantum mechanics provides not only a simple explanation for why this does not occur, why matter is actually stable, but also a marvelously detailed account of the structure of atoms.

Quantum mechanics also provides a natural explanation for a property of matter that is so deeply rooted in our consciousness that we merely take it for granted: the repetition of forms in nature. Why are identical atoms the same everywhere? On the human scale, identical objects are never exactly identical. On the atomic scale, they are. Without such exact identity, the atomic and molecular structures that underlie the physical and biological world around us could not exist. The concept of identity is inherent in the structure of quantum mechanics. Unless particles are exactly the same, they are totally different; there are no such things as "almost identical" particles. Once again, quantum mechanics defies naive common sense. To the generous eye, the world as we see it is a gift of quantum mechanics.

Another fallacious belief is that the uncertainty principle makes it impossible to know things accurately. The truth is otherwise. The precision of modern physics, in which some fundamental theories have been verified to a part in a trillion and most of the important physical constants have been measured to at least a part in a million, is in reality a gift from quantum mechanics. Although the uncertainty principle determines the scale for precision in individual measurements, it by no means limits the ultimate accuracy to which things can be measured. Rather than making things unknowable, the uncertainty principle provides the foundation for the incredible accuracy of measurements in modern physics.

The wave-particle duality is yet one more target for antiscientists. According to the principle of complementarity, physical entities can display properties that are fundamentally incompatible. To some people this is deeply disturbing because it defies—once again—common sense. The best known example is the wave-particle duality. It is a simple fact of observation that particles can exhibit wavelike properties, as in matter wave diffraction, and waves can exhibit particle-like properties, as in the photoelectric effect. There is, however, no inherent contradiction. Whether one wishes to observe particle properties or wave properties is a matter of choice, but the rule is that one cannot observe them simultaneously. This is the essence of the principle of complementarity. If one wishes, this can be viewed as a consequence of the uncertainty principle. Experimentally, there is no confusion. One can design an experiment to measure wave properties or particle properties, or, if one is more ambitious, even some combination of the two. In the latter case, however, the measurements of each property are invariably degraded.

In practice, quantum mechanics is a superb tool for studying and describing phenomena at the atomic level. Having given it this accolade, let me turn briefly to the debate on its conceptual foundations and the problems of measurement theory described elsewhere in these proceedings by Sheldon Goldstein.[3] That debate has been waged at various levels since quantum mechanics was first formulated, but the standard interpretation, the so-called Copenhagen interpretation, has stood the test of time and is widely accepted. Alternative formulations appear ultimately to yield the same description. The major advance in the past thirty years was the analysis by John Bell of possible incompleteness in the quantum description.[4] He proposed an experimental test that was carried out and which provided dramatic evidence for the standard formulation. So, unless the debate leads to new experimental tests, or provides a simpler description of known facts, it will continue to lie outside of the mainstream of physics. To me, if there is an underlying worry about the nature of quantum mechanics, it is that the theory works too well. Physical theories ultimately rest on observation, and just as observations are necessarily limited in scope, theories are necessarily limited in validity. So far, however, nobody seems to know the limitations of quantum mechanics.

To the critics who believe that physics affronts common sense and defies intuition, I pass on the advice sometimes given to students by a colleague at MIT, Arthur Mattuck, Professor of Mathematics: "***Don't use your intuition if you don't possess it!***" The tirades against relativity or quantum mechanics reflect little knowledge of the subject. Neither physicists nor philosophers need licenses to practice their professions, much less lay critics of science. These critics would command more respect, however, if they took the trouble to understand what they were criticizing. Physicists, for their part, have an obligation not to remain silent but to speak up when they encounter nonsense about physics.

## NOTES

1 Brian Appleyard, "Understanding the Present," p. 157.
2 Paul R. Gross & Norman Levitt, *Higher Superstition: The Academic Left and Its Quarrels With Science*, p. 51.
3 "Quantum Philosophy: The Flight from Reason in Science."
4 J. S. Bell, *Physics* 1 (1964): 195; and A. Aspect, "Atomic Physics 8," p. 103.

## REFERENCES

APPLEYARD, BRIAN. *Understanding the Present: Science and the Soul of Modern Man.* London: Picador, 1992.

ARONOWITZ, STANLEY. *Science as Power: Discourse and Ideology in Modern Society.* Minneapolis, MN: University of Minnesota Press, 1988.

ASPECT, A. "Atomic Physics 8." Edited by I. Lindgren, A. Rosen & S. Svanberg. New York, NY: Plenum Press, 1983.

BELL, J. S. *Physics* 1 (1964): 195.

GOLDSTEIN, SHELDON. "Quantum Philosophy: The Flight from Reason in Science." *Annals of the New York Academy of Sciences* 775 (1996): 119–125 (this volume).

GROSS, PAUL R. & NORMAN LEVITT. *Higher Superstition: The Academic Left and Its Quarrels with Science.* Baltimore, MD: Johns Hopkins University Press, 1994.

HOLTON, GERALD. *Einstein, History and Other Passions.* New York, NY: American Institute of Physics, 1995.

———. *Science and Anti-Science.* Cambridge, MA: Harvard University Press, 1993.

SNOW, C. P. *The Search.* New York, NY: Scribner, 1958.

# SCIENCE OF CHAOS OR CHAOS IN SCIENCE?

JEAN BRICMONT

IN THIS CHAPTER I try to clarify several confusions in the popular literature concerning chaos, determinism, the arrow of time, entropy and the role of probabilities in physics. Classical ideas going back to Laplace and Boltzmann are explained and defended, while some recent views on irreversibility, due to Prigogine, are criticized.

## 1 INTRODUCTION

> *We might characterize today's breakdown of industrial or "Second Wave" society as a civilizational "bifurcation," and the rise of a more differentiated, "Third Wave" society as a leap to new "dissipative structures" on a world scale. And, if we accept this analogy, might we not look upon the leap from Newtonianism to Prigoginianism in the same way? Mere analogy, no doubt. But illuminating, nevertheless.*
>
> —ALVIN TOFFLER

Popularization of science seems to be doing very well: the Big Bang, the theory of elementary particles or of black holes are explained in countless books for the general public. The same is true for chaos theory, irreversibility or self-organization. However, it seems also that a lot of confusion exists concerning these latter notions, and that at least some of the popular books are spreading misconceptions. The goal of this article is to examine some of them, and to try to clarify the situation.

In particular, I will make a critical evaluation of the various claims concerning chaos and irreversibility made by Prigogine and by Stengers, since "La Nouvelle Alliance." Several of those claims, especially the most recent ones, are rather radical: "the notion of chaos leads us to rethink the notion of 'law of nature.' "[1] For chaotic systems, "*trajectories are eliminated from the probabilistic description* . . . The statistical description is *irreducible.*"[2] The existence of chaotic dynamic systems supposedly marks a radical departure from a fundamentally deterministic worldview, makes the notion of trajectory obsolete, and offers a new understanding of irreversibility. Prigogine and Stengers claim that the classical conception was unable to incorporate time in our view of the world[3] or to account for the irreversibility of macroscopic phenomena. Boltzmann's attempt to explain irreversibility on the basis of reversible laws failed.[4]

On the basis of these physical theories, a number of speculations are put forward on the notion of "event," on the place of human beings in Nature, or even on overcoming Cartesian dualism.[5] These writings have been indeed quite influential, mostly among nonexperts. They are frequently quoted in philosophical or cultural circles as an indication that chaos, nonlinear phenomena, and the "arrow of time" have led to a profound revolution in our way of thinking.

I want to develop quite different views on most of these issues. In my opinion, chaos does not invalidate in the least the classical deterministic worldview (the existence of chaotic dynamical systems actually strengthens that view). Besides, chaos is not related in a fundamental way to irreversibility. Finally, when they are correctly presented, the classical views of Boltzmann perfectly account for macroscopic irreversibility on the basis of deterministic, reversible, microscopic laws. Part of the difficulty in understanding those views comes from some confusions about the use of the words "objective" and "subjective," associated with probability or entropy. I will try to be careful with these notions. I will also argue that most of the speculation on the new alliance between Man and Nature is misguided.

On the other hand, I believe that the ideas of Laplace and of Boltzmann are worth defending against various misrepresentations and misunderstandings. Quite independently of the work of Prigogine, there are serious confusions that are found in the literature on irreversibility, chaos, or time (some of which go back to philosophers such as Popper, Feyerabend, or Bergson). Besides, many textbooks or popular books on statistical mechanics are rather obscure, at least in the part concerning the foundations of the field (e.g., on the role played by ergodic theorems). I will try to clarify these questions too.

I wrote this paper in nontechnical language, relegating formulas to footnotes and remarks. Nothing of what I say is new.[6] In fact, everything is quite standard and old, and it is a sad fact that those ideas that were so nicely explained by Boltzmann a century ago[7] have to be reexplained over and over again.

Finally, I have to emphasize that this is in no way a criticism of Prigogine's work in general, and even less of the Brussels school. I am concerned here only with the part dealing with chaos and irreversibility (which, however, occupies the largest fraction of the popular books). I believe that a lot of interesting scientific ideas have been developed around Prigogine and that he has had an exceptional taste for discovering new directions in physics, whether in irreversible thermodynamics or in chaotic phenomena. But this does not put his views on foundational questions beyond criticism.[8]

## 2 CHAOS AND DETERMINISM: DEFENDING LAPLACE

*The concept of dog does not bark.*

—SPINOZA

### 2.1 *Determinism and Predictability*

A major scientific development in recent decades has been popularized under the name of "chaos." It is widely believed that this development implies a fundamental philosophical or conceptual revolution. In particular, it is thought that the classical worldview brilliantly expressed by Laplace in his "Philosophical Essay on Probabilities" has to be rejected.[9] Determinism is no longer defensible. I think this is based on a serious confusion between *determinism* and *predictability.* I will start by underlining the difference between the two concepts. Then it will be clear that what goes under the name of chaos is a major scientific progress but does not have the radical philosophical implications that are sometimes attributed to it.

In a nutshell, determinism has to do with how Nature behaves, and predictability is related to what we human beings are able to observe, analyze, and compute. It is easy to illustrate the necessity for such a distinction. Suppose we consider a perfectly regular, deterministic *and* predictable mechanism, like a clock, but put it on the top of a mountain, or in a locked drawer, so that its state (its initial conditions) becomes inaccessible to us. This renders the system trivially unpredictable, yet it seems difficult to claim that it becomes nondeterministic.[10] Or consider a pendulum: when there is no external force, it is deterministic and predictable. If one applies to it a periodic forcing, it may become unpredictable. Does it cease to be deterministic?

In other words, anybody who admits that *some* physical phenomena obey deterministic laws must also admit that some physical phenomena, although deterministic, are not predictable, possibly for "accidental" reasons. So a distinction must be made.[11] But once this is admitted, how does one show that *any* unpredictable system is *truly* nondeterministic, and that the lack of predictability is not merely due to some limitation of our abilities? We can never infer indeterminism from our ignorance alone.

Now, exactly what does one mean by determinism? Maybe the best way to explain it is to go back to Laplace: "Given for one instant an intelligence which could comprehend all the forces by which nature is animated and the respective situation of the beings who compose it—an intelligence sufficiently vast to submit these data to analysis—it would embrace in the same formula the movements of the greatest bodies of the universe and those of the lightest atom; for it, nothing would be uncertain and the future, as the past, would be present before its eyes."[12] The idea expressed by Laplace is that determinism depends on what the laws of nature are. Given the state of the system at some time, we have a formula (a differential equation, or a map) that gives in principle the state of the system at a later time. To obtain predictability, one has to be able to measure the present state of the system with enough precision, and to compute with the given formula (to solve the equations of motion). Note that there exist alternatives to determinism: there could be no law at all; or the laws could be stochastic: the state at a given time (even if it is known in every conceivable detail) would determine only a probability distribution for the state at a later time.

How do we know whether determinism is true—i.e., does nature obey deterministic laws? This is a rather complicated issue. Any serious discussion of it must be based on an analysis of the fundamental laws, hence of quantum mechanics, and I do not want to enter this debate here.[13] Let me just say that it is conceivable that we shall obtain, some day, a complete set of fundamental physical laws (like the law of universal gravitation in the time of Laplace), and then we shall see whether these laws are deterministic or not.[14] Any discussion of determinism outside of the framework of the fundamental laws is useless.[15] All I want to stress here is that the existence of chaotic dynamic systems do not affect this discussion *in any way.* What are chaotic systems? The simplest way to define them is through sensitivity to initial conditions. This means that, for any initial condition of the system, there is some other initial condition, arbitrarily close to the first one so that, if we wait long enough, the two systems will be markedly different.[16] In other words, an arbitrarily small error on the initial conditions makes itself felt after a long enough time. Chaotic dynamic systems are, of course, unpredictable in practice, at least for long enough times,[17] since there will always be some error in our measurement of the initial conditions. But this does not have any impact on our discussion of determinism, since we are assuming from the beginning that the system obeys some deterministic law. It is only by analyzing this deterministic system that one shows that a small error in the initial conditions may lead to a large error after some time. If the system did not obey any law, or if it followed a stochastic law, then the situation would be very different. For a stochastic law, two systems with the *same* initial condition could be in two very different states after a short time.

It is interesting to note that the notion that small causes can have big effects (in a perfectly deterministic universe) is not new at all. Maxwell wrote: "There is a maxim which is often quoted, that 'The same causes will always produce the same effects.'" After discussing the meaning of this principle, he adds: "There is another maxim which must not be confounded with that quoted at the beginning of this article, which asserts 'That like cause produce like effects.' This is only true when small variations in the initial circumstances produce only small variations in the final state of the system."[18] One should not conclude from these quotations[19] that there is nothing new under the sun. A lot more is known about dynamic systems than was known in the time of Poincaré. But the general idea that not everything is predictable, even in a deterministic universe, has been known for centuries. Even Laplace emphasized this point: after formulating universal determinism, he stresses that we shall always remain "infinitely distant" from the intelligence that he just introduced. After all, why is this determinism stated in a book on *probabilities*? The reason is obvious: for Laplace, probabilities lead to rational inferences in situations of incomplete knowledge (I shall come back later to this view of probabilities). So he is assuming from the beginning that our knowledge is incomplete, and that we shall never be able to *predict* everything. It is a complete mistake to attribute to some "Laplacian dream" the idea of perfect predictability.[20] But Laplace does not commit what E. T. Jaynes calls the "Mind Projection Fallacy": "We are all under an ego-driven temptation to project our private thoughts out onto the real world, by supposing that the creations of one's own imagination are real properties of Nature, or that one's own ignorance signifies some kind of indecision on the part of Nature."[21] As we shall see, this is a most common error. But whether we like it or not, the concept of dog does not bark, and we have to carefully distinguish between our representation of the world and the world itself.

Let us now see why the existence of chaotic dynamic systems in fact supports universal determinism rather than contradicts it.[22] Suppose for a moment that the world would be such that no classical mechanical system can behave chaotically. That is, suppose we have a theorem saying that any such system must eventually behave in a periodic fashion.[23] It is not completely obvious what the conclusion would be, but certainly *that* would be an embarrassment for the classical worldview. Indeed, so many physical systems seem to behave in a nonperiodic fashion that one would be tempted to conclude that classical mechanics cannot adequately describe those systems. One might suggest that there must be an inherent indeterminism in the basic laws of nature. Of course, other replies would be possible: for example, the period of those classical motions might be enormously long. But it is useless to speculate on this fiction, since we know that chaotic behavior is compatible with a deterministic dynamics. The only point of this story is to stress that deterministic chaos increases the explanatory power of deterministic assumptions and therefore, according to normal scientific practice, *strengthens* those assumptions. And if we did not know about quantum mechanics, the recent discoveries about chaos would not force us to change a single word of what Laplace wrote.[24]

### 2.2 *Trajectories and Probabilities*

Now I will turn to the main thesis of Prigogine and Stengers on chaotic dynamic systems: the notion of trajectory should be abandoned, and replaced by probabilities. What does it mean? Let me quote Prigogine: "We must therefore eliminate the notion of trajectory from our microscopic description. This actually corresponds to a realistic description: no measurement, no computation leads strictly to a point, to the consideration of a *unique* trajectory. We shall always face a *set* of trajectories."[25]

Let us see how reasonable it is to "eliminate the notion of trajectory" for chaotic systems by considering a concrete example.[26] Take a billiard ball on a sufficiently smooth table, so that we can neglect friction (for some time), and assume that there are suitable obstacles and boundaries so that the system is chaotic. Now suppose that we use an "irreducible" probabilistic description—that is, instead of assigning a position to the ball, we assign to it a probability distribution.[27] Consider next the evolution of that probability distribution. Since we are dealing with a chaotic system, that distribution will spread out all over the billiard table. This means that after a rather short time, there will be an almost uniform probability of finding the ball in any given region of the table. Indeed, even if our initial probability distribution is well peaked around the initial position of the ball, there will be lots of nearby initial conditions that will give rise to very different trajectories (that is exactly what it means to say that the system is chaotic). But now we can hardly take the probability distribution after some time seriously as an "irreducible" *description* of the system. Indeed, whenever we look at the system, we find the ball somewhere, at a rather well-defined position on the table. It is certainly not completely described by its probability distribution. The latter describes adequately our knowledge (or rather our ignorance) of the system, obtained on the basis of our initial information. But it would be difficult to commit the Mind Projection Fallacy more radically that to confuse the objective position of the ball and our best bet for it. In fact, chaotic systems illustrate this difference: if all nearby initial conditions followed nearby trajectories, the distinction between probabilities and trajectories would not matter too much. But chaotic systems show exactly how unreasonable is the assignment of "irreducible" probabilities, since the latter quickly spread out over the space in which the system evolves.

What the example of the billiard ball shows is that we must distinguish different levels of analysis: first of all, we may describe the system in a certain way: we may assign to the ball at least an approximate position at each time, hence an approximate trajectory.[28] Certainly the ball is not *everywhere*, as the "irreducible" probabilistic description would suggest. The next thing we can do is try to find exact or approximate laws of motion for the ball. The laws of elastic reflection against obstacles, for example. Finally, we may try to solve the equations of motion. We may not be able to do the last step. But this does not mean that one should give up the previous ones. We may even realize that our laws are only approximate (because of friction, of external perturbations, etc. . . .). But why give up the notion of (approximate) trajectories? Of course, since we are not able to predict the evolution of trajectories, one may *choose* to study instead the evolution of probability distributions. This is perfectly reasonable, as long as one does not forget that, in doing so, we are not only studying the physical system but also our ability or inability to analyze it in more detail. This will be very important in the next section.

At this point, I want to briefly discuss the classical status of probabilities in physics, i.e., of probabilities as "ignorance." This will also be very important in the next section. To quote Laplace again: "The curve described by a molecule of air or of vapour is following a rule as certainly as the orbits of the planets: the only difference between the two is due to our ignorance. Probability is related, in part to this ignorance, in part to our knowledge."[29] Let us consider the usual coin throwing experiment. We assign a probability 1/2 to heads and 1/2 to tails. What is the logic of the argument? We examine the coin, and we find out that it is fair. We also know the person who throws the coin, and we know that he does not cheat. But we are unable to control or to know exactly the initial conditions for each throw. We can, however, determine the average result of a large number of throws. This is simply because, if one throws the coin $N$ times, the overwhelming majority (for $N$ large) of the results will have an approximately equal number of heads and of tails. It is as simple as that, and there will be nothing conceptually more subtle in the way we shall use probabilities below. The part "due to our ignorance" is simply that we *use* probabilistic reasoning. If we were omniscient, it would not be needed (but the averages would remain what they are, of course). The part "due to our knowledge" is what makes the reasoning work. We could make a mistake: the coin could be biased, and we did not notice it. Or we could have a "record of bad luck" and have many more heads than tails. But that is the way things are: our knowledge *is* incomplete, and we have to live with that. Nevertheless, probabilistic reasoning is extraordinarily successful in practice, but, when it works, this is due to our (partial) knowledge. It would be wrong to attribute any constructive role to our ignorance. And it is also erroneous to assume that the system must be somehow indeterminate when we apply probabilistic reasonings to it. Finally, one could rephrase Laplace's statement more carefully as follows: "Even if the curve described by a molecule of air follows a rule as certainly as the orbits of the planets, our ignorance would force us to use probabilistic reasonings."

## 3 IRREVERSIBILITY AND THE ARROW OF TIME

> ***Since in the differential equations of mechanics themselves there is absolutely nothing analogous to the second law of thermodynamics, the latter can be mechanically represented only by means of assumptions regarding initial conditions.***
>
> —L. BOLTZMANN

### 3.1 *The Problem*

What is the problem of irreversibility? The basic physical laws are reversible, which simply means that, if we consider an isolated system of particles, let it evolve for a time $t$, then reverse exactly the velocities of all the particles, and let the system again evolve for a time $t$, we get the original system at the initial time with all velocities reversed.[30] Now there are many motions that we see without ever observing their associated "time-reversed" motion: we go from life to death but not vice versa, coffee does not jump out of the cup, mixtures of liquids do not spontaneously unmix themselves. Some of these examples taken from everyday life involve nonisolated systems, but that is not relevant.[31] I shall center the discussion below on the canonical physical example (and argue that the other situations can be treated similarly): consider a gas that is initially compressed by a piston in the left half of a box; it is then released and expands into the whole container. We do not expect the particles to go back to the left half of the box, although such a motion would be as compatible with the laws of physics as the motion that does take place. So the question, roughly speaking, is: if the basic laws are reversible, how is it that we see some motions but never their time-reversed ones?

The first point to clarify is that this irreversibility does not give rise to a *contradiction* with basic physical laws.[32] Indeed, the laws of physics are always of the form: given some initial conditions, here is the result after some time. But they never tell us how the world *is or evolves.* In order to account for that, one always needs to assume something about the initial conditions. The laws of physics are compatible with many possible worlds: there could be no earth, no life, no humans. Nothing of that would contradict the fundamental physical laws. Actually, one would have to give a rather strong argument to show that there is a conflict between the reversibility of the laws and the existence of irreversible phenomena. But no argument at all is given, beyond a vague appeal to intuition, as for example: "No speculation, no body of knowledge ever claimed the equivalence between doing and undoing, between a plant that grows, has flowers and dies, and a plant that resuscitates, becomes younger and goes back to its primitive seed, between a man who learns and becomes mature and a man who becomes progressively a child, then an embryo, and finally a cell. Yet, since its origins, dynamics, the physical theory that identifies itself with the triumph of science, implied this radical negation of time."[33] But nobody says that there is an "equivalence" between the two motions, only that both are compatible with the laws of physics. Which one, if any, occurs depends on the initial conditions. And if the laws are deterministic, assumptions on initial conditions are ultimately assumptions on the initial state of the Universe.

Once one has remarked that there is no contradiction between irreversibility and the fundamental laws, one could stop the discussion. It all depends on the initial conditions, period. But much more can be said. It is perfectly possible to give a natural account of irreversible phenomena on the basis of reversible fundamental laws, and of suitable assumptions on initial conditions. This was essentially done a century ago by Boltzmann, and despite numerous misunderstandings and misguided objections (some of them coming from famous scientists, such as Zermelo or Poincaré), its explanation still holds today. Yet, Prigogine writes: "He (Boltzmann) was forced to conclude that the irreversibility postulated by thermodynamics was incompatible with the reversible laws of dynamics."[34] This is in rather sharp contrast with Boltzmann's own words: "From the fact that the differential equations of mechanics are left unchanged by reversing the sign of time without anything else, Herr Ostwald concludes that the mechanical view of the world cannot explain why natural processes run preferentially in a definite direction. But such a view appears to me to *overlook that mechanical events are determined not only by differential equations, but also by initial conditions.* In direct contrast to Herr Ostwald I have called it one of the most brilliant confirmations of the mechanical view of Nature that it provides and extraordinarily good picture of the dissipation of energy, as long as one assumes that the world began in an initial state satisfying certain initial conditions" (italics are mine).[35] I shall now explain this "brilliant confirmation of the mechanical view of Nature" and show that all the alleged contradictions are illusory.[36]

### 3.2 *The Classical Solution*[37]

First of all, let us see which systems do behave irreversibly. A good test is to record the behavior of the system on a movie, and then to run the movie backwards. If it looks funny (e.g., people jump out of their graves), then we are facing irreversible behavior. It is easy to convince oneself that all the examples of irreversible behavior involve systems with a large number of particles (or of degrees of freedom). If one were to make a movie of the motion of one molecule, the backward movie would look completely natural. The same is true for a billiard ball on a frictionless billiard table.[38] If, however, friction is present, then we are dealing with many degrees of freedom (the atoms in the billiard table, those in the surrounding air, etc.).

There are two fundamental ingredients in the explanation of irreversibility. The first one has already been introduced: initial conditions. The second one is suggested by the remark that we deal with systems with many degrees of freedom: we *have* to distinguish between microscopic and macroscopic variables. Let us consider the phase space $\Omega$ (see Note 30) of the system, so that the system is represented by a point $\mathbf{x}$ in that space and its evolution is represented by a curve $\mathbf{x}(t) = T^t(\mathbf{x})$. Various quantities of physical interest—for example, the density or the average energy or the average velocity in a given cubic millimeter—can be expressed as functions on $\Omega$.[39] These functions (call them $F$) tend to be many-to-one, i.e., there is typically a huge number of configurations giving rise to a given value of $F$.[40] For example, if $F$ is the total energy, then it takes a constant value on a surface in phase space. But if $F$ is the number of particles in a cubic millimeter, there are also many microscopic configurations corresponding to a given value of $F$. Now let me make two statements, the first of which is trivial and the second not. Given a microscopic initial configuration $\mathbf{x}_0$, giving rise to a trajectory $\mathbf{x}(t)$, any function on phase space follows an induced evolution $F_0 \rightarrow F_t$, where $F_0 = F(\mathbf{x}_0)$, and $F_t = F[\mathbf{x}(t)]$. That is the trivial part. the nontrivial observation is that, in many situations, one can find a suitable family of functions (I shall still denote by $F$ such a family) so that this induced evolution is actually (approximately) *autonomous*. That is, one can determine $F_t$ given $F_0$ alone, and not knowing the microscopic configuration from which it comes.[41] This means that the different microscopic configurations on which $F$ takes the value $F_0$ will induce the same evolution on $F_t$. A very trivial example is given by the globally conserved quantities (like the total energy): for all microscopic configurations, $F_t = F_0$, for all times. But that is not interesting. It is more interesting to observe that all the familiar macroscopic equations (Navier-Stokes, Boltzmann, diffusion, . . .) are of that type. Actually, there are several provisos to be made here: first of all, it is not true that *all* microscopic configurations giving rise to $F_0$ lead to the same evolution for $F_t$. In general, there will only be a (vast) majority of microscopic configurations that will do that.[42] Moreover, if we want that evolution to hold strictly for all times, then this set of microscopic configurations may become empty.[43] Finally, the laws used in practice (Navier-Stokes, Boltzmann, diffusion, . . .) are usually approximations to the real laws satisfied by the corresponding $F_t$'s.

So the precise justification of a macroscopic law should be: given $F_0$, and given a (not too large) time $T$[44], there exists a large subset of the set of $\mathbf{x}$'s giving rise to $F_0$ (i.e., of the preimage in $\Omega$, under the map $F$, of $F_0$) such that the induced evolution of $F_t$ is approximately described by the relevant macroscopic equations up to time $T$. It should be obvious that it is not easy to prove such a statement mathematically. One has to deal with dynamic systems with a large number of degrees of freedom, about which very little is known, plus one has to identify limits in which one can make sense of the approximations mentioned above (a large subset, a not-too-large time $T$, . . .). Nevertheless, this can be done in some circumstances, the best known probably being the derivation of Boltzmann's equation by Lanford.[45] In the APPENDIX, I discuss a model due to Mark Kac that, while artificially simple, can be easily analyzed and shows exactly what one would like to do in more complicated situations.[46]

Let us come back to the problem of irreversibility: should we expect those macroscopic laws to be reversible? A priori, not at all. Indeed, I have emphasized in the abstract description above the role of initial conditions in their derivation.[47] The macroscopic equations may be reversible or not, depending on the situation. But their derivation shows that there is no *logical* reason to expect them to be reversible. Note that I am not discussing irreversibility in terms of entropy, but rather in terms of the macroscopic laws. After all, when we observe the mixing of different fluids, we see a phenomenon described by the diffusion equation, but we do not see entropy flowing. The connection with entropy will be made in SECTION 5.

### 3.3 *The Reversibility Objection*

Let me illustrate this explanation of irreversibility in a concrete physical example (see also APPENDIX for a simple mathematical model). Consider the gas introduced in SECTION 3.1 that is initially compressed by a piston in the left half of a box, and that expands into the whole box. Let $F$ be the density of the gas. Initially, it is one (say) in one half of the box and zero in the other half. After some time $t$, it is (approximately) one half everywhere. The explanation of the irreversible evolution of $F$ is that the overwhelming majority of the microscopic configurations corresponding to the gas in the left half, will evolve deterministically so as to induce the observed evolution of $F$. There may, of course, be some exceptional configurations, for which all the particles stay in the left half. All one is saying is that those configurations are extraordinarily rare, and that we do not expect to see even one of them appearing when we repeat the experiment many times, not even once "in a million years," to put it mildly. [48]

This example also illustrates the answer to the reversibility objection. Call "good" the microscopic configurations that lead to the expected macroscopic behavior. Take all the good microscopic configurations in the left half of the box, and let them evolve until the density is approximately uniform. Now, reverse all the velocities. We get a set of configurations that still determines a density of one half in the box. However, they are not good. Indeed, from now on, if the system remains isolated, the density just remains uniform according to the macroscopic laws. But for the configurations just described, the gas will move back to the left half, leading to a gross violation of the macroscopic law. What is the solution? Simply that those "reversed-velocities" configurations form a very tiny subset of all the microscopic configurations giving rise to a uniform density. So that, if we prepare the system with a uniform density, we do not expect to "hit" even once one of those bad configurations.[49]

Now comes a real problem. We are explaining that we never expect to get a microscopic configuration that will lead all the gas to the left side of the box. *But we started from such a configuration.* How did we get there in the first place? Well, obviously the system was not isolated: an experimentalist pushed the piston. But why was there an experimentalist? Do we expect that to happen? In fact, this is even more unlikely than the gas going to the left half of the box. We can go back step by step, human beings depend on the food they eat (and on the history of natural selection), which itself ultimately depends on the sun (through the plants and their photosynthesis).

As discussed in Penrose,[50] for example, the earth does not gain energy from the sun (that energy is reradiated by the earth), but low entropy; the sun sends (relatively) few high-energy photons, and the earth reradiates more low-energy photons (in such a way that the total energy is conserved). Expressed in terms of "phase space," the numerous low-energy photons occupy a much bigger volume than the incoming high-energy ones. So the solar system, as a whole, moves towards a larger part of its phase space while the sun burns its fuel. That evolution accounts, by far, for what we observe in living beings or in other "self-organized" structures.[51] I shall come back to this point in SECTION 6. Of course, for the sun to play this role, it has to be itself out of equilibrium, and to have been even more so in the past. We end up with an egg and hen problem and we have ultimately to assume that the universe started in a state far from equilibrium, an "improbable state" as Boltzmann called it. To make the analogy with the gas in the box, it is as if the universe had started in a very little corner of a huge box.[52]

To account in a natural way for such a state is, of course, a major open problem, on which I have nothing to say[53] (see Penrose, *The Emperor's New Mind*, for further discussion, and Figure 7.19 there for an illustration), except that one cannot avoid it by "alternative" explanations of irreversibility. Given the laws of physics, as they are formulated now, the world could have started in equilibrium, and then we would not be around to discuss the problem.[54] To summarize: the only real problem with irreversibility is not to explain irreversible behavior in the future, but to account for the "exceptional" initial conditions of the universe.

### 3.4 *Chaos and Irreversibility*

Now I come to my basic criticism of the views of Prigogine and Stengers, who argue that dynamic systems with very good chaotic properties, such as the baker's map, are "intrinsically irreversible." Let me quote from a letter of a collaborator of Prigogine, D. Driebe,[55] criticizing an article of J. L. Lebowitz[56] explaining Boltzmann's ideas. This letter is remarkably clear and summarizes well the main points of disagreement. "If the scale-separation argument were the whole story, then irreversibility would be due to our approximate observation or limited knowledge of the system. This is difficult to reconcile with the constructive role of irreversible processes. . . . Irreversibility is not to be found on the level of trajectories or wavefunctions but is instead manifest on the level of probability distributions. . . . Irreversible processes are well observed in systems with few degrees of freedom, such as the baker and the multibaker transformations. . . . The arrow of time is not due to some phenomenological approximations but is an intrinsic property of classes of unstable dynamical systems."[57]

Let us discuss these claims one by one. First of all, as I emphasized above, the scale-separation (i.e., the micro/macro distinction) is not "the whole story." Initial conditions have to enter into the explanation. Next, what does it mean that "irreversible processes are observed in systems such as the baker transformation"? This transformation describes a chaotic system with few degrees of freedom, somewhat like the billiard ball on a frictionless table.[58] For those systems, there is no sense of a micro/macro distinction: how could one define the macroscopic variables? To put it otherwise, if we make a movie of the motion of a point in the plane evolving under the baker's map, or of a billiard ball, or of any isolated chaotic system with few degrees of freedom, and run it backwards, we shall not be able to tell the difference. There is nothing funny or implausible going on, unlike the backward movie of any real irreversible macroscopic phenomenon. So the first critique of this alleged connection between unstable dynamic systems (i.e., what I call here chaotic) and irreversibility is that one "explains" irreversibility in systems in which nothing irreversible happens, and where therefore nothing is to be explained.

It is true that probability distributions for those systems evolve "irreversibly," meaning that any (absolutely continuous, see Note 27) probability distribution will spread out all over the phase space and will quickly tend to a uniform distribution. This just reflects the fact that different points in the support of the initial distribution, even if they are close to each other initially, will be separated by the chaotic dynamics. So it is true, in a narrow sense, that "irreversibility is manifest on the level of probability distributions." But what is the physical meaning of this statement? A physical system, chaotic or not, is described by a trajectory in phase space, and is certainly not described adequately by the corresponding probability distributions. As I discussed in SECTION 2.2, the latter reflects, in part, our ignorance of that trajectory. Their "irreversible" behavior is therefore not a genuine physical property of the system. We can, if we want, focus our attention on probabilities rather than on trajectories, but that "choice" cannot have a basic role in our explanations.

One cannot stress strongly enough the difference between the role played by probabilities here and in the classical solution. In the latter, we use probabilities as in the coin-throwing experiment. We have some macroscopic constraint on a system (the coin is fair; the particles are in the left half of the box), corresponding to a variety of microscopic configurations. We predict that the behavior of certain macroscopic variables (the average number of heads; the average density) will be the one induced by the vast majority of microscopic configurations, compatible with the initial constraints. That is all. But it works only because a large number of variables are involved, *in each single physical system.* However, each such system is described by a point in phase space (likewise, the result of many coin throwings is a particular sequence of heads and tails). In the "intrinsic irreversibility" approach, a probability distribution is assigned to *each single physical system*, as an "irreducible" description. The only way I can make sense of that approach is to consider a *large number* of billiard balls or of copies of the baker's map, all of them starting with nearby initial conditions. Then it would be like the particles in the box, the average density would tend to become uniform, and we are back to the standard picture. But this does not force us to "rethink the notion of law of nature." And for that picture to hold, one does not even need the motion to have strong chaotic properties (as we shall see in SECTION 4.2).

### 3.5 *Is Irreversibility Subjective?*

I will now discuss the alleged "subjectivity" of this account of irreversibility (it is due to our approximate observation or limited knowledge of the system). I shall consider the "constructive role" of irreversible processes in SECTION 6. Branding Boltzmann's ideas as "subjective" is rather common. For example, Prigogine writes: "In the classical picture, irreversibility was due to our approximations, to our ignorance."[59] But thanks to the existence of unstable dynamic systems, "the notion of probability that Boltzmann had introduced in order to express the arrow of time does not correspond to our ignorance and acquires an objective meaning."[60] To use Popper's image: "Hiroshima is not an illusion" (I shall come back to Popper's confusions in SECTION 4.4). This is only a dramatization of the fact that irreversible events are not subjective, or so it seems. The objection is that, if the microscopic variables behave reversibly and if irreversibility only follows when we "choose" to concentrate our attention on macroscopic variables, then our explanation of irreversibility is unavoidably tainted by subjectivism. I think that this charge is completely unfair, and reflects some misunderstanding of what irreversible phenomena really are. The point is that, upon reflection, one sees that all irreversible phenomena deal with these macroscopic variables. There is no subjectivism here: the evolution of the macroscopic variables is objectively determined by the microscopic ones, and they behave as they do whether we look at them or not. In that sense they are completely objective. But it is true that, if we look at a single molecule, or at a collection of molecules represented by a point in phase space, there is no sense in which they evolve "irreversibly," if we are not willing to consider some of the macroscopic variables that they determine.

However, it is true that the apparently "subjective" aspect of irreversibility has been sometimes overemphasized. Heisenberg wrote: "Gibbs was the first to introduce a physical concept which can only be applied to an object when our knowledge of the object is incomplete. If, for instance, the motion and the position of each molecule in a gas were known, then it would be pointless to continue speaking of the temperature of the gas."[61] And Max Born said: "Irreversibility is therefore a consequence of the explicit introduction of ignorance into the fundamental laws."[62] These formulations, although correct if they are properly interpreted, lead to unnecessary confusion. For example, Popper wrote: "It is clearly absurd to believe that pennies fall or molecules collide in a random fashion *because we do not know* the initial conditions, and that they would do otherwise if some demon were to give their secret away to us: it is not only impossible, it is absurd to explain objective statistical frequencies by subjective ignorance."[63] However, just after saying this, Popper gives what he calls "an objective probabilistic explanation of irreversible processes,"[64] attributed to Planck, which, as far as I can tell, is not very different from what I call the classical solution. The source of the confusion comes from two uses of the word "knowledge." Obviously, the world does what it does, whether we know about it or not. So, indeed, if "some demon" were to provide us with detailed knowledge of the microscopic state of the gas in the left half of the box, nothing would change to the future evolution of that gas. But we may imagine situations where one can *control* more variables, hence to "know" more about the system. When the piston forces the gas to be in the left half of the box, the set of available microscopic states is different from when the piston is not there, and obviously we have to take that "knowledge" into account. But there is nothing mysterious here.

I believe that statistical mechanics would become easier to understand by students if it were presented without using an anthropomorphic language and avoiding subjective sounding notions such as information, observation, or knowledge. Or, at least, one should explain precisely why these notions are introduced and why they do not contradict an objectivist view of natural phenomena (see the writings of Jaynes on this point[65]). But I also believe that the charge of subjectivity should be completely reversed: to "explain" irreversibility through the behavior of probability distributions (which *are* describing our ignorance), as Prigogine does, is to do as if the limitations of human knowledge played a fundamental physical role.

## 4 SOME MISCONCEPTIONS ABOUT IRREVERSIBILITY

*The Second Law can never be proved mathematically by means of the equations of dynamics alone.*

—L. BOLTZMANN

### 4.1 *The Poincaré Recurrence Theorem*

According to Prigogine[66] Poincaré did not recommend reading Boltzmann, because his conclusions were in contradiction with his premises. Discussing our example of a gas expanding in a container, Prigogine observes that "if irreversibility was only that, it would indeed be an illusion, because, if we wait even longer, then it may happen that the particles go back to the same half of the container. In this view, irreversibility would simply be due to the limits of our patience."[67] This is basically the argument derived from the Poincaré recurrence theorem (and used by Zermelo against Boltzmann[68]), which says that, if the container remains isolated long enough, then indeed the particles will return to the half of the box from which they started. Replying to that argument, Boltzmann supposedly said "You should live that long." For any realistic macroscopic system, the Poincaré recurrence times (i.e., the time needed for the particles to return to the left half of the box) are much larger than the age of the universe. So that again no contradiction can be derived, from a physical point of view, between Boltzmann's explanations and Poincaré's theorem. However, there is still a mathematical problem (and this may be what Poincaré had in mind): if one tries to rigorously derive an irreversible macroscopic equation from the microscopic dynamics and suitable assumptions on initial conditions, the Poincaré recurrence time will put a limit on the length of the time interval over which these statements can be proved. That is one of the reasons why one discusses these derivations in suitable limits (e.g., when the number of particles goes to infinity) where the Poincaré recurrence time becomes infinite. But one should not confuse the fact that one takes a limit for mathematical convenience and the source of irreversibility. In the Kac model discussed in the APPENDIX, one sees clearly that there are very different time scales: one over which convergence to equilibrium occurs, and a much larger one, where the Poincaré recurrence takes place. But the first time scale is not an "illusion." In fact, it is on that time scale that all phenomena that we can possibly observe do take place.

### 4.2 *Ergodicity and Mixing*

One often hears that, for a system to reach "equilibrium," it must be ergodic, or mixing. The fact is that those properties, like the "intrinsic irreversibility" discussed above, ***are neither necessary nor sufficient*** for a system to approach equilibrium. Let me start with ergodicity. A dynamic system is ***ergodic*** if the average time spent by a trajectory in any region of the phase space is proportional to the volume of that region. There are several precisions to be made here: average means in the limit of infinite time, and this property has to hold for all trajectories, except (possibly) those lying in a subset of zero volume. One says that it holds for "almost all" trajectories. This property implies that, for any reasonable function on phase space, the average along almost all trajectories will equal the average over the phase space.[69] Then, the argument goes, the measure of any physical quantity will take some time. This time is long compared to the "relaxation time" of molecular processes. Hence, we can regard it as approximately infinite. Therefore, the measured time average will approximately equal the average over phase space of the physical quantity under consideration. But this latter average is exactly what one calls the equilibrium value of the physical quantity. So if a dynamic system is ergodic, it converges towards equilibrium. This appeal to ergodicity in order to justify statistical mechanics is rather widespread[70] even though it has been properly criticized a long time ago by, e.g., Tolman, Jaynes, and Schwartz.[71]

Let us see the problems with this argument: a well-known, but relatively minor problem, is that it is very hard to give a mathematical proof that a realistic mechanical system is ergodic. But let us take such a proof for granted, for the sake of the discussion. Here is a more serious problem. Assume that the argument given above is true: how would it then be possible to observe or measure ***any nonequilibrium*** phenomenon? In the experiment with the box divided into two halves, we should not be able to see any intermediate stage, when the empty half gets filled, since the time for our measurements is supposed to be approximately infinite. So where is the problem? We implicitly identified the "relaxation time" with what one might call the "ergodic time"—i.e., the time taken by the system to visit all regions of phase space sufficiently often so that the replacement of time averages by spatial averages is approximately true. But, whatever the exact meaning of the word "relaxation time" (for a few molecules) is, the ergodic time is certainly enormously longer. Just consider how large is the volume in phase space that has to be "sampled" by the trajectory. For example, all the particles could be in the right half of the box, and ergodicity says that they will spend some time there (note that this is not implied by Poincaré's theorem; the latter guarantees only that the particles will return to the part of the box from which they started—i.e., the left half here). To be more precise, let us partition the volume of phase space into a certain number of cells, of a given volume, and consider the time it takes for a given trajectory to visit each cell, even once, let us say.[72] That, obviously, will depend on the size (hence, on the number) of the cells. By taking finer and finer partitions, one can make that time as large as one wishes. So if one were to take the argument outlined above literally, the "ergodic time" is infinite, and speaking loosely about a relaxation time is simply misleading.

At this point of the discussion, one often says that we do not need the time and space average to be (almost) equal for all functions, but only for those of physical relevance (like the energy or particle densities). This is correct, but the criticism of the "ergodic" approach then changes: instead of not being *sufficient* to account for irreversibility, we observe that it is not *necessary*. To see this, consider another partition of phase space: fix a set of macroscopic variables, and partition the phase space according to the values taken by these variables[73] (see, e.g., Figures 7.3 and 7.5 in R. Penrose, *The Emperor's New Mind*, for an illustration, and the APPENDIX here for an example). Each element of the partition consists of a set of microscopic states that give the same value to the chosen macroscopic variables. Now these elements of the partition have very different volumes. That is just a restatement of the law of large numbers. There are (for $N$ large) vastly more results of $N$ throws of a coin where the number of heads is approximately one half than throws where it is approximately one quarter (the ratio of these two numbers varies exponentially with $N$). By far the largest volumes correspond to the *equilibrium values* of the macroscopic variables (and that is how "equilibrium" should be defined). So we need a much weaker notion than ergodicity. All we need is that the microscopic configuration evolves in phase space towards those regions where the relevant macroscopic variables take their equilibrium values. The Kac model (see APPENDIX) perfectly illustrates this point: it is not ergodic in any sense, yet, on proper time scales, the macroscopic variables evolve towards equilibrium.

There is a hierarchy of "ergodic" properties that are stronger than ergodicity: mixing, K-system, Bernoulli.[74] But none of these will help us to understand, in principle, irreversible behavior any more than ergodicity.

The problem with all those approaches is that they try to give a purely mechanical criterion for "irreversible behavior." Here is the basic dilemma: either we are willing to introduce a macro/micro distinction and to give a basic role to initial conditions in our explanation of irreversibility or we are not. If we make the first choice, then, as explained in SECTION 3, there is no deep problem with irreversibility, and subtle properties of the dynamics (like ergodic properties) play basically no role. On the other hand, nobody has ever given a consistent alternative, namely, an explanation of irreversibility that would hold for *all* initial conditions or apply to *all* functions on configuration space (therefore avoiding the micro/macro distinction). So we have to make the first choice. But then everything is clear, and nothing else is needed.

Another critique of the "ergodic" approach is that a system with one or few degrees of freedom may very well be ergodic, or mixing, or Bernoulli (like the baker's transformation). And, as we discussed in SECTION 3.4, it makes no sense to speak about irreversibility for those systems. So this is another sense in which the notion of ergodicity is not sufficient.[75]

To avoid any misunderstandings, I emphasize that the study of ergodic properties of dynamic systems gives us a lot of interesting information on those systems, especially for chaotic systems. Besides, ergodic properties, like other concrete dynamic properties of a system, may play a role in the form of the macroscopic equations obeyed by the system, in the value of some transport coefficients, or in the speed of convergence to equilibrium. But, and this is the only point I want to make, ergodic properties are not very relevant conceptually for the problem of irreversibility or of approach to equilibrium.

### 4.3 *Real Systems Are Never Isolated*

Sometimes it is alleged that, for some reason (the Poincaré recurrences, for example) a truly isolated system will never reach equilibrium. But it does not matter, since true isolation never occurs and external ("random") disturbances will always drive the system towards equilibrium.[76] This is true but irrelevant.[77]

In order to understand this problem of nonisolation, we have to see how to deal with idealization in physics. Boltzmann compares this with Galilean invariance.[78] Because of nonisolation, Galilean (or Lorentz) invariance can never be applied strictly speaking (except to the entire universe, which is not very useful). Yet, there are many phenomena whose explanation involve Galilean (or Lorentz) invariance. We simply assume that the invariance is exact, and we argue that the fact that it is only approximate does not spoil the argument. It is the same thing in statistical mechanics. If we can explain what we want to explain (e.g., irreversibility) by making the assumption that the system is perfectly isolated, then we do not have to introduce the lack of isolation in our explanations. We have only to make sure that this lack of isolation does not conflict with our explanation. And how could it? The lack of isolation should, in general, speed up the convergence towards equilibrium.[79] Also, if we want to explain why a steamboat cannot use the kinetic energy of the water to move, we apply irreversibility arguments to the system boat + water, even though the whole system is not really isolated.

Another way to see that lack of isolation is true but irrelevant is to imagine a system being more and more isolated. Is irreversibility going to disappear at some point? That is, will different fluids not mix themselves, or will they spontaneously unmix? I cannot think of any example where this could be argued. And I cannot with a straight face tell a student that (part of) our explanation for irreversible phenomena on earth depends on the *existence* of Sirius.

### 4.4 *Bergson, Popper, Feyerabend (and Others)*

Here, I will discuss various confusions that have been spread by some philosophers. Bergson was a rather unscientific thinker, and many readers will wonder why he belongs here. I have myself been very surprised to see how much sympathy Prigogine and Stengers seem to have for Bergson.[80] But Bergson has been extremely influential, at least in French culture, and, I am afraid he still is.[81] In particular, he is one source of the widespread confusion that there is a contradiction between life and the second principle of thermodynamics. Roughly speaking, Bergson saw a great opposition between "matter" and "life," and a related one between intellect and intuition. The intellect can understand matter, but intuition is needed to apprehend life.[82] Bergson was not a precursor of the discovery of the DNA, to put it mildly. The second principle of thermodynamics, which he called the "most metaphysical of the laws of physics," was very important for him.[83] It reinforced his "vision of the material world as that of a falling weight."[84] Hence, "all our analyses show indeed in life an effort to climb the slope that matter has descended."[85] "The truth is that life is possible wherever energy goes down the slope of Carnot's law, and where a cause, acting in the opposite direction, can slow down the descent."[86] It is all metaphorical, of course, but Bergson's philosophy *is* entirely a "metaphorical dialectics devoid of logic, but not of poetry," as Monod calls it.[87] In any case, life is perfectly compatible with the Second Law (see SECTION 3.3).

Turning to Popper, we have already seen that he had many problems with statistical mechanics. Since Popper is generally considered positively by scientists,[88] it is worth looking more closely at his objections. He took too literally the claims of Heisenberg, Born, and Pauli on irreversibility as "subjective" (see SECTION 3.5), which he thought (maybe rightly so) were precursors of the subjectivism of the Copenhagen interpretation of quantum mechanics.[89] Besides, he was strongly opposed to determinism, and he was convinced that "the strangely law-like behaviour of the statistical sequences remain, for the determinist, ***ultimately irreducible and inexplicable.***"[90] As I discussed in SECTION 2.2, there is no problem in using probabilities, even in a deterministic universe. He then invented a rather obscure "propensity" interpretation of probabilities. He also felt that one should define "objectively" what a random sequence is. A sequence (of zeros and ones) will be random if there are (almost) as many zeros and ones, as many pairs 00, 01, 10, 11, etc. . . .[91] He did not seem to realize that this is like saying that a "microscopic configuration" (a sequence) gives to certain "macroscopic variables" (the average number of occurrences of finite subsequences) the values that are given to them by the overwhelming majority of sequences. So the difference from what he calls the "subjective" viewpoint is not so great.

Finally, Popper was very critical of Boltzmann. Although he admires Boltzmann's realist philosophy, he calls Boltzmann's interpretation of time's arrow "idealist" and claims that it was a failure. As we saw, any explanation of irreversibility ultimately forces us to say that the universe started in an "improbable" state. Boltzmann tried to explain it as follows: in an eternal and infinite universe globally in equilibrium, all kinds of fluctuations will occur. What we call our universe is just the result of one such gigantic fluctuation, on its way back to equilibrium. But this explanation does not really work. Indeed, the most probable assumption, if a fluctuation theory is to hold, is simply that my brain is a fluctuation out of equilibrium, just at this moment and in this small region of space; while none of the familiar objects of the universe (stars, planets, other human beings) exist, and all my (illusory) perceptions and memories are simply encoded in the states of my neurons (a "scientific" version of solipsism). However improbable such a fluctuation is, it is still far more probable than a fluctuation giving rise to the observed universe, of which my brain is a part. Hence, according to the fluctuation theory, that "solipsist" fluctuation must actually have occurred many more times than the big fluctuation in which we live, and therefore no explanation is given for the fact that we happen to live in the latter.[92]

Boltzmann's cosmology does not work. So what? When Popper wrote (1974), no one took Boltzmann's cosmology seriously anyway: it had long since been superseded by cosmologies based on general relativity. Besides,

Popper does not raise the objection I just made. His criticism is, rather, that this view would render time's arrow "subjective" and make Hiroshima an "illusion." This is complete gibberish. Boltzmann gives a complete and straightforward explanation of irreversible processes in which Hiroshima is as objective as it unfortunately is (when it is described at the macroscopic level, which is what we mean by "Hiroshima"). Of course, questions remain concerning the initial state of the universe. In the days of Boltzmann, very little was known about cosmology. What the failure of Boltzmann's hypothesis on the origin of the initial state shows is that cosmology, like the rest of science, cannot be based on pure thought alone.[93]

Popper was also too much impressed with Zermelo's objections to Boltzmann, based on the Poincaré recurrence theorem, and discussed above.[94] But he has even stranger criticisms: in "Irreversibility; or, Entropy since 1905" he argues that Brownian motion (where fluctuations may pull the particle against gravity) is a serious problem for the Second Law. Maxwell had already observed that "The Second Law is constantly being violated . . . in any sufficiently small group of molecules. . . . As the number . . . is increased . . . the probability of a measurable variation . . . may be regarded as practically an impossibility."[95] Going from bad to worse, Feyerabend invents a "perpetuum mobile of the second kind" (i.e., one respecting the first law but not the second) using *a single molecule.*[96] He adds that he assumes "frictionless devices" (he had better do so!). Those claims are then repeated in his popular book *Against Method*, where it is explained that Brownian motion refutes the second law.[97] This is how the general educated public is misled into believing that there are deep, open problems that are deliberately ignored by the "official science"!

Unfortunately, this is not the end of it. Contemporary (or postmodern) French "philosophy" is an endless source of confusion on chaos and irreversibility. Here are just a few examples. The well-known philosopher Michel Serres says, in an interview with the sociologist of science Bruno Latour entitled paradoxically "Eclaircissements": "Le temps ne coule pas toujours selon une ligne (la première intuition se trouve dans un chapitre de mon livre sur Leibniz, pp. 284–286) ni selon un plan, mais selon une variété extraordinairement complexe, comme s'il montrait des points d'arrêt, des ruptures, des puits, des cheminées d'accélération foudroyante, des déchirures, des lacunes, le tout ensemencé aléatoirement, au moins dans un désordre visible. Ainsi le développement de l'histoire ressemble vraiment à ce que décrit la théorie du chaos. . . ."[98] Another philosopher, Jean-François Lyotard writes: "L'idée que l'on tire de ces recherches (et de bien d'autres) est que la prééminence de la fonction continue à derivée comme paradigme de la connaissance et de la prévision est en train de disparaître. En s'intéressant aux indécidables, aux limites de la précision du contrôle, aux quanta, aux conflits à l'information non complète, aux "*fracta*," aux catastrophes, aux paradoxes pragmatiques, la science postmoderne fait la théorie de sa propre évolution comme discontinue, catastrophique, non rectifiable, paradoxale. Elle change le sens du mot savoir, et elle dit comment ce changement peut avoir lieu. Elle produit non pas du connu, mais de l'inconnu. Et elle suggère un modèle de légitimation qui n'est nullement celui de la meilleure performance, mais celui de la différence comprise comme paralogie."[99] A sociologist, Jean Baudrillard, observes that "Il faut peut-être considérer l'histoire elle-même comme une formation chaotique où l'accélération met fin à la linéarité, et où les turbulences créées par l'accélération éloignent définitivement l'histoire de sa fin, comme elles éloignent les effets de leurs causes. La destination, même si c'est le Jugement dernier, nous ne l'atteindrons pas, nous en sommes désormais séparés par un hyperespace à réfraction variable. La rétroversion de l'histoire pourrait fort bien s'interpréter comme une turbulence de ce genre, due à la précipitation des événements qui en inverse le cours et en ravale la trajectoire."[100] Finally, Gilles Deleuze and Félix Guattari understood chaos as follows: "On définit le chaos moins par son désordre que par la vitesse infinie avec laquelle se dissipe toute forme qui s'y ébauche. C'est un vide qui n'est pas un néant, mais un *virtuel*, contenant toutes les particules possibles et tirant toutes les formes possibles qui surgissent pour disparaître aussitôt, sans consistance ni référence, sans conséquence (Ilya Prigogine et Isabelle Stengers, *Entre le temps et l'éternité*, pp. 162–163)."[101] Of course, Prigogine and Stengers are not responsible for *these* confusions (in that reference, they discuss the origin of the universe). But this illustrates the difficulties and the dangers of the popularization of science. Besides, Guattari wrote a whole book *Chaosmose*, which is full of references to nonexistent concepts such as "nonlinear irreversibility thresholds" and "fractal machines."[102]

## 5 ENTROPIES

*Holy Entropy! It's boiling!*

—G. GAMOW

There is some kind of mystique about entropy. According to K. Denbigh and M. Tribus,[103] von Neumann suggested that Shannon use the word "entropy," adding that "it will give you a great edge in debates because nobody really knows what entropy is anyway." Another author says that entropy "is nothing material but is, so to speak, purely spiritual and yet it rules the world."[104] But there is a very simple way to understand the notion of entropy. Just consider any set of macroscopic variables (at a given time) and consider the volume of the subset of phase space (of the microscopic variables) on which these macroscopic variables take a given value. The *Boltzmann entropy* (defined as a function of the macroscopic variables) equals the logarithm of that volume. Defined this way, it looks quite arbitrary. We may define as many entropies as we can find sets of macroscopic variables. Furthermore, since the micro/macro distinction is not sharp, we can always take finer-grained entropies, until we reach the microscopic variables (the positions and the momenta of the particles), at which point the entropy is constant and equals zero (giving a volume equal to one to a single microstate, which is rather a quantum-mechanical way of counting).

But one should make several remarks:

1. These entropies are not necessarily "subjective." They are as objective as the corresponding macroscopic variables. Jaynes, following Wigner, calls these entropies "anthropomorphic."[105] A better word might be "contextual"—i.e., they depend on the physical situation and on its level of description.

2. The "usual" entropy of Clausius corresponds to a particular choice of macroscopic variables (e.g., energy and number of particles per unit volume for a monoatomic gas without external forces). The derivative with respect to the energy of *that* entropy, restricted to equilibrium values, defines the inverse temperature. One should not confuse the "flexible" notion of entropy introduced above with the more specific one used in thermodynamics.[106]

3. The Second Law seems now a bit difficult to state precisely. "Entropy increases"; yes, but which one? One can take several attitudes. The most conservative is to restrict oneself to the evolution of a given isolated system between two equilibrium states; then the increasing entropy is that discussed in Point 2 above. The Second Law is then a rather immediate consequence of the irreversible evolution of the macroscopic variables: the microscopic motion will go from small regions of phase space to larger ones (in the sense of the partitions discussed in SECTION 4.2). The gas in the box goes from an equilibrium state in the left half of the box to another equilibrium state in the whole box. There are many more microscopic configurations corresponding to a uniform density than there are configurations corresponding to the gas being entirely in one half of the box. But this version of the Second Law is rather restrictive, since most natural phenomena to which we apply "Second Law" arguments are not in equilibrium. When used properly in nonequilibrium situations, reasonings based on the Second Law give an extremely reliable way to predict how a system will evolve. We simply assume that a system will never go spontaneously towards a very small subset of its phase space (as defined by the macroscopic variables). Hence, if we observe such an evolution, we expect that some hidden external influence is forcing the system to do so, and we try to discover it.[107]

4. In most nonequilibrium situations, most of these entropies are very hard to compute or even to estimate. However, Boltzmann was able to find an approximate expression of his entropy (minus his $H$ function), valid for dilute gases (e.g., for the gas in the box initially divided in two of SECTION 3) and to write down an equation for the evolution of that approximate entropy. A lot of confusion is due to the identification between the "general" Boltzmann entropy defined above, and the approximation to it given by (minus) the $H$-function.[108] Another frequent confusion about Boltzmann's equation is to mix two conceptually different ingredients entering in its derivation[109]: one is an assumption on *initial conditions*, and the other is to make a particular approximation (i.e., one takes the Boltzmann-Grad limit; see H. Spohn, *Large Scale Dynamics of Interacting Particles*, in which the approximation becomes exact; in the Kac model in the APPENDIX, this limit reduces simply to let $n$ go to infinity for fixed $t$). To account for irreversible behavior, one has always, as we saw, to assume something on initial conditions, and the justification of that assumption is statistical. But that part does not require, in principle, any approximation. To write down a concrete (and reasonably simple) equation, as Boltzmann did, one uses this approximation. Failure to distinguish these two steps leads one to believe that there is some deep problem with irreversibility outside the range of validity of that approximation.[110]

5. Liouville's theorem[111] is sometimes invoked against such ideas. For instance, we read in Prigogine and Stengers: "All attempts to construct an entropy function, describing the evolution of a set of trajectories in phase space, came up against Liouville's theorem, since the evolution of such a set cannot be described by a function that increases with time."[112] What is the difference with our entropies? Here I consider *a single system* evolving in time and associate with it a certain set of macroscopic variables, to which in turn an entropy is attached. But since the values of the macroscopic variables change with time, the corresponding set of microstates changes too. In other words, I "embed" my microscopic state into different sets of microscopic states as time changes, and the evolution of that set should not be confused with a set of *trajectories*, whose volume is indeed forced to remain constant (by Liouville's theorem).[113]

6. A related source of confusion comes from the fact that Gibbs' entropy, $-\int \rho \log \rho d\mathbf{x}$, which looks more natural and more "fundamental" (since it is expressed via a distribution function $\rho$ on phase space), is indeed constant in time (by Liouville's theorem again). But why should one use this Gibbs entropy out of equilibrium? In equilibrium, it agrees with Boltzmann and Clausius entropies (up to terms that are negligible when the number of particles is large), and everything is fine.[114] When we compare two different equilibrium states all these entropies change, and the direction of change agrees with the second law.[115] The reason is that the values taken by the macroscopic variables are different for different equilibrium states. Actually, trying to "force" the Gibbs entropy to increase by various coarse-graining techniques gives then the impression that irreversibility is due only to this coarse-graining and is therefore arbitrary or subjective.[116]

7. Finally, why should one worry so much about entropy for nonequilibrium states? A distinction has to be made between two aspects of irreversibility: one is that macroscopic variables tend to obey irreversible laws, and the other is that when an isolated system can go from one equilibrium state to another, the corresponding thermodynamic entropies are related by an inequality. Both aspects are connected, of course, and they can both be explained by similar ideas. But this does not mean that in order to account for the irreversible behavior of macroscopic variables, we have to introduce an entropy function that evolves monotonically in time. It may be useful or interesting to do so, but it is not required to account for irreversibility. All we really *need* is to define suitably the entropy for equilibrium states, and that has been done a long time ago.

8. Jaynes rightly says that he does not know what is the entropy of a cat.[117] The same thing could be said for a painting, an eye, or a brain. The problem is that there is no well-defined set of macroscopic variables that is specified by the expression "a cat."

## 6 ORDER OUT OF CHAOS?

*In my view all salvation for philosophy may be expected to come from Darwin's theory. As long as people believe in a special spirit that can cognize objects without mechanical means, or in a special*

> *will that likewise is apt to will that which is beneficial to us, the simplest psychological phenomena defy explanation.*
>
> —L. BOLTZMANN

In this section, I will discuss the "constructive role" of irreversible processes.[118] But I also want to discuss the impact of scientific discoveries on the cultural environment. At least since the Enlightenment and the Encyclopaedia, scientists have communicated their discoveries to society, and, through popular books and the schools, have profoundly influenced the rest of the culture. But one has to be very careful. In his recent book on Darwin, the philosopher D. Dennett makes a list of popular misconceptions about the theory of evolution.[119] One of them is that one does not need the theory of natural selection any more, since we have chaos theory! He does not indicate the precise source of this strange idea, but it illustrates how easily people can be confused by loose talk, analogies, and metaphors.

I think that one should clearly reaffirm certain principles: first of all, no macroscopic system has ever jumped out of equilibrium spontaneously. Moreover, isolated macroscopic systems always evolve towards equilibrium. These are general qualitative statements that one can make about macroscopic mechanical systems. No violations of them have ever been found. Of course, nobody explicitly denies those principles, but if one puts too much emphasis on the constructive role of irreversible processes (or, worse, on our "dialogue" with Nature),[120] many people will be confused.[121] The fact is that the main principle that is clearly understood about irreversibility and is universally valid has nothing constructive in it: isolated systems go towards equilibrium.[122]

Of course, it has always been known that very complicated and interesting phenomena occur out of equilibrium, human beings for example. But this raises two completely different problems. One is to explain those phenomena on the basis of the microscopic laws and of suitable assumptions on initial conditions. Many progresses in this direction are made, but we are far from understanding everything; and, of course, to account for the existence of human beings, Darwin's theory *is* needed.

The other question, a much easier one, is to understand why there is no *contradiction* between the general tendency towards equilibrium and the appearance of self-organization, of complex structures or of living beings. *That* is not difficult to explain qualitatively; see SECTION 3.3 and Penrose's *The Emperor's New Mind.*

Going back to Popper (again), he wanted to solve the alleged contradiction between life and the second principle (see Note 51) by turning to Prigogine's *From Being to Becoming* and saying that "*open systems in a state far from equilibrium* show no tendency towards increasing disorder, even though they produce entropy. But they can export this entropy into their environment, and can increase rather than decrease their internal order. They can develop structural properties, and thereby do the very opposite of turning into an equilibrium state in which nothing exciting can happen any longer."[123] This is correct, provided the environment *is more ordered than the system*, where "order" is taken in a technical sense: the system plus its environment (considered as approximately isolated) is in a state of low entropy, or is in a small subset of its *total* phase space and moves towards a larger subset of that space[124] (where the subsets are elements of a partition like the one discussed in SECTION 4.2). But it is misleading to suggest that order is created out of nothing, by rejecting "entropy" in an unspecified environment.[125] It is not enough to be an "open system"; the environment must be more ordered.

Here are some examples that *may* create this confusion[126]: Prigogine considers a system of particles (on a line), which start in a disordered configuration.[127] Then "the strong interactions between those particles" will push them to form an ordered crystal. It looks like a "passage from a disordered situation to an ordered one." But is this an isolated system? Probably not. If there are interactions between the particles favoring an ordered crystal, the disordered initial configuration must have been one of high potential energy; hence, the "ordered" configuration will have a high kinetic energy, and oscillations will occur. Of course, if dissipation takes place, the "passage from a disordered situation to an ordered one" is possible. But this means that some environment absorbs the energy of the system, in the form of heat; hence, it increases *its* disorder. And the environment must have been more ordered to start with.

To give another example, Prigogine and Stengers emphasize[128] that, for the Bénard instability[129] to occur *one must provide more heat to the system.* As noticed by Meessen, "It is remarkable that the structuration is initiated by a source of heat, which is usually a source of disorder."[130] This quotation shows clearly what is confusing: heating suggests an increase of disorder, while the result is the appearance of a self-organized structure. But what is needed, of course, is a temperature *difference* between the two plates. So if one heats up from below, one must have some cooling from above. The cooling acts like a refrigerator, so it requires some "ordered" source of energy. The more one heats, the more efficient must be the cooling.

These are fairly trivial remarks, but I believe they have to be made, at least for the general public, if one wants to avoid giving the impression that processes violating the second law can occur: all the emergence of complex structures, of whatever one sees, is perfectly compatible with the universal validity of the "convergence to equilibrium," provided one remembers that our universe started (and still is) in a low entropy state.[131]

Besides, one should be careful with the issue of determinism, at the level of macroscopic laws, for example, when bifurcations occur. In many places, Prigogine and Stengers put too much emphasis on the notion of *event*: "By definition, an event cannot be deduced from a deterministic law: it implies, one way or another, that what happened 'could' not have happened."[132] Let us consider Buridan's ass. One can describe it as being "in between" two packs of food. It could choose either. But that is a macroscopic description. Maybe one of the eyes of the ass is tilted in one direction, or some of its neurons are in a certain state favoring one direction. This is an example where the macroscopic description does not lead to an autonomous macroscopic law. At the macroscopic level, things are indeterminate, and the scheme of SECTION 3 does not apply: the microscopic configurations may fall into different classes, corresponding to different future evolutions for the macroscopic variables, and no single class constitutes an overwhelming majority. Thus, when we repeat the experiment (meaning that we control the same *macroscopic* variables) different outcomes will occur, because different experiments will correspond to microscopic variables that belong to different classes.

The same thing may happen in a variety of phenomena—e.g., which way a roll in a Bénard cell will turn. But that (true) remark has nothing to do with the issue of determinism, which is meaningful only at the microscopic level: in a perfectly deterministic universe (at that level) there will always be many situations in which simple autonomous macroscopic laws can be found; hence, we shall have the illusion of "indeterminism" if we consider only the macroscopic level.[133]

One should avoid (once more) the Mind Projection Fallacy. The macroscopic description may be all that is accessible to us; hence, the future becomes unpredictable; but, again, it does not mean that Nature is indeterminate.[134]

I will conclude with some remarks on Boltzmann and Darwin, which may also clarify the relation between "subjective" evaluations of probabilities and what we call an "explanation." As we saw, Boltzmann had a great admiration for Darwin. While preparing this article, I read in "La Recherche" that "the couple random mutations-selection has some descriptive value, but not at all an explanatory one."[135] That attitude is rather common (outside of biology), but it goes a bit too far. Actually, there is an analogy between the kind of explanation given by Darwin and the one given by Boltzmann, and they are both sometimes similarly misunderstood[136] (of course, Darwin's discovery, although less quantitative than statistical mechanics, had a much deeper impact). What does it mean to explain some fact, like evolution or irreversibility? As we saw, we claim to understand some macroscopically observed behavior when, given some macroscopic constraint on a system, the overwhelming majority of the microscopic configurations compatible with those constraints (and evolving according to the microscopic laws) drive the macroscopic variables in agreement with that observed behavior.

Turning to Darwin, his problem was to explain the diversity of species and, more importantly, the *complexity* of living beings, "those organs of extreme perfection and complication," such as eyes or brains, as Darwin called them.[137] The fact is that we do not know, and we shall never know, every microscopic detail about the world, especially about the past (such as every single mutation, how every animal died, etc.). Besides, the initial conditions of the world could be just so that complex organs are put together in one stroke. To use a common image, it would be like "hurling scrap metal around at random and happening to assemble an airliner."[138] This does not violate any known law of physics. But it would be similar to various "exceptional" initial conditions that we encountered before (e.g., the particles going back to the left half of the box). And we would not consider an explanation valid if it appealed to such "improbable" initial conditions. But to say that such a scenario is "improbable" simply means that, given our (macroscopic) description of the world, there are very few microscopic configurations compatible with that description and giving rise to this scenario. And, indeed, if the world were four thousand years old, the existence of those complex organs would amount to a miracle.

To understand the Darwinian explanation, one must take into account three elements, at the level of the macroscopic description: natural selection (very few animals have offspring), variation (small differences between parents and offspring occur, at least in the long run), and time (the earth is much older than one thought). Then, the claim is that the overwhelming majority of microscopic events (which mutations occur, which animals die without children) compatible with such a macroscopic description leads to the appearance of those "organs of extreme perfection and complication."[139] Note that we do not need to assume that mutations are genuinely "random." They may obey perfectly deterministic laws, and the randomness may reflect only our ignorance of the details. Note also that one can invent exceptional microscopic configurations: it is possible for the world to be just so that all the animals that, thanks to a mutation, have a slightly more sophisticated eye or brain are hit by lightning when they are young and do not have children. But they correspond now to extremely improbable initial conditions.[140]

A final point that is common to Boltzmann and Darwin (and his successors) is that they have provided "brilliant confirmations of the mechanical view of Nature."[141] Many people simply cannot swallow mechanical and reductionist explanations. They need some vital spirit, some teleological principle, or some other animist view. Their philosophies "thrive upon the errors and confusions of the intellect." And this is probably why the theories of Boltzmann and Darwin have been constantly attacked and misrepresented. Putting philosophical considerations aside, I believe that what we understand well, we understand in mechanical and reductionist terms. There is no such thing as a holistic explanation in science. And thanks to people like Boltzmann and Darwin the "mechanical view of Nature" is alive and well, and is here to stay.

## ACKNOWLEDGMENTS

I have discussed many of the issues raised in this paper with colleagues and students, and particularly with S. Goldstein, A. Kupiainen, J. L. Lebowitz, C. Maes, J. Pestieau, O. Penrose, and H. Spohn. I have also benefitted from discussions with S. Focant, M. Ghins, N. Hirtt, D. Lambert, R. Lefevere, I. Letawe, J.-C. Limpach, T. Pardoen, P. Radelet, and P. Ruelle.

## APPENDIX
## THE KAC RING MODEL

Let me analyze a simple model, provided by Mark Kac,[142] which nicely illustrates Boltzmann's solution to the problem of irreversibility, and shows how to avoid various misunderstandings and paradoxes.

I shall describe a slightly modified version of the model and state the relevant results, referring to Kac's *Probability and Related Topics in the Physical Sciences* for the proofs (the quotations below also come from this book).

"On a circle we consider $n$ equidistant points"; $m$ of the intervals between the points are marked and form a set called $S$. The complementary set (of $n - m$ intervals) will be called $\bar{S}$.

"Each of the $n$ points is a site of a ball which can be either white ($w$) or black ($b$). During an elementary time interval each ball moves counterclockwise to the nearest site with the following proviso."

If the ball crosses an interval in $S$, it changes color upon completing the move, but if it crosses an interval in $\bar{S}$, it performs the move without changing color.

"Suppose that we start with all white balls; the question is what happens after a large number of moves." Later, we shall also consider other initial conditions.

Let us emphasize the analogy with mechanical laws. The balls are described by their positions and their (discrete) "velocity," namely, their color. One of the simplifying features of the model is that the "velocity" does not affect the motion. However the "velocity" changes when the ball collides with a fixed "scatterer," i.e., an interval in $S$. Scattering with fixed objects tends to be easier to analyze than collisions between particles. The "equations of motion" are given by the counterclockwise motion, plus the changing of colors [see eqs. (5,6) below]. These equations are obviously deterministic and reversible: if after a time $t$, we change the orientation of the motion from counterclockwise to clockwise, we return after $t$ steps to the original state.[143] Moreover, the motion is strictly periodic: after $2n$ steps each interval has been crossed twice by each ball, hence they all come back to their original color. This is analogous to the Poincaré cycles, with the provision that, here, the length of the cycle is the same for all configurations (there is no reason for this feature to hold in general mechanical systems). Moreover, it is easy to find special configurations that obviously do not tend to equilibrium: start with all white balls, and let every other interval belong to $S$ (with $m = \frac{n}{2}$). Then, after two steps, all balls are black, after four steps they are all white again, etc. The motion is periodic with period 4. Turning to the solution, one can start by analyzing the approach to equilibrium in this model à la Boltzmann:

### *Analog of the Classical Solution of Boltzmann*

Let $N_w(t)[N_b(t)]$ denote the total number of white (black) balls at time $t$ (i.e., after $t$ moves, $t$ being an integer) and $N_w(S;t)[N_b(S;t)]$ the number of white (black) balls that are going to cross an interval in $S$ at time $t$.

"We have the immediate conservation relations:

$$N_w(t + 1) = N_w(t) - N_w(S;t) + N_b(S;t)$$
$$N_b(t + 1) = N_b(t) - N_b(S;t) + N_w(S;t) \qquad (1)$$

Now to follow Boltzmann, we introduce the assumption ("Stosszahlansatz" or "hypothesis of molecular chaos"[144])

$$N_w(S;t) = mn^{-1}N_w(t)$$
$$N_b(S;t) = mn^{-1}N_b(t)\text{''} \qquad (2)$$

Of course, if we want to solve (1) in a simple way, we have to make some assumption on $N_w(S;t)$,$N_b(S;t)$. Otherwise, one has to write equations for

$N_w(S;t)$,$N_b(S;t)$ that will involve new variables and lead to a potentially infinite regress.

The intuitive justification for this assumption is that each ball is "uncorrelated" with the event "the interval ahead of the ball belongs to $S$," so we write $N_w(S;t)$ as equal to $N_w(t)$, the total number of white balls, times the density $\frac{n}{m}$ of intervals in $S$. This assumption looks completely reasonable. However, upon reflection, it may lead to some puzzlement (just as the hypothesis of "molecular chaos" does): exactly what does "uncorrelated" mean? Why do we introduce a statistical assumption in a mechanical model? Fortunately, here these questions can be answered precisely, and we shall answer them later by solving the model exactly. But let us return to the Boltzmannian story.

"One obtains

$$N_w(t+1) - N_b(t+1) = (1 - 2mn^{-1})[N_w(t) - N_b(t)]$$

Thus

$$\begin{aligned} n^{-1}[N_w(t) - N_b(t)] &= (1 - 2mn^{-1})^t n^{-1}[N_w(0) - N_b(0)] \\ &= (1 - 2mn^{-1})^t \end{aligned} \tag{3}$$

and hence if

$$2m < n \tag{4}$$

(as we shall assume in the sequel) we obtain a *monotonic* approach to equipartition of white and black balls." Note that we get a monotonic approach for all initial conditions $[N_w(0) - N_b(0)]$ of the balls.

The variables $N_w(t)$,$N_b(t)$ play the role of macroscopic variables. We can associate to them a Boltzmann entropy[145]

$$S_b = \ln \binom{n}{N_w(t)},$$

i.e., the logarithm of the number of (microscopic) configurations whose number of white balls is $N_w(t)$. Since

$$\binom{n}{N_w(t)} = \frac{n!}{N_w(t)![n - N_w(t)]!}$$

reaches its maximum value for $N_w = \frac{n}{2} = N_b$, we see that (3) predicts a monotone increase of $S$ with time. We can also introduce a partition of the "phase space" according to the different values $N_w$, $N_b$. And what the above formula shows is that different elements of the partition have very different numbers of elements, the vast majority corresponding to "equilibrium," i.e., to those near $N_w = \frac{n}{2} = N_b$.

We can see here in what sense Boltzmann's solution is an approximation. The assumption (2) cannot hold for all times and for all configurations, because it would contradict the reversibility and periodicity of the motion. However, we can also see why the fact that it is an approximation does not invalidate Boltzmann's ideas about irreversibility.

Let us reexamine the model at the microscopic level, first mechanically and then statistically. For each $i = 1, \cdots, n$, we introduce the variable

$$\varepsilon_i = \begin{cases} +1 \text{ if the interval in front of } i \in \bar{S} \\ -1 \text{ if the interval in front of } i \in S \end{cases}$$

and we let

$$\eta_i(t) = \begin{cases} +1 \text{ if the ball at site } i \text{ at time } t \text{ is white} \\ -1 \text{ if the ball at site } i \text{ at time } t \text{ is black} \end{cases}$$

Then, we get the "equations of motion"

$$\eta_i(t) = \eta_{i-1}(t - 1)\varepsilon_{i-1} \tag{5}$$

whose solution is

$$\eta_i(t) = \eta_{i-t}(0)\varepsilon_{i-1}\varepsilon_{i-2}\cdots\varepsilon_{i-t} \tag{6}$$

(where the subtractions are done modulo $n$). So we have an explicit solution of the equations of motion at the microscopic level.

We can express the macroscopic variables in terms of that solution:

$$N_w(t) - N_b(t) = \sum_{i=1}^{n}\eta_i(t) = \sum_{i=1}^{n}\eta_{i-t}(0)\varepsilon_{i-1}\varepsilon_{i-2}\cdots\varepsilon_{i-t} \tag{7}$$

and we want to compute $n^{-1}[N_w(t) - N_b(t)]$ for large $n$, for various choices of initial conditions ($\{\eta_i(0)\}$) and various sets $S$ (determining the $\varepsilon_i$'s). It is here that "statistical" assumptions enter. Namely, we fix an arbitrary initial condition ($\{\eta_i(0)\}$) and consider all possible sets $S$ with $m = \mu n$ fixed (one can, of course, think of the choice of $S$ as being part of the choice of initial conditions). Then, for each set $S$, one computes the "curve" $n^{-1}[N_w(t) - N_b(t)]$ as a function of time. The result of the computation, done in ***Probability and Related Topics in the Physical Sciences***, is that, for any given $t$ and for $n$ large, the overwhelming majority of these curves will approach $(1 - 2\frac{m}{n})^t = (1 - 2\mu)^t$, i.e., what is predicted by (3). (To fix ideas, Kac suggests that one think of $n$ as being of the order $10^{23}$ and $t$ of order $10^6$.) The fraction of all curves that will deviate significantly from $(1 - 2\mu)^t$, for fixed $t$, goes to zero as $n^{-1/2}$, when $n \to \infty$.

Of course, when I say "compute," I should rather say that one makes an estimate of the fraction of "exceptional" curves deviating from $(1 - 2\mu)^t$ at a fixed $t$. This estimate is similar to the law of large numbers, and (7) is, indeed, of the form of a sum of (almost independent) variables.

*Remarks*

1. The Poincaré recurrence and the reversibility "paradoxes" are easily solved: each curve studied is periodic of period $2n$. So that if we did not fix $t$ and let $n \to \infty$, we would not observe "irreversible" behavior. But this limit is physically correct. The recurrence time ($n$) is enormous compared to any physically accessible time. As for the reversibility objection, let us consider as initial condition a reversed configuration after time $t$. Then we know that for that configuration and *that set S*, $n^{-1}[N_w(t) - N_b(t)]$ will not be close to

$(1 - 2\mu)^t$ at time $t$ (since it will be back to its initial value 1). But all we are saying is that for the vast majority of $S$'s this limiting behavior will be seen. For the reversed configuration, the original set $S$ happens to be exceptional. The same remark holds for the configuration with period 4 mentioned in the beginning. Note also that if we consider the set of configurations for which $n^{-1}[N_w(t) - N_b(t)]$ is close to $(1 - 2\mu)^t$ for *all times*, then this set is empty, because of the periodicity.

2. We could consider other macroscopic variables, like the number of white and black balls in each half of the circle ($1 \leqslant i \leqslant \frac{n}{2}$ and $\frac{n}{2} + 1 \leqslant i \leqslant n$), and define the corresponding entropies. We could go on, with each quarter of the circle etc., until we reach a microscopic configuration (number of white or black balls at each site), in which case the entropy is trivially equal to zero (and therefore constant).

3. This model, although perfectly "irreversible," is not ergodic! Indeed, since it is periodic, no trajectory can "visit" more than $2n$ microscopic configurations. But the "phase space" contains $2^n$ configurations (two possibilities—black or white—at each site). So only a very small fraction of the phase space is visited by a trajectory. This nicely illustrates the fact that ergodicity is not necessary for irreversibility. What is used here is only the fact that the vast majority of configurations give to the macroscopic variables a value close to their equilibrium value.

### *Conclusion*

I do not want to overemphasize the interest of this model. It has many simplifying features (for example, there is no conservation of momentum; the scatterers here are "fixed," as in the Lorentz gas). However, it has *all* the properties that have been invoked to show that mechanical systems cannot behave irreversibly, and therefore it is a perfect counterexample that allows one to refute all those arguments (and to understand exactly what is wrong with them): it is isolated (the balls plus the scatterers), deterministic, reversible, has Poincaré cycles, and is not ergodic.

The result obtained in the Kac model is exactly what one would like to show for general mechanical systems in order to prove irreversibility. It is obvious why this is very hard. In general, one does not have an explicit solution (for an $n$-body system!) like (5,6), in terms of which the macroscopic variables can be expressed; see (7). It is also clear in this example exactly what is the status of our "ignorance." If we prepare the system many times and if the only variables that we can control are $n$ and $m$, then we, indeed, expect to see the irreversible behavior obtained above. This is simply because it is what happens *deterministically* for the vast majority of microscopic initial conditions corresponding to the macroscopic variables that we are able to control. We may, if we wish, say that we "ignore" the initial conditions, but there is nothing "subjective" here. Finally, I shall refer to ***Probability and Related Topics in the Physical Sciences*** for a more detailed discussion, in this model, of the status of various approximations used in statistical mechanics (e.g., the Master equation).

## NOTES

1 I. Prigogine, *Les Lois du Chaos*, p. 15. Here and below I have translated texts that were available only in French.

2 Ibid., p. 59.

3 *Order out of Chaos*, chap. 1.

4 Prigogine, *Les Lois du Chaos*, p. 41.

5 See ibid., chap. 9 and Prigogine, "Why Irreversibility? The Formulation of Classical and Quantum Mechanics for Nonintegrable Systems," p. 106.

6 On the issue of irreversibility, see R. Feynman, *The Character of Physical Law*, and R. Feynman, R. B. Leighton & M. Sands, *The Feynman Lectures on Physics*; E. T. Jaynes, *Papers on Probability, Statistics and Statistical Physics*; J. L. Lebowitz, "Macroscopic Laws, Microscopic Dynamics, Times's Arrow and Boltzmann's Entropy" and *Physics Today*; and R. Penrose, *The Emperor's New Mind.*

7 L. Boltzmann, *Theoretical Physics and Philosophical Problems.*

8 I must add that I have defended, in the past, some of the ideas criticized below. Needless to say, I am interested in the critique of ideas and not of individuals.

9 For exaggeratedly negative comments on Laplace, see, e.g., I. Ekeland, *Le Calcul, L'Imprévu*, p. 31; and J. Gleick, *La Théorie du Chaos*, p. 21.

10 Likewise, only the most radical social constructivist might object to the idea that Neptune and Pluto were following their (deterministic) trajectories before they were discovered.

11 In an often-quoted lecture to the Royal Society on the three-hundredth anniversary of Newton's *Principia*, Sir James Lighthill gave an inadvertently perfect example of how to slip from unpredictability to indeterminism: "We are all deeply conscious today that the enthusiasm of our forebears for the marvelous achievements of Newtonian mechanics led them to make generalizations in this area of *predictability* which, indeed, we may have generally tended to believe before 1960, but which we now recognize were false. We collectively wish to apologize for having misled the general educated public by spreading ideas about *determinism* of systems satisfying Newton's laws of motion that, after 1960, were to be proved incorrect. . . ." (italics are mine; quoted, e.g., by L.E. Reichl, *The Transition to Chaos in Conservative Classical Systems: Quantum Manifestations*, p. 3; and by I. Prigogine & I. Stengers, *Entre le Temps et l'Éternité*, p. 93, and *Les Lois du Chaos*, p. 41). See also S. Vauclair, *Eléments de Physique Statistique*, p. 7, where, after describing a chaotic system, one concludes that "the deterministic approach fails."

12 P. S. Laplace, *A Philosophical Essay on Probabilities.*

13 I have expressed my point of view on the foundations of quantum mechanics in "Contre la philosophie de la mécanique quantique." For related views, see D. Albert, *Quantum Mechanics and Experience* and "Bohm's Alternative to Quantum Mechanics"; J.S. Bell, *Speakable and Unspeakable in Quantum Mechanics*; D. Dürr, S. Goldstein & N. Zanghi, "Quantum Equilibrium and the Origin of Absolute Uncertainty"; and T. Maudlin, *Quantum Non-Locality and Relativity.*

14 Most of the laws that are discussed in the literature on chaos (e.g., on the weather) are actually macroscopic laws, and not fundamental or microscopic ones. This distinction will be discussed in SECTION 3.

15 Opponents to determinism are quick to point out that determinism cannot be proved. Of course, no statement about the world can literally be *proved.* But they do not always see how vacuous are their own arguments in favor of indeterminism, arguments that rely ultimately on our ignorance. In *The Open Universe. An Argument for Indeterminism*, Popper gives a long series of such arguments. In a review of this book, the biologist Maynard Smith shows a rather typical misunderstanding of Laplace: first he agrees with Popper about Laplace, because the latter's computations are impossible, and then disagrees with Popper about free will and gives, as far as I can see, a perfectly causal and Laplacian account of human actions, which, of course, are not computable either (*Did Darwin Get It Right?*, p. 244). To avoid misunderstandings, I am not trying to say that determinism is or must be true. All I say is that various arguments against determinism miss the point.

16 Here is a simple example. Consider the "phase space" to be simply the interval $I$ = [0,1[. And take as (discrete time) dynamics the map $f: x \rightarrow 10x$ *mod* 1. This means, we take a number between 0 and 1, multiply it by 10, write the result as an integer plus a number between 0 and 1, and take the latter as the result [i.e., $f(x)$]. This gives again a number between 0 and 1, and we can repeat the operation. Upon iteration, we obtain the *orbit* of $x$; $x$ itself is the initial condition. To describe concretely the latter, one uses the decimal expansion. Any number in $I$ can be written as $x = 0.a_1a_2a_3\ldots$, where $a_i$ equals 0,1,2, . . ., 9. It is easy to see that $f(x) = 0.a_2a_3\ldots$. This is a perfect example of a *deterministic* but *unpredictable* system. Given the state $x$ at some initial time, one has a rule giving the state of the system for arbitrary times. Moreover, for any fixed time, one can, in principle, find the state after that time, with any desired accuracy, given a sufficiently precise characterization of the initial state. This expresses the deterministic aspect. Unpredictability comes from the fact that, if we take two initial conditions at a distance less than $10^{-n}$, then the corresponding orbits could differ by, say 1/2, after $n$ steps, because the difference will be determined by the $n$th decimal. One of the relatively recent discoveries in dynamical systems is that simple physical examples, like a forced pendulum, may behave more or less like this map.

17 How long a time this is depends on the details of the system.

18 J. C. Maxwell, *Matter and Motion*, p. 13. As for the applications to the weather, Poincaré (*Science et Méthode*, p. 69) already noticed that the rainfalls or the storm seem to occur at random, so that people are more likely to pray for rain than for an eclipse (for an exception to this rule, but based on prior knowledge, see Hergé, *Tintin et le Temple du Soleil*, p. 59). We are not able to predict the storms, because the atmosphere may be in a state of "unstable equilibrium." It may all depend on a tenth of a degree. And he adds, "If we had known this tenth of a degree, one could have made predictions," but since our observations are not sufficiently precise, it all looks as if it were due to randomness.

19 J. Hadamard ("Les Surfaces à Courbures Opposées et Leurs Lignes Géodésiques"), P. Duhem (*La Théorie Physique. Son Objet et Sa Structure*) and E. Borel ("La Mécanique Statistique et l'Irréversibilité") made similar observations. See D. Ruelle (*Chance and Chaos*) for a discussion of that history from a modern perspective, and a good popular exposition of chaos.

20 It is interesting to read the rest of the text of Laplace. First of all, he expresses the belief that there are, indeed, fundamental, universally valid laws of nature that can be discovered through scientific investigation (the only example that Laplace had of a fundamental law was that of universal gravitation). In that respect, nothing has changed today. One of the goals of physics is still to discover those fundamental laws. His basic idea could be called universal reductionism rather than universal determinism. Since reductionism is remarkably well defended in Weinberg's book *Dreams of a Final Theory*, I shall not pursue this point. Reading a little further, we see that Laplace's goal is to use science against superstition. He mentions the fears caused by Halley's comet in the Middle Ages (where it was taken as a sign of the divine wrath) and how our discovery of the laws of the "system of the world" "dissipated those childish fears due to our ignorance of the true relations between Man and the Universe." Laplace expresses also a deep optimism about the progress of science. Again, nothing of that has been refuted by the evolution of natural sciences over the last two centuries. But one will not find any claim about the computability, by us humans, of *all* the consequences of the laws of physics.

21 E. T. Jaynes, "Clearing up Mysteries—the Original Goal," p. 7. Jaynes's criticism were directed mostly at the way quantum theory is presented, but they also apply to some discussions of chaos theory or of statistical mechanics.

22 Of course, since classical mechanics is not really fundamental (quantum mechanics is), this issue is rather academic. We nevertheless want to discuss it, because there seems to be a lot of confusion in the literature.

23 Imagine that the Poincaré-Bendixson would hold in all dimensions.

24 And, concerning quantum mechanics, see the references in Note 13.

25 See also I. Prigogine & I. Stengers, *Entre le Temps et l'Éternité*, p. 28: "As we shall see, there exists, for sufficiently unstable systems, a 'temporal horizon' beyond which no determined trajectory can be attributed to them."

26 See R.W. Batterman, "Randomness and Probability in Dynamical Theories: On the Proposals of the Prigogine School," for a related, but different, critique. Batterman says that the replacement of trajectories by probabilities is "very much akin to the claim in quantum mechanics that the probabilistic state description given by the Ψ-function is complete, that is, that underlying exact states cannot exist" (p. 259). But he notes that here, unlike in quantum mechanics, no no-hidden-variable argument is given to support that claim (for the exact status of no-hidden-variable arguments in quantum mechanics, see J. S. Bell, *Speakable and Unspeakable in Quantum Mechanics*, and T. Maudlin, *Quantum Non-Locality and Relativity*).

27 An absolutely continuous one, i.e., given by a density. If one considers probabilities given by delta functions, it is equivalent to consider trajectories.

28 I say "approximate," because I describe the system as it is seen. I do not yet consider any theory (classical or quantum).

29 P. S. Laplace, *A Philosophical Essay on Probabilities.*

30 Mathematically, the microscopic state of the system is represented by a point in its "phase space" $\Omega$. Each point in that space represents the positions and the velocities of *all* the particles of the system under consideration. So, the phase space is $\mathbf{R}^{6.N}$ where $N$ is the number of particles (of the order of $10^{23}$ for a macroscopic system), since one needs three coordinates for each position and three coordinates for each velocity. Hamilton's equations of motion determine, for each time $t$, a map $T^t$ that associates to each initial condition $\mathbf{x} \in \Omega$, at time zero, the corresponding solution $T^t\mathbf{x}$ of the equations of motion at that time. Reversibility of the equations of motion means that there is a transformation (an involution) $I$ acting on $\Omega$ that satisfies the following relation:

$$T^tIT^t\mathbf{x} = I\mathbf{x}, \tag{8}$$

or $IT^t = T^{-t}I$. In classical mechanics, $I$ reverses velocities. (In quantum mechanics, $\Omega$ is replaced by a Hilbert space, and $I$ associates to a wave function its complex conjugate. For the role of weak interactions, see R. P. Feynman, *The Character of Physical Law.*)

31 We shall discuss in SECTION 4.3 a frequent confusion that assigns the source of irreversibility to the (true but irrelevant) fact that no system is ever perfectly isolated. But let us point out here that it is easy to produce nonisolated systems that behave approximately in "time-reversed" fashion: a refrigerator, for example. Living beings also seem to violate the second law of thermodynamics. But put a cat in a well-sealed box for a long enough time: it will evolve towards equilibrium.

32 Such a contradiction is suggested by the following statement of Prigogine and Stengers: " Irreversibility is either true on all levels or on none: it cannot emerge as if out of nothing, on going from one level to another" (*Order out of Chaos*, quoted by P. Coveney, "The Second Law of Thermodynamics: Entropy, Irreversibility and Dynamics," p. 412.

33 I. Prigogine & I. Stengers, *Entre le Temps et l'Éternité*, p. 25.

34 I. Prigogine, *Les Lois du Chaos*, p. 41. Stengers goes even further: "The reduction of the thermodynamic entropy to a dynamical interpretation can hardly be viewed otherwise than as an 'ideological claim' . . ." (*L'Invention des Sciences Modernes*, p. 192). We shall see below in what precise sense this "reduction" is actually a "scientific claim."

35 Quoted in J. L. Lebowitz, *Physics Today*, Sep. 1993, pp. 32–38.

36 This is, of course, not new at all. Good references, apart from Boltzmann himself (*Theoretical Physics and Philosophical Problems. Selected Writings*), include R. P. Feynman, *The Character of Physical Law*; E. T. Jaynes, *Papers on Probability, Statistics and Statistical Physics*; J. L. Lebowitz, "Macroscopic Laws, Microscopic Dynamics, Time's Arrow and Boltzmann's Entropy," and *Physics Today*, Sep. 1993, pp. 32–38; R. Penrose, *The Emperor's New Mind*; and E. Schrödinger, "Irreversibility."

37 By classical, I mean "standard." However, all the discussion will take place in the context of classical physics. But it could be extended to quantum mechanics (replacing phase points by state vectors).

38 Of course, the billiard ball itself contains many molecules. But the rigidity of the ball allows us to concentrate on the motion of its center of mass.

39 By "functions," I mean also families of functions indexed by space or time, i.e., fields, such as the local energy density or the velocity field.

40 I am a little vague on how to "count" configurations. If I consider discrete (finite) systems, then it is just counting. Otherwise, I use, of course, the Lebesgue measure on phase space. All statements about probabilities made later will be based on such "counting."

41 Although not trivial, this fact simply means that reproducible macroscopic experiments exist and that a macroscopic description of the world is possible.

42 To make the micro/macro distinction sharp, one has to consider some kind of limit (hydrodynamic, kinetic, etc.), where the number of particles (and other quantities) tend to infinity. That is a convenient mathematical setting to prove precise statements. But one should not confuse this limit, which is an approximation, with the physical basis of irreversibility. See J. L. Lebowitz, "Macroscopic Laws, Microscopic Dynamics, Time's Arrow and Boltzmann's Entropy" and H. Spohn, *Large Scale Dynamics of Interacting Particles*, for a discussion of those limits.

43 This is due to the Poincaré recurrences; see SECTION 4.1 and the APPENDIX.

44 I mean shorter than the Poincaré recurrence time.

45 O. E. Lanford, "Time Evolution of Large Classical Systems"; *Physica*; "On a Derivation of the Boltzmann Equation."

46 It is also important to clarify the role of "ensembles" here. What we have to explain is the fact that, when a system satisfies certain macroscopic initial conditions ($F_0$), it *always* (in practice) obeys certain macroscopic laws. The same macroscopic initial conditions will correspond to many different microscopic initial conditions. We may introduce, for mathematical convenience, a probability distribution (an "ensemble") on the microscopic initial conditions. But one should remember that we are physically interested in "probability one statements," namely, statements that are independent of the microscopic configuration, as opposed to statements about averages, for example. Otherwise, one would not explain why all individual macroscopic systems satisfy a given law. In practice, those "probability one statements" will hold only in some limit. Physically, they should be interpreted as "very close to one" for finite systems with a large number of particles. To see how close to one this is, consider all the bets and games of chance that have ever taken place in human history. One would certainly expect the laws of large numbers to apply to such a sample. But this number is minuscule compared to the typical number ($10^{23}$) of molecules in a cubic centimeter.

47 This remark is of some interest for the issue of *reductionism*: higher-level laws, such as the macroscopic laws, are reduced to the microscopic ones *plus* assumptions on the initial conditions. If this is the case in statistical mechanics, where it is usually granted that reductionism works, it should clarify the situation in other fields, like biology, where reductionism is sometimes questioned. In particular, the fact that some assumptions must be made on the initial conditions in going from the microscopic to the macroscopic should not be forgotten, nor should it be held as an argument against reductionism. Another frequent confusion about reductionism is to remark that the macroscopic laws do not uniquely determine the microscopic ones. For example, many of the macroscopic laws can be derived from stochastic microscopic laws or from deterministic ones. That is true, but it does not invalidate reductionism. What is true on the microscopic level has to be discovered independently of the reductionist program. Finally, what is considered microscopic or macroscopic is a question of scale. The classical description considered here at the "microscopic" level is an approximation to the quantum description and neglects the molecular and atomic structure. And the "macroscopic" level may in turn be considered microscopic if one studies large-scale motions of the atmosphere. But

despite frequent claims to the contrary, reductionists are quite happy not to explain carburetors directly in terms of quarks (see S. Weinberg, *Dreams of a Final Theory*, for a good discussion of reductionism).

48 R. P. Feynman, *The Character of Physical Law*. See the end of Note 46.

49 To put it in formulas, let $\overline{\Omega}_t$ be the configurations giving to $F$ its value at time $t$. If we denote by $F_t$ that value, $\overline{\Omega}_t$ is simply the preimage of $F_t$ under the map $F$. Let $\Omega_t$ be the set of good configurations, at time $t$, that lead to a behavior of $F$, for later times (again, not for *too* long, because of Poincaré recurrences), which is described by the macroscopic laws. In general, $\Omega_t$ is a very large subset of $\overline{\Omega}_t$, but is not identical to $\overline{\Omega}_t$. Thus, $\overline{\Omega}_0$ are all the configurations in the left half of the box at time zero, and $\Omega_0$ is the subset consisting of those configurations whose evolution leads to a uniform density. Microscopic reversibility says that $T^t\{I[T^t(\Omega_0)]\} = I(\Omega_0)$ (this is just (8) in Note 30 applied to $\Omega_0$). A reversibility paradox would follow from $T^t[I(\Omega_t)] = I(\Omega_0)$ (one takes all the good configurations at time $t$, reverses the velocities, lets them evolve for a time $t$, and one gets the original set of initial conditions, with velocities reversed). But $\Omega_t$ is *not equal*, in general, to $T^t(\Omega_0)$ (and this is the source of much confusion). In our example, $T^t(\Omega_0)$ is a tiny subset of $\Omega_t$, because most configurations in $\Omega_t$ we not in the left half of the box at time zero. Actually, $I[T^t(\Omega_0)]$ provides an example of configurations that belong to $\overline{\Omega}_t$ but not to $\Omega_t$. These configurations correspond to a uniform density at time $t$, but not at time $2t$.

50 R. Penrose, *The Emperor's New Mind* and "On the Second Law of Thermodynamics."

51 Failure to realize this leads to strange statements, as, for example, in J. Cohen & I. Stewart, *The Collapse of Chaos*: speaking of the evolution since the Big Bang, the authors write: "For systems such as these, the thermodynamic model of independent subsystems whose interactions switch on but not off is simply irrelevant. The features of thermodynamics either don't apply or are so long term that they don't model anything interesting. Take Cairns-Smith's scenario of clay as scaffolding of life. The system consisting of clay alone is *less* ordered than that of clay plus organic molecules: Order is increasing with time. Why?" (p. 259). The explanation given afterwards ignores both the action of the sun, and the original "improbable state" discussed here. As Ruelle wrote in a review of this book, "if life violates the second law, why can't one build a power plant (with some suitable life forms in it) producing ice cubes and water currents from the waters of Loch Ness?" ("Cracks in the Glass Menagerie of Science"). A similar confusion can be found in Popper: "This law of the increase of disorder, interpreted as a cosmic principle, made the evolution of life incomprehensible, apparently even paradoxical" (*The Open Universe. An Argument for Indeterminism*, p. 172).

52 I neglect here the effect of gravity, see R. Penrose, *The Emperor's New Mind*, for a discussion of the effect of gravity.

53 See R. Penrose, *The Emperor's New Mind*, for further discussion, and Figure 7.19 there for an illustration.

54 As Feynman says: "Therefore I think it necessary to add to the physical laws the hypothesis that in the past the universe was more ordered, in the technical sense, than it is today—I think this is the additional statement that is needed to make sense, and to make an understanding of the irreversibility" (*The Character of Physical Law*, p. 116).

55 D. Driebe, Letter in *Physics Today*, November 1994, p. 13.

56 J. L. Lebowitz, *Physics Today*, September 1993, pp. 34–38.

57 In a recent textbook one reads, after a discussion of the baker's map: "Irreversibility appears only because the instantaneous state of the system cannot be known with an infinite precision" (S. Vauclair, *Eléments de Physique Statistique*, p. 198).

58 The baker map is quite similar to the map discussed in Note 16, and has the same chaotic properties as the latter, but is invertible.

59 I. Prigogine, *Les Lois du Chaos*, p. 37.

60 Ibid., p. 42. See, e.g., P. Coveney, "The Second Law of Thermodynamics: Entropy, Irreversibility and Dynamics": "Another quite popular approach has been to rele-

gate the whole question of irreversibility as illusory" (p. 412). See also R. Lestienne, *Les Fils du Temps. Causalité, Entropie, Devenir* (p. 176); and I. Prigogine & I. Stengers, *La Nouvelle Alliance. Métamorphoses de la Science*, p. 284, for similar remarks.

61 W. Heisenberg, *The Physicist's Conception of Nature*, p. 38. W. Pauli made a similar remark (see "Wahrscheinlichkeit und Physik"; quoted in K. R. Popper, *Quantum Theory and the Schism in Physics*, p. 109).

62 Max Born, *Natural Philosophy of Cause and Chance*, p. 72.

63 K. R. Popper, *Quantum Theory and the Schism in Physics*, p. 106. In his textbook *Statistical Mechanics*, S. K. Ma shows similar concerns. "In one point of view, probability expresses the knowledge of the observer. If he knows more about the system, the probability is more concentrated. This is obviously incorrect. The motion of the system is independent of the psychological condition of the observer" (p. 448). And H. Bondi wrote: "It is somewhat offensive to our thought to suggest that if we know a system in detail then we cannot tell which way time is going, but if we take a blurred view, a statistical view of it, that is to say throw away some information, then we can. . . ." ("Physics and Cosmology"; quoted in P. T. Landsberg, *The Enigma of Time*, p. 135. T. Gold expressed similar views; see P. T. Landsberg, *The Enigma of Time*).

64 K. R. Popper, *Quantum Theory and the Schism in Physics*, p. 107.

65 E. T. Jaynes, *Papers on Probability, Statistics and Statistical Physics*; "The Gibbs Paradox."

66 I. Prigogine, *Les Lois du Chaos*, p. 23.

67 Ibid., p. 24.

68 E. Zermelo, *Wiener Annalen* 57 (1896): 485.

69 In formulas, let $\Omega$ be the "phase space" on which the motion is ergodic (i.e., a constant energy surface, which is a subset of the space considered in Note 30, on which is defined the measure induced by the Lebesgue measure, normalized to one, and denoted $d\mathbf{x}$). Then, ergodicity means that

$$\lim_{T\to\infty} \frac{1}{T} \int_0^T F(T^t\mathbf{x})dt = \int_\Omega F(\mathbf{x})d\mathbf{x} \tag{9}$$

for $F$ integrable and for almost all initial conditions $\mathbf{x} \in \Omega$. The LHS is the time average and the RHS the space average. If we take for $F$ the characteristic function of a (measurable) set $A \subset \Omega$, the time average equals the fraction of time spent by the trajectory in $A$, and the space average is the volume of $A$.

70 For a history of the concept of ergodicity, and some very interesting modern developments, see G. Gallavotti, "Ergodicity, Ensembles, Irreversibility in Boltzmann and Beyond." It seems that the (misleading) emphasis on the modern notion of ergodicity goes back to the Ehrenfests's paper *The Conceptual Foundations of the Statistical Approach in Mechanics* more than to Boltzmann. A careful, but nevertheless exaggerated, interest in ergodicity and mixing is found in the work of A. I. Khinchin (*Mathematical Foundations of Statistical Mechanics*) and N. S. Krylov (*Works on the Foundations of Statistical Mechanics*). It is also found, e.g., in D. Chandler, *Introduction to Modern Statistical Mechanics*, p. 57; Y. G. Sinai, *Topics in Ergodic Theory*, p. 207; T. Hill, *Statistical Mechanics*, p. 16; S. K. Ma, *Statistical Mechanics*, chap. 26; C. J. Thompson, *Mathematical Statistical Mechanics*, app. B; and N. Dunford & J. T. Schwartz, *Linear Operators; Part 1: General Theory*, p. 657. But see J. T. Schwartz, "The Pernicious Influence of Mathematics on Science," for a self-criticism of *Linear Operators; Part 1: General Theory*; in the recent textbook of S. Vauclair, *Eléments de Physique Statistique*, one reads: "One considers that during the time $\delta t$ of the measurement, the system has gone through all the possibly accessible states, and that it spent in each state a time proportional to its probability" (p. 11); and: "Only the systems having this property (mixing) tend to an equilibrium state, when they are initially in a state out of equilibrium" (p. 197).

71 R. C. Tolman, ***The Principles of Statistical Mechanics***, p. 65; E. T. Jaynes, ***Papers on Probability, Statistics and Statistical Physics***, p. 106; and J. T. Schwartz, "The Pernicious Influence of Mathematics on Science."

72 If there is some cell that has not been visited even once, there will be some function on phase space for which the space average and the time average, computed up to that time, differ a lot: just take the function that takes value one on that cell and zero elsewhere.

73 See, e.g., Figures 7.3 and 7.5 in R. Penrose, ***The Emperor's New Mind***, for an illustration, and the APPENDIX here for an example.

74 See J. L. Lebowitz & O. Penrose, "Modern Ergodic Theory."

75 See, e.g., S. Vauclair, ***Eléments de Physique Statistique***, p. 197, where the approach to equilibrium is illustrated by the baker's transformation.

76 One can even invoke a theorem to that effect: the ergodic theorem for Markov chains. But this is again highly misleading. This theorem says that probability distributions will converge to an "equilibrium" distribution (for suitable chains). This is similar, and related, to what happens with strongly chaotic systems. But it does not explain what happens to a single system, unless we are willing to distinguish between microscopic and macroscopic variables, in which case the ergodic theorem is not necessary.

77 E. Borel ("La Mécanique Statistique et l'Irréversibilité") tried to answer the reversibility objection, using the lack of isolation and the instability of the trajectories. As we saw in SECTION 3.3, this objection is not relevant, once one introduces the micro/macro distinction. And Fred Hoyle wrote: "The thermodynamic arrow of time does not come from the physical system itself . . . it comes from the connection of the system with the outside world" ("The Asymmetry of Time," quoted in P. T. Landsberg, ***The Enigma of Time***). See also J. Cohen & I. Stewart, ***The Collapse of Chaos*** (p. 260) and (X. de Hemptinne, "La Source de l'Irréversibilité," for similar ideas.

78 See L. Boltzmann, ***Theoretical Physics and Philosophical Problems. Selected Writings***, p. 170.

79 One has to be careful here. If we shake a mixture of fluids, it should become homogeneous faster. But, of course, there are external influences that prevent the system from going to equilibrium, like in a refrigerator. Also, the time scale on which the approach to equilibrium takes place may vary enormously, depending on the physical situation. This is what is overlooked in X. de Hemptinne, "La Source de l'Irréversibilité."

80 See the references to Bergson in I. Prigogine & I. Stengers, ***La Nouvelle Alliance. Métamorphoses de la Science*** and ***Entre le Temps et l'Éternité.***

81 I remember that when I first heard, as a teenager, about the special theory of relativity, it was through Bergson's alleged refutation of that theory! He thought, probably rightly so, that there was a conflict between his intuitive views on duration and the absence of absolute simultaneity in relativity. So he simply decided that there was a "time of consciousness," as absolute as Newtonian time, and that the Lorentz transformations were merely some kind of coordinates "attributed" by one observer to the other. Running into trouble with the twin paradox, he decided that acceleration is relative, like uniform motion, and that, when both twins meet again, they have the same age! (See H. Bergson, ***Durée et Simultanéité. A propos de la Théorie d'Einstein.***) At least Bergson had the good sense, after his lengthy polemic with Einstein, to stop the republication of his book. But, and this is a remarkable aspect of our "intellectual" culture, the ***very same mistake*** is repeated by some of his admirers: V. Jankelevitch, ***Henri Bergson***, chap. 2; M. Merleau-Ponty, ***Eloge de la Philosophie et Autres Essais***, p. 319; and G. Deleuze, ***Le Bergsonisme***, p. 79; see also Deleuze's later writings. Of course, all this is explained by telling the physicists that they should stick to their "mathematical expressions and language" (M. Merleau-Ponty, ***Eloge de la Philosophie et Autres Essais***, p. 320), while leaving the deep problems of the "time of consciousness" to philosophers. For a modern attempt to make some sense of Bergson's universal time, see the first appendix of I. Prigogine & I. Stengers, ***Entre le Temps et l'Éternité.***

82 See J. Monod, *Chance and Necessity*, and B. Russell, *A History of Western Philosophy*, for critiques of his philosophy and G. Moreau, "Monsieur Ilya Prigogine 'Bouleverse' la Philosophie," for the relation between Bergson and Prigogine. The main problem with Bergson's lasting influence is well expressed by Bertrand Russell: "One of the bad effects of an antiintellectual philosophy such as that of Bergson, is that it thrives upon the errors and confusions of the intellect. Hence it is led to prefer bad thinking to good, to declare every momentary difficulty insoluble, and to regard every foolish mistake as revealing the bankruptcy of intellect and the triumph of intuition" (*A History of Western Philosophy*, p. 831).

83 H. Bergson, *L'Evolution Créatrice*, p. 264. As we shall see in SECTION 5, it is probably the least metaphysical of those laws (although I do not like this terminology), since it is not a purely dynamic law.

84 Ibid., p. 266.

85 Ibid., p. 267.

86 Ibid., p. 278.

87 J. Monod, *Chance and Necessity.*

88 See, e.g., the introduction by Monod to the French edition of *The Logic of Scientific Discovery*; also, e.g., I. Prigogine & I. Stengers, *Entre le Temps et l'Éternité*, p. 173. For philosophical critiques of Popper, see H. Putnam, "The 'Corroboration' of Theories"; and D. Stove, *Popper and After. Four Modern Irrationalists.* For a critique of his views on the arrow of time, see M. Ghins, "Popper versus Grünbaum and Boltzmann on the Arrow of Time."

89 See K. R. Popper, *Quantum Theory and the Schism in Physics.*

90 K. R. Popper, *The Open Universe. An Argument for Indeterminism*, p. 102.

91 See, e.g., K. R. Popper, *Quantum Theory and the Schism in Physics*, p. 112.

92 See R. P. Feynman, *The Character of Physical Law*, and J. L. Lebowitz, "Macroscopic Laws, Microscopic Dynamics, Time's Arrow and Boltzmann's Entropy," for a discussion of that fluctuation theory.

93 There are indications that Boltzmann did not take his fluctuation theory too seriously. For example, he wrote "that the world began from a very unlikely initial state, this much can be counted amongst the fundamental hypotheses of the whole theory and we can say that the reason for it is as little known as that for why the world is as it is and not otherwise" (*Theoretical Physics and Philosophical Problems. Selected Writings*, p. 172; compare with Note 53). In general, Boltzmann is quite opposed to unscientific speculations. In his criticism of Schopenhauer, he takes a very Darwinian (and surprisingly modern) view of mankind. He starts by observing that drinking fermented fruit juices can be very good for your health: "if I were an antialcoholic I might not have come back alive from America, so severe was the dyssentry that I caught as a result of bad water . . . it was only through alcoholic beverages that I was saved" (Ibid., p. 194). But, with alcohol, one can easily overshoot the mark. It is the same thing with moral ideas. "We are in the habit of assessing everything as to its value; according to whether it helps or hinders the conditions of life, it is valuable or valueless. This becomes so habitual that we imagine we must ask ourselves whether life itself has a value. This is one of those questions utterly devoid of sense" (Ibid., p. 197). Finally, for theoretical ideas, he observes that our thoughts should correspond to experience and that overshooting the mark should be kept within proper bounds: "Even if this ideal will presumably never be completely realized, we can nevertheless come nearer to it, and this would ensure cessation of the disquiet and the embarrassing feeling that it is a riddle that we are here, that the world is at all and is as it is, that it is incomprehensible what is the cause of this regular connection between cause and effect, and so on. Men would be freed from the spiritual migraine that is called metaphysics" (Ibid., p. 198).

94 K. R. Popper, "Autobiography."

95 J. C. Maxwell, *Nature* 17 (1878): 257; quoted in J. L. Lebowitz, *Physics Today*, September 1993, pp. 32–38.

96 P. K. Feyerabend, "On the Possibility of a Perpetuum Mobile of the Second Kind."

97 This error is repeated, with many others, in A. Woods & T. Grant, *Reason in Revolt*, p. 177.

98 M. Serres, *Eclaircissements: Cinq Entretiens avec Bruno Latour.* I have left these texts in the original language, partly because they are difficult to translate, partly because the confusion might be blamed on the translator.

99 Jean-François Lyotard, *La Condition Postmoderne: Rapport sur le Savoir.*

100 Jean Baudrillard, *L'Illusion de la Fin.*

101 Gilles Deleuze & Félix Guattari, *Qu'est-ce que la Philosophie?*

102 I recommend to people interested in *tensors* (applied to psychology, sociology, etc.) Guattari's contribution to J. P. Brans, I. Stengers & P. Vincke, eds., *Temps et Devenir.*

103 K. Denbigh, "How Subjective is Entropy?"; M. Tribus, *Boelter Anniversary Volume.*

104 W. Thirring, "Boltzmann's Legacy in the Thinking of Modern Physics."

105 E. T. Jaynes, *Papers on Probability, Statistics and Statistical Physics*, p. 85.

106 This is again the source of much confusion, just like the "subjectivity of irreversibility" discussed in SECTION 3.5. See, for example, K. R. Popper, *Quantum Theory and the Schism in Physics*, p. 111; K. Denbigh, "How Subjective Is Entropy"; S. K. Ma, *Statistical Mechanics.* In M. Gell-Mann's popular book *The Quark and the Jaguar* we read: "Entropy and information are closely related. In fact, entropy can be regarded as a measure of ignorance" (p. 219). And further: "Indeed, it is mathematically correct that the entropy of a system described in perfect detail would not increase; it would remain constant" (p. 225). This is true, but it might be useful to emphasize that one does not refer to the "usual" thermodynamic entropy.

107 See also E. T. Jaynes, "The Gibbs Paradox," for a nice discussion on apparent violations to the Second Law. And see, e.g., the constraints on the plausible mechanisms for the origin of life, due to the Second Law, discussed by A. C. Elitzur, "Let There Be Life," sec. 11. There is some similarity between this use of the Second Law and the way biologists use the law of natural selection. The biologists do not believe that complex organs appear "spontaneously." Hence, when they occur, they look for an adaptive explanation (see R. Dawkins, *The Blind Watchmaker*, for an introduction to the theory of evolution). Both attitudes are, of course, similar to elementary probabilistic reasoning: if we throw a coin a million times and find a significant deviation from one-half heads one-half tails, we shall conclude that the coin is biased (rather than assuming that we have observed a miracle).

108 As emphasized by J. L. Lebowitz in "Macroscopic Laws, Microscopic Dynamics, Time's Arrow and Boltzmann's Entropy."

109 As, for example, in: "This so-called hypothesis of 'molecular chaos' admits the absence of correlations between the velocities of the molecules in the initial state of the gas, although, obviously, correlations exists between the molecules after the collisions. The hypothesis of molecular chaos amounts to introduce, in a subtle way, the irreversibility that one tries demonstrate" (R. Lestienne, *Le Fils du Temps. Causalité, Entropie, Devenir*, p. 172). Boltzmann never said that he would demonstrate irreversibility without assuming something on initial conditions. Another, more radical, confusion is due to Bergmann: "It is quite obvious that the Boltzmann equation, far from being a consequence of the laws of classical mechanics, is inconsistent with them" (in T. Gold, ed., *The Nature of Time*, p. 191).

110 To make a *vague* analogy, in equilibrium statistical mechanics, one has the concept of phase transition. Mean field theory (or the van der Waals theory, Curie-Weiss or molecular field approximation) gives an approximate description of the phase transition. But the concept of phase transition is much wider than the range of validity of that approximation.

111 This theorem says that, if $A$ is a subset of the phase space $\Omega$, then $Vol[T^t(A)] = Vol(A)$, where $Vol(A) = \int_A d\mathbf{x}$.

112 I. Prigogine & I. Stengers, *Entre le Temps et l'Éternité*, p. 104. See X. de Hemptinne, "La Source de l'Irréversibilité," p. 8, for a similar statement. See also I. Prigogine, "Un Siècle d'Espoir": "According to the mechanical view of the world,

the entropy of the universe is today identical to what it was at the origin of time" (p. 160). Or, as P. Coveney says: "As long as the dynamical evolution is unitary, irreversibility cannot arise. This is the fundamental problem of nonequilibrium statistical mechanics" ("The Second Law of Thermodynamics: Entropy, Irreversibility and Dynamics," p. 411).

113 Let me use the notations of Note 49. By Liouville's Theorem, indeed $Vol[T^t(\Omega_0)] = Vol(\Omega_0)$. But $T^t(\Omega_0)$ is a very small subset of $\Omega_t$. Confusing the two sets leads to the (wrong) idea that $Vol[T^t(\Omega_0)] = Vol(\Omega_t)$. The evolution of $\Omega_t$ does not coincide with a set of trajectories.

114 Note that these entropies agree with (minus) Boltzmann's *H* function only when the interparticle forces are negligible (as in a very dilute gas). This is rather obvious, since the *H* function is an approximation to the Boltzmann entropy. See E. T. Jaynes, *Papers on Probability, Statistics and Statistical Physics*, p. 81.

115 Amusingly enough, this conclusion can be reached using only Liouville's theorem (see E. T. Jaynes, *Papers on Probability, Statistics and Statistical Physics*, p. 83), which is blamed as the source of all the troubles!

116 See, e.g., P. Coveney, "The Second Law of Thermodynamics: Entropy, Irreversibility and Dynamics": Irreversibility is admitted into the description by asserting that we only observe a course-grained probability" (p. 412).

117 E. T. Jaynes, *Papers on Probability, Statistics and Statistical Physics*, p. 86.

118 Note that the word "chaos" in the title is used in a somewhat ambiguous way: sometimes it has the technical meaning of SECTION 2, sometimes it means "disordered" or "random."

119 D. C. Dennett, *Darwin's Dangerous Idea*, p. 392.

120 Concerning the rather poetic and enthusiastic style of Prigogine and Stengers (see, e.g., the title of Prigogine's conference at Cerisy: "A Century of Hope"), I can only agree with the comment of J. Maynard Smith: "I do not think we should embrace scientific theories because they are more hopeful, or more exhilarating. . . . I feel sensitive on this matter because, as an evolutionary biologist, I know that people who adopt theories because they are hopeful finish up embracing Lamarckism, which is false, although perhaps not obviously so, or Creationism, which explains nothing and suggests no questions at all. If nonequilibrium thermodynamics makes poets happier, so be it. But we must accept or reject it on other grounds" (Did Darwin Get It Right?, p. 257).

121 As, for example, in J. Cohen & I. Stewart: "The tendency for systems to segregate into subsystems is just as common as the tendency for different systems to get mixed together" (*The Collapse of Chaos*, p. 259). Or: "by watching some phenomena, one is led to say that time has an arrow pointing towards a greater disorder, but by considering other phenomena, it seems that time has an arrow pointing, on the contrary, towards a greater order. Then, what does this arrow mean? If we can orient it in opposite directions, it is better not to talk about it any more" (A. Meessen, "La Nature du Temps," p. 119).

122 I am not talking here about the "heat death" of the universe, but only about a "small" subsystem, like the solar system, over the "short" period during which the sun burns its fuel.

123 K. R. Popper, *The Open Universe. An Argument for Indeterminism*, p. 173.

124 One should always distinguish this precise but technical sense of order from vague, intuitive words, such as "complexity." If we speak of the "complexity of the brain," it has of course evolved from a less "complex" structure, but those words do not have a precise meaning. Note that precise definitions of complexity, such as algorithmic complexity, do not at all capture the intuitive meaning of the word, since "random" sequences are algorithmically complex, and, whatever "complexity of the brain" means, and whatever "random" means, they do not mean the same thing (a similar problem occurred with the word "information" in the past).

125 Besides, as we saw in SECTION 5, entropy is not really a "substance" to be "exported." This is a somewhat strange terminology for a philosopher like Popper, so critical of "essentialism."

126 To avoid any misunderstanding, I emphasize that I do not criticize any explicit statement made below, but I suggest that one may unwillingly mislead the general public by putting too much emphasis on certain aspects of irreversible processes.

127 I. Prigone, "Un Siècle d'Espoir," p. 157.

128 I. Prigone & I. Stengers, *La Nouvelle Alliance. Métamorphoses de la Science*, p. 427.

129 A fluid is maintained between two horizontal plates, the lower one being hotter than the higher one. If the temperature difference is large enough, rolls will appear. See, e.g., I. Prigone & I. Stengers, *La Nouvelle Alliance. Métamorphoses de la Science* and *Entre le Temps et l'Éternité* for a discussion of the Bénard instability.

130 A. Meessen, "La Nature du Temps," p. 118.

131 Somewhat to my surprise, I found the following *theological* commentary on the constructive role of irreversible phenomena: "Each time that a new order of things appear, it is marked by the dissipation of a chaotic behaviour, by a broken form of movement, that is, by a "fractal," by non-linearity, by variations or "fluctuations," by instability and randomness. In this way the dynamics of self-organization of matter, which reaches the great complexity of consciousness, manifests itself" (A. Ganoczy, *Dieu, l'Homme et la Nature*, p. 79). This is a bit of an extrapolation, starting from the Bénard cells. The quotation appears in a section of a chapter on "God in the language of the physicists," where the author refers mostly to Prigogine & Stengers, *La Nouvelle Alliance. Métamorphoses de la Science* and *From Being to Becoming.*

132 I. Prigogine & I. Stengers, *Entre le Temps et l'Éternité*, p. 46. And even more misleading: "In a deterministic world, irreversibility would be meaningless, since the world of tomorrow would already be contained in the world of today, there would be no need to speak of time's arrow" ("Un Siècle d'Espoir," p. 166).

133 It is also a bit too fast to say, as Prigogine & Stengers do, that this kind of mechanism allows us to go beyond the "very old conflict between reductionists and antireductionists" (*La Nouvelle Alliance. Métamorphoses de la Science*, p. 234; quoted in A. Boutot, *L'Invention des Formes*, p. 274). Any reductionist is perfectly happy to admit that some situations do not have simple, deterministic, macroscopic descriptions, while I doubt that antireductionists such as Popper and Bergson would be satisfied with such a simple fact.

134 Here is another theological commentary: "Irreversibility means that things happen in time *and thanks to time*, that it could be that they did not happen, or did happen otherwise and that an infinite number of possibilities are always open." " 'Inventive' disorder is part of the definition of things. . . . Impredictibility which is not due to our inability to control the nature of things, but to their nature itself, whose future simply does not yet exist, and could not yet be forecasted, even by 'Maxwell's demon' put on Sirius" (A. Gesché, *Dieu pour Penser. 4. Le Cosmos*, p. 121). The author claims to find his inspiration on the "new scientific understanding of the Cosmos" from, among others, *La Nouvelle Alliance* (Ibid., p. 120).

135 M.-P. Shützenberger, "Les Failles du Darwinisme."

136 In a critique of several "almost mystical views of life," which deny "an evolutionary role to Darwinian selection," the biologist A. C. Elitzur observes that "such a misleading discussion of evolution is based on a complete distortion of thermodynamics" ("Let There Be Life," p. 450). Besides, I disagree, needless to say, with the comment of Prigogine and Stengers that there is an "antithesis" between Boltzmann and Darwin and that the theories of Darwin were a success while those of Boltzmann failed (*Entre le Temps et l'Éternité*, pp. 23–24).

137 Here, I mean complexity in an intuitive sense. Of course, it is somewhat related to entropy, because, if we consider the set of molecules in an eye, say, there are very few ways to arrange them so as to produce an eye compared to the number of arrangements that cannot be used for vision. But as Jaynes says (see Remark 8 in SECTION 5), as long as we do not have a well-defined set of macroscopic variables that define precisely what an eye is, we cannot give a precise characterization

of this "complexity" in terms of entropy (and, probably, such an entropy would not be the right concept anyway).

138 R. Dawkins, *The Blind Watchmaker*, p. 8.

139 Not being a biologist, I do not want to enter into any debate about the origin of life, the speed of evolution, or how far the Darwinian explanation goes. I want only to underline the similarity with the type of (probabilistic) explanation used in statistical physics.

140 This remark also illustrates the well-known (but not universally accepted) fact that Darwin's theory is not a "tautology." Having a better eye helps to survive and to have children, hence to propagate the "good" genes. But "better" here is not *defined* solely in terms of the number of children.

141 Speaking about DNA as the solution to the enigma of "life," the biologist Dawkins writes: "Even those philosophers who had been predisposed to a mechanistic view of life would not have dared hope for such a total fulfillment of their wildest dreams" (*River out of Eden*, p. 17). Not surprisingly, Popper said that molecular biology became "almost an ideology" (*The Open Universe. An Argument for Indeterminism*, p. 172). As for Bergson, he must be turning over in his grave.

142 Mark Kac, *Probability and Related Topics in the Physical Sciences*, p. 99; see also C. J. Thompson, *Mathematical Statistical Mechanics*, p. 23.

143 There is a small abuse here, because I seem to change the laws of motion by changing the orientation. But I can attach another discrete "velocity" parameter to the particles, having the same value for all of them, and indicating the orientation, clockwise or counterclockwise, of their motion. Then, the motion is truly reversible, and the operation *I* of Note 30 simply changes that velocity parameter.

144 The word "chaos" here has nothing to do with "chaos theory," and, of course, Boltzmann's hypothesis is much older than that theory.

145 The simplifying features of the model (the balls do not interact) have the unpleasant consequence that the "full" Boltzmann entropy introduced here and defined in SECTION 5 actually coincides with (minus) the Boltzmann *H*-function. But, in general, the latter should be only an approximation to the former.

## REFERENCES

ALBERT, D. "Bohm's Alternative to Quantum Mechanics." *Scientific American*, May 1994.

———. *Quantum Mechanics and Experience.* Cambridge, MA: Harvard University Press, 1992.

BATTERMAN, R. W. "Randomness and Probability in Dynamical Theories: On the Proposals of the Prigogine School." *Philosophy of Science* 58 (1991): 241.

BAUDRILLARD, J. *L'Illusion de la Fin.* Paris: Galilée, 1992.

BELL, J. S. *Speakable and Unspeakable in Quantum Mechanics.* Cambridge: Cambridge University Press, 1993.

BERGSON, H. *Durée et Simultanéité. A propos de la Théorie d'Einstein.* Paris: F. Alcan, 1923.

———. *L'Evolution Créatrice.* Paris: F. Alcan, 1907.

BOLTZMANN, L. *Theoretical Physics and Philosophical Problems. Selected Writings.* Edited by B. McGuinness, Dordrecht: Reidel, 1974.

BONDI, H. "Physics and Cosmology." Halley Lecture. *Observatory* 82 (1962): 133.

BOREL, E. "La Mécanique Statistique et l'Irréversibilité." *Oeuvres.* Tome 3, p. 1697. Paris: CNRS, 1972.

BORN, M. *Natural Philosophy of Cause and Chance.* Oxford: Clarendon Press, 1949.

BOUTOT, A. *L'Invention des Formes.* Paris: Odile Jacob, 1993.

BRICMONT, J. "Contre la Philosophie de la Mécanique Quantique." In *Les sciences et la philosophie.* Paris: Vrin, 1995.

CHANDLER, D. *Introduction to Modern Statistical Mechanics.* Oxford: Oxford University Press, 1987.

COHEN, J. & I. STEWART. *The Collapse of Chaos.* New York, NY: Penguin Books, 1994.

COVENEY, P. "The Second Law of Thermodynamics: Entropy, Irreversibility and Dynamics." Nature 333 (1988): 409.

DAWKINS, R. *The Blind Watchmaker.* New York, NY: W. W. Norton, 1986.

———. *River out of Eden.* London: Weidenfeld & Nicolson, 1995.

DELEUZE, G. *Le Bergsonisme.* Paris: PUF, 1968.

DELEUZE, G. & F. GUATTARI. *Qu'est-ce que la Philosophie?* Paris: Ed. de Minuit, 1991.

DENBIGH, K. "How Subjective is Entropy? In *Maxwell's Demon. Entropy, Information, Computing.* Edited by H. S. Leff & A. F. Rex. Bristol: A. Hilger, 1990.

DENNETT, D. C. *Darwin's Dangerous Idea.* New York, NY: Simon & Schuster, 1995.

DRIEBE, D. Letter. In *Physics Today*, November 1994, p. 13.

DUHEM, P. *La Théorie Physique. Son Objet et Sa Structure.* Paris: Chevalier et Rivière, 1906.

DUNFORD, N. & J. T. SCHWARTZ. *Linear Operators; Part 1: General Theory.* New York, NY: Interscience, 1958.

DÜRR, D., S. GOLDSTEIN & N. ZANGHI. "Quantum Equilibrium and the Origin of Absolute Uncertainty." *Journal of Statistical Physics* 67 (1992): 843–907.

EHRENFEST, P. & T. EHRENFEST. *The Conceptual Foundations of the Statistical Approach in Mechanics.* Translated by M. J. Moravesik. Ithaca, NY: Cornell University Press, 1959.

EKELAND, I. *Le Calcul, L'Imprévu.* Paris: Le Seuil, 1984.

ELITZUR, A. C. "Let There be Life." *Journal of Theoretical Biology* 168 (1994): 429.

FEYERABEND, P. K. *Against Method.* London: New Left Books, 1975. (*Contre la Méthode.* Paris: Le Seuil, 1979.)

———. "On the Possibility of a Perpetuum Mobile of the Second Kind. In *Mind, Matter, and Method.* Edited by P. K. Feyerabend & G. Maxwell. Minneapolis, MN: University of Minnesota Press, 1966.

FEYNMAN, R. P. *The Character of Physical Law.* Cambridge, MA: Massachusetts Institute of Technology Press, 1967.

FEYNMAN, R., R. B. LEIGHTON & M. SANDS. *The Feynman Lectures on Physics.* Reading, MA: Addison-Wesley, 1963.

GALLAVOTTI, G. "Ergodicity, Ensembles, Irreversibility in Boltzmann and Beyond." Rome: Preprint.

GAMOW, G. *Mr. Tompkins in Paperback.* Cambridge: Cambridge University Press, 1965.

GANOCZY, A. *Dieu, l'Homme et la Nature.* Paris: Éd. du Cerf, 1995.

GELL-MANN, M. *The Quark and the Jaguar.* London: Little, Brown, 1994.

GESCHÉ, A. *Dieu pour Penser. 4. Le Cosmos.* Paris: Éd. du Cerf, 1994.

GHINS, M. "Popper versus Grünbaum and Boltzmann on the Arrow of Time." Louvain-la-Neuve: UCL Preprint.

GLEICK, J. *Chaos.* New York, NY: Viking Press, 1987. (*La Théorie du Chaos*, Paris: Coll. Champs, Flammarion, 1991.)

GOLD, T., ed. *The Nature of Time.* Ithaca, NY: Cornell University Press, 1967.

GUATTARI, F. *Chaosmose.* Paris: Galilée, 1992.

HADAMARD, J. "Les Surfaces à Courbures Opposées et Leurs Lignes Géodésiques." *Journal de Mathématiques Pures et Appliquées* 4 (1898): 27.

HEISENBERG, W. *The Physicist's Conception of Nature.* London: Hutchinson, 1958.

DE HEMPTINNE, X. "La Source de l'Irréversibilité." KUL Preprint. 1995.

HERGÉ. *Tintin et le Temple du Soleil.* Brussels: Éd. Casterman, 1949.

HILL, T. *Statistical Mechanics.* New York, NY: McGraw-Hill, 1956.

HOYLE, F. "The Asymmetry of Time." Third Annual Lecture to the Research Students' Association, Canberra, 1962.

JANKELEVITCH, V. *Henri Bergson.* Paris: F. Alcan, 1931.

JAYNES, E. T. "Clearing up Mysteries—the Original Goal." In *Maximum Entropy and Bayesian Methods.* Edited by J. Skilling. Dordrecht: Kluwer Academic Publishers, 1989.

———. "The Gibbs Paradox." In *Maximum Entropy and Bayesian Methods.* Edited by C. R. Smith *et al.* Dordrecht: Kluwer Academic Publishers, 1991.

———. *Papers on Probability, Statistics and Statistical Physics.* Edited by R. D. Rosencrantz. Dordrecht: Reidel, 1983.

———. "Violation of Boltzmann's *H*-theorem in Real Gases." *Physics Review* A4 (1991): 747.

KAC, M. *Probability and Related Topics in the Physical Sciences.* New York, NY: Interscience Publishers, 1959.

KHINCHIN, A. I. *Mathematical Foundations of Statistical Mechanics.* New York, NY: Dover Publications, 1949.

KRYLOV, N. S. *Works on the Foundations of Statistical Mechanics.* Princeton, NJ: Princeton University Press, 1979.

LANDSBERG, P. T. *The Enigma of Time.* Bristol: Adam Hilger, 1982.

LANFORD, O. E. "On a Derivation of the Boltzmann Equation." In *Nonequilibrium Phenomena 1: The Boltzmann Equation.* Edited by J. L. Lebowitz & E. W. Montroll. Amsterdam: North-Holland, 1983.

———. *Physica* A106 (1981): 70.

———. "Time Evolution of Large Classical Systems." In *Lecture Notes in Physics*, vol. 38, p. 1. Edited by J. Moser. Berlin: Springer, 1975.

LAPLACE, P. S. *A Philosophical Essay on Probabilities.* Translated by F. W. Truscott & F. L. Emory. New York, NY: Dover Publications, 1951. (*Essai Philosophique sur les Probabilités.* rééd. Paris: C. Bourgeois, 1986. Texte de la 5ème éd., 1825.)

LEBOWITZ, J. L. "Macroscopic Laws, Microscopic Dynamics, Time's Arrow and Boltzmann's Entropy. *Physica* A194 (1993): 1.

———. *Physics Today*, September 1993, pp. 32–38. (Replies in November 1994.)

LEBOWITZ, J. L. & O. PENROSE. "Modern Ergodic Theory." *Physics Today*, February 1973.

LESTIENNE, R. *Les Fils du Temps. Causalité, Entropie, Devenir.* Paris: Éd. du CNRS, 1990.

LIGHTHILL, J. *Proceedings of the Royal Society* (London) A 407 (1986): 35.

LYOTARD, J.-F. *La Condition Postmoderne: Rapport sur le Savoir.* Paris: Ed. de Minuit, 1979.

MA, S. K. *Statistical Mechanics.* Singapore: World Scientific, 1985.

MAUDLIN, T. *Quantum Non-Locality and Relativity.* Cambridge: Blackwell, 1994.

MAXWELL, J. C. *Matter and Motion.* New York, NY: Dover Publications, 1952.

———. *Nature* 17 (1878): 257.

SMITH, J. MAYNARD. *Did Darwin Get It Right?* London: Penguin Books, 1993.

MEESSEN, A. "La Nature du Temps." In *Temps et Devenir.* Edited by L. Morren *et al.* Louvain-la-Neuve: Pr. Univ. de Louvain-la-Neuve, 1984.

MERLEAU-PONTY, M. *Eloge de la Philosophie et Autres Essais.* Paris: Gallimard, 1968.

MONOD, J. *Chance and Necessity.* London: Fontana, 1972. (*Le Hasard et la Nécessité.* Paris: Le Seuil, 1970.)

MOREAU, G. "Monsieur Ilya Prigogine 'Bouleverse' la Philosophie." *Etudes Marxistes* nr. 5, December 1989.

PAULI, W. "Wahrscheinlichkeit und Physik." *Dialectica* 8 (1954): 112.

PENROSE, R. *The Emperor's New Mind.* New York, NY: Oxford University Press, 1990. See "On the Second Law of Thermodynamics." *Journal of Statistical Physics* 77 (1994): 217.

———. *Foundations of Statistical Mechanics.* Elmsford, NY: Pergamon, 1970.

POINCARÉ, H. *Science et Méthode.* Paris: Flammarion, 1909.

POPPER, K. R. "Autobiography." In *The Philosophy of Karl Popper.* Edited by P. A. Schilpp. La Salle, IL: Open Court, 1974.

———. "Irreversibility; or, Entropy since 1905." *British Journal for the Philosophy of Science* 8 (1958): 151.

———. *The Logic of Scientific Discovery.* London: Hutchinson, 1958. (*La Logique de la Découverte Scientifique.* Paris: Payot, 1978.)

———. *The Open Universe. An Argument for Indeterminism.* Totowa, NJ: Rowman & Littlefield, 1956.

———. *Quantum Theory and the Schism in Physics.* Totowa, NJ: Rowman & Littlefield, 1956.

PRIGOGINE, I. *From Being to Becoming.* New York, NY: Freeman, 1980.
———. *Les Lois du Chaos.* Paris: Flammarion, 1994.
———. "Un Siècle d'Espoir." In *Temps et Devenir.* Colloque de Cerisy, à partir de l'oeuvre d'Ilya Prigogine. Edited by J. P. Brans, I. Stengers & P. Vincke. Patino, 1983.
———. "Why Irreversibility? The Formulation of Classical and Quantum Mechanics for Nonintegrable Systems. *International Journal of Quantum Chemistry* 53 (1995): 105–118.
PRIGOGINE, I. & I. STENGERS. *Entre le Temps et l'Éternité.* Paris: Fayard, 1988. (Paris: Coll. Champs, Flammarion, 1992.)
———. *La Nouvelle Alliance. Métamorphoses de la Science.* Paris: Gallimard, 1979. (Paris: Coll. Folio, 1986.)
———. *Order out of Chaos.* London: Heinemann, 1984.
PUTNAM, H. "The 'Corroboration' of Theories." In *The Philosophy of Karl Popper.* Edited by P. A. Schilpp. La Salle, IL: Open Court, 1974.
REICHL, L. E. *The Transition to Chaos in Conservative Classical Systems: Quantum Manifestations.* New York, NY: Springer, 1992.
RUELLE, D. *Chance and Chaos.* Princeton, NJ: Princeton University Press, 1991.
———. "Cracks in the Glass Menagerie of Science. *Physics World*, December 1994.
RUSSELL, B. *A History of Western Philosophy.* London: Allen & Unwin, 1946.
SCHRÖDINGER, E. "Irreversibility." *Proceedings of the Royal Irish Academy* A53 (1950): 189. (Reprinted in P. T. Landsberg, *The Enigma of Time.* Bristol: Adam Hilger, 1982.)
SHÜTZENBERGER, M.-P. "Les Failles du Darwinisme." *La Recherche*, Janvier 1996, p. 87.
SCHWARTZ, J. T. "The Pernicious Influence of Mathematics on Science." In *Logic, Methodology and Philosophy of Science.* Edited by E. Nagel, P. Suppes & A. Tarski. Stanford, CA: Stanford University Press, 1962.
SERRES, M. *Eclaircissements: Cinq Entretiens avec Bruno Latour.* Paris: Fr. Bourin, 1992.
SINAI, Y. G. *Topics in Ergodic Theory.* Princeton, NJ: Princeton University Press, 1994.
SPOHN, H. *Large Scale Dynamics of Interacting Particles.* Berlin: Springer, 1991.
STENGERS, I. "*L'Invention des Sciences Modernes.* Paris: La Découverte, 1993. (Paris: Coll. Champs, Flammarion, 1995.)
STOVE, D. *Popper and After. Four Modern Irrationalists.* Oxford: Pergamon Press, 1982.
THIRRING, W. "Boltzmann's Legacy in the Thinking of Modern Physics. Vienna: Preprint, 1994.
THOMPSON, C. J. *Mathematical Statistical Mechanics.* Princeton, NJ: Princeton University Press, 1972.
TOLMAN, R. C. *The Principles of Statistical Mechanics.* London: Oxford University Press, 1938.
TRIBUS, M. *Boelter Anniversary Volume.* New York, NY: McGraw-Hill, 1963.
VAUCLAIR, S. *Eléments de Physique Statistique.* Paris: Interéditions, 1993.
WEINBERG, S. *Dreams of a Final Theory.* London: Vintage, 1993.
WOODS, A. & T. GRANT. *Reason in Revolt.* London: Wellred Pubs, 1995.
ZERMELO, E. *Wiener Annalen* 57 (1896): 485.

# HEALTH

NO BRANCH OF the natural sciences is so much upbraided in our time as biomedicine. Sharing this dubious honor with basic biomedical research is, of course, its application: the art of clinical medicine. In this regard, current science criticism commits its usual solecism of failing to distinguish research within strict scientific protocols from practical methods that necessarily include rules of thumb. Even worse, it now finds endemic sins and errors, including the demon Eurocentrism, in both medical practice and the rigorous science upon which it increasingly relies. Attacks now come not only from the traditional loci of obscurantism and quackery, but also from proliferating organs of New Age and postmodernist sentiment—which include, sadly, large segments of the mass media.

If ever there was a visible flight from science and reason, it is to be found in the rapid shift (now manifest, to some extent, even within the healing professions themselves) from rejection of quackery to acquiescence and accommodation. A few years ago, only Saturday Night Live comedians would have featured a shamaness gesturing above the head of a patient in open heart surgery. Today, such a spectacle is on view at Columbia's College of Physicians and Surgeons.

In his essay *Sucking with Vampires*, Gerald Weissmann recalls James Russell Lowell's loathing of superstition. Weissmann demonstrates that today's flight into medical unreason has a long, if not honorable, history within our national culture. The flirtation of NIH (and some medical schools) with the "alternative" (as seen on public television), with the babbling of "holistic" practitioners, is the echo of a long and dubious tradition; and the arguments set forth against it more than a century ago are just as valid today.

Wallace Sampson is a physician with long and intimate experience of "alternative medicine," its claims, its crooks, its depredations, and its methodological irresponsibility. He shows how the alternative, with the aid of wordsmiths, can be made to appear democratic, innocent, and hopeful, while scientific medicine is assigned the role of clown or villain, clueless, stodgy, and viciously protective of its turf. Traditional devices for peddling this point of view are now augmented by the trendy relativism, easily at hand thanks to the postmodern academy, that is examined elsewhere in this volume.

Barbara Held's findings are perhaps the most surprising. It was once the universal assumption that, whatever the idiosyncracies of training and doctrine, the psychotherapist maintained a strong sense of the gap between delusional systems and a healthy reality principle. Nowadays, many therapists, echoing the antirealism of academic theory, hold that all narratives are equally valid, that the healer is under no obligation to correct delusions, but ought to make them more comprehensive and satisfying to the patient. It is evidence from the real world that may now be stigmatized. Held demonstrates how psychotherapy damages its patients and its own future by consenting to this.

# "SUCKING WITH VAMPIRES"
## The Medicine of Unreason

GERALD WEISSMANN

*Credulity, as a mental and moral phenomenon, manifests itself in widely different ways, according as it chances to be the daughter of fancy or terror. The one lies warm about the heart as Folk-lore, fills moonlit dells with dancing fairies, sets out a meal for the Brownie . . . and makes friends with unseen powers as Good Folks; the other is a bird of night, whose shadow sends a chill among the roots of the hair: it sucks with the vampire, gorges with the ghoul and commits uncleanliness with the embodied Principle of Evil, giving up the fair realm of innocent belief to a murky throng from the slums and stews of the debauched brain.*

—JAMES RUSSELL LOWELL, *Witchcraft*

THREE YEARS AFTER he delivered his majestic *Commemoration Ode* at Harvard, in 1868, to mark the end of the Civil War, James Russell Lowell wrote a major analysis of the Salem witch trials. The Union had been preserved, Lincoln was enshrined, and abolition a fact. But progress and reason had enemies remaining. Lowell, the Edmund Wilson of his era, warned New England against the twin errors of sect and superstition. In 1692, thirty women of Salem had been convicted of witchcraft and twenty put to death in an episode connected in some fashion to intense village factionalism. Young Ann Putnam, the poor slave girl Tituba, and a score of other defendants in the witchcraft trials confessed to out-of-body experiences, the sightings of great balls of fire, to having visions of multitudes in white glittering robes, and the sensation of "flying out of body." Their reports mirrored those of the credulous in Lowell's postwar Boston, abuzz with "rappings, trance mediums, the visions of hands without bodies, the sounding of musical instruments without visible fingers, the miraculous inscriptions on the naked flesh, the enlivenment of furniture."[1]

Word for word, the out-of-body experiences described by "witches" in Salem and psychics in Boston overlap with those prescribed as cures for patients by the alternative healers of the 1990s. And thanks to the antinomian sentiments of modern universities, some of these healers are members of medical school faculties in social medicine or psychiatry—as are their colleagues who accept the claims of abduction by aliens in spacecraft. One charismatic healer from the University of Massachusetts, son of a distinguished member of the National Academy of Sciences, describes wellness as the disembodied sensation of "riding the waves, much as if I were lying on a rubber raft on the ocean and the waves were picking me up and taking me down. Lifting and falling away. Lifting and falling away."[2] The patient would have been in the dock in Salem.

A century and a half ago, Lowell was encouraged that a stiff dose of *veritas* would overcome vulgar belief. The two aspects of credulity, fancy and terror, would, he believed, vanish among the educated classes as reason was joined to reform. He announced, prematurely it turns out, that "such superstition as comes to the surface nowadays is the harmless [enthusiasm] of sentiment, pleasing itself with a fiction all the more because there is no more exacting reality behind it to impose a duty or demand a sacrifice."[3] We might note that his critique of the sentimental was later endorsed by Lionel Trilling in *Sincerity and Authenticity.* Commenting on the curious notion of R. D. Laing, Michel Foucault, *et al.* "that madness is health, that madness is liberation and authenticity," he pointed out that those who believe in authenticity, the antinomians, "don't have it in mind to go mad, let alone insane—it is characteristic of the intellectual life of our culture that it fosters a form of assent which does not involve actual credence."[4]

In our New Age of Unreason, we seem to have gone one step further. There is, of course, precious little intellectual life left in our culture, and the credence we pay is to self. Conversations from Berkeley to the beltway, from the boutiques of Madison Avenue to the Bridges of Madison County, avoid the sincere to boast the authentic. We will brook every compromise of conscience, but not of coiffure. A recent item in the "Style" section of the *New York Times* described services offered by a fashionable hair salon—*Nymph*—in New York.

> In addition to cutting, coloring, perming and shampooing, *Nymph* offers some New Age extras. Clairvoyant counseling with Elaine Woodall ($75 for a one-hour session), astrological consultations with Shelly Ackerman ($65 an hour) and dream workshops with Ann Twitty (no charge) are available by appointment. The scalp massage, though, may be the most therapeutic of all the services.

The *elan vital* of *Nymph* and the rest of the narcissistic mind/body movement is clearly that of Mesmer's fluid energy, of animal magnetism. On this Mesmeric note, it may be useful to examine a prime document of the New Age movement. Bill Moyers' *Healing and the Mind*, book and TV series, are models of Mesmer redux—or Trilling's prediction gone mad. In fact, the Moyers tome is more a product than a book; in years past, one called these volumes "ooks." The Moyers ook nestled happily near the top of the *Times* best-seller list for some forty weeks, cheek to jowl with a Jungian analyst's *Women Who Run with Wolves* and Rush Limbaugh's *The Way Things Ought to Be.* Graven on that tablet of unreason were the left and right united, boys and girls together, tripping the light fantastic on the bookshelves of New York.

Moyers's volume consists of a series of interviews with fifteen well-respected doctors and scientists who are united in the belief that the conventional medicine of penicillin and polio vaccines has for too long denied that the mind can heal the body. Those interviewed include proponents of traditional Chinese medicine, relaxation and meditation practitioners, imaging and stress-reduction specialists, group therapists, psychoimmunologists, general internists, behavioral pediatricians, and some laboratory scientists. Most direct, or work at, "centers" and "institutes"; some of these appear reputable. On the whole, the doctors are rational, devoted, and intelligent and Moyers elicits their views in the cozy, undemanding style of the Public Broadcasting System, a network that specializes in prime-time nature study. When the more self-referent of these healers become extravagant or grandiose in their claims, he makes sure that they err in the direction of Jeffrey Masson and not of David Koresh. The index has twelve references to "empowerment," and thirteen to "herbal medicine" but not one to "science," or "physiology," and the sampling offered of the fine arts is similarly weighted in favor of the occult. The text is of such minimal expectations that it refers to "a sociologist called Goffman" or an "Italian philosopher called Bruno." The book cannot be used as evidence that medicine is—or ever was—a learned profession. The healers seem to have suffered from ski injuries but not from all-nighters with dead white European males: they refer mainly to each other's work and their own exercise rituals. The few poems and literary quotes offered by these practitioners of New Age healing reveal most of them to be wildly uncultivated. Occasionally they bend the knee to elders such as Max Lerner or Norman Cousins. Mother Teresa comes up a lot. But, a tone of sweetness and light dominates the conversation. Amiably, these folks agree without dissent that the kind of scientific medicine we practice at American teaching hospitals like the Massachusetts General Hospital or Bellevue is suddenly inadequate to the needs of a New Age.

Are they right? Does the mind influence the body? Of course it does. This is not, however, great news to any of us. Doctors and patients alike have known since Hippocrates that we get ulcers when upset, break out in a sweat when afraid, suffer headaches when angry, and blush when embarrassed. Grief kills. The real question is whether these cause-and-effect relationships between mind and body require for their explanation the orphic insights of New Age meditation. One is forced to say—and to their credit many of Moyers' healers *do* say it—that the jury is still out. Were the Food and Drug Administration to have set hurdles for the approval of these unconventional medical practices as high as those now set for treatments of the conventional kind, the unconventional practices would fail to clear. All of these "treatments" are—at best—arrested at the level of what pharmacologists would call phase II trials, i.e., *before* they have been proved safe and effective by means of randomized, double-blind studies in humans.

The level of proof that is accepted by Moyers *et alii* can be judged from the following study in Los Angeles. Professional, "method" actors were asked to mimic acutely happy or sad experiences and the psychoimmunologists counted blood cells (N-K cells), which they believe to constitute the first line of defense against disease:

> DOCTOR: Then, when the actor got into that state, we looked for changes in the immune system. For example, we asked each actor to imagine that he had been rejected for a part. He began to feel very intense sad feelings. We found that

during the intense sad state there was an increase in the number of natural killer cells [N-K cells] in the actor's bloodstream, and that these killer cells were functioning more efficiently than they were when the actor was in a neutral state.

MOYERS: So when the actor became sad, *something healthy happened* [italics added] . . . [But, you also] studied each actor as he imagined he got the part and was a big success on opening night.

DOCTOR: We found that the effect of the happy state on the immune system was very similar to what we had seen as a result of the sad state.

MOYERS: What seems certain from your research is that short-term emotions do affect our immune system.[5]

What may seem certain to the uninitiated is that method acting affects the traffic of N-K cells. Besides, most card-carrying immunologists are by no means persuaded that N-K cells constitute the first line of defense against infection. That function is served by cells called neutrophils.

But, if emotions *do* influence body functions, how do the New Age healers restore what one of them calls "right inward measure"? Moyers describes a stress-reduction session at the University of Massachusetts in the course of which he underwent a "body scan," a procedure designed to promote self-awareness or -arousal:

MOYERS: What exactly is happening during the body scan?

THERAPIST: I don't know what's happening physiologically because we haven't wired up people and watched them go through this. But the chances are that they're learning how to relax and dwell in the present moment. In the body scan, you lie on the floor, and without moving begin by directing the focus of your attention to the toes of your left foot, then, gradually, up through your leg and over to the other toes, and the other leg, and eventually through the whole body.

MOYERS: That actually happened to me both times that I experienced the body scan with you. The first time I'd been up all the previous night, and this morning I had flown ten hours the day before to get here, so I came to both sessions really dragging my body like a sack of potatoes. But something happened during the body scan. I don't know what to call it, because physically, I haven't really had that kind of experience before. *Perversely, during the sitting meditation, I got agitated, and I had to get up and leave* [italics added]. That's happened to me in meditation before. The body scan worked for me, but the meditation didn't.[6]

"Healers" would call Moyers' agitation a crisis, and Moyers explains to the therapist the reasons for his agitation.

I thought the explanation for the difference I experienced was that you've got a good bedside manner, and during the body scan, your gentle soothing voice was a friendly ally in my descent into the body. But in the sitting meditation, you withdrew, and left me sitting there by myself.[7]

Well, one doesn't require training in dynamic psychiatry—or dialectical materialism—to figure out why a premature withdrawal by a "friendly ally" in a "descent into the body" might provoke a crisis in the best of us. Moyers describes the beginning of one of these sessions:

They are your everyday, garden-variety Americans, twenty-five of them, and they are sitting in a circle on the floor, their legs tucked beneath them, or on folding chairs. Their eyes are closed. They are eating raisins. Three raisins each, one at a time. "S-l-o-w-l-y," says the man in the center. "Lift one raisin slowly to your mouth. Chew it v-e-r-y s-l-o-w-l-y." He pauses and surveys the circle. "Observe your arm lifting the raisin." Pause. "Think about how your hand is holding it." Pause. "Savor it." Pause. "Notice any thoughts, either negative or positive, that you have about raisins." Pause. "Pay attention to your salivary glands." Pause. "Your jaws." Pause. "Your teeth." Pause. "Now notice your tongue as you slowly, s-l-o-w-l-y, swallow the raisin."[8]

Now, some doctors might regard unconventional medical practices of this sort as harmless shenanigans while others might recommend them to patients suffering from functional complaints of bowel or back. But unconventional medicine is on the rise, and clearly meets the needs of those who cannot gain comfort from pills or conventional medical advice alone. In a recent study, carried out by (among others) two of Moyers' interviewees, Drs. David Eisenberg and Thomas Delbanco,[9] one in three respondents in a nationwide telephone survey reported using at least one unconventional therapy in the last year (acupuncture, chiropractic, herbal medicine, spiritual healing, relaxation techniques, etc.). In fact, the estimated number of visits to unconventional medical providers (425 million/annum) exceeded those to all United States primary care physicians (388 million). The estimated cost, $10.3 billion/annum out of pocket compares to $12.8 billion spent out of pocket annually for all hospitalization in the United States!

Bill Moyers's experience with the raisin eaters may be slightly absurd, but it is possible to argue that the whole mind/body movement has a darker precedent: a recurrent *fin de siècle* preoccupation with the supernatural that culminated in seventeenth century witchcraft trials, eighteenth century Mesmerism, and nineteenth century "psychical" research. In our century, Moyers's interviewees also seem to believe that "energy" and its focus by "mind" are the keys to healing.

DOCTOR 1: Health is a dynamic energy flow that changes over a lifetime.[10]

SCIENTIST: The mind is some kind of enlivening energy in the information realm throughout the brain and body that enables the cells to talk to each other and the outside to talk to the whole organism.[11]

DOCTOR 2: I think, ultimately, everything is a different form of energy. Even matter that seems solid as a rock is energy. We know from Einstein that energy and matter are interconvertible. Now, what I find interesting is that when you can focus energy, you gain more power, for better and for worse. . . . But we can use that same principle in a healing direction rather than in a harmful one by learning to concentrate mental energy.[12]

In that Einsteinian context, I note that our local, trendy workout club in New York features a logo composed in equal parts of a Bohr-model atom and a yin/yang circle. Beneath it, on the T-shirts of the fit airheads who worship there, is stenciled the equation $e = mc^2$. When asked what that means, the workout waifs avow that "energy equals motivated conditioning." In Moyers's book, there are seven references to yin/yang theory, twenty-two to "energy," six references to "motivation," and two whole chapters devoted to "conditioning!"

The *reductio ad absurdum* of such sentimental New Age fancies is the celebrity of Marianne Williamson. This guru of spiritual healing has written an ook entitled *A Return to Love* (twenty-nine weeks as a best seller) and then hit the top of the "Advice, How-to and Miscellaneous" best seller list with another called *A Woman's Worth.* Williamson, in *Return to Love*, advised AIDS patients to write a letter to their disease as part of the "imaging" techniques so popular on the West Coast. She has counseled the deeply troubled spirits of Elizabeth Taylor, Oprah Winfrey, Judy Collins, and Mike Nichols. As reported in the *New York Observer*,[13] she disclosed in the course of a book tour a novel remedy of unconventional medicine. Speaking to an audience that crowded *Episode*, a chic Madison Avenue clothing boutique, she allowed that: "There's something interesting about speaking in a clothing store. I love clothes and I understand the healing power of shopping." Now, there's a form of alternative medicine that should drive the health care budget northward by a bundle!

Another guru to the stars has not hit the best seller lists yet, but Ms. J. Z. Knight of Yelm, Washington did grab the attention of the front page of the *New York Times* a few years ago. It seems that Ms. Knight—"the most popular of the New Age spirit channelers"—is being sued by her fifth (!) husband who argued that he was stiffed in an alimony settlement. It's a real Perry Mason case. Indeed, were he in the fragrance mood, Perry might call this "The Case of Channeler's No. 5." At any rate, Jeffrey Knight pleaded that he helped his wife transform herself from the middle-class wife of a Tacoma dentist into "a spiritual guru who lives in a two-million-dollar ranch house." Now he is dying of AIDS, she did not heal him or help him, and he's back to conventional treatment, AZT.

Ms. Knight, onetime advisor to the most spiritual of our film stars, Shirley MacLaine, gets $1,000 per trance, collected $100,000 from one troubled woman for "books, tapes, and survival gear," and had an annual income of up to $4 million. But Ms. Knight's power to heal remains unmanaged by the feds or the Pru while her diagnostic insight is guaranteed by a 35,000-year-old warrior named Ramtha. This spirit voice began speaking through her one day in 1977 after Ms. Knight had floundered through four earlier marriages and sporadic attendance at a beauty school.

Now all this would be simple fun and games, were it not for more discouraging implications of New Age healing. Spurred on by proponents of homeopathy, meditation and spiritualism in the Congress, the National Institutes of Health (NIH) has set up an Office of Unconventional Medical Practices (UMP) with funding recommended for 1995 at $6 million per annum. The NIH called a conference to clarify the goals of this office and convened many of the same experts (sic) in the fields of ayurvedic, naturopathic, chinese herbal, and homeopathic medicine who turned out to be the interviewees of Moyers' miniseries and tie-in book. This coven of experts went through the motions of proposing "multicentered, large-scale, definitive controlled trials of UMP's." They agreed that many of the unconventional methods have "associative belief systems" (twelve references in Moyers). Well, of course they do: one calls those belief systems "religions" and I am not sure that one wants NIH in the business of judging which religions work better than others in definitive, controlled trials.

As for those multicentered large-scale trials to be planned by the UMP office, one might direct the ayurvedic folk to the morbidity and mortality tables of rural India, Africa, and China today, the church records of Salem, Massachusetts of 1693, the life expectancy of Parisians in 1788, or the bills of mortality of Cambridge, Massachusetts, 1891. Before scientific medicine interfered with traditional practices, few folks lived to be as old as the youngest United States senator. Nor can the sorry public health statistics of our inner cities be attributed to inadequate access to folk remedies or group meditation. To paraphrase Lincoln Steffens: we have seen the future of unconventional medical practices and they do not work. Meanwhile, in the most important aspect of medical practice, the public health, young doctors are sweating over research grants that remain unfunded because an alliance of homeopaths and New Age toe ticklers seems to have gotten hold of a dotty Senator or two.

The Salem of Ann Putnam, the séance parlors of Beacon Hill, and today's yuppie ashrams share a common belief. Accused and accuser, seer and seen, healer and healed are convinced that magic rules the body. Generally there is money to be made from that conviction. Healing and the body, healing and the mind, healing and. . . . One of Moyers's healers reminds us that the word healing relates to making things whole. Yes, it does, but my OED informs me that its ultimate origin is Old Teutonic and in German the salutation is *Heil*, as in *Sieg Heil!* Another motto of national socialism was *Blut und Boden*, which referred to natural, folk-based German customs based on blood's affinity with the earth, an alliance that the Nazis believed had been broken by "Western cosmopolitanism." Natural medicine, based on principles of *Blut und Boden*, has had its own history of clinical trials. In their recent study, *Cleansing the Fatherland: Nazi Medicine and Racial Hygiene* (1994), Christian Pross and his colleagues present newer evidence of "the preference of early Nazi health policy for holistic medicine and natural healing over decadent, Jewish [read scientific] medicine" to the skeptical tradition of Western reason.[14] The Nazis based their movement on a popular resentment of established political parties, appeals to national honor and "family values" fueled by an open antipathy to "cultural elites."

We are told frequently, nowadays, that the expensive, autocratic, medical science of our day has substituted its own elite values for those that would better serve one or another of our subcultures. It seems to me that I have heard that song before. It is from an old familiar score. Dr. Karl Gebhard, Supreme Clinician (sic) to the SS, told the Nuremberg Tribunals that

> what the National Socialists wanted to do was to introduce a popular medicine, [they] had little regard for scientific medicine, and they were all attracted by natural medicine. All sorts of popular drugs which were not approved by the medical profession allegedly because we did not understand them or were too conceited or were financially interested in the suppression of them, were used experimentally in concentration camps. . . . The source of these experiments was Himmler's conception of medicine as pure mysticism.[15]

Of course the politics of the spiritualists and New Age healers are not those of the national socialists, but once the restraints of reason are cut, you leave public life open to takeover by Lowell's murky throng from the slums and stews of the debauched brain. The heirs of Mesmer may not only be setting out a meal for the Brownie, but inviting the vampires of Salem to suck. Much of the mind/body movement, of the workout and meditation culture, remind one of the brownshirted *Kraft durch Freude* adventure with its aim of increasing collective bliss at the expense of social justice.

William James, the finest writer ever to have come out of the Harvard Medical School, believed at the end of his century what the New Age healers believe at the end of our own: that the "thunderbolt has fallen and that the orthodox belief in reductionist science has not only had its presumptions weakened, but the truth itself [has been] decisively overthrown."[16] Well, not really. Reductionist science in James's own field of medicine and physiology has had a decent run since its "overthrow" by the Boston psychics. The sanitary revolution of Lowell's time was followed by the bacteriologic revolution of James's, and this in turn was succeeded by the biological revolution of our day, which I have called the flowering of DNA. The results are easy to quantify: In 1920, at the end of the bacteriologic revolution—and before the discovery of antibiotics—the average life expectancy in the United States was 53.6 years for males and 54.6 for females. By 1990, the life expectancy of males had increased to 71.8 and females to 78.8. There is no evidence that between 1920 and 1990 intervention from the spirit world had increased, or—James and the spiritualists to the contrary—that the "truth itself" has been decisively overthrown.

The proper response to all this antinomian nonsense was given by Dr. Oliver Wendell Holmes, author, poet, and dean of the Harvard Medical School. Holmes retorted to the alternative medical practices of *his* day in *Homeopathy and Kindred Delusions*:

> As one humble member of [the medical] profession, which for more than two thousand years has devoted itself to the pursuit of the best earthly interests of mankind, always assailed and insulted from without by such as are ignorant of its infinite complexities and labors, always striving in unequal contest with [disease] not merely for itself but for the race and the future, I have lifted my voice against this lifeless delusion, rolling its shapeless bulk into the path of a noble science it is too weak to strike, or to injure.[17]

## NOTES

1 James Russell Lowell, *Literary Essays*, vol. 2, 397.
2 Bill Moyers, *Healing and the Mind*, p. 136.
3 Lowell, *Literary Essays*, vol. 2, p. 317.
4 Lionel Trilling, *Sincerity and Authenticity*, p. 171.
5 Moyers, *Healing and the Mind*, pp. 196–197.
6 Ibid., pp. 131–132.
7 Ibid., p. 133.
8 Ibid., p. 66.
9 David Eisenberg *et al.*, "Unconventional Medicine in the United States."
10 Moyers, *Healing and the Mind*, p. 130.
11 Ibid., p. 189.
12 Ibid., p. 105.
13 *New York Observer*, May 24, 1993, p. 19.

14 Götz Aly, Peter Chroust & Christian Pross, ***Cleansing the Fatherland: Nazi Medicine and Racial Hygiene***, p. 9.
15 R. E. Conot, ***Justice at Nuremberg***, p. 283.
16 William James, ***Essays on Faith, Ethics and Morals***, p. 142.
17 Oliver Wendell Holmes, ***Medical Essays***, p. 98.

## REFERENCES

ALY, GÖTZ, PETER CHROUST & CHRISTIAN PROSS. ***Cleansing the Fatherland: Nazi Medicine and Racial Hygiene.*** Baltimore, MD: Johns Hopkins University Press, 1994.
EISENBERG, DAVID *et al.* "Unconventional Medicine in the United States." ***New England Journal of Medicine*** 328, 4 (1993): 246–252.
FOUCAULT, MICHEL. ***The Birth of the Clinic: An Archeology of Medical Perception***, translated by A. M. S. Smith. New York, NY: Random House, 1973.
CONOT, R. E. ***Justice at Nuremberg.*** New York, NY: Carroll and Graf, 1984.
HOLMES, OLIVER WENDELL. ***Medical Essays.*** Vol. 9 of ***Collected Works.*** 13 vols. Boston, MA: Houghton Mifflin, 1892.
JAMES, WILLIAM. ***Essays on Faith, Ethics and Morals.*** New York, NY: New American Library, 1974.
LOWELL, JAMES RUSSELL. ***Literary Essays.*** Vols. 1–4 of ***Collected Prose Works.*** 6 vols. Boston, MA: Houghton Mifflin, 1899.
MOYERS, BILL. ***Healing and the Mind.*** New York, NY: Doubleday, 1993.
TRILLING, LIONEL. ***Sincerity and Authenticity.*** Cambridge, MA: Harvard University Press, 1971.

# ANTISCIENCE TRENDS IN THE RISE OF THE "ALTERNATIVE MEDICINE" MOVEMENT

WALLACE SAMPSON

PROPAGANDA, CULTURAL RELATIVISM, and other postmodern doctrines that challenge objectivity—"deconstruction," for instance—are integral to today's antiscientific thinking. They are linked in a cable of thought processes that is slowly strangling medical science. They are the software through which the hardware of ineffective remedies convinces and converts.

It may seem forced to identify propaganda, developed to its highest form by Hitler and Goebbels, with cultural relativism, which eliminates prejudice from the study of foreign cultures. But they are both responsible for much current medical mischief, from quackery and Laetrile to the creation of an Office of Alternative Medicine in the National Institutes of Health. Postmodernism, a more abstract and esoteric phenomenon, takes cultural relativism one step further into a disorienting world with no standards of accuracy or ethics.

## PROPAGANDA

The word stems from the College of Propaganda, formed under Pope Urban VIII in the 1600s to educate priests for missions—the propagation of the faith. Propaganda's secular and political applications are relatively recent. Now the term refers to the intentional and at times cynical manipulation of language to reconstruct perceived reality, thence to alter beliefs, with intent to alter and control behavior. It is the most effective and pervasive of the three phenomena I propose to analyze.

Samuel Hahnemann coined the words "homeopathy" and "allopathy" two hundred years ago. "Homeopathy" described his philosophically based method of treating disorders with materials that in larger doses produce the same (homeios) symptoms, in the hope of strengthening the body's reaction against disease. Hahnemann successfully branded all of medicine as allopathic (meaning treatment with agents that oppose the symptoms), despite the fact that the allopathic method was only part of what physicians used. Scientific medicine seems to be stuck with the term, which now carries negative connotations. These are conveyed by the slogan "treating only the symptom, not the cause," used by holistic and alternative medicine proponents since the 1970s. The slogan reduces medicine to the level of treating typhoid fever with aspirin.

In the 1960s and 1970s, Laetrile inventor Ernst Krebs, Jr. began similar wordplays in his contest with the Food and Drug Administration (FDA) and the courts over his two invented "vitamins"—Laetrile (B17) and Pangamic acid (B15). Laetrile came from apricot and other fruit pits containing cyanogenic glycosides, such as amygdalin, linamarin, and prunasin. They are five-to-six percent cyanide by weight. They occur naturally in other foods, such as maize, taro root, and cassava—staples in many parts of the world. But they cause thyroid dysfunction, blindness, and paralysis from chronic cyanide poisoning in cattle and humans, and deaths from acute cyanide poisoning.[1]

Krebs labeled these compounds "nitrilosides" (a contraction of "nitrile," the cyanide-containing moiety, and "glycosides," the sugar moiety). "Nitriloside" eliminated "cyanogenic" (cyanide-forming), which would have given anyone pause before ingesting it. He labeled the final product "Laetrile," a euphonious, appealing name.

When the FDA began seizing Laetrile and bringing fraud charges for false claims, Krebs renamed it a vitamin, "B 17," and called it a food supplement.[2] This ploy did not work with the FDA, but it did with customers and followers. As fraudulent as Laetrile is, it still flourishes in Tijuana clinics for gullible, desperate Americans and Canadians (native Mexicans rarely patronize those clinics).

More recently, Laetrile was pushed and sold by the Committee for Freedom of Choice. "Freedom of Choice" became a slogan that removed the perception of quackery from ineffective cancer remedies and introduced the principle that people should have freedom to select their own treatments. Supporters successfully fought FDA and state challenges to marketing of Laetrile on the basis of freedom of choice. So successful was the freedom of choice ploy that twenty-seven states legalized Laetrile, although with some restrictions. The states and the courts ignored laws against fraudulent claims and sales, granting freedom to defraud instead. "Freedom of choice" remains a rallying slogan for the "health foods" industry and "alternative medicine" supporters.

Other groups rallying support were the Cancer Control Society and the International Association for Cancer Victims [now Victors] and Friends. They crafted words and names to change fraud to freedom and charlatans to friends. Laetrile organizations created mistrust and ridicule of medical professionals, describing surgery, radiation, and chemotherapy as "slash, burn, and poison." Promoters also used the terms "medical establishment" or "medical monopoly" to deride scientific, effective treatment. "Go to Health, FDA," read one bumper sticker. Likewise, they used "metabolic therapy" (Laetrile plus special diet and vitamins) and "nontoxic therapy" to describe their own ineffective methods.

In the 1960s and 1970s Linus Pauling invented "orthomolecular." He defined it as "the right molecule in the right place at the right time."[3] Dr. Thomas Jukes described orthomolecular as a slogan, not a treatment. Dr. Victor

Herbert compared Pauling's definition to Humpty Dumpty's remark in *Through the Looking Glass*: "When I choose a word, it means just what I choose it to mean, nothing more nor less." Dr. Pauling had little idea which molecules should be where or when, or in what concentration. He advocated that everyone take massive doses of vitamins daily to cover all bases, finally concentrating on vitamin C.[4] Orthomolecular theories and claims were demolished by a series of research projects on the common cold, cancer, and mental illness in Canada and the United States, but they are still believed by many.

In the 1970s, advocates of the holistic movement perpetrated another exercise in linguistics. In this concept, rational biomedicine was labeled "reductionistic" and its rational thinking excessively "linear." Unscientific thinking was rechristened "nonlinear," and thus more modern, like non-Euclidean geometry. The term "holistic" was appropriated and advertised as more inclusive, inserting into the medical equation such concepts as spirit, mind, and consciousness. Adding such concepts effectively diminished the status of measurement and rationality. One can now frequently find in the popular literature the consequent demeaning of rationality and the "left brain," along with a parallel elevation of the importance of subjectivity and the "right brain."

The more recent alternative medicine movement obscures quackery and fraud with euphemisms such as "alternative" itself. Now a dictionary definition of "alternative" is "a choice limited to one of two or more possibilities." Implied is that either possibility will obtain the same or a similar result. "Alternative" has not been defined clearly by proponents, but one author recently defined "alternative medicine" as anything not taught in medical school or paid for by insurance.[5] In fact, this implies that there is a rather big difference indeed. As a general rule, medical schools try to teach students to do things that are effective.

Here is a partial list of terms used by alternative medicine pseudoscience promoters:

| | | |
|---|---|---|
| alternative | Eastern medicine | metabolic therapy |
| unorthodox | life-force | nontoxic |
| unconventional | spirit | homeopathic |
| nontraditional | healing | essence |
| complementary | power | paradigm change |
| holistic | natural | energy |
| orthomolecular | organic | |

Herein, alternative medicine has created a dictionary of the absurd. None of the above terms is used in its common, dictionary meaning. The first five are euphemisms, disguising methods that are actually unproved, disproved, erroneous, anamolous, ineffective, or simply fraudulent. The Office of Technology Assessment (OTA) commandeered "unconventional" to describe ineffective cancer therapies in its report to Congress.[6] This several-year review of what was until then considered cancer quackery gave a tenuous but legitimized reprieve to worthless methods and their promoters.

One recommendation in the OTA report was for a "best case review," in which the most impressive cases would be reviewed by unspecified evaluators to answer the question of whether or not an "unconventional" method deserves further research. The report went on to suggest that "one element that may be crucial to the success of a best case review is the active participation, or at least support, of the unconventional practitioner."[7] In other words, if one were unqualified, or a charlatan, or convicted of fraud, one could be recruited to evaluate one's own fraudulent activity. This concept has persisted in the Office of Alternative Medicine, so that, with one exception, its advisory committee is made up exclusively of quackery supporters.

When the Clinton health plan was under construction, proponents lobbied Hillary Clinton's committee heavily to integrate their methods into the program so they could be paid for their services. Supporters have now settled on the word "complementary" to avoid accusations of diverting patients from effective care. However, a tour of the relevant computer forums indicates that escape from the reality of effective care is the main reason for the popularity of "complementary" methods.

"Nontraditional" has been used to signify the opposite of its common meaning. "Traditional," properly speaking, means methods indigenous to a culture, or other than modern and scientific. Revisionists, however, use "traditional" to mean modern biomedicine and "nontraditional" as a milder, innocuous term for prescientific, ineffective methods. "Nontraditional" cleverly evades consideration of treatment effectiveness.

Another technique is to create a false dichotomy, intentionally apposing two contradictory terms, one a euphemism for pseudoscience, its opposite number a pejorative characterization of its scientific counterpart. Thus:

unorthodox/orthodox
nontraditional/traditional
unconventional/conventional
homeopathic/allopathic
holistic/reductionistic
Eastern medicine/Western medicine

"Orthodox" and "conventional" implicitly introduce a subtle jab at the Establishment. "Eastern" versus "Western" is similarly misleading; the real difference is between prescientific methods of both East and West and scientific, rational methods. Science, after all, is the same in the East as in the West.

Lerner, in a recent book on alternatives, not only redefined the words "healing" and "curing," but redefined what physicians do as well.[8] It will surprise most physicians to learn that they cure, but do not heal: "It is a distinction yet to be fully recognized . . . in mainstream American medicine." "A cure is a successful medical treatment . . . that removes all evidence of the disease." "Curing is what doctors hope to do. . . ." "Healing," in contrast, is "an inner process through which a person becomes whole." "Healing can take place at the physical level. . . . It can take place at an emotional level. . . ."[9]

These statements make a certain amount of sense, but if one were to ask the average person the difference between healing and curing, he would most likely find it difficult to specify. One also finds no difference between the two in a dictionary: "heal—to make whole or sound; restore to health; free from ailment; . . . to cure."[10] This makes no mention of emotional healing. The result of the ad hoc redefinition is to assert that only alternatives really heal. Of course, most physicians spend most of their professional time increasing the comfort and extending the lives of people with incurable diseases, while curing few. But Lerner's redefinition conveniently neglects that activity to promote the alternative treatments he supports.

When one reads the holistic and alternative literature, one encounters these redefinitions on almost every page. The reader is often befuddled, not recognizing that this makes sense only when one accepts the rewriting of dictionaries. This, however, is excellent propaganda technique.

## RELATIVISM

In the early twentieth century, social anthropologists developed the notion of analyzing other cultures within a value-free system. This allowed a culture to be seen as a functional whole, without the intrusion of observer prejudices. It permitted a more realistic description of a culture, its mores, and its history, and also resulted in a more tolerant political view of one culture or nation by another. Or so it would seem, ideally. The principle is not easy to adhere to. It was, for instance, challenged a number of years ago by a graduate student doing research in China, who aided women in avoiding state-mandated abortions and sterilizations. He was deported, but also lost his appointment at his university for having violated the ethic of the value-neutral observer. The episode brought into relief the clash of principles between two disjoint functions of the same person; that of the scientific observer of a society and that of humanitarian activist. This conflict seems not to have an easy resolution; each activity has its own social value.

A personal experience brought the paradox to my attention. In the 1970s I studied people who were taking Laetrile, which I knew to be worthless for cancer and a cover business for fraudulent stock sales in Canada and the United States.[11] All thirty-three patients interviewed shared an antagonistic attitude to their physicians and to the "medical Establishment"; believed in a conspiracy against Laetrile by the FDA, the American Medical Association (AMA), and pharmaceutical companies; and had failed to seek out information from the American Cancer Society, universities, medical organizations, or state agencies. All but one continued Laetrile until death. The situation seemed to operate like a religious cult, with cognitive dissonance playing a major role.[12]

Asked for help in developing the thesis and for further investigation, a sociology professor and his graduate students explained to me that, as sociologists, they did not concern themselves with the medical validity of the substance Laetrile, nor with the criminal backgrounds of its proponents. They were concerned only with the interaction of the patients and the system. (Did they have trusting relationships with the Laetrile proponents? Did their chosen medical system function to their satisfaction?) The sociologists' evaluation was value free. The possibility that the patients were involved in a cultlike system and were perhaps shortening their lives was not germane. That the Laetrile claims were fraudulent was something that sociologists could not evaluate because, qua sociologists, they had no more reason to believe a medical investigator than the Laetrile boosters.

The Laetrile situation was evaluated by sociologists at the annual meeting of the American Association for the Advancement of Science in 1979.[13] The authors of the resulting monograph tried to maintain an even-handedness in their treatment of the controversy, perceiving that their role would be prejudiced were they to take sides. Thus, there were comments such as "Critics of Laetrile have asserted that the use of Laetrile is . . . ineffective [and] dangerous," offset by "Laetrile advocates counter these claims with epidemiological data. . . . Hunzakuts, a . . . Pakistani tribe, . . . live to 100 years, and their longevity is attributed to a diet rich in amygdalin. . . ."[14] That Laetrile was part of a swindle, that people had died from cyanide poisoning from apricot pits, and that the Hunzakut data were unreliable could not be acknowledged for fear of bias.

Another sociologist, summing up the discussants' papers, stated, "If Laetrile is a fraud, . . . then one can hardly fault anti-Laetrile bias. But we do not know if proponents are any more fraudulent than . . . promoters of many orthodox cancer remedies. . . ."[15] Perhaps the sociologist could be forgiven his ignorance of medicine, but to take such a position, given the history of Laetrile promoters, is stretching the value-free principle beyond reason! In his conclusion, this commentator wrote: "Any analysis must carry some bias; even neutrality is a bias. . . . It seems to me that any bias will do as well or poorly as another. . . . the analyst ought to consider the degree to which his perceptions and conclusions depend on his particular bias rather than on 'objective fact.'"[16]

Although it is difficult to pinpoint direct examples of value-free analysis and cultural relativism in the advance of pseudoscience in medicine, the footprints can be seen in many places where the validity of information takes second place to emotion and comforting philosophies.

Acupuncturists are now petitioning the FDA to remove the "investigational device" classification from acupuncture needles. With drugs, FDA approval requires clear evidence of safety and efficacy. By way of contrast, FDA officials have told me that despite reported complications and deaths from acupuncture, despite numerous studies demonstrating effectiveness only as a placebo, despite lay acupuncturists being unqualified to diagnose or treat disease, the FDA will probably agree to the acupuncturists' petition. Validity of information is secondary to the philosophical (and economic) wishes of a significant segment of the population. The controlled studies—over sixty on pain alone—might as well not have been done.

Two generations of students are taking their places in society having been educated to see scientific as well as social problems in a content- and value-free way. They seem unable to distinguish fiction from fact, or to evaluate the probabilities of validity. Even when they can detect fiction or estimate probabilities, it seems not to matter much to them.

## POSTMODERNISM

The organizers of the conference on which this volume is based have described the controversies wrought by postmodern, or poststructuralist, approaches to history, politics, economics, sociology, and to much of science.[17] Postmodernism's history is recent, and it has not matured to the point where its terminology and concepts have been co-opted by medical pseudoscience. But it will probably not be long before that happens.

Postmodernism—embodied in deconstruction, among other doctrines—takes relativistic cultural analysis one step further. Value-free cross-cultural analysis states that science has validity within the community accepting its point of view, without necessarily being more valid than pseudoscience. Each position supposedly has its own criteria. Postmodernism, as I understand it, posits that there is not necessarily any validity at all to the scientific view, which is taken to represent the prejudices of a system dominated by European males! The various forms of postmodernism are apparently appealing to the young, who hold firmly and emotionally to antiestablishment views. Consequently, a new generation of students, inculcated with antagonism to science, will soon be taking its place in government, the legal system, journalism, education, and so forth.

It is not inconceivable that postmodernism may already have been influential indirectly. Along with self-interested propaganda and culturally relative social analysis, it may be responsible for much of the current distrust of rationality, a distrust so pervasive that otherwise sensible legislators now consent to funding for the Office of Alternative Medicine. The "consciousness" movement now has a vigorous life of its own, with best-sellers by Deepak Chopra, Bernie Siegel, and Larry Dossey. Despite lancing criticism from a small cadre of scientists, the popularity of such stuff continues to swell.

## MORE WORDS ABOUT ALTERNATIVE MEDICINE

Alternative medicine proponents reject or misuse most criteria for evaluating validity. Their predecessors, the holistic medicine proponents, claim that an adequate view of medicine—i.e., health and disease—must include aspects of mind, spirit, and culture to be complete. Alternative proponents claim that contemporary biomedicine does not include psychological and social matters. They intentionally sequester for analysis one aspect of biomedicine—research—and importune their readers to equate the word "reductionist" with the actual practice of medicine. But the practice of medicine has always been, by its nature, holistic. Alternative medicine advocates conveniently forget that the fields of psychiatry, psychology, social and preventive medicine, and public health are integral parts of modern biomedical practice, as is cooperation between physicians and clergy, as manifested by the presence of chaplains in most hospitals.

Concepts such as mind, consciousness, and spirit do not lend themselves to measurement, thus generating claims that cannot be corroborated or proved. Alternativists take advantage of this; their definitions of "reductionism," as against "holism," render scientific method unappealing, while simultaneously glorifying methods impervious to disconfirmation, at least so far as those who have swallowed the alternative paradigm are concerned.

Of course, supporters of alternative medicine rarely refer to evidence against their methods. For instance, overwhelming evidence shows that acupuncture has only placebo effects, and that homeopathy has no effects. Yet proponents of both alternatives blithely ignored this evidence and convinced Congress to fund a special office to evaluate them. With a single exception, none but proponents were appointed to its two advisory boards; every staff member is a supporter as well. If this happened in another branch of government, there would be an uproar.

Three million dollars per year has been appropriated to support the office and to investigate the methods. All research funds have been awarded to alternative advocates, most of whom have never done credible research. There is no provision for peer review of the work or for its publication. The awards are in essence a political payoff to the well-financed lobby behind the office's creation, and will foster its aims.

Despite having brought pressure for the government to fund research, alternative medicine proponents are now having second thoughts about carrying it out. In the recently launched journal *Alternative Therapies*, the editor inveighed as follows:

> While this double-blind method of evaluation may be applicable to certain alternative therapies, it is inappropriate for the majority of them. Many alternative interventions are unlike drugs and surgical procedures. Their action is affected by factors that cannot be specified, quantified, and controlled in double-blind designs. Everything that counts cannot be counted. To subject alternative therapies to sterile, impersonal double-blind conditions strips them of intrinsic qualities that are part of their power. New forms of evaluation will have to be developed if alternative therapies are to be fairly assessed.[18]

The author ignores the fact that double-blind, controlled studies are not the only form of scientific evidence. Single-case reports ($n = 1$) are valuable if performed properly, and lead to new discoveries. Almost all cancer therapy trials are unblinded, but are still perfectly valid because the patients are properly stratified and randomized. The studies are reproducible in the hands of multiple investigators. The author of the paragraph above goes on for nine pages about the supposed fact that blinded studies cannot measure the effects of initial conditions or of consciousness; yet he offers no substitute system for evaluation. He thus leaves us with this non sequitur: present knowledge is adequate to dismiss the utility of most alternative methods; but there are ineffable qualities that our methods cannot detect and alternatives cannot define; therefore, alternative methods must be accepted, their practitioners licensed, and their services paid for by public funds and health insurance.

## THE PRESS

The press amplifies and exacerbates the problem with its attempts at balance. Its ethic is ostensibly to present objective and balanced articles. But in reality, the technique for reporting medical pseudoscience is to find a proponent or satisfied patient, quote that source for two columns, and then "balance" that encomium by quoting a skeptical physician or scientist for one or two paragraphs. Of course, the piece concludes with a rebuttal of the skeptics by the original proponent. This formula is ubiquitous; it is widely demanded by editors and apparently taught in journalism schools. "Balanced" reporting is thus the journalistic equivalent of the "value-free" cross-cultural analysis technique, and regards respect for scientific fact as mere "bias."

## CONCLUSION

Markle and Petersen described the Laetrile successes of the 1970s thus: "Laetrile advocates . . . have created the most effective challenge to medical orthodoxy in American history."[19] That comment, written in 1979, would now have to be amended. The holistic movement and its successor, the alternative movement, characterized by an even broader, all-inclusive attack on reason, have parlayed techniques of propaganda, academic resentment of science, and dubious philosophical speculations into the most effective assault yet on scientific biomedicine. Its object is probably the appropriation of the political and social power now in the hands of scientists and physicians, and its attendant economic rewards. The battle will be widespread, with universities providing one important arena. Change may take decades to occur. Meanwhile, the scientific and medical communities will be wise to learn from the success of pseudoscience.

## NOTES

1 V. Herbert, *Nutrition Cultism*, pp. 29–32; *Toxicants Occurring Naturally in Foods*, pp. 449–450.
2 M. Culbert, *B 17. Forbidden Weapon against Cancer*, p. 92.
3 L. Pauling, *How to Live Longer and Feel Better*, p. 93; *Cancer and Vitamin C*, pp. 102–103.
4 Ibid.
5 D. Eisenberg *et al.*, *New England Journal of Medicine* 286, 4 (1993): 236–240.
6 Office of Technology Assessment, *Unconventional Cancer Therapies*, p. 232.
7 Ibid.
8 M. Lerner, *Choices in Healing*, p. 13.
9 Ibid.
10 *Random House Dictionary.*
11 *Securities and Exchange Commission vs. Biozymes Intl., Ltd.*, Andrew R. L. McNaughton (and six others), Records of the U. S. District Court, Spencer Williams, Presiding Judge, Docket Records, San Bruno, Calif. Records of the Court of Sessions of the Peace, *Her Majesty the Queen vs. Andrew J. McNaughton*, Province of Quebec, District of Montreal, No. 499-72. W. Sampson, Laetrile Administrative Rule-Making Hearing, Oral Argument, Docket No. 77N-0048, Food and Drug Administration, 1977.
12 L. Festinger, *A Theory of Cognitive Dissonance.*
13 G. Markle & J. Petersen, *Politics, Science, and Cancer*, p. 6.
14 Ibid.
15 Ibid., pp. 175–179.
16 Ibid.
17 Paul R. Gross & Norman Levitt, *Higher Superstition: The Academic Left and Its Quarrels with Science.*
18 L. Dossey, *Alternative Therapies* 1, 2 (1995): 6–10.
19 Markle & Petersen, *Politics, Science, and Cancer.*

## REFERENCES

COMMITTEE ON FOOD PROTECTION. *Toxicants Occurring Naturally in Foods.* Washington, DC: National Academy of Sciences, 1973.

CULBERT, M. *B 17. Forbidden Weapon against Cancer.* New Rochelle, NY: Arlington House, 1974.

DOSSEY, L. *Alternative Therapies* 1, 2 (1995): 6–10.

EISENBERG, D. *et al. New England Journal of Medicine* 286, 4 (1993): 236–240.

FESTINGER, L. *A Theory of Cognitive Dissonance.* Stanford, CA: Stanford University Press, 1957.

GROSS, PAUL R. & NORMAN LEVITT. *Higher Superstition: The Academic Left and Its Quarrels with Science.* Baltimore, MD: Johns Hopkins University Press, 1994.

HERBERT, V. *Nutrition Cultism.* Philadelphia, PA: George F. Stickley, 1980.

*Her Majesty the Queen vs. Andrew J. McNaughton*, Records of the Court of Sessions of the Peace, Province of Quebec, District of Montreal, No. 499-72.

LERNER, M. *Choices in Healing.* Cambridge, MA: MIT Press, 1994.

MARKLE, G. & J. PETERSEN. *Politics, Science, and Cancer: The Laetrile Phenomenon.* Boulder, CO: Westview Press, 1980.

OFFICE OF TECHNOLOGY ASSESSMENT. *Unconventional Cancer Therapies.* Washington, DC: United States Government Printing Office, 1990.

PAULING, L. *Cancer and Vitamin C.* Menlo Park, CA: Linus Pauling Institute of Science and Health, 1979.

———. *How to Live Longer and Feel Better.* New York, NY: W. H. Freeman, 1986.

*Random House Dictionary.* New York, NY: Random House, 1992.

SAMPSON, W. Laetrile Administrative Rule-Making Hearing, Oral Argument, Docket No. 77N-0048, Food and Drug Administration, 1977.

*Securities and Exchange Commission vs. Biozymes Intl., Ltd.*, Andrew R. L. McNaughton (and six others), Records of the U.S. District Court, Spencer Williams, Presiding Judge, Docket Records, San Bruno, Calif.

# CONSTRUCTIVISM IN PSYCHOTHERAPY
## Truth and Consequences[a]

BARBARA S. HELD

THE INFLUENCE of postmodernist ideas on the theory and practice of psychotherapy has been growing exponentially. That influence is typically expressed by way of the philosophical doctrine of constructivism, along with the variant of it known as social constructionism. As core components of most postmodern theories, these doctrines are founded upon the antirealist epistemological supposition that knowers do not discover objective reality as it is; rather, they make or construct their own subjective "realities" in language. These doctrines therefore pay much attention to stories and meanings. In the case of therapy, they direct those of us who are therapists to attend to the personal meanings clients find in their lives or, more true to these doctrines, the stories clients *construct* about their lives.

Although constructivism and social constructionism have received widespread attention in the humanities and social sciences, in my opinion they have not received sufficient critical attention—certainly not within psychotherapy, where their adoption has many real consequences. It is therefore my purpose to consider the true meaning for psychotherapy of constructivism/constructionism.[1] In order to show how an antirealist epistemology indeed defines those two doctrines, I begin by providing capsule definitions of realism and antirealism, after which I give evidence of the antirealist leanings of those therapists who have promoted a constructivist/constructionist basis for therapy. I then consider the implications and consequences of antirealism for therapy in particular and for knowledge in general, with attention paid to the logical problem that arises when therapists use antirealist doctrines to make claims about the real workings of therapy; I explain how the word "construct" is behind much of the confusion. Finally, I offer reasons for the meteoric rise of antirealism in therapy and maintain that the problem generating

[a] An earlier version of this paper entitled "The Real Meaning of Constructivism" appears in the *Journal of Constructivist Psychology*, 8 (1995): 305–315. In both papers I have drawn from my book *Back to Reality: A Critique of Postmodern Theory in Psychotherapy*, New York, NY: W. W. Norton, 1995.

that trend can be solved not by an appeal to antirealism, but rather by a clearer understanding of the nature and use of theoretical systems in therapy. In my conclusion, I suggest an alternative to the postmodern/antirealist trend in therapy.

## THE REALISM/ANTIREALISM DISTINCTION

Although the realist doctrine contains a spectrum of formulations, in general it claims that the knower can attain (some) knowledge of an independent reality—that is, reality that is objective in the sense that it does not originate in the knower, or knowing subject.[2] The expression "independent reality," then, refers to the independence of the thing known from the knower—more precisely, from the knower's cognitive operations and the theoretical, linguistic, or narrative constructions produced by those operations. Thus, for the realist, nothing—no theory, for instance—*necessarily* intervenes between the knower and the known, or between the knower and the (mind-independent) object of the knower's inquiry.

In those cases when the knower does need to use a theory to understand some aspect of reality (e.g., the physics of the very small, or causality in general), that theory does not automatically alter or distort the real/independent reality it purports to explain. Thus, according to the realist doctrine, *knowable* reality—indeed *reality* itself—is not merely a theoretical, cognitive, or linguistic construction on the part of the knower, a construction that tells us something about the knower and the knower's cognitive operations, but little, if anything, about the object of inquiry, that is, about the independent reality to be known. A typical realist epistemology, then, implicitly acknowledges the existence of an independent ontological reality—a real "thing in itself"—that can be known to some extent. In therapy, for example, the therapist can be certain that the person who sits there claiming to be threatened by some life experience and behaving in a way consistent with that claim is not *merely* a by-product of the therapist's own theoretical or linguistic construction made on the basis of some utterly unknowable stimulus.

According to the antirealist doctrine, which also contains a spectrum of formulations of either a more or a less radical sort, the knower cannot under any circumstances attain knowledge of a reality that is objective, independent of the knower, or how the world really *is.* Instead, knowers make, invent, constitute, create, construct, or narrate, in language, their own subjective realities, or, in the usual terminology, antirealities, or nonrealities.[3]

For most antirealists, then, a theory, language, construct, or narrative always intervenes, or mediates, between the knower and the known—that is, between the knower and the targeted independent reality that is usually presumed to exist, but inaccessibly so;[4] therefore, the knower can never have direct (unmediated by theory) awareness of an independent (of the knower) reality. Moreover, according to antirealists, the intervening (mediating) theory, language, construct, or narrative *always* precludes unaltered or undistorted awareness of that independent reality. That is because, for antirealists, the knower's own cognitive operations and the theoretical, linguistic, or narrative constructions produced by those operations always alter or distort in experience the true nature of the targeted independent reality. In short, for antirealists, all knowledge is inescapably subjective or relative.

## EVIDENCE OF ANTIREALISM IN CONSTRUCTIVIST/ CONSTRUCTIONIST THERAPIES

Therapies that are based on a constructivist/constructionist, or antirealist, doctrine go by several names that sometimes co-occur: postmodern/narrative/ constructivist/social-constructionist therapies. I distinguish the last two of these in a moment. Just here, let me say that there are a great many statements made by members of the above-named therapy movements that document the promotion of some form of antirealism, in that they claim that reality is a function of the knower, that we have no single truth but only multiple views of reality, that all knowledge is relative and so can never be true but only useful, and so on.[5] Let us review a sample of them.[6]

1. In a chapter entitled "Postmodern Epistemology of Practice," Polkinghorne states, "Our observations cannot be trusted to represent the real, nor can our rationality be assumed to mirror the order of the real. We have no sure epistemological foundation upon which knowledge can be built. Our experience is always *filtered through* [italics added] interpretive schemes."[7] He later says, "A common theme of the postmodern epistemology is that linguistic systems stand between reality and experience (Rorty, 1989). Each language system has its own *particular* way of *distorting*, *filtering*, *and constructing* experience"[8] [all italics added]. Note here that linguistic systems not only mediate between reality and the knower's experience of reality, but consistent with my definition of antirealism, they also alter or distort in that process.

2. In their "Introduction" to their book *Therapy as Social Construction*, McNamee and Gergen state, "Our formulations of what is the case are guided by and limited to the systems of language in which we live. . . . Thus, for example, one cannot describe the history of a country or oneself on the basis of 'what actually happened;' rather, one has available a repertoire of storytelling devices or narrative forms and these devices are *imposed* [italics added] on the past."[9]

3. Neimeyer, in "An Appraisal of Constructivist Psychotherapies," says, "Like the broader postmodern zeitgeist from which it derives, constructivist psychotherapy is founded on a conceptual critique of objectivist epistemology. In particular, it offers an alternative conception of psychotherapy as the quest for a more viable *personal knowledge* [italics added], in a world that lacks the fixed referents provided by a directly knowable external reality."[10]

4. In "A Universe of Stories," Parry states, "The complete adoption of a narrative paradigm for therapy is proposed . . ., especially for life in a postmodern world that lacks any objective frame of reference."[11]

5. Hoffman, in "Constructing Realities: An Art of Lenses," says, "All therapy takes the form of conversations between people and . . . the findings of these conversations have no other 'reality' than that bestowed by mutual consent."[12]

With this summary of the place of antirealism in constructivist/constructionist therapy in mind, let us now examine the distinction between constructivism and social constructionism.

## THE CONSTRUCTIVISM/SOCIAL CONSTRUCTIONISM DISTINCTION

According to the many discussions of constructivism found in psychotherapy literature, knowledge is relative to each knower. That is, subjectivity is personal or, more strongly put, uniquely personal. This means that we each distort in experience the true nature of independent reality or, more radically, we each create some "reality" or nonreality uniquely as a function of our own linguistic constructions. To put this bluntly, each knower gets it—independent reality—wrong uniquely.

By contrast, social constructionists[13] say that knowledge is relative to each social/cultural/linguistic group of knowers. Subjectivity is therefore interpersonal. This means that we all distort in experience the true nature of independent reality or, more radically, we all create some "reality" or nonreality in groups as a function of our common/consensual linguistic constructions. This process allows communication/shared understandings as a function of the social, rather than the uniquely personal, construction of nonreality. Put bluntly, each knower gets it—independent reality—wrong in common with other knowers in the same linguistic context.

Although the distinction between the nature and workings of personally vs. socially constructed knowing and knowledge is not made clear or precise enough for me, there is one clear and precise point to be made about both of these doctrines: they are both antirealist doctrines.

## IMPLICATIONS AND CONSEQUENCES OF CONSTRUCTIVISM/CONSTRUCTIONISM FOR THERAPY AND KNOWING

In previous writings I have tried to call attention to the problems therapists face when they adopt antirealism as their core theoretical construct. Here I limit my comments to two concerns.

### *Reality or Truth Claims: A Problem of Logic*

Despite their professed antirealism, constructivist/constructionist therapists make—as they must and as they should—general reality or truth claims about the effectiveness of their therapy. For instance, in "The Pluralist Revolution: From the One True Meaning to an Infinity of Constructed Ones," Omer and Strenger claim that "psychotherapy works by transforming a person's self-narrative and the self-concept embodied in the narrative. . . . Our self-narratives *become* [italics added] our lives."[14]

These are claims about therapy that transcend a merely linguistic effect. To be more precise, constructivist/constructionist therapy, as all therapy, surely changes the client's story or narrative, that is, the client's language *itself.* But it is also alleged to change the extralinguistic realities—the objective behaviors and events—of the client's life to which that story refers and to which we supposedly, according to the professed antirealism, have no access. Thus, constructivist/constructionist therapists find themselves caught in a logical inconsistency, and that is but one of the problematic consequences of adopting that epistemological doctrine.

Nonetheless, no constructivist/constructionist therapist that I know would accept this scenario: that as a result of successful constructivist/constructionist therapy a battered wife *says* she is no longer beaten; however, we cannot take her story to give us any extralinguistic truth about her battered status. That is, she may, despite her story, still be battered, and all we and she have is the story, or linguistic construction itself. Yet that is what constructivism/constructionism calls upon us to accept.

### *The Word "Construct" and the Confusion about Active vs. Passive Knowing*

The word "construct" (noun or verb), and by extension "construction," is, in my experience, behind much of the confusion. In short, the idea of an active rather than a passive knower is falsely assumed to imply antirealism. This implication of the terms "constructivism" and "constructionism" themselves is important; let me elaborate.

We all sometimes *construct* theories about how the world works, both in science and in life. The fact that knowing involves an active process on the part of the knower does not make all knowers antirealists, and all knowledge subjective. Put differently, *all theories are themselves linguistic constructions.* Constructing theories is the business of science, all science. But that fact does not make all scientists constructionists/antirealists. Those participating in the realism/antirealism debate take it for granted that all theories are constructions, an assumption that does not automatically assign anyone to one side of the controversy.

However, to then say that the theoretical construction we have just created is the *only* reality we have is to confuse two things: it confuses (a) the linguistic status of the theory itself with (b) the extralinguistic or extratheoretic reality that the theory is attempting to approximate indirectly. That is, the reality under investigation is not itself a mere linguistic construction—a nonreality, if we take seriously the proclamations of postmodernists/constructionists—either as it *exists* itself, or as it is *known* to the investigator.[15] And no therapist I know of treats it as such, despite his or her antirealist epistemological declarations.

## REASONS FOR THE ADOPTION OF ANTIREALISM AS A FOUNDATION FOR THERAPY

I divide reasons for the antirealist turn in some sectors of the field into those that seem cynical in tone and those that do not. A cynical reason focuses upon the inherent trendiness of therapy as a discipline. In short, antirealism pervades late twentieth-century humanities, sometimes in the form of so-called postmodernism. That therapy has actively cultivated this latest trend should come as no surprise, since therapists have a long history of adopting popular trends. The reason for this tendency may be less frivolous than the tendency itself. I think that reason has in part to do with therapy's real discomfort/difficulty with maintaining its own traditions. By this I mean therapists have tended to look beyond therapy for legitimate paradigms—indeed, for legitimation itself. Consider, for example, the turn to physical sciences that began at least as early as Freud and has continued more recently by way of an appeal to quantum physics to support antirealist leanings.[16] Now, in the last few years, we find an appeal to postmodern humanities to support those same leanings. However intellectually captivating those exercises may be, I am not convinced that either the world of subatomic physics or the world of literary theory, with its deconstructionist method, has much to tell us about how to do good therapy.

Nonetheless, I do find reasons for this turn to antirealism that do not have cynical overtones. These have to do with what I consider to be a sincere attempt to solve the eternal problem of achieving a *system* of therapy that maintains the unique individuality of each client. In looking carefully at the writings of many postmodern/narrative/constructivist/constructionist therapists, I have found a pervasive concern, one that strikes me as more fundamental than antirealism: how to preserve in therapy the client's unique individuality—including personal views of unique life experience. Reference again to a quotation of Neimeyer, in which he frames constructivism as the "quest for a more viable *personal* [italics added] knowledge,"[17] illustrates this concern,[18] as does the following quotation:

> Constructivism elevates the client's view of reality . . . to paramount importance in the therapeutic process. The application of constructivism to therapy makes the client's meaning system hierarchically superior to the therapist's theoretical orientation and/or personal beliefs. Constructivism, therefore, provides a strong rationale for respecting the preeminence of the client's world view. In practical terms, it emphasizes the client's idiosyncratic meaning system as the impetus for therapy.[19]

The goal of individualizing therapy, then, is typically pursued by rejecting the use of general, predetermined theories, meanings, and contents. But what most fail to realize is that this goal is confounded with antirealism—and unnecessarily.

It is my claim that the realism/antirealism distinction is completely irrelevant to the attempt to individualize therapy. That is because—and this may well be considered radical in some circles—there is no need to assume that the client's awareness (or experience) of unique personal events and interpretations of those unique, personal events/experiences are antirealist ones. Most fundamental to my argument is this: it is the *composition* of the therapy system that guides one's practice that is relevant to that pursuit. In particular, if we consider a generic model of therapy systems that contains three fundamental components,[20] my point becomes clearer.

Let us begin by assuming that therapy systems, when complete, can be characterized by the following three predetermined components: (a) one or more descriptions of what constitutes problems, pain, or, put less neutrally, pathology—e.g., depression or chronic indecision; (b) one or more theories about what causes problems, pain, or pathology—e.g., a neurotransmitter defect or early emotional trauma. These I call theories of problem causation, and they, along with the descriptions themselves, constitute the content of therapy—*what* must be considered and discussed, indeed changed, in the course of therapy; and (c) one or more theories (with attendant methods) about how to alleviate problems, pain, or pathology—e.g., challenging irrational beliefs or teaching the client new skills for coping. These I call theories-cum-methods of problem resolution, and they constitute the process of therapy—*how* therapy, or change, occurs, whatever its content.

We are now in a position to see just how those who promote the adoption of antirealism in therapy are attempting to individualize practice by eliminating predetermined content from the theoretical systems they construct and use. They seek to replace that content with unique, *client*-determined meanings. But all they have done—the *real* consequence of their activity—is to make the therapy system they use less complete, and so the practice guided by that system is less systematic, rule governed, and replicable; hence I have called their aspirations antisystematic.[21] Moreover, they have not made the system that remains—namely, the predetermined method of problem resolution, or therapeutic process—more open to an antirealist interpretation. And, again, the client's personal meanings/views/understandings need not be given an antirealist interpretation (recall the battered wife, whose *particular* battered experience is perfectly unique to her), certainly not any more than general, predetermined theories of problem causation and resolution need to be taken as realist, or extratheoretically true: the latter could simply be bad theories. Thus, what we have is a confusing of the problem of individualizing therapy (and the related problem of systematizing it) with antirealism, or subjectivity.

To reiterate: the individualization (and systematization) of therapy is not a function of the realism or antirealism of the theoretical system used to guide one's practice. It is a function of the completeness of that system in terms of the three component parts I outlined above. This is true even for those systems of therapy that are taken to be completely antirealist, by design or otherwise.

## A REALISTIC ALTERNATIVE

I hope that, in the final analysis, constructivist/constructionist therapists will spend more of their time reconsidering the nature and composition—the completeness—of the theoretical systems that guide their practice, and less time extolling the virtues of antirealism as a foundation for therapy. I also hope that a good portion of that time gets devoted to making those systems as extratheoretically true as possible, an activity that must include the systematic empirical evaluation of the effectiveness of those systems. After all, as any postmodernist will tell you, language, including the language of therapy systems, matters precisely because it affects the extralinguistic reality we all must cope with. This reality most surely matters to therapists of any philosophical preference.

## NOTES

1 I sometimes use the term "constructionism" or "constructionist" to include both doctrines. Where I use the term "constructivist" or "constructivism" alone, I mean only that doctrine, except in the title itself.

2 Note that I am deliberately excluding Kant's sense of objectivity. For Kant, objectivity was subjective in origin and therefore phenomenal. Hence, although he took scientific knowledge to be objective, he also thought that reality, as it is "in itself,"

was not attainable by the knower. For Kant, then, the terms *objectivity* and *reality* were not equivalent.

3 B. S. Held, "What's in a Name? Some Confusions and Concerns about Constructivism"; *Back to Reality: A Critique of Postmodern Theory in Psychotherapy*; B. S. Held & E. Pols, "The Confusion about Epistemology and 'Epistemology'—and What to Do about It"; E. Pols, *Radical Realism: Direct Knowing in Science and Philosophy.*

4 Those more radical antirealists who reject even the existence of any mind-independent reality usually see their own mental constructions as the only reality. For them, theory or language can therefore play no intervening or mediating—and thus no distorting—role in the act of knowing, because there is nothing out there—beyond mind—to be mediated/distorted by their knowing process. "Notice that it is a more radical claim to say . . . that *reality* does not exist independently of the knower than it is to say . . . that our *experience* of reality, being structured by ourselves, may be different from what reality in fact is" (Held & Pols, "The Confusion about Epistemology," p. 513). I direct my comments to less radical antirealists, those who posit the existence of an independent reality—albeit an inaccessible one—since in my experience they comprise the majority of constructionist therapists. But whether the antirealism is of the more or the less radical sort, it always precludes any access (direct or indirect) to an independent reality, and *that* is the point of all antirealist doctrines. So it is also the point of constructivism/constructionism itself.

5 E.g., H. Anderson & H. A. Goolishian, "Human Systems as Linguistic Systems: Preliminary and Evolving Ideas about the Implications for Clinical Theory"; S. de Shazer, *Putting Difference to Work*; J. S. Efran, M. D. Lukens & R. J. Lukens, *Language, Structure, and Change: Frameworks of Meaning in Psychotherapy*; R. T. Hare-Mustin & J. Marecek, "Gender and the Meaning of Difference: Postmodernism and Psychology"; L. Hoffman, "Constructing Realities: An Art of Lenses"; M. J. Mahoney, *Human Change Processes: The Scientific Foundations of Psychotherapy*; S. McNamee & K. J. Gergen, Introduction to *Therapy as Social Construction*; R. A. Niemeyer, "An Appraisal of Constructivist Psychotherapies"; H. Omer & C. Strenger, "The Pluralist Revolution: From the One True Meaning to an Infinity of Constructed Ones"; A. Parry, "A Universe of Stories"; D. E. Polkinghorne, "Postmodern Epistemology of Practice"; D. P. Spence, *Narrative Truth and Historical Truth: Meaning and Interpretation in Psychoanalysis*; P. Watzlawick, ed., *The Invented Reality*; M. White & D. Epston, *Narrative Means to Therapeutic Ends.*

6 See Held, *Back to Reality*, pp. 96–101 for a more comprehensive listing.

7 Polkinghorne, "Postmodern Epistemology of Practice," p. 149.

8 Ibid., pp. 149–150.

9 McNamee & Gergen, "Introduction," p. 4.

10 Niemeyer, "Appraisal," p. 230.

11 Parry, "Universe," p. 37.

12 Hoffman, "Constructing Realities," p. 4

13 E.g., K. J. Gergen, "The Social Constructionist Movement in Modern Psychology."

14 Omer & Strenger, "Pluralist Revolution," p. 253. See Held, *Back to Reality*, pp. 143–146, for more examples.

15 To call the linguistic construction itself a nonreality is misleading in one sense. Even if the linguistic construction *were* the only reality, this would still pose a problem for antirealists: the construction would then have a real existence in language *as* the linguistic construction it was formed to be, regardless of its truth status. How, then, can we know *that* (linguistic) reality? See Pols, *Radical Realism*, and Held, *Back to Reality*, for extensive discussions of that problem.

16 E.g., W. J. Doherty, "Quanta, Quarks, and Families: Implications of Quantum Physics for Family Research."

17 "An Appraisal of Constructivist Psychotherapies," p. 230.

18 See Held, *Back to Reality*, pp. 82–89, for additional evidence of this concern.

19 A. D. Solovey & B. L. Duncan, "Ethics and Strategic Therapy: A Proposed Ethical Direction," p. 55.

20 B. S. Held, "The Process/Content Distinction in Psychotherapy Revisited"; *Back to Reality.*

21 Held, *Back to Reality.*

## REFERENCES

ANDERSON, H. & H. A. GOOLISHIAN. "Human Systems as Linguistic Systems: Preliminary and Evolving Ideas about the Implications for Clinical Theory." *Family Process* 27 (1988): 371–393.

DE SHAZER, S. *Putting Difference to Work.* New York, NY: W. W. Norton, 1991.

DOHERTY, W. J. "Quanta, Quarks, and Families: Implications of Quantum Physics for Family Research." *Family Process* 25 (1986): 249–263.

EFRAN, J. S., M. D. LUKENS & R. J. LUKENS. *Language, Structure, and Change: Frameworks of Meaning in Psychotherapy.* New York, NY: W. W. Norton, 1990.

GERGEN, K. J. "The Social Constructionist Movement in Modern Psychology." *American Psychologist* 40 (1985): 266–275.

HARE-MUSTIN, R. T. & J. MARECEK. "Gender and the Meaning of Difference: Postmodernism and Psychology." In *Making a Difference: Psychology and the Construction of Gender*, edited by R. T. Hare-Mustin & J. Marecek, 22–64. New Haven, CT: Yale University Press, 1990.

HELD, B. S. *Back to Reality: A Critique of Postmodern Theory in Psychotherapy.* New York, NY: W. W. Norton, 1995.

———. "The Process/Content Distinction in Psychotherapy Revisited." *Psychotherapy* 28 (1991): 207–217.

———. "What's in a Name? Some Confusions and Concerns about Constructivism." *Journal of Marital and Family Therapy* 16 (1990): 179–186.

HELD, B. S. & E. POLS. "The Confusion about Epistemology and 'Epistemology'—and What to Do about It." *Family Process* 24 (1985): 509–517.

HOFFMAN, L. "Constructing Realities: An Art of Lenses." *Family Process*, 29 (1990): 1–12.

MAHONEY, M. J. *Human Change Processes: The Scientific Foundations of Psychotherapy.* New York, NY: Basic Books, 1991.

MCNAMEE, S. & K. J. GERGEN. "Introduction." In *Therapy as Social Construction*, edited by S. McNamee & K. J. Gergen, 1–6. Newbury Park, CA: Sage, 1992.

NEIMEYER, R. A. "An Appraisal of Constructivist Psychotherapies." *Journal of Consulting and Clinical Psychology* 61 (1993): 221–234.

OMER, H. & C. STRENGER. "The Pluralist Revolution: From the One True Meaning to an Infinity of Constructed Ones." *Psychotherapy* 29 (1992): 253–261.

PARRY, A. "A Universe of Stories." *Family Process* 30 (1991): 37–54.

POLKINGHORNE, D. E. "Postmodern Epistemology of Practice." In *Psychology and Postmodernism*, edited by S. Kvale, 146–165. Newbury Park, CA: Sage, 1992.

POLS, E. *Radical Realism: Direct Knowing in Science and Philosophy.* Ithaca, NY: Cornell University Press, 1992.

SOLOVEY, A. D. & B. L. DUNCAN. "Ethics and Strategic Therapy: A Proposed Ethical Direction." *Journal of Marital and Family Therapy* 18 (1992): 53–61.

SPENCE, D. P. *Narrative Truth and Historical Truth: Meaning and Interpretation in Psychoanalysis.* New York, NY: W. W. Norton, 1982.

WATZLAWICK, P., ed. *The Invented Reality.* New York, NY: W. W. Norton, 1984.

WHITE, M. & D. EPSTON. *Narrative Means to Therapeutic Ends.* New York, NY: W. W. Norton, 1990.

# ENVIRONMENT

ONLY IGNORANCE or duplicity permits the conclusion that, in the state of the biosphere and its physical environment, all is well. That there is nothing to worry about, or that something will turn up to deal with whatever problems there are, is the faith of the somnambulist. Yet at the other end of the spectrum of environmental viewpoints, also, there is a great deal to worry about. A current of opinion, once limited to obscure sects but now diffusing into the mainstream environmental movement, holds that not only technology and industrialism, but science itself, and the attitude of critical rationality that underlies it, are the true threats to life on earth and man's hope of finding a proper place within it.

> . . . I fully concur that the passage into the twenty-first century sees the world in a state of ecological crisis. I further agree that much of the blame must be assigned to technologies that owe their existence to the success of Enlightenment thought. But I am also convinced that a wholehearted commitment to reason and science offers the only way out of the dilemma.

Thus, Martin W. Lewis, whose essay here, like his book *Green Delusions*, scrutinizes the irrationalist core of radical ecologism. Lewis details the ways in which the resentment of science embedded in fashionable ecomania must subvert or misdirect serious and necessary environmental initiatives.

Stanley Rothman and S. Robert Lichter adopt a severely empirical approach in analyzing specific ways in which the clichés of some schools of environmental activism come to diverge from the best-warranted scientific information. They show that, notwithstanding a consonance of general political and social perspectives, scientists working in cancer epidemiology have a far different view of what constitutes a serious threat of environmental cancer than nonscientists who regard themselves as activists for environmental sanity.

In her essay, Rene Denfeld contrasts the romantic ecoradicalism endemic among "difference" or "gender" feminists with the serious and committed environmentalism that it displaces or discourages. She points to the ways in which young people—young women especially—have had deflected their spontaneously serious interest in environmental questions by the sectarianism, the woolly self-righteousness, and the disdain for science that have characterized what is now called "ecofeminism."

# RADICAL ENVIRONMENTAL PHILOSOPHY AND THE ASSAULT ON REASON

MARTIN W. LEWIS

THE RELATIONSHIP BETWEEN environmentalism and science is one of profound ambivalence. Environmentalists rely on scientific findings for their assessments of our environmental predicament, and the movement as a whole is so closely intertwined with the science of ecology that it threatens to usurp its very name; "ecology," for many, is indistinguishable from "environmentalism."[1] But most committed Greens are also wary of the scientific endeavor, viewing it as complicit in planetary destruction. As one moves toward the more radical fringes, not only science but rationality itself is denounced as lying at the root of humanity's deadly estrangement from the natural world.

Environmentalism's flight from science and reason would be little cause for concern if it were merely confined to a small group of activists at the movement's periphery. Unfortunately, hostility to science, coupled with misgivings about reason, is the norm among a sizeable and influential group of academics devoted to the study of what might loosely be called environmental philosophy. Whether teaching in departments of philosophy, political science, history, geography, or "environmental studies," scholars examining ideas about nature and their effects on the human-environmental relationship commonly regard the scientific revolution of the seventeenth and eighteenth centuries as humanity's greatest mistake. Most environmental scientists, policy analysts, and economists remain relatively little influenced by this assault on modernity and Enlightenment thought. But among the broader community of the ecologically concerned, resistance to such philosophical radicalism seems to be minimal. Concepts quickly filter out of the academic context to affect virtually the entire environmental movement, even, in a muted manner, its nonradical political core. It is now a common article of faith among the most concerned Greens that the very survival of human civilization, if not life itself, depends on a wholesale rejection of science and reason—on a repudiation of the Enlightenment project that alienated us from nature and set us on a course of accelerating destruction.[2]

As a self-professed environmentalist, I fully concur that the passage into the twenty-first century sees the world in a state of ecological crisis. I further agree that much of the blame must be assigned to technologies that owe their existence to the success of Enlightenment thought. But I am also convinced that a wholehearted commitment to reason and science offers the only way out of the dilemma. Only through scientific investigation can we know the origin and magnitude the planet's problems. But of equal significance is the fact that only through science can we devise less harmful technologies that will allow us to continue to enjoy the fruits of modernity—which twentieth-century humans will not forswear, no matter how urgent the pleading—without undercutting natural systems in the process. By attacking science and reason, ecophilosophers assail their potential allies, jeopardizing the environmental cause.

The principal aim of this paper is to outline and criticize radical environmental philosophy's principal charges against science and reason. A secondary concern is the connection between the anti-Enlightenment position of the Greens and that of literary deconstructionists. The relationship between these two camps is highly problematic; ecophilosophers are attracted to the general notion of "postmodernity," but remain wary of the nihilism associated with the poststructuralist movement. One response has been to advocate a "constructive" version of postmodernism that shares many of the premises of literary poststructuralism while diverging markedly from its main platform. In my conclusion to the present paper, I consider the political implications of this Green version of postmodern philosophy. Here I contend that for all its "progressive" posturing, ecoradicalism retains a reactionary core.

## SCIENCE AND REASON ON TRIAL

Most ecoradicals believe that human beings existed for millennia in a state of environmental grace as merely one species among a myriad in a balanced, harmonious global ecosystem. Environmental consonance was eventually to be shattered, however, when *Homo sapiens* arrogantly escaped from the interlocking networks of natural equilibrium and began clambering its way to dominance. The ultimate consequences of its eventual supremacy is apparent today in the form of impending environmental apocalypse. The task for ecophilosophy is to explain how such a total rupture could have occurred, and more importantly, to show how balance might be restored in time to save the planet from annihilation. Assuming that a harmonious existence was once humanity's birthright, ecophilosophers are eager to find the signal error—what Val Plumwood calls the "fatal flaw"[3]—that set our course astray.

The key error is often assumed to lie in the ideological realm, particularly in concepts about nature and the human position within it. This idealist orientation is hardly surprising, for it is precisely the study of ideas that preoccupies environmental philosophers. Some writers, to be sure, are wary of the charge of idealism, and suggest that underlying economic and social-structural changes determined the course of ideological development.[4] Still, even among those who might call themselves materialists, the emphasis remains on the transformation of ideas about nature. If the deadly errors we have embraced in our thinking could only be erased, the argument goes, our relationship with the natural world might be set right again.

Most ecoradicals believe that no amount of mere political reform or technological fiddling can avert ecological catastrophe, charging that our current commitment to continuous economic growth is predicated on a world view where humans necessarily *dominate* nature. If this so-called dominant paradigm could be dissolved, however, we might reestablish our lost intellectual (and spiritual) connections with the earth. Refastening those bonds will in turn fulfill the human spirit, and ward us away from intrinsically destructive behaviors. The momentous duty of environmental philosophers is thus nothing less than to correct the central defect in human thought that has brought us to the threshold of ecological disaster.

For many radical ecophilosophers, the great error was nothing less than the glorification of reason that began in Europe during the early modern era[5] and that culminated in modern scientific methodology. These developments, it is argued, brought in their train a corresponding depreciation of allegorical understanding, mythic interpretation, and spiritual values. In the most common version of ecointellectual history, Galileo Galilei, Francis Bacon, Isaac Newton, and especially René Descartes are accused of promulgating the great error by intellectually "killing" the earth.[6] Prior to the revolution in thought these men engendered, we are informed, moral constraints limited the damage that humankind could wreak upon the natural world. Francis Bacon, however, placed humanity in a Promethean position from which all ties to nature could be cast aside as the quest for total dominion commenced. Galileo, Newton, and Descartes, for their part, demolished the old view of the earth as a living entity intrinsically worthy of respect and due consideration. In its stead they erected the hard and lifeless construct of mechanism, one in which all nonhuman organisms were deemed mere machines that could be used and abused at will. The resulting "framework of reductive mechanism," explains Val Plumwood, "permits the emotional distance which enables power and control, killing and warfare, to seem acceptable."[7] The death of nature," writes Carolyn Merchant, "legitimated its domination."[8]

The new world that emerged in the Europe of the seventeenth century—and which was culturally solidified during the Enlightenment of the eighteenth—was one ruled by the arrogance of reason. Minds were now increasingly divorced from bodies, and masculinist "objective" intelligence severed all connections from the emotional and ethical domains. Science and reason now began a futile campaign to deny death[9]—and hence organic reality—and in so doing lost the attachment to Gaia that had previously underwritten human society. The supposed linkages between seventeenth-century rationality and contemporary ecocide are often portrayed as direct and unproblematic. One is led on a straight path from the scientific revolution through the Enlightenment to the industrial revolution and hence to the ecotribulations of the present day. Rupert Sheldrake, for example, takes us from Galileo's physics to the mass slaughtering of American bison in the space of two paragraphs.[10]

Denunciations of Descartes and other early modern mechanistic philosophers are a staple feature of ecoradical philosophy. But the question of whether their views of nature constituted the *key* rupture remains open; some thinkers argue that the roots of the West's malignant disaffection from the earth run much deeper. Descartes, after all, although hoping to begin philosophy anew, was a product of a long line of Western thought. Several ecophilosophers have thus located the crucial break in ancient Greece, identifying it with the development of atomism, dualism, or, most simply, reason itself. For Val Plumwood, Plato supplants Descartes as the original serpent in the garden.[11] Yet other writers insist that the West's ideological escape from a natural existence is older still, dating back to the origins of the Judeo-Christian tradition. The corrupting influence here is said to be the book of Genesis, in which humankind is "granted" dominion over all other living things.[12]

The main thrust of ecophilosophy is thus to outline a history of error in the Western world's intellectual apprehension of the earth. A concomitant project is to celebrate societies that have maintained a correct set of ideas and spiritual practices. Contemporary "tribal" societies—often called "primal" (a term with positive connotations in this usage)—are typically held up as eco-exemplars, as are Native Americans and Asian peoples of the past. Accounts of such peoples are highly generalized; one or two brief examples of harmonious relations are proffered to make the case that balance with the earth is the *essential* human condition in the absence of corrupting Western ideas and practices.

Some ecophilosophers do seek at least partially to rehabilitate the Western legacy by showing that harmony was the norm prior to the modern debasement of ideology. Thus one commonly encounters the assertion that medieval Europeans, unlike their modern descendants, intellectually lived within the framework of natural ecosystems without threatening their integrity. According to Charlene Spretnak, Europeans of the Middle Ages inhabited an "organic female universe" replete with a "sense of oneness, continuity, and organic justice."[13] Contending that the later ideology of domination infused the social no less than the ecological order, such writers often attempt to minimize the class distinctions of medieval society by stressing its organic unity and communalism.[14] A related exercise is to exhume evidence of earlier societies that seem to lack class stratification, patriarchal subjugation, and militarization altogether. Marija Gimbutas's reconstruction of Neolithic "Old European" society provides the perfect scenario here, one that is especially appealing to ecofeminists:[15] a peaceful, gender-egalitarian, classless society living fully in tune with its natural environment. This paradise would be lost when violent, patriarchal Indo-European peoples of the steppes invaded and conquered—peoples whose social structure would eventually be translated into the ideology of masculine domination associated with such men as Plato, Descartes, and Newton.

The doctrine of the "key error" is thus more complex than it might appear at first glance. The scientific revolution of the seventeenth century may represent the ultimate rupture between humanity and the earth, but this disaster was allegedly foreshadowed for several millennia. Indeed, the most radical environmental thinkers argue that the overriding break occurred earlier still—

with the Neolithic emergence of agriculture. According to this view (usually associated with male theorists), hunting and gathering is our "natural" mode of existence, and only in hunting-gathering societies can people really be personally fulfilled and truly at one with the earth. As Max Oelschlaeger argues, "No one knows for certain how long prehistoric people existed in an Edenlike condition of hunting-gathering, but 200,000 years or more is not an unreasonable estimate of the hegemony of the Great Hunt."[16]

While different authors thus tell slightly different stories, environmental philosophy as a whole presents a fairly consistent narrative of humanity's decline into its present predicament. The original humans lived in Edenic bliss, in union with Gaia. Things began to go wrong thousands of years ago when domination emerged. Still, natural systems maintained their balance as human technologies remained limited and as ethical and religious standards constrained development. Then, some 400 to 500 years ago, mechanistic philosophy finally "killed nature" and began our race toward oblivion.

It is worth noting that this historical vision is accompanied by a stock conception of world geography. The fatal journey was initiated in only one small portion of the world: the West.[17] The key ideological errors all occurred within Western philosophy, which is often viewed as irredeemably flawed. Instrumental rationality, amoral objectivism, value-free science, and, ultimately, death-dealing technology and insatiably destructive economic growth are all considered unique to the West. The rest of the world is pictured as having existed in a state of near ecological bliss, from which it would be torn when Western ways were imposed through imperialism and advertising, and when Western thought was internalized by local elites. In the rhetoric of radical environmental philosophy, the West is coincident with modernity, and is the singular origin of environmental destruction.

In the typical ecoradical rendition of world geography, the non-West—pictured as the land of intrinsic harmony—is either identified with the archetypal small-scale "primal" society or with the historical zone of Buddhist and Hindu civilization (the "Eastern" world stretching from India to Japan). Little effort is made, however, to distinguish these two divergent versions of the non-Western world, let alone to draw out their internal diversity. Moreover, huge sections of the planet—most notably the realm of Islam—disappear altogether from the ecoradical globe.[18]

## THE FAILURE OF ECORADICAL PHILOSOPHY

The standard ecoradical narrative is surely correct in selecting the seventeenth century as a crucial turning point in intellectual history, one that would eventually have profound environmental repercussions. But the causal links in the narrative from that point forward do not stand up to scrutiny. First, the signal occurrence of the seventeenth century—the scientific revolution—was by no means a direct outgrowth of Descartes's or Bacon's beliefs about the relationship between people and nature. Second, the philosophical relationships among the vilified thinkers of the period are also far more complex than the ecoradical account portrays them to be. Finally, there is little evidence that the industrial revolution of the late eighteenth and early nineteenth centuries was itself a direct consequence of the scientific revolution.[19] In fact, it is quite possible that the industrialization of the coal age, with its ecological degradation, would have occurred regardless of this intellectual shift. To link *science* conclusively with environmentally harmful technologies, one must wait for the late nineteenth century. Each of these issues deserves more extended commentary.

Francis Bacon no doubt inspired many scientists and engineers in the early modern period, but he was not a significant figure in the scientific revolution. Nor was the role of Descartes critical. The latter occupies an important place in the history of thought, both for his mathematics and for his insistence that philosophy must begin from a standpoint of utter skepticism. Most of his scientific views, however, were soon discredited. In actuality, Descartes's insistent rationalism and unaccommodating mechanism thwarted the development of genuine science, for they discouraged empirical inquiry. For this reason, many of the leading *philosophes* of the French Enlightenment vilified Descartes as a "rash metaphysician."[20] Environmental philosophers often ignore such niceties in their quest to identify the ecovillains of Western history. Thus the prominent ecofeminist Val Plumwood informs us that "the only force allowed within the mechanistic framework [of Newtonian physics] is that of kinetic energy . . . all other purported forces, including action at a distance, being regarded as occult."[21] Ironically, such a statement unknowingly echoes the charge leveled by the soon-to-be-discredited Cartesians and Leibnizians against the physics of Newton and its "occult" notion of gravity.

Newton, of course, like Galileo before him, was a pivotal figure in the scientific revolution. Moreover, his physics was certainly characterized by mechanistic explanations. But this does not mean that the scientific revolution—much less the Enlightenment—was predicated on a thoroughgoing metaphysics of dualistic mechanism, one that reduced all nonhuman living beings to the status of mere machines that could be used without ethical considerations. What was significant was rather the formalization of a view of nature as characterized by regularity and by causal relations, and, more importantly, the development of reasonably reliable methods for answering specific questions about natural structures and processes. While it may be true that a metaphysical shift at the time replaced the Aristotelian view with an atomistic and mechanistic conception of *inert matter*,[22] such a transformation was not as far-reaching as the ecoradicals would have it. Ultimate metaphysical (and even spiritual) conceptions about the nature of life and about humanity's relationship to the rest of the natural world were of little consequence to the scientific revolution, so long as they did not unduly intrude upon scientific investigations at hand.[23] Newton's own final vision of the universe, after all, was based on Unitarian and prophetic Christianity,[24] and was complemented by faith in astrology. The subsequent development of Western thought certainly saw its share of radical mechanists—from the Enlightenment's La Mettrie to this century's B. F. Skinner[25]—but mechanism in this sense is indeed rank metaphysics, not science. On such ultimate issues genuine science is silent, for it addresses only those questions whose answers can be verified empirically.[26]

In my view, the ideas implicit in the scientific revolution and, more generally, the Enlightenment, cannot be directly blamed for our contemporary environmental crisis. The problem is not that humanity embraced, at a specific time and place, a set of faulty beliefs. It is true that modern science, once it was effectively joined with technology, gave humankind tremendous power over natural systems. That such power was used, often in ways detrimental to nature, is explicable simply by reference to the common human desire to improve one's material well-being, which easily shades over into outright greed. Metaphysical justification, *pace* Plumwood, has never been necessary for people to countenance "power and control, killing and warfare." It is hardly reasonable to lay the charge of Cartesian "dualism" at the industrial revolution's inventors of machine production. Such inventors were mainly "practical men," with little interest in abstract philosophy or even science, whose milieu was still profoundly influenced by traditional Christian understanding. And to blame the slaughter of millions of bison on the American Great Plains in the nineteenth century on Descartes's mind/body dualism is simply an instance of metaphysical determinism run wild.

The ecoradical assessment of the historical record is just as tendentious, and just as ill informed, as its appraisal of Western philosophy. The belief in modernity's deadly rupture is so pervasive that anything nonmodern is automatically celebrated. Thus medieval European societies are not only portrayed as functioning harmoniously within the limits of natural systems, but as less hierarchical and exploitative than those of today. The latter allegation is comical, but even the former does not withstand scrutiny. Medieval industries may not have threatened the ozone layer, but before the population decline of the fourteenth century many European societies were approaching local ecological limits and were suffering severe consequences.[27] When one looks at deforestation and the steady diminution of wildlife, the evidence of premodern European ecological despoliation is overwhelming. Even the ecofeminist image of the European Neolithic as a time of complete social and ecological concord fails to withstand careful inspection. Indeed, two prominent *feminist* archeologists have recently demonstrated that the Gimbutas thesis simply cannot be substantiated.[28]

The ecoradical search for "primal" peoples, uncorrupted by technological advance or social hierarchy, also runs into an empirical morass. As Robert Edgerton[29] has recently shown, it is not particularly difficult to "challenge the myth of primitive harmony." Torture of animals, male oppression of females, and outright (local) ecological devastation may not have been universal conditions, but they were common enough everywhere. Even if we retreat back to the upper Paleolithic, Edenic bliss is hard to find. Not only was life in the late Pleistocene difficult and short,[30] but much evidence suggests that human beings at this time were responsible for the extinction of dozens of species of large mammals.[31] If the Pleistocene overkill hypothesis is correct, then Paleolithic humanity is guilty of causing an ecological catastrophe almost as great as the one presently being visited upon the planet.

The ecoradical vision fails on geographical grounds no less than on those of history. Recent research indicates that, well into the early modern period, western Europe was one of a series of linked civilizations in an "Afro-Eurasian ecumene"[32] that had historically moved more or less together in economic and technological evolution, and that even supported similar ideological structures.[33] While western Europe certainly controlled the "world system" by the nineteenth century, several regions of Asia were approaching their own capitalist transitions before their development was cut short by imperial force.[34] Nor can western Europe lay any special claim to rationality. Historically speaking, reason and unreason have coexisted in each civilization. Sung China in particular stands out for its development of critical thought as applied to natural phenomena. As Jacques Gernet explains:

> The immeasurable times and spaces, the intermingling of living beings—demons, animals, infernal beings, men, and gods, through their transmigrations—this whole cosmic phantasmagoria disappeared, leaving in its place only the visible world. Man became man again in a limited, comprehensible universe which he only had to examine if he wished to understand it. . . . What is strikingly manifest is the advent of a practical rationalism based on experiment, the putting of inventions, ideas, and theories to the test.[35]

It is undeniable that modern science originated in western Europe and not anywhere else. But to a certain extent, genuine science emerged as much in spite of as because of Western culture. It was vigorously opposed, after all, by the foundational institution of Western civilization, the Roman Catholic Church,[36] and a continuous line of Western thinkers—culminating with the contemporary postmodernists—has always stridently opposed scientific methodology if not rationality itself. Far from being an attribute of Western culture, scientific thought transcends cultural context.[37] All scientific claims, of course, are shot through with cultural expectations, but science itself is a self-correcting process that over time weeds out the culturally contingent for the universal.[38] It was precisely because of its universalism that modern science diffused so rapidly across the globe, creating a contemporary scientific project that is remarkably cosmopolitan. The strength of contemporary Indian and Chinese science, in other words, is by no means a simple result of Asian "Westernization." Science may have emerged within the Western cultural context, but it was never imbued with a Western cultural essence. To speak of "Western science," as both cultural conservatives and cultural leftists are wont to do, is thus inaccurate.

If there is no Western science, then there can be no Western sciences—no Western physics and no Western medicine. If one wants to contrast (traditional) Far Eastern medicine with Western medicine, one must examine the *traditional* medical practices of western Europe—in other words, the humoral theory of disease, associated with such "curative" practices as massive bloodletting. Such medical "science" was, after all, standard throughout the West until well into the nineteenth century. It would eventually yield to modern medicine, not least because its own methods were often deadly. If "Eastern medicine" survives to this day as an alternative to modern healing—in California no less than in Japan—it is largely because its "cures" are at worst rather innocuous.

The ecoradical thesis that western Europe is the unique location of humanity's ecological sins likewise fails on mundane grounds of historical evidence.[39] The destruction of nature has a long history in Asia as well as in Europe; widespread belief in the supposedly nondualistic and nature-oriented creeds of Buddhism, Taoism, and Shinto did not—despite the fondest dreams of (Western) environmental thinkers—protect the natural world when human interests contravened. Consider, for example, Peter Perdue's description of "exhausting the earth" in early modern Hunan province:

> Mountains were stripped bare and swamps drained to produce the maximum amount of cultivated land. Forests that abounded in wood, bamboo, ramie fiber, and charcoal exhausted their production as cultivated fields spread. . . . Even wild areas like Sangzhi county, whose mountains were filled with tigers and leopards when the area was first opened to Han settlement by the Yongzheng Emperor in the 1720s, reported, by the 1760s, that the wild animals had disappeared and that "all the mountains have been turned into cultivated fields."[40]

In regard to the contemporary world, one should ponder the devastating effects of *traditional* Chinese and Korean gastronomy and pharmacology on endangered species. It did not require Westernization—much less conversion to Cartesian dualism—for Chinese epicures to sanction cooking bear paws on live bears, scooping monkey brains out of live monkeys, or concocting aphrodisiacs out of the horns of rhinoceroses and the penises of tigers. Traditional eastern Eurasian medicine may not be as deadly as traditional western Eurasian medicine, but it is far more ecologically destructive.

In a word, neither modern science nor environmental degradation is a unique feature of Western culture. One might make a better case for regarding ecoradical philosophy as an intrinsically Western cultural expression. Its ideas have little salience in the East Asian or Islamic cultural spheres, although they do have some well-known proponents in South Asia.[41] Postmodernism as well essentially derives from Western philosophy.[42] It is to the complicated relationship between environmental philosophy and postmodern discourse that we now turn.

## THE TURN TO POSTMODERNISM

The ecoradical attack on reason and science was initiated within a framework of reasoned debate. Historical evidence was examined, and plausible linkages were hypothesized among developments in philosophy, science, technology, and economics—developments that were not unreasonably posited as lying at the foundation of the contemporary environmental crisis. Ecophilosophers also sought confirmation of their vision of premodern ecological harmony from the archeological and anthropological record. Moreover, they attempted to ground their entire framework in the science of ecology. Thus they commonly argued that since ecologists had demonstrated that stability follows from diversity, then the increasingly monocultural world of modernity must be rejected as overly vulnerable to disruptions of catastrophic proportions.

The problem, however, is that more careful consideration of the same lines of argument has since discredited the principal concepts of ecoradical philosophy. The roots of modern society are far more entangled and multistranded than they would have it, and the premodern world is now known to have been far less ecologically and socially benign. Even in regard to the dim reaches of the Neolithic and Paleolithic, prudent scholars are forced to conclude that the available evidence does not allow the favored reconstructions. To imagine that "Old Europeans" were intrinsically irenic and gender-egalitarian is problematic enough, but to imbue Magdalenian hunter-gatherers with a "Paleolithic consciousness"[43] that ensured ecological rapport is revealed to be pure fantasy. Even the science of ecology has failed the Greens, for it now emphasizes continuous flux and patchy distribution patterns, rather than the stability of coherent ecosystems that once underwrote the vision of harmonious relations between people and nature.[44]

One might imagine that such difficulties with evidence and theory would lead to a crisis of confidence and a questioning of assumptions. But ecoradical beliefs are often held with a religious vigor; the very existence of life on earth is thought to be at stake, and environmental philosophy itself is accorded a salvational role. Inasmuch as it is a religious world view, this position is impervious to evidence against its key tenets. But environmental philosophy is only partially religious, being in equal measure a scholarly pursuit. As scholars, Green thinkers must address the evidentiary problems outlined above. It is here that postmodernism comes in: as a ready exit from their quandary.

The overriding attraction of a postmodern attitude is that it annuls the inconvenient requirement of empirical confirmation. In more extreme versions, the notion of evidence, like the formal rules of logic, is regarded merely as a social construct that society's power holders use to maintain and justify their positions. Stories of the human past invented by an active ecoradical imagination—stories aimed at subverting existing power structures—can thus be argued to have just as much legitimacy as the reconstructions of professional archeologists and other "scientists" trapped within the confines of objectivist discourse. If anything, they have more validity because of their moral authority; in the postmodernists' world, ethics are not to be separated from matters of "fact." By the same criteria, the problems implicit in the new ecology can simply be ignored. Ecologists are merely constructing their own stories about nature, and those currently being told in the scientific journals may be regarded as suspect, for they could potentially be used to justify a modernist agenda of human-imposed environmental change. As Donald Worster writes:

> If [the new ecological models] are not mere reflections of global capitalism and its ideology, they are nonetheless highly compatible with that force dominating the earth. The newest ecology, with its emphasis on competition and disturbance, is clearly another manifestation of what Frederic Jameson has called the "logic of late capitalism."[45]

If taken to its (il)logical extreme, such a postmodern stance allows one to endorse any tales about the human past that prove convenient for one's political agenda. And, indeed, many philosophical radicals have liberated themselves almost entirely from the demands of evidence. Several prominent scholars go so far as to argue that "the historical data on goddesses should be ignored if they do not present an image that is healthful for modern feminists seeking an alternative spirituality."[46] In popular literature, even the boundary between fictional and nonfictional environmental accounts has been blurred, if not entirely "transcended," as is clearly evident in the reception of the farfetched best-seller, *Mutant Message Down Under.*[47]

In one sense, the romance between ecoradical philosophy and postmodernism is entirely natural. Postmodernism began its career with an attack on the sterile architecture of high modernism, and what, after all, could be more antithetical to the organic ideal of the Greens than Bauhaus "machines for living"? Similarly, the poststructuralist offensive against the supposedly disembodied, logocentric, objectivist, totalizing, imperialistic, Eurocentric project of Western science and rationality conforms impeccably with the ecoradical critique of Western philosophy. Concordance can also be found in the emphases placed on diversity in human cultural expression and in nature itself. Even the postmodernists' accent on playfulness finds its echoes in the Greens' favored account of natural processes.

Yet it would be a serious error to conclude that postmodernism and ecoradical philosophy share identical concerns, much less that the two movements have somehow merged. Most environmental philosophers strongly mistrust the mainstream Derridean/Foucauldian schools of postmodernism, and they commonly regard the more flamboyant displays of "po-mo" irreverence with dismay and disgust.[48] Extreme postmodernism is far too relativistic and skeptical for most Greens. Whereas poststructuralists condemn the search for the "transcendental signified" as a pointless quest, ecoradicals not only want to isolate the "transcendental signified" in the form of nature,[49] but propose literally to worship it. Moreover, the unquiet embrace of technology encountered in some postmodern quarters finds scant support in the Green movement. Donna Haraway's "Cyborg Manifesto,"[50] for example, which celebrates transgressing the boundaries between life and nonlife, the machine and the organic, and the human-created and naturally existing, is blasphemy to the true believer. Even the postmodern call for the end of "metanarrative" finds little assent in environmental circles, where the story of humanity's self-imposed exile from Eden functions as a veritable litany.[51]

Ecoradical thinkers may disdain modernity, but few are naive enough to think that we can return to a pristine premodern condition. Most hope rather for the advent of a "postmodern" world, one in which a revolution in consciousness brings an end to relentless economic growth and mind-numbing consumerism. This vision of postmodernity, however, is distinctly unlike that of most literary theorists; as Charlene Spretnak argues, the environmentalists' goal is the forging of a *constructive*, rather than a *deconstructive*, postmodernity (compare Michael Zimmerman's advocacy for a distinctly ecological version of "critical" postmodernism).[52] Waving aside the pastiche, superficiality, and cool skepticism of the scholarly *avant garde*, most ecoradicals rather seek a reassertion of religious or quasi-religious values founded upon a spiritualized ecology. They are, in other words, serious in their pursuit of "reenchantment." And many of them believe that such a New Age is indeed at hand. As Spretnak argues, "We are living through a period of spiritual searching and renewal on many parts of the planet. As the cultural grip of modernity weakens, the insights of spiritual teachings can be shared once again."[53] Seen in this light, the contrast with literary postmodernism could hardly be more pronounced.[54]

## SCIENCE AND THE POSTMODERN TURN

In the postmodern view, whether in its Green or its deconstructive versions, science is not to be dispensed with so much as domesticated. Modern science is to lose its place of privilege, with "non-Western"—and nonrational—"ways of knowing" elevated to an equal position. Yet if anything is to be "privileged," it is the so-called postmodern sciences, fields such as chaos theory that supposedly obliterate the reductionistic "old paradigm." According to ecoradical optimists, the "new sciences" will address a *living nature*, as a thoroughgoing vitalism—extending even to rocks—replaces the dismal philosophy of mechanism.[55] Modernist science, however, still has limited uses in the ecoradical program. The favored cosmology of environmental philosophy rests on the standard accounts of stellar and organic evolution, infused with a strong dose of pantheistic mysticism. And, of course, even the most reductionistic forms of science are still invoked when they warn of the thinning stratospheric ozone layer or some other potential ecological calamity.

While most specific elements of the ecocritique of standard science are of little significance, it is a matter of concern that ecoradicalism contributes to a generalized hostility toward science. As this volume demonstrates, reasoned inquiry is under attack from many sides. By spreading the message that science is no more reliable than shamanism, and especially by arguing that reason itself is the ultimate source of our environmental crisis, Green philosophers do little to enhance the public's ability to think clearly about the world and its very real problems. Earth-spirit worship may be psychologically beneficial for certain individuals, but at a societal level it is symptomatic of a dangerous tendency toward escapism.

A more particular peril is the discrediting of environmental science. Many contemporary environmentalists believe that the is/ought division must be erased as a vestige of now-discredited objectivist discourse.[56] Discarding this distinction is a moral imperative for many Greens; if one knows that a certain activity has negative ecological consequences, the argument goes, then one *ought* to oppose it. Moving thus from "is" to "ought" is a defensible, if philosophically convoluted, position. But shifting in the other direction from "ought" to "is" is absolutely fatal to the cause of knowledge. Following such a conviction, an environmental scientist may feel an obligation to "prove" that a certain suspect activity is indeed ecologically destructive, regardless of the evidence uncovered. The result of such a procedure would be bogus science, the "findings" of which could be disproved by anyone following standard "objectivist" practices.

The unsettled question here is whether practicing environmental scientists are susceptible to the ecoradical judgments undercutting the standards of their own profession. I imagine that the great majority are not, yet it is undeniable that philosophical radicalism has penetrated the entire environmental movement. Antienvironmentalists, at any rate, argue that much environmental science is tainted in precisely such a manner.[57] Such accusations may be baseless, but the statements of prominent ecophilosophers lend them a certain credibility. This is a shame, for in the struggle to preserve nature, trustworthy environmental science is essential.

## THE POLITICS OF ECOLOGICAL POSTMODERNITY

By embracing the concept of the postmodern, while infusing it with their own meaning, ecophilosophers signal their belief that the era of modernity is coming to an end. In the favored scenario, a revolution of consciousness will finish the bankrupt experiment that the West, dragging along the rest of the world, has undertaken over the past four centuries. Reason and science, it is said, not only have failed to fulfill basic human needs, but have led us to the brink of an ecological holocaust. Salvation thus requires nothing less than the repudiation of the Enlightenment foundations of modern, secular, liberal society.

But the belief that modernity is now yielding, or will soon begin to yield, to postmodernity is little more than wishful thinking. If taken at face value, the term "postmodern" actually makes sense only in relation to the artistic movement of modernism dating to the turn of this century. One can, indeed, contrast postmodernist architecture with modernist architecture—although the former may already be passé.[58] But modernism in this limited sense has only the most tenuous relationship with the broader phenomenon of modernity. And there is no evidence of which I am aware that suggests we are on the threshold of modernity's demise. The postmodern thesis, in this guise, is merely an academic repackaging of the 1960s' countercultural belief in the "dawning of the age of Aquarius." If anything, the opposite is the case. The end of the twentieth century is witnessing intensified global integration, capitalism, urbanization, and accelerated technological advance, or to put it in more abstract terms, the compounding of space-time compression.[59] Such developments are definitive of modernity. Far from being postmodern, the world we are entering is *hyper*modern.

This is not to say that all of the classical qualities of modernity are in ascendance. While economic and technological elaboration proceeds with a seemingly inexorable logic, cultural developments follow disparate pathways. Global integration sparks countervailing movements of cultural resistance, often intensifying local ethnic and religious identities. Moreover, hypermodernity—like all forms of modernity—generates profound discomfort among large segments of the population even in the most "advanced" countries. Antimodernism has perennial appeal.[60] But the postmodern versions of discontent are arguably the least significant in the world today, being limited to a few tens of thousands of (mainly Western) intellectuals, and to some extent a few million "New Age" sympathizers. More significant global challenges to modernity are associated with reactionary politics, identified with religious fundamentalism (Christian in the United States, Hindu in India, and Muslim from Morocco to Malaysia)[61] and extremist nationalism, which could easily turn into full-fledged fascism in much of the postcommunist world. Proponents of these antimodernist creeds[62] number at least in the hundreds of millions. The authoritarian and technocratic version of capitalism (epitomized by Singapore but now being recreated in China), marked by an unyielding hostility to democratic institutions and basic human freedoms, can be seen in this light as an attempt (like that of fascism) to merge elements of the antimodern with the hypermodern; the liberal politics of the Enlightenment are rejected, while science and technology are enthusiastically embraced.[63]

In the United States, religious antimodernists are now in alliance with right-wing modernists. Few seem to realize how deep the cleavages are between these two parties; fundamentally, Pat Robertson[64] has greater affinity with the leaders of Egypt's Islamic Brotherhood than he does with mainstream Republicans. But in regard to specifically environmental considerations, this gap is inconsequential; the "conservative movement" as a whole has apparently come to the conclusion over the past few years that environmental protection is a luxury we cannot afford. No matter how much scientific evidence is accumulated pointing to the dangers of continued fossil fuel addiction, the present leaders of Congress have evidently decided that no change is necessary. Remaining faithful to the Enlightenment ideal of (economic) freedom, they have lost sight of the overriding Enlightenment value of reason itself.[65] Much evidence suggests that it would require little more than changing the tax code to force the internalization of negative environmental externalities in order to initiate the age of solar power. But the vested interests of Exxon shareholders and Texan politicians call for other policies. The defense of reason in American environmental management thus lies within the left-wing branch of Enlightenment liberalism, represented by the Democratic Party. Yet even here, most political leaders (with the notable exception of Al Gore[66]) do not seem to realize the severity of the problems.[67] Only the mainstream environmental groups—which remain within the Enlightenment tradition—offer hope.

But just as with the antienvironmentalist coalition, environmentalism as a whole is composed of groups allied for the sake of convenience but rent by irreconcilable philosophical differences. The rift between ecoliberals and ecoradicals is colossal, but only those in the latter camp have the courage to recognize its existence; the unstated liberal credo seems to be that one must never counter—let alone attack—the radical leftist position for fear that doing so would only strengthen the right-wing opposition.[68] Liberals are also often morally cowed by radicals, and have been reluctant even to take credit for the many successes of their own environmental activism.[69] As a result, ecoradical philosophy is assumed by many to represent the voice of the movement as a whole. Inasmuch as their message provokes distrust (if not disgust) in many thoughtful voters, the more rabid Greens may inadvertently be strengthening the forces of antienvironmental reaction.

Ironically, radical ecophilosophy itself has deeply conservative roots. Such conservatism is most evident in the rejection of modernity and the romantic longing for a premodern existence, but it is also visible in the advocacy of localism and the rhetorical emphasis on stability, harmony, and family.[70] Anna Bramwell has traced the disturbing similarities between today's Green radicals and the eco-Nazis of the 1930s.[71] While several radical environmental philosophers, notably Michael Zimmerman,[72] have made brave and creative attempts to retain a core of progressivism rooted in Enlightenment thought while promoting a Green version of postmodernism, such an agenda is not easily advanced. Postmodernism and progressive liberalism remain diametrically opposed.

The common hope of postmodernists (whether of the skeptically deconstructive or the romantically optimistic school) that modernity is finished may be regarded as little more than fantasy. Barring an ecological apocalypse, technological progress and capitalist production seem likely to continue for a good while to come. But if we may be fairly certain that the modern world will live on for some time, we cannot assume that its exact path is mandated beforehand. This is the premise of real political action (as opposed to academic debates): attempting to guide social, economic, and technological policy within the context of a vigorous global modernity. The task of environmentalism, in my view, ought to be to direct public policy along the most "enlightened" tracks possible. To espouse a philosophy of antimodernism is to cede the terrain to those who envisage a technologically progressive but socially and environmentally stagnant or even regressive future.

## POSTSCRIPT

Five years ago I grew concerned that the increasing radicalism of Green philosophy posed a threat to the ecological movement. In *Green Delusions: An Environmentalist Critique of Radical Environmentalism*, I warned that an antiliberal,[73] antimodernist, antiprogressive, and ultimately antirational rhetoric could provoke a major backlash. In retrospect, it seems that my thesis has been at least partially vindicated. Even the most institutionalized forms of environmental protection are now under attack, and the membership of mainstream environmental organizations has declined sharply. The antiecological assault is multifaceted, but a leading line of attack is to portray the most fanatical assertions of the most strident Green as representative of environmentalism as a whole. Such a tactic may be mendacious, but that is beside the point. What is significant is that the mainstream environmental community has not been able to respond effectively, in part because it seems incapable of divorcing itself from the radical fringe. The philosophically inclined members of that fringe thus continue to enjoy tremendous intellectual clout. Remaining unwaveringly opposed to modernity, the Enlightenment project, and the tradition of liberal reformism, they have undermined the movement from within at the moment of its greatest vulnerability.

## NOTES

1 See Anna Bramwell, *Ecology in the 20th Century: A History.*

2 I have developed this general thesis at greater length in *Green Delusions: An Environmentalist Critique of Radical Environmentalism.* The present paper at-

tempts to address more precisely the connections among ecoradicalism, postmodernism, and anti-Enlightenment thought, and it relies heavily on recently published works that were unavailable when I was writing *Green Delusions.* One unavoidable problem with the present paper is that it tends to reduce all instances of ecoradical philosophy to a single voice. What I have attempted to do is to distill the core ideas of the most prominent schools; on the diversity of positions within ecoradical philosophy, see *Green Delusions.*

3 Val Plumwood, *Feminism and the Mastery of Nature*, p. 72.

4 See, for example, Carolyn Merchant, *Radical Ecology: The Search for a Livable World*, pp. 44–45.

5 Reason itself is often singled out for attack in ecoradical works. "Reason," Plumwood argues, "has been constructed as the privileged domain of the master . . ." (*Feminism and the Mastery of Nature*, p. 3). Michael Zimmerman informs us that "The patriarchal ego divinizes reason because it seems eternal in comparison with the mortal flesh" (*Contesting Earth's Future: Radical Ecology and Postmodernity*, p. 206).

6 See especially Carolyn Merchant, *The Death of Nature: Women, Ecology, and the Scientific Revolution and Radical Ecology*; Max Oelschlaeger, *The Idea of Wilderness: From Prehistory to the Age of Ecology*, pp. 76 ff.; Charlene Spretnak, *States of Grace: The Recovery of Meaning in the Postmodern Age*, pp. 150 ff.; and Jonathon Porritt, *Seeing Green: The Politics of Ecology Explained*, p. 105.

7 Plumwood, *Feminism and the Mastery of Nature*, pp. 118–119.

8 Merchant, *Radical Ecology*, p. 42.

9 On the Western drive to deny death, see Zimmerman, *Contesting Earth's Future.*

10 Rupert Sheldrake, *The Rebirth of Nature: The Greening of Science and God*, pp. 57–58.

11 Plumwood, *Feminism and the Mastery of Nature*, ch. 3. Plumwood does maintain, however (p. 112) that Descartes was a direct heir to the Platonic tradition.

12 This view, associated with Lynn White's 1967 essay "The Historical Roots of Our Ecological Crisis," does not seem to be as popular as it was two decades ago.

13 Spretnak, *States of Grace*, pp. 150–151.

14 See Merchant, *Radical Ecology*, p. 45; Brian Tokar, *The Green Alternative: Creating an Ecological Future*, p. 11.

15 See Marija Gimbutas, *The Godesses and Gods of Old Europe, 6500–3500 BC.* Gimbutas's ideas have infiltrated into environmental discourse especially through Eisler's popular rendition (*The Chalice and the Blade*).

16 Oelschlaeger, *The Idea of Wilderness*, p. 24.

17 See especially Plumwood, *Feminism and the Mastery of Nature*; see also Merchant, *Radical Ecology*, pp. 100–102.

18 One tactic is merely to merge this area into the West; Jim Mason, for example, discusses the "West's aggressive and rigidly monothesistic Judeo-Christian-Islamic Megareligion" (*An Unnatural Order: Uncovering the Roots of Our Domination of Nature and Each Other*, p. 30). While there are close linkages between these three religions, it is absurd—and insulting—to engulf the entire Islamic realm into an expanded West. (Just imagine trying to sell this idea in Algiers, Tehran, or Karachi!) Ecoradical world geography is often based on rank ignorance. Merchant, for example, believes that Asia lies in the southern hemisphere (*Radical Ecology*, p. 25), while Barry Commoner contends that most of the world's population resides below the equator (*Making Peace with the Planet*, p. 166). Lester Milbrath writes that before the coming of Europeans, North America, South America, and Africa were "populated by humans who made their livings mainly by hunting and gathering . . ." (*Envisioning a Sustainable Society: Learning Our Way Out*, pp. 14–15). Donella Meadows, Dennis Meadows, and Jorgen Randers maintain that the per capita gross national product of China and Indonesia declined during the 1980s, while that of the U.S.S.R. and other "rich" countries rose substantially (*Beyond the Limits: Confronting Global Collapse, Envisioning a Sustainable Future*, pp. 38–39; see especially the chart on page 38).

19 The modern consensus is that the industrial revolution bore little relationship to the scientific revolution (although Margaret Jacobs makes a good case for the opposite conclusion in *The Cultural Meaning of the Scientific Revolution*). It is for this reason that Lynn White, Jr. argued that one must search much farther back to locate the roots of the West's environmental problems ("Historical Roots").

20 Peter Gay, *The Enlightenment: An Interpretation. The Science of Human Freedom*, p. 147. It is also important to note the profound disparity between the thought of Bacon and that of Descartes. As Steven Toulmin—a critic of the philosophical foundations of modernity—shows, Bacon's "modest empirical methods" place him more in the Renaissance tradition of Montaigne than in that of Descartes (*Cosmopolis: The Hidden Agenda of Modernity*, p. 130).

21 Plumwood, *Feminism and the Mastery of Nature*, p. 125.

22 David C. Lindberg, *The Beginnings of Western Science: The European Scientific Tradition in Philosophical, Religious, and Institutional Context, 600 B.C. to A.D. 1450*, pp. 361–362.

23 In Descartes's case a metaphysics of mechanism based on actual physical contact did intrude, and partly as a result he did little to advance physics.

24 See Paul R. Gross & Norman Levitt, *Higher Superstition: The Academic Left and Its Quarrels with Science*, pp. 65–66.

25 Notably both of these thinkers included human beings within their mechanistic framework, thus obviating the Cartesian dualism that ecoradicals find so objectionable.

26 It is also important to note that the ecoradical charge that modern science involves a complete separation of reason from intuition and intellect from emotion is also without merit; as Gerald Holton reminds us, "science is the mobilization of the whole spectrum of our talents and longings, in the service of shaping more and more adequate world pictures." (*Science and Anti-Science*, p. 139).

27 See, for example, Jean Gimbel, *The Medieval Machine: The Industrial Revolution of the Middle Ages.*

28 See Margaret W. Conkey & Ruth E. Tringham, "Archeology and the Goddess: Exploring the Contours of Feminist Archaeology." See also Rene Denfeld, *The New Victorians: A Young Woman's Challenge to the Old Feminist Order*, for a feminist critique of the ecofeminist Goddess cult. The parallels between Denfeld's critique of radical feminism and my critique of radical environmentalism are striking.

29 Robert B. Edgerton, *Sick Societies: Challenging the Myth of Primitive Harmony.*

30 The thesis of Paleolithic affluence rests heavily on unwarranted extrapolation from hunter-gatherer societies studied in the 1950s and 1960s, particularly the !Kung "Bushmen" of the Kalahari; see, for example, the outdated and extremely problematic evidence that Oelschlaeger presents (*The Idea of Wilderness*, p. 358 note 19). More recent work (for example Melvin Konner, *The Tangled Wing: Biological Constraints on the Human Spirit*, pp. 371 ff.) has demolished the notion of "original affluence," as well as the easy and condescending suggestion that such people can be regarded as the living representatives of the Pleistocene (see also Martin W. Lewis, *Green Delusions: An Environmentalist Critique of Radical Environmentalism*).

31 See the discussion in Lewis, *Green Delusions.*

32 See Marshall G. S. Hodgson, *Rethinking World History: Essays on Europe, Islam, and World History.*

33 See the lengthy discussion in Martin W. Lewis & Karen E. Wigen, *The Myth of Continents: A Critique of Metageography.* As we argue here, the "postmodern deconstructive" perspective can be usefully employed to ferret out the ideological implications of (nonscientific) geographical (or metageographical) "constructs"—constructs such as Asia, the Orient, and the East. Such entities exist only as figments of the Western imagination. My own opinion is that postmodernism can be beneficial in limited quantities, but becomes deadly if taken to extreme positions.

34 See J. M. Blaut, *The Colonizer's Model of the World: Geographical Diffusionism and Eurocentric History.*

35 Jacques Gernet, *A History of Chinese Civilization*, p. 330. Of course, irrationalist philosophy also maintained a strong position in Sung China. It is notable, how-

ever, that Taoism—considered by many the quintessential ecophilosophy—was informed by a quest to deny death (and achieve immortality) much stronger than any found in the Western tradition. Moreover, Taoism, like European Renaissance "magic," actually contributed to the development of the foundations of scientific thought.

36 On the formation of Western civilization and its relationship with the church, see Robert Bartlett, *The Making of Europe: Conquest, Colonization, and Cultural Change 950–1350*, especially ch. 10.

37 On the idea that reason transcends culture, see Ernest Gellner, *Reason and Culture.* I do not believe that science is able to free itself *entirely* from culture simply because its concepts must, at some level, be expressed in *language*, which is necessarily culturally arbitrary. But the linguistic expression of scientific concepts—and the metaphors in which they are couched—is secondary to the whole endeavor (see Gross & Levitt, *Higher Superstition*, pp. 116, 121). Science, unlike poetry, is, after all, easily translated from language to language—with its mathematical foundations remaining entirely unaltered.

38 See Michael Soule's insightful comments ("The Social Siege of Nature," p. 154).

39 Stephen Kellert has also shown that in regard to *contemporary* attitudes, the East–West environmental divide is largely fictional ("Conceptions of Nature East and West").

40 Peter C. Perdue, *Exhausting the Earth: State and Peasant in Hunan 1500–1850*, pp. 86–87. One of the reasons why deforestation was so severe in southern and south-central China is because entire forests were repeatedly burned merely to obtain ash that could be sold elsewhere as fertilizer; see Perdue, pp. 35 and 247.

41 See, for example, Vandana Shiva, *Staying Alive: Women, Ecology, and Development.*

42 Many non-Western scholars (again, especially South Asians) do adhere to postmodern beliefs, especially in their "postcolonial form." My point here, however, is that the roots of postmodernism are to be found in the history of Western philosophy; see James F. Harris, *Against Relativism: A Philosophical Defense of Method.*

43 On the notion of a "Paleolithic consciousness," see Oelschlaeger, *The Idea of Wilderness.*

44 See, for example, Daniel B. Botkin, *Discordant Harmonies: A New Ecology for the Twenty-first Century.*

45 Donald Worster, "Nature and the Disorder of History," p. 77. Worster is actually very critical of the postmodern movement, but as the passage cited above shows, he is more than ready to employ one of its main tenets when he finds it convenient to do so. Merchant (*Radical Ecology*, p. 236) more clearly forwards the postmodern position in regard to ecological science: "ecology is likewise [i.e., like all other sciences] a socially constructed science whose basic assumptions and conclusions change in accordance with social priorities and socially accepted metaphors." For a more committed postmodern view of ecology, environmentalism, and environmental history, see David Demeritt, "Ecology, Objectivity, and Critique in Writings on Nature and Human Societies."

46 Cited in Conkey & Tringham, "Archeology and the Goddess," p. 13 in manuscript.

47 Marlo Morgan, *Mutant Message Down Under.* This book is officially on the fiction list, but the author—and her fans—portray it as factual. Actually, there is little to suggest that the author spent any time at all in the Australian outback, let alone that she lived and wandered extensively with a band of Aborigines. That such an absurd book (proclaiming, for example, that "the Real People" have perfect hearing because they allow bush flies to crawl into their ears to clean out wax and grit [p. 68]) is taken seriously in this country is itself a cause for serious concern.

48 As Paul Shephard writes, "Life is indistinguishable from a video game, one of the alternatives to the physical wasteland that the Enlightenment produced around us. As tourists flock to the pseudo history villages and fantasylands, the cynics take refuge from overwhelming problems by announcing all lands to be illusory. Deconstructionist postmodernism rationalizes the final step away from connection: beyond relativism to denial" ("Virtual Hunting Reality in the Forest of Simulacra,"

p. 25.). The one commonality of Shephard and the deconstructionists, it would seem, is utter contempt for the Enlightenment.

49 See Zimmerman, *Contesting Earth's Future*, p. 138.

50 Donna Haraway, "A Manifesto for Cyborgs: Science, Technology, and Socialist Feminism in the 1980s."

51 See Zimmerman, *Contesting Earth's Future*, pp. 184 ff. In actuality, literary postmodernism has its own metanarrative—the very term "postmodern" entails as much. As Harris writes, "The general attempt to 'delegitmate,' relativize, and 'hermeneuticize' knowledge surely amounts to one of the grandest narratives ever told about the nature of knowledge" (*Against Relativism*, p. 114).

52 Spretnak, *States of Grace*; see also Oelschlaeger, *The Idea of Wilderness*, p. 203; Zimmerman, *Contesting Earth's Future*, pp. 15–16. Zimmerman's position is uniquely synthetic: "In critical postmodernism, I discern an intriguing intersection of modernity's emancipatory goals, postmodern theory's decentered subject, and radical ecology's vision of an increasingly nondomineering relationship between humans and nonhumans" (p. 16). Young has attempted to label many of the same ideas as "postenvironmentalism." Pauline Rosenau argues that postmodernists can be divided into "skeptical" and "affirmative" variants (*Postmodernism and the Social Sciences: Insights, Inroads, and Intrusions*). Ecoradicals obviously fit on the "affirmative" side, where they find much company that is not necessarily explicitly ecological in orientation.

53 Spretnak, *States of Grace*, p. 27.

54 Several deconstructionists, not surprisingly, have argued against many of the key tenets of environmentalism, moderate and radical. Shane Phelan, for example, writes that "[Carolyn] Merchant is not alone in this failure to ask the nature of nature. The environmental movement as a whole has not done this. Whether it is the fully essentialized 'Nature' of the deep ecologists or simply the 'nature' of the National Resources Defense Council, environmentalists talk as if they know what nature is" ("Intimate Distance: The Dislocation of Nature in Modernity," p. 5). Much more combative is William Chaloupka's and R. McGreggor Cawley's assertion that environmentalists are easily transformed into "mirror-image rationalists whose discourse is shaped by a search for mastery" ("The Great Wild Hope: Nature, Environmentalism, and the Open Secret," p. 19). Of course, ecoradicals simply return the accusation.

55 This vision is most closely associated with the work of Rupert Sheldrake (*The Rebirth of Nature*). On the idea of ecological postmodern science, see Zimmerman, *Contesting Earth's Future*, pp. 70, 82.

56 See Merchant, *Radical Ecology*, pp. 79–80.

57 See Michael Fumento, *Science under Siege: Balancing Technology and the Environment.*

58 See Ada Louise Huxtable, "The New Architecture."

59 See David Harvey, *The Condition of Postmodernity.*

60 See Leszek Kolakowski, *Modernity on Endless Trial.* As he perceptively argues (p. 12), "It would be silly, of course, to be either 'for' or 'against' modernity *tout court*, not only because it is pointless to try to stop the development of technology, science, and economic rationality, but because both modernity and anti-modernity may be expressed in barbarous and antihuman forms."

61 In a brilliant little book, *Postmodernism, Reason, and Religion*, Ernest Gellner argues that three visions now contend for dominance: religious fundamentalism, postmodern relativism, and Enlightenment liberalism. Gellner seriously misinterprets religion in the United States, however (p. 5), remaining oblivious to the phenomenon of Christian fundamentalism. On the similarities among Muslim, Christian, and Jewish fundamentalists, see Bruce Lawrence, *Defenders of God: The Fundamentalist Revolt against the Modern Age.*

62 As Lawrence shows (ibid.), religious fundamentalists are opposed to only certain aspects of the modern world—others they embrace with vigor. As he pithily puts it, "Fundamentalists are moderns, but they are not modernists" (p. 1).

63 This strategy, however, is not without serious contradictions; witness, for example, the hair pulling now going on in the Singaporean government over access to the Internet. According to Samuel Huntington ("The Clash of Civilizations?"), the stance of Singapore and China merely reflects the antiliberal proclivities of a resurgent Confucian civilization. Such an idea ignores, however, the relative success of liberalism in Japan, South Korea, Taiwan, and (in a depoliticized version) Hong Kong. See Lewis & Wigen, *The Myth of Continents*, for an in-depth critique of the Huntington theses.

64 See Michael Lind, "The Rev. Robertson's Grand International Conspiracy Theory."

65 Of course, many economic conservatives argue that it is indeed rational to do little or nothing to reduce the emissions of greenhouse gases because of scientific uncertainties, high costs, and even potential benefits from global warming (see Gregg Easterbrook, "Stay Cool" and "Reading the Patterns"). Much of the thinking behind such an optimistic vision, however, rests on geographical ignorance. *The Economist*, for example, tells us that "Cold wastelands . . . might become fertile farmlands" ("Stay Cool," p. 11)—ignoring the fact that virtually all "cold wastelands" have thin and highly acidic soil.

66 See Al Gore, *Earth in the Balance: Ecology and the Human Spirit.* While Gore's environmentalism is in general moderate and reasonable, he does veer off occasionally into ecoradical fantasies, giving antienvironmentalists an easy target.

67 As Gregg Easterbrook insists, environmental protection has on the whole been successful in the United States ("Here Comes the Sun"). The same cannot be argued, however, if one considers the crucial global scale of analysis.

68 Liberal feminists like Rene Denfeld have, however, undertaken a concerted critique of radical feminism (*The New Victorians*), forming an incipient movement that gives me some hope for the future of a liberal political movement in this country.

69 See Easterbrook, "Here Comes the Sun."

70 I am indebted to Michael Freeden for pointing out the rhetorical similarities between ecoradicalism and conservatism.

71 See Bramwell, *Ecology in the 20th Century: A History.*

72 Zimmerman's effort to forge an ecological postmodernism is intellectually impressive (*Contesting Earth's Future*), and he must be given credit for attempting to retain a core of liberalism within ecoradical discourse.

73 On antiliberalism in contemporary ecoradicalism, see Richard Ellis, *The Illiberalism of Radical Egalitarianism.*

## REFERENCES

BARTLETT, ROBERT. *The Making of Europe: Conquest, Colonization, and Cultural Change 950–1350.* Princeton, NJ: Princeton University Press, 1993.

BLAUT, J. M. *The Colonizer's Model of the World: Geographical Diffusionism and Eurocentric History.* New York, NY: The Guilford Press, 1993.

BOTKIN, DANIEL B. *Discordant Harmonies: A New Ecology for the Twenty-first Century.* Oxford: Oxford University Press, 1990.

BRAMWELL, ANNA. *Ecology in the 20th Century: A History.* New Haven, CT: Yale University Press, 1989.

CHALOUPKA, WILLIAM & R. MCGREGGOR CAWLEY. "The Great Wild Hope: Nature, Environmentalism, and the Open Secret." In *In the Nature of Things: Language, Politics, and the Environment*, edited by Jane Bennett & William Chaloupka, pp. 3–23. Minneapolis, MN: University of Minnesota Press, 1993.

CONKEY, MARGARET W. & RUTH E. TRINGHAM. "Archaeology and the Goddess: Exploring the Contours of Feminist Archaeology." In *Feminism in the Academy: Rethinking the Disciplines*, edited by Abigail Stewart & Donna Stanton. Ann Arbor, MI: University of Michigan Press. In press.

COMMONER, BARRY. *Making Peace with the Planet.* New York, NY: Pantheon, 1990.

DEMERITT, DAVID. "Ecology, Objectivity, and Critique in Writings on Nature and Human Societies." *Journal of Historical Geography* 20 (1994): 20–37.

DENFELD, RENE. *The New Victorians: A Young Woman's Challenge to the Old Feminist Order.* New York, NY: Warner Books, 1995.
EASTERBROOK, GREGG. "Here Comes the Sun." *The New Yorker*, April 10, 1995, pp. 38–43.
———. "Stay Cool." *The Economist*, April 1, 1995, p. 11.
———. "Reading the Patterns." *The Economist*, April 1, 1995, pp. 65–67.
EDGERTON, ROBERT B. *Sick Societies: Challenging the Myth of Primitive Harmony.* New York, NY: The Free Press, 1992.
EISLER, RIANE. *The Chalice and the Blade.* San Francisco, CA: Harper and Row, 1988.
ELLIS, RICHARD J. *The Illiberalism of Radical Egalitarianism.* In press.
FUMENTO, MICHAEL. *Science under Siege: Balancing Technology and the Environment.* New York, NY: William Morrow, 1993.
GAY, PETER. *The Enlightenment: An Interpretation. The Science of Human Freedom.* New York, NY: W. W. Norton, 1969.
GELLNER, ERNEST. *Reason and Culture.* Oxford: Blackwell, 1992.
———. *Postmodernism, Reason, and Religion.* London: Routledge, 1992.
GERNET, JACQUES. *A History of Chinese Civilization.* Cambridge: Cambridge University Press, 1982.
GIMBUTAS, MARIJA. *The Goddesses and Gods of Old Europe, 6500–3500 BC.* Berkeley, CA: University of California Press, 1982.
GIMBEL, JEAN. *The Medieval Machine: The Industrial Revolution of the Middle Ages.* London: Penguin, 1976.
GORE, AL. *Earth in the Balance: Ecology and the Human Spirit.* Boston, MA: Houghton Mifflin, 1992.
GROSS, PAUL R. & NORMAN LEVITT. *Higher Superstition: The Academic Left and Its Quarrels with Science.* Baltimore, MD: Johns Hopkins University Press, 1994.
HARAWAY, DONNA. "A Manifesto for Cyborgs: Science, Technology, and Socialist Feminism in the 1980s." *Socialist Review* 15 (1985): 64–107.
HARRIS, JAMES F. *Against Relativism: A Philosophical Defense of Method.* LaSalle, IL: Open Court, 1992.
HARVEY, DAVID. *The Condition of Postmodernity.* Oxford: Basil Blackwell, 1989.
HODGSON, MARSHALL G. S. *Rethinking World History: Essays on Europe, Islam, and World History*, edited by Edmund Burke III. Cambridge: Cambridge University Press, 1993.
HOLTON, GERALD. *Science and Anti-Science.* Cambridge, MA: Harvard University Press, 1993.
HUNTINGTON, SAMUEL P. "The Clash of Civilizations?" *Foreign Affairs* 72 (1993): 23–49.
HUXTABLE, ADA LOUISE. "The New Architecture." *The New York Review of Books*, April 6, 1995, pp. 18–21.
JACOBS, MARGARET C. *The Cultural Meaning of the Scientific Revolution.* New York, NY: Alfred A. Knopf, 1988.
KELLERT, STEPHEN R. "Conceptions of Nature East and West." In *Reinventing Nature? Responses to Postmodern Deconstruction*, edited by Michael Soule & Gary Lease, pp. 103–121. Washington, DC: Island Press, 1995.
KOLAKOWSKI, LESZEK. *Modernity on Endless Trial.* Chicago, IL: University of Chicago Press, 1990.
KONNER, MELVIN. *The Tangled Wing: Biological Constraints of the Human Spirit.* New York, NY: Holt, Rinehart, and Winston, 1982.
LAWRENCE, BRUCE B. *Defenders of God: The Fundamentalist Revolt against the Modern Age.* San Francisco, CA: Harper and Row, 1989.
LEWIS, MARTIN W. *Green Delusions: An Environmental Critique of Radical Environmentalism.* Durham, NC: Duke University Press, 1992.
LEWIS, MARTIN W. & KAREN E. WIGEN. *The Myth of Continents: A Critique of Metageography.* Berkeley, CA: University of California Press. In press.
LIND, MICHAEL. "The Rev. Robertson's Grand International Conspiracy Theory." *The New York Review of Books*, February 2, 1995, pp. 21–25.

LINDBERG, DAVID C. *The Beginnings of Western Science: The European Scientific Tradition in Philosophical, Religious, and Institutional Context, 600 B.C. to A.D. 1450.* Chicago, IL: University of Chicago Press, 1992.

MASON, JIM. *An Unnatural Order: Uncovering the Roots of Our Domination of Nature and Each Other.* New York, NY: Simon & Schuster, 1993.

MEADOWS, DONELLA, DENNIS MEADOWS & JORGE RANDERS. *Beyond the Limits: Confronting Global Collapse, Envisioning a Sustainable Future.* Post Mills, VT: Chelsea Green Publishing, 1992.

MERCHANT, CAROLYN. *The Death of Nature: Women, Ecology, and the Scientific Revolution.* San Francisco, CA: Harper and Row, 1980.

———. *Radical Ecology: The Search for a Livable World.* London: Routledge, 1992.

MILBRATH, LESTER. *Envisioning a Sustainable Society: Learning Our Way Out.* Albany, NY: State University of New York Press, 1989.

MORGAN, MARLO. *Mutant Message Down Under.* New York, NY: Harper Collins, 1994.

OELSCHLAEGER, MAX. *The Idea of Wilderness: From Prehistory to the Age of Ecology.* New Haven, CT: Yale University Press, 1991.

PERDUE, PETER C. *Exhausting the Earth: State and Peasant in Hunan 1500–1850.* Cambridge, MA: Harvard University Press, 1987.

PHELAN, SHANE. "Intimate Distance: The Dislocation of Nature in Modernity." In *In the Nature of Things: Language, Politics, and the Environment*, edited by Jane Bennett & William Chaloupka, pp. 44–62. Minneapolis, MN: University of Minnesota Press, 1993.

PLUMWOOD, VAL. *Feminism and the Mastery of Nature.* London: Routledge, 1993.

PORRITT, JONATHON. *Seeing Green: The Politics of Ecology Explained.* Oxford: Basil Blackwell, 1985.

ROSENAU, PAULINE MARIE. *Postmodernism and the Social Sciences: Insights, Inroads, and Intrusions.* Princeton, NJ: Princeton University Press, 1992.

SHELDRAKE, RUPERT. *The Rebirth of Nature: The Greening of Science and God.* Rochester, VT: Park Street Press, 1991.

SHEPHARD, PAUL. "Virtual Hunting Reality in the Forest of Simulacra." In *Reinventing Nature? Responses to Postmodern Deconstruction*, edited by Michael Soule & Gary Lease, pp. 17–29. Washington, DC: Island Press, 1995.

SHIVA, VANDANA. *Staying Alive: Women, Ecology, and Development.* London: Zed Books, 1988.

SOULE, MICHAEL E. "The Social Siege of Nature." In *Reinventing Nature? Responses to Postmodern Deconstruction*, edited by Michael Soule & Gary Lease, pp. 137–170. Washington, DC: Island Press, 1995.

SPRETNAK, CHARLENE. *States of Grace: The Recovery of Meaning in the Postmodern Age.* San Francisco, CA: Harper San Francisco, 1991.

TOKAR, BRIAN. *The Green Alternative: Creating an Ecological Future.* San Pedro, CA: R. & E. Miles, 1987.

TOULMIN, STEPHEN. *Cosmopolis: The Hidden Agenda of Modernity.* New York, NY: The Free Press, 1990.

WHITE, LYNN, JR. "The Historical Roots of Our Ecological Crisis." *Science* 155 (1967): 1203–1207.

WORSTER, DONALD. "Nature and the Disorder of History." In *Reinventing Nature? Responses to Postmodern Deconstruction*, edited by Michael Soule & Gary Lease, pp. 65–85. Washington, DC: Island Press, 1995.

ZIMMERMAN, MICHAEL E. *Contesting Earth's Future: Radical Ecology and Postmodernity.* Berkeley, CA: University of California Press, 1994.

# IS ENVIRONMENTAL CANCER A POLITICAL DISEASE?

STANLEY ROTHMAN

S. ROBERT LICHTER

IN THE PAST TWO YEARS, several journalists and scholars have been chastising the mainstream environmental movement in a manner to which its leaders are not accustomed. In a series of articles in the *New York Times* in 1993 (March 21, 24, and 26), Keith Schneider essentially accused both environmental groups and the media of exaggerating the dangers to the environment and to persons from man-made chemicals. The articles were followed by David Shaw's series of essays in the *Los Angeles Times* (September 11, 12, and 13, 1994) and a major television documentary (ABC's "Are We Scaring Ourselves to Death?") making the same points. Similar arguments have been advanced in other articles and books by long-time supporters of environmental causes such as Gregg Easterbrook's *A Moment on the Earth* (1995) and Martin W. Lewis's *Green Delusions* (1992). Still other critiques, such as Charles T. Rubin's *The Green Crusade* (1994), have accused members of the scientific community of becoming politicized—i.e., of knowingly making judgments that do not comport with scientific evidence.

Needless to say, many in the environmental movement have sharply criticized these news stories, programs, and books, claiming that their own spokespersons accurately analyze the dangers they describe.

Are the revisionists correct? Have the cancer-causing dangers of man-made chemicals been exaggerated by environmentalists and/or scientists? If they have been, what are the reasons? We have a modest contribution to make to the discussion of these issues. In 1993, we surveyed a random sample of 400 members of the American Association of Cancer Researchers (AACR) who specialize either in carcinogenesis or epidemiology. Our response rate was sixty-five percent. The study replicated (with slight variations) a random survey of cancer epidemiologists and oncologists completed in 1984. An analysis of our

TABLE 1 Scientist Survey of Causes of Cancer: Sample

| Affiliation | | | Occupation | |
|---|---|---|---|---|
| Medical school | 46 | 63 | Faculty | 33 |
| Other academic | 17 | | Researcher | 34 |
| Government agency | 13 | | Physician | 1 |
| M.D.–private practice | 1 | | Administrator | 5 |
| Hospital | 3 | | Faculty/research | 23 |
| Industrial medical lab | 4 | | Other | 4 |
| Private industry | 7 | | | |
| Other | 9 | | | |
| | 100% | | | 100% |

1. 73% have *never* consulted for private industry; only 8% have done so more than three times.
2. 92% are currently involved in research on cancer causation or prevention.
3. 55% have published 40 or more professional journal articles.

results reveals that the revisionists do have the support of the scientific community (which does not appear to be politicized on this issue) for their views and that both journalists and environmental activists do seem to have been influenced in their perceptions by their ideological commitments. In short the answer to the question posed by the title to this essay would seem to be yes, at least to some degree.

Our sample was primarily made up of academics (see TABLE 1). Sixty-three percent of our respondents were on the faculties of a medical school or had another university affiliation. Another thirteen percent worked for a government agency. Only seven percent worked in private industry. Of those not working in private industry, seventy-three percent had *never* served as consultants for industry; only eight percent had done so more than three times. Those we interviewed were actively involved in research and publication. Ninety-two percent were currently involved in research on cancer causation or prevention. Fifty-five percent have published forty or more articles in professional journals.

Politically, the scientists tend to the liberal side of the political spectrum (see TABLE 2). Forty-eight percent classify themselves as liberal; twenty-eight percent classify themselves as moderate, and a mere seventeen percent consider themselves conservatives. At the same time a substantial majority consider themselves Democrats or lean toward the Democratic Party. The results are hardly surprising given that the scientists are predominantly academics even if some may also be M.D.s.

Given their political views, one might expect the scientists to be sympathetic to environmental concerns in this area. This turns out not to be the case. The cancer experts were asked (among a great many other things) to evaluate seventeen specific substances in terms of their contribution to human cancer rates in the United States. They rated each aspect on a scale of zero to

TABLE 2 Cancer Scientists: Marginals ($n$ = 401).

| Ideology | | Party | | | |
|---|---|---|---|---|---|
| Liberal | 48 | Democratic | 46 | Democratic | 62[a] |
| Moderate | 28 | Republican | 11 | Republican | 18 |
| Conservative | 17 | Independent | 37 | Independent | 13 |
| DK[b] | 8 | DK | 6 | DK | 7 |
| | 100% | | 100% | | 100% |

[a] Those who "lean" Democratic or Republican are assigned to the appropriate party.
[b] DK = do not know.

ten, where zero indicates that something makes no contribution to cancer rates, and ten indicates that it makes a very important contribution.

In presenting the experts' ratings on this dimension, TABLE 3 ranks each environmental agent according to the average (mean) score that it received. Also listed is the proportion of scientists who rated each choice as a "major" cause (7 to 10 on the scale), a "moderate" (4 to 6) cause, or a "minor" cause (0 to 3). The category containing a plurality of responses is indicated in boldface.

Cancer experts place tobacco in a league of its own among cancer agents. Indeed, only one out of twenty researchers rate smoking as less than a major carcinogen. Chewing tobacco is named as a major carcinogen by sixty-six percent of the scientists. That rating placed this form of tobacco well ahead of asbestos, the only other substance named as a major cancer agent by a majority (fifty-six percent) of experts.

Second-hand tobacco smoke is the only other substance deemed a major cause of cancer by a plurality of cancer experts (forty-six percent). Thus, various forms of tobacco accounted for three of the top four substances on this rating list. No other substance was rated as a major contributor to cancer rates by more than about one-third of those in the sample. Nor did any of the thirteen other substances receive a mean rating above 5.4 on the scale.

A cluster of five additional substances elicited a lower level of concern, with mean ratings ranging from roughly 4.6 to 5.4. In descending order, these included fat in the diet, the natural chemical aflatoxin, low fiber in the diet, dioxin, and alcohol. About one in three scientists rated high-fat diets, aflatoxin, and dioxin as major contributors to cancer rates, compared to one in four who expressed as much concern about low dietary fiber. But dioxin's average rating dropped below that of both dietary factors, owing to the relatively large proportion of experts (forty percent) who rated it as only a minor cause of cancer.

Beginning with dioxin, the remaining ten substances were all rated as minor causes of cancer by a plurality of those experts who expressed an opinion. Five of these were clustered within one increment on the rating scale—alcohol, EDB, radon, hormones, and DDT.

The scientists expressed the least concern about five substances whose cancer-causing potential has generated headlines, but which they consensually regarded as only minor contributors to cancer rates. Nuclear power, the

TABLE 3 Scientist Survey of Causes of Cancer: Ratings on 0–10 Scale[a]

| | | Level of Concern (%) | | | |
|---|---|---|---|---|---|
| Substance | Mean Score | Major | Moderate | Minor | DK |
| 1. Smoking tobacco | 9.19 | **95** | 4 | 1 | 0 |
| 2. Chewing tobacco | 7.34 | **66** | 19 | 14 | 1 |
| 3. Asbestos | 6.49 | **56** | 22 | 20 | 2 |
| 4. Second-hand smoke | 5.88 | **46** | 33 | 20 | 1 |
| 5. Fat in diet | 5.39 | 33 | **44** | 21 | 3 |
| 6. Aflatoxin | 4.85 | 34 | **48** | 13 | 4 |
| 7. Low fiber in diet | 4.83 | 26 | **41** | 31 | 2 |
| 8. Dioxin | 4.74 | 33 | 22 | **40** | 5 |
| 9. Alcohol | 4.59 | 17 | 33 | **48** | 2 |
| 10. EDB | 4.22 | 17 | 16 | 30 | **37** |
| 11. Radon | 4.00 | 18 | 28 | **49** | 5 |
| 12. Hormones[b] | 3.99 | 14 | 37 | **45** | 4 |
| 13. DDT | 3.83 | 21 | 20 | **52** | 7 |
| 14. Food additives, preservatives | 3.27 | 10 | 29 | **57** | 4 |
| 15. Nuclear power plants | 2.46 | 7 | 16 | **73** | 4 |
| 16. Alar | 2.18 | 6 | 12 | **64** | 18 |
| 17. Saccharin | 1.64 | 3 | 9 | **85** | 3 |
| 18. Other sweeteners | 1.19 | 1 | 6 | **83** | 10 |

[a] Major = rating of 7 to 10. Moderate = rating of 4 to 6. Minor = rating of 0 to 3. DK = do not know. Boldface = modal rating (plurality).
[b] As used in treatments.

pesticide Alar, and artificial sweeteners, such as saccharin and other sweeteners, generated mean ratings ranging from only about 1.2 to 2.5. All were regarded as minor cancer agents by large majorities, the size of which ranged from sixty-four percent (Alar) to eighty-three percent (sweeteners other than saccharin). No more than seven percent of cancer experts rated any of these substances as a major contributor to cancer rates. Tobacco and asbestos were rated by far the most dangerous substances in terms of their contribution to cancer rates.

These are almost exactly the same results as those obtained in the 1984 study conducted by Stanley Rothman and William Lunch. Thus they seem to reflect a long-term consensus rather than a recent shift of scientific opinion.[1]

We may compare these results with the positions taken by a sample of leading environmentalists from the major environmental organizations, to whom we asked many of the same questions (see TABLE 4).

Environmental leaders assigned higher risks than cancer researchers to eleven out of thirteen substances listed. Only tobacco (smoking) and sunlight attracted slightly more concern from the researchers. In the case of man-made substances, the differences in their ratings were frequently dramatic. At least twice as many activists as scientists detected "major" cancer threats from Alar, sweeteners other than saccharin, DDT, dioxin, food additives, and nuclear power plants.

TABLE 4 Activist Survey of Causes of Cancer: Ratings on 0–10 Scale[a]

| | | Level of Concern (%) | | | |
|---|---|---|---|---|---|
| Cause | Mean Score | Major | Moderate | Minor | DK |
| Smoking | 9.1 | **85** | 4 | 1 | 10 |
| Dioxin | 8.1 | **67** | 9 | 7 | 17 |
| Asbestos | 7.8 | **63** | 15 | 6 | 16 |
| EDB | 7.3 | 28 | 6 | 4 | **62** |
| DDT | 6.7 | **47** | 13 | 18 | 22 |
| Pollution | 6.6 | **40** | 37 | 2 | 21 |
| Sunlight | 6.3 | **41** | 38 | 8 | 13 |
| Fat in diet | 6.0 | **39** | 26 | 12 | 23 |
| Food additives | 5.3 | 19 | **48** | 12 | 21 |
| Nuclear plants | 4.6 | 22 | 18 | **31** | 29 |
| Alar | 4.1 | 16 | 23 | **32** | 28 |
| Saccharin | 3.7 | 12 | 22 | **42** | 24 |

[a] Major = rating of 7 to 10. Moderate = rating of 4 to 6. Minor = rating of 0 to 3. DK = do not know. Boldface = modal rating (plurality).

The differences were even greater at the other end of the scale, with scientists far more likely than activists to rate most substances as relatively "minor" causes of cancer. Thus, using the mean score of carcinogenicity, about twice as many researchers as environmentalists rated Alar and nuclear power as minor carcinogens; three times as many researchers placed DDT and asbestos at the low end of the scale; five times as many researchers regarded food additives and dioxin as minor threats; and a whopping seventeen times as many cancer specialists as environmental leaders (by thirty-four to two percent) saw pollution as a minor contributor to cancer rates. Interestingly, the ratings of scientists and activists do not differ that much on natural sources of cancer. It is on man-made potential carcinogens that the differences are really substantial.

How has this disparity between expert and activist opinion affected public perception on the issue? Specifically, how have the media communicated these debates to the general public and how accurately do the media messages represent the views of the scientific community on these issues? To ascertain this, we selected a representative sample of the most visible reports from television, news magazines, and leading newspapers. Specifically, we examined all news stories on this topic that appeared on the ABC, CBS, and NBC evening newscasts or in *Time*, *Newsweek*, and *U.S. News and World Report*, as well as stories on the front page of any section of the *New York Times*, *Washington Post*, and *Wall Street Journal*, between 1972 and 1992. This produced a total sample of 1206 news items.

The patterns of media coverage are very complicated and change over time. However, in general, as FIGURE 1 shows, the most serious sources of cancer in terms of frequency of mention in the mass media (including attribution to scientists) are quite at variance with the demonstrated views of the scientific community.

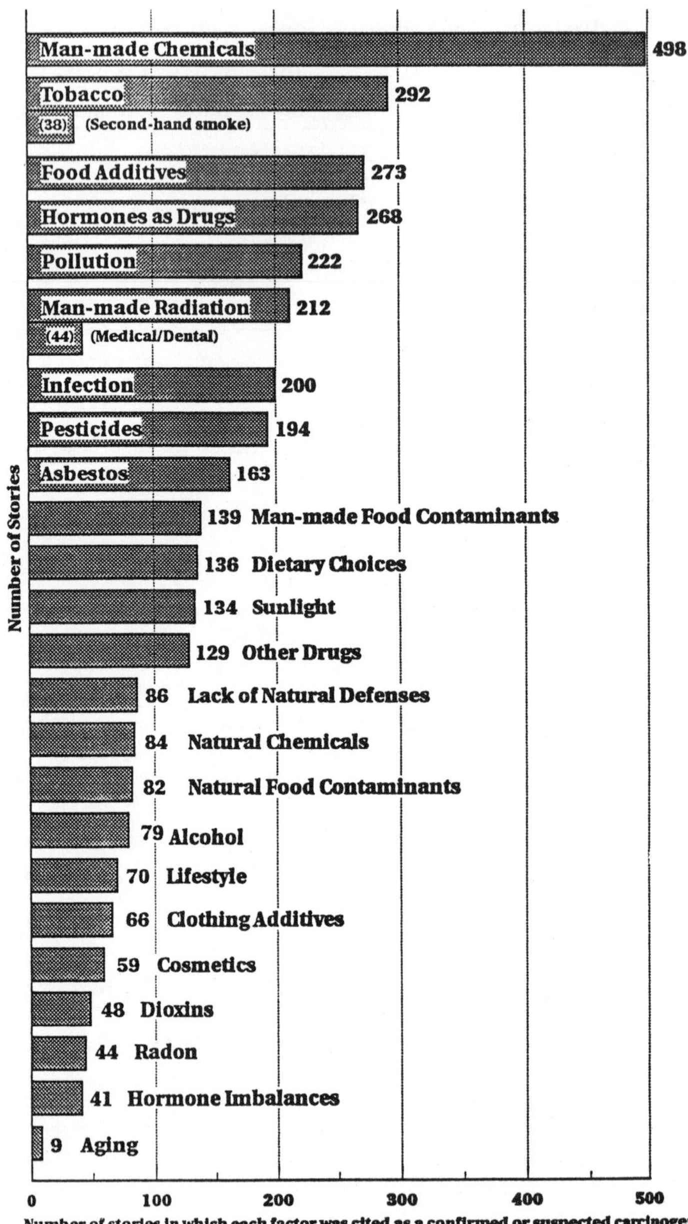


FIGURE 1 What the media say causes cancer, 1972–1992.

As we can see, man-made chemicals head the list of carcinogens mentioned by media sources, whether or not the sources are identified as expert. While tobacco is mentioned quite frequently, sunlight and dietary choices are quite low on the list of those factors covered.

Just as important, media coverage of various controversies about cancer is very much at odds with those of the scientific community and much closer to that of the environmentalist community. For example, as FIGURE 2 shows, a very large majority of media sources argue that we face a cancer epidemic. Yet less than a third of the cancer researchers surveyed believe that to be the case.[2]

The pattern is repeated on other controversial issues. For example, consider the question as to whether cancer-causing agents are unsafe at any dose. As a technical question, this involves the determination of tolerance levels for carcinogenic substances. A catchword of toxicology is the adage that the dose determines the poison, and most researchers apply this principle to suspected cancer agents. Respondents in our survey of experts disputed the assertion that carcinogens are unsafe regardless of the dose, by a margin of over two to one (sixty-four percent to twenty-eight percent, with the rest unsure).

If one turns to media coverage of the issues, however, one comes away with the opposite impression. Some sixty-six percent of media sources agree that cancer causing agents are unsafe at any dose.

Can the results of animal studies of suspected carcinogens be extrapolated to humans in order to assess the health risks associated with specific substances? This is a standard procedure for establishing carcinogenicity in accordance with federal regulations. This approach has long been controversial, because it involves giving very high doses of substances to animals and extrapolating the results to humans who are exposed to far lower doses of the same substance.

When we put this question to the expert community, the result was close to a consensual rejection of these procedures and the assumptions they involve. Only one in four cancer researchers (twenty-seven percent) endorsed the current practice of assessing human cancer risks by giving animals what is termed the "maximum tolerable dose" of suspected cancer-causing agents. More than double that number (sixty-three percent) disagreed, with the remainder unsure. According to the major media's quoted sources, however, the situation is more ambiguous, with fifty percent agreeing that these animal tests are adequate to assess the dangers of suspected carcinogens. One can, of course, argue that in the absence of good epidemiological data, we must use animal studies. That is no excuse, however, for giving the impression that such results are more solid than they actually are.

Finally, we asked scientists to evaluate the "zero risk" standard embodied in the Delaney clause, which holds that chemicals and additives must be banned from food and drugs if they are ever shown to cause cancer in any species. This principle has sparked public controversy ever since it led to the Food and Drug Administration's decision to ban saccharin in 1977, a decision that was never implemented because of public protest.[3] But the research community rejects this principle by an overwhelming seven-to-one margin (eighty-five percent to twelve percent, with the rest unsure). The impression given

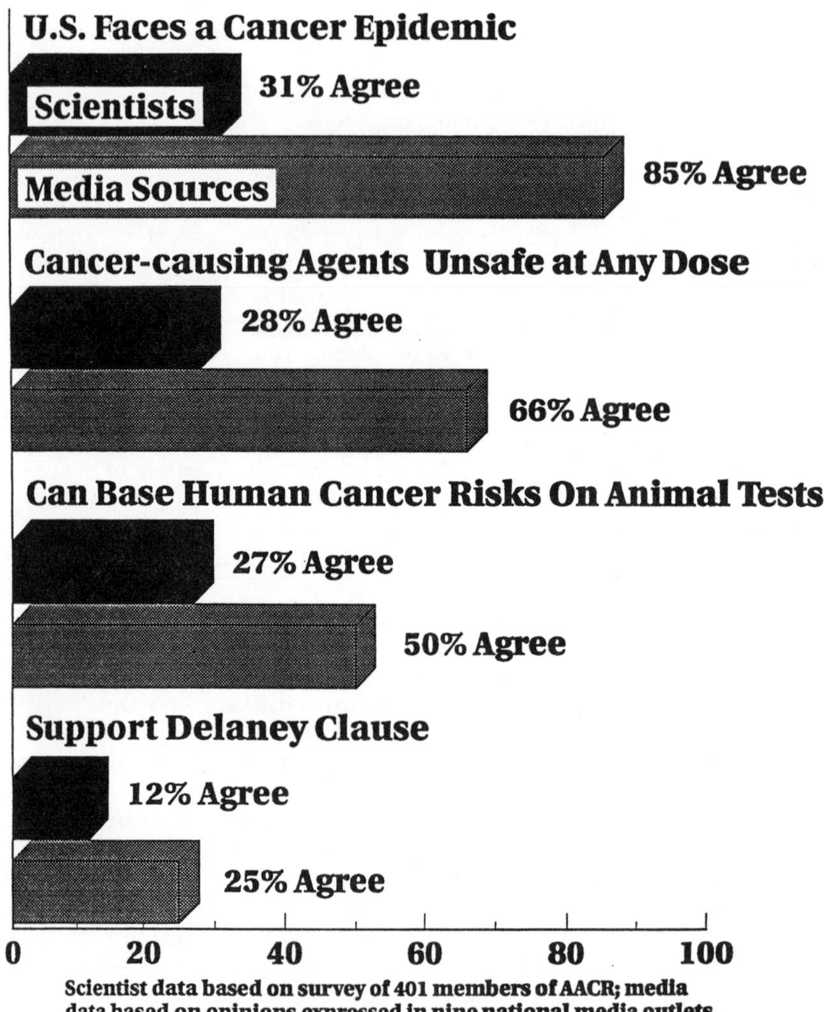


FIGURE 2 Cancer controversies: scientists vs. media.

by the media is similar, but some twenty-five percent of the experts cited by them support the Delaney clause, twice as many as do in our sample.

In summary, large majorities of the research community reject as over–risk aversive several propositions and practices that currently guide environmental cancer policy. Most cancer experts dismiss the popular notion of a cancer epidemic in the United States, and they attribute the observable rise in cancer rates to tobacco use and aging rather than the products of modern industry.

Most researchers also reject some of the principles that underlie the current regulatory approach to environmental cancer. These include the inference of human cancer risk from animal tests involving high dosages of suspected car-

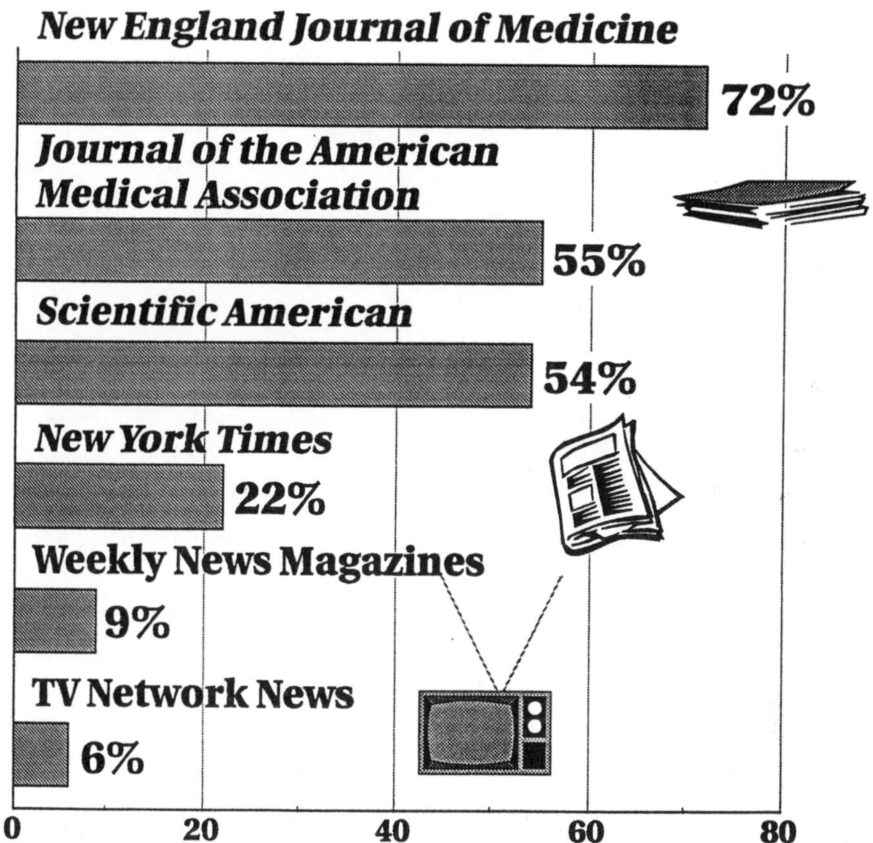


FIGURE 3 Scientists who trust cancer information from various media.

cinogens, the idea that cancer-causing agents are unsafe at any dose, and the analogous "zero risk" regulatory standard for evaluating food and drugs.

Media reports give a quite different impression. In fact, on those issues for which we have data, they are much closer to the views of environmental activists than they are to the views of the scientists we surveyed. For example, thirty-nine percent of activists support the Delaney clause; over sixty percent believe we are facing a cancer epidemic, and over fifty percent believe that cancer-causing agents are unsafe at any level.

It is little wonder that scientists regard media reporting of cancer issues as very poor. Their vote of no confidence in the media is recorded in FIGURE 3. It may come as no surprise that only one out of sixteen scientists in our sample (six percent) rate television news as a highly reliable source of information on

TABLE 5 Scientists's Reputational Ratings: Confidence in Individual Expertise on Environmental Cancer

| | Level of Confidence (%)[a] | | | |
|---|---|---|---|---|
| Individual Rated | High | Medium | Low | DK |
| Bruce Ames | 67 | 19 | 6 | 8 |
| Richard Peto | 57 | 15 | 3 | 24 |
| Richard Doll | 53 | 8 | 2 | 37 |
| Samuel Broder | 49 | 18 | 7 | 26 |
| Robert Weinberg | 47 | 18 | 11 | 24 |
| Joseph Fraumeni | 40 | 15 | 4 | 41 |
| Irving Selikoff | 30 | 15 | 4 | 51 |
| John Higgenson | 29 | 12 | 4 | 56 |
| Sidney Wolfe | 24 | 15 | 11 | 50 |
| Samuel Epstein | 24 | 20 | 17 | 40 |

[a] Ratings on zero to ten scale, where low = 0 to 3, medium = 4 to 6, and high = 7 to 10. DK = do not know.

environmental cancer, while nearly ten times as many (fifty-five percent) rate TV newscasts as unreliable sources. But the ratings were hardly any better for the weekly news magazines, despite their advantages of running longer stories on more leisurely deadlines. Only nine percent of cancer experts rate news magazines highly, compared to forty-nine percent who give them low reliability ratings.

Most striking of all, though, is the lack of scientific respect for the *New York Times*, which is renowned among journalists for its award-winning weekly science section. Fewer than one in four researchers (twenty-two percent) rate the *Times* as highly reliable in its cancer coverage. In fact, the proportion of cancer experts who rate the paper of record of the United States as unreliable exceeded the proportion who found it highly reliable by thirty to twenty-two percent. As a baseline for comparison, fifty-four percent rated *Scientific American* as highly reliable, while only eight percent call it unreliable.

Nor does the scientific community have much respect for most of those cited most frequently by the media as sources of opinion about cancer sources (see TABLE 5). Only twenty-four percent of the scientists polled express relatively high confidence in Sidney Wolfe and Samuel Epstein, two of the most widely quoted media sources, and frequently cited as experts by environmentalist organizations. On the other hand, sixty-seven percent of the scientists polled express great confidence in Berkeley biochemist Bruce Ames, who is regarded as a lackey of business by many activists. Close behind Dr. Ames in reputational ratings are Richard Peto and Richard Doll, whose conclusions about cancer causation are quite similar to those of Ames. (In fact, based on mean ratings that excluded respondents who did not rate each individual, Doll finished slightly ahead of Ames atop the list, while Epstein fell slightly behind Dr. Wolfe at the very bottom.)

Some of the major environmental groups receive similarly poor ratings. Only sixteen percent of the scientists rate the Environmental Defense Fund

TABLE 6 Scientists's Rating of Institutional Expertise on Cancer

| | Level of Confidence (%)[a] | | | |
|---|---|---|---|---|
| Institution Rated | High | Medium | Low | DK |
| National Cancer Institute | 92 | 7 | 0 | 1 |
| National Institutes of Health | 87 | 9 | 2 | 3 |
| Centers for Disease Control | 70 | 21 | 6 | 3 |
| World Health Organization | 64 | 28 | 6 | 2 |
| Center for Science and the Public Interest | 15 | 22 | 12 | 51 |
| Environmental Defense Fund | 16 | 29 | 21 | 33 |
| Tobacco Institute | 5 | 14 | 73 | 8 |

[a] Ratings on a zero to ten scale, where low = 0 to 3, medium = 4 to 6, and high = 7 to 10. DK = do not know.

highly, as compared to ninety-two percent who give high ratings to the National Cancer Institute or the eighty-seven percent who give that rating to the National Institutes of Health (see TABLE 6). Interestingly, a large proportion of scientists do not feel that they know enough about the two activist groups listed (Center for Science in the Public Interest and Environmental Defense Fund) to even rate them.

Why are the views of environmental activists so different from those of the scientific community on the dangers of man-made sources of cancer, though not on natural causes? Clearly many factors are involved, though scientific illiteracy does not seem to be a key variable. We argue that the views of some environmental activists derive in part from a broader set of political and social attitudes. The leaders of the mainstream environmental groups are not as far to the left as are the leaders of some of the more radical environmental groups, or as were an earlier generation of mainstream environmental leaders. They are, however, well to the left of the general population.[4] Our survey reveals that they are relatively secular, suspicious of business, and overwhelmingly supportive of the liberal wing of the Democratic Party. They favor extensive governmental economic intervention; they are also much more supportive of abortion rights, the rights of homosexuals, and preferential treatment for minority groups than is the population as a whole.

Thus, their belief that contemporary capitalism produces cancer is but another manifestation of a critical view of their society. Of course, the cancer researchers in our sample, which is largely composed of academics or government scientists, also hold liberal views on social issues. However, their political orientation does not carry over to their area of expertise. Further they are relatively moderate as compared to the environmental activists whose views contrast sharply with that of the general population on a wide range of subjects including environmental issues, despite the public's general environmental concern at this point in time.[5]

For example, while sixty-one percent of the general public believe that government regulation of business is harmful, only six percent of the environmentalists share that view. Similarly, forty percent of the public believes that business attempts to balance profits and the public interest—double the proportion (nineteen percent) of environmental leaders who agree.[6]

And why have reporters turned to the environmental groups for news about the environment and believed them rather than others? There are clearly many reasons for such behavior, not the least of which is the ability of environmental groups to organize effectively, and (though this can be exaggerated) journalists's penchant for "bad" news. A penchant for bad news, however, does not explain the willingness of the media in the 1950s to side with the scientific establishment on the fluoridation controversy of that period,[7] or to have reported the AIDS controversy with such care and sympathy, stressing that casual contact with an AIDS victim is not dangerous and convincing the general public of that fact.[8]

As Douglas and Wildavsky point out,[9] it is equally hard to make the case that concern about, say, pesticides is merely a function of a particular type of risk assessment engaged in by laymen as against scientists. People are more willing to take risks they conceive of as being imposed by themselves (e.g., driving an automobile) than those perceived as imposed on them by others (e.g., the use of pesticides). However this pattern should not affect *estimates* of risk, especially estimates of environmental activists who have been working in the field for some time.[10]

At least some critics have argued that media reportage can be explained, paradoxically, by the very professionalism of journalists who are trained to believe that there are two sides to every issue.[11] But it is difficult to argue that professional norms lead to giving equal credence to positions that are rejected by most experts in the field, or by reporting on issues in such a way as to lead the public to believe that a majority of experts support a given risk assessment when the experts actually do not.

We strongly suspect that one important reason for journalist's behavior is related to their own particular beliefs. A number of studies, several of which are summarized in Lichter and Rothman,[12] have demonstrated the liberalism of elite journalists, and their tendency to turn for information to public interest groups and "nonestablishment" scientists for information.[13] For journalists, political liberalism has gone hand-in-hand with support of environmental causes. In short, many journalists have taken their cues from environmentalists because they have found the views of such activists congenial on a variety of social issues. John Stossel, who reported ABC News documentary "Are We Scaring Ourselves to Death?," explained his reasons for believing environmental groups:

> We consumer reporters approached it from the bias that on the one hand is business which is greedy and has an ulterior motive and will distort the data, and on the other hand is the noble environmental group, which has no other motive than to help the public.[14]

The problem is not only that the news per se differs from expert opinion, leaving the public poorly informed about cancer risk. The deeper problem is that the public is being misinformed about the nature of expert opinion on cancer risk. The latter is not only a more serious impediment to an informed public, it is a more direct indictment of the journalistic profession. We may not know how to achieve objectivity. We can at least demand accuracy.

Why have such attitudes begun to change? That is another long story, which we can only discuss briefly here. The environmental movement in its early stage was clearly even more ideological than it is now.[15] As it has become part of the establishment and as its composition has changed, it has become more subject to criticism. After all, as Easterbrook points out,[16] the mainstream movement has won many of its battles, with results that can be seen. Beyond that, knowledge about cancer and cancer causation and the relation of both to public policy has been growing, partly stimulated by the environmental movement itself. The presence of an increasing number of scientifically trained persons in these fields has impacted on journalists who, themselves, have acquired more expertise. For example, the highly regarded Harvard Center for Risk Analysis has become an important source of expert information for journalists, as have such activist (more environmentally "conservative") groups as the American Council for Science and Health.

Finally, practicing scientists have begun to take a more active role in publicly assessing risks from various activities or substances. The recent report by the American Physical Society on the supposed relations between power lines and cancer is a good example of this development (*New York Times*, May 14, 1995).

What can we conclude? It is clear that environmental groups were, for many years (and still are to some extent), able to capture the interest and support of the media for claims about the findings of scientists that were simply inaccurate. There were many reasons for this, but the end result was that bad science outpointed good science for some time because ideology triumphed over evidence. Yes, environmental cancer, as defined by environmental groups, and the media, has partially been a political disease. It should be added that while a few scientists have lent their names to bad science, the great majority have not. That, to some, they appear to have done so is a reflection of the poor quality of media reporting.

In certain respects that trend has been turned around. However, while it would be nice to end this essay on a positive note, it is not clear how permanent the victory is. For that matter, we are not even sure that rationality has won a victory. The attention paid to revisionists like Easterbrook, Lewis, and Schneider does not mean that their views now dominate within the mass media. Things are still up for grabs. Indeed, more journalists have attacked them than have supported them, or so it seems.

Beyond that, as Gross, Levitt,[17] and others have pointed out, a wave of antiscience is now sweeping segments of the academy manifesting itself as "deep ecology," "ecofeminism," "Afrocentrism," and the like. While support for such ideologies is currently rather limited, the evidence would seem to indicate that these views are making progress among important groups in the population, and that they are relatively immune to rational criticism. Whatever the reasons for this phenomenon, it illustrates again that science is a rather fragile endeavor, despite its successes. Human beings do not naturally think scientifically.[18] The struggle against irrationality and against the bending of science to wishes of one kind or another never ends.

## NOTES

1 The complete results of both studies will be published S. Robert Lichter & Stanley Rothman, *Environmental Cancer as a Political Disease.*

2 We compared those stories in which reporters do not attribute views to anyone in particular, but simply state them as conclusions, to those in which such statements are specifically attributed to one or another expert. The results were the same. According to media stories the views of experts closely resemble those of environmental activists.

3 Aaron Wildavsky, *But Is It True?: A Citizen's Guide to Environmental Health and Safety Issues.*

4 For a discussion of the earlier generation of activists see Stanley Rothman, "Environmental Cancer and the Environmental Movement" and Martin Lewis, *Green Delusions: An Environmental Critique of Radical Environmentalism."* For discussions of the contemporary environmentalist fringe see the latter and Paul R. Gross & Norman Levitt, *Higher Superstition: The Academic Left and Its Quarrels with Science.*

5 For a recent survey see Everett Ladd & Karlyn Bowman, *Attitudes toward the Environment.*

6 The views of the environmental leaders are derived from our study; those of the public are from 1993 national surveys by Gallup, Roper, and CBS/*New York Times.*

7 Allen Mazur, *The Dynamics of Technical Controversy.*

8 *Media Monitor*; H. G. Pope & S. R. Lichter, "The Reporting of AIDS."

9 Mary Douglas & Aaron Wildavsky, *Risk and Culture.*

10 For a short review of the psychological literature on the differences in outlook regarding risk assessment between laymen and experts, see Clayton Gillette & James Krier, "Risks Courts and Agencies."

11 Michael Schudson, "How News Becomes News."

12 Robert S. Lichter & Stanley Rothman, *The Media Elite.*

13 June Goodfield, *Reflections on Science and the Media.*

14 Quoted by David Shaw in the *Los Angeles Times,* September 11, 1994, part a, p. 32.

15 Stanley Rothman, "Environmental Cancer and the Environmental Movement."

16 Gregg Easterbrook, *A Moment on the Earth: The Coming Age of Environmental Optimism.*

17 Gross & Levitt, *Higher Superstition.*

18 Lewis Wopert, *The Unnatural Nature of Science*; James Alcock, "The Belief Engine."

## REFERENCES

ALCOCK, JAMES E. "The Belief Engine," *Skeptical Inquirer*, May/June 1995, pp. 14–18.

DOUGLAS, MARY & AARON WILDAVSKY. *Risk and Culture.* Berkeley, CA: University of California Press, 1982.

EASTERBROOK, GREGG. *A Moment on the Earth: The Coming Age of Environmental Optimism.* New York, NY: Viking Press, 1995.

GILLETTE, CLAYTON P. & JAMES E. KRIER. "Risks Courts and Agencies." *University of Pennsylvania Law Review* 138 (April, 1990): 1028–1109.

GOODFIELD, JUNE. *Reflections on Science and The Media.* New York, NY: American Association for the Advancement of Science, 1981.

GROSS, PAUL R. & NORMAN LEVITT. *Higher Superstition: The Academic Left and Its Quarrels with Science.* Baltimore, MD: Johns Hopkins Press, 1994.

LADD, EVERETT CARLL & KARLYN H. BOWMAN. *Attitudes toward the Environment.* Washington, DC: American Enterprise Institute, 1995.

LEWIS, MARTIN W. *Green Delusions: An Environmental Critique of Radical Environmentalism.* Durham, NC: Duke University Press, 1992.

LICHTER, S. ROBERT & STANLEY ROTHMAN. *Environmental Cancer as a Political Disease.* In press.

LICHTER, S. ROBERT, STANLEY ROTHMAN & LINDA LICHTER. *The Media Elite.* Washington, DC: Adler and Adler, 1986.

MAZUR, ALLEN. *The Dynamics of Technical Controversy.* Washington, DC: The Communications Press, 1985.

*Media Monitor* (December 1987).

POPE, H. G. & S. R. LICHTER. "The Reporting of Aids." *Journal of the American Medical Association* 212,14 (1989): 1949–1950.

ROTHMAN, STANLEY. "Environmental Cancer and the Environmental Movement." *Ecology Law Quarterly* 21,1 (1994): 387–429.

RUBIN, CHARLES. *The Green Crusade: Rethinking the Roots of Environmentalism.* New York, NY: The Free Press, 1994.

SCHUDSON, MICHAEL. "How News Becomes News." *Forbes Media Critic* (June 1995): 76–85.

WOLPERT, LEWIS. *The Unnatural Nature of Science.* Cambridge, MA: Harvard University Press, 1993.

WILDAVSKY, AARON. *But Is It True?: A Citizen's Guide to Environmental Health and Safety Issues.* Cambridge, MA: Harvard University Press, 1995.

# OLD MESSAGES
## Ecofeminism and the Alienation of Young People from Environmental Activism

RENE DENFELD

THE ENVIRONMENT is a relatively new issue for the feminist movement. The term *ecofeminism* was coined in 1974 by the French writer Françoise d'Eaubonne to describe a feminist-based ecological movement. But it was not until the 1980s that ecofeminism began growing in popularity, both in feminist circles and the environmental movement. The growth of ecofeminism within the women's movement is the result of several different trends, including academic feminism extolling women's "difference," the emphasis on female victimization in feminist theory, and the impact of New Age, Jungian, and recovery causes on the movement. Overwhelmingly, ecofeminism is one result of the feminist retreat from political and economic activism into realms of morality. This has been happening over the last fifteen or so years, leading to a movement that bears little resemblance to the reform-minded feminism of the late 60s and 70s.[1]

Whether dealing with pornography or the environment, most feminist trends today have only dubious connection (to borrow an ecofeminist term) with delivering equality to women. Instead, current feminism is a *revolutionary* ideology. Like the Australian liberal feminist Beatrice Faust, I am using the term "revolutionary" instead of "radical," because many of the goals of equity feminism can be considered radical, whereas the current feminist outlook is one that eschews such reform in favor of social revolution.[2] Fighting for child care programs, economic equality, and other liberal feminist aims is considered by the revolutionaries to be tantamount to putting bandages on a sick system, if not colluding with the patriarchy.

Though not a leading cause, ecofeminism has had an impact throughout the movement, gaining influence in feminist organizations (the National Organization for Women, for example, includes environmental issues on its agenda) and assigned in some women's studies courses. The reading list for an introductory women's studies class at the University of New Hampshire, for instance, includes a section titled "Ecofeminism: Our Mother Earth." In other women's studies programs, students are taught ecofeminist theory under teachings on goddess worship and feminist spirituality. Gloria Steinem lauds ecofeminism in her bestseller *Revolution from Within*, for helping women "recover a lost identity."[3] Outside of a few liberal feminists, ecofeminism has gone uncriticized, gaining an implicit stamp of approval.

The purpose of this paper is to explain and criticize the ecofeminist philosophy, and put the cause in context of younger people, who are quite concerned about environmental issues as a group, but who find the messages and aims of ecofeminism—and the environmental movement that adopts it—unappealing.

The basic tenet of ecofeminism is that environmental oppression is linked to female oppression. "It is our belief that man's dominion over nature parallels the subjugation of women in many societies, denying them sovereignty over their lives and bodies," the "Declaration of Interdependence" from the Women's Environment and Development Organization (WEDO) of the Women's Foreign Policy Council declares. "Until all societies truly value women and the environment, their joint degradation will continue."[4]

According to ecofeminists, the root of environmental ills is male sexism. It is men who have dominated the earth, raped "her." They oppress the earth for the same reason they oppress women: hatred of the "feminine," or what ecofeminist and spiritualist Charlene Spretnak terms "patriarchal fear and resentment of the elemental power of the female."[5]

This linking of environmental problems with women's oppression highlights an important aspect of current feminist thought, one that needs to be addressed in order to understand ecofeminism. Just about everything in popular feminist theory boils down to male oppression of women: if we cut down rainforests, it is because men hate women; if we look at pornography, it is because men hate women; if we go to war or build weapons or pass punitive policies against the poor, it is because men hate women.[6] Whatever the problem, you can be sure that it is rooted in male hatred of women—and whatever women do, no matter how much we might feel to the contrary, we are victimized and defined by this hatred. While this egocentric attitude may generate attention, it loses credibility by ignoring the sticky realms of racism, classism, and other social ills that are often perpetuated by women—and victimize men.

In order to tie such disparate issues as the environment and women's inequality together, ecofeminist philosophy points to the "androcratic" values allegedly underlying both. Following popular feminist theory, they point the finger of blame at hierarchal thoughts and practices as the foundation of male oppression. Ecofeminist Judith Plant writes that "the idea of hierarchy has been used to justify social domination; and it has been projected onto nature, thereby establishing an attitude of controlling the natural world."[7] Plant asserts that "there are, in fact, no hierarchies in nature"—something that my dog might disagree with.

It is easy to see where this slippery slope leads. Anything that involves hierarchy, or "power-over," is thought to be male and oppressive, from meat eating to using rational thought.[8] For this reason, science, technology, reason, and other examples of what is seen as a mind/body split epitomize evil in the popular feminist view.

A class at George Washington University on feminist theory, for example, claims to challenge the "masculinist world views which have structured the public world. . . . We will initiate our inquiry by demystifying malestream 'theory,' 'reason,' and 'science.'" Students keep a journal to help them reject "the binary dualisms of male 'science' which separates thought from feeling and reason from emotion." The problem, as many current feminists see it, is that reason and technology have allowed people to become dislocated—in a spiritual, if not physical sense—from their environment. The answer, then, is to feel and think everything all at once, without any hierarchal ordering. This mulligan stew approach to life is seen as the Answer To It All.

Even technology aimed at relieving environmental problems is attacked by some ecofeminists as yet another manifestation of the real problem, which is male assumption of control. In short, ecofeminist theory is "passive" theory. It condemns the very concept of change, which requires action—and power over.

Fundamental to the ecofeminist philosophy is the notion that women have retained a connection with the natural world that men have lost. The reason women are capable of this connection—and supposedly do not abuse the environment—is those biological qualities that differentiate women from men: namely, the fact that we can give birth. Arisika Razak asserts that birth can "serve as the nucleus around which to build a paradigm for positive human interaction."[9] Female reproduction, she writes, is the model for a new age built on harmony and transcendence.

According to ecofeminism, women's reproductive systems give us special insights and consciousness. Charlene Spretnak, for instance, writes that women are "predisposed" to connect with nature through menstruation, orgasm, pregnancy, natural childbirth, and motherhood.[10] As a woman as well as a liberal feminist, I am always fascinated by the idea that getting my period is—or should be—a spiritual experience. Such assumptions are based on the idea that women are controlled by their bodies: that what happens in our uteruses determines our consciousness, our character—and our social worth.

In short, ecofeminist philosophy denounces the "male" *mind* at the same time it exalts the female *body*, and all the notions of traditional femininity associated with it: connection with the earth, spirituality, emotionalism, irrationality, and intuition.

This animalistic, biologically driven concept of women may sound extreme. Yet it is remarkably common throughout ecofeminist work. The group Ecofeminist Visions Emerging sums it up by declaring that "the patriarchal mindset is obsessed with transcending our *natural animal condition*— epitomized by a disdain for the female body which gives birth, bleeds, and lactates. Women's bodies and disabled bodies are both glaring reminders that "man" is not god" (emphasis added.)[11]

It is the "natural animal condition" of women that ecofeminists exalt. Women's bodies, not women's minds, will save the planet. Brian Swimme claims it is "primal peoples and women generally" who have an "holistic, poetic vision." This is because "women are beings who know from the inside out what it is like to weave the Earth into a new human being."[12] Presumably sterile women are bereft of this consciousness.

This gross reduction of women to our reproductive organs is enough to alarm any thinking person. Ecofeminist Carolyn Merchant admits that an analysis that "makes women's essence and qualities special ties them to a biological destiny that thwarts the possibility of liberation."[13] She is right. This is precisely what ecofeminism does. It is politics of the underdog, wherein women cannot gain until we exalt powerlessness. We cannot win for losing.

Implicit in the ecofeminist analysis is the notion that women are not *capable* of male thoughts and practices. The ecofeminist vision does not just revive traditional views of women as the moral saviors of the world, it revives notions of women as less intelligent than men, and inherently incapable of wielding power.

The ecofeminist attack on new reproductive technologies is a good example of where this line of thinking can lead. Many ecofeminists (and quite a few other feminists) believe that new technologies, such as *in vitro* fertilization, are yet another form of male oppression. Some attack almost all medical intervention in pregnancy, such as Cesarean sections, and others even skirt close to condemning all birth control. Maria Mies and Vandana Shiva, coauthors of *Ecofeminism*, claim that the separation of sex from procreation in liberal philosophy makes women "experience themselves as passive and alienated from their own bodies, their procreative capacities and from any subjectivity."[14]

Here the biological essentialism of ecofeminism collapses completely into sexism. The ability (or need) for a woman to separate herself from procreation is now considered *antifeminist.* Given that control over reproduction is so necessary for the majority of working American women—because of our economic situation—this stance is not just remarkable for its sexism, but its classism as well. It seems unthinkable to ecofeminists that allowing women to separate from the reproductive elements of their bodies might not lead to passivity and alienation from themselves, but to *increased* connection with their sexual elements. After all, the uterus is not the clitoris.

In the ecofeminist view, whatever positive results women gain through reproductive technology are really our loss. Shiva, for instance, writes that the new reproductive technologies negate women's "intimate link" and "direct organic bond" with their fetuses. In this way women become "inert containers." Birth control, *in vitro* fertilization, and surrogate motherhood are all technologies that can be abused, such as racism in mandated Norplant. The assumption, however, that such technologies are inherently abusive diminishes the experiences of individual women, who may feel anything but violated by their use.

Mies and Shiva do not seem to worry about the future of birth control in the ecofeminist vision. They claim that "wise women" of the past knew "many" methods to control reproduction, but unfortunately these wise women were annihilated in alleged mass witch killings in preindustrial

Europe, a much-beloved feminist theory which is also much disputed. Still, according to Mies and Shiva, once we understand that sexual intercourse is "a caring and loving interaction with nature" we will naturally "find birth control methods which do not harm women."

What is most remarkable about ecofeminism is not what it says, which women have heard for centuries now, but what it does not say. It does not say that women have gained an incredible amount of ground in the past several decades; it does not say that women do have power in our society; and it does not say that poverty and other economic troubles clearly play a major role in environmental problems. To give serious analysis to these issues would be tantamount to admitting that often the problem is not technology or so-called male thinking, *but lack of it*, especially in those areas and cultures in which poverty is rampant.

It would also be an admission that women play a role in environmental ills—and not just as unwitting dupes of or collaborators with men. Ironically, in order to make the environment gender related, ecofeminists must downplay women's actions and stress instead a group identity. The result is women's needs are made secondary to saving the environment. In a conflict between equality and the environment—whether in women joining trade unions or becoming scientists—the ecofeminist philosophy finds itself opposed to women's rights.

Outside of their title, ecofeminists have little to do with the original aims of modern feminism. In my mind, the feminist rejection of rationality and science is actually a rejection of equality, gussied up in fancy academic discourse and New Age trappings. What ecofeminists are truly rejecting is the public realms of economics, the workplace, politics, and the sciences: the very realms that can deliver full equality to women.

Ecofeminism is more religion than environmental activism. Feminist spiritualism and goddess worship form the base of many leading ecofeminist writers and activists. In fact, there is much crossover between goddess worship, feminist witchcraft, and ecofeminism. Carol Christ, for instance, is a leading feminist spiritualist who claims the crisis facing the earth is "at root spiritual." "The preservation of the Earth," she enthuses, "requires a profound shift in consciousness: a recovery of more ancient and traditional views that revere the connection of all beings in the web of life."[15]

The vagueness of Christ's recipe for earth preservation is standard in ecofeminism. Few ecofeminists ever draw a clear picture of just what future they envision, or how we will ever arrive there. Instead, their vision of the future is an echo of the past. Ecofeminists harken glowingly back to a fanciful, if inaccurate, rendition of prehistoric societies as utopian, goddess-worshiping Edens.

According to many feminist writers, such as Marija Gimbutas, Riane Eisler, Marilyn French, Starhawk, and others, prehistoric Europe was home to a continent-wide goddess cult. From the Upper Paleolithic period (the Old Stone Age) through the Neolithic (the New Stone Age, with the development of agriculture), people supposedly lived in harmony with their environment, worshiping the fertility and sacredness of women. It was an idyllic existence, free of violence, warfare, and power-over.[16]

That is, until men ruined everything. The saga of the goddess has a villain, and naturally that villain is male. Starting around 4400 BC, or so the theory goes, a tribe of vicious male-god worshiping horsemen known as Kurgans (presumably it was an all-male society; at any rate, women are not mentioned) staged a series of assaults on the peaceful goddess worshipers. Within a few thousand years, they conquered the goddess cults, and the downfall of women began.[17]

Archaeologists, however, strongly dispute the likelihood of a monolithic goddess religion, especially over thousands of years and profound transitions from foraging to farming, from crude huts to organized towns. Despite feminist convictions that these various peoples were peace-loving and egalitarian, the evidence indicates warfare, human sacrifice, and violence. Further, the idea that violent horsemen conquered these supposed goddess-worshipers also does not hold up under investigation. In fact, some evidence suggests that the change in culture came about because of deforestation and environmental misuse, not mysterious invaders.[18]

Yet this dramatic tale of paradise lost is frequently cited by ecofeminists as evidence of the turning point of humankind, when hierarchy reared its ugly male head. It has gained an amazing amount of credence within feminist circles, even filtering into mainstream culture. The evidence is not the important part to most ecofeminists: the message is. Put women in charge, and presto—we will once again cavort happily with mother earth.

The religious myths of ecofeminism do not really challenge traditional religious beliefs. They only replace them. The same dichotomies, the same struggles between good and evil—between the female body and the male mind—and the same rejection of counterevidence make these "new" forms of spirituality more similar to religious fundamentalism than the liberal basis of modern feminism.

The fundamentalism of ecofeminism is apparent in the striking similarity between the ecofeminist theory of prehistory and Nazi theory. As anthropologist David W. Anthony writes,[19] both ecofeminists and the German Nazis have relied upon distorting evidence to support their notion of a peaceful prehistoric Europe invaded by patriarchal IndoEuropeans, the "Kurgans."

The only difference is that while ecofeminists present the invasion as the original sin, leading to all the world's woes, the Nazis paraded it proudly as the beginnings of the Aryan race. Of course, both theories are equally false—grounded in wishful thinking, and dismissive of objective science. In fact, without the Nazis, who laid out the groundwork notion of migrating Aryans, it is possible that ecofeminist theory on the Kurgan invasion would not exist.

It is a history that speaks not only to the conservatism[20] of ecofeminism, but hints at its racist and classist undertones. "Perhaps we—White women," ecofeminist Catherine Keller claims, "can only begin to regain the wisdom and power of relation as we move into contact with non-White, nonpatriarchal, and nonmodern models of connection with the physical world."[21] This remarkably patronizing characterization of all "non-White" people as little savages, charmingly free from the taint of modernity, is unfortunately common in ecofeminist work. Notice that Keller advocates coming into "contact" with non-Whites, not making a society free of such distinctions. While ecofeminists romanticize an undifferentiated category of "primal" and "non-White" peoples (as if they are all one and the same), they seem unconcerned about difficulties facing minorities and the poor today, except as they can be used as abject examples of white male oppression.

As a writer on modern feminism, I cannot offer an expert analysis on the state of the environmental movement today. It does seem, however, that ecofeminism is gaining popularity within environmental circles. With their strictures against hierarchy—whether in group leadership or using rational thought—ecofeminist groups are noteworthy for not being able to accomplish much of anything.

The future of the environmental movement, like the future of feminism, is dependent on younger people getting involved. In order to do this, the movement must include and address the perspectives of younger generations. It is my belief that, as a model for environmental activism and theory, ecofeminism fails miserably at doing just that.

Studies show that people under age 35—the first "postfeminist" generation—are at the heart of a massive change in values. In general, younger people today, far more than older generations, are rejecting "traditional" values. The narrow definitions of gender, punitive moral codes, and sexual puritanism that marked past generations are diminishing among the younger.[22]

This value shift is the result of economic as well as cultural changes: young women today have to work, and this need has led to immense changes in the way that girls (and boys) are raised, educated, and come to view their position in society. Women under the age of twenty are completely acclimated to the idea of equality in action, having grown up with female role models throughout society, from the workforce to politics.

A recent British study illustrates the changes that are taking place.[23] Younger women surveyed were much more assertive and optimistic than older women. They rate high in self-esteem, independence, empathy, internationalism, and sociability. Only a minority believe a woman needs a stable relationship to feel fulfilled. Work is increasingly a source of identity and confidence for young women—stay-at-home mothers rate lowest in self-esteem.

Overall, young women score high on "masculine" traits. They value excitement, complexity, and risk, and are much more likely to take up activities like sky diving or traveling alone to foreign countries. They tend to feel much more comfortable with sex, and are closing in on younger men in levels of sexual activity.

They are also more likely to question conventional institutions. While still interested in families, younger generations no longer need marriage as an *economic* institution. Delayed age of marriage and increased rate of out-of-wedlock births may be bemoaned as a sign of moral decay on the part of conservatives, but the fact is both are signs of women's economic independence.

Young men also demonstrate a massive shift in values. They tend to see equality not as a threat, but as a good in and of itself. In general, they hold very similar values to women their age. Both young women and men rate low in New Age beliefs like nonmaterialism and "connectedness" with the universe. They are more pragmatic, politically skeptical, and less nationalistic. They are concerned about the environment. In one American survey in 1992, seventy-three percent of young people cited "damage to the environment" as a threat to the American way of life.[24]

What does this all mean for ecofeminism, and the environmental movement that adopts it? It is my belief that ecofeminism is far too revolutionary, morally rigid, and based in traditional mores for most younger people. Younger people are more interested in concrete, pragmatic actions they can take (many, for example, are practically rabid about personal recycling) than the spiritual dogma and gender stereotyping offered by ecofeminism. Attacks on the ideals of the Enlightenment are equally unappetizing to the "Just Do It" generations. They are far more likely to become involved if tantalized with direct results or economically rewarding careers or positions.

They are also put off by the antimale messages of ecofeminism. Young women have much in common with their fellow male coworkers and friends. While some younger people do get involved in the movement—especially through campus activism—the ideals of femininity found in ecofeminism are foreign to the bulk of younger women, and openly hostile to younger men.

The technophobic aspect of ecofeminism is likely to strike younger generations as extremely bizarre, if not offensive. The condemnation of reproductive technologies, for instance, is counterproductive for a generation that must fit having children into their careers. Unlike women of older generations, younger women have grown up more comfortable discussing and using birth control. High school girls practice putting condoms on cucumbers, and young couples are more likely to incorporate diaphragms and other methods into their lovemaking. Birth control is a necessity, and as more and more women and couples put off having children until later years, reproductive technology will become increasingly important. The ecofeminist view of such technology as invasive is quite foreign to generations who see it as economically necessary and individually empowering.

The message of ecofeminism, and of the environmental movement that adopts it, is unlikely to ever gain widespread support among upcoming generations.

## NOTES

1 I examine current feminist trends, and their impact on the movement, in my book *The New Victorians: A Young Woman's Challenge to the Old Feminist Order.*

2 Beatrice Faust, *Backlash? Balderdash: Where Feminism is Going Right*, p. 11. Faust, the founder of Australia's Women's Electoral Lobby, warns that extremist "wimp feminism" is being imported into Australia, which has had a reform-minded movement. The revolutionary strain of feminism so virulent here has led to political inertia and a preoccupation with victimhood, whereas Australia's reform feminism has gained much ground in areas such as child care, maternity leave and political representation.

3 Gloria Steinem, *Revolution From Within: A Book of Self-Esteem*, pp. 294–295.

4 "A Women's Declaration of Interdependence," *Woman of Power*, Spring 1991, p. 30.

5 Charlene Spretnak, "Ecofeminism: Our Roots and Flowering," p. 11. I use the term "feminine" because it is often used in ecofeminist work, highlighting the popularity of Jungian notions of gender differences.
6 This approach is illustrated in works such as the 1991 *Backlash: The Undeclared War Against American Women*, in which author, Susan Faludi, depicts everything from high fashion to cosmetic surgery as part of a male-designed war against women.
7 Judith Plant, "Searching for Common Ground," p. 156. All following quotes from Plant are from this essay.
8 See, for instance, Carol J. Adams, *The Sexual Politics of Meat: A Feminist Vegetarian Critical Theory.*
9 Arisika Razak, "Toward a Womanist Analysis of Birth," p. 166.
10 Charlene Spretnak, *The Politics of Women's Spirituality: Essays on the Rise of Spiritual Power Within the Feminist Movement*, introduction.
11 *Ecofeminists Visions Emerging* (EVE) *Newsletter*, May 1993.
12 Brian Swimme, "How to Heal a Lobotomy," pp. 17 and 21.
13 Carolyn Merchant, "Ecofeminism and Feminist Theory," p. 102.
14 Maria Mies & Vandana Shiva, *Ecofeminism.* All following quotes are from this book.
15 Carol Christ, "Rethinking Theology and Nature," p. 314.
16 See, for instance, Marija Gimbutas, *The Language of the Goddess*; Merlin Stone, *When God was a Woman*; and Riane Eisler, *The Chalice and the Blade: Our History, Our Future.* The theory of the age of the goddess has been well accepted throughout the feminist movement. Marilyn French, for instance, flatly stated in her 1992 book *The War Against Women*, p. 9, that "archeological remains from about ten thousand years ago reflect goddess-worshiping communities living in egalitarian harmony and well-being."
17 Just about any feminist work on the "age of the goddess" advances this theory. For example, see Riane Eisler, *The Chalice and the Blade.*
18 I fully examine the evidence disputing the existence of goddess cultures and their alleged annihilation by "Kurgans" in *The New Victorians.*
19 David W. Anthony, "Nazi and Eco-feminist Prehistories: Ideology and Empiricism in Indo-European Archeology." I am indebted to Anthony for his critique of ideologically motivated interpretations of the past.
20 My thanks to Martin Lewis for illustrating how ecofeminism has more in common with conservatism than liberalism.
21 Catherine Keller, "Women Against Wasting The World: Notes on Eschatology and Ecology," p. 258.
22 For attitudes towards equality, see "High School and Beyond, Third Follow-up, 1986," U.S. Department of Education; and Nancy Gibbs, "The Dreams of Youth." For sexual activity and attitudes, see "Youth Risk Behavior Survey, 1990," from the Centers for Disease Control; and Noni E. MacDonald *et al.*, "High Risk STD/HIV Behavior Among College Students."
23 Helen Wilkinson, *No Turning Back: Generations and the Genderquake.*
24 Alan Deutschman, "The Upbeat Generation," p. 52.

## REFERENCES

ADAMS, CAROL J. *The Sexual Politics of Meat: A Feminist Vegetarian Critical Theory.* New York, NY: Continuum, 1990.

ANTHONY, DAVID W. "Nazi and Eco-feminist Prehistories: Ideology and Empiricism in Indo-European Archeology." Paper presented at the symposium Nationalism, Politics and the Practice of Archeology, Chicago, IL, November 1991.

CENTERS FOR DISEASE CONTROL. *Youth Risk Behavior Survey, 1990.*

CHRIST, CAROL. "Rethinking Theology and Nature." In *Reweaving the World: The Emergence of Ecofeminism*, edited by Irene Diamond & Gloria Feman Orenstein. San Francisco, CA: Sierra Club Books, 1990.

DENFELD, RENE. *The New Victorians: A Young Woman's Challenge to the Old Feminist Order.* New York, NY: Warner, 1995.

DEUTSCHMAN, ALAN. "The Upbeat Generation." *Fortune*, July 1992, p. 52.

*Ecofeminists Visions Emerging* (EVE) *Newsletter*, May 1993.

EISLER, RIANA. *The Chalice and the Blade: Our History, Our Future.* New York, NY: Harper Collins, 1988.

FALUDI, SUSAN. *Backlash: The Undeclared War against American Women.* New York, NY: Doubleday Anchor Books, 1991.

FAUST, BEATRICE. *Backlash? Balderdash: Where Feminism is Going Right.* Sydney, Australia: University of New South Wales Press, 1994.

FRENCH, MARILYN. *The War against Women.* New York, NY: Ballantine Books, 1992.

GIBBS, NANCY. "The Dreams of Youth." In *Time* special issue *Women: The Road Ahead*, Fall 1990, pp. 11–14.

GIMBUTAS, MARIJA. *The Language of the Goddess.* New York, NY: Harper & Row, 1989.

KELLER, CATHERINE. "Women against Wasting the World: Notes on Eschatology and Ecology." In *Reweaving the World: The Emergence of Ecofeminism*, edited by Irene Diamond & Gloria Feman Orenstein. San Francisco, CA: Sierra Club Books, 1990.

MACDONALD, NONI E. *et al.* "High Risk STD/HIV Behavior among College Students." *Journal of the American Medical Association* (June 1990): 3155–3159.

MERCHANT, CAROLYN. "Ecofeminism and Feminist Theory." In *Reweaving the World: The Emergence of Ecofeminism*, edited by Irene Diamond & Gloria Feman Orenstein. San Francisco, CA: Sierra Club Books, 1990.

MIES, MARIA & VANDANA SHIVA. *Ecofeminism.* London: Zed Books, 1993.

PLANT, JUDITH. "Searching for Common Ground." In *Reweaving the World: The Emergence of Ecofeminism*, edited by Irene Diamond & Gloria Feman Orenstein. San Francisco, CA: Sierra Club Books, 1990.

RAZAK, ARISIKA. "Toward a Womanist Analysis of Birth." In *Reweaving the World: The Emergence of Ecofeminism*, edited by Irene Diamond & Gloria Feman Orenstein. San Francisco, CA: Sierra Club Books, 1990.

SPRETNAK, CHARLENE. "Ecofeminism: Our Roots and Flowering." In *Reweaving the World: The Emergence of Ecofeminism*, edited by Irene Diamond & Gloria Feman Orenstein. San Francisco, CA: Sierra Club Books, 1990.

———."Introduction." *The Politics of Women's Spirituality: Essays on the Rise of Spiritual Power within the Feminist Movement*, edited by Charlene Spretnak. New York, NY: Doubleday Anchor Books, 1992.

STEINEM, GLORIA. *Revolution from Within: A Book of Self-Esteem.* New York, NY: Little, Brown, 1992.

STONE, MERLIN. *When God Was a Woman.* San Diego, CA: Harvest Books, 1976.

SWIMME, BRIAN. "How to Heal a Lobotomy." In *Reweaving the World: The Emergence of Ecofeminism*, edited by Irene Diamond & Gloria Feman Orenstein. San Francisco, CA: Sierra Club Books, 1990.

UNITED STATES DEPARTMENT OF EDUCATION. "High School and Beyond, Third Follow-up, 1986," In *Digest of Education Statistics.* Washington, DC: United States Department of Education, 1988.

WILKINSON, HELEN. *No Turning Back: Generations and the Genderquake.* London: Demos, 1994.

WOMEN'S ENVIRONMENT AND DEVELOPMENT ORGANIZATION, Women's Foreign Policy Council. "A Women's Declaration of Interdependence." *Woman of Power* (Spring 1991): 30.

# SOCIAL THEORIES OF SCIENCE

IN RECENT YEARS, the field of sociology of science, once represented by the distinguished Robert S. Merton, has been subjected to a palace coup such that "Mertonian" is now a term of belittlement. Formerly, it was acknowledged that however embedded in society and history science may be, it is a unique phenomenon because of its ability to produce accurate accounts of nature. Nowadays, many scholars disdain that truism, choosing instead to think of science as just another socially constructed belief system, no more or less true than other narratives or myths. They join, thus, the post-modernist parade that winds through intellectual life, opting for the satisfactions of relativism and the creed that "truth" is a delusion or a form of swindle by those who hold power.

With the lucidity and economy one has come to expect from her, Susan Haack here urges an alternative to the current enthusiasm for a relativizing "sociology of scientific knowledge." She dissects the fashionable levities of social constructivism and diagnoses its dangers. She thinks that a "sober" sociology of science is not only possible, but also necessary and valuable to scientists as well as to social theorists.

Noretta Koertge, too, examines the widely practiced and well-rewarded academic ritual of declaring science to be a mere social construct. She notes that it takes as axiomatic that, since all inquiry is unalterably social, it must be viewed as politics by others means. Since many of the constructivists, notably the feminists among them, are themselves ideologues, this view leads to the further—disastrous—conviction that the content and findings of science not only *are*, but *ought to be* dictated by political agenda.

Stephen Cole, a noted sociologist of science to whom the dominant trends in the field appear catastrophic, sets forth the means and events by which social constructivism became the preeminent, indeed the "hegemonic" doctrine of sociology of knowledge. He shows that, monopolistic as its ambitions are, it is more than simply a theoretical position. It is also a victorious interest group, appropriating the prestige of earlier science studies, and attempting to expropriate the remaining prestige of science.

The fourth contributor, Oscar Kenshur, analyzes the theories of Bruno Latour, leader of the Paris School of sociology and anthropology of science. The controversial Latour has not only been influential among social theorists, but also appeals to literary theorists and other humanists who aspire to engage in "cultural criticism" of science. Kenshur explains Latour's broad popularity, which is puzzling given Latour's idiosyncratic, not to say opaque, discursive style. The appeal lies in theory that not only declines to choose between philosophical realism and the various antirealisms, but insists that such a choice would be meaningless. Attractive as is this ambivalence to some scholars, Kenshur shows that it is at root deeply problematical.

# TOWARDS A SOBER SOCIOLOGY OF SCIENCE[a]

SUSAN HAACK

> *[Self-styled "sociologists of knowledge"] are people who have so far succeeded in transcending the cognitive limitations of their own "class-situation" as to be in a position to inform the rest of us that no-one can ever transcend the cognitive limitations of his class-situation.*
>
> —DAVID STOVE

DAVID STOVE'S shrewd observation[1] identifies exactly what is wrong with some recently dominant trends in sociology of knowledge. But I don't believe that sociology of knowledge must, in the nature of the case, be the "stupid and discreditable business" it has of late too often been.[2] So my purpose in what follows is to articulate what distinguishes good from bad sociology of knowledge, and to show how that distinction parallels another, between true and false interpretations of the claim that knowledge and inquiry are social. I shall restrict myself here, however, to speaking of sociology of science; not, I hasten to add, because I think scientific knowledge is all the knowledge there is, but simply to keep the task within manageable bounds. I hasten to add, also, that nothing I shall say precludes the possibility of *bad* good sociology of knowledge, i.e., sociology of the right kind but poorly executed.

Briefly and roughly, the difference is that good sociology of science recognizes, and bad sociology of science denies or ignores, the fact that science is not *simply* a social institution like banking or the fashion industry, but a social institution *engaged in inquiry*, attempting to discover how the world is, to devise explanatory theories that stand up in the face of evidence. To put this more precisely requires a distinction between questions of warrant—how good is the evidence for the theory?—and questions of acceptance—what is the standing of the theory in the relevant scientific community? The good, sober kind of sociology of science accommodates considerations of warrant

[a] This paper was prepared for publication with the help of NEH Grant #FT-40534-95. I wish to thank Mark Migotti for helpful comments on a draft.

as well as acceptance; the bad, self-defeating kind refuses considerations of warrant, or transmutes them into considerations of acceptance. Bad sociology of science is thus *purely* sociological, whereas good sociology of science, acknowledging the relevance of evidential considerations, is not.[3] Good sociology of science, in consequence, requires some grasp of scientific theory and evidence, while bad sociology of science does not. (A cynic might suspect that this partially explains why there is so much bad sociology of science about.)

Bad sociology of science, ignoring or denigrating the relevance of evidential considerations, is invariably ***debunking*** in tendency. But, though proponents of the bad kind like to suggest otherwise, good sociology of science is not invariably ***legitimating*** in tendency. Acceptance and warrant may *or may not* be appropriately correlated; and a sober sociology of science will ask, not only what the mechanisms are by which they are appropriately correlated, but what goes wrong when they are not. And when, as a sober sociology of science will acknowledge sometimes happens, a scientific theory gets accepted though the evidence for it is weak, or gets rejected though the evidence for it is strong, the explanation of its acceptance or rejection will be more purely sociological, the more acceptance and warrant are disconnected.

It is because it is invariably debunking that bad sociology of science is invariably self-undermining. Claiming or suggesting that the real explanation of the currency of a scientific theory is always something about the historical or social circumstances of its origin, never a matter of its having been recognized that there is good evidence for it, bad sociology of science undermines its own pretensions to supply *warranted* explanations of the currency of this or that scientific theory.

Good sociology of science rests on a correct understanding of the sense in which it is true, and epistemologically important, that science is social.

An adequate theory of scientific knowledge involves two distinct, though related, projects: an articulation of what constitutes good evidence for or warranted acceptance of a theory, and of what constitutes good procedure, how to conduct inquiry.[4]

The first epistemological project is often construed in purely logical terms, as a question of the relation of evidence $E$ to theory $T$. This oversimplifies. The project has causal, personal, and social, as well as logical, dimensions. A causal dimension is necessary, because "the evidence" with respect to any scientific claim must include some experiential evidence, and there can be only causal, not logical, relations between a person's belief and his experiences (e.g., between his seeing a black swan and his believing that not all swans are white). A personal dimension is necessary, because it is individuals who have experiences, who see where the needle on the dial points, what pattern appears in the bubble chamber. And a social dimension is also necessary because, in view of the role of experiential evidence, "how warranted theory $T$ is" must be taken as elliptical for "how justified a scientific community is, at a time, in accepting $T$"; which depends in a complex way on how justified an individual who possessed all the evidence known to each member of the community would be in accepting $T$, discounted by some index of how justified each member of the community is in believing the others to be reliable.

The structure of justification is like a crossword puzzle, with experiential evidence the analogue of the clues, background beliefs of already-completed intersecting entries. How justified a belief is depends on how well it is supported by experiential evidence and background beliefs (analogue: how well a crossword entry is supported by its clue and already-completed intersecting entries); how justified those background beliefs are, independent of the belief in question (analogue: how reasonable those intersecting entries are, independent of the entry in question); and how comprehensive the evidence is (analogue: how much of the crossword has been completed).

The second epistemological project has often been construed as seeking an articulation of "the scientific method." This also oversimplifies. In the narrow sense in which the phrase purportedly refers to a set of rules that can be guaranteed to produce true, or probably true, or progressively more nearly true, results, there is no such thing as "the scientific method." And in the broader, vaguer sense in which the phrase refers to making conjectures, developing them, testing them, assessing the likelihood that they are true, though certainly there is such a thing, not only scientists, but also historians, detectives, investigative journalists, and the rest of us, use "the method of science" in that sense. What is distinctive about inquiry in the sciences is, rather: systematic commitment to criticism and testing, and to isolating one variable at a time; experimental contrivance of every kind; instruments of observation from the microscope to the questionnaire; sophisticated techniques of mathematical and statistical modeling; *and* the engagement, cooperative and competitive, of many persons, within and across generations, in the enterprise of scientific inquiry.

Since doing science is more like doing a huge crossword puzzle than shelling an enormous quantity of peas or carrying a very heavy log, this is no simple matter of "many hands make light work." It is a matter, at least, of specialization *and* overlapping of competencies; of cooperation *and* competition; of institutionalized mutual criticism *and* the institutionalized authority of well-warranted results. Ideally, like inquiry generally, scientific inquiry should combine creativity and carefulness; it should encourage bold, fruitful conjectures, but also keep acceptance and warrant appropriately correlated. And whether, or to what degree, science approaches this ideal depends significantly on social matters—its internal organization and its environment in society at large.

The organization and environment of science may, at different times and places, be more *or less* conducive to good, creative, honest, scrupulous inquiry. Potential hindrances include: a volume of publications so large as to impede rather than assist communication; pressure to find evidence supporting a politically desired conclusion or to ignore questions perceived as socially disruptive; pressure to solve problems perceived as socially urgent; the necessity to spend large amounts of time and energy on obtaining resources, and to impress whatever body provides the funds, in due course, with one's success; dependence for resources on bodies with an interest in the research coming out a certain way, or in rivals' being denied access to the results.

The right kind of sociology of science can help us understand what features of its internal organization and of its external environment encourage, and what discourage, successful science. One thinks, for example, of Polanyi's reflections on how to organize science so as to keep authority and criticism in balance, Campbell's on how to combine division of labor with overlapping competencies, Hull's on the workings of the peer review system, Rauch's on the relation of freedom of thought and the progress of inquiry, Beyerchen's studies of Nazism and science, Soyfer's of Lysenko and "the tragedy of Soviet biology."[5]

And then, by contrast, one thinks of the Edinburgh School's "strong program" in sociology of science, allegedly revealing the "threadbare fabric of . . . traditional philosophical accounts"; of Collins's or Gergen's assurances that the world has little or no role in the construction of scientific knowledge; of Latour's or Woolgar's insistence on approaching science as a process of producing inscriptions and thereby constructing facts; of Hubbard's or Bleier's or Nelson's or Longino's announcements that all inquiry is biased by the inquirer's gender, class, or racial perspective; of the proponents of "democratic epistemology" and the alleged "strong objectivity" of multiple standpoint theory;[6] and so on.

The alert reader, recalling my earlier acknowledgment of the possibility of bad good sociology of science, may by now be wondering about the possibility of good bad sociology of science. In any instance in which acceptance and warrant were quite disconnected, a purely sociological account of the acceptance of the theory would be appropriate; and, in such a case, the correct, purely sociological, account of how the theory came to be accepted conceivably might be given by a proponent of the bad kind of sociology of science. And this, I suppose, could fall under the rubric, "good bad sociology of science." But that our proponent of bad sociology of science got it right for once would be more by luck than judgment; for bad sociology of science is conducted under the influence of mistaken conceptions of what science is and does. If the good kind of sociology of science is well described as "sober," the bad kind might with equal justice be described as "intoxicated"—intoxicated by one or another of various misunderstandings of the thesis that science is social.

One misunderstanding is that the warrant status of a scientific claim is "just a matter of social practice." Warrant is social in the sense that talk of how warranted a scientific claim is, is elliptical for talk of how justified a scientific community is in accepting it; but how justified they are in accepting it does not depend on how justified they ***think*** they are, but on how good their evidence is.

A second misunderstanding is that "how good," here, can only mean "how good relative to the standards of community C," since standards of evidence, it is supposed, vary incommensurably among communities or paradigms. But the supposed community—or paradigm—relativity of evidential standards is not real incommensurability, but deep-seated disagreement in background beliefs, producing disagreement about what counts as relevant evidence. The crossword analogy sheds useful light here. You think, given your solution to 7 across, that the fact that a solution to 2 down ends in an "E" is evidence in its favor; I, given mine, that the fact that it ends in an "S" is evidence in its favor. You think this candidate for the job should be ruled out on the grounds that his handwriting indicates that he is not to be trusted; I think graphology is bunk and scoff at your "evidence." Our judgments of relevance of evidence depend on our background beliefs, but relevance of evidence is, nevertheless, objective.

A third misunderstanding moves from the true observation that scientific inquiry is a social enterprise to the ambiguous conclusion that scientific knowledge is "socially constructed," which is then given a false interpretation, that scientific knowledge is nothing more than the product of processes of social negotiation. Scientific inquiry is a social process, and the scientific knowledge we now possess has been the work of many persons, within and across generations. But to describe the processes concerned as a matter of "social negotiation" is thoroughly misleading; the processes by which scientific knowledge is achieved are processes of seeking out, checking, and assessing the weight of evidence.

At this point a fourth misunderstanding kicks in. The processes concerned *must* be essentially a matter of social negotiation, it is supposed, since theories are underdetermined by data, so that "social values" have to take up the slack. Evidence never *obliges* us to accept this claim rather than that, the thought is, and we have to accept something; so acceptance is always affected by something besides the evidence. But we do not "have to accept something." Not all scientific claims are accepted as definitely true or rejected as definitely false, nor should they be; indeed, keeping warrant and acceptance appropriately related requires, inter alia, that, when the evidence is insufficient, we acknowledge that we don't know. Perhaps some have been misled by the thought that sometimes we have to act, and so have to accept some theory as the basis on which to act; but this is easily enough counteracted by distinguishing accepting a theory as true from deciding, without committing oneself to its truth, to act as if it were true.

Some, committing the same kinds of confusion twice over, maintain that the objects of scientific knowledge are socially constructed. Scientific theories are of course devised, articulated, developed by scientists; theoretical concepts like *electron*, *gene*, *force*, and so forth, are, if you like, their creation. And the entities posited in true scientific theories are real. But it does not follow, and neither is it true, that electrons, genes, forces, etc., are brought into existence by the intellectual activity of the scientists who create the theories that posit them. True, as science proceeds, instrumentation and theory get more and more intertwined, and one increasingly encounters claims that refer, not to natural, but to what one might call "laboratory," phenomena. But that such phenomena are created in the laboratory does not mean that they are made real by scientists' theorizing.

Some proponents of this misunderstanding may have been misled by the fact that social institutions and categories (marriage, banking, gender, . . .)—

the objects of sociological theories—are socially constructed in the sense that, if there were no human societies, they would not exist. But even these are still not made real by scientists' theorizing. As for those who are led to believe that reality is socially constructed by reflection on the proposition that you can't describe the world without describing it, I shall say only that, since this is a simple tautology, any inference to a startling social-constructivist conclusion is manifestly a non sequitur.

If the "construction" of scientific knowledge *were* a matter simply of social negotiation, calls for a more democratic epistemology—for the inclusion of previously excluded parties to the negotiation—would be appropriate. But it isn't, so they aren't. True, freedom of thought and speech are important conditions for scientific inquiry to flourish; and it may be that some who urge "democratic epistemology" have confused the concept of democracy with the concept of freedom of speech. If so, the only reply needed is that these are distinct concepts. And as for those who argue that since scientific knowledge is nothing but a social construction, the physical sciences must be subordinate to the social sciences, the only reply needed is that it isn't, so—thank goodness!—they aren't.

I have been unremittingly negative for several pages; so perhaps it is prudent to conclude by reminding you that the point is not that there can be no useful, sober sociology of knowledge, but that it is vital to distinguish this useful enterprise from the bad, self-undermining kind. At any rate, that is what I have tried to do.

## NOTES

1 David Stove, "Epistemology and the Ishmael Effect," p. 62.

2 The description of sociology of science as a "stupid and discreditable business" comes from the first page of Stove's "Cole Porter and Karl Popper: The Jazz Age in the Philosophy of Science." Shortly thereafter, however, Stove acknowledges that sociology of knowledge is not necessarily and in principle self-undermining, and suggests a position quite close to the one I shall be defending.

3 This is a bit subtler than either of two more familiar distinctions: between "externalism" and "internalism" (the former taking into account only "external," i.e., purely sociological, factors, the latter also taking into account the content of the theories concerned); and between the "strong" and the "weak" programs in sociology of science (the former treating true and false theories strictly alike, the latter treating them differently). By my lights, reference to content is necessary but not sufficient for good sociology of science; and it is the distinction warranted/unwarranted, not true/false, that is crucial.

4 Here and in the next several pages I am drawing upon my *Evidence and Inquiry: Towards Reconstruction in Epistemology*; "Puzzling Out Science"; and "Science as Social?—Yes and No." The reader is referred to these for further details of the arguments sketched here.

5 Michael Polanyi, "The Republic of Science"; Donald T. Campbell, "Ethnocentrism of Disciplines and the Fish-Scale Model of Omniscience"; David Hull, *Science as a Process*; Jonathan Rauch, *Kindly Inquisitors: The New Attacks on Free Thought*; Alan Beyerchen, *Scientist Under Hitler* and "What We Now Know About Nazism and Science"; and Valery N. Soyfer, *Lysenko and the Tragedy of Soviet Science.*

6 David Bloor, "Sociology of Knowledge," p. 486; Harry Collins, "Stages in the Empirical Programme of Relativism," p. 3; Kenneth Gergen, "Feminist Critique of Science and the Challenge of Social Epistemology," p. 37. Bruno Latour & Steve Woolgar,

*Laboratory Life: The Social Construction of Scientific Facts*; Bruno Latour, *Science in Action*; Ruth Hubbard, "Some Thoughts about the Masculinity of the Natural Sciences"; Ruth Bleier, "*Science* and the Construction of Meanings in the Neurosciences"; Lynn Hankinson Nelson, *Who Knows? From Quine to a Feminist Empiricism*; Helen Longino, "Can There be a Feminist Science?" and *Science as Social Knowledge*; Sandra Harding, *Whose Science? Whose Knowledge?* and "After the Neutrality Ideal: Science, Politics and 'Strong Objectivity'."

## REFERENCES


BEYERCHEN, ALAN. *Scientists Under Hitler.* New Haven, CT: Yale University Press, 1977.

———. "What We Now Know about Nazism and Science." *Social Research* 59 (1992): 616–641.

BLEIER, RUTH. "*Science* and the Construction of Meanings in the Neurosciences." In *Feminism within the Science and Health Care Professions: Overcoming Resistance*, edited by Sue Rosser. New York, NY: Pergamon Press, 1988.

BLOOR, DAVID. "Sociology of Knowledge." In *Companion to Epistemology*, edited by Jonathan Dancy & Ernest Sosa. Oxford: Blackwell, 1992.

CAMPBELL, DONALD T. "Ethnocentrism of Disciplines and the Fish-Scale Model of Omniscience." In *Interdisciplinary Relationships in the Social Sciences*, edited by Muzafer Sherif & Carolyn W. Sherif. Chicago, IL: Aldine, 1969.

COLLINS, HARRY. "Stages in the Empirical Programme of Relativism." *Social Studies of Science* 11 (1981): 3–10.

GERGEN, KENNETH. "Feminist Critique of Science and the Challenge of Social Epistemology." In *Feminist Thought and the Structure of Knowledge*, edited by M. M. Gergen. New York, NY: New York University Press, 1988.

HAACK, SUSAN. *Evidence and Inquiry: Towards Reconstruction in Epistemology.* Oxford: Blackwell, 1993.

———. "Puzzling Out Science." *Academic Questions* 8.2 (Spring 1995): 20–31.

———. "Science as Social?—Yes and No." In *A Dialogue on Feminism, Science, and Philosophy of Science*, edited by Jack Nelson & Lynn Hankinson Nelson. Dordrecht: Kluwer. In press.

HARDING, SANDRA. "After the Neutrality Ideal: Science, Politics and 'Strong Objectivity.'" *Social Research* 59 (1992): 567–587.

———. *Whose Science? Whose Knowledge?* Ithaca, NY: Cornell University Press, 1991.

HUBBARD, RUTH. "Some Thoughts about the Masculinity of the Natural Sciences." In *Feminist Thought and the Structure of Knowledge*, edited by M. M. Gergen. New York, NY: New York University Press, 1988.

HULL, DAVID. *Science as a Process.* Chicago, IL: University of Chicago Press, 1986.

LATOUR, BRUNO. *Science in Action.* Cambridge, MA: Harvard University Press, 1987.

LATOUR, BRUNO & STEVE WOOLGAR. *Laboratory Life: The Social Construction of Scientific Facts.* Beverly Hills, CA: Sage Library of Scientific Research, 1979.

LONGINO, HELEN. "Can There Be a Feminist Science?" In *Women, Knowledge and Reality*, edited by Ann Garry & Marilyn Pearsall. Boston, MA: Allen Hyman, 1989.

———. *Science as Social Knowledge.* Princeton, NJ: Princeton University Press, 1990.

NELSON, LYNN HANKINSON. *Who Knows? From Quine to a Feminist Empiricism.* Philadelphia, PA: Temple University Press, 1990.

POLANYI, MICHAEL. "The Republic of Science." In *Knowing and Being*, edited by Marjorie Grene. Chicago, IL: University of Chicago Press, 1969.

RAUCH, JONATHAN. *Kindly Inquisitors: The New Attacks on Free Thought.* Chicago, IL: University of Chicago Press, 1992.

SOYFER, VALERY N. *Lysenko and the Tragedy of Soviet Science.* Newark, NJ: Rutgers University Press, 1994.

STOVE, DAVID. "Cole Porter and Karl Popper: The Jazz Age in the Philosophy of Science." In *The Plato Cult.* Oxford: Blackwell, 1991.

———. "Epistemology and the Ishmael effect." In *The Plato Cult.* Oxford: Blackwell, 1991.

# WRESTLING WITH THE SOCIAL CONSTRUCTOR[1]

NORETTA KOERTGE

IN HIS *Novum Organum*, Francis Bacon presented a method of scientific inquiry that he hoped would root out "the idols and false notions which are now in possession of the human understanding."[2] Bacon argued that these sources of systematic delusion would continue to cause trouble "unless men, being forewarned of the danger, fortify themselves as far as may be against their assaults."[3] As the founders of the Royal Society began to design an institutional base for Bacon's dream of a "Great Insaturation of the Sciences," they emphasized the importance of excluding the discussions of politics, religion, and what we today would call ideology from the conduct of the professional affairs of science.

In the intervening three centuries, scientists and philosophers have criticized and refined the details of Bacon's inductive theory of scientific method, but few have questioned the wisdom of his distrust of the Idols of the Tribe, Cave, Marketplace, and Theatre. Few, that is, until today, when we find within the academy calls for feminist and ethnic sciences as well as demands that science be guided by political commitments. It is now claimed that science not only *is*, but *should be*, "politics by other means."[4] On this view, the proper aim is not value-free science; rather the goal must be science that is infused with the "correct" political values constructed within the framework of the "correct" ideology!

*Social construction* or *constructivist* epistemology are trendy terms that, while they signal a certain sympathy towards nouveau ideas, have no precise referent.[5] They do, however, direct attention to those properties of a phenomenon that depend on culture and are therefore presumed to be amenable to change. Some of these claims are plausible and insightful. Thus when Joel Best in his book *Threatened Children* argues that the problem of the kidnapping of children by strangers is "socially constructed," what he means is that our society's perception of the seriousness of this problem is more influenced by media attention and the rhetorical strategies of activists than by empirical data demonstrating the actual dimensions of the problem.

A contrasting, but equally unobjectionable, use of the term occurs in an article that raises ethical questions about the "social construction" of dairy cows. Certain breeds have undergone such an intense process of artificial selection, guided by the financial interests of dairy farmers, that they can now hardly survive without being hooked up to milking machines. In this case what "social construction" designates is really a process of *genetic* construction designed to serve social purposes.

In each of the above cases, it is quite appropriate to emphasize the causal and constructive roles of social factors; nor do the authors deny that biological or other material factors also play a determining role—some children *are* forcibly removed from their parents. It is only our feelings about and interpretations of these states of affairs that are open to social negotiation. And although we may have chosen some of the characteristics of dairy cows, there remain all sorts of physical and biological constraints on our abilities to design viable organisms.

However, there are also extreme social constructionists who engage in a form of "biodenial" whereby they deliberately downplay or even totally ignore the role of nonsocial elements. In *Professing Feminism*,[6] we described feminists who begin by correctly pointing out the conventional or socially constructed nature of many aspects of gender stereotypes or gender roles, but end up by denying that biology has any relevance to social arrangements. In one absurd episode, young women's studies students insisted that the pain of childbirth was a "construction" of patriarchal society that would not be an issue in Amazonia. (Talk about "blaming the victim"! This anecdote illustrates how easy it is for feminists to discredit totally the experience of other women if it happens not to be politically correct!)

Unfortunately, it is not just college sophomores who believe that we can reshape the natural world to make it fit a new social consensus. Some feminists have argued that biological sex is just as socially constructed as gender. Thus in "The Five Sexes" Fausto-Stirling claims that in viewing *Homo sapiens* as a two-sexed species biologists have overlooked a purportedly large number of hermaphroditic or intersexed children. This is a popular theme among feminists. After all, gender could not possibly be anything more than a social construct if sex itself were also socially constructed! Feminists are also quick to accuse their opponents of "dualistic" thinking—e.g., of trying to force the rich complexity of a sexual spectrum into the artificially tidy and mutually exclusive boxes of biological male vs. biological female.

The way Fausto-Stirling argues for her proposal is quite typical of a social constructionist approach that purports to be politically progressive: she cites no empirical data indicating the nature and distribution of intersexed humans. The four percent figure comes from an informal estimate by John Money, who, as a specialist on hermaphroditism, gender dysphoria, and paraphilias, might be expected to guess on the high side. Neither is there any discussion about how the "five sex" proposal would mesh with the rest of biological science, such as evolutionary theory, theories of reproduction, comparative anatomy, etc. (Should we also declare there to be five sexes of chimpanzee?) Rather the argument hinges on the assumption that intersexed people would be better off if they were not pressured to conform and accommodate to a two-sexed world. If we as a society were simply to reconceptualize biological sex, so the speculation goes, these people would no longer have a need for plastic surgery on their genitals or hormone therapy, and they would presumably have less trouble finding accepting sexual partners. Surely, the argument goes, it is more humane to alter our concepts than to drive people into unnecessary surgery! Although this particular attempt at social reconstruction will probably have little impact on biology, it is cause for worry because it is part of a wide-ranging systematic attempt to change the way science is done. It is a concrete example of what is sometimes called the "sociological turn" in epistemology.

There are many deep and interesting questions to ask about the social dimensions of knowledge. For example, John Searle in *The Construction of Social Reality* analyzes the nature of the complex group commitments that are prerequisite for "social facts" such as money or stop signs to operate. Merton and his students have studied the institutional arrangements that foster scientific inquiry. And various decision theorists have struggled with questions about the relationship between individual rationality and group rationality. There is no question that social epistemology can be a valuable enterprise. Unfortunately much of what goes by that name today consists of barely disguised ideological initiatives. One attempt begins with a discussion of the role of values in science and then attempts to incorporate political values into the construction of science. Let us now analyze this line of attack[7] so that we, like Bacon's readers, can be "forewarned of the dangers. . . ."

If we use as a framework the traditional triumvirate of Truth, Beauty, and Goodness (see FIGURE 1), it would seem obvious that the overriding professional goal of science is to discover truth, and thus the scientist's value system should properly center around such criteria as empirical adequacy, precision, and generality. However, scientists also place great stock in such aesthetic desiderata as simplicity, symmetry, and coherence. And some philosophers

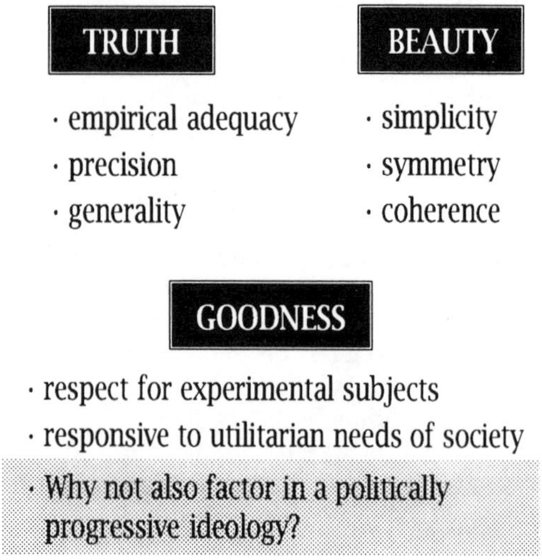

FIGURE 1 The traditional triumvirate of values.

would argue that since theory choice is empirically underdetermined, such factors must always be invoked. So, for example, we draw simple, smooth curves to summarize arrays of data points.

In addition, the scientist has various ethical responsibilities above and beyond the obvious requirements of contributing to and not interfering with the search for truth. There are now all sorts of codes dealing with the responsible treatment of human and animal subjects. Our standards of what is acceptable practice have generally tightened up over the years (although recall that human dissections were illegal for much of history). But there have always been some limits on experimental interventions and hence some conflicts between the high value placed on gaining truth and ethical concerns. Since science often justifies its existence and requests for public financial support in terms of its potential contributions to the utilitarian needs of society, there are also priority conflicts between what Bacon described as the search for the "light" of understanding and the development of the "fruits" of technological application.

It is precisely at this point that those who favor a politically progressive science try to make an opening for deliberate ideological intervention. So, their argument goes, you admit that the quest for pure scientific truth is already restricted by ethical prohibitions on harming experimental subjects and tempered by a responsiveness to the utilitarian needs of society at large. Why should we not also harness science to the task of making our society more just and more politically progressive? On the face of it, this suggestion has an attractive humanistic ring to it. Who can be against justice and political progress? But let us look at the details of the proposal. How exactly would it impact on the scientific process?

The flow chart in FIGURE 2 presents a simplified sketch of the traditional account of the process of scientific inquiry. Within what Reichenbach called

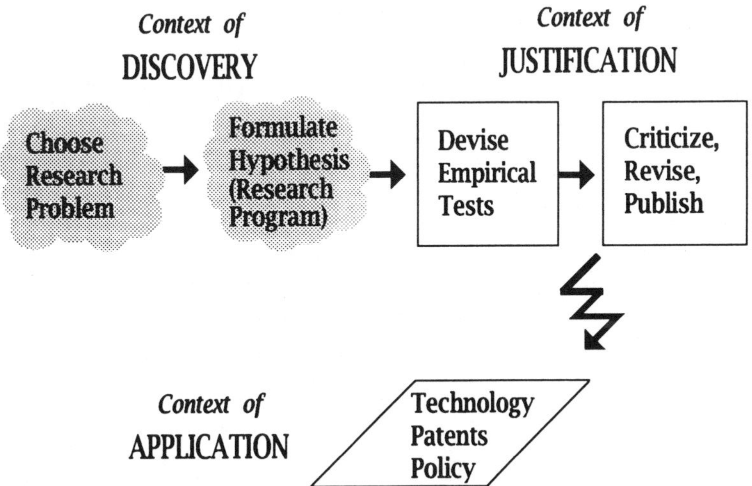


FIGURE 2 Flow chart showing the traditional account of the process of scientific inquiry.

the "context of discovery"[8] scientists choose research problems and develop strategies for formulating tentative solutions to them. This is followed by the "context of justification," in which the results of the scientist's informal or private ruminations are submitted to public empirical testing and the critical scrutiny of the scientific community. Reichenbach believed that all sorts of personal idiosyncrasies, interests, and commitments might well be influential within the context of discovery—this domain could be illuminated by the psychologist or historian. It was the procedures within the context of justification that were to guarantee the objectivity of scientific results, and these could best be described by logicians and epistemologists of science. The context of application lay outside the domain of science proper, and it was here as well as in the choice of funding priorities that social values and policies took their rightful place. Or so goes the traditional account. The proponents of politically progressive science, however, would argue that science has always been impregnated with social values. Their agenda is *not* to try to make science more value neutral. Rather, it is to inject the "correct" values at all stages.

Let us review their specific proposals for the modification of the above account, beginning with the context of application. Everyone would agree that extreme care should be taken in *applying* scientific results that are either still preliminary or could easily cause harm. These are the sorts of technology assessment issues that the Food and Drug Administration and product liability experts wrestle with all the time. However, the social constructionists would go much further, arguing that some results, *even if true*, would be so politically dangerous that they should never be published—and hence never studied in the first place. Hence, the moves to suppress publications or research conferences on possible biological correlates of crime, intelligence, or sexual orientation.[9]

Social constructionists would also introduce political or ideological considerations directly into the heart of the context of justification. Some would argue that a respect for human subjects requires the anthropologist not only to double-check his facts with his informants, but also to get their approval of any interpretations or theoretical analyses before publishing. The social constructionist has also been successful in introducing government policy requiring that the pool of experimental subjects for drug trials should be stratified by political categories, *not* according to the likely biological variability in response to the drug.[10] And AIDS activists temporarily succeeded in modifying requirements for the use of control groups for testing HIV drugs. But the subtlest—and hence the most potentially dangerous—form of ideological intrusion would take place within the so-called context of discovery, a stage of scientific inquiry that has only recently been scrutinized by philosophers. The line of attack is by now familiar: look for the role of nonepistemic values, and then try to add one's own political desiderata to the list.

Consider, for example, the factors that might legitimately influence an individual scientist's (or scientific community's) choice of research questions. Even in the purest ivory tower scenario, one will consider not just the intellectual interest of the problem—i.e., the extent to which its solution will deepen our understanding of the world—but also its technical feasibility. Is the problem "ripe" for investigation at this time? Do the requisite experimental instruments and mathematical tools already exist? Is funding available for such research? Our advocate of politically sensitive science will merely add that scientists should also consider the political ramifications of even asking certain questions while ignoring others. So, the argument goes, we should study the social construction of sexism, not sex differences, and analyze the origins of homophobia instead of the etiology of homosexuality.

Such an ideological filter could introduce a serious truncation of the topics that could be investigated and damage the ethos of free inquiry so treasured in science, but would not in itself directly distort the content of scientific findings. However, serious problems would arise if ideology were to affect the process of hypothesis formation. As many philosophers and scientists who comment on the development of science have pointed out, the process of constructing tentative solutions to a research problem is not one of totally random trial and error. Rather, scientists typically structure their search by what Lakatos[11] calls a scientific research program—i.e., they work, utilizing a variety of heuristic principles, within a metaphysical or theoretical framework that has proved fruitful in the past and that they believe will lead them to the most plausible candidates to a solution to their research problems. Furthermore, when deciding which among a potentially infinite number of paths to pursue, they deem it reasonable to let pragmatic factors such as ease of theoretical manipulation or ease of experimental investigation play a role in the decision as to which hypothesis to investigate first.

Now, says our inveterate ideologist, if plausibility estimates and pragmatics are permitted to guide our search, why not also screen potential problem solutions in terms of their political progressiveness? As Longino puts it: ". . . I am suggesting that a feminist science practice admits political considerations as relevant constraints on reasoning . . ."; ". . . if faced with a conflict between [political] commitments and a particular model of brain-behavior, we allow the political commitments to guide the choice."[12] And although Feyerabend has been castigated by feminists for sexist imagery, he, too, believes we need to subjugate science to our desires: " . . . [We should] change science from a stern and demanding mistress into an attractive and yielding courtesan who tries to anticipate every wish of her lover"; " . . . it is up to us to choose either a dragon or a pussycat for our company."[13]

Feyerabend wanted to make science subservient to hedonism and deeply resented those who would criticize the scientific status of astrology or parapsychology, thereby diminishing the pleasure of those who believe in it. Today's feminist social constructionists also want to subordinate science, but to a more severe agenda, whereby the scientific acceptability of a claim would be contingent on its perceived political expediency. History should have taught us by now the futility of such a strategy—the act of banning Copernicanism hurt the Catholic Church much more than heliocentrism could ever have done. Lysenkoism did nothing to help peasants improve their agricultural practices while it delayed the adoption of hybrids developed in the West. Women will not benefit in the long run from attempts to block the study of biological differences. And already we have evidence that being credulous about the phenomenon of so-called "recovered memories" not only harms innocent family members but also inflicts great damage to the victim herself. To violate a person's ability to distinguish fact from fantasy is the epistemological equivalent of rape. Yet the ideology of social constructionism and politically correct science is becoming increasingly influential in universities, schools of education, and even science policy organizations such as the American Association for the Advancement of Science.

So what is to be done? As friends of science and reason, we need to become more sophisticated about the nature of scientific inquiry. After taking my first high school science course, I believed that science proceeded linearly by the steady accumulation of facts, that science always brought us "better things for better living (through chemistry)" and that electron spin was strictly analogous to the earth's diurnal motion. We want students today to be less naive about the complex nature of science. We want them to realize that the scientist is a sojourner in an ever-changing intellectual world, that scientists use their imaginations to construct models that are sometimes inaccurate and always incomplete, and that in the process of correcting and refining the details of these representations, scientists often disagree vehemently even while sharing a vast store of background information, methods, and values.

We must provide students and the general public with a sophisticated account of science lest they be overwhelmed the first time they encounter relativist arguments about the underdetermination of theories by data, the theory ladenness of experience, or the social construction of concepts. And we must also speak more clearly and honestly to students about *all* the values that operate within science. We live in cynical times, and as intellectuals many of us are much more comfortable making wry remarks about the value that scientists place on credit and the resultant priority disputes, stratagems to get funding, and academic politics. These mundane values are, of course, part of scientific life—and increasingly students and the public will not let us forget it. But there are other values operating, too. All of us know more significant stories about scientific life, anecdotes that illustrate the role of intellectual honesty and the courage to acknowledge errors publicly, stories that describe the constant striving for objectivity and the ability of scientists to abandon central dogmas and favored theoretical commitments. We must talk about the intellectual virtues that are constantly guiding science and not cede the moral high ground to ideologues.

## NOTES

1 This title was inspired by a talk given by Wayne Dynes at a conference entitled "Homosexuality: Which Homosexuality?" that was held at the Free University of Amsterdam in 1987. The debate about social constructionism within Gay and Lesbian Studies is particularly heated because each side claims their own position to be politically advantageous.

2 Francis Bacon, *Novum Organum*, Aphorism 38.

3 Ibid.

4 S. Harding, *Whose Science? Whose Knowledge? Thinking from Women's Lives*, p. 10.

5 For a useful survey, see B. S. Held, "The Real Meaning of Constructivism."

6 D. Patai & N. Koertge, *Professing Feminism: Cautionary Tales from the Strange World of Women's Studies.*

7 For a related critique see N. Koertge, "Ideology, Heuristics and Rationality in the Context of Discovery."
8 H. Reichenbach, *The Rise of Scientific Philosophy.*
9 On this latter point see R. Horton, "Is Homosexuality Inherited?"
10 S. Satel, "Science by Quota: P. C. Medicine."
11 I. Lakatos & A. Musgrave, *Criticism and the Growth of Knowledge.*
12 H. Longino, *Science as Social Knowledge*, p. 191.
13 P. Feyerabend, *Killing Time*, p. 149.

## REFERENCES

BACON, FRANCIS. *Novum Organum.* New York, NY: The Liberal Arts Press, 1960.

BEST, JOEL. *Threatened Children: Rhetoric and Concern about Child-Victims.* Chicago, IL: University of Chicago Press, 1990.

DYNES, WAYNE. "Homosexuality: Which Homosexuality?" Paper presented at the Free University of Amsterdam, 1987.

FAUSTO-STIRLING, A. "The Five Sexes: Why Male and Female Are Not Enough." *The Sciences* March/April 1993, pp. 20–25.

FEYERABEND, P. *Killing Time.* Chicago, IL: University of Chicago Press, 1995.

HARDING, S. *Whose Science? Whose Knowledge? Thinking from Women's Lives.* Ithaca, NY: Cornell University Press, 1991.

HELD, B. S. "The Real Meaning of Constructivism." *Journal of Constructivist Psychology.* In press.

HORTON, R. "Is Homosexuality Inherited?" *New York Review of Books* 42, 12 (1995): 36–41.

KOERTGE, N. "Ideology, Heuristics and Rationality in the Context of Discovery." In *Correspondence, Invariance and Heuristics*, edited by S. French & H. Kamminga. Dordrecht: Kluwer Academic Publishers, 1993.

LAKATOS, I. & A. MUSGRAVE. *Criticism and the Growth of Knowledge.* Cambridge: Cambridge University Press, 1970.

LONGINO, H. *Science as Social Knowledge.* Princeton, NJ: Princeton University Press, 1990.

PATAI, D. & N. KOERTGE. *Professing Feminism: Cautionary Tales from the Strange World of Women's Studies.* New York, NY: Basic Books, 1994.

REICHENBACH, H. *The Rise of Scientific Philosophy.* Berkeley, CA: University of California Press, 1951.

SATEL, S. "Science by Quota: P. C. Medicine." *The New Republic*, February 27, 1995.

SEARLE, JOHN R. *The Construction of Social Reality.* New York, NY: The Free Press, 1995.

# VOODOO SOCIOLOGY
## Recent Developments in the Sociology of Science

STEPHEN COLE

UP UNTIL THE 1970s, sociologists of science did not examine the actual cognitive content of scientific ideas, as they believed that these were ultimately determined by nature and not a product of social processes and variables. Beginning in the late 1970s and early 1980s, a group of European sociologists, adopting a relativist epistemological position, began to challenge this view. At first they called themselves "relativist-constructivists" and later, more simply, "social constructivists." Their numbers were few, but within the short time span of roughly one decade, this group has come to completely dominate the sociology of science and the interdisciplinary field called the social studies of science. Although some like to deny this dominance because ideologically they do not like to see themselves as the power elite, their control of all the major associations and specialty journals is clear to anyone participating in the field. This dominance may easily be seen in the recently published *Handbook of Science and Technology Studies*,[1] published by the Society for the Social Study of Science. Virtually all the contributors are either constructivists or political allies of the constructivists. Tom Gieryn, a former student of Merton, but now a convert to constructivism, in his contribution to this handbook claims as an aside, "If science studies has *now convinced everybody* that scientific facts are only contingently credible and claims about nature are only as good as their local performance. . . ."[2]

It is very important to point out that social constructivism is not simply an intellectual movement, a way of looking at science, but it is an interest group that tries to monopolize rewards for its members or fellow travelers and exclude from any recognition those who question any of its dogma. In fact, the editors of the *Handbook* discussed above refused to include a chapter on the important topic of social stratification in science (in the most prestigious general journals of sociology like the *American Sociological Review*, more articles have been published on this topic than any other), claiming at first that they could find no one willing to write such a chapter. When I volunteered, an invitation to contribute was never forthcoming.

In the leading monographs that established the social constructivists—the two most important being the laboratory studies by Bruno Latour and Steve Woolgar, *Laboratory Life*, and Karin Knorr-Cetina's *Manufacturing Knowledge*, published in 1981—the epistemological position taken was highly relativistic. Scientific facts were not constrained by nature but were socially constructed or made up in the laboratory by the scientists. Harry M. Collins, another leading constructivist, stated that the "*natural world has a small or non-existent role in the construction of scientific knowledge.*"[3] Some constructivists, including Collins, later claimed that he "really didn't mean this;" it was just being used in a "polemic" or a "programmatic" statement or to describe "methodological relativism" as opposed to "epistemological relativism." Well that is fine; because if constructivists do not *really* subscribe to epistemological relativism, then there is nothing very radical about their work and no reason why it cannot be integrated with other nonrelativistic work in the social studies of science. But clearly he and they did and do really mean it, as it is the core belief of the entire program.

Some of the constructivists, influenced by symbolic interactionism and ethnomethodology, argued that what is a scientific fact was a result of social negotiations among scientists in the laboratory. Others, with a more neo-Marxist orientation, argued that social and economic interests determined the content of scientific ideas. Thus, David Bloor argued[4] that Boyle's law was influenced by his conservative political beliefs and his desire to maintain the status quo in order to protect his vast Irish land holdings. This argument was strikingly similar to that made in the 1930s about Newton's work by the Soviet sociologist of science Boris Hessen.[5]

Where did these views come from? Although similar movements were occurring in other disciplines at the same time, literary theory, for example; there is little evidence that the constructivists in science were directly influenced by developments in other fields.[6] The single most important influence on the development of social constructivism was clearly the classic book by Thomas S. Kuhn, *The Structure of Scientific Revolutions.* In light of the growing popularity of relativistic views in the more humanistic areas of the social sciences (not the dominant quantitative areas that were then and still are today committed to an old fashioned positivism), Kuhn's book seemed to give warrant to the view that social consensus determined "nature" rather than nature determining scientific consensus.

Politics also played a role in the emergence of constructivism. There were many young politically left people who entered sociology in the 1960s, a substantial portion being neo-Marxist or sympathetic to such views, who considered the then dominant approach to sociology—the functionalist view of Robert K. Merton and his students—to be politically conservative and welcomed any vehicle that allowed them to attack this group.

As the relativist-constructivist approach to science began to gain steam in the early 1980s, it began to draw some serious critical attention. Perhaps the earliest people to attack the constructivist approach were philosophers of science, with Larry Laudan,[7] Ronald Giere,[8] and David Hull[9] being prominent among them. With a group of eminent philosophers of science as contributors to this volume, I will not attempt to discuss the objections that philosophers had to the constructivist approach. Historians also began to raise questions about the accuracy of the constructivist portrayal of the development of science. Among the critics were Stephen Brush,[10] Martin Rudwick,[11] Peter Galison,[12] and ultimately Kuhn himself. In Kuhn's 1992 Rothschild Lecture he stated:

> "the strong program" [another term for the relativist-constructivist approach] has been widely understood as claiming that power and interest are all there are. Nature itself, whatever that may be, has seemed to have no part in the development of beliefs about it. Talk of evidence, or the rationality of claims drawn from it, and of the truth or probability of those claims has been seen as simply the rhetoric behind which the victorious party cloaks its power. What passes for scientific knowledge becomes, then, simply the belief of the winners. *I am among those who have found the claims of the strong program absurd: an example of deconstruction gone mad.*[13]

It is interesting that many members of the constructivist school do not see some of the scholars mentioned above, such as Rudwick and Galison, as being opposed to their position. In fact these two are frequently positively cited by constructivists. This is because all of the people I have mentioned have rejected the stereotyped view of positivism that the constructivists have set up as a straw man and therefore on some issues can be seen as having the same views as the constructivists. Virtually everyone writing about science today, the constructivists and their critics (including myself), reject the overly rationalized and idealized view of science that was prominent prior to the 1970s. But although scholars like Galison and Giere will reject the same stereotyped positivism that Kuhn rejected, they are also just as opposed to the relativism that is at the heart of the constructivist program.

Although virtually all of the leaders of the social constructivist school are or claim to be sociologists, criticisms by sociologists have been almost nonexistent. In 1982 Thomas Gieryn, a Merton student, published a criticism of the relativist-constructivists, arguing (I believe incorrectly) that all the major insights of the constructivists had been made by Merton.[14] Soon after, Gieryn, sensing the increasing power of the constructivists, had a conversion and today is proudly displayed as a member of that school as converts were displayed during the cold war. In fact, it was not until 1992 with the publication by Harvard University Press of my book *Making Science: Between Nature and Society* that there was any significant critique by a sociologist of the constructivist school.

Why were sociologists of science so late in coming to the party? First, as I pointed out above, many sociologists disliked the so called "Mertonians" for political reasons and were glad to see the rise of any school that would challenge Merton's perceived control of the specialty. Although Merton is widely respected by nonsociologists and sociologists in many countries abroad,[15] many of his ideas are currently "out of fashion."

Second, many correctly perceived the constructivist approach as an attack on the natural sciences and were pleased to see these sciences, which have long lorded it over the social sciences, knocked off their pedestal. The most eminent leader of the constructivists, Bruno Latour, readily admits that the deligitimization of the natural sciences was one of their goals. For example, in a recent debate with Collins, Callon and Latour argue that:

> The field of science studies has been engaged in a moral struggle to strip science of its extravagant claim to authority. Any move that waffles on this issue appears unethical, since it could also help scientists and engineers to reclaim this special authority which science studies has had so much trouble undermining.[16]

And in the same piece they go on to say: "we wish to attack scientists' hegemony on the definition of nature, we have never wished to accept the essential source of their power: that is the very distribution between what is natural and what is social and the fixed allocation of ontological status that goes with it."[17] And a few pages later when they list their goals, the first listed is "disputing scientists' hegemony."[18]

Third, most American sociologists of science lacked the philosophical training of the European constructivists and either did not really understand much of the constructivist work or were afraid to engage in a battle that would necessarily involve philosophical argument—an area where they felt distinctly disadvantaged. It is my bet that more than fifty percent of those citing the work of Latour could not give a coherent explanation of that work if asked to do so. People have jumped on a bandwagon that they do not really understand.

Fourth, Merton himself has a well-known dislike for controversy and has never made a public statement on the social constructivist school. Thus, the "Mertonians" were left without a leader in the debate. And finally, up until the last five years, the constructivists had very little influence in the United States and thus could safely be ignored as a European phenomenon that had no effect on the ability of nonconstructivist sociologists of science (a group dwindling in size all along) to do their quantitative more traditional work and have that work published in mainstream journals where the rewards for sociologists lie. American sociologists of science, for the most part, considered themselves to be primarily interested in sociology, with science used as a research site. The European constructivists were primarily interested in science and in many (not all) cases were abysmally ignorant of more general sociological issues and concerns.

Let me now briefly summarize my critique of the social constructivist approach to science. If a sociologist wants to show that social variables influence the cognitive content of science, she must be careful in specifying exactly what about the cognitive content is being influenced. There are three different ways in which social factors could influence cognitive content. The first has been called the foci of attention or what problems scientists choose to study. There is no question that problem choice is influenced at least to some significant extent by social factors. This was illustrated well by Merton in his classic study of seventeenth-century English science[19] in which he showed how practical military and economic problems of the day played a strong role in determining what problems scientists attempted to solve. A second way to look at how social factors influence the cognitive content is to look at the rate of advance. How does the social organization of science and the society in which it is embedded influence the rate at which problems will be solved. The centers of scientific advance have changed over time; how can we explain this? This was also analyzed by Merton in his study of seventeenth-century science when he asked the question of what influenced the dispersion of talent within a society. The question has also significantly been addressed by the late Joseph Ben-David[20] and Derek de Solla Price.[21] Tom Phelan, a former student of mine, and I are currently investigating this problem using nation states as the unit of analysis and the number of highly cited papers written as a measure of the dependent variable.

The third way to look at cognitive content is to look at the actual substance of solutions to specific scientific problems. For example, in *Laboratory Life*, Latour and Woolgar try to show how the discovery of the chemical sequence of thyrotropin releasing factor (TRF) made by R. Guillemin and A. V. Schally was socially constructed. The scientific community came to believe that TRF was made up of the sequence Pyro-Glu-His-Pro-$NH_2$ rather than some other sequence. It is this latter sense of cognitive content that the social constructivists are interested in. And they claim that since science is not constrained by nature, the solution to the chemical structure of TRF could have been different and that the specialty of neuroendocrinology would have progressed to the same degree or perhaps an even greater degree had some other structure been identified.

I argue in my book that there is not one single example in the *entire constructivist literature* that supports this view of science. In order to demonstrate the credibility of their view, one must show how a specific social variable influences a specific cognitive content. In all of their work there is always at least one or another crucial piece missing. In some of their work they illustrate well how social processes influence the doing of science; but they fail to show how they have had a significant effect on what I call a knowledge outcome or a piece of science that has come to be accepted as true by the scientific community and thereby entered the core knowledge of that discipline.

In order to demonstrate this, it is necessary to do a very close reading of the texts produced by the constructivists. First, I will give an example of how they discuss the social processes influencing scientists as they go about their work; but fail to show how these processes influence any scientific outcome. In *Laboratory Life*, probably the most heavily cited and influential work done by constructivists, Latour and Woolgar present a long description of the social negotiation taking place between two scientists, Wilson and Flower.[22] They succeed in showing that scientists engage in social negotiations about their work in the same way as any other people negotiate about other aspects of their interaction. But what they have decisively failed to show is how this social negotiation influenced any aspect of science. *They have in effect "black boxed" the science; the very thing they so indignantly accused the "Mertonians" of doing.*

This is not just one example. Much of the work in which the constructivists talk about how scientists engage in social negotiations or are "human" fails to show how such negotiation or "humanness" influences any piece of communally held science in any way. Knorr-Cetina does the same thing when she describes social negotiation between a Watkins and a Dietrich[23]—but she never says how it influenced the scientific outcome. Also, in her famous analysis of how scientists negotiated fifteen different versions of a paper they are writing, she does not say whether it made any difference. Was the last published draft a significantly different piece of science than the first draft? And even if it were, if no one paid attention to this particular paper (if, in other words, it was a trivial piece of science she was analyzing), what difference would all the negotiation make for the scientific community?

In other cases, constructivists analyze scientific conflicts but fail to show how the resolution of these conflicts were influenced by social variables. It is a sad commentary on the constructivist program that the example they like to cite most frequently is the now well-known work by Andrew Pickering, *Constructing Quarks.* In this book and several earlier papers, Pickering discusses the debate theoretical high-energy physicists had over two theories—one was called "charm" and the other "color." The former theory won out; but Pickering fails to show that this was a result of social factors. His own analysis leads to the opposite conclusion that it was data from experiments that led to the resolution of the "conflict." And research recently done by some students of mine in an undergraduate seminar suggests that there may never have been a "conflict" at all. There were almost no citations to the "color" theory other than by its two proponents. If there ever really was a conflict, we would expect to see some citation to both theories and then, when the conflict was resolved, a drop in citation to the "loser." Whom, we may ask, ever believed the "color" theory, or did Pickering manufacture this "conflict" because it would be an easy one to deal with? The primary point, however, is that if there was or was not a conflict, it was resolved, as Pickering himself points out by empirical evidence.

In what the constructivists would call the naive and outdated language of positivist sociology, they either fail to have an adequate dependent variable, an adequate independent variable, or to demonstrate the link between these variables. Even the constructivists frequently show that they are aware that their theoretical approach cannot explain what they empirically observe. For example, consider another case from *Laboratory Life.* Latour and Woolgar tell us that Schally was about to publish the formula that eventually turned out to "be" TRF; but he believed in Guillemin's work more than his own and held back in publication. Guillemin was at that time arguing that TRF might not even be a peptide. Essentially what Latour and Woolgar are saying is that Guillemin's authority was so great that it served to delay the "discovery" of TRF by several years. But what they decisively fail to say is that the structure of TRF would have been anything other than Pyro-Glu-His-Pro-$NH_2$ whether or not this discovery would have been made several years earlier or at the time that it was. In other words, they have sneakily changed the dependent variable to be the rate of advance rather than the cognitive content of knowledge.

As a general rule, readers of the work of the social constructivists should always ask (1) have they identified a real social independent variable? and (2) have they shown that it has influenced the actual cognitive content of some piece of science rather than the foci of attention or the rate of advance? By influence, we mean that the cognitive content (as it was accepted by the scientific community) turned out one way rather than another *because* of some social process. Of course, processes like those described by Latour and Woolgar effect how long it takes for a particular discovery to be made; but they have failed to give a single example where the social processes influence the content of such a discovery.

A frequently used constructivist rhetorical trick is to argue that it is impossible to separate the technical from the social; that all science is inherently social. This turns their entire argument into a tautology. If science is inherently social, this means that the technical aspects of scientific discoveries by necessity must be determined socially. This is indeed the question we are examining—the extent to which the technical aspects of science have social determinants. If we take as an assumption one answer to the question we are researching, then we might as well all pack up our bags and go home because there would be nothing more to research.

If there is only one correct outcome or solution to a scientific problem, like the structure of TRF, *before the actual discovery is made*, then this gives no room for social factors to influence it. If the constructivist position is correct, this means that it would have been possible that some other structure of TRF could have been accepted as fact and that the discipline in both its pure and applied aspects would have proceeded with just as great success. But even Latour and Woolgar show their skepticism about such a belief. In discussing the story of TRF, they point out that at one time it looked like the scientists might have been forced to abandon the program by inability to obtain enough research material:

> It was then feasible that partially purified fractions would be continued to be used in the study of modes of action, that localization and classical physiology could have continued, and that Guillemin would merely have lost a few years in working up a blind alley. TRF would have attained a status similar to GRF or CRF, each of which refers to some activity in the bioassay, *the precise chemical structure of which had not yet been constructed.*[24]

Note again that they are not saying that the chemical structure of TRF would have been any different; but only that the problem would not have been solved.

They fail to explain why some local "productions" (as researchers like Knorr-Cetina likes to call the results of laboratory science) are successful in the larger community and others are not. Latour's discussion of strategies and power clearly fails to explain cases like DNA in which the discovery was accepted into the core almost overnight. Its authors were both unknown, and their opponents were leaders of the field. What made opponents such as Pauling, Wilkens, and Chargaff enroll in the Watson and Crick bandwagon?

My book is full of many examples, based upon detailed readings of other constructivist texts, that show how in each and every case they fail to do what they claim. An examination of what happened to this book is good evidence of how the constructivists treat criticism. First, all reviews by constructivists were harshly negative including one by Shapin in *Science* and one by Pickering in the *Times Literary Supplement.* Fuller actually wrote two negative reviews in two different journals. All the reviews of the book in mainstream American sociology journals that I have seen were moderately to strongly positive, including an extremely positive review by Mary Frank Fox in *Contemporary Sociology.* But the most noticeable aspect of the constructivists' reaction to the book was to ignore it. Where they have to review it, they will give it a good bashing, but where they have any control, they feel the best course of action is to keep the book unknown. Thus the book has gone unreviewed in the two leading specialty journals in the field, the *Social Studies of Science* and *Science, Technology, and Human Values.* It is quite probable that considerably more than half of the members of the Society for the Social Studies of Science do not even know of the book's existence.

Besides out and out distortion of what is said in the book and using the review as an occasion for general "Mertonian" bashing, the most frequent tactic taken is what I call the "we never said that (or meant that)" tactic. Most of the constructivist leaders criticized in my book are not stupid; far from it. They know that what they say cannot hold water and, when pushed to its real foundations, is logically absurd. Therefore, the only way to defend themselves is to say that they never said what I said they said (or if they said it they did not mean it). There are two answers to this rebuttal. First, sit down with my book and their texts and read both closely and then determine whether they did or did not say what I said they said. For example, a recent review[25] says that my criticism "relies on misreadings" of the constructivist works. "He insists that the constructivist research program is premised on denial that the realities of nature play any part in scientists' deliberations, whereas his antagonists merely presume that these realities cannot be abstracted from the theories and technologies that frame them."[26] Statements such as these make me wonder how the reviewer ever got out of the eighth grade. Can she or can she not read the direct quote from Collins above in this article? Or does she want to accept the claim that he did not really mean it? To the extent that constructivists do not "really mean" their statements of relativism, then there is nothing contradictory between their beliefs and my own.

Shapin is another one who is notorious for putting a sugar coating on the constructivist pill.[27] For example, in a recent piece on the sociology of scientific knowledge, Shapin describes Latour's work as showing that there were more "politics" within the walls of scientific workplaces than there were outside and that to secure the support of other scientists for a scientific claim was a thoroughly social process.[28] There is no "Mertonian" sociologist of science who would disagree with such conclusions. In the long review article, he fails to deal at all with the relativism that is at the core of the constructivist program. Shapin himself, who used to write polemics supporting the constructivist program,[29] is now doing fairly traditional social history of science on topics that Mertonians studied more than twenty years ago. In his most recent book,[30] he essentially asks what social processes were involved in establishing authority in seventeenth-century science. He emphasizes the importance of the "gentleman" as an individual who could be trusted. There is no discussion, anywhere in the book, of how any social process influenced the content of science. The science is virtually black boxed throughout the book. He is arguing against the rationalistic view of science that rejects trust and authority as mechanisms that influence belief as opposed to direct observation. Consider the following quote:

> According to the classical view of the history and philosophy of science, consensus is determined by the empirical phenomena themselves. Theories supported by empirical observation would become part of the consensus; theories at odds with observable "facts" would be discarded. . . . Once we accept the notion that consensus does not automatically spring from nature, we are forced to pay more attention to the sociological processes through which consensus is developed, maintained, and eventually shifted. One of the primary mechanisms through which consensus is maintained is the practice of vesting authority in elites.

There is one problem here; this quote is not from Shapin's book. Rather it is from a book published in 1973 by Jonathan and Stephen Cole, entitled *Social Stratification in Science.*[31] This latter book is not cited by Shapin. Now the point is not to say that we have priority on this matter. Clearly we never did the type of detailed and admirable historical work conducted by Shapin to show how the mechanisms of authority may have developed in a particular society at a particular time. It is, however, to point out that without the relativism, there is no great gap between contemporary work done in the social studies of science and the past work of the misguided "Mertonians."

What is the current state of constructivist sociology of science? It is a field that is in intellectual disarray but stronger than ever in its political control of organizations, journals, and science studies programs within universities. After the constructivists's unexpected and amazingly easy takeover of the sociology of science, when it was no longer fun to flog the "Mertonian" bad boys, they fell out among themselves and split into a bunch of warring clans.

One problem always faced by the constructivists was that of reflexivity. If in the natural sciences facts were not based upon empirical evidence from the external world, then how could they be said to be so based in the sociology of science? Why should anybody bother reading the works of Latour *et al.* if they only represented an attempt to push a point of view by power? Some constructivists such as Woolgar and Ashmore and to some extent Mulkay went off in the direction of taking their own work as the subject of analysis.[32] In the mean time Bruno Latour, who had become the demigod of the constructivist movement, and his French sidekick Michel Callon, began to recognize the problems inherent in their relativist position and turned on a dime and claimed they are not now and never have been relativists. Instead, he and Callon are now developing what they call actor-actant network theory. The most interesting part of this work is that what used to be considered the object of study; quarks, for example, now become equal to the humans. Scientific, and indeed social, outcomes are a result of interaction among a network of scientists, practitioners, other people, and things.

In a famous paper by Callon in which he analyzes some applied science on scallop fishing,[33] he concludes his story by saying that the reason that the experiment failed was because the scallops would not cooperate. In a vicious and amusing polemic between Latour and Callon on one side and Harry Collins and Steven Yeareley on the other, Collins correctly accuses Latour of abandoning the relativist-constructivist program. As Collins and Yearley say: "The crucial final quotation [in Callon's article on scallops] is: 'To establish that larvae anchor, the complicity of the scallops is needed as much as that of the fishermen.' "[34] It does not take very much insight to see that this is a nifty way to bring nature back into the analysis. Latour and Callon have no answer to this, although they do successfully make some criticisms of Collins's work which privileges sociology while trying to attack the privileging of the natural sciences. For example:

> That scallops do not interfere at all in the debate among scientists striving to make scallops interfere in their debates—is not only counterintuitive but empirically stifling. ***It is indeed this absurd position that has made the whole field of SSK*** [social study of scientific knowledge—another term for constructivism] ***look ridiculous*** and lend itself to the "mere social" interpretation.[35]

An intelligent reader of this polemic between Collins and Yearley on the one side and Callon and Latour on the other can do nothing other than agree with both sides. Their work if taken seriously is nothing other than absurd or voodoo sociology. The sociology of science with its many potentially interesting questions has gotten lost in a tautological mess of philosophical arguments that, as many of even the constructivists have now seen, lead nowhere.[36]

Latour's latest book[37] is an obscure philosophical essay (which he proudly proclaims has no examples because of his "Gallic" tradition) that does not answer the pressing problems facing the sociology of science today and goes little beyond Callon and Latour[38] in developing actor-actant network theory. In both pieces, Latour argues that modernists have tried to locate things along a nature-society polarity and this has made it impossible for them to deal with "hybrids" or with phenomena that move back and forth between the nature end of the continuum and the social end of the continuum as they develop. Thus, he would argue there is no way to deal with a phenomenon like TRF, which is social as the scientists struggle to define it and becomes a result of nature once it has been defined. Latour suggests throwing in another dimension, which is time. But this does not solve any of the problems because it does not tell us why certain objects become stabilized and others do not and why some become stabilized at the nature end of the pole and others at the social. It is perhaps because of the deep obscurity of Latour's latest book that his followers are defining it as his "best contribution yet."

I find it exceedingly strange when my book can be criticized for addressing itself to the following type of questions:

> Can the rate of scientific innovation in a nation be counted a linear function of the sheer number of scientists it manages to sustain (and, by implication, is scientific advance somehow linked to national development)? Can non-scientists whose decisions affect the direction of research be provided with quantitative measures of the quality of scientific work? Do universalistic criteria inform routine procedures for evaluating proposed projects, so that the most deserving applicants are funded? Do status differentials within disciplinary communities im-

> pede the free flow of communication so necessary to the growth of knowledge, possibly to a degree that precludes recognition of outstanding work done by persons who toil in professional obscurity? Do scientists' creative powers diminish as they age, so that an individual of relatively advanced age should, *ceteris paribus*, be presumed an unworthy recipient of support? And, even if there is no association between creativity and age, should younger grant applicants receive preferential consideration, lest the scientific community become moribund through failure to promote young talent (and the population as a whole thereby cease to enjoy the fruits of scientific progress)?[39]

Are not these indeed the type of questions that sociologists of science should be interested in and representative of a type of work that has literally been wiped out by the dominance of the new social constructivists?

In my book, *Making Science*, I call for a rapprochement between the social constructivists and the more traditional sociologists of science. Much of the work of the constructivists has been useful in pointing out that science is not the type of rational endeavor as depicted in the introductory philosophy of science chapters in science courses. It clearly is not easy to determine what is true or false, worthwhile or trivial. In my own work, I make a distinction between the frontier and the core. The core consists of a small set of theories, analytic techniques, and facts that represent the given at any particular point in time. Core knowledge is that which is accepted by the scientific community as being both true and important. The other component of knowledge, the research frontier, consists of all the work currently being produced by all active researchers in a given discipline. The research frontier is where all new knowledge is produced. On the research frontier, science is characterized by a lack of consensus. In fact I have shown that at the frontier, there is no more consensus in the natural sciences than there is in the social sciences. The lack of consensus is so great that whether or not one receives a National Science Foundation grant is fifty percent a result of the quality of the proposal and fifty percent a result of luck—which reviewers out of the pool of eligibles happen to be sent your proposal.[40]

Clearly social factors play an important role in the evaluation of new knowledge; but so does evidence obtained from the natural world. The sociologist of science should study how social factors and such evidence interact in the evaluation process. This is perhaps the most crucial question of the discipline. And if one abandons the frequently programmatic and polemic relativism of the constructivists, their work is of use in answering this question. In fact, I like to think of myself as a realist-constructivist. Yes science is socially constructed, but yes how it is constructed is to various degrees and extent constrained by nature.

My work in the sociology of science has led me to strongly reject the conclusion that the natural sciences are entirely socially constructed; but my life in the social sciences has made me more amenable to the possibility that these sciences may indeed be entirely socially constructed. Ideology, power, and network ties seem to determine what social scientists believe; evidence is frequently entirely ignored. I have recently begun to address this problem in an article entitled "Why Sociology Doesn't Make Progress Like the Natural

Sciences." That social science is completely or almost completely socially constructed helps explain how the social constructivist view of science could have become so powerful in the absence of any good supporting evidence and in the face of such devastating empirical critiques as those found in books like Peter Galison's *How Experiments End.*

## NOTES

1 S. Jasonoff, G. D. Markle, J. C. Peterson & T. Pinch, eds., *Handbook of Science and Technology Studies.*
2 Ibid., p. 440 (emphasis added).
3 Harry M. Collins, "Stages in the Empirical Program of Relativism," p. 3 (emphasis added).
4 David Bloor, "Durkheim and Mauss Revisited: Classification and the Sociology of Knowledge."
5 Boris Hessen, "The Social and Economic Roots of Newton's *Principia.*"
6 For a critique of the social constructivists and a discussion of similar attacks made in other fields of scholarship see Paul Gross & Norman Levitt, *Higher Superstition: The Academic Left and Its Quarrels with Science.*
7 Larry Laudan, *Science and Values: The Aims of Science and Their Role in Scientific Debate*; *Relativism and Science.*
8 Ronald Giere, *Explaining Science.*
9 David Hull, *Science as Process.*
10 Stephen Brush, "Should the History of Science Be Rated X?"
11 Martin Rudwick, *The Great Devonian Controversy: The Shaping of Scientific Knowledge among Gentlemanly Specialists.*
12 Peter Galison, *How Experiments End.*
13 Thomas S. Kuhn, Rothschild Lecture, pp. 8–9 (emphasis added).
14 Thomas Gieryn, "Relativist/Constructivist Programmes in the Sociology of Science: Redundance and Retreat."
15 So much so, that he probably would have to be considered the most eminent living sociologist.
16 M. Callon & B. Latour, "Don't Throw the Baby Out with the Bath School! A Reply to Collins and Yearley," p. 346.
17 Ibid., p. 348.
18 Ibid., p. 351.
19 R. K. Merton, *Science, Technology, and Society in Seventeenth-Century England.*
20 Joseph Ben-David, "Scientific Productivity and Academic Organization in Nineteenth-Century Medicine."
21 Derek de Solla Price, *Little Science, Big Science . . . and Beyond.* For a review of literature on the rate of scientific advance see Stephen Cole, *Making Science: Between Nature and Society*, ch. 9.
22 Bruno Latour & Steve Woolgar, *Laboratory Life: The Construction of Scientific Facts*, pp. 154–158.
23 Karin Knorr-Cetina, *The Manufacture of Knowledge: An Essay on the Constructivist and Contextual Nature of Science.*
24 Latour & Woolgar, *Laboratory Life*, p. 128 (emphasis added).
25 H. Kuklick, "Mind over Matter?"
26 Ibid., p. 370.
27 Unfortunately, because as the deadline for this article approached and passed my reply to Shapin's review of my book in *Science* remained somewhere in transit between Stony Brook and Queensland, I will be unable to include here details of his misinterpretations. But certainly one of the most important is to say that constructivists do not really mean their relativist manifestos.
28 S. Shapin, "Here and Everywhere: Sociology of Scientific Knowledge."
29 S. Shapin, "History of Science and Its Sociological Reconstructions."

30 S. Shapin, *A Social History of Truth: Civility and Science in Seventeenth-Century England.*
31 Jonathan Cole & Stephen Cole, *Social Stratificaiton in Science*, pp. 77–78.
32 See Andrew Pickering, ed., *Science as Practice and Culture*; S. Shapin's review article on the sociology of scientific knowledge in *Annual Reviews of Sociology*; S. Jasonoff, G. D. Markle, J. C. Peterson & T. Pinch, eds., *Handbook of Science and Technology Studies* for discussions of recent trends in social constructivism.
33 Michel Callon, "Some Elements of a Sociology of Translation: Domestication of the Scallops and the Fishermen of St. Brieux Bay."
34 Harry M. Collins & Steven Yearley, "Epistemological Chicken," p. 314.
35 Michel Callon & Bruno Latour, "Don't Throw the Baby Out with the Bath School! A Reply to Collins and Yearley," p. 353 (emphasis added).
36 Andrew Pickering, *Science as Practice and Culture.*
37 Bruno Latour, *We Have Never Been Modern* (originally published in French in 1991).
38 Callon & Latour, "Don't Throw the Baby Out."
39 H. Kuklick, "Mind over Matter?" p. 369.
40 Stephen Cole, *Making Science: Between Nature and Society*, ch. 4.

## REFERENCES

BEN-DAVID, JOSEPH. "Scientific Productivity and Academic Organization in Nineteenth-Century Medicine." *American Sociological Review* 25 (1960): 828–843.

BLOOR, DAVID. "Durkheim and Mauss Revisited: Classification and the Sociology of Knowledge." *Studies in History and Philosophy of Science* 13 (1982): 267–297.

BRUSH, STEPHEN. "Should the History of Science Be Rated X?" *Science* 183 (1974): 1164–1172.

CALLON, MICHEL. "Some Elements of a Sociology of Translation: Domestication of the Scallops and the Fishermen of St. Brieux Bay." In *Power, Action, and Belief: A New Sociology of Knowledge?*, edited by John Law, pp. 196–229. London: Routledge and Kegan Paul, 1986.

CALLON, MICHEL & BRUNO LATOUR. "Don't Throw the Baby Out with the Bath School! A Reply to Collins and Yearley." In *Science as Practice and Culture*, edited by A. Pickering. Chicago, IL: University of Chicago Press, 1992.

COLE, JONATHAN & STEPHEN COLE. *Social Stratification in Science.* Chicago, IL: University of Chicago Press, 1973.

COLE, STEPHEN. *Making Science: Between Nature and Society.* Cambridge, MA: Harvard University Press, 1992.

———. "Why Sociology Doesn't Make Progress Like the Natural Sciences." *Sociological Forum* 9 (1994): 133–154.

COLLINS, HARRY M. "Stages in the Empirical Program of Relativism." *Social Studies of Science* 12 (1981): 3–10.

FOX, MARY FRANK. "Realism, Social Constructivism and Outcomes in Science." *Contemporary Sociology* 22 (1993): 481–483.

GALISON, PETER. *How Experiments End.* Chicago, IL: University of Chicago Press, 1987.

GIERE, RONALD N. *Explaining Science.* Chicago, IL: University of Chicago Press, 1988.

GIERYN, THOMAS. "Relativist/Constructivist Programmes in the Sociology of Science: Redundance and Retreat." *Social Studies of Science* 12 (1982): 279–297.

GROSS, PAUL R. & NORMAN LEVITT. *Higher Superstition: The Academic Left and Its Quarrels with Science.* Baltimore, MD: Johns Hopkins University Press, 1994.

HESSEN, BORIS. "The Social and Economic Roots of Newton's *Principia.*" In *Science at the Crossroads: Papers Presented to the International Congress of the History of Science and Technology.* London: Frank Cass, 1971.

HULL, DAVID. *Science as Process.* Chicago, IL: University of Chicago Press, 1988.

JASONOFF, S., G. D. MARKLE, J. C. PETERSON & T. PINCH, eds. *Handbook of Science and Technology Studies.* Thousand Oaks, CA: Sage Publications, 1995.

KNORR-CETINA, KARIN. *The Manufacture of Knowledge: An Essay on the Constructivist and Contextual Nature of Science.* New York, NY: Pergamon Press, 1981.

KUHN, THOMAS S. Rothschild Lecture. 1992.

——— *The Structure of Scientific Revolutions.* Chicago, IL: University of Chicago Press, 1970.

KUKLICK, H. "Mind over Matter?" *Historical Studies in the Physical and Biological Sciences* 25 (1995): 361–378.

LAUDAN, LARRY. *Relativism and Science.* Chicago, IL: University of Chicago Press, 1990.

———. *Science and Values: The Aims of Science and Their Role in Scientific Debate.* Berkeley, CA: University of California Press, 1984.

LATOUR, BRUNO. *We Have Never Been Modern.* Cambridge, MA: Harvard University Press, 1993.

LATOUR, BRUNO & STEVE WOOLGAR. *Laboratory Life: The Construction of Scientific Facts.* Princeton, NJ: Princeton University Press, 1986.

MERTON, R. K. *Science, Technology, and Society in Seventeenth-Century England.* New York, NY: Howard Fertig, 1970.

PICKERING, ANDREW. *Constructing Quarks: A Sociological History of Particle Physics.* Edinburgh: Edinburgh University Press, 1984.

———. *Science as Practice and Culture.* Chicago, IL: University of Chicago Press, 1992.

PRICE, DEREK de SOLLA. *Little Science, Big Science . . . and Beyond.* New York, NY: Columbia University Press, 1986.

RUDWICK, MARTIN J. S. *The Great Devonian Controversy: The Shaping of Scientific Knowledge among Gentlemanly Specialists.* Chicago, IL: University of Chicago Press, 1985.

SHAPIN, S. "Here and Everywhere: Sociology of Scientific Knowledge." In *Annual Review of Sociology*, edited by John Hagan & Karen S. Cook, pp. 289–321. Palo Alto, CA: Annual Reviews, Inc., 1995.

———. "History of Science and Its Sociological Reconstructions." *History of Science* 20 (1982): 157–211.

———. *A Social History of Truth: Civility and Science in Seventeenth-Century England.* Chicago, IL: University of Chicago Press, 1994.

# THE ALLURE OF THE HYBRID
## Bruno Latour and the Search for a New Grand Theory

OSCAR KENSHUR

*Yes, the scientific facts are indeed constructed, but they cannot be reduced to the social dimension because this dimension is occupied by objects mobilized to construct it. Yes, those objects are real but they look so much like social actors that they cannot be reduced to the reality "out there" invented by the philosophers of science. The agent of this double construction—science with society and society with science—emerges out of a set of practices that the notion of deconstruction grasps as badly as possible. The ozone hole is too social and too narrated to be truly natural; the strategy of industrial firms and heads of state is too full of chemical reactions to be reduced to power and interest; the discourse of the ecosphere is too real and too social to boil down to meaning effects. Is it our fault if the networks are* simultaneously real, like nature, narrated like discourse, and collective, like society?

—BRUNO LATOUR

THIS PASSAGE from *We Have Never Been Modern* will give a sense both of Latour's undertaking and of the vehemence with which he sets it forth. Latour presents his project in terms of the avoidance of two extremes, represented by alternative approaches to the study of science: the more traditional one that examines science in terms of its adequacy to the real workings of the natural world; and the currently fashionable approach that interprets science as the projection of social or cultural forces. Those who consider these alternatives to be mutually incompatible, and who embrace one of them, while vehemently rejecting the other, seem to assume that despite his protestations to the contrary, Latour must really align himself with one of the extremes.

But the humanists who are attracted to Latour take him at his word. Indeed, his popularity among humanists derives, I believe, precisely from a widespread wish to avoid reductivist methodologies and paradoxical theories. After all, humanist scholars who are interested in denigrating science, or in demystifying its claims to knowledge, need only choose from an array of available theories that have come into fashion during the past twenty-five years—theories ranging from deconstruction to scientific constructivism—that have had the effect of relativizing knowledge claims in general and of treating science as a mere social construct. The humanists who turn instead to Latour wish to be able to study scientific texts and practices as symbolic structures, but without accepting the counterintuitive and intellectually irresponsible proposition that in order to treat a theory or hypothesis as ideology, or culture, or rhetoric, they are obliged to deny that it is really science and tells us something about the world. At the same time, these humanists need to reject the proposition that because science tells us about the world, it is therefore immune from consideration in ideological or cultural terms.

But since the unpalatable extreme views, including, very prominently, the view that reduces science and reality to mere texts, have come into prominence on grand theoretical wings, humanists who wish to treat science as the object of responsible inquiry—but without treating the natural world as a mere construct—seem to feel that they need to be armed with a better theory, a more nuanced alternative to deconstruction and constructivism.[1] And for many, Latour's rejection of deconstruction and of constructivism, and his positing of entities that are neither purely social nor purely natural—entities that he calls hybrids, or quasi-objects—appears to provide just such a theoretical alternative.

To be able to denounce paradoxical extremes, however, is by no means tantamount to providing a coherent alternative, and what one person calls nuance another may call waffling. However much I sympathize with my fellow humanists who yearn for an adequate way of understanding and justifying their study of science as a symbolic structure, my aim here is to suggest that they do not need Latour and are better off without him.

My discomfort with Latour arises in the first instance from the way the hybrid as an object of analysis fits so snugly with the disciplinary practice that emerges triumphantly from his analysis of inadequate disciplines. This critique of disciplines, in turn, is intimately bound up with a historical narrative that begins with the Enlightenment and culminates in Latour himself and the disciplinary practice that he champions.[2]

Latour's historical narrative is not a history of disciplinarity per se. Indeed, at first glance, it resembles the genre of Enlightenment-bashing that serves as a quasi-historical justification for much contemporary theory in the humanities. Once again, the starting point is the epistemological wrong turn that is identified with the rigorous separation of the subjective and objective realms. In Latour's version, however, Descartes is not cast as the villain of the piece. The wrong turn, characterized by Latour as a "purification," is epitomized, with the help of Schaffer and Shapin's *Leviathan and the Air Pump*, by the dispute between Hobbes and Boyle over the epistemological validity of laboratory science. The ensuing separation of the realm of the human or social, symbolized by Hobbes's commonwealth in *Leviathan*, from the natural realm summoned up by Boyle and his experimental followers results in a kind of double vision. On the one side there is the science of society: "The Leviathan is made up only of citizens, calculations, agreements or disputes . . . of nothing but social relations."[3] On the other side is nature, which "has always existed and has always already been there," but whose secrets can now be discovered by scientists. It is scientists who allow the mute objects of the natural world to speak through their mediating power.[4]

The cunning stratagem of modernity, however, according to Latour, is not simply the establishment of this separation, but the paradoxical undermining of it in ways that go unrecognized or unacknowledged. Latour's claim that the demarcation between the natural and the social or subjective realms has been insisted upon in the official dogmas of modernity and even ostensibly policed, but that this demarcation has necessarily been violated in actual practice, is the source of his title, *We Have Never Been Modern.*

As long as the paradoxes and contradictions are occulted, however—as they have been until very recently—the demarcation allows the natural sciences to pretend to reveal nature in its purity—by means of one emerging scientific discipline after another—as well as to provide the Enlightenment with the basis for critiques of the obscurantism and superstition that had tainted premodern modes of thought.[5]

Despite its inception under the provocative tutelage of Shapin and Schaffer, and despite its diagrams, its revelations of paradoxes, Latour's characterization of the Enlightenment thus ends up on familiar ground—which, for all its familiarity, is not immune to objection. But I will defer my substantive objections until I complete my outline of Latour's historical schema.

The first Enlightenment is followed by a second, nineteenth-century one, marked by the rise of social science and the critique of ideology:

> With solid support from the social sciences, it became possible to distinguish the truly scientific component of the other sciences from the component attributable to ideology.[6]

Latour's history of the exfoliation of post-Enlightenment disciplines pays much more attention to developments in the social sciences than to those in the natural sciences or the philosophy of science, and his sharpest irony seems to be directed at those tendencies in the history of sociology from which he is most anxious to distinguish his own method. Thus the tone of his account of the emergence of the second Enlightenment is much less cool and detached than his account of the first one:

> . . . a succession of radical revolutions created an obscure "yesteryear" that was soon to be dissipated by the luminous dawn of the social sciences. The invincible

moderns even found themselves able to combine the two critical moves by using the natural sciences to debunk the false pretensions of power and using the certainties of the human sciences to uncover the false pretensions of the natural sciences, and of scientism. Total knowledge was finally within reach.[7]

This animus toward social science as it follows out the logic of the Enlightenment seems in large part based on a scorn for the science/ideology dichotomy of classical marxism, and the concomitant ideological debunking of the nonscientific or less scientific by the more purely scientific. Later, in an arresting discussion of sociologists after Durkheim, Latour accuses them of a kind of elitism abetted by a paradoxical double vision. Ordinary people, he tells us,

> imagine that the power of gods, the objectivity of money, the attraction of fashion, the beauty of art, come from some objective properties intrinsic to the nature of things.[8]

The sociological demystification of these beliefs reduces the purported objects to a "surface for the projection of our social needs and interests."[9] However, when ordinary people indulge in naive beliefs about free will and human agency, according to Latour, the same sociologists switch from sociological reductivism to mechanistic reductivism, a switch that elicits some of Latour's heaviest sarcasm:

> . . . fortunately, social scientists are standing guard, and they denounce and debunk this naive belief in the freedom of the human subject and society. This time they use the nature of things—that is, the indisputable results of the sciences—to show how it determines, informs and moulds the soft and pliable wills of the poor humans.[10]

The problem with these sociologists is not just that they are arrogant and inconsistent, but that they recklessly continue to embrace the modern separation of the subjective and objective realms. To quote Latour again,

> They too "see double." In the first denunciation [that is the treatment of religion and art] objects count for nothing; they are just there to be used as the white screen on to which society projects its cinema. But in the second, they [objects] are so powerful that they shape the human society while the social construction of the sciences that have produced them remains invisible.[11]

Latour's critique of sociology as the inheritor of the Enlightenment's double vision culminates in a critique of the constructivist sociology of science. When sociology finally applies to science itself the tools that it had previously applied to religion and to beauty and to other phenomena that social scientists, according to Latour, "happen to despise," the result is not a happy consistency, but rather the last paradoxical gasp of the Enlightenment paradigm, the final crisis that reveals the bankruptcy of dualism and that allows us to see that we were never modern.

This final turn is represented for Latour by the Edinburgh school of science studies.[12] By turning the "critical repertoire that was reserved for the 'soft' parts of nature to debunk the 'harder' parts"—that is, to turn against the science that had been the bedrock of sociology the weapons previously employed against religion and esthetics and other things that the sociologists did not believe in anyway—the Edinburgh daredevils, as Latour calls them, "deprived the dualists and themselves . . . of half of their resources":

> Society had to produce everything arbitrarily including the cosmic order, biology, chemistry, and the laws of physics. The implausibility of this claim was so blatant for the hard parts of nature that we suddenly realized how implausible it was for the "soft" ones as well.[13]

In other words, it is only when the denunciations carried out in the spirit of Enlightenment dualism reach their final extreme stage, in Edinburgh, that we can recognize "how badly constructed were the social theory as well as the epistemology that went with those denunciations."[14] This recognition, in turn, allows us to recognize that the objects of our inquiries

> are much more social, much more fabricated, much more collective than the "hard" parts of nature but [that] they are in no way the arbitrary receptacles of a full-fledged society. On the other hand, they are much more real, nonhuman and objective than those shapeless screens on which society needed to be projected.[15]

Thus science studies, as represented by the Edinburgh school, by trying the impossible task of providing purely social explanations for hard scientific facts, force us to recognize that dualism is dead and that all the objects of our inquiries, both "hard" and "soft," are actually quasi-objects, or hybrids.

At this point sociology seems to give way to the discipline that is ideally suited to the investigation of hybrids that are neither purely social nor purely natural, the discipline under whose banner Latour carries out his own science studies—namely, anthropology. But I think that it would be more accurate to say that sociology gives way to Latour. That is to say, Latour's history is set forth in such a way as to suit his disciplinary purposes, which involve the legitimation of his own work as an antidote to the failings of the entire tradition that preceded him. My ultimate purpose, however, is not to accuse Latour of using a self-serving rhetoric or of having a vigorous ego. Indeed, like Latour, I would like to see a more balanced and responsible alternative to competing reductivisms. I also share his mistrust of the science/ideology dichotomy because, as I have argued elsewhere, in a given instance the same symbolic structure may be analyzable both in ideological terms—that is, with respect to the ways in which it serves sectional interests—and in scientific terms—that is, in terms of its explanatory adequacy according to the norms of appropriate disciplines.[16] But the lesson that should be derived from the recognition that the same symbolic structures may be alternately amenable to analysis qua ideology and analysis qua science is an antiessentialist one: a given scientific theory, for example—no matter how great its explanatory power, and no matter how great its ideological usefulness—is not essentially science or essentially ideology. Latour, on the contrary, seems to be postulating an alternative essentialism, one that renounces metaphysical dualism in favor of what appears to all the world to be a metaphysical hybridism. Every scientific theory, in his view, appears to be, in essence, half ideology and half science.

One could of course object that Latour is more concerned with methodology than with metaphysics, and that the point of talking about hybrids is not to postulate composite essences but to set forth a model that permits us to treat entities as if they were hybrids. But such a defense would have no real bearing on my critique. For what concerns me is not a fine point of metaphysics, but precisely the consequences of the hybrid model—regardless of the ontological commitments one attached to it—for the way we approach and study a great array of cultural objects. To see what is wrong with "hybridism" as a methodological principle, we may begin by looking more closely at Latour's critique of the double vision that hybrids are supposed to supplant.

Let us return to Latour's rhetorically seductive description of the elitist sociologists who scorned the beliefs of ordinary people, but did so in allegedly contradictory ways. Some beliefs, such as those of religion and esthetics, it will be recalled, are reduced by these sociologists to social projections, whereas others, such as the belief in moral agency, are debunked by appeals to external nature, that is, to mechanistic determinism. In one instance, according to Latour, the social scientists demystify naturalization, and in the second they themselves naturalize the mechanistic reality that determines belief in free will. They are constructivists when it comes to religious beliefs and naturalists when it comes to belief in the mechanistic universe that determines beliefs. In one case they are reductive in one direction, and in the other case they are reductive in the other direction.

But Latour's neat formulations turn out, on closer inspection, to be quite misleading. Those who explain religious beliefs as projections do not necessarily deny natural mechanistic explanation for those beliefs. Hobbes, for example, whom Latour identifies as an inventor of the separation between the social and the natural, clearly ties the two sorts of explanation together. Far from pretending to exclude the natural, Hobbes claims that the religiopolitical projections of the sectarian fanatics are explicable by the fact that those fanatics are machines whose stated motives are reducible to the crude mechanisms of self-interest. By the same token when Hobbes and later social scientists "debunk [the] naive belief in the freedom of the human subject" in favor of a deterministic model, they are not switching from social to natural explanation. After all, the belief in the freedom of the human subject could be understood as a projection that serves particular social or cultural purposes. The explanations in both examples contain both social and natural elements that are by no means mutually incompatible. In fact the two cases are remarkably similar to the degree that both include natural causes and social projections, and it would appear that it is Latour's rhetoric, rather than the method of classical sociology, that suppresses the natural in one case, the social in the other case.

While Latour's examples are grossly misleading, they do provide an insight into his animus against previous sociology and science studies. In both examples just discussed, there are beliefs that the investigators treat as projections or illusions; namely, the religious or esthetic beliefs in one case, and the belief in the freedom of the human subject in the other case. The sociological explanation that Latour is attacking explains these beliefs as caused, as it were, from below, by social or psychological mechanisms, rather than from above, by the efficacious reality of God or beauty or human agency. Latour seems to condemn such debunking or demystifying analysis because he sees it as tied to the science/ideology dichotomy that forces the investigator to come down on one side or the other. Each belief is either caused by a correct understanding of external nature, and hence is scientific; or it is caused by our needs and desires and hence is ideological. Latour, quite rightly, wants to avoid such dichotomizing. He does not wish to treat scientific beliefs, such as the existence of the ozone layer or of quarks, as mere social projections, since he wants scientific beliefs to be caused, at least partially, by the actual facts or objects; but he also does not wish to treat such beliefs as explicable purely in terms of external nature. Accordingly he seems to feel the need to reject any modes of explanation in which there are beliefs that are not taken seriously. Thus, paradoxically, Latour's avoidance of reductive or one-sided explanations results in his refusal to recognize that not all phenomena require the same sort of explanation.

Latour's problem is not that he refuses to accept the possibility of differences between the sociology of religion and the sociology of science; the problem is that he is unwilling or unable to acknowledge that our judgment as to which beliefs need to be explained in social terms, which need to be explained in natural terms, and which need to be explained as both social and natural is indeed a judgment and depends on the nature of the case. Consider, for example, the following two cases of religious or quasi-religious beliefs: (1) People living in various parts of the United States claim that Elvis Presley, despite official reports to the contrary, is alive and well and shopping at a K-Mart in Kansas. (2) People living in the vicinity of Port Moresby, New Guinea have built their version of a runway that they hope or expect will attract gods similar to those that land on the real runway at the real airfield in Port Moresby and disgorge treasures. Now, in both cases there might be an element of social projection in the explanation, and there might also be a naturalistic explanation of what causes people to make such projections. But the anthropological investigator may not feel obliged, in the case of the Elvis sightings, to do empirical research in order to ascertain whether the belief in the K-Mart "Elvis" was caused, as it were, from above, by the real presence of Elvis Presley. If, however, the anthropologist who investigated the Port Moresby cargo cult failed to consider whether this belief was related to the real existence of the empirical airplanes that landed on the real airstrip, then we might have reason to question the investigator's professional competence.

The difference is not that the one investigator is more reductive than the other, or that one investigator is so enlightened as to deal with hybrids. Both are attending to what is relevant or interesting or in need of explanation in the given case, but the two cases are different from one another, and require investigators to attend to different sorts of things.

This is not to deny that many investigators in every discipline approach situations with methodological blinkers that impel them to ignore relevant factors and hence to be reductive. But this results not from the underlying contradictions of post-Enlightenment thought, but rather from a quasireligious belief in the magical applicability of a given model or methodology or diagram to every case. And this belief is one that Latour shares with those whom he most persuasively attacks. His hybrids are neither unpredictable nor idiosyncratic, but are consistently balanced entities that always lend themselves to a single specific mode of analysis, his own.

I said at the outset that Latour's theory is attractive to humanistic practitioners of science studies partly because it allows them to treat science from the perspective of their home disciplines without requiring them to deny that there is a reality out there that science talks about and even explains. The burden of my argument has been to show not only that they are wrong to think they need a theory like Latour's, but that they are wrong to think they need any theory at all—in the sense of a general law that stipulates how and why each part of a symbolic structure has ontological status or claims to our attention.

Ideological criticism, I would argue, is the study of particular cases. If one embraces a theory that stipulates that ideology and science are mutually exclusive, then one has embraced a bad theory. Latour and his followers are right to reject such theories. But our recognition that the science/ideology dichotomy deserved to be abandoned does not require us to replace it with a new grand theory that stipulates that every phenomenon is half ideology and half science, and that fails to allow us to respond to the nature of the particular case. We need to be able to decide in each case whether there is an interesting ideological dimension to a scientific theory, or whether a belief that patently serves ideological interests nonetheless has scientific plausibility apart from its ideological usefulness. We need to be able to judge which is more interesting, more in need of explanation—the images on the screen, the mechanisms that produce them, or the world that they are striving to represent.

[*Notes and references begin overleaf.*]

## NOTES

1 Quentin Skinner has observed that it is both paradoxical and apt to characterize as "grand theory" contemporary versions of skepticism that debunk traditional metaphysical systems and global explanations (*The Return of Grand Theory in the Human Sciences*, pp. 12–14). Indeed, it may be impossible to deny the possibility of making large claims without implicitly or explicitly making large claims of one's own. The same paradox may be found in traditional formulations of absolute skepticism, where doubt is justified on the ground that there is equal evidence on each side of each issue, and hence where the denial of certainty is transmuted, as Stephen Pepper pointed out (*World Hypotheses: A Study in Evidence*, p. 7), into the grand metaphysical claim that "the universe is infinitely divisible into dichotomous parts *pro* and *con.*"

2 For Latour, as for others who theorize about the way facts are not merely unproblematic givens, there always turn out to be realms of inquiry where facts can be dealt with unproblematically. For Latour, the realm of history and the realm of texts seem to be completely untouched by the problems that apply to science. In this respect, he shares with constructivists the tendency to allow pure facts to exist in ancillary disciplines that are used to problematize the status of facts in his object discipline. On this tendency in scientific and literary constructivism, see Oscar Kenshur, "The Rhetoric of Incommensurability."

3 Bruno Latour, *We Have Never Been Modern*, p. 28.

4 Ibid., p. 30.

5 Ibid., p. 35.

6 Ibid.

7 Ibid., p. 36.

8 Ibid., p. 51.

9 Ibid., p. 52.

10 Ibid., pp. 52–53.

11 Ibid., p. 53.

12 Latour is referring to the "strong programme" at the University of Edinburgh Science Studies Unit, which came into prominence in the seventies. In a book that has served as a kind of manifesto for the Edinburgh group, David Bloor argued that for a sociologist to hold back from treating science in the same sociological terms as he or she treated religion or pseudoscience was to exhibit a "lack of nerve and will" (*Knowledge and Social Imagery*, p. 4).

13 Latour, *We Have Never Been Modern*, pp. 54–55.

14 Ibid., p. 55.

15 Ibid.

16 Oscar Kenshur, *Dilemmas of Enlightenment: Studies in the Rhetoric and Logic of Ideology.*

## REFERENCES

BLOOR, DAVID. *Knowledge and Social Imagery.* 2d edit. Chicago, IL: University of Chicago Press, 1991.

KENSHUR, OSCAR. "The Rhetoric of Incommensurability." *The Journal of Aesthetics and Art Criticism* 42 (1984): 375–381.

———. *Dilemmas of Enlightenment: Studies in the Rhetoric and Logic of Ideology.* The New Historicism: Studies in Cultural Poetics, no. 26. Berkeley, CA: University of California Press, 1993.

LATOUR, BRUNO. *We Have Never Been Modern.* Translated by Catherine Porter. Cambridge, MA: Harvard University Press, 1993.

PEPPER, STEPHEN C. *World Hypotheses: A Study in Evidence.* Berkeley, CA: University of California Press, 1942.

SHAPIN, STEVEN & SIMON SCHAFFER. *Leviathan and the Air-Pump: Hobbes, Boyle, and the Experimental Life.* Princeton, NJ: Princeton University Press, 1985.

SKINNER, QUENTIN, ed. *The Return of Grand Theory in the Human Sciences.* Cambridge: Cambridge University Press, 1985.

# HISTORY, SOCIETY, POLITICS

THE EFFECT OF postmodern nihilism on the social sciences has been anything but uniform. Some precincts of sociology and anthropology have been as deeply awash in it as any literary studies department, while other areas—mathematical economics, for instance—know it only as a rumor. The contributors to this section are all students of society, or rather, of societies, ancient and contemporary. They reflect on the corrosive effects of unreason in our society and on several of the disciplines that study society.

The section begins with an exchange between Wellesley College classicist Mary Lefkowitz and Howard University Egyptologist Ann Macy Roth on the popularity, especially among black students and intellectuals, of revisionist history of the ancient Mediterranean, which not only exalts Egypt as the fountainhead of all classical culture and learning, but depicts it as a prototypical black African society. Lefkowitz recounts her struggle to challenge these tendentious and often bizarre accounts by identifying their lapses. She stresses the ultimate duty of the scholar to promote the untiring use of evidence, coherent argument, and common sense in dealing with history, and faults not only enthusiastic Afrocentrists, but also academics in other disciplines who steer fastidiously clear of the argument, not only because they accept enhanced self-esteem as a remedy for the handicaps borne by some black students, but also because they have absorbed the ambient epistemic relativism.

Professor Roth, on the other hand, finds something hopeful in the thirst of black students to retrieve some sense of history, even if this means, initially, the celebration of distorted or fictitious versions of ancient Egypt, among other things. She realizes how problematical this embrace of appealing myths can be, but tries to explore methods for bringing such students to a sense of intellectual responsibility, and for enabling them to internalize the canons of sound historical argument. It is her faith that the capacity to form reasoned judgments, once launched, is a powerful antidote to even the most seductive chauvinism.

It is of no small interest that in all the human sciences there was, for a time, a conviction that work should be "scientific," meaning that it should be disciplined by rules of logic and evidence. But the relativistic impulse always lies just beneath the surface. In cultural anthropology it has again broken forth and now decries reason as a species of cognitive imperialism. Anthropologist Robin Fox gives a historical account, deeply informed by personal experience, of how this came to be, noting the irony that an intellectual current now associated with would-be progressives had its source in reactionary antimodernism.

Political scientist Simon Jackman finds that while political liberalism and scientific realism are, indeed, properly coupled as the twin children of the

Enlightenment, they do not completely proscribe relativism of a certain useful kind. It is simply that a rational relativism—perhaps better named, as it sometimes is, pragmatism—carries no necessary denial of the value of evidentiary standards, which are not only required for, but effective in, the honest assessment of society.

How honest such assessments are, however, when political necessity and social pressure, reinforced by well-meaning clichés, impinge on the effort to determine how things are, is unpredictable. Christina Sommers, who has been subjected to violent attack for raising such questions, demonstrates how readily "science"—that is to say, logic and evidence—can be set aside when the sociopolitical resonance energy is large. She analyzes the modes of myth making, including feeble methodology and ungrounded psychologizing, that have been used to justify a picture of our primary and secondary schools as dens of sexist iniquity.

# WHATEVER HAPPENED TO HISTORICAL EVIDENCE?

MARY LEFKOWITZ

WHATEVER HAPPENED to historical evidence? Perhaps it is naive to ask this question, but in this essay I would like to offer a real-life demonstration of where the new historicism, or history-without-facts, is leading us. My subject is the effect of factlessness on the study of ancient Greece, and I will try to explain why disregard of evidence has made it possible and even fashionable to claim that many of the great achievements that have always been attributed to the ancient Greeks were borrowed or even stolen from Egypt.

I cannot pretend to know all the reasons, but I can at least identify a few of the most important ones. Several apply to all subject areas, and have been described with great effect in *Higher Superstition.*[1] In general, the acquisition of specific knowledge has become *déclassé*. It has apparently become pedagogically chic to say that we aim to teach students how to think, and that therefore we need to concentrate on method rather than on content. Some students seem to be insulted when I ask them to learn (and spell) the names of the principal Greek gods. They complain if I test them on Friday about what they were supposed to know Wednesday. They seem not to realize that specific knowledge, although burdensome at first, is in the end liberating. Ultimately, I suppose, John Dewey is responsible for this state of affairs. In 1891 he spoke enthusiastically of the lecture as means of education: "it has helped to dispel those vicious methods of *rote* study" based on text books.[2] As a result, we have—at least in theory—placed a higher value on method than on facts. We are implying, even if we do not mean to, that facts are less important.

But I think that there is another, and fundamentally much more troubling reason why many of our students, and their instructors, tend to place less and less emphasis on the acquisition of factual information. It is that they do not really believe that there are such things as facts, or to put it the other way round, they think that facts are meaningless because they can be manipulated and reinterpreted. If it is true (and I think it probably is) that no historical work can be written without bias of some sort, none can be trusted to give an entirely accurate picture of what the writer is seeking to describe.

Of course historians (and their readers) have always been aware that they can and do write with an evident bias—the first-century AD. Roman historian Tacitus tells us at the beginning of his *Annals* (1.1.3) that he proposes to write the history of the emperors from Augustus to Nero *sine ira et studio*, "without anger or partisanship." He knew that his audience would know that he meant the exact opposite of what he said. But recently many historians have been concentrating on another type of bias, this time unconscious: the blinkers put on everyone's vision by the values of their particular societies. These scholars insist that history is always composed in conformity or response to the values of the society in which it is produced, and for that reason it can be regarded as a cultural projection of the values of that society, whether individual writers are aware of it or not.

Such beliefs, if carried to their logical extreme (and they often are, particularly by academics in this country), make it possible to say that all history is by definition fiction. If history is fiction, it is natural to deny or to minimize the importance of all historical data. Instead, these writers concentrate instead on cultural *motives*. Historians, in their view, write what they *are.*[3] This way of looking at history produces some astounding results, such as modern Holocaust denial. Holocaust deniers argue that the extent of the slaughter was exaggerated or even falsified, because the history of the Holocaust has been written by Jewish historians and other people with reason to be angry at the Nazis.

Once motive becomes more important than facts, the author of any work of history, or for that matter, any work of literature becomes a subject of greater interest than the books or articles he produces. That trend in itself would not be quite so harmful if it meant that the study of history were simply being replaced by biography, the history of individuals rather than of geographic or governmental entities. But the trouble is that in biography as well as in history there is a new concentration on cultural motives.

Cultural motives are not factors like personal animosities or family connections. They are rather the kind of impulses or habits of thought that are dictated by sex, ethnicity, and nationality. As a result, a modern writer about the *Iliad* might not be concerned with Homer the individual, about whom in any case virtually nothing is known, but with Homer as a European male in a society that allowed few rights and privileges to women and that tolerated slavery. It is almost unnecessary to read the text of the *Iliad*, or at least to read it very closely, before condemning it for its patriarchal values and Eurocentric intent.

I should emphasize that looking at history and literature in this way has an immediate appeal. It allows the looker-historian to take a stance of considerable moral superiority, because *we* (the historians) know (as Homer and so many ancient people did not) that women should be valued as highly as men and that slavery is evil. Concentrating on *cultural* motivations (however inaccurately defined) allows us to form judgments without the careful amassing of detail characteristic of traditional research or even the learning of foreign languages.

The inevitable result is a portrait of the past painted with broad strokes and bright colors of our own choosing. It is almost as if we removed all the Raphaels in the museum and replaced them with Mondrians in order to study the history of the Renaissance. We are left with a vivid history of the concerns of our own society. We can now see in the past not the issues that they thought were important, whatever these might have been, but a biased history written to the dictates of dead white European males, and a literature largely insensitive to the needs and aspirations of women and cultural minorities.

We ought to have seen right from the start that this "new historicism" has some serious shortcomings. But in fact most of us are just beginning to emerge from the fog far enough to see where history-without-facts can lead us, which is right back to the fictive history developed to serve the Third Reich. This is the era not just of Holocaust denial, but of Aristotle bashing. Aristotle bashing is one of the more remarkable developments in the process of deconstructing Ancient Greece.

Although the study of Greek literature and Greek philosophy occupies only a small place in the university curriculum, it has nonetheless been treated with respect because Greek literature and philosophy are regarded as the foundation of Western Civilization. Their role as founders makes the ancient Greeks liable to attack—ironically—for *both* their virtues *and* their vices. Feminist scholars concentrate on their vices, such as their attitude toward women and slaves. Here, for example, is a recent pronouncement on the value of ancient history by a classicist, Professor Eva Keuls, controversial author of *The Reign of the Phallus*: "Why continue to study a historical record that is incurably male-centered and has been purged over the ages of all vestiges of accomplishments that were not [incurably male-centered]?"[4]

Meanwhile, Afrocentrist historians argue that the concentration on ancient Greece and Rome has prevented the study of other, even more ancient African civilizations, such as those of Egypt and Nubia, from occupying a place in the curriculum. If one counters this argument by observing that, while we can study Egyptian art and architecture, we owe to the Greeks and Romans modern scientific thought, governmental, and literary structures (not to mention technical language), we are told that the Greeks stole their philosophy from Egypt, where it originated.

It is not at all remarkable that this claim, and others like it, has been readily acceptable to large numbers of people, many of whom are otherwise well educated. Many people would like it to be true, because they would like an African civilization to get the credit for some of the greatest discoveries in human history. What is surprising is the readiness of educated people to believe in the notion that Greek culture was stolen from Africa. I doubt that such a belief would have carried much conviction as recently as thirty years ago. Its credibility can only be a direct result of the new historicism, or history-without-facts.

When I observed at a Wellesley College faculty meeting that Aristotle could not have stolen his works from the library of Alexandria because (a) the library was not built until after his death, and (b) there is no evidence that Aristotle ever went to Egypt, a colleague in the Political Science Department said that he didn't see why it mattered who stole what from whom. When I mentioned this incident recently at another university, a political scientist there explained that people in his field were not interested in facts, only in how to make generalizations.

One hopes that this sample of political scientists is not representative. But even if it is, we need to ask why the story about Aristotle's plagiarism seemed plausible to them. Was it that it now seems natural to assume that Aristotle would have had predatory motives? Undeniably Aristotle was a European male, which is to say, a representative of a colonial power inclined to look down on Africa and on women. No use in my pointing out that there would not have been much point for Aristotle to steal books from the library of Alexandria, even if it had existed during his lifetime, because most of the books in it were in Greek, and presumably available to him in Athens.[5]

Once cultural plausibility and cultural wish fulfillment become the principal criteria for determining what is true, it is not difficult to understand why American faculties have become increasingly tolerant of nonfactual, or to put it more positively, fictional addenda to their history offerings. A case in point is Wellesley College's "Africans in Antiquity" course, which is intended to present ancient history from an "Afrocentric Perspective." The principal text for this course is not Professor Frank Snowden's authoritative *Blacks in Antiquity*, but *Stolen Legacy* by Professor George G. M. James, a book few classical scholars have ever even heard of.

James was a professor of Greek and mathematics at several colleges in Arkansas. Although his book was published only after his death in 1954, it has had since its publication wide influence among African-Americans. In *Stolen Legacy* James asserts, with frequent citations to ancient and modern sources, that Aristotle stole his philosophy from the library at Alexandria and passed it off as his own when he returned to Greece. According to James, Aristotle had ample opportunity to learn the Egyptian language when he was a student in Egypt, studying the Egyptian mystery system with Egyptian priests; the final proof that Aristotle stole his philosophy is that no one person could know enough to write on so many different topics.

As I have suggested, in the American academic environment today such fictions are only too easily understood as plausible alternative interpretations. When Classicists and Egyptologists try to explain why the notion of a "stolen legacy" is wrong, their arguments are dismissed on the grounds that they stem from "Eurocentric motives," that is, the cultural motivations of (white) peoples of European descent, who naturally (it is alleged) wish to disregard the achievements of African civilization. When four years ago, in an article in *The New Republic*,[6] I tried to show why James's central thesis was wrong, one of the leading Afrocentrists, Prof. Molefi Kete Asante of Temple University, dismissed my whole discussion by attacking my cultural motives: ". . . Lefkowitz and those who share her views are not interested in understanding Afrocentricity. Their intention is fundamentally the same projection of Eurocentric hegemony that we have seen for the past five hundred years."[7]

Asante also takes me to task for being surprised when a student told me she had been upset by my failure to point out in a second-year Greek course on Plato's *Apology* that Socrates was black. She had been taught by an instructor at Wellesley that the flat nose and thick lips in his ancient portraits suggest that he might have been of African origin. I simply had not known that anyone had made the suggestion. It is of course extremely unlikely that Socrates (being an Athenian citizen) had a foreign origin. The comic poets would never have let him forget it. In any case his portrait sculptures are inspired by jokes about his being flat-nosed and bald and on the caricature of him in Plato's *Symposium*, where Alcibiades compares him to a silenus and a satyr (215b).[8] I did not believe that Socrates was black just because it is conceivably possible that Socrates (or any other Greek) might have had an African ancestor. But here is how Asante interprets the incident: "Lefkowitz's response to the student and use of the student's alleged statements demonstrates one of the major issues involved in the attacks on Afrocentricity: white racism."[9]

If what Asante alleges were true, I would have great sympathy with his arguments. But it is hardly fair to Classicists today to say that they are trying to ignore or disregard the African and Near Eastern Civilizations that influenced ancient Greece or have any wish not to give credit where credit is due. On the contrary, such connections as there are have been pointed out, and their importance even exaggerated. But the actual state of classical studies has been ignored in favor of an imaginary portrait, where classicists appear as racist historians seeking to maintain a hegemony over the curriculum that they have not had (at least in the United States) since before the First World War.

Certainly little progress can be made in the debate if we insist that whatever has been said by anyone has been predetermined by cultural motives. Perhaps because scholars like Asante are themselves writing history from a cultural perspective, they think that everyone else is doing the same. James clearly wrote *Stolen Legacy* for a purpose: to liberate black people, in part by eliminating the false praise traditionally given in all Western educational institutions to the Greeks: "the term Greek philosophy, to begin with is a misnomer, for there is no such philosophy in existence."[10] "The Greeks were not the authors of Greek philosophy, but the Black people of North Africa, The Egyptians."[11]

If we look at how James uses his evidence and what sort of evidence he used, it is of course not difficult to show that not only are the main contentions of the book mistaken, but some are probably deliberately fraudulent. But the work is interesting for another reason as well, and that is as a type of fiction that could be called myth-history. As such it provides an excellent illustration of the process I have been trying to describe: what happens when it becomes possible to abandon the idea of warranted evidence and to write history as one would like it to have been? The problem is not hypothetical. The portrait of the Greeks presented in *Stolen Legacy* has been accepted as authoritative by many Americans, and study materials based on it have been prepared (and used) by some public school systems.

Schools have adopted this idiosyncratic version of ancient history primarily for cultural reasons, as part of a program designed to make students aware of the importance of the African heritage. But why do so few people object when it is pointed out, with full documentation, that many of the claims of the Afrocentric curriculum are simply false? To a large degree, the answer is that, as I have suggested, people are ceasing to believe in the notion of valid evidence. Another contributing factor is basic general ignorance. University faculties are becoming increasingly specialized, and not everyone has a broad acquaintance with even the basic outline of European history. If one knows nothing about the ancient world, and the type of evidence that Egyptologists and classicists must deal with, why would such a person imagine himself qualified to deny that Greek philosophy was heavily influenced by Egyptian philosophy? On the other hand, he would not be so reluctant to judge cultural motives, and it would seem perfectly plausible to him that classicists might not be willing to admit that Greek philosophy had its origins in Egypt, either out of ignorance or prejudice.

How should one respond to this challenge? I do not think we can simply ignore it and hope that it will go away by itself. It might do that, eventually, but not before considerable damage is done. First, one needs to demonstrate that we actually do not have the motives that are attributed to one's "culture" or discipline. Perhaps the easiest way to do that is to point out how little cachet there is in being an anti-Semite or racist, or (for that matter) insisting on warranted evidence. Why would anyone these days be eager to maintain that the Egyptian influence on Greek thought was minimal, since they could win much more general approval by insisting that it was "borrowed massively from Egypt," as Martin Bernal has claimed?[12] Once I can show that there is nothing in it for me, my critics may be willing at least to listen to my side of the story.

Here (in brief) is what I would tell them.[13] I would begin by asking them to describe the corpus of Egyptian philosophy that formed the basis for the Greek philosophy. If the great Greek philosophers had stolen their ideas from the Egyptians, as James asserts, we would expect someone to provide parallel texts showing frequent verbal parallels. As it is, he can point only to some general similarities between Egyptian religious ideas and Greek theories. As James observes, Aristotle wrote a treatise *On the Soul*; the Egyptians believed in the immortality of the soul. But there the similarity ends. James admits that there is no close resemblance, because Aristotle's theory is only a "very small portion" of the Egyptian "philosophy" of the soul, as described in the Egyptian *Book of the Dead*.[14] But anyone who looks at a translation of the *Book of the Dead* can see that it is not a philosophical treatise, but rather a series of ritual prescriptions to ensure the soul's passage to the next world. Nothing could be more different from Aristotle's abstract consideration of the nature of the soul.

I would then look at the other material James provides in support of his contention. One of his sources is a passage in the Church father, Clement of Alexandria, a Greek who lived in the second century AD. Clement describes a procession of Egyptian priests who carry forty-two treatises containing what he calls "all of Egyptian philosophy": their subject matter includes hymns, astrology, cosmography, temple construction and provisions, sacrifice, priestly training, and various branches of medicine.[15] Even if we ignore the problem of chronology and assume that the works Clement lists in the second century AD are copies of traditional ancient writings, it is important to note that by "philosophy" Clement meant not what we now call philosophy, but learning in general, and in this particular case a body of knowledge that had little or no connection with anything Greek.[16]

Another possible source of the notion that Greek philosophy derived from Egyptian thought comes from the Egyptians themselves. In the early centuries AD, hundreds of years after the deaths of Plato and Aristotle, treatises were written in Greek that purported to have been composed at the beginning of time by Hermes Trismegistus, grandson of the god Hermes, or Thoth (his Egyptian analogue). In fact these writings (the so-called *Hermetica*) are much influenced by later thought, including Plato, Aristotle, and their followers, and the Gnostics.[17] There is apparently no record of any Egyptian-language original from which they were derived, and in fact the treatises could not have been composed without the conceptual vocabulary and rhetoric of Greek philosophy.

There is, finally, a third source of the notion that the Greeks learned from the Egyptians rather than vice versa, and that is the ancient Greeks themselves. James and other Afrocentrists are impressed by this ancient evidence. But their assurance is based on a false assumption, namely, that if what a historian says is ancient, it is also accurate. Ancient Greek historians have biases and limitations, just like modern ones. Also, the accuracy of what they record depends upon the quality of their information. So we need to look critically at the stories they tell about the visits made to Egypt by Greek philosophers and poets.

The Greeks had a profound respect for the antiquity of Egyptian culture. The fifth-century historian Herodotus was so impressed with Egypt that he was keen to discover possible connections. He tried to match up the Greek gods with their Egyptian counterparts. He even went so far as to claim that the names of the Greek gods came from Egypt, but the few examples he produces do not stand up to modern linguistic analysis. He pointed out that Greek myth suggests that parts of Greece were colonized by Egyptians, or at least by Egyptians descended from Greeks who had emigrated there. But such vague and imaginative correspondences, even if they could be confirmed by archaeological discoveries, do not amount to any kind of proof that Greek philosophy was stolen from Egypt.

The fullest account of the visits of Greek philosophers to Egypt is given by Diodorus of Sicily, a Greek writer who lived in the first century BC, several hundred years after Plato and Aristotle. Diodorus says that the Egyptian priests of his day relate that various Greek poets and philosophers came to visit Egypt.[18] He cites as evidence for their visits statues, houses, and inscriptions with their names, and offers illustrations of what each admired and transferred from Egypt to their own country.

The similarities between Greek and Egyptian culture cited by the priests are at best superficial, and do not stand up to close examination. For example, the priests pointed out that both Egyptian and Greek myths tell of a dwelling place of the dead located beyond a body of water; here perhaps their Egyptian notions may have had some influence on the formation of early Greek myth, but their beliefs about the fate of the soul after death and their burial customs are widely divergent. It is clear from these and other instances cited by the priests that they were determined to make the most of such resemblances as there were between the religious observances of two cultures. Since they had no information about religious rites as they had been practiced at the times when Pythagoras and Plato visited Egypt, they were compelled to make their deductions on the basis of the rituals practiced in their own times, after several centuries of Greek occupation and influence.

The Egyptian priests in Diodorus's account are even less explicit about the Egyptian influence on what we would now call philosophy. They claim that Lycurgus, Plato, and Solon "transferred many instances of Egyptian practices into their law codes," but cite no examples. In fact the only recognizable similarities are that both Egyptians and Greeks had laws. On that basis it would be possible to conclude that any earlier civilization "influenced" any later civilization, even if they had little or no opportunity for contact with one another. Using the same methodology, Jews living in Alexandria in the second and first centuries BC claimed that Plato studied with *Moses.*[19]

There are also significant problems with some of the priests' other claims about what Greek philosophers learned in Egypt. According to the priests Pythagoras took from Egypt his teachings about religion, geometry, number theory, and the transmigration of souls. Although we know that the Greeks based their mathematical theories on the arithmetical calculations of both Babylonians and Egyptians, there is in fact nothing in Egyptian religion that resembles Pythagoras's theory of the transmigration of souls; if he had to get it from some other religion, and did not simply invent it himself, it would have come from India. The priests seem not to have been aware that astrology was primarily a Babylonian invention, brought to Egypt by Greeks after the conquest of Alexander. The Greeks could have learned about *astronomy* from the builders of the pyramids, but about that the priests were silent.

Did the great philosophers whose works still survive ever go to Egypt? Plutarch in the second century AD and other late biographers claim that Plato himself studied in Egypt, and even name his teachers. But the earliest biographical information we have about him says nothing about it. Where did the later biographers get the idea that he went there? Apparently, from his own writings.[20] Plato talks about Egyptian customs, religion, and legends in some of his dialogues. Although his references do not suggest that he had a first-hand knowledge of Egypt, they were enough to lead later biographers to believe that he went to Egypt as a young man. Foreign travel was their way of explaining why writers included references to foreign customs and geography in their works.[21] The story of Plato's trip in Egypt was invented by later biographers to explain his interest in Egypt, and to provide physical "proof" of the importance of Egyptian culture that (as we have seen) the Egyptian priests in later antiquity were eager to establish.

None of the accounts of the lives of Socrates or Aristotle says anything about their travels there. Socrates is, in fact, recorded by a close contemporary, Plato, as saying that during his lifetime he hardly ever left Athens and never went outside of Greece.[22] James argues that the Greek writers' silence about Socrates and Aristotle "proves" that the Greeks conspired to conceal the extent of their debt from posterity. But, of course, the same evidence of silence has led other scholars to the natural conclusion that none of them actually ever went there.

At this point one of my critics might well be prepared to ask why—if there was no formal school of philosophy in Egypt—James speaks of an Egyptian mystery system. Here I would explain that the notion of an Egyptian mystery system is a relatively modern and almost completely European fiction. The earliest descriptions of academies for Egyptian priests, with large libraries and art galleries, occur not in any ancient text, but in an eighteenth-century French work of historical fiction, the novel *Séthos* by the Abbé Jean Terrasson, first published in 1731. Terrasson's novel was widely read; it had a profound influence on portrayals of Egyptian religion in later literature, such as Mozart's *Magic Flute.*[23] In particular, initiation of Terrasson's hero into the Egyptian priesthood served as the inspiration for Masonic rituals.[24]

Terrasson was compelled to reply for his description of Egypt on Greek and Latin literature. Egyptian sources were not available to him, or to anyone before the 1830s, when hieroglyphics were deciphered on the basis of the bilingual Rosetta Stone.[25] So Terrasson based his account on Roman sources—Virgil's description of the hero Aeneas's visit to the lower world in the *Aeneid* (1st century BC) and a famous account of an initiation into the Greco-Roman cult of Isis, written in Latin by Apuleius of Madauros in North Africa in the second century AD. The conversion follows the pattern of the journey of discovery and initiation characteristic of Greek hero myths.

Terrasson develops the initiation ceremony into a complex series of tests and trials, including a descent into darkness, and walking between blazing coals.[26] The only truly Egyptian element in the initiation is the procession of priests, which he took not from an Egyptian source, but from the Church father Clement.[27] In fact, there were no "mystery" or initiation cults in Egypt until the Greeks settled in Alexandria in the third century BC, after Alexander's invasion.

It is certainly understandable that Terrasson was unable to distinguish Greek rituals from indigenous Egyptian rituals. He could not read any inscriptions or papyri that described ancient Egyptian rites and beliefs, since they were written using Egyptian symbols, such as hieroglyphics or hieratic script, which no one at the time could read or understand. It would also be unreasonable to suppose that the Masons, who do not pretend to be serious scholars, would have sought to revise their rituals and notions of their own history in the light of the new information about Egypt that became available after hieroglyphics could be read. James, of course, knew better. If he had intended to write a serious academic book (rather than a myth-history), he should have taken recent discoveries about Egypt into consideration. Instead, he seems to base his views on the anachronistic notion of "Egyptian Mysteries" preserved in Masonic ritual; he speaks of Egyptian "Grand Lodges," another distinctive feature of the Masonic Order, and cites Masonic literature.

But instead of concentrating on what is now known about Egyptian myth and ritual, James cites *Anacalypsis* (or "Revelation") by Godfrey Higgins, a book published in 1833 that is completely out of date. Higgins argued that Egyptian writing could never be deciphered because it was a *secret* system.[28] That of course is nonsense, as decipherment of hieroglyphics has shown. In *Stolen Legacy* James likewise insists that no records (in any language) of the

Egyptian mystery system have come down to us because it was *secret*. Because it would not suit his purpose, James does not mention the other, and more obvious explanation for the absence of records, which is, of course, that *no such system ever existed.*

Thus, most ironically, the "Egyptian Mystery System" described by James is not African, but essentially Greek, and in its details specifically European. James has in effect accused the Greeks of borrowing from themselves, and said nothing about the real and distinctively Egyptian ideas that influenced the Greeks during their long contact with each other. The best evidence for the interchange of ideas between Greece and Egypt, of course, comes from the period after Alexander's conquest, when Egypt was ruled by the Ptolemaic dynasty.

In conclusion, if we consider all the surviving evidence in its historical context, it is difficult to discover exactly what the Greek philosophers were supposed to have learned in Egypt, if they ever went there; it is much easier (though I do not propose to do so now) to show how their work derives from their Greek predecessors, some of whom clearly were familiar with near-Eastern accounts of the creation of the world. And I think that although we can sympathize with James's mission and the cultural reasons for the composition of his mythistory, we cannot accept as an actual historical fact that (as Martin Bernal says) "Greek science and philosophy [were] borrowed massively from Egypt."[29]

In this essay I have attempted to show why this kind of history writing is being approved of and produced in American universities today, and why I think we need to call it "mythistory" and treat it as such. If we continue to confuse it with history, certainly we will be in serious trouble, because each department, or possibly each person, will be teaching that version of the past that best suits his or her cultural agenda, and all challenges will be dismissed because their motivations will be perceived as different or inferior.

Clearly it is time to stop imagining that each culture can live by its own particular sets of values and myths (and "facts"), without regard for any of the others. Humanists and social scientists need to join with scientists in acknowledging that there is a common body of knowledge (and nonknowledge). For once, classicists have the qualifications to take the initiative. Our training requires us to know about a variety of academic disciplines, and to pay close attention to specific details of language. We are in a good position both to describe what is happening, and to suggest practical remedies. Study of ancient Greek civilization in this case as in others is a positive advantage. After all, it was the ancient Greeks who invented the notions of history and of writing about the various cultures of the known world with sympathy, and with a view to something other than their own aggrandizement.

## NOTES

1 Paul R. Gross & Norman Levitt, *Higher Superstition: The Academic Left and Its Quarrels with Science.*

2 John Dewey, *The Early Works, 1882–1898*, vol. 3, p. 147. My thanks to Jonathan Imber for this reference.

3 For an excellent discussion of this tendency (with bibliography), see P. Cantor, "Stephen Greenblatt's New Historicist Vision."
4 E. Keuls, *American Historical Review* 10 (1993): 1216.
5 See Mary Lefkowitz, "Afrocentrism Poses a Threat to the Rationalist Tradition." This article is dismissed as "an eloquent testimonial to the power of white Jewish skin privilege" by T. Martin in *The Jewish Onslaught: Despatches from the Wellesley Battlefront*, p. 63.
6 Mary Lefkowitz, *Not out of Africa.*
7 Molefi Kete Asante, "On the Wings of Nonsense: The Attack on Afrocentricity."
8 Cf. K. J. Dover, *Plato: Symposium*, p. 166.
9 Asante, "On the Wings of Nonsense," p. 39.
10 G. G. M. James, *The Stolen Legacy*, p. 1.
11 Ibid., p. 158.
12 Martin G. Bernal, *Black Athena: The Afroasiatic Roots of Classical Civilization*, vol. I, p. 38.
13 For more details, see Mary R. Lefkowitz, "The Myth of a 'Stolen Legacy'" and my book *Not out of Africa.*
14 James, *The Stolen Legacy*, p. 125.
15 Clement, *Stromata*, 6.4; cf. G. Fowden, *The Egyptian Hermes: A Historical Approach to the Late Pagan Mind*, pp. 54–59.
16 Cf. Fowden, *The Egyptian Hermes: A Historical Approach to the Late Pagan Mind*, pp. 62–63.
17 For details, see B. Copenhaver, *Hermetica*, pp. xiii–xlv.
18 Diodorus Siculus, 1. 96–98.
19 Cf. Lefkowitz, "Ethnocentric History from Aristobulus to Bernal."
20 T. Hopfner, *Plutarch über Isis und Osiris*, vol. 2, pp. 85–90.
21 Mary Lefkowitz, *The Lives of the Greek Poets*, pp. 43–45.
22 Plato, *Crito* 52b.
23 Cf. E. Iversen, *The Myth of Egypt and Its Hieroglyphs in European Tradition*, p. 122; and K. Thomson, *The Masonic Thread in Mozart*, pp. 26, 155.
24 Bernal, *Black Athena: The Afroasiatic Roots of Classical Civilization*, vol. I, p. 180.
25 For a brief account of the decipherment, cf. Iversen, *The Myth of Egypt*, pp. 138–146.
26 J. Terrasson, *The Life of Sethos*, vol. 1, pp. 155–196.
27 See Note 15.
28 G. Higgins, *Anacalypsis*, p. 483.
29 See Note 11. Cf. A Preus, *Greek Philosophy: Egyptian Origins.*

## REFERENCES

ASANTE, MOLEFI KETE. "On the Wings of Nonsense: The Attack on Afrocentricity. *Black Books Bulletin*: Wordswork 16.1–2 (1993–94): 38.

BERNAL, MARTIN G. *Black Athena: The Afroasiatic Roots of Classical Civilization*, vol. 1. New Brunswick, NJ: Rutgers University Press, 1987.

CANTOR, P. "Stephen Greenblatt's New Historicist Vision." *Academic Questions* 6.4 (1993): 21–36.

COPENHAVER, B. *Hermetica.* Cambridge: Cambridge University Press, 1992.

DEWEY, J. *The Early Works, 1882–1898*, vol. 3. Carbondale, IL: Southern Illinois University Press, 1972.

DOVER, K. J. *Plato: Symposium.* Cambridge: Cambridge University Press, 1980.

FOWDEN, G. *The Egyptian Hermes: A Historical Approach to the Late Pagan Mind.* Cambridge: Cambridge University Press, 1984.

GROSS, PAUL R. & NORMAN LEVITT. *Higher Superstition: The Academic Left and Its Quarrels with Science.* Baltimore, MD: Johns Hopkins University Press, 1994.

HIGGINS, G. *Anacalypsis.* Chesapeake, NY: ECA Associates, 1991.

HOPFNER, T. *Plutarch über Isis und Osiris*, vol. 2. Darmstadt: Wissenschaftliche Buchgessellschaft, 1967.

IVERSEN, E. *The Myth of Egypt and Its Hieroglyphs in European Tradition.* Princeton, NJ: Princeton University Press, 1992.

JAMES, GEORGE G. M. *Stolen Legacy.* New York, NY: Philosophical Library, 1954.

KEULS, EVA. *The Reign of the Phallus.* American Historical Review 10 (1993): 126.

LEFKOWITZ, MARY. "Afrocentrism Poses a Threat to the Rationalist Tradition." *Chronicle of Higher Education* 5/6 (1992): 52.

———. "Ethnocentric History from Aristobulus to Bernal." *Academic Questions* 6.2 (1993): 12–20.

———. *The Lives of the Greek Poets.* Baltimore, MD: Johns Hopkins Press, 1981.

———. "The Myth of a 'Stolen Legacy.'" *Society* 31.3 (1994): 27–33.

———. "Not out of Africa." *The New Republic*, February 10, 1992, pp. 29–36.

———. *Not out of Africa.* New York, NY: Basic Books, 1996.

MARTIN, T. *The Jewish Onslaught: Despatches from the Wellesley Battlefront.* Dover, MA: The Majority Press, 1993.

PREUS, A. "Greek Philosophy: Egyptian Origins." Binghamton, NY: Research Paper of the Institute for Global Cultural Studies of Binghamton University, 1992.

SNOWDEN, FRANK. *Blacks in Antiquity.* Cambridge, MA: Harvard University Press, 1970.

TERRASON, J. *The Life of Sethos.* Translated by Thomas Lediard. London: J. Walthoe, 1732.

THOMSON, K. *The Masonic Thread in Mozart.* London: Lawrence and Wishart, 1977.

# BUILDING BRIDGES TO AFROCENTRISM
## A Letter to My Egyptological Colleagues[a]

ANN MACY ROTH

"WHAT COLOR WERE the ancient Egyptians?" This is a question that strikes fear into the hearts of most American Egyptologists, since it so often presages a barrage of questions and assertions from the Afrocentric perspective. Few of us have devoted much thought or research to the contentions of the Afrocentric movement, so we nervously try to say something reasonable, and hope that the questioner won't persist and that we won't end up looking silly or racist or both.

In late 1993, I received a temporary appointment to the faculty of Howard University and began teaching Egyptological subjects to classes that were almost entirely African-American. As a result, I have been dealing with Afrocentric issues on a regular basis, and have spent a good deal of time and energy thinking and talking about them. Since my appointment, many of my Egyptological colleagues at other universities have asked me about Afrocentric sentiment at Howard and my strategies for teaching traditional Egyptology to the students who espouse it. The tone of these inquiries has demonstrated to me both the curiosity and the discomfort that American Egyptologists feel about Afrocentrism. This attempt to write an account of my impressions is partly inspired by such questions, which I have had difficulty answering cogently in short conversations. More importantly, however, I have come to believe that the Afrocentric movement has a great potential to advance or to damage our field. Which of these directions it takes will depend upon the degree to which traditionally trained American Egyptologists can come to understand and adapt to its existence. This essay is my attempt to speed that process.

"Afrocentric Egyptology," as practiced today, has an international scholarly literature behind it. (The movement is, if anything, more prominent in France than it is here, to judge from the numerous displays of Afrocentric books and journals I saw in Paris bookshops last summer.) In America, however, Afro-

[a] Reprinted with permission from the *Newsletter of the American Research Center in Egypt*, nos. 167–168, September and December 1995.

centric Egyptology is less a scholarly field than a political and educational movement, aimed at increasing the self-esteem and confidence of African-Americans by stressing the achievements of African civilizations, principally ancient Egypt. As such, it is advocated in popular books, textbooks, and even educational posters sponsored by major breweries. It has apparently thus far enjoyed considerable success in its educational aims. As a result, it is being taught to students from grade school through the university level all over America, and its tenets are frequently cited as established fact by the media and the educational establishment. Coming to Howard as part of a tentative Egyptological experiment, I was amazed at the quantity of Egyptology that was already being taught in courses ranging from drama to mathematics to philosophy. (An Afrocentric work by Ivan van Sertima on Egypt is included in the recommended reading for freshman orientation.) The movement continues to grow in importance and influence, and, whatever one thinks of its content, it has an increasing degree of popular acceptance by a large audience.

This kind of Egyptology has little to do with the Egyptology that we professional Egyptologists practice, and many of us currently regard its incursions upon our field as a nuisance. We see it only when its exponents ask aggressive and seemingly irrelevant questions in classes and public lectures, or make extravagant claims about ancient Egyptian achievements (the harnessing of electricity, the conquest of large parts of southern Europe), citing authors of dubious credibility and outdated theories and translations (often by E. A. W. Budge). Especially annoying are those who combine Afrocentrism with the age-old mystical-crackpot approach to our field, claiming for the Egyptians fantastic lost skills and secret knowledge. In most cases, our reaction to Afrocentrism is avoidance: we deal with the issue by dismissing it as nonsense, by disparaging the knowledge of its proponents, and by getting back to "real" Egyptology.

By doing this, however, we are both ignoring a danger and missing an opportunity. The number of African-Americans who are taught this material is growing, and we will increasingly have to deal with its inaccuracies and exaggerations simply in order to teach our students. This gap between our field and the Afrocentric version of it is not going to go away; if we ignore it, it will surely widen. And by setting ourselves against the whole phenomenon in an adversarial and often condescending way, we make it impossible for the responsible educators involved in the movement (and there are many) to tap our expertise and improve the accuracy of the materials they teach.

At the moment, however, we have the opportunity to narrow the gap by taking a more positive direction. By granting that an Afrocentric perspective may have something to offer our field, we can exorcise the defensiveness and hostility that is so often engendered by the assertions of Afrocentrists. By making our classes more hospitable to those with Afrocentric views, we take the first steps towards training a new generation of Afrocentric scholars in the traditional methods of our field. They will then be able to correct and improve the argumentation of Afrocentric scholarship so that the content of their movement benefits from traditional Egyptology's decades of research and hard-won conclusions. Afrocentric Egyptology need not necessarily conflict with traditional Egyptology; it seems to be possible to combine the two, to the benefit, perhaps, of both.

First, however, it is necessary for traditional Egyptologists to understand the underpinnings of Afrocentric Egyptology. Its contentions, as I have encountered them, fall under four rough rubrics: (1) that the ancient Egyptians were black, (2) that ancient Egypt was superior to other ancient civilizations (especially that of the ancient Greeks, which is seen to be largely derivative), (3) that Egyptian culture had tremendous influence on the later cultures of Africa and Europe, and (4) that there has been a vast racist conspiracy to prevent the dissemination of the evidence for these assertions. Most traditional Egyptologists recognize these contentions but do not understand the motives behind them, and so deal with them in a counterproductive way. I will address them one by one.

1 *The contention that the ancient Egyptians were black.* Like most Egyptologists, I had never thought of the ancient Egyptians as being any color in particular. Neither black nor white seemed an appropriate category—they were simply Egyptian. This view, in fact, is probably the one held by most Egyptians themselves, both ancient and modern. As we know from their observant depictions of foreigners, the ancient Egyptians saw themselves as darker than Asiatics and Libyans and lighter than the Nubians, with different facial features and body types than any of these groups. They considered themselves, to quote Goldilocks, "just right." These indigenous categories are the only ones that can be used to talk about race in ancient Egypt without anachronism. Even these distinctions may have represented ethnicity as much as race: once an immigrant began to wear Egyptian dress, he or she was generally represented as Egyptian in color and features. Although there are occasional indications of unusually curly hair, I know of no examples of people with exaggeratedly un-Egyptian facial features, such as those represented in battle and tribute scenes, who are represented wearing Egyptian dress, though such people must have existed.

As for indigenous categories in modern Egypt, I have been told by most of the modern Egyptians with whom I have discussed the question that, if they had to use the categories of the modern Western world, they would describe themselves as white. (There are some exceptions, but few would describe themselves as black.) As evidence of this, one can point to the consternation that was produced in Egypt when it was announced that the black actor Lou Gossett would portray President Anwar Sadat in a biographical film. There exist terms in modern colloquial Egyptian Arabic to describe skin color, most commonly "white," "wheat-colored," "brown," and "black." In practice, however, these terms are frequently applied inaccurately, so that people are (flatteringly) described as lighter in color than they actually are. The term "black" is viewed almost as a pejorative, and is rarely used. This categorization of the modern population is only partly relevant to the question, although it contributes to the reluctance of Egyptologists working in Egypt to describe the ancient Egyptians as "black."

I have encountered arguments that the ancient Egyptians were much "blacker" than their modern counterparts, owing to the influx of Arabs at the time of the conquest, Caucasian slaves under the Mamlukes, or Turks and French soldiers during the Ottoman period. However, given the size of the Egyptian population against these comparatively minor waves of northern immigrants, as well as the fact that there was continuous immigration and occasional forced deportation of both northern and southern populations into Egypt throughout the pharaonic period, I doubt that the modern population is significantly darker or lighter, or more or less "African" than their ancient counterparts. It should be noted, however, that we really do not know the answer to this question. More research on human remains needs to be, and is being, done.

But what of scientific racial categories? What of the three races we learned about in grade school? In talking to several physical anthropologists, I have learned that these three races have no clear scientific meaning. Anthropologists today deal with populations rather than individuals, and describe ranges of characteristics that occur within a population as being similar to or different from the ranges of characteristics of another population, usually expressing the degree of affinity with a percentage. There is no gene for blackness or whiteness, and nothing that can allow a scientist to assign a human being to one or the other category, beyond the social definitions of the culture in which the scientist is a participant. While anthropologists sometimes describe people in terms of the traditional three races, this is not a result of applying objective criteria based on clear biological distinctions, but is instead a shorthand convenience. Such judgments work backwards from the social categories to arrive at an identification that would be recognized by a member of society. For example, when a forensic anthropologist gives the race of an unidentified dead body as "white," it is simply a prediction that the missing person form with which it will be compared probably described the person that way. Scientific determinations are thus just as dependent upon social categories as more impressionistic judgments are.

Even comparative studies can be biased by the assumptions that underlie them. Some "Eurocentric" criteria for race acknowledge the wide variety of physical characteristics found in Europe, and define as black only those populations that differ markedly from all European populations. As a result, populations that resemble any European population are excluded from the category "black." This is often what happens when scientists are asked about the remains of ancient Egyptians, some of whom closely resembled southern Europeans. By this model, only Africans living south of the Sahara desert, which separates them more markedly from European gene pools, are defined as black. The categorizations arrived at by reversing the same procedure are equally extreme. If the range of physical types found in the African population is recognized, and the designation "white" is restricted to those populations that have none of the characteristics that are found in any African populations, many southern Europeans and much of the population of the Middle East can be characterized as "black." This method was at one time adopted by "white" American schools and clubs, which compared applicants to the "white" physical types of Northern Europe, and found that many people of Jewish or Mediterranean heritage did not measure up. Neither of these ways of determining "race" can result in a definitive division between "black" and "white," because those are not in fact distinct categories but a matter of social judgment and perspective. What is a continuum in nature is split into two groups by our society. (The terms "African" and "European," although easier to distinguish because of their geographic basis, are no less subjective and problematic as cultural categories.)

Race, then, is essentially a social concept, native to the society in which one lives. It is anachronistic to argue that the ancient Egyptians belonged to one race or another based on our own contemporary social categories, and it is equally unjustifiable to apply the social categories of modern Egypt or of ancient Greece or any other society, although all of these questions are interesting and worthy of study on their own. The results tell us nothing about Egyptian society, culture, and history, which is, after all, what we are interested in.

This is not, however, what the Afrocentrist Egyptologists are interested in. They want to show that according to modern Western categories, the ancient Egyptians would have been regarded as black. This approach is not invalidated by the cultural limitations of racial designations just outlined, because it is an attempt to combat a distinct modern, Western tradition of racist argument, a tradition which has the effect of limiting the aspirations of young African-Americans and deprecating the achievements of their ancestors. This argument contends that black peoples (that is, people that we would describe as black) have never achieved on their own a satisfactory civilization and by extension can never achieve anything of much value. "Look at Africa today," argue the adherents of this notion, ignoring the added burdens imposed by economic exploitation, cultural imperialism, and a colonial past on most African nations, and ignoring the African states that do not appear regularly in the newspapers. "Look at history," they add, discounting Egypt as part of the Near East and ignoring (generally through ignorance) the other great African cultures.

These misconceptions are argued in many parts of American society. President Richard Nixon was quoted as making several of these arguments in the recently released diaries of his chief of staff, H. R. Haldeman. Similar assertions were made occasionally in the more intemperate discussions of the Los Angeles riots. And I understand that the Pennsylvania chapters of the Ku Klux Klan give each new member a leather-bound book with the gilded title *Great Achievements of the Black Race*, which is filled entirely with blank pages. Is it any wonder that the members of this maligned group want to inscribe on those blank pages the Great Pyramid and the Sphinx, the gold of Tutankhamun, the Asiatic conquests of Thutmose III, and the fame and political acumen of Cleopatra?

At this juncture, however, many Egyptologists miss the point. "Why not use Nubia?" I have been asked, "or any of the other great African civilizations? Why can't they leave Egypt alone?" The answer is that these other civilizations did not build pyramids and temples that impressed the classical writers of Greece and Rome with their power, antiquity, and wisdom. Nor have most modern Americans and Europeans heard of the civilizations of Nubia, Axum, Mali, Ife, Benin, and Zimbabwe. Hannibal is famous enough to be worth claiming, but few other non-Egyptians are. The desire to be associated with historical people who are generally acknowledged to be "great" by the Western cultural canon accounts for the frequent and (to Egyptologists) puzzling contention that Cleopatra was black, despite the fact that she was demonstrably descended from a family of Macedonian generals and kings who married their sisters, and therefore had little claim to either a black or an African origin (although one of my Classicist colleagues at Howard tells me that her paternal grandmother is unknown, and might have been Egyptian). The reason she is identified as black is that among modern Americans she is probably the best known ancient Egyptian of them all. Shakespeare and Shaw wrote plays about her, her life has been chronicled in several popular films, and her name is regularly invoked in our popular culture to signal the exotic, the luxurious, and the sexy. In this sense, "Afrocentric" Egyptology is profoundly Eurocentric, and necessarily so: it plays to the prevalent cultural background of its intended audience.

If the question of the race of the ancient Egyptians is entirely subjective and political, then, why does it bother Egyptologists at all? Why would we rather the Afrocentrists "used Nubia"? I think our reasons are largely related to the tenuous place our field holds in academia. Afrocentrists see Egyptologists as a strong, academically supported, establishment force; but despite, and perhaps even partly because of, the popular fascination with its contents, Egyptology tends not to be taken quite seriously by people who study other parts of the ancient world. Already many noted departments of Near Eastern studies with extensive faculty in ancient Mesopotamia and the Levant do not feel it necessary to teach or support research in Egyptology at a similar level. We fear, perhaps, that if we endorse the view that ancient Egypt was a "black civilization," we will further cut ourselves off from our colleagues who study other civilizations contemporary with ancient Egypt. At the same time, there is no place for us in African studies departments, which generally tend to address questions related to modern history and current political and social problems. While anthropologists working in Africa may offer us insights and models, the methods and concerns of our field require more, rather than less, contact with scholars studying other ancient Mediterranean and Near Eastern cultures. We have been too isolated for too long as it is.

The politics of the situation, as well as the requirements of course topics such as archaeology, make it important for us to deal with the question of the race of the ancient Egyptians in our university classes. My own method, developed long before coming to Howard, is to be very explicit about my own views on the question. I give a lecture on the land and the people of Egypt, normally very early in the semester, before the question is brought up by students; and I try to present the question neutrally, without defensiveness or antagonism. I explain the social nature of racial categories and the categories used by the Egyptians themselves, their representation of foreigners, and the frequency of foreign (Asian and African) immigration to Egypt in all periods of its history, extending back into the Paleolithic. Discussions of geography and language are also useful here. It is also necessary to address the political question. In doing so, I often make use of Bruce Williams's observation (which really goes to the heart of the matter) that few Egyptians, ancient or modern, would have been able to get a meal at a white lunch counter in the American South during the 1950s. Some ancient Egyptians undoubtedly looked very much like some modern African-Americans, and for similar historical reasons. Very few, if any, of them looked like me. I also explain the politics of the question in modern Egypt. Finally, I explain the irrelevance of the political question to the subject I will be teaching, a circumstance that allows me to respect the students' political convictions (which I treat rather as I might treat a religious conviction), and should allow them to learn about Egyptian culture in my class without violating their beliefs. By making my position clear at the outset, I forestall the Afrocentric students' speculations and attempts to "trap" me into committing myself to the exaggeratedly "Eurocentric" views that they might otherwise assume I espouse. It also reassures students that they can come to me with questions about their Afrocentric readings, or their own Afrocentric questions about course materials; the topic is no longer taboo. It is impossible to build bridges if we discourage discussion.

2 *The contention that the Egyptians were the greatest civilization in history.* Contrary to the expectations of most Afrocentrists, most Egyptologists are less bothered by the contention that the Egyptians were black than by the exaggerated claims made about the achievements of Egyptian civilization. These claims, including attribution to the Egyptians of great mathematical, scientific, and philosophical sophistication, are often based on misinterpretations or exaggerations of the evidence, and in some cases pure fantasy and wishful thinking. Many of the arguments advanced show a complete ignorance of (or disregard for) the facts of chronology—for example, the contention that the Greeks "stole" their philosophy from the library at Alexandria and then burned it down to cover their theft, or the claim that the architecture of Greek peripheral temples was borrowed from the eastern mamisi at Dendera.

Paradoxically, while it is in the details of this contention that Egyptologists find the most grounds for outrage and dismissal of the entire movement, this is also the area where we can do the most help the Afrocentrists move towards a more rigorous and respectable scholarship. In principle, few Egyptologists would deny that ancient Egypt was a great civilization, and that the ancient Egyptians achieved wonderful things and made unique contributions to history and global culture. It in no way detracts from these contributions that they had terrible difficulties adding fractions because of a ludicrously clumsy system of notation, or that they did not understand the importance of the brain, or that they may have borrowed the idea of writing from Sumerian civilization. On these points the Afrocentrists need to develop a better appreciation of where the strengths of Egyptian civilization really were. Most Afrocentrists do not want to be in the position of teaching their children things that are not true. However, because of the political desire to find great Egyptian achievements in areas that the West values, and because of the limited material available to them and their limited familiarity with the culture, they often misinterpret the evidence and seize upon unsubstantiated ideas that fit their agenda.

The way we can help here is not, however, to argue against these misunderstandings and mistaken ideas individually. There are too many of them, and the arguments tend to be both unpleasantly adversarial and futile.

"See, this is a model of an ancient Egyptian glider-plane."

"Actually, it's a Late Period model of a bird. If the Egyptians could fly gliders at that period, don't you think Greek and Egyptian sources would have mentioned it?"

"But it's aerodynamically perfect!"

"Well, of course it is; it's a bird."

"But it's different from all the other bird models. Besides, what do you know about aerodynamics?"

This sort of argument gets us nowhere. The only strategy that is effective is more fundamental. We must familiarize students with the evidence and the way one argues from it. Students who have read translations of ancient Egyptian literature and other texts and discussed how social and cultural deductions can be drawn from primary sources will generally not stand for assertions about ancient Egypt that are blatantly contradicted in these texts. Likewise, students who have read about the forms of pyramids and the theories about their construction, or who have become familiar with Egyptian tomb iconography, will not believe claims that do not correspond to the evidence they have seen. (There will, of course, be ideologues who will hold on to their groundless convictions in the teeth of the evidence, but most of them will have dropped the class after the initial discussion of the race of the ancient Egyptians.) Teaching students a more source-based, critical approach not only will improve their ability to evaluate the contentions of Afrocentric Egyptology, but should help them deal with other subjects as well, and lays the foundation for academic and other work that will give them pride in their own achievements as well as their heritage. Moreover, an explicitly source-based approach has the added advantage of forcing us to reexamine our own basic assumptions.

When Afrocentrists base their conclusions on the evidence, the results can serve their purposes without violating the sensibilities of scholars. The validity of the evidence also lends authority to the ideological position being argued. One example that goes some distance towards this goal is an Afrocentric poster given me by one of my students, designed and produced by a group called the Melanin Sisters, for grade-school children. The poster is decorated with hieroglyphs and urges the reader to adopt behavior in accordance with the ancient Egyptian concept of Ma'at. As a guide to the requirements, the Negative Confession is quoted (albeit with some substitutions for the weird bits). Another student showed me a book called *Hip-Hop vs. Maat: A Psycho-Social Analysis of Values* (by Jawanza Kunjufu, Chicago: African American Images, 1993), which again uses the Negative Confession, as well as selections from Egyptian wisdom literature, to construct a system of morality that the author contrasts favorably with the street ethics prevalent among some young

African-Americans. The use of actual Egyptian evidence in developing Afrocentric materials could be encouraged and made more authentic if Egyptologists took a less adversarial attitude toward its creators.

If we teach Afrocentric students to find evidence for their assertions and to construct convincing arguments, there will always be the possibility that they will use these tools to argue points that we find uncongenial to our pictures of Egyptian civilization. At a conference some years ago, I praised an innovative and provoking argument to a colleague, and his reply was, "Yes, I suppose it was interesting, but just imagine what they will do with it." To use such fears of exaggeration in the popular sphere (regardless of whether they are justified) as an excuse for suppressing arguments that contradict our own reconstruction of the past is unjustifiable and unscholarly. Political bias is unavoidable, so the current wisdom goes, and we will find it more difficult to accept some arguments than others, depending upon our own previous ideas or our feelings about the person making the argument. But such predispositions are something that we all deal with frequently, and should have learned to set aside. We are scholars, and we should not be afraid of the truth, whatever it turns out to be.

3 *The contention that Egyptian civilization had extensive influence on Europe and Africa.* This argument really has two parts, which are in some ways symmetrical, but which have two entirely different motivations. The argument for Egyptian influence in Europe is an extension of the argument for the overall superiority of Egypt to other cultures: by rooting Greek and Roman civilizations in Egypt, Africa can be seen as the source of the civilization we find most impressive: our own. The argument for the influence of Egypt on other African civilizations, in contrast, is intended to allow modern African-Americans (who are in most cases the decendents of people abducted from non-Egyptian parts of Africa) to claim the Egyptian cultural heritage as their own.

The half of this question that has been most discussed of late is the claim that Egypt colonized Greece, and that classical Greek culture is essentially Egyptian. Greece is traditionally viewed by Western culture as the source of beauty and reason, so (again, for political reasons) it is felt especially important to show that ancient Egypt was extremely influential in its development. *Black Athena*, Martin Bernal's work on the question, has been at the center of the recent debate on this claim, and has given it a degree of prominence and respectability in the non-Afrocentric scholarly community. Despite this, I feel strongly that Bernal's books do an ultimate disservice to the cause he is trying to advance. In the short term, of course, they have brought both the issue and Bernal himself to the forefront of public consciousness. However, his arguments are so chosen and presented that they cannot serve as a solid foundation for the academically credible Afrocentric Egyptology that he hopes to create.

In many cases, Bernal has either intentionally misled his readers by his selection of evidence, or he has neglected to investigate the full context of the evidence on which he builds his arguments. He routinely cites late Classical traditions that support his argument and ignores the Egyptian evidence that does not. A good example of these problems is his discussion of the connections of Egypt with bull cults on Crete (vol. 2, pp. 22–25, and more fully in chapter 4, especially pp. 166–184). After an initial foray proposing dubious connections between Min, bulls, Pan, and the Minoan king Minos, Bernal connects Minos to Menes and the name of Memphis, *Mn-nfr*, because of their phonetic similarity and their connection with the bull cult of Apis. (*Mn-nfr*, of course, comes from the name of the mortuary temple of Pepi I and has nothing to do with Menes, who is called the founder of the Apis cult only by a late Roman writer.) The name of the Mnevis bull also contains the magic letters *mn* in the Classical sources. The fact that the name was consistently written *Mr-wr* by the Egyptians is not mentioned in the summary, while in the fuller argument it is dismissed as "confusion among the three biconsonantals *mr*, *mn* and *nm*" in words referring to cattle (possibly due to onomatopoeia). The fact remains that the Mnevis bull is only rarely called anything but *Mr-wr.* The "winding wall" sign in *Mr-wr*, which is also used in *mrrt*, "street," is connected in his summary with the labyrinth of the Minotaur.

The result of these arguments is a "triple parallel": the connection of a bull cult in both Egypt and Crete "with the name *Mn*, the founding pharaoh, and a winding wall." But in Egypt neither the name *Mn* nor the founding king was clearly connected to the Apis cult; and the connection of the "winding wall" sign with the Mnevis bull was probably purely phonetic. The triple parallel reduces to a single coincidence: the founding king of Egypt and the most famous king of the Minoans both had names with the consonants "Mn." This relationship, as Bernal points out, has been discussed by previous scholars. That both countries had bull cults, like most other ancient Mediterranean cultures, is hardly worthy of remark. The following discussion of "the bull Montu" is even more tenuous, since Montu is generally characterized as a falcon, and is no more to be equated with the Buchis bull with which he shares a cult place than the sun god Re is to be equated with the Mnevis bull. That these arguments are flawed does not prove Bernal's conclusions wrong, of course; but such arguments can never prove him right, and in the meantime they obscure the debate.

The connections and contacts between Egypt and the Greek world have long been recognized, and Bernal misrepresents the degree to which modern scholars suppress evidence for them. Certainly the influence of Egyptian statuary on Archaic Greek *kouroi* is widely accepted, among classicists as well as Egyptologists, although the differences in their function and execution are obviously of importance too. In arguing for an Egyptian colonization of Greece, however, Bernal and his followers disregard the extensive Egyptian textual tradition (surely if Thutmose III had conquered southern Europe and set up colonies there he would have mentioned it in his annals, for example), as well as the arguments of the scholars who have been investigating these questions for decades. Most of Bernal's arguments, interestingly, rest on the Greek textual tradition, which was of course a product of its own culture's cultural and political situation and requirements, and often made use of the Egyptians' antiquity and reputation for wisdom. By crediting the Greek evidence over the Egyptian, European over the African, Bernal takes advantage of the fact that his Western audience is more familiar with (and more inclined to credit) the Classical tradition than the Egyptian. That few of the myriad reviews of Bernal's books have been written by Egyptologists is an obvious indication of the European provenience of his evidence.

If we are honest, most Egyptologists would admit that we would like nothing better than to find indisputable evidence that all Western culture derived from Egypt; such a discovery would make us far more important, more powerful, and wealthier than we are today. Because of this bias, we are justifiably cautious in making such claims.

The other half of this contention, that Egyptian civilization had a wide influence in the rest of Africa, is argued most prominently in the writings of Sheikh Anta Diop. Many turn-of-the-century scholars made such a claim, and they are widely and reverently quoted in the Afrocentric literature to support the more recent contentions. Interestingly, their motivation was essentially racist. The invention of the "Hamitic" racial group, defined as a population essentially "white" in skeletal features, but with the peculiar anomaly of dark skin, allowed some early Egyptologists to categorize the Egyptians and the Nubians as "white." Then, working on the racist assumption that "blacks" were incapable of higher civilization, they attributed anything that looked like civilization in the reminder of Africa to "ancient Egyptian colonization." While there is a rather pleasant poetic justice in the fact that the flawed conclusions resulting from these racist assumptions are currently being used to argue for the connection of all Africans and African culture with the glories of ancient Egypt, the evidence for these conclusions is hardly acceptable from a scholarly point of view. As with the European conquests and colonies hypothesized by Bernal, African conquests and colonies beyond Upper Nubia are unlikely because of the silence of the Egyptian records, although other kinds of contact are not impossible.

These two contentions of Egyptian influence outside of Egypt are among the most difficult Afrocentric claims to deal with. Unlike the question of race, these are not subjective judgments, and yet like the question of race they are yes–no questions that lie at the heart of the Afrocentric hypothesis. In particular, to deny the claim that all Africans are descended culturally and genetically from the ancient Egyptians is seen as an attack on African-Americans' right to claim the ancient Egyptian heritage as their own. At the moment, these claims have neither been definitively proved nor disproved, so it is probably wisest to take an agnostic position regarding them. The nature and extent of Mediterranean connections with ancient Egypt are worthy of further study, and may offer scope to arguments more truly Afrocentric than those propounded by Bernal. In Africa, too, there clearly were connections of some kind with areas beyond Nubia, as we know from the depiction of trade goods; and the degree of contact with Western Africa through Libya and the Oases has not been exhaustively studied. All of these areas have been receiving more attention in recent years, and it may be that there was more contact between Egypt and the rest of Africa, or between Egypt and Europe, than our current interpretations allow. If there was, let those who would argue it argue from evidence rather than authority.

4 ***There has been a scholarly conspiracy among Eurocentric Egyptologists to suppress evidence about the blackness of the ancient Egyptians, their greatness, and their influence on European and other African civilizations.*** This is probably the most offensive manifestation of Afrocentrism we encounter, implying as it does that Egyptologists as a group have routinely abandoned their scholarly integrity, simply in order to further some racist agenda. (As an epigrapher, I find the charge that we have recarved the faces of Egyptians represented in tomb reliefs particularly ludicrous.) Its most frequent manifestation is the Napoleon-knocked-the-nose-off-the-Sphinx-so-no-one-would-know-it-was-black contention, a silly argument that demonstrates the movement's unattractive paranoia. For the evidence against this, incidentally, I refer the reader to a fascinating article by Ulrich Haarmann, "Regional Sentiment in Medieval Islamic Egypt," which records that according to Makrizi, Rashidi, and other medieval Arab authors the face of the Sphinx was mutilated in 1378 A.D. (708 A.H.) by Mohammed Sa'im al-Dahr, whom Haarmann describes as "a fanatical sufi of the oldest and most highly respected sufi convent of Cairo."

Although some Afrocentrists may have found individual Egyptologists uncooperative, for reasons made clear above, we are hardly likely to deny the achievements of the Egyptians. In one sense, we are far more Afrocentric than the Afrocentrists, since we try, where possible, to study Egyptian civilization on its own terms, rather than comparing it to our own culture. Most of us have developed a great respect for the skills of the Egyptians: their abilities and sophistication as sculptors, writers, diplomats, theologians, painters, architects, potters, bureaucrats, builders, warriors, and traders will not be denied by those who have studied the results of their work. Even greater skill is apparent in the suitability of these achievements to the needs of the ancient culture as a whole, and this suitability is better appreciated the better one understands the cultural context in which the achievement occurred. To yank a building or a statue or a poem from its indigenous cultural milieu in order to compare it with its Western counterparts is decidedly Eurocentric, especially when one uses the Western products as the standard against which the Egyptian are to be judged; and yet, for political reasons, this is the most common approach of the Afrocentrists.

In another sense, however, the contention that Egyptologists are Eurocentric has at its center a kernel of truth. Any Egyptologist who proposes to do something constructive about the Afrocentric movement must admit that, in its origins and to some extent in its current preoccupations, Egyptology is a Eurocentric profession. It was founded by European and American scholars whose primary interest was in confirming the Classical sources and in confirming and explicating the Old and New Testaments for the furtherance of Christianity. A look at the earliest Egypt Exploration Society publications illustrates the way that early scholars "sold" their work by connecting it to familiar Classical and (especially) Biblical names and places: ***The Store City of Pithom*** *and the Route of the Exodus* (1885), *Tanis* (1885), *Naukratis* (1886 and 1888), *The Shrine of Saft el Henneh and the Land of Goshen* (1887), *The City of Onias and the Mound of the Jew* (1890), and *Bubastis* (1890). Furthermore, the fact that the cultures to the north and east of Egypt provide texts that we can use to correct and augment the Egyptian evidence, while those to the south and west do not, provides a third reason for concentrating our research on foreign relations to the northeast. Insofar as Nubian cultures have been studied, they have until recently been seen as distorted and somewhat comical attempts to replicate their great neighbor to the north. Because of these circumstances (the Classical focus of Western culture, Christianity, and the distribution of writing), as well as the often unconscious racism of early scholars that has affected the shape of our field, Egyptologists have too often ignored the rest of Africa.

This ignorance has not been complete. As a result of the birth of cultural anthropology around the turn of the century, there was a great interest in finding the origin of Egyptian traditions in those of "other primitive cultures," i.e., the societies of contemporary Africa, which were taken as models for what Egypt was like "before civilization." This rather weird perspective led to such anachronisms as the claim that the ancient Egyptian jubilee ceremony "derived" from the alleged eighteenth-century African practice of killing a king who became too old to rule effectively.

Despite the nature of the underlying assumptions, this early work in anthropological comparisons contains many interesting ideas. (I have found the work of A. Blackman especially rich.) Such similarities between cultures, reviewed and reworked to accord with current scholarly standards, may help explicate some of the puzzling elements in Egyptian culture. It must be remembered, however, that similarity does not prove influence, or even contact. As the archaeology and cultural anthropology of Africa becomes better known, and as Egyptologists, Afrocentric and traditional, become more familiar with and sophisticated about African cultures, it may be that patterns of such similarities can be identified, categorized, and traced with sufficient scholarly rigor to show routes of contact. These are important questions, and represent an area where the Afrocentric perspective might make substantial contributions not just to the education and self-esteem of African-Americans but to the international scholarly field of Egyptology as well. Such discoveries would add immeasurably to the resources of the entire field of Egyptology, widening our horizons and broadening our understanding of Egyptian culture.

Afrocentric Egyptology, properly pursued, has the potential to achieve important political goals: improving the self-image of young African-Americans and enhancing their belief in their own potential for achievement, by combating the racist argument that no one from Africa or with a dark skin has ever achieved anything worthwhile. The less exaggerated and the more rooted in accepted scholarly argument its teachings are, the more authority the curriculum will have. As the movement grows more sophisticated and better grounded, and as mainstream Egyptologists grow commensurately more accepting of its perspectives, it will be possible, I hope, to do away with the defensiveness that so often characterizes Afrocentric teachings currently. Instead of learning a doctrine on faith, teachers of Afrocentrism should encourage students to investigate the primary evidence and refine our knowledge of Egypt and other African civilizations on their own, truly Afrocentric, terms. Teachers should not worry that students will find that ancient Egypt was not a great civilization after all—on the contrary, the deeper one goes into its cultural productions, the more one comes to appreciate the ingenuity of the Egyptians.

At the same time, Afrocentric scholars with traditional training can serve as a useful corrective to the European vantage point inherent in traditional Egyptology, by focusing on questions that it might not occur to traditional Egyptologists to ask. We all ought to help train these scholars. The level of interest and enthusiasm about ancient Egyptian culture is amazingly high in the African-American community. When I first arrived at Howard University, I was stunned by the enthusiasm I met with, both from my own students and from students outside of my classes (not to mention the prevalence of Egyptian-themed clothing and jewelry). At Howard, Egyptology is not a peripheral field in which one might take an elective as a novelty or to add an exotic line to one's law school application; Egyptian culture is seen as a heritage to be proud of, and something worth learning more about. Whether or not one agrees with the premise that inspires this enthusiasm (and, as I've said, this is largely a matter of faith and definition), there is a real potential for the expansion of our field among these students. While some Afrocentric students will lose interest once they get past the political questions, others will remain fascinated by the culture. A few of these may go on to become Egyptologists, whether with an Afrocentric agenda or not. Others will enter other professions, enriched by an appreciation for a culture other than their own, but to which they feel some connection.

In a time when university administrators talk endlessly of bottom lines and judge the validity of scholarly fields by the number of students they attract, we cannot afford as a field to ignore such an audience for the material we want to teach. In view of the growing influence of Afrocentrism in the educational and larger community, we cannot afford to maintain our adversarial attitude towards it and to refuse to contribute to its better grounding in Egyptological evidence and research. Most importantly, as scholars and teachers, we cannot afford to ignore enthusiastic, talented students with new perspectives that have the potential to expand both our academic field and our understanding of ancient Egypt.

## REFERENCES

BERNAL, MARTIN. *Black Athena: The Afroasiatic Roots of Classical Civilization.* 2 vols. to date. New Brunswick, NJ: Rutgers University Press. 1987–

HAARMANN, ULRICH. "Regional Sentiment in Medieval Islamic Egypt." *Bulletin of the School of Oriental and African Studies* 43 (1980): 55–66.

KUNJUFU, JAWANZA. *Hip-Hop vs. Maat: A Psychosocial Analysis of Values.* Chicago, IL: African American Images, 1993.

# STATE OF THE ART/SCIENCE IN ANTHROPOLOGY[a]

ROBIN FOX

I HAVE BEEN CHALLENGED to explain an apparent inconsistency in various published pronouncements on the issue of humanism and science. On the one hand I appear as a champion of science and the scientific method in the evermore acrimonious debate on the status of anthropology as a "science."[1] On the proverbial other hand,[2] I appear as a gloomy critic of the "academic/scientific enterprise" in its entirety.[3] Now I could just claim with Emerson (no, not Winston Churchill, although he loved to use the phrase) that a foolish consistency is the hobgoblin of little minds, and slither past this one. But the great Ralph Waldo did say a *foolish* consistency, and no one wants to admit to that who claims either scholarly or scientific status, much less both.

No, I am afraid I must protest my innocence here and hence claim not to have been inconsistent at all. I would not bother, except that I think perhaps there is a little lesson to be learned in understanding why, a lesson very relevant to the current debate. So let us back up a little and take an autobiographical peep into the Golden Age of Anthropology. I admit I came in on the tail end of the shining epoch—the boring '50s. But I was just in time to be socialized into the notion that there was no other game in town but science if one wanted academic and epistemological respectability. People called their books *The Science of Culture* (White) or *The Natural Science of Society* (Radcliffe-Brown), or even *A Scientific Theory of Culture* (Malinowski); anthropology was still unashamedly "the science of man." They might have quarreled about whether the subject matter of the science was indeed culture or society, or custom or behavior (was that ever settled?), but they did not quarrel about the science bit.

The symbolism of the quest for scientific status was marked: the archetypical image was not clad in academic robes but in a white lab coat, preferably with a slide rule (remember the slide rule?) sticking out of the pocket. Everyone wanted "laboratory" status. I worked at Harvard in both the Laboratory of Social Relations and the Laboratory of Human Development, and

[a] This is an expanded version of a paper published as "Scientific Humanism and Humanistic Science" in *Anthropology and Humanism*, Volume 19, Number 1, 1994.

even in France the burgeoning structuralists at the Collège de France instituted the Laboratoire de l'anthropologie sociale—still there and still flourishing under the same title. Both Britain and the United States then had a Social Science Research Council, and would have settled for nothing less. The vindictive Tory government in the United Kingdom has recently stripped the *Science* from its research council title—surely to avenge themselves on the pesky Fabians from the London School of Economics. Signs of the times indeed. We anthropologists understand these symbolic gestures, no? But back then it was science or bust. The terror of being excluded from scientific grace was palpable. The last thing a Ph.D. candidate wanted to hear was that he was being "unscientific." Words of doom.

Just what constituted being scientific was in turn much debated. At Harvard it raged between the statistical crowd (analysis of variance qualified you for scientific heaven) and the Freudians. But even many of the latter, who were in turn anthropologists, tried manfully (as we used to say) to "operationalize" psychoanalysis and so wring science out of it. This meant mostly that "Freudian hypotheses" were "tested" by statistical methods—Yale to the rescue waving the Human Relations Area Files and flurry of *t* tests and Chi-squares and assorted other measures of association. Triumphantly, ancestral claims to legitimacy were established through a once-forgotten article by Sir E. B. Tylor, and everyone breathed a collective sigh of scientific relief. The strict ethnographers, not having hypotheses to test or anything to predict, strove to claim at least the status of the "observational sciences." They were more like natural history or astronomy than physics, but they were sciences nonetheless. As Popper taught us—and taught me personally at the London School of Economics—it was method, not subject matter, that distinguished science, and insofar as we were dealing with "public" data, objectively gathered and both confirmable and refutable, then we were doing science at however humble a level.

Again, simply doing science—while the only road to truth (what else was there, theology, metaphysics?)—did not mean that one was automatically right. It is true that in the first flush of Popperian enthusiasm this seemed to be forgotten. We were all so concerned with the *status* of simply doing science that whether what we were doing was worth the effort was a question rarely raised. Thus there could be both wrong science and trivial science, but this mattered less than being science. After all, it was part of the price one paid for being on the track of truth. It meant, however, that a great deal of pseudoscience got through unnoticed except by a few curmudgeonly critics like Pitirim Sorokin. We tended to forgive anything so long as it was, at the very least, conducted in the scientific spirit.

Now this might seem cynical but it isn't meant to be. I am wholly in favor of proceeding in the scientific spirit at however low a level, rather than abandoning the effort simply because a lot of it is trivial. Let that be clear. On one shoulder sits the spirit of Popper demanding science in the form of falsifiable hypotheses, and on the other sits the ghost of Dean Swift parodying the crazy scientists of the Royal Society trying to extract light from cucumbers. I love Swift's humor, but I defer to Popper's judgment: at the very least we should try to emulate the virtues of science. If we do not, then we are either merely expressing opinions—however erudite and insightful—or, worse, doing metaphysics. Metaphysical statements, let us remember, can be *verified* but not falsified: this was the basis of the Popperian criticism of Logical Positivism. Thus "all events are ideas in the mind of God" could be verified by producing God and showing the correlation between events and his ideas. But in the absence of God and his ideas, or in the presence of alternative hypotheses ("all events are the products of previous events"), the metaphysician can still hold fast to his belief in his mind-of-God hypothesis, and he cannot be proved wrong.

He cannot be proved right either; that was Popper's clincher. To be truly scientific the hypothesis had to be vulnerable to disproof; otherwise, as David Hume said, commit it to the flames. If there was a lot of trivial science, then there was even more trivial metaphysics and worthless opinion. The only other game in town in those days that commanded a lot of attention was Existentialism. (Actually I found they had barely heard of it at Harvard.) And some of us remember A. J. Ayer's perfunctory dismissal of it as "a simple-minded misuse of the verb to be," or even better that "nothing is not the name of something." (I believe it was also Sir Alfred who said "existence is not a predicate.") As for the ancestors of Existentialism—Heidegger, Husserl, and Kierkegaard—they represented the worst excesses of continental Idealist metaphysics and, according to Ayer and the others of his persuasion, were talking literal nonsense. It is amusing to me, therefore, to find almost forty years later that I am expected to come to terms with these monsters and be held accountable to their excesses of unreason.

No thank you. Even trivial science had the virtue of vulnerability, and since there were a great many aspirant scientists and not many good ideas, triviality was the price we had to pay for the few gems that made it into the permanent record. It used to be an old joke that the strictures of the National Science Foundation made it impossible to get a grant for anything that had not already been done anyway. Rats were run silly in an attempt to achieve "replicability." At the same time, any daring new idea was almost bound to be turned down, since it was novel and hence its "replicability" was in question.

But here we must draw a very necessary distinction between the ideals of science and the way science is conducted. And an even more necessary distinction between the results of science and the uses to which they are put. And a yet more necessary distinction between those same scientific ideals and the failure of many scientists to live up to them.

The conduct of science can lead to boring triviality. Even great results can be used to evil ends. Scientists, being children of Adam, can be fools and charlatans or even just blind and biased. Indeed, my own despair at the "academic/scientific enterprise" which raised the initial question is a despair over the inevitability of human frailty, not over the ideals of scientific discovery. Since science has no value agenda of its own it is always subject to hijacking by fanaticism and idealism. Thus, when I am told that some particular theory—Mendelian genetics, for example—has been used to bad ends by eugenicists, racists, sexists, and fascists, I am depressed but not surprised. It is part of the basic "design failure" of human nature that leads to the initial despair.[4] But this does nothing to shake my faith in the truth of Mendelian genetics. This will be established by the testing of Mendelian hypotheses, and if these are falsified, fine; we are still doing science and the result will be, as it was after Morgan and de Vries discovered mutations, a better genetics. And again, and perhaps most important of all, one should not confuse the valid results of science with the *provenance* of those results.

I seem to hear repeatedly today that science is somehow disreputable because it is the province of European white bourgeois males (or something such).[5] Mendel was such, he was even an Augustinian monk, but he got it right about the wrinkled peas; and it would not have mattered if he had been a black handicapped Spanish-speaking lesbian atheist. These incidental facts about *scientists* tell us a lot about the history and sociology of science. They are indeed, and always have been, the province of the sociology of knowledge. Many critics of science today in anthropology speak as though they have just discovered the idea that knowledge is relative to class, race, sex, religion, etc. This idea was in fact at the basis of modern sociology as seen in Weber's question about the role of religion in the development of occidental rationality (and, for that matter, Marx and Engels on the nature of "superstructure"). But the next erroneous step was one that Weber never took: to say that the *truth* of propositions so generated was relative in the same way as the generation of them. (Marx and Engels unfortunately did take that step, and announced that revolutionary proletarian truth was the only truth. But it is worth noting that they never doubted that the issue was truth or that their truth was "scientific"—indeed they made much of this distinction in their pursuit of "scientific socialism.") Truth is independent of the source. Bias and prejudice and the like there will always be, there *has to be* in order for ideas to be generated at all. We do not think in a rational vacuum. As David Hume, again, said, "Reason is, and ought to be, the slave of the passions."

My despair sprang from a belief that human beings can never use their reason for disinterested ends. I still believe this. The sociology of knowledge, the very mechanisms of the brain, and all the complaints, for example, of the feminists and deconstructionists, confirm this belief for me. But at the same time I do not accept that bias in the generation or use of science affects the status of scientific propositions themselves. There is, in other words, no such thing as "feminist science" any more than there is "Aryan science" or "Jewish science." When feminists claim that there is "Eurocentric white male science," what they are talking about is bias in the choice of subjects or the conduct of experiments or observations, or even more in the use of metaphors to popularize science. All this may well be true. But here is the nub: *the very ideals of science itself are the only real antidote to its misuse.* To insist on putting "feminist science" in place of "male chauvinist science" makes no more sense than putting "Aryan science" in place of "Jewish science," if what we are interested in is the truth of the scientific propositions themselves. Here the only solution is to put "good science"—that is, science conducted according to objective, rational empirical procedures aimed at eliminating as much observer bias as possible—in place of "bad science."

Even so, scientific hypotheses have to come from somewhere. In this sense there always has to be "bias." The metaphor of "bias" derives from the fact that according to quirks of its construction, a freely rolling ball may tend to run in one direction rather than another. But if it were not for such biases we would never take a direction. Let me return to the issue of wrong (and/or trivial) science in order to cull an example. In the era I was describing, the dominant paradigm (as we have come to call it post-Kuhn) of behavioral science was Behaviorism (disregarding for the moment the Marxists and "scientific" socialism). No line of enquiry escaped its influence—whether philosophy, linguistics, anthropology, or anything else—and even communist doctrine via Pavlov and Lysenko was happily Behavioristic (indeed Pavlov was one of its founding fathers.) Even if its influence was indirect, as in a lot of anthropology, it nevertheless helped to reinforce and underline the prevailing environmentalism of the subject, and its metaphors were all-enveloping.

Cultural determinism was in fact a form of Behaviorism in which a blank-slate organism, the "culture carrier," was molded through "enculturation" (read "conditioning") without any regard to what happened in the "black box" of the organism's brain and consciousness. In many culture-and-personality versions the behaviorism was quite overt and its language and theories used to "explain" cultural traits and behavior.

Now there was no question that Behaviorism was "scientific." It passed all the tests. It sought disconfirmation of hypotheses.[6] It worked on the logico-inductive method. It was objective, and its results were replicable. It had only one slight fault: as a total theory of human behavior and culture it was wrong. It explained some things about human behavior, but when it tried to push into other areas—chiefly, for example, in language—it simply failed to account for phenomena that it claimed to explain—language learning by children, for example. Alternative theories became available, such as those of the ethologists, which could account for animal behavior where Behaviorism failed. Experimenters began to produce falsification of the theory of "reinforcement schedules" by showing how animals resorted to "instinctive" behavior patterns under repeated trials. So great was the Behavioral bias that these initial results were simply not believed. One editor of a leading psychology journal, confronted with the findings of John Garcia and his colleagues, announced that he would believe the results when they found bird shit in cuckoo clocks. Well, the bird shit cometh, and indeed hitteth the fan. Chomsky demonstrated how children could not possibly learn language by conditioning. And so it went. The paradigm of Behaviorism had pushed too far and had run up against phenomena it could not explain. Its hypotheses were disconfirmed. Back to the drawing board.

The Behaviorists went kicking and screaming. Scientists are human. They put vast investments of time, money, prestige, and ego into their work, and they do not like to see it faulted. What is more, they had a three-hundred-year tradition of *tabula rasa* philosophy and psychology behind them to give strength to their prejudices. But the opposition equally had its biases. Chomsky derived his from Cartesian linguistics, Garcia from the natural behavior of animals, the ethologists from Darwinian evolution, and the social science fellow travelers initially from the ethologists, but ultimately from a plain dislike of the Behaviorist/Environmentalist paradigm as evidenced in, for example, the shocking case of the scientific tyranny of Lysenko and Stalin among other things.

In other words, a whole mixture of information and "bias" went into the questioning of the Behaviorist wisdom, and hence the search for information and the design of experiments to disconfirm the Behaviorist hypotheses. In the process, other biases than those motivating the Behaviorists (both capitalist and communist versions) became enshrined in the new natural sciences of behavior. These in turn will undoubtedly be challenged and if they push too far, disconfirmed, by people with other biases. The robust findings of Behaviorism and Ethology will be retained, but science will push in new directions where they could never go. Indeed, many would claim that it was the very failure of Ethology to explain its own firm observations on the basis of "group selection" theory, that led to the development of the alternative of "sociobiology" based on individual selection. The debate still rages, as it should, but it is a debate that rages within the confines of a basic set of assumptions about how such issues should be settled in science. It is, in other words, a truly scientific debate, not an ideological or metaphysical one.

The point then that science can be "wrong" is beside the point. It is the business of science to be wrong. That is one way we know it is science.

The point that science can be "biased" is equally beside the point. We have to have biases to motivate us to look beyond established paradigms.

The point that science can be "trivial" is also beside the point. We have to have a lot of trivial science to keep scientists employed, for out of the trivia some pure gold will emerge, often serendipitously, and since we do not know in advance where or how it will emerge we have to put up with the trivia in the meantime.

The point that science can be used for evil purposes is beside the point. Art and music can be used for evil purposes, but no one proposes abandoning either. Anything can be used for evil purposes. I am not going to stop listening to Wagner just because Hitler liked him.

All these objections are beside the point if the point is the *truth* of science. The truth of scientific propositions is independent of these objections, including my own despair at our human inability to use science wisely.

It was my own "bias" against the "inhumanity" of Behaviorism as exemplified in the Skinner box and *Walden II*—my own deep-rooted organistic Burkean conservatism if you like—that led me to search for other and more plausibly "human" theories of human behavior, and has taken me from Ethology through the evolution of behavior to neurosociology, cognitive science and the developing breakthrough in the philosophy of mind and human consciousness. My "bias" was against socialist totalitarianism (just like Orwell's—although he thought it was socialism gone wrong, whereas I thought socialism was wrong to start with). I freely admit to the bias. I am proud of the bias. I am glad to have lived to see the day when I can turn round to lifelong critics on the political and the scientific front and say: "I told you so." But the scientific truth of the theories I was driven to explore as alternatives to the totalitarian ideology of Behaviorism will have to stand on its own regardless of my bias, even if intellectual historians might find the connection interesting.

To return to the original issue then, of whether or not I was inconsistent, the answer should be clear: I was talking not of the truths of science, but of the uses of science when I despaired of science. I was, in short, taking a "humanistic" position with regard to science and aligning myself with H. G. Wells at the end of his days, with George Sorel, with Albert Camus, with George Orwell or William Golding or D. H. Lawrence or T. S. Eliot or a host of humanists who have deplored the uses of science in the twentieth century in particular (although the seeds of this attitude were firmly planted by Ruskin and Carlyle, Dostoevsky and Mary Shelley, and others in the nineteenth century as well).

There has been a chorus of dismay about the consequences of scientific hubris. But what I want to argue here, and what this build-up has meant to establish, is that this humanistic despair is misplaced if it is directed to science itself as opposed to the uses, or the biases in the generation of, science and technology. Indeed, it is the marriage of the former to the latter, when technology turns sour, that really frightens the humanists. What frightens them is not Einstein's discovery of the relation of energy to mass and the speed of light squared, but the translation of this into atomic weapons. It is not electronics and wave theory, but the use of this to produce television and mechanical control of thought. It is not Mendelian ratios as such, but their use to prove that some races are inferior, and not natural selection as such but its use in arguments for exterminating less-fit races.

It is true that the humanists do not always make this distinction. They do indeed blame science as such for the mess. But these are usually the religious humanists for whom science challenges the truth-claims of religion, and so is to blame for separating us from God. Few current anthropological humanists fall into that category. Yet they too seem to be making the same mistake. They blame the gardener because the tree has poisonous fruits. Again they are too often confusing science and technology, which is not altogether their fault since we all tend to equate them. But there was technology long before there was science. Science is a way of knowing: an epistemological system. It has in fact nothing to do directly with technology, which can flourish without it, as it did in China. Edison was not really a scientist in the strict sense. He was a technological *bricoleur* who operated by trial and error and hunch and know-how and cumulative successes. But science and advanced technology are inextricably linked, in fact if not in theory, since advanced technology is dependent on the findings of science.

But a disillusionment with the fruits of technological advance (weapons of mass destruction, worldwide pollution, thought control and totalitarianism, consumer materialism, etc.) really has nothing to do with the question of whether or not anthropology should or should not be a science. This question, which I answer in the affirmative, has to do with the claims of anthropology to be able to answer real questions about the real world: to establish scientific truths. It has nothing to do with how such truths might be misused or the provenance of their discovery.

In theory. In practice, if I am consistent, I will have to say that of course they will be misused. This is part of my "humanistic" complaint: that we are fated to misuse knowledge by the very nature of our devotion to ideas.[7] This hypothesis, I must admit, verges on the metaphysical. It could be falsified only by observing an indefinitely long period of human progress free from the misuse of science and the fanatical devotion to ideas. No one lives long enough to test such hypotheses. So far, human history roundly confirms it, but I cannot speak for the future. I would dearly like to think I am wrong—at least I would know I am doing science and the spirit of Popper would be appeased. But this is why I insist that this is a humanistic observation, an opinion, a judgment call, and not a scientific hypothesis. Because it seems to me we have no real alternative. Unless we are to abandon the search for "truth"—which is evidence about the real world established by the testing of falsifiable hypotheses within the framework of a revisable theory connecting them—then we are stuck with "science" and we have to go with it, whatever our nervousness about its possible misuse. We have to use our humanistic imaginations to try to second guess the misuse and so avoid it as much as possible. This does not amount to controlling or censoring science as such, but to constructing the possible scenarios for its abuse and seeking to avoid these.

This is what the humanist–scientist Michael Crichton did in his brilliant morality tale *Jurassic Park*, and what, indeed, many of the best of the science fiction writers do, which is why they turn out to be better social scientists than we are much of the time. Here I cite Frank Herbert's wonderful *Dune* series and the collected works of Asimov, Heinlein, Brin, Clarke, and Bradbury. This area has been too little explored by "humanistic" anthropology—let us call it "the anthropology of the future"—and this is a pity, since it is a perfect area where humanistic imagination and sensitivity and scientific knowledge (and technology) best come together. The possibilities of scientific hubris are here accepted, but so is science as a way of knowing. The ideal, often unstated but always there, is the melding of humanistic and scientific ideals in the service of mankind, not their bitter opposition. In this, science fiction at its best transcends the gloomy pessimism of the Sorelian antiscience tradition, while keeping a level head about the possibilities of a scientific utopia. My point is that it recognizes the distinction I am urging between the truths of science and the biases and uses of science, and searches for humanistic ends without ditching the search for scientific truth.

But, some of you have been itching to interrupt, has the whole issue of "objective scientific truth" not been called into question?" Is this not the point of the "humanistic" objection after all? What is the point in trying to seek a rapprochement if there is no scientific "truth" and all truths are relative? Are we not then left only with the humanistic modes of interpretation as practiced in the humanities proper? For this type of objection (having its roots of course in "deconstruction" and neo-Idealist philosophy) it is not the fruits of science (although these may have been the source of this particular "bias") that are at issue, but the status of scientific truth itself.

Actually, in practice, the arguments get all muddled. Thus, the argument goes, there are no absolute truths, since all knowledge is relative to the social condition of the knower: the Marxist and sociology-of-knowledge position. In the latter-day version, however, the social condition of the knower has been expanded from social class, technically defined, to "gender" (an egregious solecism meaning, in essence, sex),[8] ethnicity, race, religion, class (widely defined as position in the social dominance system), historical period, ideological position (which should be a product but has become a producer), generation, and Lord knows what else.

Thus we can get such monsters as "Eurocentric white male heterosexual bourgeois Protestant science."[9] This "science" has no claim to universal absolute truth, the position holds, *because* of this provenance. It is only "true" in this "context" just as, for example, in the Middle Ages, the pre-Copernican theory of the universe was "true" in the context of a society governed by Catholic theology. But there is no universal truth of science outside these relative truths of particular sciences. There is much invoking of Einstein on relativity and quantum physics and the Heisenberg principle—often in startling ignorance of the real principles involved in each—and of course Derrida, Foucault, Feyerabend, and sometimes Dilthey (by the better educated).

The hapless Tom Kuhn (who is horrified by this particular mangling of his theory of paradigms) and Richard Rorty are invoked like gods to justify an ultimately totally relativistic epistemology, as is Willard Quine, despite his being an uncompromising materialist.[10]

These arguments have been so well ventilated now, and by people better able to deal with them than I, that I am not here going to fight the battle all over again.[11] Obviously I do not agree with the relativist position. While accepting the connection between social position in all the above senses and the generation and use of scientific ideas, I clearly do not accept that this relativity affects the truth value of scientific propositions per se. A proposition, so long as it is in the form of a falsifiable hypothesis, is not invalidated by being placed in a different social context. This is where the Heisenberg principle is so often and so strangely misunderstood and misused by such critics. The proposition in question might never have arisen in another context, or might have arisen in a very different form, and this is in itself an interesting question for the sociologists of knowledge to pursue. Indeed it was the starting point for Weber's brilliant, and soundly scientific, comparative sociology of knowledge based on the principle of concomitant variation first propounded by John Stuart Mill. Strange how Weber is so often cited then by the relativists as though he were a founding father of their movement! This is based on a misunderstanding of his principle of *Verstehen,* loosely translated as "understanding," and which is certainly concerned with the subjective states of the actors. (Basically Weber was asking "what did they think they were doing?") But he never thought this removed analysis from the burden of objectivity and proof.

At the risk of being boring I repeat: the truth value of a proposition, even a proposition about subjectivity, is not affected by context. This is the whole point of science; the whole point of the revolution in thinking that Weber set out to analyze under the heading of "rationality." And he saw it applying across the board—to music, mathematics, business, theology, law, and religion, as well as science. (How many "humanistic" anthropologists so free with Weber's name read him on the rational evolution of music in the West? Not many to my knowledge.) And it happened once in history in one particular place, but—and here is the revolutionary thing—unlike every previous system of thinking, its truths were potentially universal in their application to reality. Unlike religious or magical beliefs, its truths were totally independent of social and cultural context. We will get nowhere trying to control the world with the principles of sympathetic magic as practiced by a Siberian shaman, and indeed they will be intelligible only in their cultural context; but the same shaman will do very well with the observation of Boyle's law, or by following genetic principles in his breeding of reindeer—and so would his Zulu counterpart.

It was because of this principle that Weber was able to conduct his "scientific" investigation of comparative civilizations and ideological systems. If Weber were the relativist he is made out to be, he would have had to dismiss his own lifework as inherently false. Weber would have seen quite clearly the impossibility of the relativists' position: if it were true, then it must be false. They are caught like the Cretan liar (who said that all Cretans were liars): if all truths are indeed epistemologically relative and have no universal application, then the proposition that all truths are epistemologically relative is itself relative and has no universal application, and we have no reason to accept it. It is the product of its own context, biases, social conditions, etc.

Indeed, it is. This brings us back to the question we started with about anthropology and the quest for scientific status in the heyday of the scientific paradigm. At this point most anthropologists would have been what it is fashionable to call "value relativists." This was an often incoherent position, but at its most general it said that "we" could not judge other societies on a scale with ourselves at the top. All societies were ethically equal in this view. The great sin was "ethnocentrism." This position itself is not logically sound, but leave that for a moment. It was essentially a humanitarian attempt to oppose the view of the "natives" as "savages" and to plead for a deeper understanding of customs that appeared at first, to the "ethnocentric" observer, as cruel or disgusting. But I know of no anthropologists who extended this to epistemological relativism—to the view discussed in the previous paragraph. There was no way they could do this and maintain a "scientific" status. Paradoxically, it was argued that value relativism was more "scientific" than "absolutism" or other nonrelativistic positions in ethics, and the Logical Positivists (and their linguistic philosophy successors) were often invoked as philosophical backup for this view: ethical statements were "emotive" not "descriptive," and hence there could be no absolute ethical standards, etc. The world of anthropology was thus kept safe for science. The "natural science of society" for Radcliffe-Brown was essentially what we now know as "structural-functionalism," and he (mistakenly) thought that this meant it should look for "general laws" of social functions on the model of "general laws" of physics. None were ever found, nor should they have been; science does not proceed by looking for general laws, which are, in any case, always provisional hypotheses in real science as opposed to pseudoscience (for example, the "evolutionism" of Herbert Spencer—the "development hypothesis" as he misleadingly called it).

Scientific "truth" is indeed not fixed and absolute "out there" in the world waiting to be discovered, but is ***a special kind of relationship between the knower and the known.*** The nature of this relationship, however, as we have seen, is unique and confers a unique status on the propositions (hypotheses) that result. The Functionalists did not understand this. Radcliffe-Brown himself declared that he was an evolutionist of the school of Spencer and his "science" was a nineteenth century mechanistic version that never caught up even with Popper (and despite my loyalty to my old teacher, I have to admit that we have progressed in the philosophy of science since!).[12]

Thus, in Britain and France, a reaction set in against the scientism of Functionalism. In these early days it was seen essentially as a shift initiated by Evans-Pritchard and Lévi-Strauss from explanation to interpretation, from cause to meaning, from science to symbolism, from social structure to mental structure. It saw itself as reviving the historical division between the natural and cultural sciences (after Dilthey, after Kant) and coming down in favor of ***Kulturwissenschaft.*** It therefore saw itself as moving away from "science" as such, since "science" was associated with the discredited Functionalism. For the French, the paradigm of linguistics (de Saussure, Jakobson) was first invoked as an alternative; later hermeneutics and rhetoric got their turn, and the rest is history. In the United States Clifford Geertz (who also ended up down the road from me in Princeton) led a group of young resistance fighters against Functionalism in its Culturalist versions and in particular in its evolutionary or ecological materialist varieties; and "symbolic anthropology" here, too, lined up against "science" and with the new European symbolic and structuralist movements.

In its origin, this move from "function to meaning," from science to humanism, was anything but radical. On the contrary, some of its manifestations were positively reactionary, or at least seemed so to the scientific rationalists of the time. In the United Kingdom, at Oxford, it was an affair conducted largely by Roman Catholic (some converted) anthropologists (Evans-Pritchard, the Lienhardt brothers, Turner, Douglas, etc.), who were in a frank reaction against the positivist-rationalist tradition of Sir James Frazer and his admirer, Bronislaw Malinowski, at the London School of Economics.[13] (For the record, I too was reared in that tradition, and in some sense still consider myself a Malinowskian social anthropologist.) I remember the suspicion that this latter-day "Oxford Movement" engendered. Sir Raymond Firth, Malinowski's successor, groaned deeply and shook his head sadly when Turner decided to call various stages of Ndembu rituals "stations." And the super-positivist Max Gluckman of the rigidly empiricist Manchester school (which he created, of course), referred to the whole movement as "the oratory." Leach at Cambridge responded with a series of "structuralist" deconstructions of the Old Testament, just to keep them on the defensive. Cambridge as a whole was still in good rationalist hands with Fortes, Goody, and Leach, who, like most of their generation (including Evans-Pritchard), hailed from the London School of

Economics graduate program. (During World War II the London School of Economics was evacuated to Cambridge, so a natural affinity existed: if anything, Cambridge was more left wing than the London School of Economics, despite the latter's reputation.) The Cantabrigian countermovement was to come from young and as-yet-unknown sociologists who reacted equally against the Fabian empiricism of the still-dominant London School of Economics, and who were eventually to kick off the "cultural studies" movement, aligning themselves with Paris and Frankfurt.

In the United States, we associate this trend largely with Geertz and Schneider, and perhaps with Bellah in sociology—all, like myself, products of the Harvard Social Relations Department, and all reacting against it. Again, they were seen at the time as more reactionary than radical, especially by the Left and very especially by the Marxists. When I was there in the late 1950s, there were always one or two people around busily quoting Suzanne Langer or Kenneth Burke or Alfred Schutz, but no one took them seriously. We were wrestling with Talcott Parsons's grand synthesis, and this was the point of departure, pro or con. Actually, Geertz seemed to start in the direction of trying to build a "general theory of social action," to use the jargon of the time, but then backed off into "thick description" and "interpretation" and the like and away from grand Parsonian generalizations. He never seems to have given a coherent reason for this switch; it seems more a matter of taste than anything else.

But my point is that none of these initial movements were "radical" either intellectually or politically. They were even just the opposite. They predated the "spirit of '68" and the philosophical revolutions in continental Europe. And, despite the currently fashionable conspiracy theories of knowledge, I am inclined to see them as genuine, perhaps even predictable reactions to the overlong dominance of Behaviorism and Functionalism, positivism and empiricism. They were in this sense genuine intellectual movements. They were also genuine movements in the direction of a "humanistic" as opposed to a "positivistic" view of the role of the social "sciences." But at least initially they did not attack or denigrate science as such. They simply quietly differentiated themselves and what they did from it. The Oxford Catholics, for example, saw no future in attempts to "explain" religion according to reductionist psychological schemes such as those of culture-and-personality anthropology, or to "laws of development" of a Frazerian or Comtean kind, or according to a Malinowskian theory of "needs," or a Functionalist theory of social utility. Religious symbols, rituals, and beliefs could not be so "explained"; they could only be "interpreted" in terms of what they meant to the believers. The humanist-Catholics and their nonreligious humanist brethren were, across the board, remarkably undogmatic about it all, even in their most programmatic statements such as Evans-Pritchard's *Social Anthropology* of 1951 and his *Theories of Primitive Religion* of 1965. They preferred on the whole to make their points by making superior demonstrations rather than by claiming a superior epistemological status.

What has happened since is that these movements, which, as I keep insisting, were genuine intellectual resistances initiated by the postwar generation to what appeared to be a barren Functionalist heritage, have been hijacked by ideologically motivated, blatantly political movements of the anti-Vietnam baby-boom generation. The "spirit of '68" infuses them; the women's rights movement in the latest avatar of "feminism" has climbed aboard; the fashionable movements in philosophy of knowledge in Frankfurt and Paris (deconstruction, hermeneutics) lend strength to them; and, following on the genuine achievements of the civil rights movement, various groups claiming "empowerment" have plugged in, with the demands for "multiculturalism" and the overthrow of "Western civilization" that have become so depressingly familiar and so politically oppressive. The rather bewildered leftover Marxists who have seen their real and intellectual worlds crumble are trying to accommodate, however clumsily. What was a shift in emphasis in the social sciences has become a revolutionary, relativistic, antiscientific political ideology, with a frightening tendency, in the United States, at least, to harness the worst forces of puritanical fanaticism, forces that seem so eager to burst out and have their day, in a new wave of campus totalitarianism that threatens with academic gulags and thought reform those who do not accept the moral absolute of the cultural relativists. (Logic has been the most obvious loser in the whole sorry history.)

The sadness of this for me—and I write as a humanist, in the broad sense, for humanists—is that the majority of "humanistic anthropologists" seem to feel it necessary to identify with the hijackers, and hence with their antiscientism. There is not only no need for this, but I would argue that it is a dangerous and, in the end, futile road to take. In reacting against Functionalism in any of its versions, we were reacting essentially against a misconceived science. Because the ideas of science most humanists hold are as outdated as the ideas of the functionalists themselves, they see their revolt as a necessary rejection of science as such.[14]

But let me slip into autobiography again. I, too, revolted against Functionalism as early as the "symbolists"; but I did not throw over science, since I saw that what I was rejecting in Durkheim's or Boas's versions of social science was not in itself very good science (although it was in the scientific spirit, and I will come back to that). Remember our earlier point: just to be doing science in some way or another is not good enough; one has to be doing it right. My reaction was to equate Functionalism with "inadequate science" (and cultural anthropology, in fact, with outworn ideology) and to seek for a more adequate scientific approach to human society—one that eschewed the *tabula rasa* and the Durkheimian separation of individual and social, for example, and proceeded within the framework of a theory (e.g., natural selection) that would produce testable hypotheses about proximate mechanisms in human social behavior.[15]

I, personally, found it in Ethology, as it was then called. Originally a science of animal behavior based on observation growing out of Darwinian "natural history," it was introduced to experiment by Lorenz, von Frith, and Tinbergen, and expanded to human behavior by a growing group of interested social scientists with varying degrees of scientific usefulness, and indeed a few wild and woolly exercises thrown in. Ethology needs its own history, but now, commonly known as "sociobiology" after a coinage of E. O. Wilson's, it is thoroughly established as workable science (normal science in Kuhn's terms), with its branches and its schools and its infighting—typical of all young sciences, where youngsters out to make a name constantly reinvent the wheel and call it a vehicular motion-facilitation device.[16]

I do not mean here to defend this "human ethological" approach in detail, for that would be out of place. In any case I get tired of having to explain basic processes of natural selection, with which I expect all freshmen to be familiar, to senior colleagues of the humanistic persuasion. It is embarrassing. I simply mean to point out that the route out of Functionalism was not *necessarily* the "symbolic" or "interpretative" one; that one did not have to ditch science and opt for some other mode of knowing in order to escape the trap. For that way leads to the absurdities of epistemological relativism and even more dangerous ideological traps. The way out of bad science is to find good science, not to ditch science altogether and embrace various forms of opinion mongering that masquerade as knowledge while denying its possibility. (Logic loses again.) Others who were disillusioned with Functionalism did not necessarily go the ethological route; they chose other routes like cognitive anthropology[17] or cultural (historical) ecology[18] or a more cybernetic approach[19]—all of which stuck to science while rejecting its teleological Functionalist version.

But, again, you might be itching to interject, we *are* humanists. We want to interpret, not explain. We want to look for meaning, not for cause. We do see what we do as "literary" not as "science." And we like it that way. To which I say: bunkum! (Actually, I wrote something else, but this is a polite academic publication, so I changed it.) You want to be *believed.* When you insist that something is the case either about human behavior in general or about some local behavior in particular, you want your reader to accept it as "true." You do not in fact say: well, this is just my opinion, this is just a story like *Gulliver's Travels*, and you can take it or leave it. If you have asserted that the X do Y about Z, you will be peeved if another observer says no, they do W about Z, not Y. You will want to show that you are right in your "interpretation" and that it was not just a whimsical invention. There may by "multivocal" interpretations, but you will be prepared to admit that some are simply off the wall and others "make sense." You must therefore appeal to some criteria of judgment—some things we would all as rational observers agree on—to decide the matter. You would want, in other words, to frame a falsifiable hypothesis and test it at however low a level.

The last phrase is significant since it goes back to what I said and put on one side about "in the spirit of science" and all that. You may mistakenly think that science is what happens in physics or biology classes (a lot of "humanists" seem to think this way) and that it must, for example, always involve quantification and statistics. But this is not the case. Science is a mode of knowing. If we have a disputed line in the work of a French troubadour that scholar A insists is genuine and scholar B says was inserted later, we can settle the matter if evidence is available. Say the line refers to an artifact that was not in existence when the troubadour wrote, then we can all accept that it must have been inserted later (unless there is some other, nonmetaphysical, hypothesis that is better, of course).

We are here accepting the "spirit of science" as much as if we quantify and use statistics. These are only relevant to certain kinds of hypotheses. If you are indeed interested in the "truth"—and whatever you might think, you really are—then you *must* use the scientific mode, at however low a level, to arrive at it. Evolution would not work otherwise. We would not be here to discuss the matter today. In our everyday thinking we are constantly testing and confirming and falsifying hypotheses; this, more than "conditioning," explains how we behave as we do. We are natural natural scientists; we have no other choice. I am here invoking Popper's powerful argument that our perception of the world, and our decision making about it, work on the basic principles of hypothesis testing and refutation, and that "scientific method" therefore is simply the extension of basic cognitive principles.[20]

This is why the current antiscientific relativism makes no sense to humanists. It is simply a kind of throwing in the towel—a confession of intellectual cowardice. The sins of functionalism lay in the false notion of science as a pursuit of general laws (which were largely teleological truisms).

The virtue of Functionalism lay in a devotion to the idea of rational enquiry at the empirical level; to at least an attempt to adhere to objective standards in fieldwork, the anthropological mode of gathering data.[21] This did not necessarily involve quantification and statistics; it all depended on the kind of question posed and the kind of answer sought. But at the very least there was a commitment to a descriptive objectivity that was in principle "replicable" by other fieldworkers. Truly, when it came to interpretation and judgment, there were differences; but it was accepted that somewhere in the conflicting accounts, the different styles of writing and presentation, there was a possibility of truth. Of course there was sloppy work, and sometimes there did seem to be irreconcilable differences. But these were rationally (if not always reasonably) argued, and the source of bias or the nature of the different interpretations examined in an attempt to see just what was bias (what, for example, resulted from incompatible prior assumptions) and what was fact. Anthropologists, in other words, were held accountable in a way that tourists and journalists and novelists were not. And that is how it should be. It is tough, but no one said anthropology was easy. If you want it easy, then be honest: join the creative writing program.

There was, in fact, a pretty good agreement as to what constituted evidence for a statement of "fact" in ethnography, and in teaching fieldwork seminars I always stressed the need for evidence to back up generalizations. I was always leery of ethnographies that simply stated the "customs" of the so-and-so about, say, land tenure, in the absence of any detailed maps and evidence of inheritances. When I came to do my own fieldwork, inspired by the examples of Malinowski on Kiriwina, the meticulous data gathering of the Manchester school, Firth on Tikopia, and Leach on Pul Eliya (among others), I made sure that I documented my "interpretation" of Tory Island land holding with as rich a database as possible, and one that was open to objective scrutiny by any other interested anthropologists who might wish to contest my version of the facts.[22]

Land tenure is "hard" data, I suppose, but the principle applies just as much to interpretations of religious symbolism or magical rites: if you say the X do Y because of Z, I want to see the data that support this view as opposed to the view that they do it because of W. If you say that the X "mean" Z when they do Y, then I want a good reason to suppose that this is a fact and not just your fancy. And indeed, as you know very well, fieldwork and interpretation could not proceed and be convincing if this were not the case. After all, in some sense, all studies of cultural rules and customs are studies of "meaning." What did it "mean" for the Tory Islanders to say that all children should be provided for from the land when at least half the children never got any land? One could not answer this question of meaning by writing confessional poetry or deconstructing the concept of land tenure, but only by gathering empirical data to test various hypotheses about what the meaning could possibly be. This is the "spirit of science" at the fairly low observational level at which we practice it. Most of the time we do not notice it because it is, as I have said, the normal human way of processing and testing knowledge anyway. But when we come to want our interpretations accepted by the community of scientists/scholars, then we have to become self-conscious about them and play by the elementary rules of science whether we like it or not.

In short: ***If you wish to be believed, you must accept the burden of falsifiability.*** You must accept that your statements are hypotheses that are in principle subject to refutation. If you refuse to accept this burden, on any grounds whatsoever, then there is no reason why we should pay any further attention to anything you say, since you could just as well utter complete nonsense or gibberish; it would make no difference. The same goes for so-called deconstruction as an intellectual activity. The critical analysis of concepts in order to reconstruct them as better hypotheses is very necessary to science. Again, I have done more than my fair share of this critical service. But while the deconstruction of concepts as an end in itself, and, if I understand Derrida rightly (and I am convinced that part of his program is that he should not be understood), as a never-ending self-cancelling activity may satisfy some cloudy demands of Husserl, Heidegger, and the "phenomenologists" (who have as far as I can see no relevance to social science whatsoever), it is useless to those of us concerned with the assessment of empirical reality. Of course, you will respond, the existence of empirical reality is what these theories hold to be moot (or at least they question the possibility of our knowing it). To which I can only reply: let me hear you say that when told you need a difficult operation to save your life, or the life of one of your children. Christian Scientists are at least consistent on this issue; academics who hold these ridiculous theories are simply hypocrites.

I have done "science" at all levels from the purely descriptive to the quantifiable and statistical. I also think of myself as a "humanist" in the broad sense: I approach my fellow human creatures as being in a deep sense the "same" as I am, and I "interpret" their differences from me against the measure of this sameness. This is what we do about all other people all the time, starting with the most familiar and working out to the seemingly unfathomable other. The real poet, like any artist, tries all the time to see the general in the particular. In this he is no different from the scientist. They are siblings under the skin. In its original, Renaissance, meaning, "humanist" referred to someone who took man as the measure (as opposed to God or angels). It did not differentiate between scientists and artists in this respect, and the greatest of the humanists was himself the greatest artist and scientist of his day, who saw no conflict between the two ways of knowing. If for Leonardo they were one, why not for us? Later ages, which split off "humanities" from "sciences" (beginning, I think, in the seventeenth century with the use of the word humanist to mean a student of the classics), started a rot of which we are the ultimate heirs. The tragedy of anthropology is that it is the perfect discipline to unite the two again, and thus to be a light to enlighten the gentiles and the glory of "humanism." The "science of mankind" is not a science that would or could ignore art and poetry. How would this be possible since these are two of mankind's most distinguishing achievements? But it would try to deal "scientifically" with them in the sense I have outlined above, not just reiterate their own structures, but envelop them in the fold of humane scientific examination, which is its own kind of poetry for those who have the ear.

Let me end by reaffirming that I am indeed committed, as an anthropologist, to the furtherance of humanistic studies as currently understood. I am, after all, a humanist manqué, and would have been a composer, guitarist, poet, or playwright in a perfect world. Nothing I have said about the necessity of scientific method in the pursuit of truth need alarm anyone who is devoted to the study of the art, poetry, music, or drama of native peoples (or anyone else). Go ahead with my enthusiastic blessing. All I am asking is that you do not join the fashionable science bashing that politically and ideologically motivated groups and individuals seem to think is necessary to their positions. It is not necessary to humanistic anthropologists, who, I maintain, if they are doing a good job as such, will not violate the rules or the spirit of science anyway. Let us return ***humanism*** to its original meaning (the meaning that led Sartre to insist that Existentialism was a humanism), and let anthropology be the shining example. Humanistic insight and scientific objectivity are not and never should be opposed: a devotion to humanistic values will lead to a more insightful science, and an equal devotion to scientific values will lead to a more convincing humanism. We are equal partners in the task of achieving a better understanding of mankind. Let us cease the useless warfare and conclude a fruitful peace. Both anthropology as a discipline and mankind as a species will be the better for it.

## NOTES

1 Robin Fox, *The Search for Society*; *The Challenge of Anthropology: Old Encounters and New Excursions.*

2 Who was it who defined an intellectual as someone whose on-the-other-hand did not know what his on-the-other-hand was doing? I have heard that it was Pierre Trudeau, but I have also heard this challenged.

3 Robin Fox, *The Violent Imagination.*

4 See the essay of that title in Fox, *The Violent Imagination.*

5 I was once accused in print of being a "bourgeois establishment scientist." I immediately rushed off a copy to my mother, so she would have printed proof that I had made it.

6 Actually Kuhn's criticism of normal science paradigms is relevant here, since what it, in fact, constantly sought was confirmation of hypotheses. The possibility of disconfirmation was always taken care of by the null hypothesis, of course; but negative results were simply discarded and the premises reworked until confirmation was achieved. In the end, however, it did succumb to disconfirmation, however unwillingly.

7 See again Robin Fox, "Design Failure," in *The Violent Imagination.*

8 "Gender, n., is a grammatical term only. To talk of persons or creatures of the masculine or feminine gender, meaning of the male or female sex, is either a jocularity (permissible or not according to context) or a blunder" (H. W. Fowler, *A Dictionary of Modern English Usage*, p. 211).

9 In 1957 I was on a student committee for the relocation of Hungarian refugee students after the uprising. It goes without saying that we were very unpopular with the student Left. We interviewed a candidate who told us he had been studying at a Marxist-Leninist institute. His subject was "proletarian philosophy." He told us he wanted to continue his studies at Oxford. We asked what he wanted to study there. Without hesitation he replied, "bourgeois philosophy." We figured he would survive just fine.

10 I remember almost thirty years ago being one of the few people in Princeton willing to talk at length with Dick Rorty about his enthusiasm for Idealist philosophy, which no one else seemed to share. He put me on to reading Royce, and I fired him up over Bradley. I might add that I was not too enthusiastic about their theories of reality, but I was interested in their social theories. Tom Kuhn, also in Princeton, and I discussed, while feeding his pet monkeys, the "paradigm shifts" in behavioral science I have discussed above. Thus do the wheels of history turn in strange and crooked ways.

11 See Paul R. Gross & Norman Levitt, *Higher Superstition: The Academic Left and Its Quarrel with Science*, for the best summary. But see also the interesting critiques of the "Dallas School" of humanists, who do not accept the mainstream antiscience position of their colleagues (Frederick Turner, *The Culture of Hope*; Alexander Argyros, *A Blessed Rage for Order*). A massive tour de force on these same lines is Joseph Carroll, *Evolution and Literary Theory.*

12 See Paul Arthur Schilpp, ed., *The Philosophy of Karl Popper.*

13 Malinowski's admiration for Frazer actually seems to stem from his permanent residence in England and applies to his writings in English. I gather his early writings in Polish are quite critical of Frazer. But it was *The Golden Bough* that drew him into anthropology.

14 This essay, in an earlier version, was originally directed at humanist anthropologists in one of their own journals, so I did not need to explain to them who they were or too much about their "interpretative community." I can only say here, to the nonanthropological reader, that what I describe is now utterly pervasive in cultural anthropology. The reader can pick up the catalogs of the university presses and see the hundreds of books that pour out each year based on assumptions that, indeed, are rarely even argued any more but are taken as given. It has become impossible, for example, to talk to most cultural anthropology graduate students—always desperate to be up-to-date—except in this dreadful sublanguage. I told one student that I did not think the Wenner-Gren Foundation would like his grant proposal, since it was far too empirical. He explained that the proposal he would actually submit would frequently mention "hegemony" and "patriarchy" as well as "signifiers" and "others," so it should be all right.

15 See Robin Fox, *The Search for Society*, ch. 3–5.

16 Robin Fox, "Sociobiology."

17 See the recent excellent history by Roy D'Andrade, *The Development of Cognitive Anthropology.*
18 E.g., Carole L. Crumley, ed., *Historical Ecology.*
19 Paul Bohannan, *How Culture Works.*
20 See Sir Karl Popper, *Objective Knowledge.* I do not mean that this argument fully accounts for the principles on which our cognition works. I have written at length on our use of intuition, probability, stereotyping, matching, and representability, for example (see Robin Fox, *The Challenge of Anthropology,* ch. 14; and *The Search for Society,* ch. 8). I am simply stressing the Popperian component as basic and necessary.
21 I am here addressing the humanistic anthropologists and so concentrate on fieldwork, which is their métier. Obviously I do not need to address these remarks to physical and biological anthropologists, who as a matter of course adhere to the scientific method, as do most archaeologists—although there is some wavering here (see Robin Fox, "One World Archaeology").
22 Robin Fox, *The Tory Islanders.*

## REFERENCES

ARGYROS, ALEXANDER. *A Blessed Rage for Order.* Ann Arbor, MI: University of Michigan Press, 1991.

BOHANNAN, PAUL. *How Culture Works.* New York, NY: Free Press, 1995.

CARROLL, JOSEPH. *Evolution and Literary Theory.* Columbia, MO: University of Missouri Press, 1995.

CRUMLEY, CAROLE L., ed. *Historical Ecology.* Santa Fe, NM: School of American Research Press, 1994.

D'ANDRADE, ROY. *The Development of Cognitive Anthropology.* Cambridge: Cambridge University Press, 1995.

FOWLER, H. W. *A Dictionary of Modern English Usage.* Oxford: Clarendon Press, 1926.

FOX, ROBIN. *The Challenge of Anthropology: Old Encounters and New Excursions.* New Brunswick, NJ: Transaction Publishers, 1994.

———. "One World Archaeology: An Appraisal." *Anthropology Today* 9, 5 (1993): 6–10.

———. *The Search for Society.* New Brunswick, NJ: Rutgers University Press, 1989.

———. "Sociobiology." In *The Social Science Encyclopedia*, edited by A. Kuper & J. Kuper. London: Routledge, 1995.

———. *The Tory Islanders.* Cambridge: Cambridge University Press, 1978.

———. *The Violent Imagination.* New Brunswick, NJ: Rutgers University Press, 1989.

GROSS, PAUL R. & NORMAN LEVITT. *Higher Superstition: The Academic Left and Its Quarrel with Science.* Baltimore, MD: Johns Hopkins Press, 1994.

POPPER, SIR KARL. *Objective Knowledge.* Oxford: Clarendon Press, 1979.

SCHILPP, PAUL ARTHUR, ed. *The Philosophy of Karl Popper.* La Salle, IL: Open Court Press, 1974

TURNER, FREDERICK. *The Culture of Hope.* New York, NY: Free Press, 1995.

# LIBERALISM, PUBLIC OPINION, AND THEIR CRITICS
## Some Lessons for Defending Science[1]

SIMON JACKMAN

*Let her and Falsehood grapple; who ever knew Truth put to the worse in a free and open encounter?*

—MILTON, *Areopagitica*

*The light of the public is the light of the Enlightenment, a liberation from superstition, fanaticism, and ambitious intrigue.*

—SCHMITT, *The Crisis of Parliamentary Democracy*

*As though it was not man who invented science but some superhuman ghost who prepared this world of ours and only, through some incomprehensible obliviousness, forgot to change man into a scientific animal; as though man's problem were to conform and to adjust himself to some abstract niceties. As though science could ever be more than man; and, consequently, as though such a gap between scientific and social knowledge could ever be more than wishful thinking.*

—ARENDT, *Essays in Understanding*

LIBERALISM AND SCIENCE are both products of the Enlightenment. Perhaps even more strongly, taken together liberalism and science *define* what we refer to as the Enlightenment. This essay is a defense of both science and liberalism, but I begin by apparently ceding an important point to critics of these Enlightenment projects. Note that in references to "liberalism" and "science" we often use the phrase I have employed here: "product of the Enlightenment." That is to say, liberalism and science are "socially constructed" (to put it broadly)—a charge that is in many ways one of the strongest and least equivocal points made by critics of liberalism and science. I concede, therefore, to a relativism that is often seen as mortal to Enlightenment projects.

But far from damning either science or liberalism, I demonstrate here that both are best pursued and derive much of their strength via this concession to relativism. I do this by means of several related arguments. First, I review the familiar connection between liberalism and science in Enlightenment thinking. Second, I show that a large component of what we refer to as political science is driven by liberal concerns. In particular, the study of public opinion is a central component of any discipline claiming to be a science of liberal politics. I briefly survey the uses of statistics in measuring public opinion, to argue that quantification and surveys (central tools in the study of public opinion) serve the liberal aims of political science especially well. I review and rebut criticisms of this position, some of which have been around for a long time and others that have a definite postmodern ring to them. In spite of what might be called a social construction of reality, I argue that there are some more or less durable components of the phenomena we refer to as American public opinion; that these components are real, measurable, are reasonably robust, and, moreover, have political content. Postmodern critics of positivist social science underestimate or ignore the extent to which a "normal (social) science" like public opinion research is relevant if not indispensable to studying democratic politics in the wake of the information revolution.

I conclude by returning to the links between liberalism and science. Those defending science against postmodern critiques can learn from the resilience of liberalism. As the history of the study of public opinion shows, liberalism can accommodate shifting opinions, relativism, and even the charge that there is a social construction of reality. Just as this concession does not fatally wound liberalism or the science of liberalism (political science), it need not seriously threaten science more generally.

## SCIENCE AND REASON IN POLITICAL SCIENCE

Liberalism's foundational texts present an argument linking reason, science, and political order that define the modern conception of democracy. The Enlightenment's political theorists were impressed by science as a model of knowledge generation. Mysticism impeded and suppressed not only scientific innovations like Copernicus's heliocentric astronomy but also developments in political thought and practice. In its various guises—papal infallibility, the divine right of kings, Aristotelian essentialism, the immortality of the soul, original sin, and so on—mysticism confounded science per se, including a nascent science of politics.

The Enlightenment's reaction against mysticism and the embrace of science drives Hobbes's *Leviathan* (1651), an important precursor to classical liberalism (if not canonical itself) and one of the first self-consciously scientific approaches to politics of the modern era. Hobbes's influences are easily discerned: Hobbes was a close friend of Francis Bacon, the founder of English empiricism; while on his third journey to the European continent (1634–7) he made a "pilgrimage" to Florence to meet Galileo.[2] In *Leviathan* Hobbes argues for a political science informed by reason and observation, free of the strictures of religious faith or the artificial distinctions in the metaphysics of Aristotelian "School-Men."[3] Hobbes's method is an application of Galileo's:

first, society is reduced to its constitutive parts—individual people, in Hobbes's view. After understanding the "motions" of individual people as a product of separate forces, human behavior and interaction (and, in particular, politics) emerge as the aggregation of these various forces.

Hobbes and other liberals purport to make very few a priori assumptions as to the moral content of these forces. In this way liberalism claims to be normatively "thin": liberalism is less a prescription for "good politics" than a *method* for "good political science." Beyond self-preservation (for Hobbes), liberals make few explicit assumptions about human nature. This is the source of liberalism's attractiveness as both a political theory and a model for organizing societies ranging from England in the late seventeenth century, to the United States in the late eighteenth century, and to the Czech Republic in the late twentieth century.

In this way liberalism exhibits the characteristics of what we have come to recognize as good science. Just as good science is nothing more than a method, so, too, is good politics, according to liberals. Good science, which subsumes good political science, does not presuppose its substantive findings: rather, the demands science makes of us are methodological. Science, like liberalism, is a means rather than an end. Liberalism, like science, is a method premised on weak assumptions. Liberal methods for politics, like scientific methods, both purport to be and have shown themselves to be generalizable, and readily applicable to new problems and contexts.

These connections between the emergence of scientific method and scientific attempts to study politics are familiar to us. Three centuries later, we struggle to recall if liberalism (broadly conceived) gave rise to science, or vice versa.[4] Moreover, it is our concern over the current status of these connections that has given rise to this volume. In drawing out these historical and theoretical connections between liberalism and science, I aim to show that assaults on scientific method simultaneously assault liberalism. Liberalism, under assault for some time now, may be unlike science in this regard. Indeed, the appeal to science was a powerful weapon in the hands of liberalism's critics.[5]

## LIBERALISM, LIKE SCIENCE, IS PRAGMATIC AND RELATIVIST

In the mid-nineteenth century, John Stuart Mill elaborated the relationship between science and liberal political thought, guided in good measure by Bentham's utilitarianism. The outcome is a relativism of sorts, or at least an understanding that knowledge is generated through social processes, that the nature of those processes (and hence, the store of knowledge itself) varied historically, and that liberal principles and the generation of knowledge complemented one another in ways that also displayed variability through history.[6]

In a sense, a relativism had always been an element of liberalism: in substituting human experience and reason for dogma and superstition, earlier liberals had already taken an important first step towards a pragmatic, epistemological relativism, if only by acknowledging that knowledge is cumulative. But Mill advances this position further, as the following passage from *On Liberty* demonstrates:

> . . . I forgo any advantage which could be derived to my argument from the idea of an abstract right, as a thing independent of utility. I regard utility as the ultimate appeal on all ethical questions; but it must be utility in the largest sense, grounded on the permanent interests of man as a progressive being.[7]

The last phrase, "the permanent interests of man as a progressive being," is important. Here Mill is subjugating his utilitarian pragmatism to a liberal *telos*: "progress" is measured by the extent to which people are free to maximize their happiness. But learning is also an important component of Mill's notion of "man as a progressive being," as it is for other liberals. Mill writes that man ". . . is capable of rectifying his mistakes, by discussion and experience."[8] The pursuit of happiness is aided by better understanding of nature, the economy, political institutions, and so on. In this way, science itself is part of the justification for liberalism. Free speech, a key liberal guarantee, is cherished not as a right in itself, but because it promotes the generation of knowledge. The overall picture is rather sanguine and even familiar: experience and reasoned discussion generate knowledge, and since knowledge makes for greater enjoyment of the world, political institutions should be designed so as to ensure vigorous, wide-ranging, untrammeled discussion.

Importantly, there are strong leanings towards pragmatism and relativism underlying Mill's position. Mill asserts that "[t]here is no such thing as absolute certainty"; all we have is "assurance sufficient for the purposes of human life."[9] Accordingly, under a wide set of circumstances, "all silencing of discussion is an assumption of infallibility"[10] and risks obscuring the truth or progress towards the "sufficient assurances" mentioned above. Discussion is vital to bringing about learning, since experience is insufficient or is incomprehensible without interpretation: "There must be discussion, to show how experience is to be interpreted."[11] While "[w]rong opinions and practices gradually yield to fact and argument,"[12] facts and arguments do not simply present themselves to us. "Very few facts are able to tell their own story, without comments to bring out their meaning."[13] To the extent that anyone is more or less certain of an opinion, it is because they have kept their mind open to criticism;

> because it has been his practice to listen to all that could be said against him; to profit by as much of it as was just, and expound to himself, and upon occasion to others, the fallacy of what was fallacious. . . . No wise man ever acquired his wisdom in any mode but this . . .[14]

In short, free speech is essential in a liberal polity, but not because of any romantic preoccupation with inalienable, self-evident, natural rights or other enduring truths. Rather, Mill's advocacy of free speech was absolutely driven by practical considerations. Put simply, free speech permits discussion, makes possible the reconciliation of differences of interpretation, and leads us to better discern the truth on important questions. And liberal political institutions (e.g., guarantees of free speech, a free press, parliamentary sovereignty) are mechanisms for allowing the experience and reason of a particular society to be effectively applied to the problems it faces. For Mill, what discussion yields is none the weaker in the face of contingent truths.

## POLITICAL SCIENCE IS A LIBERAL SCIENCE

In no small measure, political science has concerned itself with the study of liberal political institutions. Questions having to do with the formation of liberal states dominate the subfield of comparative politics,[15] spurred on recently by the demise of the Soviet Union, democratic transitions in Latin America, and regional and ethnic conflicts in the Balkans, Africa, and Russia. Relatively stable, mature democracies have provided ample opportunities for political scientists to study liberal political institutions such as political parties, election campaigns, electoral systems, legislatures, political psychology, and public opinion more generally.[16]

For the most part, studies of liberal democracies have adopted liberal postulates. Methodological individualism is almost unquestionably accepted, despite vigorous challenges from "systems analysis,"[17] structuralism (in its various Marxist and functionalist forms), and network analysis: for most students of politics, ***individuals*** are the causal actors of interest.[18] Accordingly, the aims of political science have a decidedly liberal cast: without appearing flip, one could summarize large swaths of political science as the study of preferences (the study of public opinion, broadly conceived) and the interaction and aggregation of people and their preferences via political institutions (e.g., trade unions, political parties, campaigns and elections, electoral systems, legislatures, constitutions).

This political science is "liberal" because a priori substantive assumptions about citizens' preferences are kept to a minimum. We presume to know little about the details of what citizens want; these preferences are themselves a matter for speculation and discovery by students of public opinion. Just as classical liberals ascribed self-preservation to their subjects, so, too, most contemporary social scientists uncontroversially imbue citizens of mature, liberal, democracies with preferences for peace and prosperity, *ceteris paribus.*[19] But beyond these weak assumptions, substantive rationality is a matter of some controversy in contemporary political science. Assumptions of nontrivial substantive rationality are typically confined to analyses of relatively small, specialized, well-defined groups of political actors,[20] and most political scientists are less troubled by the assumption of instrumental rationality.[21]

Of course, social scientists are well aware of the circumstances in which rationality fails,[22] and it is not my goal here to survey social science's responses to these shortcomings. Instead, I want to focus on the liberal character of the study of public opinion—citizen's preferences—and some recent attacks on it.

## THE LIBERAL STUDY OF PUBLIC OPINION

Central to the liberal political science envisaged by Hobbes, Locke, and Mill is the notion that while making few a priori assumptions about citizens' substantive preferences, it is, nonetheless, possible (and, indeed, highly desirable) to obtain knowledge of those interests. As a political practice, liberalism demands that these interests and opinions be made known. Learning about citizens' interests and opinions motivates the design of institutions that permit and encourage the vigorous interplay of these interests and opinions. A good deal of the substance and method of what is now called political science was anticipated by Hobbes in the following passage from *Leviathan*:

> The Skill of making, and maintaining Common-wealths, consisteth in certain Rules, as doth Arithmetique and Geometry; not (as Tennis-play) on Practise onely; which Rules, neither poor men have the leisure, nor men that have had the leisure, have hitherto had the curiosity, or the method to find out."[23]

The challenge posed by Hobbes was taken up enthusiastically by one of his students, William Petty, who invented the phrase "political arithmetick" in describing his enumerations of population and land usage in late seventeenth century England. Quantification became central to the development of a science of public administration; in turn, the development of "statistics" was in large measure driven by the demands of liberal states presiding over the Industrial Revolution for information regarding "geography, economics, agriculture, trade, population, and culture."[24] The word itself—statistics—has a controversial etymology, "although most writers of the early nineteenth century agreed that it was intrinsically a science concerned with states, or at least with those matters that ought to be known to the 'statist.'"[25] Interestingly, in the term's first English usage, in Sir John Sinclair's twenty-one–volume *Statistical Account of Scotland* (1791–1799), a distinction was made between the German *Statistik* (more directly "statist," dealing with "political strength" and "matters of state") and the idea of statistics as a means of ascertaining "the 'quantum of happiness' enjoyed by the inhabitants of a country, 'and the means of its future improvement.' "[26] The connections between emergent scientific techniques and the prerequisites of a utilitarian liberalism seem stark here. Notes Porter:

> Implicitly, at least, statistics tended to equalize subjects. It makes no sense to count people if their common personhood is not seen as somehow more significant than their differences. The Old Regime saw not autonomous persons, but members of estates. They possessed not individual rights, but a maze of privileges, given by history, identified with nature, and inherited through birth. The social world was too intricately differentiated for a mere census to tell much about what really mattered.[27]

In time, though, the application of statistics to matters of government drifted away from the ostensibly liberal concern with measuring citizens' individual happiness and more towards the study of aggregates. Prominent in this shift was Adolphe Quetelet, a Belgian mathematician and astronomer, who came to be highly influential in European scientific communities in the first half of the nineteenth century. Quetelet was impressed by the application of Laplace's work on probability to problems in astronomy and geodesy. Quetelet struggled to apply results from probability and mathematics to statistics at a time when the latter was regarded with some suspicion as the "province of physicians and amateur reformers."[28] Quetelet is remembered more for his efforts than his successes in this regard,[29] and for his farsightedness: it was the British statisticians of the second half of the nineteenth century who successfully merged probability with the budding social science of statistics.[30] Quetelet's powerful command of metaphor, more than any substantive insight, attests to his influence.[31] In addition to attempting to cast the empirical regularities he observed in demographic data in terms of natural or mathematical laws, Quetelet introduced the abstraction of the "average man" (*l'homme moyen*), a concept so influential it continues to powerfully shape ideas about public opinion in a liberal polity.

For Quetelet, the average man was in large part a rhetorical device for conveying what he was learning from his application of elementary probability models to demographic data. His audience found the description of their society in terms of its more moderate members to resonate with liberal, egalitarian ideals. The average citizen is made up of little bits of each of us, and so we can trust what she says. Her preferences reflect our preferences, and so her biases and prejudices seem less outrageous. Due to the "miracle of aggregation," her preferences and biases tend to lie well toward the center of the more radical or reactionary voices in the polity. Likewise, random sampling ensures that all citizens have an equal chance of having their opinion count in the determination of "average" opinion. In the wake of the political upheaval through Europe in the late eighteenth and early nineteenth century, the moderate quality of Quetelet's average man was welcomed, as it has been in liberal polities ever since. Given the liberal insistence on freedom of expression, the average citizen's opinions served as a calming and even familiar voice amid a sometimes confused din.

The power of this argument is not lost on politicians, lobby groups, or scholars. Claiming public opinion as an ally is a common tactic in a liberal, democratic polity, and politicians typically tread lightly when they perceive themselves to be at odds with public opinion. And it is no exaggeration to claim that modern, democratic politics is in large measure about the competition for the support of Quetelet's average man, or, as it is known to contemporary political scientists, the "median voter."[32]

## CRITICISMS OF SURVEY RESEARCH

With so much at stake politically, it comes as no great surprise that some truly stupendous violations of scientific method have been committed under the auspices of "the study of public opinion." Nonrandom sampling and provocative or naive question wordings in surveys typically underlie the more egregious cases; the attendant biases have come to be well understood and cataloged by public opinion professionals of various backgrounds (statisticians, sociologists, political scientists, psychologists, geographers and demographers, and, more recently, journalists, educationalists, and even historians). The highly partisan "straw polls" that were a prominent feature of United States politics in the late nineteenth century and early twentieth century are today dismissed as historical curiosities, though no doubt they once had important symbolic meaning.[33] The idea of a random sample is familiar to the mass public, and public reactions to surveys have become increasingly sophisticated; for instance, people are more likely to question the methodology of a public opinion poll if the poll's findings are inconsistent with their prior beliefs,[34] as a well-trained scientist might carefully scrutinize the methodology underlying a seemingly novel finding. Published poll results are increasingly accompanied by information regarding sampling techniques, sample size, response rate, and margins of errors on sample proportions.

Criticisms of survey research as "unscientific" are fairly easily accommodated by students of public opinion: in almost all cases the remedy is to do "better science," be it drawing "more random" samples or being more attuned to the effects of question wording, question sequence, interview context, and so on. And in no small measure, this methodological progress is *the* story of the study of public opinion in the twentieth century. Progress towards greater rigor and professionalization can be clearly discerned over time in journals like *Public Opinion Quarterly* (dating from 1937, and later to become the flagship journal of the American Association for Public Opinion Research) and journals from disciplines like political science that rely extensively on survey research.[35] By the late 1940s roughly a quarter of the published articles in one survey of sociology journals employed survey data; the corresponding proportion in the late 1970s was around 56%. In political science, the percentages are 2.6% in the late 1940s and 35% in the late 1970s.[36] These increases have been accompanied by a growing awareness of the shortcomings of survey-based research: secondary users of surveys are typically well aware of the problems posed by nonrandom selection and nonresponse,[37] and various "method" effects stemming from question format and question sequence.[38]

The lesson of this rushed intellectual history is that contests, discussions, and investigations of public opinion were about an abstraction and how to describe it, and not an attempt to unmask an enduring truth. It is widely acknowledged that the success of Quetelet's metaphorical average man is due to its political utility. But scientific arguments about the interests, preferences and even the prejudices of this average man are no less possible because he is an abstraction. As I show below, one of the more well-known criticisms of public opinion is ultimately answered with recourse to science.

### *The "Old-Left" Critique*

As the study of public opinion became more professionalized and ostensibly scientific, a counterargument from the Left also emerged. However, an important feature of this criticism is that it is not antipositivist or based on any strong epistemological quarrel with liberal science. Rather, it is more concerned with the ignored or unrealized political and ideological consequences of mass opinion research. For these critics there is more to politics than that which is open to a liberal field of vision, let alone a liberal wearing statistical eyeglasses. This line of criticism has its origins in Marxism, which, like liberalism, asserts that political opinions have their origins in material interests. But, of course, the point of departure for the Marxist is the assertion that class power blinds citizens to their objective interests; false consciousness helps account for the repeated finding that working class people are more supportive of capitalist values than one would expect on the basis of the interests that analysts might objectively ascribe to them.

In this view, which we may as well dub an old Left perspective, the study of public opinion is part of a dehumanizing capitalist apparatus, stripping away the (possibly radical) subjectivity of a worker to essentialize her as a consumer, a housewife, or a voter, and employing statistics to tabulate the analyst's colorless abstraction, the views of the average citizen. Views of this type we typically associate with Frankfurt School radicals such as Marcuse[39] and critics of modernity such as Arendt.[40] For instance, in *The Human Condition* Arendt sees the uses of quantitative techniques in the social sciences stemming from the rise of "the modern science of economics" as "the social science par excellence":

> The application of the law of large numbers and long periods to politics or history signifies nothing more than the wilful obliteration of their very subject matter, and it is a hopeless enterprise to search for meaning in politics or significance in history when everything that is not everyday behavior or automatic trends has been ruled out as immaterial. . . .
>
> Statistical uniformity is by no means a harmless scientific ideal; it is the no longer secret political ideal of a society which, entirely submerged in the routine of everyday living, is at peace with the scientific outlook inherent in its very existence.[41]

Likewise for Ginsberg, who sees public opinion polling as submerging "individuals with strongly held views in a more apathetic mass public,"[42] thereby "transforming public opinion from a spontaneous assertion to a constrained response"[43] and providing the state with a vehicle for monitoring dissident opinion without having to formally acknowledge or negotiate with "putative spokespersons"[44] of groups in civil society; "in essence, polling intervenes between opinion and its organized or collective expression."[45] And perhaps more sinisterly,

> Polling erodes one of the major competitive advantages that has traditionally been available to lower-class groups and parties—a knowledge of mass public opinion superior to that of their middle- and upper-class opponents.[46]

While highly critical of social science's use of survey research, this "old Left" critique nonetheless remains sympathetic to positivism. The critics just cited all believe that people do have opinions and preferences and engage in observable, political acts, and that there does exist something called public opinion. The more compelling problem for these critics is that survey research obscures or even destroys the capacity for social science to study radical opinions or potentially transformative political acts. National random probability sampling abstracts the citizen from social contexts that powerfully shape political dispositions and political behavior such as family, neighborhoods, race, social class—and, most tellingly, from other citizens. To the extent that survey research informs political debate and policy making, the net effect is to delegitimate group identities in political consciousness—a "feedback" effect, of sorts. Since the social group is the (theoretical) locus of political transformation, this feedback effect of public opinion polling in turn stifles social change itself.

Note that nothing in this line of criticism depends crucially on the claim that public opinion research is epistemologically suspect. Rather, the point is that public opinion researchers are naive or ignorant as to the political and ideological consequences of their scholarly "interventions" into the polity. Put slightly differently, old Left critics in effect urge public opinion researchers to be better scientists. Unsurprisingly, social scientists have been able to respond to these charges with what might be called methodological innovation. By amending and enhancing the techniques and practices of survey research, public opinion researchers can respond to the criticism identified by the old Left. There is no reason why surveys cannot be used to measure radicalism or political deviance; today, most self-conscious survey researchers are well aware of the limitations of full, national probability samples for studying attitudes or behavior more localized in scope. Small-area studies, target-group oversamples, participant-observer research, focus groups, controlled experiments, and panel studies are just some of the different ways empirical social science has gone about studying not just the regularities of political and social life, but also radicalism or deviance. Notable examples include an international survey-based study of social class and class consciousness in the 1980s;[47] participant-observer research of worker organization and resistance in industrial settings in Africa and Eastern Europe;[48] a unique blend of political biography and participant-observer research in analyzing members of Congress' relations with their constituents;[49] survey-based studies of black public opinion and political behavior using over-sampling techniques, or sampling solely from the target population;[50] and experiments on the effects of television on political attitudes.[51]

### *Public Opinion as Power-Knowledge*

However, the old Left critiques have more recently been buttressed by an antipositivism. This line of argument attacks not just public opinion research, but cuts more deeply to the epistemological foundations of positivist social science. Foucault is the generally accepted starting point for the argument that knowledge *is* power.[52] To see the two concepts as distinct is an error:

> We should admit rather that power produces knowledge . . .; that power and knowledge directly imply one another; that there is no power relation without the correlative constitution of a field of knowledge, nor any knowledge that does not presuppose and constitute at the same time power relations.[53]

This position amounts to nothing less than the collapsing of the fact-value distinction. And the prescription for social-scientific practice that typically results is *not* to do "better science," but to give up on the endeavor altogether. Public opinion research, in particular, does nothing but reproduce the biases of pollsters. Opinions do not exist independently of the apparatus designed to measure them—i.e., survey questions and the professional or ideological interests of pollsters. Claims Pierre Bourdieu:

> What needs to be questioned is the very notion of "personal opinion". . . . The idea of "personal opinion" perhaps owes part of its self-evident character to the fact that . . . it expressed from the very beginning the interests of the intellectuals, small, self-employed opinion producers whose role developed parallel to the constitution of a specialized field of knowledge and a market for cultural products, then of a sub-field specializing in the production of political opinions (with the press, the parties and all the representative bodies).
>
> The act of producing a response to a questionnaire on politics, like voting, . . . is a particular case of a supply meeting a demand.[54]

### *The Incoherence of American Public Opinion*

Somewhat ironically, evidence to support for this radical critique position is readily drawn from the findings of public opinion researchers themselves. Since Walter Lippman, students of public opinion have observed a vast disjuncture between public opinion among "political elites" and mass opinion: it is sometimes claimed that the latter is so diffuse that "measured" public opinion is really nothing but a reflection of elite opinion.[55] The first decades of scholarship employing large-scale social surveys "made sport of civics-book accounts" of American public opinion. Far from being a nation of Aristotles or "omnicompetent citizens"[56] 1950s and 1960s behavioral social science found

> . . . that the citizen was nonideological to the point of intellectual disorganization, uninformed about government policies and the parties' positions vis-à-vis those policies, oblivious to congressional politics, and unable to comprehend, if not hostile to, basic democratic freedoms. . . . Again and again the electorate fell short of academically promulgated standards that supposedly were preconditions of responsible government.[57]

In the words of a famous 1964 study of ideology in the American electorate, mass public opinion seems free of ideological "constraint." Citizens appeared to have little idea about "what goes with what," which is to say that a large section of the American public seemed to possess policy preferences that did not cohere in an ideologically consistent fashion.[58] When respondents were interviewed in a series of polls (a panel study), their opinions displayed over-time instability that seems consistent with "guessing," giving rise to the label "nonattitudes."[59]

Foucauldian "power-knowledge" and the bleak findings of 1950s behavioralism combine to pose a formidable challenge to social scientific investigations of mass politics. Not only do social scientific investigations of mass opinion merely echo the categories and priorities of the knowledge elite, but for the most part this is all that social scientists can *recognize* in the opinions of the mass public. Public opinion—the cornerstone of the liberal ideal of free-ranging, reasoned, deliberation—can hardly be said to exist in any meaningful or independent fashion; opinions exist only to the extent that pollsters create them. If so, then no serious student of democratic politics could take survey-based investigations seriously.

### *Public Opinion and Postmodernity*

If public opinion was in this kind of deep trouble a generation or so ago, then one might reasonably suspect that postmodernists have public opinion on its last legs today. And so they do. Enhancing an already damning line of argument against public opinion is the postmodernist understanding of the media (television, in particular) as a source of what might be loosely called "social meaning." To the extent that people rely on television for politically relevant information, then it would seem that public opinion reflects in large measure the daily hum and clutter of the news, as critics such as McLuhan and Baudrillard suggested. These writers argue that the distinction between reality and the media's representation of reality has "imploded." The social world is that which appears on television, or, to play on McLuhan's famous (1964) phrase, "the medium is the message."[60] Representation has become indistinguishable from reality and sign indistinguishable from signifier, so much so that "we no longer know what effects the media have on the masses and how the masses process the media."[61]

For Baudrillard, public opinion is part of the media "cyberblitz" itself and is a key part of the "implosion" postmodernists dwell on. Since reports about public opinion figure so prominently in the media, it is this aggregate-level abstraction—a postmodern version of Quetelet's average man—that is politically and ideologically efficacious, more so than the underlying individual opinions themselves (which have ceased to exist in any meaningful way). Public opinion—the public's own opinions, tallied, aggregated, and sanitized—are reflected back to the public via the media; we are no longer able to ascribe causal primacy to either opinion or its representation in surveys. Put differently, there is no longer any "specificity" to public opinion. Under these conditions the study of mass politics amounts to nothing but an empty statistical sorting of mush from slush, an attempt to pass back through a "black hole" to somehow undo the implosion of public opinion and its representation. As Baudrillard puts it, ". . . it is now impossible to isolate the process of the real, or to prove the real."[62]

As with most nostalgic utterances, Baudrillard's depends on a fantastic recreation of the past. We need not accept Baudrillard's implication that "the real" is a prerequisite for science, or, indeed, that it *ever* existed. Liberal science for the most part does not turn on a notion of "reality." Liberals like Mill and important precursors like Hobbes described themselves as defining the conditions for an orderly and political progressive interpretation of social conditions, rather than unmasking prior truths. To be sure, though some modern students of politics may have lost sight of this fact, liberals find contingency at the root of scientific method.

In this way, Quetelet's convenient abstraction—the average man—was no less an abstraction for being convenient. Liberal thinkers and citizens could applaud, utilize, and even fight over this abstraction without ever insisting that the average man was real. The real issue concerned what would be attributed to this admitted abstraction. Similarly, contemporary political discussions of public opinion that unabashedly traffic in abstractions are no less scientific for it. To further my case that liberalism, science, and relativism are consistent with one another, I turn to an example from contemporary public opinion.

## A POLITICAL ECONOMY OF POLITICAL INFORMATION

At the same time as postmodernism was gaining momentum (largely outside of the social sciences), students of public opinion were engaged in a thorough reevaluation of the bleak reports of 1950s and 1960s survey research (surveyed above in the section, THE INCOHERENCE OF AMERICAN PUBLIC OPINION). The idea that "political information" is costly slowly made its way into political science from economics, with a key bridging work in Downs's *An Economic Theory of Democracy*. Many students of public opinion now acknowledge that under certain conditions, political ignorance is rational.[63]

The argument here is very simple: "information" about political issues, candidates' positions, and even seemingly elementary facts about American political institutions is relatively costly to acquire for most Americans, relative to the benefits that information will bring.[64] It may surprise some members of a so-called knowledge elite to learn that most citizens are not avid consumers of political information. For most people, politics and political knowledge is not highly sought after in its own right, but is more likely a by-product of their other activities. Notable exceptions aside, decisions made in Washington seldom impinge on their lives so directly, frequently, or with consequences so dire as to make investment in detailed political information worthwhile. "Rational ignorance" comes as less of a shock once a mildly sophisticated view of public opinion and information acquisition is adopted.

In this view, partisanship,[65] political ideology, and even emotional or affective responses to political symbols (including candidates) are understood as "heuristics," or economizing devices that provide "information shortcuts" for citizens evaluating sometimes complicated issues.[66] In short, a detailed, well-organized, ideologically coherent "map" of the parties' platforms is not necessary in order to make sense of American politics. Voters do not carry such maps around in their heads, nor, by any reasonable understanding of where their interests might lie, should they. If anything, we should ask why it is that citizens know as much as they do about politics.

The implication of this line of research is easily summarized. Contemporary research shows that citizens can reason about politics even when they are presumed to know very little. The successful conduct of liberal politics does not depend on the successful apprehension of "reality."

### *Economics and Elections*

Information about economic conditions is a special case of "political information." Unlike political information per se, information about one's own economic situation is relatively costless; we cannot help but learn something about economic circumstances in the course of everyday living. Information about the macroeconomy is relatively plentiful, too; a casual glance at a headline or the television news will often be enough to gain a quick impression of the health of the national economy, beyond what one learns from one's social network (family, co-workers, friends, neighbors, etc.). For instance, while in 1985 only 4% of Americans could name the Japanese prime minister, and less than one-third knew what form of government Japan has, 87% knew that the United States runs a trade deficit with Japan, and only 6% thought the United States ran a surplus with Japan.[67] Likewise, guesses about important economic indicators are generally not too far out, though political ignorance remains high: about fifty percent of the population guessed the unemployment rate within a percentage point or two in 1980; but in 1985, after George Schultz had served four years as secretary of state, only 25% of the population could recall his name; and in 1987, after seven years of public debate about Contra aid, only about one-third of the American public knew that Nicaragua is in Central America.[68]

Accordingly, there are good reasons to believe that economic information is of immense political consequence; and, at least since Marx, social scientists have believed that politics is about conflict over material conditions. Indeed, the relationship between economic growth and electoral outcomes is a pillar of social science; incumbents are generally well rewarded for good economic performance (and the promise of more), while they are fairly harshly punished for economic downturns.[69]

My own research[70] shows that the aggregate American public—itself an abstraction—is a reasonable forecaster of economic conditions. I find a small systematic bias towards optimism in the aggregate public's economic forecasting over a variety of economic indicators. But in general it appears that the "real" and the "perceived" economies do not stray far from one another. While there is clearly some "social construction of [macro-] economic reality" taking place, it seems to be a fairly benign construction. I trace out the political implications of the "equilibrium relationship" between the real and perceived economies by examining the consequences of each for presidential approval. Out-of-equilibrium economic forecasts typically result in oversanguine evaluations of the president, but these dissipate relatively quickly over time as new economic information becomes available. I consider several plausible sources of out-of-equilibrium economic beliefs, including so-called rally-round-the-flag events (when presidents' approval typically soars in response to a foreign policy triumph or disaster, the most notable recent example being the Gulf War) and media reports of the economy. In general, I am impressed by the over-time robustness of the relationships between "political" variables like presidential approval, the perceived economy, and objective economic indicators.

In summary, there is some coherence to the American political economy, and some justification for studying it. First, a politically naive and disinterested public is not inconsistent with a reasoning, liberal polity, and indeed may even be demanded by a postulate of rationality. But because economic information is relatively cheap to acquire, the normative consequences of a rationally ignorant public are not as dire as one might first presume. To the extent that economic information is used by voters and is reasonably uncontaminated by the buzz and clutter of the media (or at least, not contaminated for long), there appears to be a firm link between objective reality and political outcomes.

Contrary to postmodern ruminations on the fluidity of political interests and identities, many Americans' political concerns are *not* easily manufactured by an all-powerful knowledge elite. There *are* measurable, objective realities confronted by Americans every day that continue to powerfully shape political life. And these are worthy of our continued attention.

## IN DEFENSE OF LIBERALISM AND SCIENCE

The history of the study of public opinion I have sketched here reveals two recommendations that I see may be generalized beyond the context of the social sciences. First, for the most part, critics of science have *not* been epistemologically motivated, and so a mutually acceptable response to left-leaning critics of positivist social science has been to do "better" social science. Self-consciousness, introspection, and methodological improvements—all good scientific practices—have helped bring the tools of positivist social science to bear on a wider array of research agendas than first thought possible, and have helped overcome the ideological qualms of positivism's critics. The lesson here is that science's best response to its critics may well be to do "better" science, be it generalizing theorems to broader classes of cases, testing additional, plausible rival hypotheses, or providing even clearer statements of assumptions, procedures, findings, and so on. Returning to my liberal theme, this recommendation amounts to the advice that if your opponent is a liberal, then you can be sure that reason and discussion are at your disposal, and that there is a very good chance at least one of you will emerge wiser from the encounter.

This advice would be fine *if* all our critics shared a commitment to the liberal values of reason and dialogue. My second point is to reemphasize the distinctiveness of the postmodern attack on science. In this case it is not at all clear that our opponent shares our liberal values—doing "better science" will not satisfy the postmodernist; "science" itself is the problem for some of our more trenchant critics. What is the appropriate response under these circumstances? What does a liberal do when faced with an opponent apparently hateful of her precisely *because* of her commitment to liberal values? Consider when a liberal's opponent sees "capitalist oppression" instead of "democracy" and "progress," or a dehumanizing bureaucratization instead of scientific rigor? What is the liberal response?

Even liberals radically committed to free speech recognize that at times we would be forced to repress, silence, or expel "noxious creatures" in our midst. Mill's commitment to a progression towards truth via vigorous, free, public debate came at a price: acting on beliefs that one held. In *On Liberty*, among the most celebrated defenses of free speech ever penned, Mill notes circumstances in which free speech must give way to "sure" opinions:

> It is the duty of governments, and of individuals, to form the truest opinions they can; to form them carefully, and never impose them upon others unless they are quite sure of being right. But when they are sure . . . it is not conscientiousness but cowardice to shrink from acting on their opinions, and allow doctrines which they honestly think dangerous to the welfare of mankind, . . . to be scattered abroad without restraint. . . . We may, and must, assume our opinion to be true for the guidance of our own conduct: it is assuming no more when we forbid bad men to pervert society by the propagation of opinions which we regard false and pernicious.[71]

I am not yet persuaded that the response to illiberal critics of science is to repress them as one might the "bad men" Mill described. We might see some of them as "perverting" society with "false and pernicious" opinions, but, at least as far as the social sciences are concerned, I do not see postmodern critiques amounting to, say, yelling fire in a crowded theater. I can well imagine things being different in the medical sciences, environmental sciences, and so on.

A better strategy for now, I think, is to engage the postmodern critic, and perhaps "bring him around" to the liberal view. As I argued earlier (LIBERAL-

ISM, LIKE SCIENCE, IS PRAGMATIC AND RELATIVIST), liberalism is more usefully thought of as a methodology than as an end—as procedural rather than normatively prescriptive. That "normal" social scientists (modernists) and postmodernists can read one another's work and talk to each other suggests that there are at least some shared values and assumptions upon which one might base a liberal conversation, although wading through Baudrillard's later writings does approach the upper limit of my methodological pluralism.[72]

One strategy I find appealing is to question what it is we really learn about society from postmodern critiques of social science. The postmodern implosion of social reality into social representation is, like many bad social theories, *too strong*. By being a theory about everything, it becomes a theory about nothing. How does an implosion of reality and representation help us answer questions about racial inequality, the causes of war, why the Equal Rights Amendment failed, why there is no universal health care in the United States, the prospects for peace in the Middle East, the political economy of sub-Saharan Africa, European integration, to name just a few? Questions of importance even to science's critics are not as well served by the hard-line anti-positivism implied by the Baudrillardian version of postmodernism. For instance, how are we to answer questions like "whose science?" and "whose knowledge?"[73] in a world where meaning itself is lost? Those of us with transformative political projects in mind (as have many a social scientist) will find little to grasp hold of here. The complaint of a well-known feminist is relevant in this regard:

> Post-Lacan, actually post-Foucault, it has become customary to affirm that sexuality is socially constructed. Seldom is specified what, socially, it is constructed of, far less who does the constructing or how, when, or where. . . . Power is everywhere therefore nowhere, diffuse rather than pervasively hegemonic. "Constructed" seems to mean influenced by, directed, channeled, as a highway constructs traffic patterns. Not: Why cars? Who's driving? Where's everybody going? What makes mobility matter? Who can own a car? Are all these accidents not very accidental?[74]

But to concede that reality is socially constructed is not to throw the (positivist) baby out with the bath water. If anything, this position should be an impetus towards social science, not an excuse for critique and endless introspection.[75] "Whose social construction?" and "what are the implications?" are questions that ought to immediately follow the oft-heard but empty assertion that a particular topic of analysis is "socially constructed." Likewise, statements such as "the meaning of *x* is *contingent*" amount to nothing more than saying "*x* warrants study." The vacuous assertion that something is "contingent" begs (if not screams) the question "contingent on *what*?"

Even if we accept that politics is a great contest for "meaning" (say, if we employ Foucault's notion of power-knowledge), then surely it matters who the winners and losers turn out to be in particular settings. And to the extent that it is not always easy to observe who are the victors in this contest, then a powerful set of observational and analytical tools might be required—a fortiori if one happens to be interested in predicting future contests or explaining past contests. To the extent that these tools tend not to bias our investigations in one way or another, or could be applied to different contexts (say, if generalization or theory building happened to be our goal), then we might label these tools *science.*

## CONCLUSION

I began this essay by linking both science and liberalism as complementary products of the Enlightenment. I remarked that both science and liberalism are relativist and pragmatic—liberals and scientists make few presumptions about absolute truth. The arguments within social science that I survey here show that liberalism is actually fairly accommodating of "conditionality" or "contingency," concepts dear to postmodernists, and, as it turns out, concepts that are actually at the heart of science as well. The admission of a contextualized basis for knowledge is not an abandonment of science, but rather an acknowledgment of the richness of the world that is, if anything, an invitation to inquiry. This admission was the mutual origin of both science and liberalism, is the source of their resilience, and will ensure their safe passage through the postmodernist "storm."

## NOTES

1 My colleague Lynn Sanders helped me organize my thoughts on the topics pursued here, in addition to pointing me toward numerous useful references. Students in my undergraduate class "Liberalism and Its Critics" also bore the brunt of some of my thinking out loud. Errors and omissions remain my own responsibility.

2 This pilgrimage was made while Hobbes was resident in Paris and a member of the circle of scientists centered around the mathematician Marin Mersenne; Mersenne's correspondents included Fermat, Descartes, Pascal, and Galileo. See C. B. MacPherson's "Introduction" to ***Leviathan***.

3 For Hobbes, religion is tantamount to superstition, with its roots in a "Feare of things invisible"; in turn, these fears can be exploited, formed into "Lawes" by some people, so that they may "govern others, and make unto themselves the greatest use of their Powers" (***Leviathan***, p. 168).

4 For instance, see Paul R. Gross & Norman Levitt, ***Higher Superstition: The Academic Left and Its Quarrels with Science***, pp. 16–23. In particular, "[B]y the time of the French Revolution . . . [t]he empiricism and rigor of the sciences were emulated in the analytic strategies of political thought. It is of course possible, and tempting, to speculate whether a similar system of scientific discourse might have arisen in an entirely different social context. . . . Or could science have matured only upon a substrate of subtly congenial social ideals and institutions, like those found in seventeenth-century Europe?" (pp. 19–20).

5 I have in mind here Marx's historical materialism, and the appeal to science so prominent in the structuralist Marxism of Althusser and his followers.

6 "Liberty, as a principle, has no application to any state of things anterior to the time when mankind have become capable of being improved by free and equal discussion" (John Stuart Mill, ***On Liberty and Other Essays***, p. 15).

7 Ibid.

8 Ibid., pp. 24–25.

9 Ibid., p. 24.

10 Ibid., p. 22.

11 Ibid., p. 25.

12 Ibid.

13 Ibid.

14 Ibid.

15 E.g., Robert D. Putnam, ***Making Democracy Work: Civic Traditions in Modern***

*Italy*; and Gabriel A. Almond & Sidney Verba, *The Civic Culture: Political Attitudes and Democracy in Five Nations.*

16 E.g., G. Bingham Powell, *Comparative Democracies.*

17 E.g., Fernando Cortes, Adam Przeworski & John Sprague, *Systems Analysis for Social Scientists.*

18 For instance, Elster asserts: "The elementary unit of social life is the individual human action. To explain social institutions and social change is to show how they arise as the result of the action and interaction of individuals. This view, often referred to as methodological individualism, is in my view trivially true" (*Nuts and Bolts for the Social Sciences*, p. 13). Green and Shapiro briefly survey the triumph of methodological individualism over turn-of-the-century "group theorists" in studies of American politics (*Pathologies of Rational Choice: A Critique of Applications in Political Science*). And the "analytical Marxist" movement of the 1980s brought a methodological individualist framework to questions about class and political economy. See, for instance, Jon Elster, *Making Sense of Marx*; Adam Przeworski, *Capitalism and Social Democracy*; Adam Przeworski & John Sprague, *Paper Stones: A History of Electoral Socialism*; and John E. Roemer, ed., *Analytical Marxism* and *Free to Loose.*

19 This point was put somewhat bluntly not by a political scientist, but by an economist, George Stigler: "Prosperity is even more uncontroversial than motherhood . . ." ("General Economic Conditions and National Elections," p. 166). Contrast Holmes's argument that contemporary political science and economics overstate the extent to which classical liberalism relied on postulates of a self-interested, substantive rationality (*Passions and Constraint: On the Liberal Theory of Democracy*, ch. 2). Hobbes himself noted that "the Passions of men, are commonly more potent than their Reason" (*Leviathan*, p. 241).

20 E.g., vote-maximizing legislators (see, e.g., Keith Krehbiel, *Information and Legislative Organization*), lobbyists, special interest groups, or political entrepreneurs maximizing revenues from membership or rents from the public purse (Mancur Olson, *The Logic of Collective Action*; Lawrence S. Rothenberg, *Linking Citizens to Government: Interest Group Politics at Common Cause*; David Austen-Smith and John R. Wright, "Counteractive Lobbying"), budget-maximizing bureaucrats (William A. Niskanen, *Bureaucracy and Representative Government*), or a group of actors, each representing the state as a whole, as is common in studies of international conflict, maximizing net gains from war or threats of aggression (see, e.g., James D. Fearon, "Domestic Political Audiences and the Escalation of International Disputes"; Bruce Bueno de Mesquita & David Lalman, *War and Reason*). Downs's assumption that voters maximize a "stream of benefits" in choosing between political parties is a reasonably rare extension of assumptions of a substantive rationality to the mass electorate (*An Economic Theory of Democracy*).

21 For instance, in an important review of political science applications of rational choice, Lalman, Oppenheimer, and Swistak see rationality in instrumental terms: "Rationality, in itself, has nothing to say about whether the desires, or preferences, of an individual are benevolent or evil. Theories of rational action do not explain where preferences come from" ("Formal Rational Choice Theory: A Cumulative Science of Politics"). On the distinction between substantive and instrumental rationality, and criticisms of rational choice applications in political science, see Donald P. Green & Ian Shapiro, *Pathologies of Rational Choice: A Critique of Applications in Political Science*, pp. 17–18, though even these authors agree that "thin [instrumental] rationality" keeps "controversial assumptions about human goals and motivation to a minimum."

22 Elster (*Sour Grapes: Studies in the Subversion of Rationality* and *Nuts and Bolts for the Social Sciences*) summarizes some common and notable "subversions" of rationality. See also the essays in Karen Schweers Cook & Margaret Levi, eds., *The Limits of Rationality.*

23 Hobbes, *Leviathan*, p. 261.

24 Theodore Porter, *The Rise of Statistical Thinking*, p. 25.

25 Ibid., pp. 23–24.
26 Sir John Sinclair, quoted in Porter, *The Rise of Statistical Thinking*, p. 24.
27 Ibid., p. 25.
28 Ibid., p. 43.
29 Quetelet's attempt to forge a "social physics" was notable more for its promise than its insights. In his most famous work, *Sur l'homme et le développement de ses facultés, ou essai de physique sociale* (1835), Quetelet attempted to infer law-like propositions about society that would stand with astronomy's achievements from the previous century (Stephen M. Stigler, *The History of Statistics: The Measurement of Uncertainty before 1900*, p. 170) and aimed to be the "Newton of statistics" (Porter, *The Rise of Statistical Thinking*, p. 46). The phrase *physique sociale* itself is actually Comte's (Porter, p. 41).
30 Stephen M. Stigler, *The History of Statistics: The Measurement of Uncertainty before 1900*, ch. 8).
31 Quetelet had a strong secondary interest in literature while pursuing his doctorate in mathematics, and prior to 1823 he had written "a libretto to an opera, a historical survey of romance, and much poetry (Stigler, *History of Statistics*, p. 162).

In addition, another of Quetelet's important contributions was a professionalization of statistics. Stigler (*The History of Statistics: The Measurement of Uncertainty before 1900*, p. 162). claims Quetelet was "active in the funding of more statistical organizations than any other individual in the nineteenth century." These include the (Royal) Statistical Society of London, the Statistical Section of the British Association, the International Statistical Congresses, to name only those outside Quetelet's native Belgium.
32 Anthony Downs, *An Economic Theory of Democracy*; James Enelow & Melvin J. Hinich, *The Spatial Theory of Voting: An Introduction*; Peter Coughlin, *Probabilistic Voting Theory.*
33 Susan Herbst, *Numbered Voices: How Opinion Polling Has Shaped American Politics*, ch. 4.
34 Daniel M. Merkle, "The Impact of Prior Belief and Disclosure of Methods on Perceptions of Poll Data Quality and Methodological Discounting."
35 E.g., the *American Political Science Review* (American Political Science Association), the *American Journal of Political Science* (Midwestern Political Science Association), and the *Journal of Politics* (Southern Political Science Association). Converse (*Survey Research in the United States: Roots and Emergence 1890–1960*) provides a detailed history of the professionalization of survey research in the United States, starting from pioneering efforts in the interwar years, through the formation of AAPOR in 1947 and the development of the large academic survey research centers (SRC at Michigan and NORC at Chicago).
36 Stanley Presser, "The Use of Survey Data in Basic Research in the Social Sciences"; John Brehm, *The Phantom Respondents*, p. 15.
37 Brehm, *The Phantom Respondents.*
38 E.g., Robert B. Eubank & David John Gow, "The Pro-Incumbent Bias in the 1978 and 1980 National Election Studies"; more generally, see Howard Schuman & Stanley Presser, *Questions and Answers in Attitude Surveys: Experiments on Question Form, Wording, and Context.*
39 Herbert Marcuse, *One-Dimensional Man.*
40 This line of criticism relies in no small measure on Weber's themes of "rationalization," "bureaucratization," and "modernization." Herbst (*Numbered Voices: How Opinion Polling Has Shaped American Politics*, pp. 12–20) outlines the influence of Weber on Frankfurt School understandings of the rise of statistical thinking in Western thought.
41 Hannah Arendt, *The Human Condition*, p. 42–43.
42 Benjamin Ginsberg, *The Captive Public: How Mass Opinion Promotes State Power*, p. 65.
43 Ibid., p. 63.
44 Ibid., p. 73.

45 Ibid., p. 74.
46 Ibid., p. 76.
47 E.g., Erik Olin Wright, *Classes.*
48 Michael Burawoy, *The Politics of Production: Factory Regimes under Capitalism and Socialism.*
49 Richard Fenno, *Homestyle: House Members in Their Districts.*
50 Michael Dawson, *Behind the Mule: Race and Class in African-American Politics*; Patricia Gurin, Shirley Hatchett & James S. Jackson, *Hope and Independence: Blacks' Response to Electoral and Party Politics.*
51 E.g., Shanto Iyengar & Donald R. Kinder, *News That Matters.*
52 Though the Weberian roots of this position are easy to discern; e.g., Susan Herbst, *Numbered Voices: How Opinion Polling Has Shaped American Politics*, pp. 20–24.
53 Michel Foucault, *Discipline and Punish: The Birth of the Prison*, p. 27.
54 Pierre Bourdieu, *Distinction: A Social Critique of the Judgement of Taste*, p. 399.
55 Susan Herbst, *Numbered Voices: How Opinion Polling Has Shaped American Politics*, p. 46; Pierre Bourdieu, "Public Opinion Does Not Exist."
56 Walter Lippman, *Public Opinion.*
57 Morris P. Fiorina, *Retrospective Voting in American National Elections*, p. 3.
58 Philip E. Converse, "The Nature of Belief Systems in Mass Publics," p. 213.
59 Philip E. Converse, "Attitudes and Non-attitudes: Continuation of a Dialogue."
60 Marshall McLuhan, *Understanding Media: The Extensions of Man.*
61 Douglas Kellner, *Jean Baudrillard: From Marxism to Postmodernism and Beyond*, p. 69.
62 Jean Baudrillard, *Simulacra and Simulation*, p. 21.
63 In addition to Downs's seminal work, see also Gary S. Becker, "Competition and Democracy"; Samuel L. Popkin, John W. Gorman, Charles Phillips & Jeffrey A. Smith, "Comment: What Have You Done for Me Lately? Toward an Investment Theory of Voting"; Richard Nordhaus, "Alternative Approaches to the Political Business Cycle," p. 39; and the review in Morris P. Fiorina, "Information and Rationality in Elections."
64 "For example, the health of the national economy may in fact have a greater effect on voters than whether their next vacation is fabulous or merely good; but time spent deciding where to travel leads to better vacations, whereas time spent evaluating economic policies leads not to better policies but only to a better-informed vote" (Samuel L. Popkin, *The Reasoning Voter*, p. 22).
65 Morris P. Fiorina, *Retrospective Voting in American National Elections.*
66 Samuel L. Popkin, *The Reasoning Voter.*
67 Ibid., p. 27.
68 Ibid., p. 35.
69 E.g., Edward R. Tufte, *Political Control of the Economy*; Helmut Norpoth, Michael Lewis-Beck & Jean-Dominique Lafay, eds., *Economics and Politics: The Calculus of Support*; Michael S. Lewis-Beck & Tom Rice, *Forecasting Elections.* Cf. Jay P. Greene, "Forewarned before Forecast: Presidential Election Forecasting Models and the 1992 Election."
70 Simon Jackman, "Perception and Reality in the American Political Economy."
71 John Stuart Mill, *On Liberty*, pp. 23–24.
72 E.g., "There is no more hope for meaning. And without a doubt this is a good thing: meaning is mortal. But that on which it has imposed its ephemeral reign, what it hoped to liquidate in order to impose the reign of the Enlightenment, that is, appearances, they, are immortal, invulnerable to the nihilism of meaning or of non-meaning itself. This is where seduction begins" (Baudrillard, *Simulacra and Simulation*, p. 164).
73 Susan Harding, *Whose Science? Whose Knowledge? Thinking from Women's Lives.*
74 Catherine A. MacKinnon, *Towards a Feminist Theory of the State*, p. 131. See also Richard Rorty, "Feminism, Ideology, and Deconstruction: A Pragmatist View."

75 Social scientists do their best work when they stop worrying about epistemology; philosophers do this better than we can anyway. And, indeed, social scientists may only be able to *start* work once they exercise their erstwhile epistemological consciences.

## REFERENCES

ALMOND, GABRIEL A. & SIDNEY VERBA. *The Civic Culture: Political Attitudes and Democracy in Five Nations.* Princeton, NJ: Princeton University Press, 1963.

ARENDT, HANNAH. *The Human Condition.* Chicago, IL: University of Chicago Press, 1957.

AUSTEN-SMITH, DAVID & JOHN R. WRIGHT. "Counteractive Lobbying." *American Journal of Political Science* 38 (1994): 25–44.

BAUDRILLARD, JEAN. *Simulacra and Simulation.* Ann Arbor, MI: University of Michigan Press, 1994.

BECKER, GARY S. "Competition and Democracy." *Journal of Law and Economics* 1 (1958): 105–109.

BOURDIEU, PIERRE. *Distinction: A Social Critique of the Judgement of Taste*, Cambridge, MA: Harvard University Press, 1984.

———. "Public Opinion Does Not Exist." In *Communication and Class Struggle*, edited by A. Mattelart and S. Siegelaub. New York, NY: International General, 1979.

BREHM, JOHN. *The Phantom Respondents.* Ann Arbor, MI: University of Michigan Press, 1993.

BUENO DE MESQUITA, BRUCE & DAVID LALMAN. *War and Reason.* New Haven, CT: Yale University Press, 1992.

BURAWOY, MICHAEL. *The Politics of Production: Factory Regimes under Capitalism and Socialism.* London: Verso, 1985.

CONVERSE, JEAN M. *Survey Research in the United States: Roots and Emergence 1890–1960.* Berkeley, CA: University of California Press, 1987.

CONVERSE, PHILIP E. "Attitudes and Non-attitudes: Continuation of a Dialogue." In *The Quantitative Analysis of Social Problems*, edited by Edward R. Tufte. Reading, MA: Addison-Wesley, 1970.

———. "The Nature of Belief Systems in Mass Publics." In *Ideology and Discontent*, edited by David E. Apter. New York, NY: Free Press, 1964.

COOK, KAREN SCHWEERS & MARGARET LEVI, eds. *The Limits of Rationality.* Chicago, IL: University of Chicago Press, 1990.

CORTES, FERNANDO, ADAM PRZEWORSKI & JOHN SPRAGUE. *Systems Analysis for Social Scientists.* New York, NY: Wiley, 1974.

COUGHLIN, PETER. *Probabilistic Voting Theory.* Cambridge: Cambridge University Press, 1992.

DAWSON, MICHAEL. *Behind the Mule: Race and Class in African-American Politics.* Princeton, NJ: Princeton University Press, 1994.

DOWNS, ANTHONY. *An Economic Theory of Democracy.* New York, NY: Harper and Row, 1957.

ELSTER, JON. *Making Sense of Marx.* Cambridge: Cambridge University Press, 1985.

———. *Nuts and Bolts for the Social Sciences.* Cambridge: Cambridge University Press, 1989.

———. *Sour Grapes: Studies in the Subversion of Rationality.* Cambridge: Cambridge University Press, 1983.

ENELOW, JAMES & MELVIN J. HINICH. *The Spatial Theory of Voting: An Introduction.* Cambridge: Cambridge University Press, 1984.

EUBANK, ROBERT B. & DAVID JOHN GOW. "The Pro-Incumbent Bias in the 1978 and 1980 National Election Studies." *American Journal of Political Science* 27 (1983): 122–139.

FEARON, JAMES D. "Domestic Political Audiences and the Escalation of International Disputes." *American Political Science Review* 88 (1994): 577–592.

FENNO, RICHARD. *Homestyle: House Members in Their Districts.* Boston, MA: Little, Brown, 1978.

FIORINA, MORRIS P. "Information and Rationality in Elections." In *Information and Democratic Processes*, edited by John A. Ferejohn & James H. Kuklinski. Urbana, IL: University of Illinois Press, 1990.

———. *Retrospective Voting in American National Elections.* New Haven, CT: Yale University Press, 1981.

FOUCAULT, MICHEL. *Discipline and Punish: the Birth of the Prison.* New York, NY: Vintage, 1979.

GINSBERG, BENJAMIN. *The Captive Public: How Mass Opinion Promotes State Power*, New York, NY: Basic Books, 1986.

GREEN, DONALD P. & IAN SHAPIRO. *Pathologies of Rational Choice: A Critique of Applications in Political Science.* New Haven, CT: Yale University Press, 1994.

GREENE, JAY P. "Forewarned before Forecast: Presidential Election Forecasting Models and the 1992 Election." *PS: Political Science and Politics* 26, 1 (1993): 17–21.

GROSS, PAUL R. & NORMAN LEVITT. *Higher Superstition: The Academic Left and Its Quarrels with Science.* Baltimore, MD: Johns Hopkins University Press, 1994.

GURIN, PATRICIA, SHIRLEY HATCHETT & JAMES S. JACKSON. *Hope and Independence: Blacks' Response to Electoral and Party Politics.* New York, NY: Russell Sage Foundation, 1989.

HARDING, SUSAN. *Whose Science? Whose Knowledge? Thinking from Women's Lives.* Ithaca, NY: Cornell University Press, 1991.

HERBST, SUSAN. *Numbered Voices: How Opinion Polling Has Shaped American Politics.* Chicago, IL: University of Chicago Press, 1993.

HOBBES, THOMAS. *Leviathan.* Introduction by C. B. MacPherson. Harmondsworth, Middlesex: Penguin, 1968.

HOLMES, STEPHEN. *Passions and Constraint: On the Liberal Theory of Democracy.* Chicago, IL: University of Chicago Press, 1995.

ISYENGAR, SHANTO & DONALD R. KINDER. *News That Matters.* Chicago, IL: University of Chicago Press, 1987.

JACKMAN, SIMON. "Perception and Reality in the American Political Economy." Ph.D. diss., University of Rochester, 1995.

KELLNER, DOUGLAS. *Jean Baudrillard: From Marxism to Postmodernism and Beyond.* Stanford, CA: Stanford University Press, 1989.

KREHBIEL, KEITH. *Information and Legislative Organization.* Ann Arbor, MI: University of Michigan Press, 1991.

LALMAN, DAVID, JOE OPPENHEIMER & PIOTR SWISTAK. "Formal Rational Choice Theory: A Cumulative Science of Politics." In *Political Science: The State of Discipline II*, edited by Ada W. Finifter. Washington, DC: American Political Science Association, 1993.

LEWIS-BECK, MICHAEL S. & TOM RICE. *Forecasting Elections.* Washington, DC: Congressional Quarterly Press, 1992.

LIPPMAN, WALTER. *Public Opinion.* New York, NY: Macmillan, 1922.

MACKINNON, CATHERINE A. *Towards a Feminist Theory of the State.* Cambridge, MA: Harvard University Press, 1989.

MARCUSE, HERBERT. *One-Dimensional Man.* Boston, MA: Beacon, 1964.

MCLUHAN, MARSHALL. *Understanding Media: The Extensions of Man.* New York, NY: McGraw-Hill, 1964.

MERKLE, DANIEL M. "The Impact of Prior Belief and Disclosure of Methods on Perceptions of Poll Data Quality and Methodological Discounting." Paper presented at the annual meeting of the Association of Education in Journalism and Mass Communication, Boston, MA, 1991.

MILL, JOHN STUART. *On Liberty and Other Essays.* Oxford: Oxford University Press, 1991.

NISKANEN, WILLIAM A. *Bureaucracy and Representative Government.* New York, NY: Aldine and Atherton: 1971.

NORDHAUS, RICHARD. "Alternative Approaches to the Political Business Cycle." *Brookings Papers on Economic Activity* 2 (1989): 1–68.

NORPOTH, HELMUT, MICHAEL LEWIS-BECK & JEAN-DOMINIQUE LAFAY, eds. *Economics and Politics: The Calculus of Support.* Ann Arbor, MI: University of Michigan Press, 1991.

OLSON, MANCUR. *The Logic of Collective Action.* Cambridge, MA: Harvard University Press, 1965.

POPKIN, SAMUEL L., JOHN W. GORMAN, CHARLES PHILLIPS & JEFFREY A. SMITH. "Comment: What Have You Done for Me Lately? Toward an Investment Theory of Voting." *American Political Science Review* 60 (1976): 640–654.

———. *The Reasoning Voter.* 2d edit. Chicago, IL: University of Chicago Press, 1994.

PORTER, THEODORE. *The Rise of Statistical Thinking.* Princeton, NJ: Princeton University Press, 1986.

POWELL, G. BINGHAM. *Comparative Democracies.* Cambridge, MA: Harvard University Press, 1982.

PRESSER, STANLEY. "The Use of Survey Data in Basic Research in the Social Sciences." In *Surveying Subjective Phenomena*, edited by Charles F. Turner and Elizabeth Martin. New York, NY: Russell Sage Foundation, 1984.

PRZEWORSKI, ADAM. *Capitalism and Social Democracy.* New York, NY: Cambridge University Press, 1985.

PRZEWORSKI, ADAM & JOHN SPRAGUE. *Paper Stones: A History of Electoral Socialism.* Chicago, IL: University of Chicago Press, 1986.

PUTNAM, ROBERT D. *Making Democracy Work: Civic Traditions in Modern Italy.* Princeton, NJ: Princeton University Press, 1993.

ROEMER, JOHN E., ed. *Analytical Marxism.* New York, NY: Cambridge University Press, 1986.

ROEMER, JOHN E. *Free to Loose.* Cambridge, MA: Harvard University Press, 1988.

RORTY, RICHARD. "Feminism, Ideology, and Deconstruction: A Pragmatist View." In *Mapping Ideology*, edited by Slavoj Zizek. London: Verso, 1994.

ROTHENBERG, LAWRENCE S. *Linking Citizens to Government: Interest Group Politics at Common Cause.* New York, NY: Cambridge University Press, 1992.

SCHUMAN, HOWARD & STANLEY PRESSER. *Questions and Answers in Attitude Surveys: Experiments on Question Form, Wording, and Context.* New York, NY: Academic Press, 1981.

STIGLER, GEORGE J. "General Economic Conditions and National Elections." *American Economic Review* 63 (1973): 160–167.

STIGLER, STEPHEN M. *The History of Statistics: The Measurement of Uncertainty before 1900.* Cambridge, MA: Harvard University Press, 1986.

TUFTE, EDWARD R. *Political Control of the Economy.* Princeton, NJ: Princeton University Press, 1978.

WRIGHT, ERIK OLIN. *Classes.* London: Verso, 1985.

# PATHOLOGICAL SOCIAL SCIENCE
## Carol Gilligan and the Incredible Shrinking Girl

CHRISTINA HOFF SOMMERS

AN ARTICLE in the November 1994 issue of the *Atlantic Monthly* exposed as highly questionable the widely credited proposition that electrical power lines are significantly linked to cancer.[1] The author, Gary Taubes, a contributing correspondent for *Science*, puts the findings on electromagnetic fields in the context of the history of science, which is "littered with examples of what Irving Langmuir, the 1932 Nobel prize winner in chemistry, called 'pathological science' or the 'science of things that aren't so.'" Pathological science occurs when researchers report the existence of an illusory effect. A recent celebrated example is cold fusion. According to Langmuir:

> These are cases where there is no dishonesty involved but where people are tricked into false results . . . being led astray by subjective effects, wishful thinking, or threshold interactions.[2]

Pathological science, as Langmuir noted, never issues in a fruitful research program. No new discoveries are made; pathological science either dies out or else just goes "on and on."

Today, "pathological *journalism*" must be added to the mix. It was not until *New Yorker* writer Paul Brodeur published an alarming three-part article in 1989[3] that people took electromagnetic fields to be a serious danger. Since then trepidation has been high and pervasive. The American Physical Society recently issued a statement informing the public that it could find no evidence linking electromagnetic fields to cancer, adding that "more serious environmental problems are neglected for lack of funding and public attention."[4] According to a July 1992 article in *Science* by H. Keith Florig, "the cost to society of public anxiety about EMF now exceeds $1 billion annually."[5]

I shall devote most of my paper to commenting on one important and influential kind of pathological social science. Within the past decade there has been a profusion of research on women's issues. Much of it concerns "women's ways of knowing." Related to this is the research that came out with the sensational finding that the nation's adolescent girls are suffering a severe loss of self-esteem. Self-esteem research is a paradigm of what can happen when pathological science is nourished by pathological journalism.

The scare over the allegedly diminishing self-esteem of American schoolgirls was created in large part by the media (in particular the *New York Times*) which promoted and gave credence to the views of Carol Gilligan. In 1983 Carol Gilligan, a professor at the Harvard University Graduate School of Education, wrote a controversial book called *In A Different Voice* in which she claimed that women have special ways of dealing with moral dilemmas; she argued that, being more caring, less competitive, less abstract and morally more sensitive than men, women speak "in a different voice." According to Gilligan, women's culture of nurturing and caring and their habits of peaceful accommodation could be the salvation of a world governed by hypercompetitive males and their habits of abstract moral reasoning.

Though her views tap into quite a few old stereotypes about men and women, her empirical theories in moral psychology and adolescent development suffer for want of evidence. Independent research tends to *disconfirm* Gilligan's thesis that there is a substantive difference in the moral psychology of men and women. Lawrence Walker of the University of British Columbia has reviewed 108 studies on gender difference in solving moral dilemmas. He concludes, "sex differences in moral reasoning in late adolescence and youth are rare."[6] William Damon (Brown University) and Anne Colby (Radcliffe College) point out that "There is very little support in the psychological literature for the notion that girls are more aware of others' feelings or are more altruistic than boys. Sex differences in empathy are inconsistently found and are generally very small when they are reported."[7]

Even other feminist research psychologists have taken to dismissing Gilligan's findings. Faye Crosby, a psychologist at Smith College, finds Gilligan's anecdotal approach methodologically defective:

> Gilligan referred throughout her book to the information obtained in her studies, but did not present any tabulations. Indeed she never quantified anything. The reader never learns anything about 136 of the 144 people from [one of her three studies], as only 8 are quoted in the book. One probably does not have to be a trained researcher to worry about this tactic.[8]

Gilligan herself seems untouched by any of the criticism; in any case she shows little sign of tempering her theories or changing her anecdotal style of research. In recent years, Gilligan's research is no longer comparative but concentrates solely on girls about whom she makes a second, equally sensational, claim: that our "male-voiced culture" discourages and "silences" girls. According to Professor Gilligan, young girls are spontaneous, forthright and truthful, only to be betrayed in adolescence by an acculturation that diminishes their spirit and induces in them a kind of "self-silencing." In her 1990 book *Making Connections* she claims to have found that eleven-year-old girls possess a natural sagacity about social reality and, if asked, they will openly share their insights with you. They do not tell you what they think *you* want to hear. "Eleven-year-olds are not for sale," says Gilligan.[9]

By ages twelve or thirteen, things begin to change for the worse. According to Ms. Gilligan, the girls come up against the "wall of Western Culture," and they learn that the knowledge they possess—in particular their insights into human relations—is subversive in our patriarchal society. To protect themselves, they begin to hide their knowledge from others. Many bury it so deep inside themselves that they lose touch with it. Says Gilligan:

> Interviewing girls in adolescence . . . I felt at times that I was entering an underground world, that I was led in by girls to caverns of knowledge, which then suddenly were covered over, as if nothing was known and nothing was happening.[10]

In January 1990, Gilligan was featured in the cover story of the *New York Times Magazine.* The author, feminist novelist Francine Prose, cites these oft-quoted words of Gilligan's:

> By 15 or 16 . . . [girls] start saying, I don't know, I don't know, I don't know. They start not knowing what they had known.[11]

Prose acknowledges that Gilligan's research "has provoked intense hostility on the part of academics," and has generated "cult-like veneration" among her followers. But she herself finds Gilligan's discoveries evocative and revolutionary.

> [Gilligan's] observation may cause many women to feel an almost eerie shiver of recognition, and inspire them to rethink that period in their lives. . . . *Making Connections* may well oblige traditional psychology to formulate a more accurate theory of female adolescence.[12]

I belong with the academics. I do not get an "eerie shiver of recognition" when I read Gilligan. On the contrary, I find her descriptions of adolescent girls overly sentimental and widely off the mark. What exactly are these little girls supposed to "know"? As an example of their startling clarity and openness, Gilligan tells of a twelve-year-old girl who was asked to complete this sentence: "What gets me in trouble is . . ." The girl wrote "chewing gum and not tucking my shirt in" and she added, parenthetically, "but it's usually worth it." This parenthetical remark electrified Ms. Gilligan and her colleagues for its candor and boldness. They cited it as typical of the "outrageously wonderful statements" preteen girls can make—before they go "underground" at adolescence. This "telling" anecdote duly made it into Ms. Prose's reverential *New York Times Magazine* article.

Since Gilligan no longer studies boys, we are left with a number of questions. Do preteen boys also have hidden "caverns of knowledge"? Do they also make "outrageously wonderful statements"? Or does Gilligan believe that, unlike girls, eleven-year-old boys are "for sale"? We are given to understand that preteen girls are cognitively *special.* If boys do *not* "know" what the girls "know," then Gilligan has, indeed, made a discovery that would revolutionize the field of developmental psychology. But where is the evidence? To establish such a claim we should need to embark on a massive, carefully designed study of many thousands of American boys and girls. Gilligan's methods cannot begin to meet the case.

I know I am belaboring the obvious, but when the obvious is being ignored or rejected, we are forced into belaboring it. By confining her attention to girls, Gilligan fails to offer the slightest shred of evidence for the thesis that eleven-year-old boys do not have the same natural wisdom and forthrightness that she finds in girls. She does not even offer anecdotal evidence.

Though Gilligan would appear to deny it, we have to consider that preteen boys may be as astute as preteen girls. If so, there are two possibilities to consider. Perhaps adolescent boys are also "silenced" and "forced underground." But if that were the case, then Gilligan's sensational claim that girls are at special risk appears to be false. Alternatively, it may be that it is only girls who "hit the wall of Western Culture" and are forced to "sell out." In that case, it is only the girls who become inarticulate and conformist little Stepford teens. Adolescent boys remain sage and honest interpreters of social reality. This does not seem right; and surely Gilligan would reject any alternative that "valorizes" boys.

A third and more plausible possibility is to abandon Gilligan's perspective altogether and say that well-balanced girls and boys do not differ significantly in respect to astuteness and candor and that both pass from childhood to adolescence, becoming less narcissistic, more reflective, and less sure about their grasp of the complex world that is opening up to them. Leaving junior high school they emerge from the "know-it-all stage" to a more mature stage in which they begin to appreciate that there is a vast amount they do not know. If so, it is not true that "girls start not knowing what they had known" but that older children of both sexes go through a period of realizing that what they thought they knew may not be true at all—and there is a lot out there to learn.

In a more recent book, *Meeting at the Crossroads*, Gilligan and her coauthor Lyn Mikel Brown repeat points made in *Making Connections* with even greater urgency. They claim that "women's psychological development within patriarchal societies and "male-voiced" cultures is inherently traumatic."[13] Speaking in characteristically sweeping terms, they say that American girls undergo various kinds of "psychological foot-binding" caused by being constantly told that "People . . . did not want to hear what girls know."[14] They lose their candor and begin to cultivate a superficial "niceness and kindness."

It should be noted that Gilligan and Brown are fully aware that other scholars will challenge them to produce evidence for their claims. To deflect this challenge, they take the bold tack of saying that the demand for conventional evidence is misplaced. In the first chapter of *Meeting at the Crossroads*, Gilligan and Brown point out that they are studying "girls' responses to a dominant culture that is out of tune with girls' voices and for the most part uninterested in girls' experiences."[15] To adhere to "proper" methods of inquiry and proof would be to fall into the masculinist trap of using "a method of psychological inquiry appropriated from this very system." Their object is to free women from the strictures imposed by what some feminist theorists call "patriarchal constructions of knowledge." Why should they betray their purpose by applying the methods of the "dominant culture"? So, right at the outset, Gilligan and Brown commit what the radical feminist Mary Daly (approvingly) calls "methodicide." They announce that they are deliberately abjuring masculinist methods of inquiry in favor of a more relational, dialogical approach to their subjects. As if in justification for abandoning all pretense to providing us with anything like standard scientific proofs for their large claims about girls and boys, they quote the late poet Audre Lorde's admonition: "The master's tools will never dismantle the master's house."[16]

But Ms. Lorde's much-quoted remark is no more than clever. Even if one's research project is part of a larger antipatriarchal project that aims at "dismantling the master's house," why *not* take advantage of tools that were used to build it? Gilligan and Brown's "feminist" justification for deserting scientific method in establishing their claims does not withstand critical scrutiny. They offer no proof of their claims. Their excuse that the demand for proof is "masculinist" comes down to saying, "We don't *have* to play by the rules; the boys wrote them."

Reading Gilligan and Brown, one finds oneself asking: is anything they are doing and claiming science? Is it even respectable as informal social commentary? The late Christopher Lasch was considerably less impressed by them than was the *New York Times*, which honored *Meeting At the Crossroads* as a "*New York Times* Notable Book of the Year." Lasch, in a review for the *New Republic*, points out that Gilligan and Brown's idealized view of female children as noble, spontaneous, and naturally virtuous beings who are progressively spoiled and demoralized by a corrupting socialization has its roots in Jean-Jacques Rousseau's theory of education.[17] Rousseau, however, sentimentalized boys as well as girls. Lasch argues that both Rousseau and Gilligan are wrong. In particular, real girls do not change from a Rousseauian ideal of natural virtue to something more muted, conformist, "kind and nice." On the contrary, when researchers look at junior high school girls without preconceptions, they are often struck by a glaring *absence* of "niceness and kindness." Of Gilligan and her associates, Lasch says:

> They would have done better to remind themselves, on the strength of their own evidence, that women are just as likely as men to misuse power, to relish cruelty, and to indulge the taste for cruelty in enforcing conformity.[18]

I think Lasch has it right. Things may have changed, but when I think back to the girls in my Southern California junior high school "niceness and kindness" are not the first words that spring to mind. I recall rigid popularity hierarchies, reminiscent of a severely stratified medieval monarchy. There were two reigning princesses, and there was their court of privileged female consorts: then there are the masses, and the untouchables. What was the boy's role in all this? They floated around, trying to figure out their assigned place. It was the girls who were in charge.

But perhaps it *has* all changed and the Gilligan-Brown picture is now the accurate one. That is doubtful. There are indications that my junior high school days are being repeated in other schools today. Matawan Regional High School in Aberdeen, New Jersey recently offered students a workshop on sexual harassment. The "gender equity facilitator" asked the students, "In this school, who are the powerful people?" "The cheerleaders," was one answer. "Girls," another. The facilitator was clearly unprepared for these replies and tried to get the students to say it was the powerful male elites.[19] The kids did not buy it. Neither do I. That is the picture that Gilligan and her colleagues keep missing.

Gilligan's shortcomings as an empirical psychologist seem not to matter to the large number of journalists and feminist activists who continue to be thrilled by her findings. The feminist literary critic and mystery writer Carolyn Heilbrun reviewed *Meeting at the Crossroads* for the *New York Times.*[20] She shares with Francine Prose the negative attribute of having no special expertise in adolescent development or academic psychology. That is often the way the media reports on advocacy research. In particular, when it comes to dealing with controversial feminist research claims, the media, and in this case *The New York Times*, seem studiously to avoid writers who might pose critical challenges to the findings. Ms. Heilbrun, like Ms. Prose, does not stint in her praise of the importance of what Gilligan and her coworkers had found. She calls the book "Original and incisive . . . revolutionary." According to Heilbrun,

> They discovered the startling veracity of young female thought and expression, and the even more amazing submergence of that veracity as adolescence proceeded. The girls were now seen to have moved from "authentic into idealized relationships," sacrificing truth on the altar of niceness.[21]

Heilbrun does concede that Gilligan's research has been challenged by scholars, but she assures readers that Gilligan's contribution remains a "landmark in psychology."

That Gilligan's work marks a landmark in the history of psychology is doubtful in the extreme. It does however represent a kind of landmark in the history of advocacy research. By popularizing Gilligan's views and promoting her as a matron saint of a new psychology of women, the media placed her at the very center of the flourishing self-esteem movement.

In particular Gilligan's claims caught the attention of the American Association of University Women (AAUW). In 1990, the AAUW leaders were "intrigued and alarmed" by Gilligan's early findings that "pointed to the onset of adolescence as a crisis in girls' lives."[22] The "AAUW wanted to know more" and commissioned a scientific study to prove that Gilligan was right. As the (then) AAUW president Sharon Schuster candidly told the *New York Times*, "We wanted to put some factual data behind our belief that girls are getting shortchanged in the classroom."[23]

The AAUW chose Carol Gilligan to help the Greenberg Lake polling firm develop a self-esteem questionnaire.[24] In effect, Gilligan cooperated in designing a study that was supposedly offering independent confirmation of her claims. Nancy Goldberger and Dr. Janie Victoria Ward are the two other researchers who helped develop the questionnaire and interpret the results.[25] Ward had written her dissertation under Gilligan at the Harvard Graduate School of Education. Goldberger, a psychologist at the Fielding Institute in Santa Barbara, is a coauthor of *Women's Way of Knowing*, the bible of feminist epistemology. Carol Gilligan is mentioned in the acknowledgement as someone who "inspired and informed" the authors. If anyone could be counted on to do a good job "putting some factual data" behind Ms. Schuster's belief that girls are being "shortchanged," it would be Ward, Goldberger, and Gilligan. As it turned out, the team proved themselves more than equal to the task.

In January of 1991 the *New York Times* announced the distressing results of the AAUW's study:

> Girls emerge from adolescence with a poor self-image, relatively low expectations from life and much less confident in themselves and their abilities than boys, a study to be made public today has concluded.[26]

The *Times* interviewed only one expert on adolescent psychology in its news article: Carol Gilligan. No effort was made to see if the AAUW findings were consistent with other studies. Alluding to Gilligan's research, the *Times* spoke of the AAUW Study as "confirming earlier studies that were smaller and more anecdotal."[27]

Still, we may ask, was Gilligan's diminished self-esteem hypothesis actually confirmed by the AAUW study? And the answer is: not at all. Soon after its self-esteem study began to make headlines in 1991, a little-known but respected magazine called *Science News* ran a story reporting the skepticism of leading researchers about the AAUW's findings.[28] William Damon, Director for the Center for the Study of Human Development at Brown University, had also looked carefully at the "purported finding . . . that girls in our society 'lose' their self-esteem when they reach adolescence." He reports his conclusion that "the scientific support . . . is as shaky as jello."[29] Joseph Adelson, editor of the *Handbook on Adolescent Psychology*, said this about the AAUW findings: "When I saw the report I thought, 'This is awful. I could prove it is awful, but it's not worth my time.'"[30]

Once I got my hands on the original survey and the full data report, I too could see why the experts were unimpressed. In fact, showing that the AAUW results are wrong is not as time consuming as Joseph Adelson imagined it to be.

The AAUW had used a self-report survey technique and had found that when you ask schoolboys to respond to the statement "I am happy the way I am" or "I am good at a lot of things" a large percentage will say "Always true," which is to say "I am always happy the way I am, always good at a lot of things."

Forty-six percent of high school boys checked "always true" to "I am happy the way I am" compared to only twenty-nine percent of girls.[31] In what the AAUW called "a crucial measure of self-esteem," the AAUW announced a seventeen-point self esteem gap favoring the boys. But what the AAUW never made clear and what the press reports left out is that the survey had offered five possible responses: "always true," "sort of true," "sometimes true sometimes false," "rarely true," and "never true." To the suggestion that they were happy as they are, girls favored the second and third responses, "sort of true" or "sometimes true sometimes false." If you count the first three answers as indicating normal or healthy levels of self-esteem, the gap between boys and girls drops from seventeen points to four points (eighty percent–eighty-four percent). But in its public announcements, the AAUW ignored all but the first response; in effect any girl who failed to check the "always true" option was classified as lacking in self-esteem.

Nor does the AAUW consider the possibility that forty-two percent of boys who checked "always true" to "good at a lot of things" or "happy the way I am" may be showing a lack of maturity or reflectiveness or a want of humility. It is not necessarily a mark of insecurity or low self-esteem to admit to feeling not prodigiously proficient some of the time. Is it the girls we should be worrying about here?

Also unpublicized was the very awkward finding that African-American boys, who are educationally most at risk, score ***highest*** of all of the AAUW's self-esteem indexes: seventy-eight percent of African-American boys said they were "always" "happy the way I am" compared with thirty-four percent of white girls. Black girls too are well ahead of white boys on the self-esteem scale. These results undermine either the link the AAUW claims between self-esteem and academic performance or the study's controversial methodology of measuring "self-esteem" by self-reports.

The AAUW explains that they did not highlight their findings on the African-American boys because the sample was too small to be statistically reliable. But surely it was not too small to have alerted them that something might have been seriously amiss in their methodology. Given the fact that the AAUW researchers had found unusually high levels of self-esteem in a statistically significant sample of African-American girls, they had to consider the strong possibility that the high scores of the black boys were not a fluke. They were duty-bound to investigate the hypothesis, so devastating to their entire project, that "self-esteem" as they were measuring it was not correlated with academic achievement.

Consider this major piece of evidence adduced by the AAUW to highlight the difference in boys' and girls' aspirations for success:

> Self-esteem is critically related to young people's dreams and successes. The higher self-esteem of young men translates into bigger career dreams. . . . The number of boys who aspire to glamorous occupations (rock star, sports star) is greater than that of young women at every stage of adolescence, creating a kind of "glamour gap."[32]

A glamour gap? Most kids do not have the talent and drive to be rock stars. The sensible ones know it. (The number one career aspiration of the girls, by the way, was lawyer.) What the responses of the children suggest, and what many experts on adolescent development will tell you, is that girls mature earlier than boys who at this age, apparently, suffer from a "reality gap."

The AAUW study did find areas where girls show similar or higher levels of self-confidence, but they do not mention these findings in their "shortchanged girl" brochures, newsletters, press reports, or documentary. In a section of the AAUW study called "classroom experience," we learn that eighty percent of high school girls and seventy-one percent of boys say the "teachers think girls are smarter." Among high school students, seventy-six percent of girls and seventy-four percent of boys say teachers prefer girls.

What of Gilligan's claim that girls' personalities become more and more hidden as they reach adolescence? That too is gainsaid by the AAUW's own results. When asked to respond to the sentence, "People don't know the real me," the AAUW survey found that only fourteen percent of the high school girls said "always true," compared to nineteen percent of the boys! What is more, the girls seemed to feel *better* understood, not less, as they grew older. (In elementary school, twenty-one percent of girls say "always true" to the statement that people do not really know them. By middle school it drops to seventeen percent, and in high school drops further to fourteen percent.)[33] The AAUW and Dr. Gilligan do not explain how these findings square with their thesis of neglected, hidden, "silenced," and progressively demoralized girls.

Though a surprising number of intelligent people now believe the myth of the nation's diminished shortchanged schoolgirls, the facts belie that girls are worse off than boys. The data on boys from the United States Department of Education's *Digest of Education Statistics* and *Condition of Education* indicate that far more boys than girls suffer from learning disabilities, delinquency, alcoholism, and drug abuse; five times as many boys as girls commit suicide. Girls get better grades; more girls graduate from high school and college. Even the frequently cited claim that girls score lower on standardized tests is misleading. In the 1992 National Assessment of Educational Progress Test of seventeen-year-olds, boys outperformed girls by four points in math and ten points in science but girls more strikingly outperformed boys by twelve points in reading and seventeen points in writing. Girls are catching up in math and science; boys continue to lag far behind in reading and writing.

When the announcement about the nation's diminished girls, suffering from the endemic gender bias of the nation's "patriarchal" school systems, was made, the feminist establishment embraced it. Popular journalists like Anna Quindlen, Judy Mann, Anita Diamant, and Peggy Orenstein were bowled over. We soon had a spate of books and articles carrying the bleak tidings about America's threatened female teens.

In *Failing at Fairness: How America's Schools Cheat Girls*, Myra and David Sadker predict the fate of a six-year-old girl on top of a playground slide:

> There she stood on her sturdy legs, with her head thrown back, and her arms flung wide . . . full of energy, self reliance and purpose. She feels confident about what she can do and who she can become. . . . If the camera had photographed the girl . . . at twelve instead of six . . . she would be looking at the ground instead of the sky; her sense of self-worth would have been an accelerating downward spiral.[34]

Writing in *Mirabella*, Jane O'Reilly, (a founder of *Ms.* magazine) contrasts two groups of girls playing at a lake shore, the younger "ardent, vital and confident," the older girls "mannered, anxious and doubtful."[35] O'Reilly can "practically see the older girls' self-esteem draining away with the rivulets of lake water running down their legs." Anita Diamant says that she went on "red alert searching for ways to protect my daughter's God-given sparkle and snap."[36]

A feminist holiday for children was launched. Anna Quindlen, Gloria Steinem, Naomi Wolf, Marlo Thomas, and Callie Khouri (the scriptwriter of *Thelma and Louise*) joined with the *Ms.* Foundation to inaugurate "Take Our Daughters to Work Day." Quindlen insisted that the holiday must be exclusively for girls because "a survey by the AAUW . . . showed that self-esteem, confidence and expectations of girls go south during adolescence."[37]

The AAUW intensive lobbying campaign quickly prompted Congress to attach a "Gender Equity Package" to the current Elementary and Secondary Education Act. The package provides millions of dollars for gender equity programs, workshops, and materials. Representative Patricia Schroeder expressed her appreciation for the AAUW research, "Today we know that little girls as young as 11 years old suffer from low levels of self-esteem."[38]

The Gender Equity Bill diverts resources badly needed for coping with the very real learning gap that separates American and foreign children. On one recent international math test, American boys lagged behind Korean girls by a gap more than fifteen times larger than the gap between our boys and girls.[39] Yet, American students express far more confidence in their math and science proficiency than their more proficient peers in Taiwan or Korea. Among advanced industrial nations, American children rank near the bottom—but they're "happy the way they are."[40]

---

Pathological science is expensive. The public will soon be spending countless millions to address a fake self-esteem crisis. In another field, Berkeley scientist Bruce Ames reminds us of the enormous costs to the public caused by the food toxin scares. And I have mentioned the millions being spent to combat the dubious dangers of electromagnetic rays. Few readers of this would deny that in all such cases the resources diverted by pathological science could be put to better use.

But pathological science is not just economically draining. Blatantly bad science is an integral part of the assault on truth, objectivity and plain good sense. Oxford biologist Richard Dawkins was recently asked his opinion about the fashionable New Age Gaia principle (that the earth is a living organism). He replied, "The idea [of Gaia] is not dangerous or distressing except to academic scientists who value the truth."[41] That is right, but it understates the harm.

In the recent past, scientists have let others fight their fight against the "methodicides." They wrongly believed that the struggle was internal to the humanities and that the sciences are largely unaffected. Even internally, responsible scientists did little or nothing to expose and discredit the practitioners and promulgators of pathological science. The pseudoscientists have been allowed to go their merry way, contributing to a climate of suspicion that gives all of science a bad name.

There is reason to hope that this is now changing. Scientists have finally begun to recognize science itself is at risk and that they really cannot afford to be neutral and aloof. This is the first volume of the New York Academy of Sciences devoted to the assault on reason, common sense, and traditional standards as it explicitly impacts on science and the teaching of science in our society. For the first time scientists and intellectuals who are concerned about the threat to science have gathered to take counsel on how to meet it. I for one am very proud to have a part in this endeavor.

## NOTES

1 Gary Taubes, "Fields of Fear."
2 Ibid., p. 107.
3 Paul Brodeur, "Annals of Radiation."
4 William Broad, "Cancer Fear Is Unfounded, Physicists Say."
5 H. Keith Florig, "Containing the Costs of the EMF Problem."
6 Lawrence Walker, "Sex Differences in the Development of Moral Reasoning: A Critical Review," p. 681.
7 William Damon & Anne Colby, "Listening to a Different Voice: A Review of Gilligan's *In a Different Voice*," p. 475.
8 Faye Crosby, *Juggling: The Unexpected Advantages of Balancing Career and Home for Women and Their Families*, p. 124.
9 Quoted by Francine Prose in "Confident at 11, Confused at 16," *New York Times Magazine*, January 7, 1990. p. 23.
10 Carol Gilligan, Nona Lyons & Trudy Hanmer, eds., *Making Connections: The Relational Worlds of Adolescent Girls at Emma Willard School*, p. 14.
11 Prose, "Confident at 11, Confused at 16," p. 23.
12 Ibid.
13 Lyn Mikel Brown & Carol Gilligan, *Meeting at the Crossroads*, p. 216.
14 Ibid., p. 218.
15 Ibid., p. 10.
16 Ibid.
17 Christopher Lasch, "Gilligan's Island."
18 Ibid., p. 38.
19 Reported by Cathy Young in "The Frontiers of Flirting," p. 109.
20 Carolyn Heilbrun, "How Girls Become Wimps."
21 Ibid., p. 13.
22 Elizabeth Debold, Marie Wilson & Idelisse Malave, *Mother Daughter Revolution: From Betrayal to Power*, p. 9.
23 Suzanne Dealy, "Little Girls Lose Their Self-Esteem on Way to Adolescence, Study Finds."
24 Ibid.
25 American Association of University Women, "Summary," *Shortchanging Girls, Shortchanging America*, p. 17.
26 Dealy, "Little Girls Lose Their Self-Esteem."
27 Ibid.
28 Bruce Bower, "Does Adolescence Herald the Twilight of Girls' Self-Esteem?"
29 William Damon, *Greater Expectations: Overcoming the Culture of Indulgence in America's Homes and Schools*, p. 75.
30 Quoted by Christina Hoff Sommers in *Who Stole Feminism? How Women Have Betrayed Women?*, p. 145.
31 American Association of University Women, "A Call to Action," *Shortchanging Girls, Shortchanging America*, p. 24.
32 American Association of University Women, "Summary," *Shortchanging Girls, Shortchanging America*, p. 8.
33 See the AAUW/Greenbert-Lake Full Data Report, question 35.
34 Myra Sadker & David Sadker, *Failing at Fairness: How America's Schools Cheat Girls*, pp. 77–78.
35 Jane O'Reilly, "Lost Girls," p. 118.
36 Anita Diamant, "For Our Daughters," p. 72.
37 Anna Quindlen, "Take Her with You."
38 Testimony of Rep. Patricia Schroeder on the Gender Equity in Education Act before the House Education and Labor Committee Subcommittee on Elementary, Secondary, and Vocational Education, April 21, 1993. The Gender Equity in Education Act was incorporated into the 1994 $12 billion Elementary and Secondary Education Act (ESEA). It establishes a "Special Assistant for Gender Equity" and provides federal funding for "equity policies, programs, activities and initiatives" in our

public schools. The American Association of University Women and other feminist lobbying groups have succeeded in getting girls categorized as an "historically underserved population." This means that, under the ESEA, many more millions than were available under the original Gender Equity in Education Act will now be allocated for "gender equity" programs. For example, Title I in ESEA (which has a $7.5 billion appropriation) makes provisions for the "incorporation of gender equitable methods and practices" in the K–12 schools.

39 International Assessment of Mathematics and Science (IAMP).

40 Harold W. Stevensen, "Children Deserve Better than Phony Self-Esteem," pp. 12–13. See also Harold W. Stevensen & James W. Stigler, *The Learning Gap: Why Our Schools Are Failing and What We Can Learn from Japanese and Chinese Education.*

41 Quoted by John Brockman in *The Third Culture: Beyond the Scientific Revolution*, p. 86.

## REFERENCES


AAUW/Greenburg-Lake Full Data Report. Washington, DC: American Association of University Women, 1991.

AMERICAN ASSOCIATION OF UNIVERSITY WOMEN. *Shortchanging Girls, Shortchanging America.* Washington, DC: American Association of University Women, 1991.

AMES, BRUCE *et al.* "Oxidants, Antioxidants, and the Degenerative Diseases of Aging." *Proceedings of the National Academy of Sciences* 90 (1993): 7915–7922.

BOWER, BRUCE. "Does Adolescence Herald the Twilight of Girls' Self-Esteem?" *Science News*, March 23, 1991, pp. 184–186.

BROAD, WILLIAM. "Cancer Fear is Unfounded, Physicists Say." *New York Times*, May 14, 1992, pp. 34–39.

BROCKMAN, JOHN. *The Third Culture: Beyond the Scientific Revolution.* New York, NY: Simon and Schuster, 1995.

BRODEUR, PAUL. "Annals of Radiation." *New Yorker*, June 12, 19, 26, 1989.

BRODY, JANE E. "Scientist at Work." *New York Times*, July 5, 1994.

BROWN, LYN MIKEL & CAROL GILLIGAN. *Meeting at the Crossroads.* New York, NY: Ballantine Books, 1992.

COLBY, ANNE & WILLIAM DAMON. "Listening to a Different Voice: A Review of Gilligan's *In a Different Voice.*" *Merrill-Palmer Quarterly* 29, 4 (October 1983): 475.

CROSBY, FAYE J. *Juggling: the Unexpected Advantages of Balancing Career and Home for Women and Their Families.* New York, NY: Free Press, 1991.

DAMON, WILLIAM. *Greater Expectations: Overcoming the Culture of Indulgence in America's Homes and Schools.* New York, NY: Free Press, 1995.

DEALY, SUZANNE. "Little Girls Lose Their Self-Esteem on Way to Adolescence, Study Finds." *New York Times*, January 9, 1991, p. B6.

DEBOLD, ELIZABETH, MARIE WILSON & IDELISSE MALAVE. *Mother Daughter Revolution: From Betrayal to Power.* Reading, MA: Addison-Wesley, 1993.

DIAMANT, ANITA. "For Our Daughters." *Parenting*, April 1994, p. 72.

EDUCATIONAL TESTING SERVICE. *International Assessment of Mathematics and Science* (IAEP). Princeton, NJ: Educational Testing Service, 1992.

FLORIG, H. KEITH. "Containing the Cost of the EMF Problem." *Science*, July 1992, pp. 468–469, 490, 492.

GILLIGAN, CAROL. *In a Different Voice: Psychological Theory and Women's Development.* Cambridge, MA: Harvard University Press, 1983.

GILLIGAN, CAROL, NONA LYONS & TRUDY HANMER, eds. *Making Connections: The Relational Worlds of Adolescent Girls at Emma Willard School.* Cambridge, MA: Harvard University Press, 1990.

GOLDBERGER, NANCY, MARY FIELD BELENKY *et al. Women's Way of Knowing: The Development of Self, Voice, and Mind.* New York, NY: Basic Books, 1986.

GROSS, PAUL R. & NORMAN LEVITT. *Higher Superstition: The Academic Left and Its Quarrel with Science.* Baltimore, MD: Johns Hopkins Press, 1994.

HEILBRUN, CAROLYN. "How Girls Become Wimps." *New York Times Book Review*, October 4, 1992, p. 13.
LASCH, CHRISTOPHER. "Gilligan's Island." *The New Republic*, December 7, 1992, pp. 34–39.
O'REILLY, JANE. "Lost Girls." *Mirabella*, April 1994, p. 118.
PROSE, FRANCINE. "Confident at 11, Confused at 16." *New York Times Magazine*, January 7, 1990, p. 23.
QUINDLEN, ANNA. "Take Her with You." *New York Times*, March 28, 1993, section 4, p. 15.
SADKER, MYRA & DAVID SADKER. *Failing at Fairness: How America's Schools Cheat Girls.* New York, NY: Charles Scribner's Sons, 1994.
SOMMERS, CHRISTINA HOFF. *Who Stole Feminism? How Women Have Betrayed Women?* New York, NY: Simon and Schuster, 1994.
STEVENSEN, HAROLD W. "Children Deserve Better than Phony Self-Esteem." *Education Digest* 58, 4 (December 1992): 12–13.
STEVENSEN, HAROLD W. & JAMES W. STIGLER, *The Learning Gap: Why Our Schools Are Failing and What We Can Learn from Japanese and Chinese Education.* New York, NY: Summit Books, 1992.
TAUBES, GARY. "Fields of Fear." *Atlantic Monthly*, November 1994, p. 107.
UNITED STATES DEPARTMENT OF EDUCATION. *Condition of Education.* Washington, DC: Office of Educational Research and Improvement, 1995.
———. *Digest of Education Statistics.* Washington, DC: Office of Educational Research and Improvement, 1995.
WALKER, LAWRENCE J. "Sex Differences in the Development of Moral Reasoning: A Critical Review." *Child Development* 55 (1984): 681.
YOUNG, CATHY. "The Frontiers of Flirting." *Men's Health*, October 1994, p. 109.

# FEMINISMS

At the end of the twentieth century, the American feminist movement no longer speaks with one voice. In the academy, a particularly influential version reigns, declaring that women not only speak "in a different voice" but encounter the world through a different "way of knowing." This view is now coming under challenge from other feminists, among them the contributors to this section, to whom this seems a thinly disguised regression to the belief that women have innate intellectual and social limitations.

Philosopher Janet Radcliffe Richards provides a close analysis of the fashionable practice of "feminist epistemology." Leaving aside the vexed question of whether its contradictions and obscurities are consistent with naming it a philosophical position at all, Radcliffe Richards makes the case that, within any reasonable definition, feminism as such provides little justification for epistemological radicalism, and that it faces practical dangers from the versions popular on many campuses.

Noretta Koertge, a philosopher of science with long experience within the women's movement, also finds that feminist epistemology and the spirit that endorses it present a threat to the intellectual ambitions of women scholars. The present popularity of one particular brand of feminist philosophy puts pressure on women philosophers to conform to the style, and to present themselves in the job market as adherents and expositors even if their interests—indeed, their convictions—are markedly different.

Meera Nanda, who works in a science studies program and has a strong background in biology, questions the assumption, now nearly ubiquitous in women's studies and in the radical wing of science studies, that the rationalist traditions of the Enlightenment are inherently oppressive to non-Western peoples, particularly to women of the Third World. She rallies to Ernest Gellner's view that Enlightenment liberalism is indispensible to the emancipation of women in traditional societies. Moreover, the scientific viewpoint, per se, is a key ingredient of the frame of mind needed to challenge the myriad forms of non-Western obscurantism that keep women in thrall all over the world.

Mathematician Mary Beth Ruskai, to whom "feminism" has always meant women's right to intellectual respect on the basis of full intellectual parity, finds that some of the more recent developments in feminist theory present a grave danger to this position. In particular, they demean and marginalize women who have succeeded, in sometimes-hostile environments, in the professions of science. Even worse, the "gender theorists" discourage or misdirect younger women with the talent and ambition to become scientists.

# WHY FEMINIST EPISTEMOLOGY ISN'T[1]

JANET RADCLIFFE RICHARDS

TWENTY YEARS AGO, when feminism was younger and greener, it would have been much easier to set about discussing the movement from the point of view of flights from reason and science. To start with, there would have been relatively little dispute that such a flight was going on. Many feminists themselves claimed to be taking wing from all such male devices for the oppression of women, as well as from morality, which was another of them. And, furthermore, the position was a relatively easy one for the skeptical outsider to attack. Unless feminists could say such things as that the present treatment of women was morally wrong, or prevailing ideas about their nature false or unfounded, or traditional reasoning about their position confused or fallacious, it was difficult to see on what basis they could rest the feminist case. And of course as they did say such things, all the time, it was obvious that any systematic attempt to reject ethics and rationality was systematically undercut by feminists' own arguments.[2]

In these more sophisticated times, however, the issue is much more complicated. The language, at least, has changed, and few feminists now can be heard to say that reason or science should be abandoned altogether. What they say instead is that particular, traditional accounts of these things must go, to be replaced by new, feminist conceptions of them; and this makes matters very different.

Here, for instance, is Elizabeth Grosz in an essay on feminist epistemology, describing the work of another feminist, Luce Irigaray[3]:

> Irigaray's work thus remains indifferent to such traditional values as "truth" and "falsity" (where these are conceived as correspondence between propositions and reality), Aristotelian logic (the logic of the syllogism), and accounts of reason based upon them. This does not mean her work could be described as 'irrational, 'illogical,' or 'false.' On the contrary, her work is quite logical, rational, and true in terms of quite *different criteria*, perspectives, and values than those dominant now. She both combats and constructs, strategically questioning phallocentric knowledges without trying to replace them with more inclusive or more neutral truths. Instead, she attempts to reveal a *politics* of truth, logic and reason.

Statements of this kind obviously make matters much more difficult for the critic who is concerned about flights from reason and science in general, and suspects contemporary feminism of being at the forefront of the rush. If feminists claim that all they are doing is offering new and improved conceptions of these things, it no longer seems possible to object in principle to the whole project. Investigating the foundations of science, epistemology, logic, ethics, and all the rest is a perfectly respectable philosophical activity, in which feminists seem as entitled to join as anyone else.

Presumably, therefore, any complaint about irrationality must be directed to the particular conceptions of science and epistemology put forward under the name of feminism. But here things become extraordinarily difficult, because anyone of even vaguely familiar epistemological views who has tried to tackle feminist works in these areas will know how quickly there comes the sensation of being adrift in uncharted seas, with no familiar landmarks in sight. There is nothing so simple as particular claims that, by more conventional standards, seem mistaken. In so many ways that they defy representative quotation, much of what is said seems already so laden with revisionary theory that the innocent reader is left with the uncomfortable feeling of having missed the story so far. What are "knowledges"?—which presumably, if the term is not just a perverse neologism, are intended as something other than just different things that are known. What is it for them to have, or for that matter fail to have, such astonishing qualities as phallocentricity? What is the significance of their being said to be "produced,"[4] rather than acquired, or by that production's being "intersected by gender,"[5] or by women's asserting "'a right to know' independent of and autonomous from the methods and presumptions regulating the prevailing (patriarchal) forms of knowledge"?[6] What, even, is "feminist knowledge"?[7] To the uninitiated it soon begins to look as though it would take whole chapters or books to unpick and contend with the presuppositions of even a single sentence, even if there were any obvious way to set about doing it.

Now in a sense this cannot reasonably be complained about. Feminist epistemology is supposed to seem strange to the traditionalist; if it were not at odds with more familiar ideas, there would presumably be no point in doing it. But on the other hand if there is no common ground, how can the debate be tackled at all? The first difficulty seems to be to find a way of even approaching the problem.

This is the issue I want to address, and the point from which I propose to start is this. Feminist epistemology may indeed look pretty bizarre, and completely detached from familiar ways of thinking, to anyone who plunges straight in; but its recommenders cannot intend there to be no connection at all with more familiar ground. The term is not supposed to be an arbitrary label adopted by people who go around wrenching words from their accustomed forms and contexts for the fun of it. Feminists who advocate feminist epistemology did, after all—or at least their foremothers did—grow up with the kind of epistemological view they now claim to have overthrown, so they must themselves have had reasons, presumably connected with their feminism, for rejecting it. The idea seems to be that once you have a proper feminist view of things, you begin to see the limitations of the "standard"[8] views you started with, and can take off in new and enlightened directions. And if so, it should be possible to explain to others the reasons for doing so.

We can, then, focus the problem by concentrating on the question of what arguments might persuade someone of such standard epistemological and scientific views to abandon them for the unfamiliar world of feminist epistemology; and, in particular, on the question of why a *feminist* should think of making the change, which—the idea seems pretty obviously to be—she should. This question can be regarded as the landmark to which it is always possible to turn when there is nothing else recognizable in sight.

Consider, then, the situation of someone whose epistemological views are still of a fairly commonsensical sort: someone who holds such unremarkable opinions as that there are some things we know, some we do not know, and others of which we are unsure; that different people know different things; that we often think we know what we turn out not to have known; and probably also—what Alcoff and Potter call a "philosophical myth"—that philosophical work can be good only "to the extent that its substantive, technical content is free of political influence."[9] And think of this standard epistemologist also as a feminist—whom I shall, to avoid complications irrelevant to the matter in hand, take to be a woman—on the shores of this somewhat alarming sea, dipping in the occasional tentative toe and wondering what reasons the mermaids can offer for her abandoning the familiar landscape behind and casting off.

## THE FOUNDATIONS OF FEMINISM

It looks as though the first problem must be to say what can be meant by the claim that the feminist on the shore is indeed a feminist, especially since, *ex hypothesi*, she has not yet embraced the various ideas and approaches that go by the name of feminist science and epistemology. It is well known, and notorious, that there is no generally accepted definition of feminism, and I certainly do not want to send things awry at the outset by adopting a prescriptive definition that other feminists would reject. Fortunately, however, there is no need to. It will be enough to take an absolutely minimal account, and say that whatever the details of what she thinks, a feminist is, at the very least, someone who thinks that *something* has been seriously wrong with the traditional position and treatment of women.

Now of course this could hardly be more minimal, since it allows for great differences between feminists in their ideas about what is wrong and how to put it right. Nevertheless it is enough for my purposes here, because if a feminist is someone who is making a complaint about the present state of things, there are various things that must be generally true about her starting position, no matter what the details are.

In the first place, she must obviously have a view about the way things *are*, or she could not think there was anything wrong with it; and she must also have some ideas about what possibilities there are for change, or she would not be able to say that things should be otherwise. She must, in other words, have a range of first-order beliefs about the world: the kind of belief that is supported by empirical, often scientific, investigation. Beliefs of this kind also imply that she has other beliefs about second-order questions of epistemology and scientific method, since in reaching conclusions about what to believe about what the world is like and how it works she has, however unconsciously, depended on assumptions about how these things can be found out, and how to distinguish knowledge from lesser things. These assumptions will become more explicit if any part of her feminism involves (as it is pretty well bound to) accusing the traditionalist opposition of prejudice, or of perpetrating or perpetuating false beliefs about women.

Similar points apply to questions of value. In order to make any complaint whatever about the way things are, a feminist must at least implicitly appeal to standards that determine when one state of affairs or kind of conduct is better or worse than another; and if her complaint takes a moral form rather than a simply self-interested one—if, like virtually all feminists, she expresses her complaints in terms of such things as injustice and oppression and entitlements to equality—she must be appealing to moral standards of good and bad or right and wrong, of which she thinks the present state of things falls short. And if she has such normative, first-order, standards, that in turn will imply something about her attitudes to the higher-order questions of metaethics, whether or not she thinks of them as such.

To say this is, once again, to say nothing at all about the content of such beliefs and standards. The claim so far is only that for feminism to get going at all, in any form, there must already be in place ideas about the way things are, and standards by which they are found wanting. Different people may have different beliefs and standards, and so reach quite different conclusions about what is wrong. However, since the specific problem being addressed here is of how someone could get from *familiar* ideas of science and epistemology to the kinds that are claimed as feminist, it is most useful to start by assuming fairly ordinary kinds of belief about both facts and values, and to consider how anyone starting from that kind of position can have been led to feminism at all.

This may in itself seem to present a problem; for how, it may be asked, can anyone be both a feminist—of any kind—and a holder of traditional views? If feminism is essentially a challenge to received beliefs and attitudes, as it is, its starting point must be the idea that these views are in some way wrong. There must therefore be *some* differences between what even the most cautious feminist accepts, and what is standardly believed by people who have not yet reached this degree of enlightenment.

And of course this is true; by its very nature, feminism must challenge some received ideas. But feminism began as a movement—as it probably does for most individuals—not with some sudden éclaircissement that led its supporters to reject all familiar standards, and to embrace instead new ones according to which prevailing ideas about women appeared as wrong from the foundations. Rather, it began in effect with the recognition that familiar ideas about women were *anomalous*, in that they could not be justified by the standards that holders of these ideas quite routinely accepted in other contexts. The original point was that traditional standards of evidence and argument in science and ethics *themselves* did not support traditional conclusions about women.

At the very simplest level, consider the early feminist challenges to received beliefs about the nature of women. Mill, for instance, pointed out that since women and men had since records began been placed in different social situations and given systematically different kinds of education (as everyone knew, since that was the status quo the insisters on women's difference were trying to defend), none of the observed psychological or intellectual differences between the sexes could reliably be attributed to nature.[10] This was an objection to established beliefs, but it involved the adoption of no new standards of epistemology or scientific procedure; prevailing beliefs were challenged by appeal to the standards that would be applied in any other scientific context. The feminist claim was essentially that there was what might be called a *sex-connected incoherence* in the current view of things. On the basis of the most fundamental current views about the nature of knowledge and standards of evidence, some less fundamental beliefs could be shown to be unfounded.

The same kind of thing happened with early feminist challenges to moral values. Of course some traditional moral values—about the propriety of women's remaining subservient to their husbands and away from public life—were challenged by feminists. But the original challenges were made not by reference to completely new standards of moral assessment that transmuted traditionalist right into feminist wrong, but by arguments showing that familiar general ideas of morality, such as most people professed most of the time, were incompatible with traditional ideas about the treatment of women. Even if it were assumed that most women were congenitally unsuited to the kind of occupation reserved for men (which anyway, from the previous argument, there was no adequate reason to assume), widely held views about open opportunities and letting people rise by their own efforts were incompatible with a wholesale exclusion of women that did not allow them even to try. Or even if it were conceded both that women were systematically inferior to men in strength and intellect, and that the weak needed protection, it still took a pretty remarkable twist of reasoning to reach the accepted conclusion that this provided a justification for making women weaker still, by placing them in social and legal subordination to men.[11]

So the arguments through which traditional feminism reached its first conclusions involved no departure from familiar standards of evidence and argument in ethics, epistemology, and science, but actually presupposed them. It was *by appeal* to these standards that the position of women was first claimed to be wrong. And notice that all arguments of this kind also depend on absolutely ordinary logic. It is *because* the traditional conclusions do not follow from the traditional premises, or *because* traditional beliefs are incompatible with traditional standards of assessment, that the challenge to the received view in its own terms is possible.[12]

Here, then, are the beginnings. Although feminism, as a critical movement, necessarily challenges parts of the status quo, it typically does so, at the outset, by appealing to other, more fundamental parts that it holds constant.

Feminism as a movement started with the broad standards of moral and empirical investigation and argument that most other people accepted at the time, and the recognition that these could not support familiar, supposedly commonsensical, ideas about women and their position. Most individual feminists probably begin in more or less this way as well.

## FEMINIST PROGRESS

The fact that feminism must start with appeals to existing standards, however, does not imply that it can never escape them. Any aspect of belief can be rethought at any time; and since even in its earliest stages feminism considerably affects the way the world appears to its converts, it is likely that once these first changes have been made, other adjustments will soon be found necessary. What must now be considered is how, starting with the kind of first- and second-order beliefs most people have, and having reached some kind of feminism on their basis, the new feminist might be led *by her feminism* to reject them, and eventually adopt radically different ideas about reason and science.

Consider then the situation of the novice feminist, who has come to recognize that traditional ideas about women cannot be justified by traditional standards. She will of course immediately recognize the need to work for political change, but that will not be all. An equally important consequence of her new view of the world will be an increasing awareness of new questions that need to be answered. Once what was previously accepted as knowledge has been thrown into doubt, the problem inevitably arises of what should now be put in its place; and, as happens with any new perspective or information, there also comes the recognition of questions that people simply never thought of asking before.

For instance, once feminists realize that positive obstacles have always been placed in the way of women's achievements, they may start wondering whether the attitudes that brought about this state of affairs could also have led to the overlooking of what women actually did do. They may start searching historical records for evidence that, in spite of the obstacles, women achieved a good deal more than was traditionally thought. Or if, having brought about apparently equal treatment of the sexes in some area, feminists find women still doing less well than men, they may suspect the existence of more subtle obstacles, and set up experiments to investigate that possibility. They may try such things as swapping round the names of men and women on academic articles, to see whether this affects readers' assessment of their merits,[13] or making controlled observations to see whether girls who are interested in science are actively discouraged by their teachers or other children.[14]

So feminism can open up a range of inquiries that would probably otherwise have remained closed, and any of which may lead to further changes in beliefs about what the world is like and how it works. And of course every new discovery will in turn have its own implications. Some will lead to an expanded political program. If, for instance, it turns out that people are unconsciously biased against women even when they think they are being impartial, feminism will have a different, and much more complex, problem on hand from the one that arises only from awareness of overt discrimination. Most discoveries will also themselves suggest further questions, which, again, would not have arisen but for the feminist awareness that started the inquiries in the first place.

It is, however, essential to see exactly how feminism connects with changing beliefs of these kinds. The sorts of investigation I have been discussing are of a kind that arise directly from feminism: without feminism to suggest where to look, nobody might have thought of launching them. And when they are finished, feminists may well find their feminist agenda widened: they may recognize more scope for feminist research and more need for feminist action. But the extent to which this happens *depends on how the inquiries turn out*, and that has nothing to do with feminism at all. In the kind of investigation considered so far, there is no point at which feminism provides the *justification* for any change of belief. Feminism does not determine what counts as a proper inquiry, or what counts as a result one way or the other, or what the results should be. In conducting inquiries of this kind, feminists are still using the standards of evidence and argument that brought them to feminism in the first place. If they come to belief that girls are disregarded or discouraged in science classes, or that academic articles thought to be written by women are systematically underrated, that is because the evidence shows this to be so. It is not because feminist principles demand that it must be.

During these early stages of feminist progress, therefore, the situation is still essentially the same as at the beginning. The feminist's background beliefs and standards themselves provide the justification for her expanding feminism, not the other way round. And, for that reason, her conclusions should be demonstrable to any impartial investigator who shares her basic standards and will look. Even though whatever changes in belief result from these investigations would probably not have occurred but for feminism, the new beliefs are not feminist in the sense of there being any reason for a feminist to hold them that a non-feminist—someone who has not yet recognized that women have grounds for complaint—has not.

It is important to stress this, because although the point is simple and obvious once seen, it seems to be widely overlooked. If so, this is probably at least partly a result of the ambiguity of "because" between cause and justification. You may change your views *because of feminism* in the sense that you would not otherwise have embarked on the inquiries that led to those changes, but that does not mean you change them *because of feminism* in the sense that feminism provides the justification for the change. Various well-known advances in science were (reputedly) made because their begetters soaked in baths, or reclined in orchards, or dreamt of snakes with their tails in their mouths; but no one thinks these causes of inspiration provided any part of the justification for accepting the resulting theories of displacement or gravitation or benzene rings. Feminism may set investigators on the track of new discoveries, and the feminist program may expand as a result of them, but that provides no more reason to count the new *beliefs themselves* as feminist than to count the others as bathist or appleist or snakeist.

So, to relate all this again to the lingerer on the shores of feminist epistemology, we can see that her feminism may well have made considerable progress since it first began, but nevertheless the *justification* for any changes in her view of things has so far had nothing to do with her being a feminist. The novice has exactly the same kinds of reason for progress within feminism as she had for becoming a feminist in the first place. Her mind has been changed only because of what the evidence has shown, and this has involved an appeal to her old familiar standards of epistemology and scientific method.

## INTERMEDIATE STANDARDS

One way of expressing all this is that the discussion so far has been of feminism only as an applied subject. Feminism and feminist progress have been shown as emerging from standards of science and rationality that are not themselves feminist. The view of feminists in the thick of feminist epistemology, however, is that this is only the beginning:

> Feminism made its first incursions into philosophy in a movement from the margins to the center. Applied fields, most notably applied ethics, were the first areas in which feminist work was published. . . . But from the applied areas we moved into more central ones as we began to see the problems produced by androcentrism in aesthetics, ethics, philosophy of science, and, finally and fairly recently, the "core" areas of epistemology and metaphysics. . . . the work of feminist philosophers is in the process of producing a new configuration of the scope, contours, and problematics of philosophy *in its entirety* [original emphasis].[15]

And when feminism reaches this new ground, quite different questions seem to arise. Presumably nothing that has been said about the progress of feminism as an applied subject, within the traditional framework, can be assumed to apply to feminist challenges to that framework itself.

Nevertheless, it is important to have discussed the less radical kind of change, because it is essential for clarifying the general issue of feminist challenges to accepted standards. Standards come in hierarchies, and a good many changes can be made at superficial or intermediate levels long before the fundamentals of rationality, philosophy of science, and epistemology are even approached. And in particular, it is essential to recognize that changes in first-order beliefs—about what the world is like and how it works—always in themselves amount to potential changes in standards, because well-entrenched first-order beliefs are automatically and necessarily used as the basis for assessing others regarded as less well established.

This is a fundamental fact about the way we ordinarily reason—and must reason—that is obvious as soon as it is thought about. Consider, for instance, the Phoenicians in Herodotus, who returned from their voyage to circumnavigate Africa claiming that as they had sailed westward round what we know as the Cape, the sun had lain on their right, to the north. Since this account was incompatible with contemporary views about the relationship of the earth to the sun, Herodotus, not unreasonably, did not believe them. It seemed much more likely that returning travelers should spin fantasies than that the sun should change its course. Now, however, our changed beliefs about astronomy have changed the standards by which we assess the story; and we regard it not only as true, but also as providing the best possible evidence that the Phoenicians really had circumnavigated Africa.[16] This kind of thing happens in every aspect of life. In forensic medicine we decide guilt by reference to our fundamental beliefs about blood groups or DNA, whereas once we might have decided it by whether the accused sank or floated in water. If some fringe medicine makes claims that are incompatible with well-entrenched scientific theory, that theory will be used as proof that the fringe claims must be wrong, and any anecdotal evidence in their favor will be explained away in other ways.[17] But if some previously unknown causal mechanism is eventually found to exist (as was briefly claimed in the case of homeopathy a few years ago[18]), attitudes may change and the fringe be incorporated, wholly or at least in part, into the main stream.

How likely such changes are to happen will depend on how well entrenched, how comprehensive, and how vulnerable any particular range of beliefs is; and this makes it likely that feminism, once begun, will lead eventually to extensive changes of standards at this level. This is essentially Mill's point again. Understanding of the world comes through observation of its constituents under varying circumstances; and since ideas about women and men have developed while women have been seen only in rather limited situations, it is to be expected that proper investigation will dislodge many traditional beliefs connected with them. And when changes do occur—either positively, in the acceptance of new beliefs to replace or supplement the old, or negatively, in the recognition that old beliefs are insecure—standards for the assessment of other beliefs and ideas will necessarily change with them.

This is already familiar from more or less everyday life. If the idea is challenged that women's nature allows them to find happiness only through making husbands and children the main focus of their attention, this will result in fundamentally different approaches to the assessment of individual women who are chronically discontented about the course of their lives. If it is accepted that sexual abuse of girls by their fathers is widespread, allegations that this has happened will become more likely to lead to criminal investigations of fathers than psychological probings of women's oedipal delusions. But the same general point is also potentially relevant to many of the issues discussed in the broad context of feminist epistemology and science.

For instance, suppose feminists are right in claiming that many beliefs typically held by women—passed down the generations, perhaps—have traditionally been dismissed as old wives' tales. If this has happened because of conflict with entrenched scientific theories, then the dislodging of those theories—perhaps as a result of feminist inquiries—would remove the basis on which the women's beliefs had been dismissed; and if enough independent evidence could be accumulated in favor of the women's beliefs, they might even themselves become the basis for rejecting the established theories. Or perhaps—another feminist idea—women may have ways of investigating the world that have been disregarded because they are different from the ones currently regarded as paradigmatic of good scientific procedure. But if feminist-inspired investigation eventually showed not only that such female techniques did exist, but also that they were just as scientifically effective, by ordinary criteria for scientific success, as the ones currently used as the touchstone for the worth of research and the promise of researchers, that would support feminist demands for changes in the structure of the scientific establishment and the standards by which aspirant scientists were assessed.[19] Whether or not such changes turn out to be justified, there is no theoretical problem about the possibility that they might be.

Just because of this relevance to matters of science and knowledge, however, it is essential to stress again that when changes in standards of these kinds do occur, they have nothing to do with changing standards of epistemology or scientific method. Changes in first-order beliefs of the kinds just discussed provide the basis for corresponding changes only in ***superficial*** or ***intermediate*** standards for the assessment of knowledge claims. They still give no reason for changes in epistemology or fundamental attitudes to science. Quite the contrary, in fact, because whether or not the feminist-inspired research actually justifies the relevant changes in first-order beliefs ***depends***, once again, on the acceptance of more fundamental standards of science and epistemology.

This point is of great importance to the traditional feminist on the epistemological shore, because unless she takes great care to distinguish these different levels of standard, she may well slip into feminist epistemology by accident. Inevitably, in the course of her developing feminism, she will encounter innumerable traditional knowledge claims she regards with suspicion; and it would be a serious mistake to respond by saying that if these were what traditional epistemology counted as knowledge, there must be something wrong with traditional epistemology. This would not only be much too precipitate; it would also give far too much credit to patriarchal man. It would by implication concede that whatever he had claimed as knowledge must, by traditional standards, really *be* knowledge, to be dealt with only by complicated revolutionary epistemologies through which it could be shrunk into mere phallocentric knowledge, or otherwise emasculated. It is usually much simpler—and, you would think, much more feminist—to start with the assumption that what has been *claimed* or *accepted* as knowledge may not be knowledge of any kind, even phallocentric, but, by patriarchal man's very own epistemological standards, plain ordinary (frequently patriarchal) ***mistakes.***

It is difficult to say how much of the impulse towards feminist involvement in epistemology and other fundamental parts of philosophy and science arises from the blurring of this distinction. My suspicion is that a great deal of it does.[20] Fortunately, however, this is not a matter that needs to be investigated here, since it would make no difference to the arguments of this paper whether the answer were all or none. All that matters for my purposes is to make the distinction clear. This is why it has been important to discuss feminist progress in first-order knowledge, ***within*** traditional views of science and epistemology, and to show how this does in itself bring about changes in standards for the assessment of other knowledge claims, or of abilities to make advances in knowledge. Only when that issue is out of the way is it possible to make an uncluttered assault on the real question, of how the feminist on the shore should approach the problem of genuine epistemological change.

## EPISTEMOLOGY PROPER

There is of course no problem of principle about the inquiring feminist's taking her feminist awareness into the study of epistemology: in this context as in others she may want to raise new questions, or check that female-connected ideas have not been overlooked or given inadequate consideration. There is also no problem about her considering new epistemological ideas, since that is something anyone can do at any time; and there is no reason, at least in advance of detailed consideration, to rule out the possibility of her deciding to abandon her old views for the kind she sees advocated by feminist epistemologists. The specific question here, however, is not of whether the move to these ideas can be justified at all, but of whether any part of the justification can be provided *by feminism.*

In the cases so far discussed, feminism has not been involved in the justification of new beliefs. All the changes in our feminist's view of the world have been justified in terms of the same epistemological and scientific standards as she appealed to when becoming a feminist in the first place. Those standards, however, are now themselves at issue. The question therefore arises again of whether her feminism—her commitment to the pursuit of proper treatment for women—gives her reasons to abandon her old ideas and change to new ones.

Something on these lines does seem to be widely implied, and not only in the term "feminist epistemology" itself. Alcoff and Potter, for instance, in the introduction to their anthology, say:

> The history of feminist epistemology itself is the history of clash between the feminist commitment to the struggles of women to have their understandings of the world legitimated, and the commitment of traditional philosophy to various accounts of knowledge—positivist, postpositivist and others—that have consistently undermined women's claims to know.[21]

And later, in a comment on one of the essays in the collection, they refer to "Alcoff and Dalmiya's concern that traditional epistemology has reduced much of women's knowledge to the status of 'old wives' tales'."[22] Both these comments suggest that one purpose of feminist epistemology is to find an account of knowledge that would result in women's knowledge claims' being accepted rather than dismissed.

Alcoff and Potter also make broader claims about the aims of feminist epistemology:

> For feminists, the purpose of epistemology is not only to satisfy intellectual curiosity, but also to contribute to an emancipatory goal, the expansion of democracy in the production of knowledge. This goal requires that our epistemologies make it possible to see how knowledge is authorized and who is empowered by it.[23]

They also say that what the essays in their collection have in common is (nothing more than) "their commitment to unearth the politics of epistemology,"[24] and that "feminist work in philosophy is unashamedly a political intervention"[25]; and, referring to the work of one of their contributors, that "feminist epistemologies must be tested by their effects on . . . practical political struggles."[26] And another contributor says of one particular (nonfeminist) account of epistemology:

> Critics must ask for whom this epistemology exists; whose interests it serves; and whose it neglects or suppresses in the process.[27]

All this suggests that there are feminist, emancipatory, oppression-resisting *reasons* for taking up these approaches to epistemology, and that the adequacy of any candidate theory must be judged by the extent to which it contributes to that emancipation.

So how should the feminist on the shore respond to this? Should she adopt this new approach to epistemology on the grounds that her present, traditional, theories are themselves part of the apparatus by which women have been oppressed? Consider first the suggestion that traditional epistemology has been responsible for the relegation of women's knowledge to the status of old wives' tales, and should be abandoned *for that reason*. Could she be led to the conclusions of feminist epistemology by this route?

The first point to notice here is that the claim that familiar ideas of epistemology have "consistently undermined women's claims to know"[28] cannot just be slipped into the argument as if it were obviously true. Since the challenge is to demonstrate to the traditional feminist the inadequacy of her present position, and show why she should adopt instead the ideas of feminist epistemology, this is something of whose truth she needs to be persuaded. She will need to be shown that there really are such substantial and systematic differences in the kinds of knowledge claim made by men and women, that the ones most often ruled out by her present standards are women's, that it is actually her epistemology, rather than her first-order beliefs, that leads to their being ruled out, and that the recommended change in epistemology would lead to their being recognized as knowledge.

These are not small matters; and if the traditional feminist is like most of the uninitiated, all this will present her would-be persuaders with considerable problems. Even the simplest part of the claim, about the sexes' making systematically different knowledge claims that are differently treated, may well strike her as implausible; and there is likely to be even more difficulty about the idea that whatever beliefs she does discount are the result of her traditional epistemology, since that is something she is likely not only to reject, but even to have difficulty in understanding.[29] So problems of this kind might well be enough on their own to put an end to any real prospect of converting the traditional feminist to feminist epistemology by an argument of this kind.

Even if there were some reasonable prospect of their being overcome, however, there would still remain a more fundamental and completely intractable problem. Suppose our feminist agreed to accept, at least for the sake of argument, that the traditional epistomological standards she still accepts really did discount a good deal of what women had traditionally claimed to know. How exactly is that supposed to justify the change to a new epistemology?

It is significant that in the quotations above Alcoff and Potter say first that it is women's *claims to know* that have been undermined by traditional epistemology, and later that it is women's *knowledge.* Now of course if our feminist's traditional theory actually ruled out women's *knowledge*, it would indeed be wrong and should be changed; that is analytically true. But the whole problem is that she does not yet accept that what she is dismissing *is* knowledge. What she is dismissing are knowledge *claims*, which, by her present standards, *really do* amount to nothing more than old wives' tales. To accept the crucial premise that these claims did represent genuine knowledge, she would *already have to accept* the new epistemology that the argument is supposed to be justifying. So she cannot be persuaded to change by an argument of this kind, because if the premise is taken to be about knowledge claims, it provides no reason at all for any change in epistemology; and if it is taken to be about knowledge, it is flagrantly question begging.

And, furthermore, *until* she has seen reason to change her epistemology, the still-traditional feminist must also conclude that it is the advocates of feminist epistemology who are treating women wrongly. Her feminist principles combine with her present epistemological views to suggest that the proper way to treat any women who make these misguided knowledge claims is to give them a proper education and bring them out of their ignorance. To offer them instead an epistemology that passes their ignorance off as knowledge is only to cheat them into collusion with their own deprivation, and this is obviously something she must regard as a scandalous perpetuation of the traditional wrongs of women, to be fought with all the feminist energy she can muster.

Of course this particular argument, about getting women's knowledge properly acknowledged, is only one among many possible lines of feminist argument to the conclusion that epistemological change is needed, and to show that this one does not work is still to allow for the possibility that others might. But in fact a version of the same problem arises whenever the emancipation of women is used as part of an argument for change in fundamental standards of epistemology or science, even when the substance is quite different.

Suppose, for instance, our still-traditional feminist were urged by feminist epistemologists to recognize that prevailing epistemological standards had been put in place by men, and that women could never be free from oppression as long as they were judged by male standards, epistemological or otherwise. She would, once again, have to be persuaded that there really were such differences between male and female standards (which would certainly present problems, since she herself accepts the ones said to be male). She would also have to be persuaded that women's knowledge and abilities were bound to fare badly as long as male standards prevailed (which might be equally difficult, since what she has seen of the standards recommended by feminists is unlikely to make her yearn to be judged by them). But even if those difficulties could be overcome, the more fundamental problem would remain. The idea that any group's knowledge claims cannot be properly assessed by the standards of another group *is itself* the epistemological theory being advocated, *opposed* to the one the inquirer now holds, and therefore cannot be invoked as any part of an argument that her present view should change. And, again, from the point of view of her present ideas of epistemology, any move towards judging each group by its own standards would itself constitute a wrong to women, in inducing them to mistake their real deprivation for inappropriate attitudes of the privileged.

The same thing happens if it is claimed that knowledge is at root a matter of politics: that feminist epistemology is a matter of revealing "a *politics* of truth, logic, and reason," and that until this is understood women will be misled by standards that pretend to objectivity, but are really nothing but manifestations of unjust male power. Once again, the idea that knowledge is a matter of politics is *itself* the epistemological position being defended, and therefore the claim that women are wrongly treated by epistemologies that deny it cannot be used as part of the argument in its defense. And *until* the traditional feminist has been persuaded to change her mind, she will continue to think that attempting to persuade women of its truth is, once again, to delude them into thinking that nothing but politics is needed to transform their ignorance into knowledge, and so obstruct their acquisition of the real knowledge that is needed for effective political activity of any kind.

The problem cannot be escaped even by a retreat to the most blatantly political position of all: the argument that we must adopt the epistemological ideas claimed as feminist because until we do, things will be worse for women; that the progress of women depends on making this change. Even that, if taken as a defense of a serious epistemology (as opposed to one professed in public for political reasons but denied in private) presupposes the idea that epistemology is logically secondary to ethics, which is itself an epistemological theory.[30]

Because the root of this matter is a logical one, it makes no difference how many variations of detail are tried. The point is essentially the one that was made at the beginning of this piece, about the foundations of feminism, that criteria by which proper treatment and assessment can be recognized need to be in place before it can be said that the present state of things is falling short of them. The claim that some set of epistemological and scientific standards results in inappropriate treatment of women and their ideas cannot be used as an argument against those standards, because, necessarily, to accept those standards *is* to accept that women should be treated according to them. But anyone who is suspicious of such succinct generalities can easily test arguments individually by recalling the image of the feminist on the shore, who must be offered reasons to abandon her current views for those of feminist epistemology. It will be found for any argument she might be offered that if the reasons given for the recommended change from traditional to feminist epistemology are themselves *feminist*—if they depend in any way on the idea that current epistemology wrongs women, or is bad for women in any way—they will turn out to depend not only on highly contentious empirical premises, but also on revisionary epistemological claims that presuppose the conclusion and therefore beg the question.[31]

Now of course all this shows only that *feminist* justifications of what is claimed as feminist epistemology cannot work. It still leaves open the possibility that the feminist on the shore might find other, non-feminist, reasons for changing her epistemological ideas, just as she earlier found non-feminist justifications for changes in various first-order beliefs about the world. Philosophers have for centuries been producing arguments about epistemology that have nothing to do with feminism, and some such non-feminist argument might persuade her to accept the approaches now claimed as feminist. Furthermore, if she did become convinced that these new epistemological ideas were right, ***and*** that women would do better under these than under the old ones, she would also—necessarily—conclude that women were wrongly treated by traditional epistemology, and might well see it as part of her feminist politics to develop, and persuade others to adopt, the epistemology she now regarded as right.[32]

This is pretty obviously what has happened in the case of feminist epistemologists. They have been persuaded by particular approaches to epistemology—typically the kind that derive from ideas about the sociology of knowledge and science, and stress the idea that dominant groups set the standards—and have made these ideas the foundation both of their future inquiries and of the form their feminist politics takes. And that, as far as it goes, is fine in principle, but it must not be mistaken for there being any feminist reasons for accepting that, or any other, approach to epistemology.

So what all this means is that the situation is just the same for the feminist on the shores of feminist epistemology as it was in the early stages of her feminist inquiry. Feminism may, perhaps,[33] prompt her to raise particular epistemological questions, and if the answers to her questions lead her to epistemological change, that change will affect both her politics and the course of future inquiries. But still her feminism cannot itself be the determinant of the answers.

Feminism, in other words, can never escape its beginnings as an applied field.[34] Conclusions about what should be done by feminists for women—irrespective of whether they want what is just or right for women, or merely what is good for them—are ***at all stages*** essentially derivative, and dependent on more fundamental ideas. No beliefs about matters of fact, and no theories of epistemology or science, can be required by feminism, because feminist conclusions depend on them.

This means that to attach the label "feminist" to particular theories of epistemology or anything else is completely arbitrary.[35] In no sense that is not seriously misleading can there be any such thing as feminist epistemology.

## THE FLIGHT FROM SCIENCE AND REASON

Now this must all be tied to the theme of this volume: the flight from science and reason.

The most immediately obvious way into the analysis of what goes by the name of feminist epistemology is to take the specific claims, presuppositions, and lines of argument claimed by their advocates as feminist, and subject them to critical analysis. That, however, is not what I have been doing here. This paper has been concerned only with the more fundamental problem of how feminism fits into these inquiries at all; and the essential conclusion is that although it certainly has a place, that place is limited. To try to go beyond it is to run into incoherences far more damaging to the idea of feminist epistemology than any criticisms of the details of its content.

To risk an analogy no doubt much too frivolous for such solemn matters, but salutary for just that reason, the place of feminism in scientific and philosophical inquiry has emerged as strikingly similar to that of James ("The Amazing") Randi[36] in the Uri Geller investigations. When scientists started investigating Geller's telepathic and spoon-bending exploits, they of course thought they were conducting a careful inquiry that eliminated the possibility of fraud. Scientists, however, do not know about conjuring. When Randi was brought in, his practiced eye went immediately to what had been made invisible to the lay observer, and the Geller tricks were exposed. And this did, in its modest way, affect the course of science. Anyone whose view of the world had been influenced by the Geller phenomena now had to eliminate these apparent data from their calculations, and rethink their view of the world—and perhaps even their ideas about scientific methodology—on a different basis. But even if Randi's contribution had been a thousand times greater—even if everyone engaged in the inquiries had been busily conjuring, and all the data had had to be scrapped—he still would have been doing nothing that could possibly be described as conjurist science. His contribution lay entirely within the familiar framework of scientific investigation. He put the scientists in the way of eliminating certain misleading information, but they themselves had to be able to confirm that the suspected tricks were actually going on: if Randi had claimed that the tricks must remain invisible to anyone who lacked his conjurist insights, nobody would have been in the least interested. And after this purge of misleading information, the scientists went on just as before, influenced by conjuring only to the extent of being aware of that kind of possibility for deception. It would have been absurd for Randi—at least qua conjurer—to say anything about what data should be taken into consideration after the spurious ones had been eliminated, or which theories were most promising, or what direction future research should take, let alone for him to have offered anything claimed as conjurist approaches to science as a whole.

Notwithstanding obvious differences between the two cases, the contribution of feminism to academic inquiry has emerged as much the same in kind. Feminists come to academic inquiry of all sorts with a particular interest—and, after a while, with some accumulated expertise—that makes them look where others had not thought to look, and that frequently results in their discovering what had previously been unknown and finding anomalies where all had been presumed smooth. But all these discoveries must be visible to anyone who is willing to consider them; and once they have been made, the question of how science and philosophy should proceed is no more the concern of feminists than of everybody else. There can be no *feminist reasons* for adopting either first- or second-order beliefs of any kind.

The proposal of this paper is that this simple point provides the most effective way of coping with the phenomenon of so-called (as I must now insist) feminist epistemology. It combines two great advantages, of being relatively simple and easy to demonstrate, and of considerable power once established.

Consider first the simple aspects that make the point relatively easy to demonstrate.

First, the argument depends *not at all* on what the content of feminist epistemology is supposed to be. This is a great advantage, because anyone who takes on the details not only has an enormous task on hand, but also runs the perpetual risk of wrangles about misrepresentation. None of the foregoing arguments depends on the details of what any feminist thinks, so it makes no difference, for instance, whether or not what is claimed as feminist epistemology is accurately characterized by Alcoff and Potter as being specifically concerned to "unearth the politics of knowledge," or whether or not I am right in my speculation that much of the impetus to feminist epistemology arises from confusions of level. The essential point is a logical one about the relationship of feminism to any theory of epistemology, or, for that matter, any other type of theory claimed as feminist.

Second, the claim is itself simple and straightforward, involving none of the appalling complications waiting to ensnare any critic brave or unwary enough to start from inside feminist epistemology and try to find a way out. Although developing and illustrating the argument may take some time in the first instance, all that is really involved is a simple logical point: that feminism can provide no justification for holding one theory rather than another. It is also relatively easy to demonstrate, since even if the general argument is thought to be in some way suspect, any particular argument that is attempted will provide an illustration. The important point to keep in view—the landmark, when everything else has vanished into the fog—is the question of why the feminist with still-conventional views of epistemology *should change* to the views claimed as feminist; and it can quickly be shown that if the argument offered has anything to do with her feminism, it will run into question begging or self-contradiction. She cannot be persuaded to change her epistemological views on the grounds that her present ones discount women's knowledge, for instance, since until she has changed those views she will not accept that what she is discounting *is* knowledge.

And finally (though Mill might have described this as "resembling those celebrations of royal clemency with which . . . the king of Lilliput prefaced his most sanguinary decrees"[37]), the argument has the advantage of being essentially mild and unprovocative, because it implies no criticism of the *substance* of what is claimed as feminist epistemology. *All* it does, as such, is insist that answers to questions of epistemology and science are presupposed in any arguments about the proper treatment of women, and therefore cannot themselves be required by feminism. It shows that if the feminist on the shore is to be persuaded to embrace the theories claimed as feminist she must be offered non-feminist reasons for doing so, but that does not imply that such reasons could not exist, or that the whole thing is nonsense.

So the case being presented here is relatively simple and relatively uncontentious, and all this sweetness and light may, perhaps, suggest that its implications cannot be very far reaching. It may even seem to leave the heart of the issue untouched, in allowing for the possibility that what goes by the name of feminist epistemology may be good epistemology even though not feminist. But although that may be technically true, the case argued here has direct implications that are almost frighteningly out of line with the prevailing culture of academic politics, and indirectly makes all the difference that any skeptic needs.

The most important direct implication is no doubt obvious. It is that a commitment to feminism—to righting the wrongs of women—gives ***not the slightest*** presumption in favor of any theory of set of beliefs that happens to have labeled itself feminist. Once it is recognized that such theories can have no feminist justification, and that the name is arbitrary and misleading, it becomes clear that they should all be treated exactly as if the name were not there. Decisions about their appropriateness for teaching in universities, or for publication by serious publishers, should positively not be distorted by the mistaken idea that women's past oppression must entitle whatever calls itself feminist to special consideration. And in fact the case is even stronger, because the argument applies equally to the question of whether some area of theory is worth even detailed preliminary study. If an initial skim of any part of the literature suggests that its content is weak or confused or misguided, then—in a world of far too many books, where deciding to read one means not reading others—even the most committed feminist can, with a limpid feminist conscience, decide to go no further. She may, of course, have made a mistake, but that is true of all the other books she has no time to read, and she would ***certainly*** be making a mistake if she allowed the spurious association with feminism to influence her decision.

The second heretical implication of these arguments is that no expertise whatever in these knowledge-connected subjects—epistemology and the sciences—comes of being a feminist. A feminist awareness that sex-connected anomalies may come up in particular areas, or that hitherto unnoticed questions may arise in them, ***does not constitute expertise in these areas.*** In fact it is rather the other way round. Until she has enough of a grip on a particular subject, a feminist cannot be adequately aware of the ways in which sex-connected anomalies may lurk within it, or where to look for undiscovered facts that might be of feminist significance. Once again the conjuring analogy is useful. If Randi had been summoned to seek out conjurist fraud among scientists dealing with esoteric parts of modern physics, he would have had to learn enough of the physics to see where lay the possibilities for deceit by conjuring. No matter how conscious a feminist may be of women's oppression, she will not spot subtle mistakes in patriarchal argument unless she has enough understanding of logic to understand how good arguments work, nor recognize inadequate evidence in any part of science unless she knows what adequate evidence looks like.[38]

And what this means, schematically speaking, is that neither science nor epistemology can be properly studied in women's studies departments, or taught by feminists qua feminist. These subjects must be taught—by all means in the company of people who are aware of the need to keep on the lookout for sex-connected anomalies—in departments of science and philosophy, by people who have enough background in the area to understand how to conduct inquiries and seek out anomalies. There is no problem of principle about this, because the arguments that show that no theories or beliefs can be feminist also show, by implication, that they cannot be patriarchal or phallocentric either; if they cannot be inherently emancipatory, neither can they be inherently oppressive.[39] And when such departments make new appointments, they are seriously misguided if they think that concerns for women oblige them to appoint specialists in what goes by the name of feminist epistemology or feminist anything else. They can advance the cause of women by looking for excellent scientists or philosophers who have some special awareness of the way feminist issues may arise in those areas, but that is quite a different matter.

Even this does not suggest that there is anything wrong with the content of what is claimed as feminist epistemology. All the theories and approaches misleadingly claimed as feminist could, in principle, survive the loss of the name, and be regarded as worth teaching and publishing whether thought of as feminist or not. But the association with feminism has provided a hothouse within which nonsense has had every chance to rampage, and the test for the theories claimed as feminist is to see how well they can survive a draft of cooler air.

It is obvious, in the first place, that the name of feminism—the apparent seal of authenticity—makes all the difference in the world to the way theories claimed as feminist are approached. Absorbers of feminist epistemology are not epistemological surfers who happened to be entranced in passing by the substance of these theories, only to find later that they were claimed as feminist. It is feminism that has drawn the crowds, and their attitudes to feminism will inevitably extend to what they see as feminist. Committed feminists will be more inclined to attribute obscurity to their own confusion than to confusion in what is claimed as feminist theory, and to accept on trust conclusions whose supporting arguments they have too little time or skill to assess. Fellow travellers will uneasily presume that there must be something in what is going on, and will be reluctant to resist as baffling appointments and courses of invisible merits proliferate around them. But take away the label, and feminists will see that they need positive reasons for venturing into these waters rather than excuses for staying out, and feminist sympathizers will be no more willing to endure the severe cognitive dissonance many of them now suffer than they would for the sake of flat earthers or crop circlers.

The protection provided by the name, furthermore, stretches even further when the related idea takes root that science, epistemology, or anything else should be approached through feminism, rather than the other way round. Just about anything can be made plausible to people who approach their subject from the far side of the relevant academic disciplines, unequipped with the techniques of detailed criticism that are the basis of all real progress in both science and philosophy, and in no position to identify as caricatures whatever silly or simplified ideas may be attributed to the opposition. Astrology can seem well founded to people whose scientific background is too vague to allow criticism of plausible generalities ("science has shown that the stars do have an influence on the earth"), but it cannot survive even a minimal acquaintance with post-Newtonian physics; the inconsistencies hidden by the generalities glare in the details. Creationism can flourish among people who start with the Bible, keep themselves entirely surrounded by creationists, and limit their acquaintance with paleontology and scientific method to selected odds and ends, but they could never have reached creationism from paleontology. Whether or not the epistemological ideas claimed as feminist have any merits, they can hardly fail to look plausible if approached from the direction of feminism, through sweeping ideas about the imposition of alien standards on an oppressed group ("men have had the power, and have used their patriarchal standards to dismiss women's knowledge," or whatever), and kept in a self-reinforcing huddle that sees all outside criticism as irrelevant because patriarchal. The test for the theories is to see to what extent they can survive an approach from the other direction, by people who are familiar with the relevant techniques of detailed criticism, and who do not have to rely for their understanding of "standard" epistemology on the rather surprising accounts that sometimes appear in the feminist literature[40]—many of which are quite enough on their own to make change seem a matter of urgency.

My own view is that not much would survive. If the whole idea of feminist epistemology rests on a mistake, that in itself bodes ill for the details, and enough has already been said in passing—for instance, about confusions between epistemological and first-order standards, knowledge and knowledge claims, and the politics of knowledge and the politics of epistemology—to suggest a range of serious problems. And to the extent that Alcoff and Potter are right in seeing feminist epistemology as "an unashamedly political intervention," there is also the fundamental logical problem of the idea that politics even can, let alone should, be at the root of things. Politics is a matter of manipulating other people to bring about a desired set of ends, and nobody can start making political calculations until they think they know more or less how the relevant parts of the world work. That not only means that first-order beliefs must precede politics; it also presupposes an epistemology that has nothing to do with politics. Anyone who tried to think seriously and in detail about how to go in for politics on the basis of an epistemology that took power to be at the root of what knowledge *actually was* would soon be stopped by dizziness.

However, to demonstrate this would take a good deal of detailed work that would go far beyond the intended scope of this paper, and might also overreach what was worth doing at all. Once so-called feminist epistemology is approached with the recognition that the name is spurious, and that writings presented under its name should be assessed as if the name were not there, it becomes clear that anything that looks like gobbledygook may reasonably be presumed such until shown otherwise, and disregarded accordingly.[41] The only reason for going into the details of anything that did look hopelessly unpromising would be to try to demonstrate to its practitioners the incoherences of their current position; but there would be no point in even that unless there were some reasonable prospect of success, and I doubt that anyone unpersuaded by the general argument given here would be more likely to be persuaded by the details. Since the deep irrationality of thinking that there can be any connection between feminism and particular theories of epistemology is the root of most of the other irrationalities, that is the one to concentrate on.

## THE POINT OF ALL THIS

Finally, all this can be connected with a general problem that may seem to confront this jamboree of science and rationality. It is all very well to argue that there are deep irrationalities in current feminism and other popular ideas that are supposed to go with politically enlightened attitudes, but what, it may be asked, is the point of an enterprise like this, where a cluster of people who like to think of themselves as rational and scientific get together and sigh about the others out there who are not? We already believe in all this, and need no persuading; and reason is not going to get us very far with people who are against it altogether, or at least think ours is the wrong sort. They are holding conferences and collecting anthologies of their own—hundreds or thousands to our one—in which arguments like ours will be dismissed as manifestations of the conceptions of science and rationality they are protesting about, and we shall all be exactly where we started.

Obviously, however, we think there are purposes in what we are doing; and one final use of the foregoing arguments is to suggest what some of these may reasonably be.

First, then, even for us—the already persuaded—these essays are not just a matter of reaffirming what we all think already, as if we were bearing witness at a revivalist meeting. Apart from the obvious fact that we need to hear from experts in other fields than our own, there is the more fundamental point that it is quite possible to have a general commitment to rationality, and an unease about the way things are going, without having pinned down exactly what the problem is, or even having established with any certainty that there is one. In many cases—certainly all matters concerning feminism—trying to work out precisely where things are going wrong is a grueling and time-consuming enterprise. To the extent that this is the kind of thing we are doing, we can think of ourselves as involved in one of those "towards" projects of which we hear so much these days—which is is particularly appropriate for people of our intentions, since if there is a single distinguishing characteristic of commitments to rationality and science, it is a refusal to take unrefined intuitions as the end of any matter.[42]

Second, we need not presume that all the people who seem to have been swept away on the tide of popular irrationality are actually beyond rescue. Much may depend on how far they have gone, and why they let themselves go in the first place. Not much can be done about any who meet arguments with rhetoric, or (and who can blame them?) resist the idea that the subjects in which they take themselves to be expert do not exist. But many others—most of us most of the time, outside our own subjects—hold confused or wrong beliefs, not because we are twisting arguments and fudging facts to justify what we are determined to believe anyway, but because we have actually been *misled* by bad arguments, or by false or misleading claims about matters of fact.[43] This does not (necessarily) mean that we are gullible or stupid; the point is, once again, that the problems lie in the details, and no one can be an expert in everything. Much of the time we have no choice but to depend on other people's authority, and we face the perennial problem of how to recognize experts without being experts ourselves.[44] But if the mistakes in argument are the cause of the mistaken conclusions, rather than the other way round, then at least when the mistakes are demonstrated we have no difficulty in changing our minds.

And finally, this is all important not just for the purposes of persuasion, but also for stiffening the moral resolution of the already persuaded. This is directly relevant to feminism, but it has implications for all our concerns.

It is extremely difficult—especially for morally sensitive men—to maintain a firm resistance to the purveyors of feminist this or that against the tide of fashionable academic and political opinion. The assumption that professing feminists must be the guardians of women's legitimate interests, combined with vague ideas about the entitlements of oppressed groups to define their own oppression, means that people who are rash enough ot protest about feminist epistemology or feminist anything else are likely to be accused of patriarchy, phallocentricity, androcentrism, and probably devouring their young. These pressures are not easy to withstand, especially now that most of the accused are sufficiently guilt ridden to wonder uneasily whether there may be something in the accusations; but it may become easier if it is recognized that tolerance of the various approaches that have taken on the name of feminism is not only not required by concerns for the emancipation of women, nor even an irritating but intrinsically harmless background to efforts to achieve it, but a positive obstruction that actually tends to perpetuate and entrench the very oppression of women that most people take feminism to be opposing.

This is another consequence of the essential derivativeness of feminism. To reach the conclusion that women are in some way oppressed, you must appeal to standards of ethics and rationality that are not being met in the treatment of women. ***Whatever*** your standards, it would be manifestly absurd to support something that ***by those standards*** actually tended to perpetuate the oppression of women, just because it called itself feminist.

The point will be most obvious to anyone who has approached feminism from the direction of fairly ordinary ideas of science and epistemology. Once you have reached such conclusions as that traditional claims about the nature and position of women have always been inadequately supported by the evidence and are often mistaken, that women have had various direct and subtle obstacles placed in the way of their acquisition of knowledge, and that their abilities have never been impartially assessed or their achievements acknowledged, these conclusions show the direction your feminist politics must take. You must therefore positively oppose any movement that encourages women to disdain as patriarchal the very ideas in philosophy and science you have been trying to ensure they are given full opportunity to learn, or that demands recognition as special ways of knowing for the contrived ignorance you have been struggling to remedy, or that seems to be trying to smooth into mere patriarchal knowledges the traditional knowledge claims you have been trying to expose as unsupported travesties. And if such a movement calls itself feminist, that only provides a reason for redoubling your efforts to thwart it.

But irrespective of whether your epistemological ideas are of these traditional kinds, or whether you think the content of the ones claimed as feminist may have something to be said for them, there is still the matter of allowing any particular view within either science or philosophy, or for that matter ethics (to return to that), to count as feminist at all. If you come to feminism through the idea that women have been seriously deprived of access to knowledge and science, and kept in a situation that has entrenched misunderstandings of their nature and potential, what appears as the most important issue is not which particular theories happen to be generally accepted at any time, but the kind of education—and, more broadly, intellectual culture—that makes the proper discussion of all these things possible. This is something that the claiming of *any* particular set of views as feminist will actively work against, for all the reasons already discussed. Women cannot learn philosophy or science as long as they see these things as aspects of feminism, rather than feminism as raising questions within them.

It is hard to imagine anything better calculated to delight the soul of patriarchal man than the sight of women's most vociferous leaders taking an approach to feminism that continues so much of his own work: luring women off into a special area of their own where they will remain screened from the detailed study of philosophy and science to which he always said they were unsuited,. teaching them indignation instead of argument, fantasy and metaphor instead of science, and doing all this by continuing his very own technique of persuading women that their true interests lie elsewhere than in the areas colonized by men. And, furthermore, outdoing even his own contrivances, in equipping them with a sophisticated, oppression-loaded, all-purpose rhetoric that actually obstructs any serious attempt at analysis.

It is not easy to resist the rhetorical fuzziness of fashionable conceptions of what is needed for the emancipation of downtrodden groups; but if these arguments are right, this is one of the clearest moral and intellectual necessities going. Our positive duty to women—not to mention such matters as truth, academic standards, and other things we normally take ourselves to be committed to—absolutely demands an unremitting opposition to all these confusions. If the simple point is held on to that there is no intellectual or moral theory that anyone should accept in virtue of being a feminist, and that the achievement of what most of us think women are entitled to will actually be held back as long as it is thought that there is, perhaps it may be easier to summon up the necessary moral courage.

## NOTES

1 This title was borrowed from Larry Alexander's "Fancy Theories of Interpretation Aren't," an uncompromising gem of analysis with which I am pleased to have even this tenuous association.

2 See, e.g., Janet Radcliffe Richards, *The Sceptical Feminist*, ch. 1, passim.

3 "Bodies and Knowledges: Feminism and the Crisis of Reason," p. 209.

4 E.g., in Alcoff & Potter, eds., *Feminist Epistemologies*, Introduction, p. 13: "For feminists, the purpose of epistemology is not only to satisfy intellectual curiosity, but also to contribute to an emancipatory goal: the expansion of democracy in the production of knowledge." For convenience, most of the illustrative quotations in this paper will come from the Alcoff and Potter anthology. This seems a pretty comprehensive and representative collection, but nothing in the argument presented here depends on whether or not this is so.

5 E.g., Elizabeth Potter, "Gender and Epistemic Negotiation," p. 172: ". . . claims put forward by feminist scholars that gender strongly intersects the production of much of our knowledge. . . ."

6 Grosz, "Bodies and Knowledges," p. 187: ". . . if the body is an unacknowledged or an inadequately acknowledged condition of knowledges, and if the body is

always sexually specific, concretely 'sexed,' this implies that the hegemony over knowledges that masculinity has thus far accomplished can be subverted, upset, or transformed through women's assertion of a 'right to know' independent of and autonomous from the methods and presumptions regulating the prevailing (patriarchal) forms of knowledge." This passage is perhaps as good an illustration as any of the general point about uncharted seas. Readers of P. G. Wodehouse may find themselves reminded of Bertie Wooster's encounter with the improving literature prescribed by one of his passing financées: "I opened it [*Types of Ethical Theory*], and I give you my honest word this was what hit me. . . ." (from "Jeeves Takes Charge").

7 E.g., in Lynn Hankinson Nelson, "Epistemological Communities," p. 122: ". . . for more than a decade feminists have argued that a commitment to epistemological individualism would preclude reasonable explanations of feminist knowledge; such explanations . . . would need to incorporate the historically specific social and political relationships and situations, including gender and political advocacy, that have made feminist knowledge possible."

8 I do not want to concede at any point that there really is any such thing as "standard" epistemology (hence the distancing inverted commas), let alone that it has the kinds of characteristic that are sometimes claimed by feminist epistemologists; but for the limited purposes of this essay, and for the sake of argument, it will do no harm here to allow the point to pass.

9 Alcoff & Potter, Introduction, p. 13. To people who accept this myth, they say, "feminist work in philosophy is scandalous primarily because it is unashamedly political intervention."

10 J. S. Mill, *The Subjection of Women*, pp. 23–24.

11 See, e.g., Janet Radcliffe Richards, "Traditional Spheres and Traditionalist Logic," pp. 319–338.

12 There is no space to deal here with feminist challenges to logic, but the broad conclusions of this paper will be seen to apply to those as well. There will also be no further discussion of feminism and ethics, but for everything that is said here about epistemology, arguments about moral and other value judgments run in parallel.

13 P. Goldberg, "Are Women Prejudiced against Women?" pp. 28–30; quoted in Ann Oakley, *Subject Women*, p. 126.

14 See, e.g., Allison Kelly, ed., *The Missing Half*, passim.

15 Alcoff & Potter, *Feminist Epistemologies*, pp. 2–3.

16 Herodotus, *The Histories*, pp. 283–284.

17 See Wallace Sampson, "Antiscience Trends in the Rise of the 'Alternative Medicine' Movement"; and Gerald Weissmann, "'Sucking with Vampires': The Medicine of Unreason," this volume.

18 Claims about the efficacy of homeopathic medicines have usually been rejected by scientists as obviously absurd, because the dilutions recommended are sometimes so extreme as to leave not a single molecule of the original drug. At one time some researchers claimed to have discovered that molecules could leave behind impressions of themselves in their absence, making it seem that there might after all be come causal mechanism by which the medicines might work. I believe this apparent finding has now itself been rejected, so the dilutions are back where they were.

19 Many feminists have taken up (without her full concurrence) Evelyn Fox Keller's work on Barbara McClintock, whose "feel for the organism" they claim as exemplifying women's approach to science, and as having been resisted by the scientific establishment. There are many possible grounds for controversy here, about whether the approach really is specifically female and whether it was really rejected by the establishment (see, e.g., Fox Keller, "The Gender/Science System"); but even if the claims were right, that would show the need for changes in standards only at an intermediate level, themselves justifiable in terms of more fundamental ideas about the nature of scientific success. These points are discussed further in Note 29.

20 This is what seems to be going on, for instance, where Lorraine Code (in "Taking Subjectivity into Account," pp. 15ff) is criticizing what she calls "S-knows-that-*p*"

epistemologies, and in doing so mentions a sociologist's claims to have proved scientifically (and therefore to know) that "orientals as a group are more intelligent, more family-oriented, more law-abiding and less sexually promiscuous than whites, and that whites are superior to blacks in all the same respects." She gives reasons for doubting this claim, and then goes on to say:

> . . . the "Science has proved . . ." rhetoric derives from the sociopolitical influence of the philosophies of science that incorporate and are underwritten by S-knows-that-*p* epistemologies. . . . The implicit claim is that empirical inquiry is not only a neutral and impersonal process but also an inexorable one; it is compelling, even coercive, in what it turns up to the extent that a rational inquirer *cannot* withhold assent.

But nobody I have ever heard of holds epistemological views according to which if someone *claims* that science has proved this or that, a rational inquirer cannot withhold assent. That is what the claimant wants us to think, of course, but as rational inquirers we can, and frequently do, distinguish between S's claiming to know that something has been proved and S's actually knowing it. We typically challenge the claim—*within* the framework of familiar epistemology—precisely by casting doubt on the first-order knowledge claims produced as evidence. The fact that claimed proofs can be mistaken—which no non-lunatic epistemology could possibly deny—does not even begin to show that there is something fundamentally flawed about traditional ideas of propositional knowledge.

A more general indication that confusions of level may be a source of problems is the huge range of topics typically raised in feminist writings about feminist science and epistemology. These may have just about anything to do with women—or some individual woman—and any aspect of science or its applications, or anything whatever to do with knowledge. This does not matter in itself, but it does matter that there is usually no systematic discussion of how the different kinds and levels of discussion relate to each other. I have often been struck in practice, for instance, by the way feminists who are (quite reasonably) angry about the male takeover of obstetrics describe the insensitive use of gadgets as "subjecting women to male science," then go on to take this as indicating some kind of global, woman-oppressing maleness of every aspect of the scientific enterprise.

21 Alcoff & Potter, *Feminist Epistemologies*, Introduction, p. 2.

22 Ibid., p. 11.

23 Ibid., pp. 13–14.

24 Ibid., p. 3. Notice that this is one step further on than usual. Many epistemologies concern the idea of a politics of *knowledge*; the idea that there is a politics of epistemology suggests that there are political reasons for adopting one *epistemology* rather than another.

25 Ibid., p. 13.

26 Ibid., p. 14.

27 Lorraine Code, "Taking Subjectivity into Account," p. 23.

28 Note in passing, though I shall not go into them, the problems inherent in using empirical evidence based on the assumptions of one epistemology as part of the argument for establishing a quite different one, which may well undermine the original evidence. There is no difficulty in accepting that a feminist may gradually change the epistemology she started with, but she cannot do so *and* keep earlier conclusions that were actually based on the rejected epistemology.

29 There is a real problem about the idea that it is epistemological standards, rather than particular first-order beliefs, that underlie the rejection of women's knowledge claims, which is difficult to explain because any case that fulfills the conditions seems bound to look absurd.

Consider again, for instance, the familiar feminist idea that much traditional knowledge of midwives has been dismissed as nonsense because it conflicts with established scientific views. Suppose that in some such case careful, feminist-inspired study revealed that a particular group of midwives had more success, in terms of well being of mothers and their children, than some corresponding group of male doctors who based their practice on current scientific theories. Individual doctors or the scientific community might, perhaps, go on insisting that the midwives were simply ignorant and should be disregarded; but no standard *epistemology* would support such an attitude. Any reasonable scientist would take the midwives' success as evidence that they were on to something (though they might well be wrong about what it was) and that there must therefore be some inadequacy in the scientific theory. Such a case would therefore show no need for revisionary epistemology, but only for changes in ideas about which first-order claims to use as the standard for judging others.

To find a case that required genuine epistemological change to turn the midwives' ignorance into knowledge, it would be necessary to move to something much more bizarre, and postulate a situation where their practices were not only at odds with received scientific theory, but also *less successful* than those of the doctors, resulting in *worse* statistics of maternal and child welfare (because if they were successful, ordinary epistemology would admit that they were raising problems for the received theory), and where feminists would argue that we must change epistemological standards until *these* practices were counted as demonstrating knowledge. It is difficult to imagine either what such standards would be, or that any feminist would want to recommend any such thing. And unless the traditional feminist can be brought to understand how a change in epistemology might result in changes in the assessment of women's knowledge claims, she obviously cannot be persuaded to make the change for that reason.

Similar problems arise with feminist ideas about the need for radical change in fundamental approaches to science, to accommodate women's ways of setting about understanding the world. There is no problem in principle with the supposition that women might have systematically different ways of doing things, or that these ways might be systematically more successful than men's (though I know of no serious evidence that either of these is actually true), but to the extent that this is what is claimed by feminists, it does not call for any changes in fundamental conceptions of science. If women were successful in this way, ordinary standards of scientific success (such as reaching successful theories more quickly than men) would show this to be so (see above, pp. 393–394). To show that more fundamental changes were needed in the criteria for scientific success it would be necessary to imagine women's being *unsuccessful* by current standards—having theories that tests kept showing were getting nowhere, making predictions that were usually unfulfilled, and so on—and then saying that scientific standards should be changed to count *this* as good science. It is, again, difficult to imagine either what such standards would be, or that any feminist would recommend them. All this provides further reason for suspecting that many feminist claims about the need for epistemological change may really be about the need for change in the first-order beliefs that provide intermediate standards for judgment.

30 And, of course, a pragmatically self-refuting one. If we can tell what is going to benefit women, we must think we know something about how the world works, and therefore must presuppose an epistemology other than the one we are supposed to be defending.

31 Or perhaps (though this possibility has not been discussed here) to depend on traditional epistemological claims that contradict the conclusion.

32 In fact it would be stretching things a bit to count this as part of *feminist* politics, since it would be for the benefit of *anyone* who was disadvantaged by the present sort. This aspect of the arbitrariness of counting a particular kind of epistemology as feminist is in effect noted by Alcoff & Potter (p. 4), though differently expressed and understood. There is no sign of their being aware of the other problems involved in claiming particular theories as feminist.

33 But that may be less likely than in the case of first-order inquiries.

34 See above, beginning of INTERMEDIATE STANDARDS, p. 392.

35 It may be objected that "feminist" can legitimately be used to mean (more or less) "done in a characteristically female way," and that in this sense of the word it is not arbitrary to claim particular approaches as feminist. This sense of "feminist" is

formally repudiated by most feminists, though it is obviously entrenched in popular usage. My own impression (which of course needs fuller justification) is that when the world is used to characterize first-order activities—scientific practice, normative ethics—it tends to mean "female," and when it characterizes second-order questions of metaethics, philosophy of science, and epistemology, it concerns getting recognition for what is supposedly female at the other level. "Feminist epistemology" never seems to mean "female epistemology." Even if it did, however, this would not help. There would, first, be the serious empirical question of whether women did really do things in the way claimed as feminist—which the many women who repudiate the ideas of feminist epistemology would say was an unfounded slur. But even if they did, to claim these ideas as *feminist*, rather than just female, is to imply an endorsement of them (the term"feminist" would not be used by a feminist who thought these female ideas nonsense). That presupposes an epistemology to supply the endorsement; and that, by the arguments presented here, cannot itself be feminist.

36 James Randi, *The Truth about Uri Geller.*

37 Mill, *The Subjection of Women*, p. 44.

38 It would be irrelevant here to ask, rhetorically, "by whose standards?" or otherwise raise questions about the standards of logic and scientific method being used. This argument is neutral between different possible standards. Resolve the fundamental problems of these matters any way you please, even reaching conclusions that are skeptical or relativist, and the argument about feminism goes through in the same way: until you understand how to apply whatever standards you do accept, you cannot see whether the treatment of women is wrong by those standards. (And anyone who does go for relativism or radical skepticism, or anything else too far from familiar standards, is likely to run into difficulties in finding anything wrong with the situation of women or anything else.)

39 There may of course be continuing anomalous *treatment* of women in such departments, or an unwillingness to address woman-connected anomalies, and these are a continuing cause for feminist concern. But that must not be confused with there being anything essentially patriarchal about particular theories.

40 See, e.g., note 31, but there are many more. If feminists think the rest of the world goes around with epistemology like this, it is no wonder they think change is needed—though of course even if it is, that does not mean that it is needed for feminist reasons.

41 This is *not* to suggest that everything that appears under the heading of feminist epistemology is gobbledygook. The very fact—already mentioned—that there is such a confusion of levels, and so many different things going on, in what is claimed as feminist epistemology or science means that the quality may be thoroughly mixed. Although I do think there is not much hope for any of the genuinely epistemology claimed as feminist, much that appears in the collections may come into the category of genuinely useful feminist criticism.

42 This is the most appropriate reply to the complaint that came from the floor of the conference, that we were "hypocritical" (why hypocritical I have no idea) in not having representatives of the *other side.* It is much more to the point than the *tu quoque* response about that other side's having conferences at which we would be unwelcome.

43 A week after Mary Lefkowitz's lecture (this volume), I heard a BBC discussion in which someone was seriously talking about the discovery that Greek culture had come from Africa. She had read the claims, and since she had no other information on the subject, she had no reason to suspect they were quite unfounded.

44 This is why it is important whenever possible to find some clear and simple way—without going into details that most people have no time to follow—in which to demolish the claims of spurious experts. It is also why it is essential to try to persuade universities and publishers to take their responsibilities seriously. People generally assume, as they should be entitled to, that university appointments and publication by respectable publishers are guarantees of opinions worth listening to.

## REFERENCES


ALCOFF, LINDA & ELIZABETH POTTER, eds. *Feminist Epistemologies.* London: Routledge, 1993.

ALEXANDER, LARRY. "Fancy Theories of Interpretation Aren't." *Washington University Law Quarterly* 73, no. 3 (1995): 1081–1082.

CODE, LORRAINE. "Taking Subjectivity into Account." In *Feminist Epistemologies*, edited by Linda Alcoff & Elizabeth Potter. London: Routledge, 1993.

FOX KELLER, EVELYN. "The Gender/Science System." *Hypatia* 2, no. 3 (Fall 1987).

GOLDBERG, P. "Are Women Prejudiced against Women?" *Transaction* 5, no. 5 (1968): 28–30.

GROSZ, ELIZABETH. "Bodies and Knowledges: Feminism and the Crisis of Reason." In *Feminist Epistemologies*, edited by Linda Alcoff & Elizabeth Potter. London: Routledge, 1993.

HERODOTUS. *The Histories.* Edited by A. R. Burn. Translated by Sélincourt. London: Penguin, 1972.

KELLY, ALISON, ed.. *The Missing Half.* Manchester: Manchester University Press, 1981.

MILL, J. S. *The Subjection of Women.* Edited by Susan M. Okin. Indianapolis, IN: Hackett, 1988.

NELSON, LYNN HANKINSON. "Epistemological Communities." In *Feminist Epistemologies*, edited by Linda Alcoff & Elizabeth Potter. London: Routledge, 1993.

OAKLEY, ANN. *Subject Women.* Oxford: Martin Robertson, 1981.

POTTER, ELIZABETH. "Gender and Epistemic Negotiation." In *Feminist Epistemologies*, edited by Linda Alcoff & Elizabeth Potter. London: Routledge, 1993.

RADCLIFFE RICHARDS, JANET. *The Sceptical Feminist.* 2nd edit. London: Penguin, 1994.

———."Traditional Spheres and Traditional Logic." In *Empirical Logic and Public Debate, Essays in Honour of Else M. Barth*, edited by Erik C. W. Krabbe, Renée José Dalitz & Pier A. Smit, pp. 319–338. Poznan Studies in the Philosophy of the Sciences and the Humanities 35. Amsterdam: Rodopi, 1993. (Reprinted in *The Sceptical Feminist*, 2nd edit., pp. 358 ff, passim.)

RANDI, JAMES. *The Truth about Uri Geller.* Buffalo, NY: Prometheus Books, 1982.

# FEMINIST EPISTEMOLOGY
## Stalking an Un-Dead Horse

NORETTA KOERTGE

FEMINIST EPISTEMOLOGY consists of theories of knowledge created *by* women, *about* women's modes of knowing, *for* the purpose of liberating women. By any reasonable standard, it should have expired in 1994. Working independently, Gross and Levitt in *Higher Superstition*, Sommers in *Who Stole Feminism?* as well as Patai and Koertge in *Professing Feminism* all identified fatal flaws in the feminist epistemological program. More detailed analyses appeared in *Feminist Epistemology: For and Against*, a special issue of the *The Monist*, edited by Haack. The simple bottom line of all these critiques is succinctly expressed by Pinnick in a 1994 issue of *Philosophy of Science:* "No *feminist* epistemology is worthy of the name, because such an epistemology fails to escape well-known vicissitudes of epistemic relativism. The central thesis of this article is that *feminist* epistemology should not be taken seriously."[1]

There is a long history of cogent criticisms of feminist epistemology—recall, for example, Radcliffe Richards's beautifully argued book, *The Sceptical Feminist*, which appeared in 1981. And at a symposium in 1980, where Harding and Hartsock were already decrying Bacon's alleged rape metaphors, I vigorously criticized their "standpoint" epistemology: "One final polemical remark: If it really could be shown that patriarchal thinking not only played a crucial role in the Scientific Revolution but is also necessary for carrying out scientific inquiry as we know it, that would constitute the strongest argument for patriarchy that I can think of! I continue to believe that science—even white, upperclass, male-dominated science—is one of the most important allies of oppressed people."[2]

But whatever its cognitive deficits, feminist epistemology is sociologically very successful. (To give just one indication, the program for a recent conference on gender and science lists 90 speakers.) So for me the most pressing question is to understand *how* feminist epistemology functions within educational institutions today and *why* it is viable. Is this just the latest example of Gresham's Law at work in the humanities, whereby simplistic bad ideas drive out the more complex good ones, or is something more unusual going on?

Feminist epistemology is but one of literally dozens of new specialties that have recently sprung up in American universities. These "paradisciplines" as I call them are no longer confined to women's studies programs. One can find them offered as alternatives or "correctives" within regular university departments. Thus we find books, journals, conferences, and college courses devoted not only to feminist epistemology, but also black epistemology and queer epistemology. There is lesbian ethics and feminist morality, feminist aesthetics and feminist musicology.

Paradisciplinary initiatives are even taking root within the sciences. Psychology of women, black psychology and biology of women have now been joined by feminist economics and feminist geography. Opposition to the most central methods and tools of science is fostered in the paradisciplines of ethnomathematics, Afrocentric science, and feminist methodology. We thus are faced with a profusion of new academic specialties that not only claim to *complement* traditional scholarship but also to replace or "reinvent" it in radical ways. How did so many of these oppositional subjects get established so quickly? Here I will briefly discuss only two of the contributing factors, academic separatism and the ethos of affirmative action.

In describing the emergence of new species, one mechanism invoked by evolutionary biologists is the "founder effect." Since small samples are generally unrepresentative of the whole population, if a small group of organisms should become geographically isolated from the rest, as the small group inbreeds, the idiosyncrasies of the founders become dominant and a new species may emerge in a relatively short time. This is exactly the situation that obtained in women's studies programs. As we describe in detail in *Professing Feminism* there was in the beginning a deliberate attempt to isolate feminist scholarship from the rest of the academy. Some authors would cite only women in their footnotes; since men were thought to be biased, only women were considered competent to referee articles for publication; men were sometimes even excluded from attending conferences and were rarely invited to speak. The policy of restricting participation in allegedly academic discussions to people of the appropriate "identity" was sometimes also used to filter out people on the basis of race, ethnicity, and sexual orientation. By severely limiting the influence of outside commentary and by aggressively promoting each others' work, the seminal (ovular?) works within these various alternative disciplines quickly gained the trappings of scholarly success. To be blunt, how can one deny tenure to someone whose book receives rave reviews in (feminist) journals and whose book jacket sports blurbs from (feminist) professors at Berkeley, Columbia, or M.I.T.? So it is easy to understand how feminist and other paradisciplines got off to a roaring start, but it remains a puzzle as to how they have become so widely accepted as part of ordinary disciplinary offerings, *even by critics who find their substantive claims unpersuasive.* To answer this question we need to look at the interaction between the ethos of affirmative action and the formation of these alternative disciplines.

Women and minorities have always tended to cluster within certain academic specialties—women are more likely to be pediatricians than surgeons, harpists rather than percussionists, ethicists instead of logicians. Some of the clustering follows gender stereotypes, so, for example, a disproportionate number of women study child development, sociology of the family, botany, etc. In other cases, the patterning is probably best explained in terms of mentoring chains and role models. One thinks, for example, of the extraordinary number of female primatologists and the famous women in X-ray crystallography such as Dame Kathleen Lonsdale and Rosalind Franklin. One purpose of second wave feminism in general and university affirmative action plans in particular was not just to increase the number of women professors, but also to expand the range of disciplinary possibilities by breaking down the stereotypes that women naturally belonged in "soft" fields.

However, one major result of these political initiatives in the university has been the creation of new "pink collar" ghettos! Feminist activists, as we saw above, have practiced deliberate segregation. But in addition, anyone who begrudges the initiatives to bring more women or minorities into the university might also be happy to see "them" shoved off into an academic ghetto where "they" will not interfere with business as usual. And the middle-of-the-road, well-meaning, guilty liberal white male has uncritically promoted the new paradisciplines, mainly from afar so as not to compete with the women and minorities who "own" these new fields. What is the effect on young women and minorities of this strange synchrony of support? I will illustrate my concern with two incidents that have come to my attention recently (I have changed identifying details):

A young woman finishing a dissertation on the foundations of statistics opined that perhaps she should do an independent reading course on feminist epistemology. When I remarked, "Oh, I didn't know you were interested in studying those issues," she replied that she was not really, but thought that since she was a woman applying for a job in philosophy departments, people would expect her to be able to teach such things. I recommended that she stick to her own research interests and not try to second-guess the market, but upon glancing through the *Jobs in Philosophy* afterwards and noting all of the interest in feminist philosophy, I wondered if I had given her good strategic advice. (Note that as long as only (or mostly) women do feminist studies, then by recruiting in such a field, it will almost always turn out that the very best candidate really is a woman. By applying affirmative action criteria to fields, one no longer need apply them to individual candidates!)

A second example came to my attention during a job search for a philosopher of biology. We received an application from an African American male who had completed a Ph.D. at a good school working with one of the leaders in the field, but whose publications all dealt with African philosophy. "I was advised that it would be easier for me to publish quickly in African philosophy," he wrote, "but I really want to get back into philosophy of science," and sure enough, one of his letters of recommendation (from a person in no position to judge the quality of his dissertation) happily took credit for the applicant's switching fields.

These cautionary tales exhibit the pressures on women and minority students to choose academic areas that are supposedly "appropriate" to their "identities." So much for expanding career possibilities and so much for providing new role models in a diversity of academic specialties! Paradisciplines, as I have defined them, are not intended just to introduce new perspectives on or vital additions to the traditional disciplines. Rather, they stand in explicit and wholesale opposition to the received approaches. So, for example, the history of women in science need not be a paradiscipline according to my definition. Learning about the careers of "forgotten" women scientists or African American scientists can certainly add an important dimension to our understanding of scientific institutions as well as social perceptions of science. Such a history may chide science for not being as open to all talented people as Merton's norms would suggest, but unless one starts redefining "scientist" to include midwives, herbalists, and scullery maids (as some would have us do), the history of women in science does not detract from our understanding of science, but enriches it. Feminist epistemology, on the other hand, stands in a sharply antithetical relationship to the core values of science. A dramatic way of summarizing the conflict is to look at feminist commentary on standard accounts of scientific norms.

Let me begin with what is intended to be a noncontroversial summary of the "received view" of scientific ideals. Talcott Parsons[3] listed four basic norms of *scientific knowledge:*

a. Logical clarity or precision
b. Logical consistency among claims
c. Generality of principles
d. Empirical validity

To foster the search for this kind of knowledge, Robert Merton[4] noted that scientific institutions need to promote:

a. Organized skepticism
b. "Universalism" (scientific contributions should not be judged on the basis of the race, religion, national origin, etc. of the scientist)
c. Disinterestedness (science should not serve a particular social/political agenda)
d. Communality (scientific results should be freely shared)

Although the credibility of science relies heavily on institutional features, such as peer review and the cross-checking of experimental results, it also depends strongly on the personal integrity of *individual scientists.* A complex of such traits can be summed up under the norm of *objectivity:*

a. Data reported and the conclusions drawn should ideally be completely *independent* from the personal preferences or idiosyncrasies of the individual scientist.
b. A good method for removing subjective elements from scientific findings is to *detach* one's own feelings or wishes from the process of scientific inquiry.
c. Although intuition and *Fingerspitzengefühl* play an essential role in the process of scientific discovery, they should have no effect on the acceptance of scientific results. Thus, while individual or local points of view may be very important in suggesting scientific strategies, the knowledge obtained eventually applies everywhere.

d. A high value is also placed on the individual scientist's *curiosity* and *intellectual fascination* with discovery and puzzle solving. (These traits are intimately connected with objectivity because ideally the only answers that scientists find pleasing are correct ones!)

Now it is quite appropriate and reasonable for feminist scholars to point out various ways in which the actual practice of science sometimes fails to live up to these norms. For example, medical or psychological theories have sometimes not been adequately tested on female subjects and hence may lack the empirical validity and generality prescribed by Parsons. And despite the Mertonian norms of universalism and communality, women scientists sometimes find that their work is not taken as seriously as that of comparable male colleagues and they may not be included in informal communication networks.

It could also be argued, given that no one can be perfectly objective, that as long as science is done primarily by males, male perspectives might indeed influence to some extent the direction of scientific research. Science itself would benefit from the input of women (and others) who might not only bring in new heuristically valuable points of view, but could also provide additional sources of critical scrutiny. Science can only be improved by such exhortations to live up to its own ideals. However, when we turn to radical feminist critiques of science based on feminist epistemology, we find a repudiation of the ideals themselves. Here is an overview[5] of their opposition to the traditional norms:

None of Parsons's norms are acceptable: logic is a patriarchal device for browbeating nonlinear thinking; since all knowledge is contextual, the search for generality is a form of imperialism; empirical validity must be tempered by moral and political appraisals.

Communality of a nonhierarchical sort is acceptable, but the rest of Merton's norms must go: a humane community would be based on trust, not skepticism; universalism should be replaced by standpoint theory, which says that reports are always to be understood as a product of the culture, gender, ethnicity, class of the observer who made them; no activity can be or should be disinterested. Quite the contrary, a commitment to correct political and social goals is to be encouraged.

Although the term "objectivity" is sometimes retained by radical feminists, the values now denoted by it are antithetical to the traditional meaning: observers should always remain emotionally connected to what they are studying; the richness of subjective experience should not be stripped away in the vain search for a lowest common denominator of objectivity; intuition should not play second-fiddle to abstract, cold rationality/objectivity; knowledge is always perspectival and tied to local context, and the attempt to find an objective or "God's eye" point of view always ends up privileging the powerful. Thus the playful curiosity so characteristic of so-called pure science must be replaced by an attitude of caring and commitment.

As I remarked at the beginning of this paper, as a philosophical system the various tenets of feminist epistemology have been decisively discredited.

Nevertheless, it is having a growing influence on science education. At Indiana University, for example, one required textbook for future science teachers is *Women's Ways of Knowing*,[6] winner of a distinguished publication award from the Association of Women in Psychology, a book that argues that girls and women are more comfortable with a "connected" style of learning as opposed to the "separated" style preferred by men—and scientists! The empirical evidence for these alleged differences is very weak—for one thing, the authors did not include any males in their study. But even if it were true that some little girls (or little boys) were uncomfortable with the sort of reasoning required to do science, I would draw quite a different conclusion from feminist epistemologists and educators, namely, that these children should expand their cognitive repertoires, not that science should abandon its modes of reasoning.

There are many ways of thinking about and learning about the world—some are antithetical to science. But even within science, we talk about algebraists vs. geometers, "lumpers" vs. "splitters," those who are good at synthesis vs. those who delight in details. A primary purpose of education at any level is to help students become better thinkers, more sophisticated, more critical, and more self-conscious about their methods of inquiry and belief formation. What a pity, if in the name of liberating women, feminists should now encourage women and members of various ethnic groups to stay comfortably within the habits of thought that conform to traditional gender and cultural stereotypes. One of the joys of liberal education in either the arts or sciences is the challenge to learn how to think differently. How patronizing to tell young women that the ways of logic, statistics, and mathematics are not women's ways—that all they need to do is to stay connected.

## NOTES

1 C. L. Pinnick, "Feminist Epistemology: Implications for Philosophy of Science."
2 N. Koertge, "Methodology, Ideology and Feminist Critiques of Science."
3 Talcott Parsons, *The Social System.*
4 Robert Merton, *The Sociology of Science: Theoretical and Empirical Investigations.*
5 This summary is a composite of the views that are carefully analyzed in the writings mentioned in the first paragraph of this paper.
6 M. F. Belenky, B. M. Clinchy, N. R. Goldberger & J. M. Tarule, *Women's Ways of Knowing: The Development of Self, Voice, and Mind.*

## REFERENCES

BELENKY, M. F., B. M. CLINCHY, N. R. GOLDBERGER & J. M. TARULE. *Women's Ways of Knowing: The Development of Self, Voice, and Mind.* New York, NY: Basic Books, 1986.

GROSS, PAUL & NORMAN LEVITT. *Higher Superstition: The Academic Left and Its Quarrel with Science.* Baltimore, MD: Johns Hopkins Press, 1994.

HAACK, S. *Feminist Epistemology: For and Against. The Monist.* La Salle, IL: Hegeler Institute, 1994.

KOERTGE, N. "Methodology, Ideology and Feminist Critiques of Science." In *PSA 1980: Proceedings of the 1980 Biennial Meeting of the Philosophy of Science Association*, edited by P. Asquith & R. N. Giere, vol. 2, pp. 346–359. East Lansing, MI: Philosophy of Science Association, 1980.

MERTON, R. K. *The Sociology of Science: Theoretical and Empirical Investigations.* Chicago, IL: University of Chicago Press, 1973.

PARSONS, T. *The Social System.* New York, NY: The Free Press of Glencoe, 1951.

PATAI, D. & N. KOERTGE. *Professing Feminism: Cautionary Tales from the Strange World of Women's Studies.* New York, NY: Basic Books, 1994.

PINNICK, C. "Feminist Epistemology: Implications for Philosophy of Science." Philosophy of Science 61 (1994): 646–657.

RADCLIFFE RICHARDS, J. *The Sceptical Feminist: A Philosophical Enquiry.* London: Routledge and Kegan Paul, 1981.

SOMMERS, C. H. *Who Stole Feminism? How Women Have Betrayed Women.* New York, NY: Simon and Schuster, 1994.

# THE SCIENCE QUESTION IN POSTCOLONIAL FEMINISM

MEERA NANDA

HIS FEMINIST SYMPATHIES are not something that the late Ernest Gellner—a philosopher and a social anthropologist—is best known for. In fact, "women" receive two, passing references and the "feminist movement" exactly one in the books that inspired this paper.[1] And yet, this Indian feminist finds a deep personal resonance in Gellner's affirmation of reason[2] as an ally of all those struggling to break free from the margins that their natal cultures have consigned them to.[3]

Gellner answers some of the doubts I have long entertained about the implications of the currently fashionable *hermeneutic* egalitarianism for the project of *political* egalitarianism. Accepting for a moment the postmodernist idea that what and how we know depends on our social location, can the assumed equality of knowledge systems of all social groups, irrespective of their location in the hierarchies of power, realistically hope to alter the balance of power in the favor of the relatively powerless, including those who have been historically denied full participation in the social order—the colonized people, the people of color, the majority of working people, and in all these cases, only doubly so, women? In other words, what will serve the emancipatory interests of those on the margins of power—an insistence on the presumed cognitive validity (and indeed superiority) of their diverse local, traditional knowledges, as has become common in the contemporary critiques of Western modernity? Or a chance to acquire and participate in the dominant knowledge systems, which in the contemporary world are closely associated with modern scientific rationality? Gellner offers an interesting and original argument for why the subjugated need to overcome the duality between the local traditions of knowledge and the "Western" scientific rationality that has become nearly universal in its reach: they need to appropriate the latter in order to enter a creative dialogue with the former. Thus, a repudiation of oppressive power structures and ideologies does not require a repudiation of reason or, in Gellner's terms, *you can negate Kipling without having to renounce Descartes.*[4]

Women may not be foremost on Gellner's agenda, but I wish to appropriate his defense of reason for women's struggle for autonomy, equality, and dignity in postcolonial (Third World) societies. I argue against the tendency of some postcolonial feminists to decry Descartes—and through him, the entire legacy of the Enlightenment and the Scientific Revolution—as "Western" and "patriarchal." Instead, I try to show why women in postcolonial societies need the spirit of Descartes to fight not just Kipling, but our own Manu[5] as well. I suggest that scientific rationality can serve to generate a critical stance toward the cultural discourses that constitute the self-identities of women living under traditional patriarchal arrangements in most postcolonial societies. Reason can become a part of Third World women's "situated knowledges" and provide a set of nonarbitrary criteria against which the claims of the arbitrary authority of the father, the family, and the community can be judged. Thus, with Gellner's help, I intend to go against the grain of contemporary feminist and postcolonial critiques of science and argue that postcolonial women can indeed use the "master's tools to dismantle the master's house."[6]

Such an explicit defense of reason has become heretical in these postmodern[7] times, especially when it follows from the work of a "humble adherent of Enlightenment Rationalist Fundamentalism" as Gellner calls himself.[8] After the humbling reason has taken lately, Gellner's ideas may strike some as quaint (or dangerous) anachronisms—an intellectual equivalent of meeting a card-carrying Bolshevik who insists on calling St. Petersburg, Leningrad. Ernest Gellner, of course, is no Bolshevik, epistemic or political. As one of the best known contemporary British philosophers, Gellner is not only familiar with, but has actively shaped the debate over, the plausibility of a notion of truth that can transcend culture, gender, race, and class. Neither is he, as we shall see, incognizant of the disasters—colonization, the Gulags, and the death camp—that have followed in the wake of each attempt to reorder societies in the image of some transcendent truth. If Gellner were simply asserting the superiority of a transcendent reason as a matter of faith, reading these books would not be worth the effort. But he provides a persuasive defense of the liberating potential of the *ethos* of scientific rationality.

This makes Gellner especially relevant for those of us for whom reason and science have come prefixed with "Western," a qualifier with connotations not only of foreign origin, but also of a history of material exploitation and cultural oppression associated with colonialism. Consequently, women from postcolonial societies have an exceptionally "uneasy alliance" (to borrow a phrase from Seyla Benhabib[9]) with both reason, and with the postmodernist repudiation of it. The Enlightenment reason and science came to the postcolonial world not as harbingers of progress—however contradictory and ambivalent it has been for our sisters in the West[10]—but as a part of the apparatus of the empire meant to administer the "natives," which included a rationalization of local patriarchal practices. This explains why the postmodernist critique of the Enlightenment and reason have found a tremendous resonance among some postcolonial intellectuals, including feminists, some of whom tend to embrace a nostalgic and nationalistic antimodernism.[11]

But if Third World feminists have not entirely lost faith in the "Western" idea of emancipation, with all that it promises—self-determination and personal autonomy,[12] equality in the public and private sphere—we have to combat *all* relations of patriarchy, Western *and* ancestral. Granted, the two are overlaid on one another, each reinforcing the other through a complex and historical complicity between them. Yet, ancestral oppressions do not become any less cruel just because they are "our own," and alien ideas do not exhaust themselves just because they are "not ours." This reasoning obviously represents the worst kind of genetic fallacy and can only lead to a suffocating, reactionary nationalism which has never been anything but oppressive for women. Indeed, as Seyla Benhabib points out,[13] there are times when in order to free ourselves from the chains "our own" culture has forged for us, a "social critic [has to] become the social exile."[14] As Third World feminists, then, we cannot unambiguously join the postmodernist and nationalistic critics in bidding farewell to modernity and reason, for reason has been both a source of oppression *and* a resource for emancipation for women. Gellner's special talent lies in the way he compels his readers to remember the often forgotten emancipatory impulse of reason.

Emancipatory. That is exactly how *I* experience the training I received in India in the field of molecular biology—I have a Ph.D. in this embodiment of "reductionist," "patriarchal," and "Western" science. Learning and doing science in my otherwise tradition-bound community opened up a whole new world for me: it gave me the intellectual resources to define a sense of self and to demand a modicum of autonomy from the claims of my family and community. Ironically, the very struggle for the right to follow my own project changed my relationship with my project itself: I could not continue to do science as a mere career, I wanted it to illuminate my whole world. I became active in science for the people movements and worked as a science writer for a newspaper in India. Looking back, I count my years in the laboratory as the most transformative period in my life.[15] That is why I am never fully convinced by the arguments of my Western friends who decry science as deeply gendered and misogynist. Arguments of my old friends from India, similarly, who have become increasingly vocal in denigrating scientific rationality as an agent of imperialism and "Western patriarchy"[16] fail to sway me.

---

Is it any surprise, then, that I find myself cheering Gellner when he writes, "autonomy requires reason and reason requires autonomy"?[17] I submit that, postmodern skepticism regarding the idea of the autonomous individual notwithstanding, we must pay attention to the historic, autonomy-enhancing role of reason.

What does autonomy consist of, and what does reason have to do with it? In philosophical terms, Enlightenment rationalism was driven by the quest for a solid and autonomous sense of the self—a self capable of constructing the world on foundations that are wholly one's own and not simply taken over from an unexamined cultural inheritance. The quest for rational self-creation—to trust nothing that you have not made and tested yourself—is what separates the moderns from the ancients. To be modern is to be compelled to face reality, unaided by the consolations of traditions, religion, or metaphysics—and unhindered by the constraints of arbitrary authority.

These are problematic statements. I am fully aware that the postmodern academy has long since declared the "death" of such an autonomous subject. Far from an idea to strive for, the idea of a sovereign, self-determining subject has been declared an arrogant fiction of liberal humanism, which has sought to universalize the Western male concept of reason over all other modes of being, perceiving, and living. According to the critics, what the Enlightenment wrought was not emancipation from convention, superstition, and arbitrary authority, but the beginning of a new totalitarianism of Western rationality. Thus, they wish to replace this humanist hubris with a sense of radical situatedness and relativeness of all human knowledge.

This is where the contemporary critics of Enlightenment rationality will have to reckon with Gellner, a life-long defender of the universal cognitive power of science. Gellner accepts that any straightforward, mimetic representation of reality in which our knowledge can be grounded is impossible. The Promethean aspiration of the Enlightenment for certain knowledge, based only on self-produced and self-tested mental/sensory resources, was bound to fail because we cannot *but* fashion our cognitive tools from the cultural resources available to us: the social relations, the language, the myths, and the rituals. But while Gellner accepts the constructed nature of scientific knowledge and rationality, he refuses to read it as announcing the "death" of reason, objective knowledge, etc. as the postmodern thinkers tend to.

If all knowledge, including scientific knowledge, is socially and culturally constructed, is it possible at all *not* to take the postmodernist and postcolonial redoubt which proclaims the cognitive equivalence of all knowledge claims? In this respect, Gellner seems to be asserting two contradictory theses: yes, reason is bound by culture; *and yet*, knowledge beyond local cultures and morality—epitomized in natural sciences—is not only possible, it is a central feature of our shared human condition in the twentieth century. This tension is palpable throughout Gellner's work, though it is finally resolved in *Reason and Culture*. In *Postmodernism, Reason and Religion*, Gellner puts up a spirited defense of the existence of "knowledge beyond culture and morality"[18] against the postmodernists (especially anthropologists among them who posit a symmetry and cognitive equivalence among all knowledge systems), and against the religious fundamentalists (who put revelation above evidence).[19] In *Reason and Culture*, Gellner leads his readers to a journey through the history of scientific rationality, starting with Descartes–Hume–Kant to Durkheim, Weber, and the intellectual progenitors of the postmodern challenge to reason, namely, Kuhn and Wittgenstein.

---

I find Gellner's resolution of this contradiction quite persuasive: rationality cannot transcend culture but it can—and has—engendered *a new kind of culture*. In this new culture, scientific rationality with its methodological imperatives of exposing our beliefs to empirical evidence becomes our tradition, our "custom and example." Gellner's own words can best convey the flavor of this new culture:

> The Cosmic Exile, the opting out of culture, is impracticable. But it constitutes the noble and wholly appropriate charter or myth of a new kind of culture, a new system of a distinctively Cartesian kind of Custom and Example. Custom was not transcended: ***but a new kind of custom altogether was initiated*** . . . [a culture] built on wholly new principles. All the same, it was *a* culture, rather than a transcendence of all cultures, as Descartes had supposed. It had its own and distinctive compulsions, and they too had their social roots, as Weber taught.[20]

While Gellner is an unabashed partisan of this new culture, he is by no means unaware of the bloody history and the spiritual/moral poverty of all attempts to engineer a rational social order. We shall shortly examine how he resolves *this* set of contradictions. But before we do that, it will be worth our while to flesh out the history and the contents of this scientific culture that Gellner wants to defend, more so because the patron saint of this new culture, René Descartes, has been lately out of favor with the feminist and postcolonial critics of the Enlightenment.

In Gellner's eloquent and sympathetic account, cognitive autonomy was what Descartes was after when he set out to systematically doubt all "truth of which I had been persuaded merely by *example and custom*."[21] For Descartes, "example and custom"—that is, inherited culture—cannot provide secure enough foundations for one's beliefs, for there is no way of ascertaining that the received wisdom is free from error, or that it is the best possible answer.[22] As Gellner describes it, the battle lines were clearly drawn between individual reason and collective culture. Descartes believed that "truth can be secured only by stepping outside prejudice and accumulated custom . . . by means of proudly independent, solitary Reason."[23] The collective and customary can be overcome only by the inner compulsion of clear and distinct ideas, the classic example of which, of course, is the argument Descartes is best known for: *I think, therefore I am.* The indubitability of this argument gave Descartes grounds for believing that cognitive trustworthiness was available. He then went on to infer the existence of a benevolent God (no less!) from the existence of the thinking self, and to conclude that the clear and distinct ideas underwritten by God are fit to guide us toward certain knowledge.

Some feminist critics have interpreted Descartes's association of reason with transcendence (i.e., overcoming) of the circumstantial, the given and the sensory—all attributes historically associated with the feminine—as inaugurating a masculine rebirth of knowledge that led to the gradual separation of women and rationality.[24] One of the major weaknesses of Gellner is that he simply does not engage with the considerable feminist literature on this subject. But in my reading, Gellner is not completely at odds with the feminist reading, although in the end, as we shall see, he does not give gender, nationality, or any other biographical feature of the knower much weight in the acquisition and appreciation of scientific knowledge.

Descartes's was only the beginning of the rationalist program. Hume and Kant—and through them, the eighteenth century Enlightenment—inherited Descartes's quest for knowledge that could be validated without falling back on the dreaded "custom and example." In a very eloquent (though not terribly original) interpretation, Gellner emphasizes the autonomy-enhancing aspects of Kant's "anti-Copernican Counter-Revolution." He presents Kant as "having used philosophy to restore centrality to mankind [sic]. He made the structure of the human mind, rather than the structure of the world, pivotal and fundamental."[25] In other words, the order that makes things knowable is not inherent in those things themselves, or backed up by a guarantee by a benevolent god: it inheres instead in the manner in which our mind handles and classifies them. We *cannot* think of the world any differently, and it is these universally shared inner compulsions that make us human.

The denouement came with Durkheim and Weber in the early twentieth century. Not satisfied with Kant's assumption of a rational order that simply inheres in all human minds, across history and across cultures, Durkheim went on to actually observe human rationality in the real world. In the process, he opened the door once more to culture and example, not as a source of error but much more fundamentally, as the very source of our rationality itself. The Kantian order does not simply inhere in the human mind, but is instilled there through *ritual*; thus, when we think, our culture thinks through us.

On a Durkheimian reading, the Cartesian flight from the merely conventional and customary was not only not possible, the scientific rationality it sought was itself the voice of the society—and not society as such, but the early-modern Western society with its own peculiar cultural-religious ethos. It was Max Weber who showed that it was a "special mutation of the sacred," i.e., the rise of Protestantism, that instilled a new kind of compulsion in Descartes and his followers: "to accept only conclusions imposed in a state of sobriety whilst contemplating lucid, clear and distinct, uncontentious notions . . . [and to reject] the old reliance on the local, unsymmetrical accidents of history, on the arbitrary rituals of this or that culture."[26] The Enlightenment codified this new kind of compulsion in the formal structure of modern science, and in the process obscured its social roots. Gellner follows the return of the repressed—all that is cultural, contingent—right through the historicism of Collingwood and Kuhn, Schopenhauer, Nietzsche and Freud and later Wittgenstein (in *Reason and Culture*) to the contemporary cultural anthropologists who have followed Clifford Geertz's "interpretive turn" (in *Postmodernism, Reason and Religion*). But in Gellner's account, all the later social and cultural constructionists appear as pale shadows of Durkheim and Weber who struck the first blow against the Enlightenment's hubris of transcendence.

Thus far, Gellner's reading is not at odds with the feminist and other post-philosophy critiques of scientific rationality. If what passes as scientific reason is only a peculiar, seventeenth century European variant of the Protestant ethos, it cannot possibly claim to transcend time, culture, and biography. How can such a cultural construct *not* bear the signature of its pioneer's and practioners's gender–the most primordial cultural category that gives meaning to most of our metaphysical categories? Once scientific rationality is placed in its cultural context, the gender of the knower can no longer be taken as irrelevant to what is known. Feminist suspicion of the masculine bias of reason is well grounded in the historical fact that the ideal type of the thinking and deliberating subject, as constructed in Western rationalist tradition from Plato onward, has coincided with that of a "man." As we know through the work of well-known feminist historians of the Scientific Revolution (notably Carolyn Merchant and Susan Bordo), the increasing value on rationality in the early modern Europe coexisted with a historically high level of gynophobia. This provides grounds for the often heard critique of Descartes as having initiated a radical separation of women and rationality. The postcolonial critics of reason similarly argue that white, bourgeois cultural ideal of reason was imposed upon the colonized people as a privileged way of knowing the "objective" reality.

But even though the feminist and postcolonial critiques of science can be read into Gellner's account of the denouement the Cartesian project has faced in the hands of anthropologists, sociologists, and present day postmodernists, Gellner himself has no sympathy for these critiques.[27] He accepts the cultural boundedness of cognition, but refuses to label scientific reason as masculine, Western, or as "the ventriloquist's dummy of a given social order."[28]

As hinted above, he maintains his apparently contradictory position by treating reason not as *transcending* culture, but as *constituting* a distinct culture in itself. In what he calls his "doctrine of the Big Ditch,"[29] Gellner believes that the new culture heralded by the Enlightenment and the subsequent Scientific Revolution in the West marks a great discontinuity in the intellectual history of humankind in the sense that a new form of knowledge came into existence that "surpasses all others, both in its cognitive power and in its social iciness."[30] Even though human cognition cannot transcend culture, yet, when scientific rationality itself provides a new cultural context, the result is knowledge that can transcend culture—knowledge that passes under the name of science. Science, Gellner holds, provides "an idiom capable of formulating questions in a way that answers are no longer dictated by the internal characteristics of the idiom of the culture carrying it, but by an independent reality."[31]

Gellner makes a strong case that in our times, scientific rationality has indeed become a part of our cultural ethos, not just in the West, but all over the world. The Cartesian quest for certain knowledge was futile, but it has become inescapable and mandatory, because:

> this aspiration defines us, even though it cannot be fulfilled. We are what we are because our ancestors tried so hard, and that effort has entered our souls and pervaded our cognitive custom. *We are a race of failed Prometheuses. Rationalism is our destiny. We are not free of culture, of Custom and Example: but it is of the essence of our culture that it is rooted in the rationalist aspiration*[32] (emphasis added).

Suspending all value judgment for a moment, there *is* ample evidence to support the proposition that scientific rationality has become "our destiny." For better or for worse, and not minimizing the impact of colonialism on demolishing the self-confidence of non-Western societies in their own knowledge systems, it is an empirical fact that modern science has acquired a near universal appeal. One need not look beyond the desperate eagerness of Third World societies to emulate the success of technologically advanced societies (often with disastrous results) to be reminded of the deep appeal this cognitive style has come to exert. While one can have reservations about its desirability, the fact that we live in a world where one cognitive style is being sought and adopted by diverse cultures and peoples can hardly be denied.

---

Accepting the empirical fact of the emergence of scientific reason as a new kind of culture with universal pretensions and appeal is the easier part. The harder part lies in answering the questions that follow. To begin with, what is it about scientific rationality that gives it the grounds to claim universality for its knowledge claims? More importantly for the political project of feminism in postcolonial societies, what makes scientific rationality a source of empowerment against the constraints of "custom and example" which often are antithetical to the idea of individual autonomy. Gellner provides answers that I find convincing, for the most part.

Ever since Foucault, the answer to the first question has become: power. Scientific reason universalizes itself because it has become indissociable from all regimes of disciplinary power. The cognitive content of science and the truth or falsity of its claims are no longer even conceivable apart from the power of science as an institution and a cultural practice. Gellner dismisses this whole tradition that stretches beyond Foucault to all species of social constructionisms in a one-liner: "the contrary view is often heard, but is simply false."[33] He substantiates this rather imperious move by arguing against the thesis of theory ladenness on which the generic social constructionist critique of science rests. Indeed, "Prometheus Perplexed," his concluding chapter of *Reason and Culture*, contains the most pithy argument against theory ladenness of observation that I have ever come across.

Theory ladenness of perception has been the Achilles heel of all attempts to ground scientific rationality in objective reality. Because we gain access to the objects of scientific investigation only through the languages and concepts in which we think, there is no way to get outside our beliefs, or to step outside language, in order to compare our claims about the object of investigation with bare, unmediated reality. We cannot thus, in principle, expose our hypotheses to the "real" world. Because nature and society are not distinct, all knowledge about the objective world (natural or social), and therefore the scientific method itself, is social. As anyone familiar with contemporary critiques of science will be able to attest, this has become the dominant view in the humanities, including most women's studies departments.

Gellner, needless to say, finds this critique highly unsatisfactory. He offers a defense of the scientific method that ties up neatly with his conception of reason *as* culture. The cultural ethos of the scientific tradition initiated by Descartes and carried over by Hume, Kant, and later empiricists makes it possible to keep theory saturation down to acceptably low levels. Unlike prescientific beliefs, which are imposed on those born into them as "package deals" and are protected from critical scrutiny, the ethos of science makes it imperative that we open the packages, that we "break up what is actually experienced, and turn it into the final court of appeal of theories."[34] These theories are themselves testable by exposing them to data *not* under the control of interpretations based on the theories being tested. This escape from theory saturatedness is possible because the interpretations that accompany perceptions are *themselves* repeatedly scrutinized over time from divers viewpoints—as the scientific method holds, at least as an ideal. Thus, *science as a collective and cumulative enterprise* continually breaks up all theoretical assumptions into experimental propositions and thus succeeds in bringing "truth under the control of nature" rather than society.[35] If nature can control our truth claims, then the existence of transcultural knowledge becomes unproblematic.

What Gellner seems to be claiming is that while the historic *origin* of reason was culturally bound (à la Durkheim and Weber), scientific rationality, understood as a sensibility, a temperament, a style of thought, can be adopted by diverse cultures situated in different geographical, historical, and linguistic spaces. The scientific temper is universally miscible with other cultures—with often unexpected and transformative results—because it is based on one clear and distinct idea that the rational mind cannot ignore, namely, "that anything which is in conflict with independently, symmetrically established evidence, cannot be true" and must be refused.[36] As shown above, Gellner believes that through collective and relentless questioning of all assumptions underlying our theories, science *can* arrive at just such independent evidence. Thus, in scientific rationality, we as a *species* have found an idiom capable of formulating questions about nature in such a way that answers are no longer dictated by the internal characteristics of the idiom or the culture carrying it but by *an independent reality*—a phrase Gellner uses often and without the apologetic quotation marks that have become almost mandatory these days.

Gellner freely admits that he has not solved the problem that has exercised philosophers of science for a long time, namely, how science works. He offers only pragmatic and fairly commonplace observations to support his conviction that culture-transcending knowledge does exist—for instance, the fact that the propositions and claims of science, in their purely technical content, are translatable without loss of efficacy into any culture and any milieu and that the new knowledge actually makes the real world obey its laws, as is obvious in technological advances all around us. But this recourse to consequences does not invalidate Gellner's argument. Indeed, it is equally possible to turn the tables on the postmodernist critics for they have *failed* to account for the pragmatic success of science.

---

So transcultural knowledge is possible. But is it desirable? What kind of knowledges are more empowering for postcolonial women struggling to gain a fuller selfhood and autonomy: purely local and experiential, as our postcolonial critics of science call for? Or the scientific as the sole privileged and correct source of knowledge, as the colonial powers and the modernizers maintained? These are the only two mutually exclusionary options postcolonial critics seem to be able to think of, and both are obviously rather stark and difficult choices in a world where the boundaries between the local and the cosmopolitan are becoming fuzzier by the day.

But Gellner strongly suggests that the critics may be posing a false antithesis between the local and the cosmopolitan. Although he does not cite Donna Haraway (or any other feminist science critic, for that matter), Gellner comes close to her idea of "situated knowledge"[37] in which the local and the universal exist in a creative tension with each other, each interrogating the other and providing the knower with a partial but reliable and verifiable knowledge that is suitable for living in freedom and dignity in this fast globalizing world in which "we have no clearly demarcated communities; [but only] fluid and unstable ones."[38] Thus according to Gellner when women with my kind of cultural background learn the idiom of modern science in our local settings, we should not be perceived as being (re)entrapped in a Western, imperialistic episteme. Instead, it is possible, according to Gellner (and again in close approximation with Haraway's notion of situated knowledge) to see postcolonial women remaking themselves through scientific rationality into Haraway's cyborgs, the carriers of "opposition consciousness," who are trying to cross the boundaries between "our own" and "their" traditions (and oppressions).

It is important to be clear about what is being actually claimed here. Science, according to Gellner, absolutizes no substantive convictions but only some formal, procedural principles of knowledge.[39] After all is said and done, the strengths assessed and weaknesses critiqued, this procedural requirement—*to accept nothing that contradicts independently established evidence* (which is not impossible, as Gellner tries to show)—is what remains as the most precious legacy of the Enlightenment. Gellner, however, deploys this procedural demand in a cultural rather than a formal sense. The scientific method appears in his works more like an ethos, a temper, an attitude rather than the scientific method, whose essence is still a matter of interminable debates among philosophers, historians, and, lately, sociologists of science.

What makes this method, understood as an ethos, a source of empowerment, for all those who have been denied their full personhood in all *anciens regimes*? The short answer is its deep *leveling quality*. The scientific ethos mandates that:

> there are no privileged or a priori *substantive* truths. All facts and observers are equal. There are no privileged Sources or Affirmations, and all of them can be queried. In inquiry, all facts and all features are separable: it is always proper to inquire whether combinations could be other than what had previously been supposed. In other words, the world does not arrive as a package deal—which is the customary manner in which it appears in traditional cultures—but piecemeal. Strictly speaking, though it *arrives* as a package deal, it is dismembered by thought.[40]

This, combined with his stress on the possibility of observations not dependent on our preconceived theories, leads to this basic formulation of the scientific ethos: separation of all questions and issues, the subjection of *all* claims to tests not under our own control, and the logical principle that no generalization incompatible with the results of these tests may be accepted.

These methodological imperatives indeed are potent enough to "place culture on trial" (1992a:169). No arbitrary authority—of God, the father, the "Custom and Example"—can survive this kind of scrutiny-by-doubt. The insistence of good reasons—which can be put to test—is fundamentally antithetical to acceptance of authority that is not supported by reason. Here Gellner is reaffirming (though in a far more optimistic vein) what Max Weber described as "disenchantment" of the world: a world where now there are no privileged individuals, institutions, or occasions, a world without miracles or saviors. No wonder, then, that the traditionalist forces everywhere—the cultural nationalists and religious fundamentalists alike—vent their fury at science.

I submit, and I am confident that Gellner would agree, that women from postcolonial societies have a stake in such a process of dethronement of arbitrary authority. *All* traditional, largely agrarian cultures (including that of the West before the Industrial Revolution) have at best offered a mixed deal to women. The warmth and closeness of the "community," so much in vogue these days among critics of modernity, have historically exacted a disproportionate price from women. In theory, science may not be able to validate its own methods and ground its answers in the indubitability of clear and distinct ideas or observations, as Descartes and the empiricists had hoped for. But in practice, the spirit of scientific inquiry can—and having lived through it, I can attest that it does indeed—enable women and all "others" to demand good reasons for the conduct required of them by their natal cultures. Only on the premises afforded by the scientific ethos—that is, to accept nothing that contradicts with independently supported, nonarbitrary facts—can one begin to see that "authority unjustified by reason is tyranny, and when supported by reason, is redundant, for reason alone should suffice."[41] This recognition is the necessary first step toward selfhood.

It is not difficult at all to shoot down Gellner's paean to Reason as the great iconoclast: indeed, in the present intellectual context, *defending* it has become much more iconoclastic. Even a cursory glance at the stifling conformism and commodity worship that prevail in the scientifically advanced cultures of the West is sufficient to lead one to doubt the role of the Great Emancipator Gellner has cast reason into. Gellner seems to credit reason with virtues it has never possessed. Women, for instance, have very good reasons to know that all knowers have never been considered equal in scientific institutions, nor have all facts and assumption been questioned with equal intensity, as Gellner claims in the passage cited above (this section). There *are* privileged metaphysical assumptions that derive their cognitive force from the cultural practices of privileged social groups. To some extent, the scientific method is able to keep them from being injected into all observations, but the play of these cultural, pregiven, and unquestioned sources of authority cannot be completely denied in science.

Gellner would agree with his critics that he has, indeed, drawn an idealized image of reason. But that is exactly what distinguishes him from his critics:

he wants to preserve the ethos of Enlightenment reason as an ideal to struggle toward, while the assorted postcolonial, postmodern critics are so disenchanted with the project of Enlightenment that they want to establish the undesirability of the very idea of science, reason, and modernity. The problem with letting go of reason even as an ideal is that then there is nothing left in whose name arbitrary authority of patriarchal institutions and practices can be challenged. (There remain the local narratives, as postmodernist critics of metanarratives are prone to point to but as Seyla Benhabib[42] and Norman Geras[43] have argued persuasively, in the absence of some nonarbitrary guidelines, any ranking of local beliefs would remain a matter of arbitrary preference which is always determined by those with power.) Moreover, if we accept that historically, women have made more substantial gains toward personal autonomy in those societies where all traditional institutions have had to justify themselves before the tribunal of reason as compared to those societies that still recognize a "Fount of Authority, whether it be a Person, a Text or an Institution"[44]; it becomes hard to willingly bid farewell to "their" scientific ethos.

---

Finally, there is only one minor, though critical, issue left: how does one reconcile the emancipatory potential of this idealized reason with its actual history of complicity in the worst kinds of oppressions, from the colonial exploits of the West to slavery and the Gulags? How can one justify and advocate faith in a scientific temper in the face of the rising chorus of voices in contemporary postcolonial societies that justifiably point to the crimes of reason against their self-identities? Moreover, following Gellner's argument to its local conclusion, by giving reason the status of a new, cosmopolitan culture that is allowed to set standards for ranking the traditional mores of the non-Western societies, are we not falling in the same old Orientalist trap Edward Said and others following him have been warning against?

What protects Gellner from charges of scientism—that is, of deifying science and reason as the final arbiters of all aspects and all forms of social and cultural life—is the fact that he fully accepts that the claims of science provide an intellectually limited and spiritually impoverished substratum to build a social order upon. He readily accepts that:

> whereas error can define a society, truth cannot. Truth does indeed corrode old coherent visions, but fails to replace them with anything permanent, concrete, rounded off and morally sustaining. The valid style of inquiry generates neither stability nor normative authority.[45]

Thus Gellner ascribes the gulags and the Nazi death camps to be inevitable results of *any* attempt to extract a concrete, definite social order from the overall vision of the Enlightenment. Thus in his scheme, he makes room for what he calls a "ritual theater,"[46] which, quite like the institution of the constitutional monarchy in Gellner's Britain, will continue to provide spiritual and moral solace and meaning, while "when dealing with serious matters, when human lives and welfare are at stake, the only kind of knowledge that may be legitimately used and invoked is that which satisfies the criteria of the Enlightenment."[47]

This is one part of Gellner's thesis with which I have serious disagreements. The clean boundary line he draws between the ritual and other "serious" matters belies his empiricist bent of mind and his total lack of engagement with the vast literature from various kinds of social studies of science—by feminists, sociologists, historians, and anthropologists—that shows how closely the scientific cognition is tied with the social/cultural meanings. Moreover, by proposing a separation of powers between science "for serious things" and ritual for matters of faith, spiritual meanings (by implication, nonserious things?!), Gellner is justifying the existence of the divide between cognition and values. This divide in my view is to a large part responsible for the spiritual and existential crisis of modernity. Condoning it and even recommending its deepening is unconscionable on Gellner's part.

So if the clean separation of reason and faith will not work, and reason as we know cannot furnish us with all the resources needed to live socially meaningful lives, why not simply stop worrying about it, as the postmodernists recommend? Perhaps, eventually, when more and more of the human race comes to live their lives by their own lights, reason would have fulfilled its emancipatory mission and it could be allowed to become merely one aspect of our social life. But that day has not yet dawned, definitely not in postcolonial societies, and most definitely not for women in these societies.

## ACKNOWLEDGMENTS

I wish to thank Langdon Winner, who in various ways encouraged me to acknowledge how strongly I felt about the emancipatory possibilities of science in the Third World. Critical comments by David Hess and Kim Laughlin were helpful in shaping this paper. And as always, Deborah Johnson provided encouragement and insights.

## NOTES

1 Ernest Gellner, ***Reason and Culture: The Historic Role of Rationality and Rationalism***, p. 193; and ***Postmodernism, Reason and Religion***, p. 108. Gellner passed away on November 5, 1995 in Prague, where he was serving as Director of the Center for the Study of Nationalism.

2 In keeping with the current practice in social theory, Gellner uses "Reason" not as a reference to the generic ability of all humans to reason, but to the specific rationality that emerged in Western Europe in the course of the Enlightenment and the Scientific Revolution, the two historic landmarks of modernity. Thus, modernity is characterized by a faith in the ability of scientific reason to discover universally applicable theoretical and practical norms upon which systems of thought and action could be built and society could be restructured. The rage against reason one finds common in postcolonial writing, which it shares with the postmodern spirit in general, is directed against this universalization of the Western idea of reason as the foundation of all systematic knowledge and as a source of progress and emancipation for all of humanity. Following Gellner, reason and science will be used as synonyms in this paper.

3 I became fully aware of how strongly I felt about the emancipatory possibilities of science in an informal conversation with Langdon Winner.

4 Gellner, *Postmodernism*, p. 30.
5 Manu, among the most influential ancient lawgivers in all of Hinduism, is best known for his *Manu Smriti* (Laws of Manu) which codified extremely misogynist laws for controlling women's sexuality, including laws about early marriage and proper conduct for wives etc. Though *Manu Smriti* was compiled between the second and the first centuries BC, it was accepted as *the* Hindu personal law by the British in consultation with the Brahmin elites. To enormous detriment to women, Manu's laws are still cited frequently by Indian judges and lawmakers. See Sucheta Mazumdar, "Women, Culture and Politics: Engendering the Hindu Nation."
6 Audre Lorde, "The Master's Tools Will Never Dismantle the Master's House."
7 The term "postmodernism" is used here to signify a *Zeitgeist*, a sensibility, that challenges the Enlightenment's faith in the possibility of progress through a rational understanding and redesign of our natural and social worlds. Philosophically, it challenges the belief that a knowing and active subject can, in principle, gain access to objective reality. What most concerns us in the context of postcolonial feminism, however, is the claim that "the very criteria demarcating true and false, as well as science and myth or fact and superstition, were internal to the traditions of modernity and could not be legitimized outside of those traditions" (Linda Nicholson, "Introduction," p. 4). In other words, science's claim to provide a "God's eye view" of the world is a mere ruse to hide its historical, political and cultural situatedness in the largely upper class, white, male Protestant culture of seventeenth–eighteenth-century Western Europe. Such skepticism about the possibility and (in most cases) desirability of culturally transcendent criteria of truth has become widespread among feminists and other social critics. Indeed, the search for feminist epistemology is premised on the assumption that rationality cannot transcend (i.e., stand apart from, come prior to or overcome) the gender of the knower. It is the postcolonial critics of science, however, who have carried this logic to the farthest extreme. For instance, in an attempt to show why there cannot be any scientific reasons to oppose female circumcision, Stephen Marglin writes: "there is no way of assessing the truth or falsity of organic discourses apart from people's beliefs. There is not only no objective truth in this realm, there is no objective falsehood either" (*Dominating Knowledge: Development, Culture and Resistance*, p. 15). Or again, Vandana Shiva, who has emerged as the leading Third World ecofeminist writes: "meaning and validity are controlled by the social world of the scientists and not by the natural world. These new accounts of modern science have left no criteria to distinguish between myths of traditional thought and the metaphors of modern science, between supernatural entities presupposed by traditional communities and theoretical entities presupposed by modern scientists" (*Staying Alive: Women, Ecology and Survival*, p. 32). Statements like these are commonplace in postcolonial critiques of science. In most instances, such statements are supported by statements by postmodernist and poststructuralist theorists from the West, including Foucault and Derrida and their followers.
8 Gellner, *Postmodernism*, p. 80.
9 Selya Benhabib, "Feminism and Postmodernism: An Uneasy Alliance."
10 See Rita Felski, "Feminism, Postmodernism and the Critique of Modernity" for a good review of Western feminism's ambiguous relation with modernity.
11 For a selection, see Frederique Marglin & Stephen Marglin, eds., *Dominating Knowledge.* A more recent statement of the antipathy of the postcolonial intellectuals to the idea of development can be found in Ashish Nandy, "Culture, Voice and Development: A Primer for the Unsuspecting." The problematic nature of the embrace of the "authentic" local traditions by the postcolonial critics of modernity is evident in the fact that these learned scholars, who are completely at home with the most avant-garde postmodern social theory, often find themselves in the same political camp occupied by the home-bred religious fundamentalists. As a specially disturbing instance, see the defense of the barbaric practice of *sati* put forward by Ashish Nandy, India's most prominent postcolonial critic of science ("The Human Factor").

12 I use self-determination and autonomy to mean, following Judith Grant (***Fundamental Feminism: Contesting the Core Concepts of Feminist Theory***, p. 183), "[the ability] to choose among the full range of human traits, without having to choose based on gender."

13 Benhabib, "Feminism and Postmodernism: An Uneasy Alliance," p. 27.

14 Benhabib writes: "the vocation of social criticism might require social exile, for there might be times when the immanent norms and values of a culture are so reified, dead or petrified that one can no longer speak in their name. The social critic who is in exile does not adopt the 'view from nowhere' but the 'view from outside the walls of the city,' wherever those walls and those boundaries might be. It may indeed be no coincidence that from Hypatia to Diotima to Olympe de Gouges and to Rosa Luxemburg, the vocation of the feminist thinker and critic has led her to leave home and the city walls" ("Feminism and Postmodernism," p. 29).

15 If it had not been for a fortunate chance that led me to study microbiology after high school, I would have most likely ended up in the role reserved for women in my traditional and a (barely) middle-class family in Northern India: marriage—arranged and early. I shall never forget the thrill I experienced as I became familiar with some concepts of molecular biology, especially the structure of DNA and the mechanism of protein synthesis. It helped me develop the intellectual resources to argue against what I intuitively understood to be arbitrary and unjust, especially the ideas of caste and certain Hindu rituals that girls and women had to observe. My intellectual growth gave me enough sense to a self that I could dare defy my family's incessant pressure to marry me off. I moved to New Delhi, obtained a Ph.D. from a prestigious laboratory, and later found work as a science correspondent for a national newspaper. After working for many years as a journalist, I have finally returned to academics. (I am now working on a research project in science and technology studies in the United States.) After all my ups and downs with science, there is one thing I am sure of: I gave up doing science not because I found it oppressive as a woman, but because I was afraid of turning it into an elitist and—in India's context—a socially unproductive activity.

16 Shiva, ***Staying Alive.***

17 Gellner, ***Reason and Culture***, p. 157.

18 Gellner, ***Postmodernism***, p. 54.

19 Indeed, Gellner's ***Postmodernism, Reason and Religion*** is almost entirely devoted to establishing the falsity of cognitive relativism that denies that knowledge beyond culture is possible. Statements like these repeat themselves at a sometimes annoying frequency: "I am not sure whether indeed we possess morality beyond culture, but I am absolutely certain that we do indeed possess knowledge beyond both culture and morality. The existence of transcultural and amoral knowledge is ***the*** fact of our lives. I am not saying that it is ***good***; but I am absolutely certain that it is a fact" (p. 54).

20 Gellner, ***Reason and Culture***, pp. 160–161.

21 Quoted by Gellner in ***Reason and Culture***, p. 2.

22 "Thus it is mere custom and example that persuades us than certain knowledge, and for all that, the majority opinion is not a proof worth anything for truths that are a bit difficult to discover . . . I could find no one whose opinions, it seemed to me, ought to be preferred over the others, and I found myself constrained to try to lead myself to my own" (Descartes, ***Discourse on Method***, p. 10).

23 Gellner, ***Reason and Culture***, p. 8

24 For a critical review of the feminist repudiation of Descartes, see Margaret Atherton, "Cartesian Reason and Gendered Reason" and Martha Nussbaum, "Feminists and Philosophy"; also, the author's unpublished paper, "I Think (like a man), Therefore I am (a Philosopher)."

25 Gellner, ***Reason and Culture***, p. 26.

26 Ibid., p. 50.

27 Indeed, Gellner says some rather unflattering things about his anthropologist colleagues who have taken the interpretive turn. He ends his ***Postmodernism, Reason***

*and Religion* thus: "To the relativists, one can only say—you provide an excellent account of the manner in which we choose our menu or our wall paper. As an account of the realities of our world and a guide to conduct, your position is laughable" (p. 96).
28 Ibid., p. 52.
29 Gellner, *Postmodernism*, p. 50.
30 Ibid.
31 Ibid., p. 75.
32 Gellner, *Reason and Culture*, p. 159.
33 Ibid., p. 167.
34 Ibid.
35 Ibid., p. 147.
36 Ibid.
37 Donna Haraway, "Situated Knowledges: The Science Question in Feminism and the Privilege of Partial Perspective," p. 187.
38 Gellner, *Postmodernism*, p. 90.
39 Ibid., p. 80.
40 Ibid.
41 Gellner, *Reason and Culture*, p. 137.
42 Benhabib, "Feminism and Postmodernism."
43 Norman Geras, "Language, Truth and Justice."
44 Gellner, *Reason and Culture*, p. 65. The complete statement reads as follows: "The claim that *either* experience *or* reasoning constitutes the final court of appeal for cognitive claims is equally in conflict with the attribution of terminal authority to some sacred person, tradition, event or some other channel of revelation. A society that recognizes and enforces a Fount of Authority, whether it be a Person, a Text, an Event, or an Institution (or some combination of these) is profoundly different from one which recognizes only some faculty, whatever it be, located in principle in all men."
45 Gellner, *Postmodernism*, p. 88.
46 Ibid., p. 91.
47 Ibid., p. 92.

## REFERENCES

ATHERTON, MARGARET. "Cartesian Reason and Gendered Reason." In *A Mind of Her Own: Feminist Essays on Reason and Objectivity*, edited by Louise M. Anthony & Charlotte Witt. Boulder, CO: Westview, 1993.

BENHABIB, SEYLA. "Feminism and Postmodernism: An Uneasy Alliance." In *Feminist Contentions: A Philosophical Exchange*, edited by Seyla Benhabib, Judith Butler, Durcilla Cornell & Nancy Fraser. New York, NY: Routledge, 1995.

DESCARTES, RENÉ. *Discourse on Method.* 3d edit. Translated by Donald A. Cress. Indianapolis, IN: Hackett Publishing Company, 1993.

FELSKI, RITA. "Feminism, Postmodernism and the Critique of Modernity." *Cultural Critique* (Fall 1989): 33–56.

GELLNER, ERNEST. *Postmodernism, Reason and Religion.* New York, NY: Routledge, 1992.

———. *Reason and Culture: The Historic Role of Rationality and Rationalism.* Oxford: Blackwell, 1992.

GERAS, NORMAN. "Language, Truth and Justice." *New Left Review* 209 (1995): 110–135.

GRANT, JUDITH. *Fundamental Feminism: Contesting the Core Concepts of Feminist Theory.* New York, NY: Routledge, 1993.

HARAWAY, DONNA. "Situated Knowledges: The Science Question in Feminism and the Privilege of Partial Perspective." In *Simians, Cyborgs and Women: The Reinvention of Nature.* New York, NY: Routledge, 1991.

LORDE, AUDRE. "The Master's Tools Will Never Dismantle the Master's House." In *Sister Outsider: Essays and Speeches.* Trumansburg, NY: Crossing Press, 1984.

MARGLIN, FREDERIQUE & STEPHEN MARGLIN, eds. ***Dominating Knowledge: Development, Culture and Resistance.*** Oxford: Clarendon Press, 1990.

MAZUMDAR, SUCHETA. "Women, Culture and Politics: Engendering the Hindu Nation." ***South Asia Bulletin*** 12, 2 (1992): 1–24.

NANDA, MEERA. "I Think (Like a Man), Therefore I Am (a Philosopher)?" Unpublished manuscript, 1993.

NANDY, ASHISH. "Culture, Voice and Development: A Primer for the Unsuspecting." ***Thesis Eleven*** 39 (1994): 1–18.

———. "The Human Factor." ***The Illustrated Weekly of India***, January 17, 1988, pp. 20–23.

NICHOLSON, LINDA. "Introduction." In ***Feminism/Postmodernism***, edited by Linda Nicholson. New York, NY: Routledge, 1993.

NUSSBAUM, MARTHA. "Feminists and Philosophy." ***New York Review of Books***, October 20, 1994, pp. 59–63.

SHIVA, VANDANA. ***Staying Alive: Women, Ecology and Survival.*** New Delhi: Kali for Women, 1988.

# ARE "FEMINIST PERSPECTIVES" IN MATHEMATICS AND SCIENCE FEMINIST?[a]

MARY BETH RUSKAI

IN RECENT YEARS the term "feminism" has often been used in a sense so different from the traditional one of "a theory of equality of the sexes" or "advocacy of equal rights"[1] that one wonders if even Susan B. Anthony would be considered a "feminist." The phrases "feminist critiques" and "feminist theories" are often applied to a rather narrow set of views that include an emphasis on differences between the sexes or genders. I dislike this terminology because it tends to color the discussion so that anyone who disagrees is labeled as non-feminist, even if that person has worked hard for equity and advancement of women. Therefore, I will use the term "gender difference theory" to describe the viewpoint promulgated by such writers as Sandra Harding, Sue Rosser, and Sherry Turkle. Since I have made specific criticisms of some of their work elsewhere,[2] I will concentrate here on some general issues.

Not only do I frequently find errors and inconsistencies in the writings of the gender difference theorists, but I feel that they are "asking the wrong questions." Here, I do not mean "wrong" in the sense of morality or some kind of "objective truth." Rather, I believe that the gender difference theorists are focusing on certain issues in ways that, instead of contributing to the goal of increasing the contributions of women to science and mathematics, may exacerbate the problem. I will begin my development of this theme with a personal anecdote.

After graduate school, I held a postdoctoral position in theoretical physics in Switzerland in 1969, the year the Swiss finally gave women the right to vote in federal elections. However, it would be another ten years before women were allowed to join the Swiss Alpine Club. Although I found a number of other (dual-gender) groups to ski, hike, and climb with, I also joined the Club

[a] Based on a talk given in the panel on "Feminist Perspectives" at the International Commission on Mathematical Instruction Study on Gender and Mathematics Education, Höör, Sweden, October 7–12, 1993. (An earlier version appeared in the proceedings of that conference.)

Suisse du Femmes Alpiniste and went with them on a week-long ski mountaineering trip. At that time, I was extremely well trained in terms of stamina and endurance, although I did not have the same mountaineering and ski skills as the Swiss women, who had been doing this all their lives. Our two male guides set a brisk pace up the mountain, as we followed laden with skis and gear for a week. After about an hour, I was the only one still keeping up with the guides, who stopped to let the others catch up. Almost fifteen minutes passed before the last two, who were somewhat older, came huffing and puffing up the mountain and practically collapsed as they dropped their packs. When they had finally recovered, both of them looked over at me and said with great concern, "Are you all right?" As the week progressed, the male guides and the other women accepted me as part of the group, realized my strength in stamina and my weakness in mountaineering skills, and gave me help and encouragement when I needed it, so that I could really enjoy the trip. However, despite blatant evidence to the contrary, these two elderly women remained convinced of—and regularly commented upon—my immanent collapse. This is an absolutely classic illustration of prejudice. Although bias by elderly Swiss women against visiting Americans is not a serious societal problem, this incident exhibited all the elements inherent in racial and sexual discrimination. I was being judged on the basis of their ideas of what American tourists are like and not on my own merits.

This, I believe, is the fundamental problem: our tendency to judge people by categories instead of by their individual attributes. Gender difference theory aggravates this problem because it focuses on specific gender characteristics within a group and gender differences between groups, rather than on people as individuals. It is entirely irrelevant whether or not, in particular circumstances, the attributes that are associated with a group are statistically valid, or whether the attributes are thought to be of biological or cultural origin. It is far more important to judge people as individuals. Despite our best efforts, most of us have hidden biases, not only about people's race, religion, and ethnicity, but about their height, marital status, hair style,[3] etc. Focusing on group attributes is not the way to overcome our prejudices.

I realize this may sound suspiciously like something that many of us used to believe in but have now been disillusioned by—namely, that we should be completely gender neutral and do everything on the basis of pure merit. However, our biases are simply too ingrained for that. Instead, we have come to accept the need for some forms of affirmative action. For example, I have learned that you will rarely have women speakers at mathematics conferences unless you make an active effort. But when you do make that effort, you usually find half a dozen women who are at least as good as the men speakers, so that you can invite them on their merits after all.

Let me give another example to illustrate the importance of treating people as individuals. In recent years I have changed my own teaching style, so that I lecture less and the students participate more. This was partly a response to reports that women were much more successful in "cooperative learning" environments. My own approach is only to insist that the students do some work in class, although I encourage them to do this in groups. What I discovered should not have surprised me, but it did. Many of the women dislike this intensely, because they feel, no matter how low-key I try to make it, that they are being put on the spot and tested immediately. However, even though this approach has not been very successful with female students, I have become so convinced that the students learn more, that I will not abandon it. Instead, I am trying to find ways to make it more palatable and effective for women. On the other hand, it has been more successful with the (extremely small number of) black students in my classes. In particular, two men who might very well have failed if taught in the traditional way earned strong B's. These two black men were quite different, the reasons why they might have failed were different, and the reasons why they responded positively were different. The only common thread was that because I interacted with students working in class, I was able to view each of them—not as "yet another poorly prepared black male sitting in the back of the classroom failing exams"—but as unique individuals with strengths and weaknesses. This led to interactions that did seem to work, in the sense that the students responded very positively and succeeded.

My attitude is that you should treat people as individuals. It may well be true that more girls need encouragement than boys. But our attitude should be to identify and encourage those students whose confidence level is such that they need it. If it turns out to be 95% girls, fine; if it is 60% girls, fine; if it is only 20%, fine. We should respond to the needs of the student, keeping aware that what works for most people in one group does not necessarily work for everybody.

One point on which I agree with the gender difference theorists is that the background and viewpoint from which one comes to a subject is important. For example, Karen Johnson made a convincing case[4] that Maria Mayer's Nobel prize work on the nuclear shell model was a direct consequence of her previous work in atomic and molecular calculations, which gave her a different viewpoint from most of the (male) nuclear physicists. In my own work, I was recently involved in a multinational collaboration in which I was the only mathematical physicist interacting with probabilists, statisticians, and mathematical biologists. I was able to "smell" the solution to one of our unsolved problems in a paper that I found rather technical and hard to understand. In response to my needs, one of the probabilists translated the relevant material into operator theory language so that our entropy equivalence question was reduced to one about the equivalence of eigenvalues of certain matrices. From this he felt it was intuitively obvious that they were not equivalent and that we should search for a numerical counterexample. As an operator theorist, I took one look at his notes and in two more lines used the max-min principle to prove that they actually were equivalent. Our backgrounds were very important to our different insights. But in both of these examples, it was scientific background, not gender, that made the difference. I have not yet seen a convincing case for gender-based perspectives providing important scientific insight except in those cases, such as hormone research, in which gender is itself intrinsic to the subject being studied.

There is another point to be made here. As soon as I communicated my observation, my colleague changed his position and agreed. This would appear to support the perception (often the result of deficient elementary education) that mathematical problems have only one correct solution. By contrast, both mathematics educators and some gender theorists have begun to emphasize that some mathematics problems have more than one correct answer, and most problems can be solved in more than one way. What may be confusing to nonmathematicians is that recognizing the validity of multiple approaches is not inconsistent with conviction that questions have right and wrong answers. Creativity and rigidity are not incompatible. A similar situation arises in music: the fact that everyone tunes their instruments to a standard pitch does not restrict musical creativity, but allows musicians to play together. That most questions have definite answers does not make mathematics unfeminine or uncreative, either. On the contrary, many women mathematicians find this feature of their discipline an equalizer. I do not know of any cases in mathematics in which someone has alleged that a theorem is wrong because a woman proved it. If someone wants to downgrade a woman's contributions, they will question the importance of her result or her role in the proof. But if a result is right, it is right.

Let me conclude by returning to the point that characterization by gender reinforces conformity, while what is needed is an appreciation of diversity. In some of the gender-difference literature the arguments seem to me particularly flawed because they define feminine in terms of certain well-known, common characteristics. They also define science in terms of other characteristics (sometimes of questionable validity, such as noncreative and nonintuitive) that they consider masculine. If one points out that there are many examples of women who do not fit this stereotype, the response is that the paradigm is cultural rather than biological, so that the women who do not fit are masculine and the men who do fit are feminine. To a mathematician this sounds suspiciously like "assuming what you're trying to prove."

One of the oft-told stories about the great mathematician Emmy Noether is that because she was so successful at mathematics, which in her culture and generation was considered unfeminine, she was often referred to as "der Noether," using the German masculine pronoun. Sometimes when I read the "feminist theory" or "gender difference" literature, particularly the stereotypes and the remarks about women scientists who work in science in the traditional way, I feel that they are simply calling us all "der."

## NOTES

1 *Webster's New Collegiate Dictionary.*

2 Mary Beth Ruskai, "Gender and Science"; "How Stereotypes about Science Affect the Participation of Women"; "Why Women are Discouraged from Studying Science."

3 Ruskai, "Stereotypes about Science."

4 K. Johnson, "Maria Goeppert Mayer: Atoms, Molecules and Nuclear Shells."

## REFERENCES

JOHNSON, K. "Maria Goeppert Mayer: Atoms, Molecules and Nuclear Shells." *Physics Today* 39 (Sept. 1986): 44–49.

RUSKAI, MARY BETH. "Gender and Science." (Based on an invited talk for the Association for Women in Mathematics [AWM] panel on Gender and Science at the AMS-MAA joint summer mathematics meeting, University of Utah, August 1987.) *Association for Women in Mathematics Newsletter* 17, 6 (Nov.–Dec. 1987): 5–10.

———. "How Stereotypes about Science Affect the Participation of Women." Unpublished manuscript. (Based on an invited talk in the Committee on the Status of Women in Physics [CSWP] Panel "Women in Physics: Why So Few?" AAPT/AAPT/AAAS meeting, San Francisco, CA, January 1989.)

———. "Why Women are Discouraged from Studying Science." *The Scientist* 4, 5 (March 5, 1990): 17, 19. Reprinted in *CSWP Gazette* 10, 2 (June, 1990): 2–4.

*Webster's New Collegiate Dictionary.* Springfield, MA: G. and C. Merriam Co., 1981.

# HUMANITIES

THE CRAZE for postmodern theory in the humanities—particularly within university literature departments—is often blamed as a source of the current, broader discontent with reasoned inquiry. However, many humanist scholars have lately become vocal in criticism of the excesses of "theory" and the philosophical conceits that sustain it.

This section begins with an exchange between two such humanists, both professors of English and both critics of the faddish nihilism that has made itself at home in their profession. This, however, does not rule out interesting disagreements. Paul A. Cantor, of the University of Virginia, scrutinizes the new Oxford Edition of Shakespeare's works and the editorial philosophy that lies behind it. He sees Oxford's decision to present *King Lear* in two distinct versions as an intent to degrade the notion of transcendent value in art, and of the artist as moral spokesman, by reducing Shakespeare's greatest play from a searing vision to a petty battle of contending texts.

On the other hand, George Bornstein, of the University of Michigan, asserts that the best way to undermine the mystifications of the deconstructionists is to defend the idea that a text is a product of a particular mind at a particular moment, not a free-floating assemblage of arbitrary signifiers. He thinks that students ought to learn literature through careful attention to the concrete circumstances of its production. Thus, in particular, he views the new Oxford Shakespeare as helpful, precisely because of its "two text" approach to *King Lear.*

In the following essay, Frederick Crews analyzes the continuing susceptibility of humanists to a doctrine far older than deconstruction and, in his view, comparably irrationalist. He notes that, despite its fall from grace among scientists, physicians, and philosophers, Freudian psychoanalytic theory continues to fascinate many academics, particularly a certain variety of literary scholar. He analyzes the reasons for such continuing appeal, while pointing out the historical and philosophical reasons for regarding Freudian dogma with the gravest suspicion.

Gerald Weissmann's piece points out that antirationalism is not a new phenomenon in American cultural history. The gothic figure of Edgar Allan Poe, in particular, presents the intellectual historian with an example of the antirational and antiscientific nostalgia that marked one strain of nineteenth-century American Romanticism. Contemplating this story, one is reminded that not much is new under the sun, and that the excesses of today's "postcontemporary" academy represent yet another recurrence of an old theme.

# ON SITTING DOWN TO READ *KING LEARS* ONCE AGAIN
## The Textual Deconstruction of Shakespeare

PAUL A. CANTOR

THE IRRATIONALITY and politicization of literary studies today must by now be clear to any reasonable person outside the field, especially those in the natural sciences who still know what standards of logic and objectivity are. But it is tempting for outsiders to remain unconcerned about current trends in literature departments. They assume that no matter how far literary theory may go in undermining our cultural heritage, there are limits to how much damage it can really do. After all, one might argue, literary critics may say whatever they want about the great books of the Western tradition, but at least the texts remain to contradict them. And indeed, turning to one of Shakespeare's plays after hearing one of today's critics discourse on the way it reflects capitalist hegemony can be a liberating and mind-cleansing experience. But not everyone can still be open to the wonder of a Shakespeare play after being taught to read it in a New Historicist or a feminist fashion; and of course one of the most baleful effects of contemporary literary study is its obsession with altering the canon, changing which texts our students read. But I want to talk about an even more basic problem, the way contemporary criticism is quietly changing the very texts themselves. In the long run, this trend may have the most profound and the most harmful consequences, for it will be difficult to appeal from contemporary criticism to literary texts if it succeeds in remaking those texts in its own image.

The way contemporary criticism is altering how texts are edited has gone largely unnoticed by people outside the field. In other disciplines, great controversies have arisen when theoretical conceptions have ended up having a concrete impact on the material form works of art take. I think of the heated and often acrimonious debate over the restoration of the Michelangelo frescoes in the Sistine Chapel. Whatever side one took in this controversy, it was clear that the plan to clean and restore these masterpieces rested on highly theoretical premises and arguments about the nature of painting. Some insisted that the true state of a painting is its original state—the way the artist left it when the paint began to dry. Others countered that an artist can anticipate how his work will age and thus actually may have intended and planned it eventually to look different from the way it did when he laid on his last brushstroke. I believe that it is now widely accepted that the Sistine Chapel restoration is a success, but the art historians who argued against it complained bitterly that if it was being undertaken on faulty theoretical grounds, the damage to one of the world's greatest works of art would be irreparable and posterity would forever be denied a chance to experience Michelangelo's frescoes in the way he intended.

While throughout the 1980s the whole art world seemed to be holding its collective breath over the fate of the Sistine Chapel, perhaps the only Renaissance artwork of comparable scale and grandeur—Shakespeare's *King Lear*—was undergoing a similar process of restoration, on similarly debatable theoretical grounds, and yet no one outside a handful of literary scholars even knew that it was happening, let alone cared about the results. If anything, the restoration of *King Lear* was a more radical process, for while the Italian conservators were merely, so to speak, dusting off and touching up the Michelangelo frescoes, an eager young group of editors and bibliographers was tearing *King Lear* apart, deconstructing it literally to pieces. How many people outside literary studies are aware that there are now two *King Lears* by Shakespeare? And I am not just talking about two *King Lears* in the imagination of cutting-edge Shakespeare scholars, the younger generation who now dominate the field and who brought about the decomposition of *Lear* under the banner of the same kind of rhetoric that spurred the Michelangelo restoration: "We must clear away centuries of encrustrations and get back to *King Lear* as Shakespeare originally created it." At first this point remained on the level of rhetoric, embodied solely in theoretical pronouncements, but today the idea is fully enshrined in the prestigious Oxford University Press single-volume edition of the complete Shakespeare, which indeed contains two different *King Lears.* To be sure, the old unitary *King Lear* is still widely available in all sorts of editions, and unlike the situation with the Sistine Chapel, no matter how literary scholars may choose to cut up Shakespeare's play, one feels reassured that the whole of *Lear* will survive and continue to be experienced as generations of readers have encountered it. But we should not be complacent about such editorial developments. The fact is that if the Oxford Shakespeare were to become the preferred teaching text in colleges and universities, at least one generation of students would, in my opinion, end up not reading the whole of *King Lear.*

But before trying to evaluate this situation, I need to explain how it came about, and it is a very complicated controversy, with, I want to stress, legitimate arguments on both sides of the debate. Very few people outside of Elizabethan studies have any idea of how dreadful is the state in which the texts of Shakespeare's plays have come down to us. Here, for example, is the text of a familiar speech in the earliest version we know of:

> To be, or not to be, I there's the point,
> To Die, to sleepe, is that all? I all:
> No, to sleepe, to dreame, I mary there it goes,
> For in that dreame of death, when wee awake,

And borne before an euerlasting Iudge,
From whence no passenger euer returnd,
The vndiscouered country, at whose sight
The happy smile, and the accurssed damn'd.[1]

Obviously a great deal of editorial work is needed to get from texts such as this to the ones we have become used to in the modern editions we read. In the case of "To be or not to be," fortunately, it is easy to get the speech right, because we have two other early texts of *Hamlet* which on all sorts of grounds are clearly to be preferred. But in the case of *King Lear*, the situation is murkier. The play has come down to us in two principal texts. Unfortunately they differ considerably and neither is clearly superior to the other. The first text is the so-called Quarto volume of 1608, probably an unauthorized publication designed to capitalize on the success of the play on the stage. Most scholars think that this text is somehow derived from Shakespeare's manuscript, though it betrays distinct signs of some form of oral transmission. The second text is from the First Folio volume of 1623, published seven years after Shakespeare's death by his theater associates. Most scholars think that this text is derived from a theatrical prompt-book and generally it seems cleaner and better edited than the Quarto volume. The main problem is that just under 300 lines of text appear in the Quarto volume that do not appear in the Folio and just over 100 lines appear in the Folio that do not appear in the Quarto. And we are not talking about trivial passages here; many of these lines are among the most important and most moving in the play. The two texts of *King Lear* thus confront scholars and especially editors with a difficult choice: to prefer one over the other would evidently require sacrificing some of Shakespeare's best poetic and dramatic writing.

For hundreds of years, the way editors have chosen to deal with this difficulty is to conflate the Quarto and the Folio texts into a single version which incorporates virtually all the lines of both and thus is longer than either.[2] This is the text of *King Lear* we all know. It must be stressed that this is an *ideal* text, an editorial construction; it corresponds to no document that we have actually inherited from the past. As such, the long accepted text of *King Lear* is deeply problematic, and during the past two decades it has increasingly come under attack. The conflated text was based on the assumption that in the vagaries and vicissitudes of transmission, a single text composed by Shakespeare was somehow mangled and split into two forms that editors must put back together as best they can. The most prominent challenge to this view is the thesis that the Quarto and the Folio texts in fact represent two distinct versions or stages of *King Lear*, and specifically that the Folio is Shakespeare's revision of the Quarto version. Reviving speculations that have cropped up before in the long history of Shakespeare editing, recent critics have argued that the differences between the Quarto and the Folio texts can be traced to the fact that Shakespeare himself had second thoughts about *King Lear*, leading him to delete some passages, add new ones, and fiddle with the wording throughout.[3] In particular, these critics claim that the Folio text is theatrically superior to the Quarto, that the revisions were intended to streamline the action by eliminating wordy passages, while at the same time filling in some details to clarify certain plot developments. Critics have also claimed that the Folio version reflects Shakespeare's rethinking some of the details of plot and characterization in *King Lear.* For example, the roles of Albany, Kent, and Edgar seem to be altered in the Folio version, and, given the cuts, the role of the French in the invasion of England is nearly eliminated. Notice the force of these claims—these critics do not treat the diminished role of the French in the Folio text as the accidental result of some form of cutting; they view it as Shakespeare's deliberate reconception of the play, a conscious effort to minimize the issue of foreign invasion in the revised version.

On the basis of the revision theory of *King Lear*, these scholars have recommended a whole new approach to editing the play. They claim that a conflated text fails to correspond to any version Shakespeare ever wrote and seriously distorts his intentions in revising his work. He tried to streamline *King Lear*, and here modern editors have come up with a version longer than either the Quarto or Folio texts. The *Lear* revisionists recommend publishing separately edited texts of the Quarto and Folio versions (exactly what Oxford has done); if one text must be chosen, they recommend the Folio, as representing Shakespeare's final thoughts on *King Lear*; some add the argument that the Folio text will play better on the stage. In the most important single volume edition of *Lear* to appear recently, edited by Jay Halio in the prestigious New Cambridge Shakespeare series, this practice is followed; Halio uses the Folio text as the basis of his edition, and relegates all the passages unique to the Quarto to an appendix.[4] One might argue that this approach does not sacrifice anything, since it still allows readers to familiarize themselves with the Quarto text. To speak in practical terms, however, any teacher who knows how difficult it is to get students today to do the assigned reading will be skeptical about the chances of getting them to do reading that they will inevitably regard as supplementary. I can already hear my students asking: "If it's in the appendix, will it be on the final?" To relegate the Quarto passages to an appendix will in the long run almost certainly consign them to oblivion, at least for pedagogical purposes.

For reasons such as this, I still on balance prefer working with a conflated text of *King Lear*, but I grant that the revisionist critics have much to say for their viewpoint. For example, I think that they have in a few cases demonstrated convincingly that the attempt to conflate a single speech out of variant Quarto and Folio passages produces a garbled text, which becomes internally contradictory when it tries to combine into one what are in fact alternate versions of a single moment on the stage.[5] But such cases do not provide a solid argument against a conflated text as such; they only demonstrate that the act of conflating has hitherto at times been done poorly; they prove only the need to rework the conflated text in specific passages. In short, much remains unproven in what may at first appear to be persuasive arguments for the revisionist thesis. Even if one accepted the idea that the Folio text is a revision of the Quarto, that would not clinch the case for preferring the Folio over the Quarto. If the Quarto text is somehow closer to Shakespeare's original manuscript and hence his original intentions, much could be said for preferring it precisely on these grounds, as being closer to his original inspiration. Perhaps

Shakespeare was forced into the Folio revisions for reasons that have nothing to do with his own artistic vision. Censorship may have brought about the reduction of the role of the French in the invasion, or adventitious theatrical circumstances, and not Shakespeare's considered artistic judgment, may be responsible for the cuts (perhaps the need to reduce the number of actors for a company tour in the provinces).

In short, given the limited knowledge we have, it is very difficult to interpret and evaluate the mere presence of disparities between the Quarto and the Folio texts. We have absolutely no external or material evidence to prove or even to suggest that the Folio text represents Shakespeare's deliberate revision of the Quarto.[6] Our knowledge of the facts in these cases is pitiful, and in their eagerness to fill in the huge gaps, Elizabethan scholars are forced into all sorts of hypothetical reconstructions. For example, much of the theory of Shakespeare editing rests on the supposed distinction between the author's manuscripts and theatrical prompt-books as the ultimate source for texts. Editors theorize that prompt-books, given their function, would supply fuller and more accurate details about such stage business as entrances and exits. The problem is that not a single one of Shakespeare's manuscripts has survived, nor has a single contemporary prompt-book of any of his plays come down to us. Natural scientists can appreciate the difficulty literary scholars have when they are constantly theorizing about objects that no longer exist; they have no reality check on their speculations. In fact the few manuscripts and prompt-books that have come down to us from Renaissance dramatists have tended *not* to confirm the theories of Shakespeare editors. The expected characteristic differences between manuscripts and prompt-books are simply not clear in the few samples we have.[7] Unfortunately, such empirical evidence has barely put a dent in the heavily armored textual apparatus of my colleagues.

Some editors may pretend that their procedures are scientific and their conclusions arrived at logically, but in the absence of the crucial evidence needed to support their arguments, their theories rest largely on conjecture and surmise, and usually they are relying on esthetic judgments, which are largely subjective. Since we have no way of knowing for sure that the Folio text is Shakespeare's revision of the Quarto, editors are forced into making arguments for the ways in which it is esthetically superior. Scholars talk about the Folio text as more effective in the theater, as if we had undisputed universal standards of theatricality. The ahistorical character of these arguments is especially evident in the work of Steven Urkowitz. He argues that in the Folio version, Shakespeare tended to cut reflective passages at the end of scenes in order to make the transitions from scene to scene faster and thus to sharpen certain dramatic contrasts.[8] What is interesting here is that Urkowitz's model of theatricality on examination turns out to be the cinematic fast cut, virtually a cliché of modern movie making. Indeed, faced with powerful competition from a rival medium, theater directors have increasingly worked to achieve cinematic effects on stage. Modern Shakespeare productions are often characterized by attempts to create the effect of fast cutting on stage, usually through the use of blackouts and sudden illuminations. But there is no evidence whatsoever that Elizabethan productions sought to achieve similar stage effects. In the absence of scenery, Elizabethan staging was fluid, but blackouts are very hard to achieve in outdoor theaters; moreover, there is evidence to suggest that at some of the points at which Shakespeare's fast cuts are supposed to have occurred, there was instead an intermission for the benefit of the audience in Elizabethan productions.[9]

Urkowitz also applauds the Folio text for cutting long passages that detail the motives of characters, leaving their psychology mysterious and hence dramatically more intriguing.[10] Once again, Urkowitz seems to be applying modern standards of drama anachronistically to Shakespeare. If anything, Elizabethan drama seems to revel in exploring the motivations of its characters in long, reflective speeches (remember "To be or not to be"). It is modern drama that prefers a pregnant pause or a prolonged silence to a fully articulated thought.[11] The more I reflect on the sudden championing of the Folio text by contemporary scholars, the more I have come to this explanation: they prefer the Folio text because, for whatever reasons, it more closely resembles the kind of heavily cut, fast-paced productions of Shakespeare currently in fashion with a generation of directors who have been heavily influenced by the cinema. In sum, although the revisionist critics are always accusing traditional editors of operating with unspoken and unexamined assumptions, they themselves are just as captive of mental habits. As Urkowitz examines the theatricality of *King Lear*, he seems to be automatically thinking in terms of contemporary productions and making little or no effort to imagine how the plays would have been staged in Shakespeare's day.

Thus in disputes over the text of *King Lear*, though scholars may pretend that they are arguing about technical matters of type compositors and press runs, in fact they are usually invoking assumptions about esthetic form, which covertly govern their judgment. That is why sorting out the textual situation in *Lear* is so complicated. The Quarto and the Folio texts differ in literally hundreds of ways on the level of individual words, but variations of larger significance occur repeatedly as well. For example, in the Quarto text, the last words are spoken by Albany, while in the Folio virtually the same words end the play but they are now spoken by Edgar. To be sure, it is possible to exaggerate the difference between the two versions.[12] Nobody claims that the variants result in versions that are fundamentally at odds, as if one *King Lear* were tragic and the other comic. In the eighteenth century, *King Lear* was almost always staged in a version created by Nahum Tate for which he in fact shaped a happy ending: Lear survives and Cordelia marries Edgar. Now *that* is a fundamentally different *King Lear*. The Quarto and the Folio texts do not differ to that extent. To put their divergence in perspective, my intuitive sense is that the two texts do not differ from each other more than any two productions of the play are likely to differ today in our era of highly creative directors. The Quarto and the Folio texts certainly do not differ as much as most film or television versions of the play do, say, Peter Brook's and Laurence Olivier's. No matter what text editors settle on, directors are always going to pick and choose among its passages, thereby bringing out certain aspects of the play and suppressing others. This fact of theatrical life is one practical argument for sticking with a conflated text and thereby offering as much of Shakespeare's *King Lear* as we have, recognizing that the process of directorial distillation is always going to reduce the play anyway.

I do not wish to reject or even minimize the achievement of the revisionist critics. They have forced us to reconsider the basic question of the text of the play, and anybody who wishes to understand *King Lear* cannot afford simply to ignore their arguments. Even those who choose to stick with a conflated text will probably be able to come up with a better conflation as a result of the critique of existing editions by the revisionists. I find many of their arguments persuasive, but in the end the case against them is settled for me by one decisive pair among the many additions and subtractions in the Folio text. One passage the Folio includes that the Quarto omits is the seminonsensical prophecy the Fool speaks at the end of Act III, scene ii, the one that begins "When priests are more in word than matter." Meanwhile one passage the Folio omits that the Quarto includes is the mock trial Lear conducts of Goneril and Regan in Act III, scene vi. Here I must seriously question whether Shakespeare's artistic consciousness can be invoked to explain this divergence between the two texts. If I were sitting down to cut *King Lear* for theatrical performance, my first reaction would not be: "Well, obviously that dull moment when Lear madly puts his daughters on trial has got to go, but no matter what, I am holding on to that wonderful prophecy of the Fool." I have seen many productions of *King Lear*, but I have never seen one in which the mock trial of Goneril and Regan was cut. But I have seen several productions in which the Fool's prophecy was quietly omitted (some editors have even maintained that the passage is spurious and not even by Shakespeare).[13] I am not claiming that the Fool's prophecy is worthless and should be banished from our text of *Lear.* Among others, the champions of the integrity and superiority of the Folio text have produced interesting arguments for the function of the Fool's prophecy in the logic of the play.[14] All I am saying is that if one had to choose between the Fool's prophecy and the mock trial, I do not think that anybody with taste and good judgment would prefer the former to the latter. If it were not for the textual anomaly, it would never occur to anyone to doubt that the mock trial is an indispensable part of *King Lear.* It is a brilliantly effective moment on stage; anyone who has seen a great Lear like Morris Carnovsky play this scene would acknowledge that it is one of the greatest moments in all of drama.

Moreover, the trial of Goneril and Regan is absolutely integral to the structure of *King Lear* as a whole. Shakespeare develops a profound series of parallels and contrasts between the mock trial and the real trial of Gloucester that occurs in the very next scene.[15] Perhaps the most amazing facet of the whole textual debate over *King Lear* is the spectacle of grown-up critics trying to educe reasons why the play is better off without the mock trial.[16] Here we witness a kind of reverse of the old New Criticism. Shakespeare scholars used to devote themselves to showing how his plays fit together into perfect wholes, how each moment is necessary to the integrated and overarching design of the play. Now they devote themselves to showing how the text does not hang together, how passages long thought to be miracles of artistic design are in fact extraneous and better left out. In the case of the mock trial scene, the strained character of these arguments—let me be more direct, their specious and sophistical character—is the weakest point of the position of the textual revisionists. I find myself wavering on many points in the debate, but on this one I can never be shaken. If the argument for the Folio version dictates a text of *King Lear* that omits the mock trial passage, then no matter what my general reservations, I must opt for some form of conflated text of the play. I frankly admit that I do not know why the mock trial is missing from the Folio, but I do feel that I can say with whatever certainty is possible in literary study that Shakespeare cannot be responsible for the omission. *King Lear* is not *King Lear* without the mock trial of Goneril and Regan.

As complicated as all these issues are, I do not want to dwell any longer on the specific case of the editing of a single play, even if we are talking about a work that many regard as Shakespeare's greatest achievement and hence perhaps the supreme product of the human imagination. I want to use the strange case of the two *King Lears* to illustrate a larger point about contemporary criticism, for I believe that recent developments in the editing of *King Lear* are emblematic of everything that has gone wrong in literary studies today. I am not talking about a crude assault on Shakespeare's artistry or a fundamental distortion of his aims. The editorial revisionists have not been able to exploit the divergence between the Quarto and the Folio texts of *King Lear* to pull a fast one, and, say, magically produce a feminist *Lear*, in which Goneril and Regan are vindicated and all the problems can be traced to the old king's phallogocentrism (they can leave that task to stage directors). The effect of current trends in editing *King Lear* is much subtler than this, but in the long run perhaps all the more insidious for just that reason. Indeed one has to ask oneself: why has this new approach to editing come along at just this moment in literary studies? There are of course professional reasons why such new theories and practices appear periodically. Young scholars need to build reputations and what better way to establish one's credentials as an editor than to show that for hundreds of years we have been laboring under false assumptions about the most basic facts concerning the text of *King Lear*? Moreover, one can see an economic impulse behind the revisionist theories. Publishers constantly wish to bring out new editions of Shakespeare but how can they do so if the plays remain the same? If one wonders about the swift acceptance of such a seemingly daring theory as that of the two *King Lears*, an acceptance institutionally embodied in the way that such a pillar of the academic establishment as Oxford University Press quickly bought into the theory, one might point to what a wonderful marketing ploy this thesis provides: "Be the first kid on your block to own a complete Shakespeare with two *King Lears* in it." In short, one can find material motives working behind the genesis and above all the quick acceptance in some circles of this new approach to editing *King Lear.*

But I am more struck by the way the new editorial approach conforms to the general mood in literary studies today. Is it any wonder that precisely the generation of critics raised on deconstruction as a critical theory has chosen to decompose the greatest single masterpiece of Western literature? In a narrow sense, deconstruction as a movement in literary theory shows signs of having reached a dead end. But in a larger sense deconstruction has changed the whole landscape of literary study and survives and indeed flourishes in such transformations as the New Historicism, which is really a kind of deconstructed Marxism. Above all, deconstruction has succeeded in transforming the basic attitude of critics toward a text, from a kind of respectful awe that issues in a desire to learn from the work to a barely suppressed hostility that reflects a desire to control and dominate it. During the reign of the New Criticism, critics had a respect, almost a blind faith, in the integrity of literary texts, what was often referred to as their organic wholeness. Critics used to give any apparently stray or odd element in a text the benefit of the doubt, always trying to figure out how it could be shown to fulfill a role in the work as a whole. If anything, the New Criticism erred on the side of finding wholeness in texts even in dubious situations. Traditional attitudes toward editing were in harmony with the New Critical approach to interpretation. In particular, the urge to come up with a single, unified text of *King Lear* reflected an underlying admiration for Shakespeare's artistic genius and a desire to see it reflected in the perfect form of an integral version of the play.

All this has changed in the postdeconstruction era, in the wake of relentless assaults on the wholeness of works of literature and in particular the systematic debunking of the idea that the author's consciousness stands behind and undergirds the text, supplying it with the integrity and coherence of an intentional artifact. A generation of critics has been brought up on the deconstructive principle that a work of literature always can be shown to mean something other than its author intended, perhaps reflecting some unconscious pressure on him or a buried division in his psyche or a contradiction inherent in his subject and/or his medium. An approach to editing that fragments works into multiple versions is the appropriate editorial counterpart of this brand of criticism. At first sight I may appear to be contradicting myself. After all, I have shown that the revisionist editors invoke the principle of authorial intention. They speak of Shakespeare consciously revising *King Lear*, for example, to improve its theatricality. In fact, for all the seeming daring of the textual revisionists, there is something faintly archaic about their position as we hear them constantly speaking of a man called Shakespeare, when in truly fashionable critical circles the author is supposed to be under erasure, as the French say. But if the revisionists' reliance on the concept of Shakespeare's intentions makes them sound old-fashioned, the end result of their efforts is as unsettling and as decentering to Shakespeare's authority as those of any Parisian poststructuralist. Deconstructionists use sophisticated reading strategies to destabilize the meaning of texts; the revisionist editors have a much simpler and more direct way to undermine the stability of Shakespeare's text. They claim to prove that there are now two texts of *King Lear* where there used to be only one, and then are able in effect to play the two texts off against each other, using one to call attention to the limitations of the other. If we step back for a moment and assess the impression created by this revolution in the editing of *King Lear*, it really is quite telling: where we used to think that Shakespeare wrote a single great play, we are now led to believe that he in fact produced two defective ones. The undermining of Shakespeare's authority is all the more insidious for being accomplished under the banner of what appears to be an ideologically neutral activity like textual editing.

To be sure, the revisionists still speak of Shakespeare's artistry when examining the way he streamlined the Folio version of *King Lear.* But notice the overall tendency of this approach—to confine Shakespeare's artistry to local effects and above all to show him operating within the constraints of ordinary notions of theatrical practicality. Think of the image of Shakespeare that emerges cumulatively from this editorial approach—he is no longer the universal genius he has long been thought to be, whose achievements transcend his era as well as the conventional limits of dramatic art. The Shakespeare of the revisionists is in general a practical man of the theater, and not all that different from other dramatists.[17] Like all authors, he was capable of making mistakes in his first drafts, and thus likely to have second thoughts about the details of his plays. The revisionists show us a Shakespeare capable of improving his works but the real thrust of their approach is to establish an image of Shakespeare as a fallible author in the first place. There must of course be at least a grain of truth in this view of Shakespeare; as much as I admire him, I am willing to grant that the revisionists have probably identified a few isolated cases where Shakespeare may have initially erred and later rethought details of *Lear.*

But I still reject the larger implications of this approach, which like all forms of historicism, works to assimilate Shakespeare into his environment and to downplay his difference from other artists. In their legitimate concern with the minutiae of the text of *King Lear*, the revisionists make it too easy for us to lose our sense of the larger picture and hence the deepest level of Shakespeare's artistry, which does not manifest itself in individual details of the text but rather in his ability to shape a meaningful artistic whole. Much of my understanding of *King Lear* has come from meditating long and hard on the significance of the mock trial scene within the whole of the play and especially its relation to other moments, such as the real trial of Gloucester. But anyone brought up on the revisionist view of the text is unlikely to think very long about the profundity of this moment in the play. Why waste time on it, after all, if Shakespeare himself decided to discard it in his final version? Neglect of the mock trial scene does not necessarily follow from its assignment to an earlier version of the play, but in practical terms I cannot imagine that the scene will continue to figure prominently in interpretations of the play if the Folio text comes to stand on its own and this passage is relegated to an appendix.

Whatever the revisionists may say about their trying to restore a true sense of Shakespeare's artistry (almost always a merely theatrical artistry, by the way), the real effect of their arguments is to cut Shakespeare down a peg or two. By stressing, admittedly with some validity, the messiness of the textual situation in *King Lear*, they leave us inevitably wondering: how much of a coherent masterpiece can this play really be if one day the mock trial was in the text and the next day it was out? As I have shown, in an inversion of traditional Shakespeare criticism, these revisionists are now devoting themselves to showing why moments like the mock trial scene do not truly belong in *King Lear.* Rather than showing how the text holds together, which directs us to the issues Shakespeare was raising, these critics concentrate on how the text falls apart, thereby foreclosing sustained enquiry into its overall logic. That is why, for all my technical doubts about the practice, I still opt for a conflated text of *King Lear.* It seems to me that chopping up the text is ultimately a way of chopping Shakespeare down to size. A conflated text may be an artificial editorial construct, but in the end it strikes me as the only way of fully respecting Shakespeare's achievement in *King Lear* and learning to explore its depths.

More generally, the revisionist approach to editing reflects the currently fashionable tendency toward a hegemony of the critic over the author and the literary text. The more fragmented the text, and the messier the textual situation, the more leeway the critic gains to exercise control over the literary work of art, to make it say what he wants it to say. The critic gets to take apart the text, to put it back together, and if need be, to take it apart again. The wholeness of the literary work begins to appear totally artificial if not arbitrary, the product of the way we view it, the product of the way we edit it, and not the result of the author's underlying and overarching vision, which we must learn to seek out and understand. In my view, the result of today's revisionist editors will be to present students, not with predigested and prepackaged texts, but with predeconstructed texts. The texts will have fallen apart even before they get to the students, who will have very little impetus to try to put them back together in any meaningful way. In short, having failed to undermine our literary heritage on the rarefied level of theory, deconstruction may yet succeed in doing so on the more basic level of textual editing. That is why I sound this warning against what may at first sight appear to be the innocuous activities of textual editors, activities so arcane in their consequences that they appear to be of no concern to anybody but specialists in Elizabethan literature. If we are not careful, we may awake one morning to find that, although the Sistine Chapel is still there in a new kind of pristine glory, *King Lear* as we know it has disappeared.

## NOTES

1 This is the text of Hamlet's soliloquy in the so-called Bad or First Quarto of 1603. See Michael J. B. Allen & Kenneth Muir, eds., *Shakespeare's Plays in Quarto*, p. 592.

2 For a good explanation and defense of this procedure, see G. Blakemore Evans, "Shakespeare's Text," in his edition, *The Riverside Shakespeare*, especially pp. 36–39. This whole essay gives an excellent introduction to the general problems of editing Shakespeare's plays. See also David Bevington's defense of a conflated text in his edition, *The Complete Works of Shakespeare*, p. A–17.

3 The central work in the attack on the traditional editing of *King Lear* is *The Division of the Kingdoms: Shakespeare's Two Versions of "King Lear"*, ed. by Gary Taylor & Michael Warren. Other important works in the attack are Steven Urkowitz, *Shakespeare's Revision of "King Lear"* and P. W. K. Stone, *The Textual History of "King Lear."*

4 See Jay L. Halio, ed., *The Tragedy of King Lear.* In his introduction, Halio provides a lucid summary of the debate concerning the text of *King Lear* (see pp. 58–80). He prints many parallel passages from the Quarto and the Folio, making comparisons of the two texts easy (see pp. 81–89, 265–289).

5 See, for example, Urkowitz (*Shakespeare's Revision*, pp. 67–79) and Halio (*The Tragedy of King Lear*, pp. 269–270), both of whom explain the problems in conflating the Quarto and Folio versions of Kent's long speech at the beginning of Act III, scene i. Urkowitz also scores some good points on p. 111 against conflated texts, but he overstates his case when he calls them "relatively shapeless hybrids."

6 For example, Stone (*Textual History*) thinks that the Folio does constitute a revision of the Quarto, but he attributes the work of revising not to Shakespeare but to Philip Massinger, a minor playwright of the period (see pp. 114, 127–128).

7 See Bevington, *Complete Works*, pp. lxxxv–lxxxvii. As Bevington points out, since author's manuscripts were sometimes used as promptbooks, the whole distinction begins to break down.

8 See, for example, Urkowitz, *Shakespeare's Revision*, p. 54.

9 See Halio, *The Tragedy of King Lear*, p. 275.

10 See, for example, Urkowitz, *Shakespeare's Revision*, p. 100: "the Quarto version of this passage yields a more comprehensible, straightforward reading of Albany's character. But the Folio provides better drama. The Quarto gives the audience the reason for Albany's choice; the Folio offers the audience only the painful moment of the choice itself." To say the least, this is not a noncontroversial notion of what constitutes "better drama"; in fact, it seems to turn Shakespeare into a modernist.

11 For Urkowitz's preference for silence over Shakespeare's text, see *Shakespeare's Revision*, p. 50: "The Quarto text indeed provides a choric commentary, a quiet ending, . . . but these qualities are not demonstrably or theoretically superior to a tableau of servants silently obedient as they remove a corpse." We see here at work the instincts of film directors, who cut the text of Shakespeare mercilessly in order to make room for their endless panning or tracking shots or other bits of cinematic self-indulgence. Like many filmmakers from Welles to Branagh, Urkowitz frequently seems to think that Shakespeare's text gets in the way of the drama; if only there weren't so many words. . . .

12 In his *William Shakespeare: King Lear*, Kenneth Muir writes that choosing one text over the other "would alter marginally our assessment of Albany and Edgar, but there would be no radical change in our understanding of the meaning of the play" (p. 117). Even Urkowitz, one of the staunchest proponents of the revisionist thesis, provides a detailed analysis of Folio revisions and then casually admits that they do "not concern the overall 'meaning' of *King Lear*" (*Shakespeare's Revision*, p. 25; see also p. 27). In the end one might conclude that the whole controversy over the text of *King Lear* turns out to be a tempest in a teapot, but in fact the textual disputes have had serious consequences by diverting attention from the more important task of trying to understand the meaning of the play.

13 See, for example, E. K. Chambers, *William Shakespeare: A Study of Facts and Problems*, vol. 1, p. 466. Even one of the revisionists criticizes the Fool's prophecy; see Stone: "It has little poetic merit and absolutely no dramatic relevance." Recall that Stone does not think that Shakespeare did the revisions in the Folio; thus he is often searching for signs of an inexpert hand in the Folio text. Hence he says in a charming understatement that the Fool's prophecy "is most unlikely to have recommended itself to a reviser of any literary pretentions" (*Textual History*, p. 111).

14 See, for example, Urkowitz, *Shakespeare's Revision*, p. 44 and Halio, *The Tragedy of King Lear*, p. 282.

15 It is even possible to argue that the whole of *King Lear* is structured around a series of trial scenes, broadly conceived. See Charles Moseley, "Trial and judgment: The Trial Scenes in *King Lear*." With regard to the textual issue, Moseley writes: "To us, to delete this scene [the mock trial] seems an extraordinary decision to have taken, and its importance to the thematic structure of the play and to Lear's development is one of the strongest arguments for regarding a conflation as the nearest we shall get to Shakespeare's ideal of his play" (pp. 65–66). I discuss the importance of the mock trial in the overarching structure of the play in my essay "Nature and Convention in *King Lear*."

16 See, for example, Roger Warren, "The Folio Omission of the Mock Trial: Motives and Consequences," pp. 45–57. Incredibly, Warren bases his argument in part on the point that the elimination of the mock trial avoids a duplication of Lear's mock justice in Act IV, scene vi. Normally, this kind of parallel would be taken precisely as evidence that the mock trial *belongs* in *King Lear*; these two dramatic moments would be seen as paralleling and reinforcing each other, developing a pattern, working out a theme. Critics were once ingenious in finding evidence for the unity of the text of *Lear*; now that some have come along with a theory of its disunity, they are equally resourceful in finding evidence for their new view. Just as one can always find reasons why a passage belongs in a work, one can always find reasons why it does not. The question is whether the reasons are on balance plausible. To his credit, Halio acknowledges in his edition that the omission of the mock trial is a grave problem with the Folio text; see p. 270. For further discussion of the Folio omission, see Gary Taylor, "Monopolies, Show Trials, Disaster, and Invasion: *King Lear* and Censorship," especially pp. 88–101.

17 For a good statement of this view, see *William Shakespeare: The Complete Works*, ed. by Stanley Wells & Gary Taylor, pp. xxxvi–xxxvii. For the *reductio ad absurdum* of this position, see Urkowitz's astonishing claim about Shakespeare: "Probably the greatest sources of his artistic fecundity and confidence were his daily association with the players in his company and his performances with them throughout his productive life" (*Shakespeare's Revision*, pp. 145–146). This is an excellent example of the creeping Marxism at work in contemporary criticism. Critics are so obsessed with rejecting the Romantic idea of individual genius that they would just as soon turn Shakespeare the playwright into an affable co-worker in the Globe Theater, nourished by his solidarity with his fellow actors. In viewing drama as a social production, this position tries to make Shakespeare appear as ordinary as possible. The problem for this view is of course those plays he wrote. If "daily association with the players in his company" was the "greatest source" of Shakespeare's "artistic fecundity," why did Ben Jonson or John Fletcher or Philip Massinger not produce works on the level of *Hamlet* and *King Lear*? (Of course Stone thinks Massinger did have a hand in *Lear*.)

## REFERENCES

ALLEN, MICHAEL J. B. & KENNETH MUIR, eds. *Shakespeare's Plays in Quarto.* Berkeley, CA: University of California Press, 1981.

BEVINGTON, DAVID, ed. *The Complete Works of Shakespeare.* New York, NY: Harper Collins, 1992.

CANTOR, PAUL A. "Nature and Convention in *King Lear.*" In *Poets, Princes, and Private Citizens: Literary Alternatives to Post-Modern Politics*, edited by Joseph Knippenberg & Peter Lawler. Lanham, MD: Rowman and Littlefield, 1995.

CHAMBERS, E. K. *William Shakespeare: A Study of Facts and Problems.* Oxford: Clarendon Press, 1930.

EVANS, G. BLAKEMORE. "Shakespeare's Text." In *The Riverside Shakespeare.* Boston, MA: Houghton Mifflin, 1974.

HALIO, JAY L., ed. *The Tragedy of King Lear.* Cambridge: Cambridge University Press, 1992.

MOSELEY, CHARLES. "Trial and Judgment: The Trial Scenes in King Lear." In *Critical Essays on "King Lear,"* edited by Linda Cookson & Bryan Loughrey. Harlow, Essex, UK: Longman, 1988.

MUIR, KENNETH. *William Shakespeare: King Lear.* London: Penguin, 1986.

STONE, P. W. K. *The Textual History of "King Lear."* London: Scolar Press, 1980.

TAYLOR, GARY. "Monopolies, Show Trials, Disaster, and Invasion: *King Lear* and Censorship." In *The Division of the Kingdoms: Shakespeare's Two Versions of "King Lear,"* edited by Gary Taylor & Michael Warren. Oxford: Clarendon Press, 1983.

TAYLOR, GARY & MICHAEL WARREN, eds. *The Division of the Kingdoms: Shakespeare's Two Versions of "King Lear."* Oxford: Clarendon Press, 1983.

URKOWITZ, STEVEN. *Shakespeare's Revision of "King Lear".* Princeton, NJ: Princeton University Press, 1980.

WARREN, ROGER. "The Folio Omission of the Mock Trial: Motives and Consequences." In *The Division of the Kingdoms: Shakespeare's Two Versions of "King Lear"*, edited by Gary Taylor & Michael Warren. Oxford: Clarendon Press, 1983.

WELLS, STANLEY & GARY TAYLOR, eds. *William Shakespeare: The Complete Works.* Compact edit. Oxford: Clarendon Press, 1988.

# CONSTRUCTING LITERATURE
## Empiricism, Romanticism, and Textual Theory

GEORGE BORNSTEIN

IN THEIR PROVOCATIVE BOOK *Higher Superstition: The Academic Left and Its Quarrels with Science*, Paul Gross and Norman Levitt use the example of natural science to illuminate the stance of contemporary theory in the humanities and related social sciences toward broad issues of the construction, verification, and status of what we used to call "knowledge."[1] The particular poststructuralist pattern of strong cultural construction whose manifestations they map across a broad spectrum includes aspects of cultural studies, feminism, environmentalism, and scientific critique. Those areas display a common aversion toward the twin checks of internal consistency and external falsifiability familiar to empirical standards of verification. But that is just the point, since empiricism itself is under attack as merely a cultural construct. As the French cultural theoretician Pierre Machery noted in his A *Theory of Literary Production* in a remark that might have made more grist for the mill of Gross and Levitt, "A rigorous knowledge must beware all forms of empiricism, for the objects of any rational investigation have no prior existence but are thought into being."[2] Such attitudes lead quickly to the position of leftist literary scholars like Catherine Belsey, whose "claim is not that such a history . . . is more accurate, but only that it is more radical."[3]

*Higher Superstition* traces the genealogy of those attitudes in literary romanticism, as does Susan Haack in a recent article.[4] I dissent not so much from their analysis of the current scene as from their genealogy of it. The notion of romanticism as irrational, emotional, idealist, or antiscientific has a long history in the twentieth century, touching such diverse figures as T. S. Eliot or the original New Critics who saw romanticism as lacking rigor, irony, or "wit," or historians like Peter Viereck who saw it as leading directly to Fascism. Gross and Levitt adopt a more nuanced and accurate view when they write of romantic opposition to "the *narrowly* empirical and the *strictly* rational" (italics mine). But the caution of those qualifiers drops away as they go on to write of Wordsworth as a "self-satisfied old Tory," of Blake's Newton as a "figment, not the preeminent mathematician and physicist of his time" (p. 105), of trendy studies of gender and science as "Goethe's—and

Wordsworth's and Whitman's—Romantic idealism in this year's Paris original" (p. 142), or of the "monumental figures" of Romantic individualism as "notable for their rejection of the worldview suggested by the orthodox science of their day" (p. 223). The common thread of these characterizations concerns an alleged Romantic hostility towards reason and empiricism as arbiters of experience.

I would like to argue not so much for calling such constructions of Romanticism "wrong" but rather for recognizing that a different construction of Romanticism is possible and even defensible, one in which the Romantics object not to reason and empiricism but to their undue dominance, and seek instead a reintegrated human psyche with reason as a major but not sole part. Such a view might begin with Blake's mythological figure of "Urizen," the pun in whose very name indicates which of the prime human faculties he represents. Blake memorably depicted him in chains in *The Book of Urizen* to indicate his fallen state, but also described in *The Four Zoas* (Night the Ninth) Urizen's joyous recovery of his original glory as soon as he renounces his desire to dominate all other powers. Correspondingly, in the last book of his epic *The Prelude*, Wordsworth called Imagination "but another name for absolute power / And clearest insight, amplitude of mind, / And Reason in her most exalted mood" and wanted future generations to think of himself and his friend Coleridge as "sanctified / By reason and by truth."[5] And Coleridge himself had announced in the very first chapter of his great *Biographia Literaria* that "no authority could avail in opposition to Truth, Nature, Logic, and the Laws of Universal Grammar." He found poetry attractive because it seemed more, not less, logical than science: "Poetry . . . ha[s] a logic of its own, as severe as that of science; and more difficult, because more subtle, more complex, and dependent on more, and more fugitive causes."[6] Shelley of course studied the science of his day carefully and went through a phase of considering himself a materialist. And even though a Romantic like Keats could write in moments of doubt that "I have never yet been able to perceive how any thing can be known for truth by consequitive [sic] reasoning" he would then immediately add "and yet it must be"[7] Those instances, which could easily be multiplied manyfold, would yield a different portrait of Romanticism and of the roots of current attitudes. The Romantic critique of "*narrowly* empirical" and "*strictly* rational" would then be no more antiscientific than that of Gross and Levitt themselves when they criticize contemporary strong constructionism in these terms: "Cultural constructivism, at least in the full-blooded version of ideologues like Aronowitz, is a relentlessly mechanistic and reductionistic way of thinking about things. It flattens human differences, denies the substantive reality of human idiosyncracy, and dismisses the ability of the intellect to make transcendent imaginative leaps" (p. 56).

My point is not to argue that Gross and Levitt are themselves Romantic, but rather that different constructions of romanticism are possible and would lead to different conclusions. How, then, do we judge between them? One way of judging, which I favor recuperating in a new form, would apply traditional empirical standards and ask how well each view fits the evidence, how much evidence contradicts each view, and how internally consistent is each theory. Another view, often more in favor in the humanities today, would choose according to which view is more likely to promote a desired political agenda. As the well-known theorist Frank Lentricchia bluntly states, "This sort of theory seeks not to find the foundation and the conditions of truth but to exercise power for the purpose of social change. It says there is no such thing as eternally 'true' theory. I conceive of theory as a type of rhetoric."[8] That view, of course, grows largely out of French poststructuralist thought, particularly that associated with Foucault and various French materialist critics. It was able to succeed so well in literary circles partly because the traditional antiempiricism and antirationality associated with one view of Romanticism had already cleared the way for it by diminishing humanistic reliance on reason and empiricism as checks. That was particularly so because even those enemies of Romanticism, the New Critics, had still construed literary study as an alternate form of knowledge to science. Those willing to challenge the new French theories on grounds of empiricism and reason were few on the ground by the late sixties, and in any case reason and empiricism were the very faculties now under attack as oppressive and constraining of attempts to establish a more just social order.

A typical position paper of a Continentally influenced theorist of the postmodern critique of objectivity runs like this one, in which I have changed only the names of the countries to make a rhetorical point:

> In France relativism is an exceedingly daring and subversive theoretical construction. In America, relativism is simply a fact. Everything I have said and done in these last years is relativism by intuition. If relativism signifies contempt for fixed categories and men who claim to be the bearers of an objective, immortal truth . . . then there is nothing more relativistic than [our] attitudes and activity. From the fact that all ideologies are of equal value, that all ideologies are mere fictions, the modern relativist infers that everybody has the right to create for himself his own ideology and to attempt to enforce it with all the energy of which he is capable.[9]

Today such critiques carry heavy claims of progressive attitudes, and seek to intimidate scrutiny and challenge by painting opposition as reactionary and oppressive. The problem is, the critique that I have just read is taken verbatim from Mussolini, and forms part of his 1921 positioning of Fascism as an advanced philosophical position in the article "Relativismo e Fascismo."[10] This should not surprise us in view of the actual politics of many poststructuralist luminaries, from the key philosopher Martin Heidegger's support of the Nazis when university rector to the influential popularizer of deconstruction Paul De Man's wartime newspaper propaganda in support of Nazi designs. Their gravitation toward authoritarian governments consorts oddly with the rhetoric of indeterminism, to say the least. Indeed, the French poststructuralists themselves regularly supported murderous regimes such as Mao's China, which they glorified just after the only years in recorded history during which the population of China actually dropped because of governmental policies underlying the Great Leap Forward.[11] Indeed, although the academic left regularly conflates elitism with capitalism and with modernism, every Marxist revolution of our century has been led by an educated elite claiming to speak in the name of a less advanced "people." The problem may be not that literature is not political, but that simply claiming that it is political tells us nothing of the humanity or viciousness of the politics of the claimant.

The notion of contemporary "theory" as progressive carries over into the essays in the influential current book *Critical Terms for Literary Study*, published by University of Chicago Press.[12] Less a handbook or dictionary than a set of brief essays on twenty-eight key terms or topics, the widely circulated volume functions as a window on state-of-the-art literary practice. The list of which topics have been privileged by inclusion repays attention. Terms like representation, discourse, determinacy/indeterminacy, culture, canon, gender, race, and ethnicity dominated the first edition (1990). To them the just-published new edition (1995) adds six more—popular culture, diversity, imperialism/nationalism, desire, class, and (improbably) ethics. Those are meant to clarify what the back cover describes as "the growing understanding of literary works as cultural practices." Both editions are as notable for what they omit as for what they include. This construction of the turn toward cultural practices does not feature such terms as evidence, consistency, or empiricism, which some might think important for analyzing the social embedding of literary activity. Nor is the currently fashionable label "desire" matched with any of the faculties such as reason or imagination that would have attended its Blakean balance in the human form divine. The book's canon also excludes "esthetics" or any related term, in accord with the current recoil from esthetic approaches to literature and (following Walter Benjamin's influential formulations) association of them with Fascism. Yet the volume is strangely silent on the other pole as well, the text as material object, and offers its quotations from previous authors as though the physical texts themselves were unproblematic, transparent lenses rather than themselves constructions. Instead of all those, *Critical Terms for Literary Study* repeatedly offers a partial selection featuring one set of terms while marginalizing or suppressing other sets that might contribute toward a fuller view. To the extent that the book accurately reflects "advanced" humanistic study, it also reflects the presuppositions that call forth responses like *Higher Superstition*. In remaining silent on material aspects of the text in particular, it misses an important opportunity to bring advanced criticism together with recent advances in editorial theory and textual construction in ways that might rejoin what Blake would regard as the fallen body of current theory.

A surprising number of terms that contemporary theory uses metaphorically have literal senses in contemporary textual construction. Among them are *text*, *margin*, *gap*, *erasure*, *production*, *instability*, and *inscription*. Those literal usages offer both an empirical check on some of the more extreme aspects of current theory and an opportunity for a fuller critical enterprise. That is especially true of the newer paradigms of editorial activity introduced over the past dozen years. In place of an older editorial enterprise aiming to recover an alleged ideal, pure, unitary form of a text corrupted in transmission, the newer paradigm accepts the validity of multiple authorized versions of texts that resemble processes as much as they do products. On the one hand it sees the "eclectic" construction of texts as creating ahistorical hybrids never before seen on land or sea, and on the other it tends to regard "texts" as comprising more than merely the words of an individual work on the page. Shakespeare and Yeats offer ready examples of what I mean.

Any reader, layman or scholar, wishing to sit down and read "Shakespeare" cannot do that with a text that is unproblematic or transparent at the level of its own construction. That is, we read Shakespeare's plays not in editions that simply transcribe faithful authorial manuscripts, for there are no such manuscripts to transcribe, but rather in editions that scholars construct out of various imperfect textual "witnesses" or surviving textual versions known to be corrupt in major ways. The general procedure is for the editor to select one or another version as "copy-text," a technical term denoting the version that the editor follows when unable otherwise to decide what decision to make from the conflicting evidence, and in Anglo-American editing particularly associated with so-called accidentals like punctuation (for this reason, editors sometimes suggest a "divided authority" in copy-text). The editor then usually emends the copy-text according to evidence from other textual witnesses deemed to carry authority in a particular case. Such editions are called "eclectic" not necessarily in that the editors freely mix and match, but rather in that the resultant texts are comprised of parts chosen or inferred from various distinct documents as emendations of the copy-texts. As D. C. Greetham summarizes in his recent and now standard *Textual Scholarship: An Introduction*, "By the third quarter of the twentieth century, it would be fair to say that the dominant mode of Anglo-American textual criticism, institutionally and academically, was the copy-text school of eclectic editing designed to produce a reading clear-text whose features were [allegedly] a fulfillment of authorial intentions by the selection of authorially sanctioned substantive variants from different states of the text."[13] As Greetham points out, this type of edition has become known as the "text that never was" but which "by implication, *ought* to have been, in the best of all possible worlds, since it constructed authorial intention in despite of the testimony of individual documents." Such substantive emendations are usually made according to criteria that Peter Shillingsburg has neatly schematized as either esthetic, historical, authorial, or sociological.[14]

Besides deliberate emendation, one other major factor enters into the texts of Shakespeare that most of us read—the modernization of spelling. This might at first seem a harmless procedure, simply for the convenience of the reader translating antiquated Elizabethan spelling into standard modern form. But our notions of standardized spelling (and punctuation, for that matter) are only a little over a century old. The Elizabethans did not share them. Turning the protean varieties of Elizabethan orthography into orthodox modern renderings regularly loses ambiguities and richnesses made possible by the earlier text. A brief and simple example occurs in the famous mock-judgment scene (usually Act III, Scene vi in modern editions) from *King Lear*, when the mad King addresses the disguised Edgar as judge, calling him a "robed man of justice." But in the sole surviving witness text to this particular scene, Lear refers to a "robbed" man of justice, and the pun on robbed/robed carries rich overtones lost in the silent modernization. Indeed, the modern texts that have Lear addressing Edgar at this point are themselves eclectic interpolations, since the original quarto does not indicate to whom Lear is speaking. This sort of thing happens with great frequency in the text. The result is that contemporary editions are multiple constructions both as eclectic copy-text editions and as modernized spelling (in contrast to "old spelling") ones. The implications of that situation are only beginning to make themselves known.

Nearly all current theorizing and critique of Shakespeare's *King Lear*, whether older or newer in orientation, utilizes widely used eclectic current editions like the Riverside or Penguin, which employ modernized spelling as well. The fact that such editions have little documentary validity but rather are based on conflations made nearly a century after composition affects adherents of both the "older" and "newer" *King Lears.* On the one hand, resisters of the "version" theory cannot simply advocate returning to the "old" or "true" text of *Lear*, because whatever text they mean by that is itself a constructed artifact, and usually a conflated one as well. On the other hand, the recuperation of versions is particularly damaging to newer readings emphasizing cultural practice and historical embedding. Surely the "text" that we would want to relate to Elizabethan cultural practices would be some variant of an Elizabethan text, rather than of a modernized construction that would better serve relation to *our* culture. The main witness documents to *King Lear* are two, the so-called Pied-Bull Quarto of 1608 and the famous "First Folio" of 1623 (there is also a "bad Quarto" of 1605). Seventeenth-century productions and texts of the play regularly followed the folio version. Not until 1709 did one Nicholas Rowe prepare the first edition of Shakespeare that we would call "critical," in that it conflated texts of both the quarto and the folio tradition into a new hybrid purporting to be closer to the Shakespearean original, which was conceived as unitary and stable. That tradition continued all the way until 1986, when the controversial Oxford Shakespeare text prepared by Stanley Wells and Gary Taylor appeared.[15] Their edition followed the arguments of scholars in *The Division of the Kingdoms* collection that the two sources—quarto and folio—represented two different *versions* of the play, which Shakespeare as practicing playwright had revised. The Oxford Shakespeare, accordingly, produced edited separate texts of the folio and quarto versions, rather than conflating them. In so doing they provide the "entire" play by printing edited texts of both the quarto and folio version, in contrast to the older eclectic or conflated texts which necessarily leave some material out.

What difference does it make? All the difference in the world. For example, Cordelia is stronger in one version than in the other, even to the point of apparently leading an army in act four. Lear's death speech differs, implying acceptance of defeat in the quarto text but continued illusion in the folio; and the quarto's famous mock-judgment scene on the heath does not appear in the folio at all. Our modern texts are *constructions* in the literal sense, and the recoil from reason and empiricism helps explain why so many critics and theorists remain ignorant of the extent to which choices made by editors affect critical possibilities. Yet the choices—quarto, folio, or hybrid—are not infinite, but circumscribed. Few would go so far as to advocate following the text of Nahum Tate, who rewrote the play in 1681 to give it, among other things, a happy ending! Incredibly, that version held the stage for a century and a half, all the way through the Romantic period. Not even the most anti-empirical theorist today advocates its return. Yet most contemporary critics still adhere to modernized conflated texts when discussing the political or social meanings of the play.

W. B. Yeats offers a more modern instance from our own century. I pass over quickly here the tendency of contemporary theorists to denounce Yeats's politics by, say, unwittingly using the 1924 revisions of texts written thirty years earlier when discussing Yeats's own early social views. They do so because the widely available collected editions of the poetry follow his own practice in keeping earlier quasi-chronological arrangements of volumes and sequences even while incorporating drastically revised texts under the earlier dates. The versions of "Dedication to a Book of Stories Selected from the Irish Novelists" or "Lamentation of the Old Pensioner" printed in the collected *Poems* in fact reproduce the drastic revisions of the early 1920s even while keeping the 1893 heading for the "Rose" section in which they appear. And even noting the change in title of a later poem like "September 1913" from the earlier "Romance in Ireland (On reading much of the correspondence against the Art Gallery)" might enable critics better to place the poem in its original context as part of the original campaign for a municipal art gallery in Dublin during the great labor battles of 1913. In those controversies, Yeats sided with the workers during the strike, just as they and their leaders sided with him on the gallery; he even published an article on their behalf, for which he received public thanks from fiery leaders like the socialist James Connelly and militant James Larkin. Knowledge of how and why the text was constructed might cause critics to reconsider the charge of elitism sometimes leveled against that poem and those surrounding it.

Because the discussion of *King Lear* has already illustrated the hazards of multiple versions, I focus here instead on the hostile reception accorded recently to Yeats's famous lyric "A Prayer for My Daughter," especially for what it can tell us about a different aspect of cultural construction—about the literary context that the writer may create for his work, and even about what constitutes the boundaries of a literary work in the first place. Especially in the case of short lyric poems, the tendency to take the poem itself as unit has led to badly flawed results. "A Prayer for My Daughter" has become a particular whipping boy for one brand of feminist critic, especially in its beautiful sixth stanza:

> May she become a flourishing hidden tree
> That all her thoughts may like the linnet be,
> And have no business but dispensing round
> Their magnanimities of sound,
> Nor but in merriment begin a chase,
> Nor but in merriment a quarrel.
> O may she live like some green laurel
> Rooted in one dear perpetual place.[16]

And here is Joyce Carol Oates's influential attack on that passage, taken from the lead essay of the journal in which it appeared: "This celebrated poet would have his daughter an object in nature for others'—which is to say male—delectation. She is not even an animal or a bird in his imagination, but a vegetable: immobile, unthinking, placid, "hidden". The activity of her brain is analogous to the linnet's song—no distracting evidence of mental powers . . . the poet's daughter is to be brainless and voiceless, *rooted.* So crushingly conventional is Yeats's imagination. . . ."[17] To say the least, Oates's reading is highly improbable as well as bad botany (a laurel tree is not a vegetable). Not Yeats's poem but rather Oates's reading is "crushingly conventional." It contradicts most of what we know about Yeats's actual attitude toward his daughter, including the fact that he sent her not to a local Irish school but rather to the same Swiss boarding school to which he sent his son. He did that because he thought that the Irish schools taught girls conformity and prepared them for nothing but marriage (the daughter, Anne, went on to become one of the leading Irish artists of her generation and to praise her father's attitude toward her education).[18] More importantly for our purposes, Oates's attack is based on an isolation of the individual lyric typical not only of her own school of criticism but also of the original New Critics against whom contemporary theoreticians like to define themselves. Yet as Hugh Kenner and others have pointed out, Yeats did not write individual lyrics, he wrote books of poems, and he spent considerable time arranging and ordering those individual volumes.[19] "A Prayer for My Daughter," for example, comes from a book called *Michael Robartes and the Dancer.* That volume begins with the comic title poem suggesting ways in which women can enlighten men, and it ends with a short lyric about Yeats restoring a house for his beloved and strong-minded wife, Anne's mother. "A Prayer for my Daughter" occurs more precisely between the ominous poem "The Second Coming," with its vision of a "blood-dimmed tide" of World War One sweeping over the modern world, and "A Meditation in Time of War," which invokes the specific setting of the Irish "troubles" or period of rebellion and civil war. In that context, "A Prayer for My Daughter" looks quite different, as a father's loving expression of the desire to protect from growing civil violence his first-born child, a baby of four months at the time. Oates would be unlikely to condemn such a poem emanating from contemporary Bosnia or Rwanda, especially if the civil war there were inscribed in the volume itself. In the Irish context, too, the proscription of "opinion" means not that Anne is not supposed to have opinions, but that her thoughts should be the opposite of mere "opinion," Yeats's code word for the opinionated ideologies that he saw bringing bloody destruction to the Ireland around him and which he began attacking with the very first line ("Opinion is not worth a rush") of the volume. In contrast, Anne's "thoughts" are to display "magnanimities" surpassing and correcting the narrowly intolerant opinions he saw around him. She is not to be "unthinking," as Oates claims, but rather a superior thinker of wholeness, crowned with the laurel of poetry and victory. Oates's construction is a misconstruction, formed by dismantling Yeats's own broader vision and substituting her own more limited one instead, in an uncanny reenactment of the kind of "opinion" which the poem challenges.

"A Prayer for My Daughter" illustrates one final aspect of literary construction appropriate for a newly empirical blend of textual scholarship with contemporary theory. Current editorial theorists distinguish between the "linguistic code" of a work, its words, with the "bibliographic code" or physical features of the text, such as layout, spacing, or design.[20] An important feature of the bibliographical code of "Prayer for My Daughter" is venue of publication. Like every new book of Yeats's poetry to appear in the twentieth century, the original volume of *Michael Robartes and the Dancer* bore on its title page the imprint of Cuala (earlier Dun Emer) Press, as well as a revealing colophon usually at the back. Dun Emer was the small fine-arts press founded along the lines of William Morris' Arts and Crafts principles by Yeats's sisters and Evelyn Gleeson shortly after the turn of the century as part of Dun Emer Industries. The name Dun Emer means Emer's Fort. It signified both the nationalist and the feminist impulses of the press, founded partly to promote artistic Irishness and partly to give employment outside the home to young Irish women. Dun Emer's first book was Yeats's collection *In the Seven Woods* (1903), and the ability to provide the first print run of his successive volumes (before the wider distribution offered by the later Macmillan editions) helped keep the press financially afloat. Publication under such auspices does not make Yeats a feminist, of course, and offered him some advantages in design and audience. But the decision to publish his volumes under such an imprint—nationalist, feminist, and with all the work done by women—ought at the very least to give pause to those who want to view his work from "Adam's Curse" through "Prayer for My Daughter" and "Politics" as antithetical to modern feminisms. Neither the women of Dun Emer/Cuala nor the author of the poems was so simplistic. Attention to the bibliographical code and to the literal construction and distribution of the work opens our eyes to some of the complexities of mire and blood here, and of the original social embedding of the poetry as serving a largely feminist enterprise.

In tracing these various constructions—of romanticism, of theory, of *King Lear*, and of "Prayer for My Daughter"—I have not meant to challenge the notion of construction itself, at least in what Gross and Levitt call its "weak" form (p. 44). Rather, I have meant to suggest that the idea of construction is itself a construction, as are the objects of its activity. Yet to say that historical labels, or theories, or works of art are "constructed" is not to say that all constructions are equal. I hope to have made clear that some constructions are better than others, because some take fuller account of the evidence available, are more internally coherent, and are grounded in a marriage of empirical procedures and theoretical inquiry. The joining of textual construction and literary theory need not produce either readings or views that are old-fashioned on the one hand or merely trendy on the other. To the contrary, they can produce views suited to our modern climate, correcting the excesses of both past and present. Doing that depends, too, upon a recuperation of one of the great targets of poststructuralism and of cultural studies—personal agency. It is that deprecated but necessary agency that allows us to say with Wallace Stevens in his great poem "Esthetique du Mal":

And out of what one sees and hears and out
Of what one feels, who could have thought to make
So many selves, so many sensuous worlds,
As if the air, the mid-day air, was swarming
With the metaphysical changes that occur,
Merely in living as and where we live.[21]

## NOTES

1 Paul R. Gross & Norman Levitt, *Higher Superstition: The Academic Left and Its Quarrels with Science.* The page number of further references will be cited inside parentheses within the text itself.
2 As quoted in Richard Levin, "The New Interdisciplinarity in Literary Criticism," p. 21.
3 Ibid., p. 23.
4 Susan Haack, "Puzzling Out Science," p. 28 (n. 6).
5 William Wordsworth, *The Prelude: 1799, 1805, 1850*, p. 469.
6 S. T. Coleridge, *Biographia Literaria*, vol. 1, pp. 14 and 4.
7 John Keats, *Selected Poems and Letters*, p. 258.
8 Frank Lentricchia, *Criticism and Social Change*, p. 12.
9 Translation quoted in Jerry Martin, "The Postmodern Argument Considered," p. 653.
10 In *Diuturna*, pp. 374–377.
11 Roderick MacFarquhar & John K. Fairbank, eds. *The Cambridge History of China*, vol. 14, p. 370f.
12 Edited by Frank Lentricchia & Thomas McLaughlin.
13 D. C. Greetham, *Textual Scholarship: An Introduction*, pp. 334–335.
14 Peter Shillingsburg, *Scholarly Editing in the Computer Age*, p. 19 and passim.
15 Wells and Taylor also prepared an original-spelling edition for Oxford. See also *The Division of the Kingdoms*, edited by Gary Taylor & Michael Warren.
16 W. B. Yeats, *The Poems*, p. 189.
17 Joyce Carol Oates, "'At Least I Have Made a Woman of Her': Images of Women in Twentieth-Century Literature," p. 17.
18 Jane York Bornstein, "More Than a Name," pp. 7–9.
19 Hugh Kenner, "The Sacred Book of the Arts," pp. 9–29. See, too, Hazard Adams, *The Book of Yeats's Poems*, especially chapter one; and George Bornstein, "What is the Text of a Poem by Yeats?" pp. 167–193.
20 The most influential statement of this distinction is in Jerome J. McGann, *The Textual Condition*, ch. 2.
21 Wallace Stevens, *Collected Poems*, p. 326.

## REFERENCES

ADAMS, HAZARD. *The Book of Yeats's Poems.* Tallahassee, FL: Florida State University Press, 1990.

BORNSTEIN, GEORGE. "What Is the Text of a Poem by Yeats?" In *Palimpsest: Editorial Theory in the Humanities.* Edited by George Bornstein & Ralph Williams. Ann Arbor, MI: University of Michigan Press, 1993.

BORNSTEIN, JANE YORK. "More Than a Name." *Michigan Today*, April 1989, pp. 7–9.

COLERIDGE, S. T. *Biographia Literaria.* Edited by J. Shawcross. Oxford: Oxford University Press, 1973.

GREETHAM, D. C. *Textual Scholarship: An Introduction.* New York, NY: Garland, 1992.

GROSS, PAUL R. & NORMAN LEVITT. *Higher Superstition: The Academic Left and Its Quarrels with Science.* Baltimore, MD: Johns Hopkins University Press, 1994.

KEATS, JOHN. *Selected Poems and Letters.* Edited by Douglas Bush. Cambridge, MA: Riverside, 1959.

HAACK, SUSAN. "Puzzling Out Science." *Academic Questions* 8 (Spring 1995): 28, n. 6.

KENNER, HUGH. "The Sacred Book of the Arts." In *Gnomon: Essays on Contemporary*

*Literature*. New York, NY: McDowell, Obolensky, 1958. Reprinted in *Critical Essays on W. B. Yeats*. Edited by Richard J. Finneran. Boston, MA: G. K. Hall, 1986.

LENTRICCHIA, FRANK. *Criticism and Social Change*. Chicago, IL: University of Chicago Press, 1983.

LENTRICCHIA, FRANK & THOMAS MCLAUGHLIN, eds. *Critical Terms for Literary Study*. 2d edit. Chicago, IL: University of Chicago Press, 1995.

LEVIN, RICHARD. "The New Interdisciplinarity in Literary Criticism." In *After Poststructuralism: Interdisciplinarity and Literary Theory*, edited by Nancy Easterlin & Barbara Riebling. Evanston, IL: Northwestern University Press, 1993.

MACFARQUHAR, RODERICK & JOHN K. FAIRBANK, eds. *The Cambridge History of China*. Vol. 14. Cambridge: Cambridge University Press, 1987.

MARTIN, JERRY. "The Postmodern Argument Considered." *Partisan Review* 60, no. 4 (1993): 653.

MCGANN, JEROME J. *The Textual Condition*. Princeton, NJ: Princeton University Press, 1991.

MUSSOLINI, BENITO. "Relativismo e Fascismo." In *Diuturno*. Milan: Casa Editrice Imperia, 1924.

OATES, JOYCE CAROL. " 'At Least I Have Made a Woman of Her': Images of Women in Twentieth-Century Literature." *Georgia Review* 37 (Spring 1983): 17.

SHAKESPEARE, WILLIAM. *The Complete Works*. Edited by Stanley Wells & Gary Taylor. Oxford: Clarendon Press, 1986.

SHILLINGSBURG, PETER L. *Scholarly Editing in the Computer Age*. Athens, GA: University of Georgia Press, 1986.

STEVENS, WALLACE. *Collected Poems*. New York, NY: Alfred Knopf, 1978.

TAYLOR, GARY & MICHAEL WARREN, eds. *The Division of the Kingdoms*. 1983. Reprint. Oxford: Clarendon Press, 1986.

WORDSWORTH, WILLIAM. *The Prelude: 1799, 1805, 1850*. Edited by Jonathan Wordsworth, M. H. Abrams & Stephen Gill. New York, NY: Norton, 1979.

YEATS, W. B. *The Poems*. Rev. edit. Edited by Richard Finneran. New York, NY: Macmillan, 1989.

# FREUDIAN SUSPICION VERSUS SUSPICION OF FREUD

FREDERICK CREWS

PERCEPTIONS OF SIGMUND FREUD and of the movement that he founded are so various that I cannot expect, in brief compass, to do more than sketch a position that will appear self-evident to some readers but impossibly strange to others. Some may be offended that I could stigmatize a revered discoverer of fundamental psychological truths as Exhibit A in "The Flight from Science and Reason." Others, however, will have lost any trust in psychoanalysis so long ago that they consider its very mention an anachronism. And many will want to know why they should worry about Freud and his brainchild when the major current threat to our rational and empirical ethos stems from an overt politicizing of intellectual discourse—a phenomenon that looks quite remote from the objective claims, however debatable, of a psychological system.

The only easy portion of my task is to show that Freud's work does remain alive and capable of exerting considerable influence. To be sure, psychoanalysis has suffered a continual and precipitous delegitimation within its primary domains of psychiatry on the one hand and academic psychology on the other. Yet this development has scarcely begun to dislodge Freudian lore from its pedestal in the popular mind—including the minds of myriad psychotherapists, counselors, and social workers, many of whom do not even realize that they are Freud's legatees.

An appalling number of them, in fact, have recently been facilitating "recovered memories" of early sexual abuse in precisely the way that Freud himself claimed to be doing with his patients in the mid-1890s, just before launching psychoanalysis as we now know it.[1] While modern recovered memory therapy is often decried as antithetical to the true psychoanalytic spirit, it depends crucially on some of the deepest assumptions of classical analysis—for example, that the infantilization of a patient within an unequal and volatile relationship is not a hazard to stability but a necessary stage on the path to maturity and freedom; that we must dredge up and abreact repressed memories of early events if we are to break neurotic patterns; and that etiologically determinative memories can be accurately reconstructed in therapy without fear of contamination by the therapist's biases and hunches.

Absent the prior diffusion of those Freudian notions, the grotesque and sometimes deadly fad of recovered memory therapy could not have taken root as it did a decade ago.

Unfortunately, that is not the only connection between psychoanalysis and our epidemic of "repressed abuse." Incredible as it may sound, recent books and articles leave no doubt that some contemporary analysts have abandoned their former emphasis on infantile sexuality and the Oedipus complex and are now, themselves, reverting to Freud's prepsychoanalytic practice of cajoling their patients into believing that they were molested as children.[2] Such a development leaves one disposed both to marvel at the elasticity of Freud's tradition and to put on hold any assumption that its days of producing massive social harm are over.

But since the audience of this paper is largely academic, and since I have repeatedly discussed the recovered memory movement elsewhere, I will give the rest of my attention to the survival of Freudianism within American universities. This, too, is no trivial phenomenon. Biographical studies by Frank Cioffi,[3] Malcolm Macmillan,[4] Max Scharnberg,[5] and others showing that Freud's "discoveries" rested on misconstrued or nonexistent data have as yet had curiously little effect on the eagerness of many humanists and some social scientists to adhere to the Freudian outlook or to regard Freud personally as our deepest modern thinker.

When Stanford University, for example, famously expanded its "Western Civilization" requirement, at the outset of this decade, to eight alternative tracks of "Culture, Ideas, and Values," Freud was awarded a place in all eight tracks—thus putting him into a tie for top cultural authority with Shakespeare and the Bible and ranking him ahead of Homer, Plato, Aristotle, Sophocles, Virgil, and Voltaire. Again, the *Chronicle of Higher Education* reported in 1992 that 38% of all literature professors in America were teaching psychoanalytic theory to undergraduates—a percentage greatly surpassing those for poststructuralism, Marxism, and what the *Chronicle* called "minority approaches." What is true on the undergraduate level is surely truer for methodologically minded graduate courses. And merely by opening an academic journal, one can be assured that psychoanalysis prospers among us, not in a few tasteful allusions to the unconscious but in full theoretical strength, with reliance on Freud's most ambitious propositions about the determining power of the repressed.

To begin to indicate why this is so, I want to direct your interest to a recent essay entitled "Freud on Trial." The author, Eugene Goodheart, is a well-respected professor of English at Brandeis University, justly known for his discernment in matters of literary theory. He is also, it now appears, a committed Freudian, though not a dogmatic one. The reasons he supplies for salvaging certain articles of the psychoanalytic faith reflect, I believe, something approaching a consensus among those academic humanists who feel that thoroughgoing critiques of psychoanalysis are to be discouraged and deplored.

Goodheart acknowledges that scientific objections have been raised against psychoanalysis, but he deflects them in a way that is by now routine within literature departments. "Philosophers of science," he writes,

> do not hold that scientific theories necessarily rise or fall when statements within the theories are falsified. Theoretical survival depends upon the character and degree of falsification. The issue is further complicated if one takes seriously Thomas Kuhn's theory of paradigms in which the demarcation between science and nonscience loses its sharpness.[6]

Although in actuality Kuhn's *Structure of Scientific Revolutions* scarcely touches on the demarcation problem, Goodheart has a point here. Typically, Kuhn holds, a scientific paradigm may remain dominant throughout a protracted crisis during which bothersome anomalies continue to accumulate. But another way of making the same point would be to say that paradigms do crumble when their unfruitfulness has become overwhelmingly evident. Therefore, we might reasonably expect Goodheart to assess whether, in the case of psychoanalysis, "the character and degree of falsification" has by now proved worrisome or perhaps even terminal.

Instead, however—and all too characteristically for an academic Freudian apologist—Goodheart shifts ground and introduces a quite different framework of evaluation. Now he sympathetically appeals to Michel Foucault's distinction between mere scientists and "founders of discursivity," of whom Freud is a leading example. In Foucault's reckoning as summarized by Goodheart, founders of discursivity "provide the master texts that determine truth claims."[7] Such thinkers "are not required to conform to the canons of science. Their own discourse constitutes the canon that determines its truth value."[8] If so, Foucault has supplied Freudians with a safe-conduct pass that they can produce whenever their epistemic credentials are challenged.

Goodheart appears at first to welcome this advantage, but he does not feel altogether comfortable with Foucault's blatant stacking of the epistemic deck. Founders of discursivity, he admonishes, always provoke both idolatry and iconoclasm, impulses that are equally worthy of our mistrust. Thus, in Goodheart's view, my own "ferocity"[9] against Freud can be discounted as a typical overreaction to Freud's privileged cultural status. According to Goodheart, a wiser middle course would be to demote the idol Freud to a mere "body of knowledge or knowledge claims, which we may both consult and criticize."[10]

Freud, then, remains enough of a paragon to Goodheart for his propositions to be exempted from scientific canons of validity, but at the same time the literary critic feels himself at liberty to decide intuitively which Freudian notions to adopt and which to spurn. He remains untroubled by the thought that each psychoanalytic proposition falls squarely within the domain of psychology and should thus, as Freud himself at least formally maintained, be subjected to the same empirical criteria as are brought to bear on every other psychological claim. Like many fellow humanists, Goodheart prefers instead to sample the Freudian smorgasbord on a frankly heuristic basis, appropriating only those ideas that strike him as possessing what he calls "suggestiveness."[11] If there are epistemic pitfalls lurking within that subjective criterion, Goodheart has not yet begun to worry about them.

The sheer number of reasons that Goodheart offers for removing Freud from scientific accountability makes one wonder how confident he feels about any one of them. To the halfhearted appeal to Kuhn and Foucault and the more general appeal to heuristics he appends three others: a claim that Freud's sheer "power of narration"[12] provides a kind of emotional truth that we could ill afford to forgo; an assertion that "Freud's achievement occurs in the company of the great masters of modern literature," who "seemed to have arrived at Freud's insights independently";[13] and a final claim that "[t]he evidence for psychoanalytic discovery is to be found . . . in case histories of particular persons . . .,"[14] and specifically in "the confirming response of the analysand"[15] to the analyst's interpretations.

It is clear from this last line of argument that Goodheart has drastically misconstrued one of the philosophers of science whom he cites as a properly appreciative and balanced commentator on Freud, Adolf Grünbaum. The burden of Grünbaum's classic critique *The Foundations of Psychoanalysis* is that, thanks to the pervasive threat of suggestion and the sheer improbability of gleaning temporally remote etiological inferences from a patient's free associations, Freudian tenets and interpretations cannot be reliably corroborated on the couch. By invoking Grünbaum, Goodheart has stumbled back into the scientific killing fields from which he had hoped to rescue Freud.

Because Goodheart's stated justifications are so unimpressive when examined one by one, they point to a sheer will on his part to cherish Freud and Freudianism, come what may. In his case and in that of countless other academics, this impulse bears a detectable connection to what one might call Freudian Whiggishness, or a special gratitude to Freud for having made us the sage moderns that we now perceive ourselves to be. "In exploring the unconscious and its relation to conscious life," Goodheart avers, Freud "thought he was defining the nature of man, and, to the extent that he succeeded—which is to say, to the extent that we recognize ourselves in the account that he provides, Freud must be the commanding figure of our culture in a way that Weber and Durkheim are not."[16] Goodheart is not saying here that anyone at all can look in his heart and determine just how commanding a figure Freud should be taken to be. It is we literati, we who have honed our sophistication not only on the difficult insights of modernist fiction and poetry but also on Freud's own writings, who constitute the jury and, in part, the very evidence for Freud's majestic stature.

This line of reasoning enhances self-regard in a segment of the populace that may already be burdened with overconfidence. By the same token, it bears the intellectual demerit of shutting off inquiry into the contingent and biased character of our present *Zeitgeist*. That failure of reflexive consciousness, however, is by no means unique to Goodheart. On the contrary, it thrives within the literary-academic "theory class," which has notoriously applied a withering perspectivism to all certitudes except its own. Goodheart himself, in an admirably critical book about the poststructuralist wave that has been sweeping through our universities for the past two decades and more, has displayed a keen ironic awareness of this inconsistency.[17] It is all the more remarkable, then, that he joins the theory mongers in treating some of the universals of Freudian doctrine as if they were imperishable laws of nature.

Many other humanists resemble Goodheart in observing no firm distinction between the cause of psychoanalysis and that of high artistic endeavor. Freudianism has graced our cultural landscape for so many decades that the prospect of its removal is perceived as a kind of ecological crime, akin to paving a rain forest to make way for theme parks and drive-in theaters. Yet the humanists' treasuring of Freudian texts *as* culture does not prevent those texts from being further treasured as *explanatory of* culture. For Goodheart in particular, the very act of exalting early modernist literature entails an endorsement of the Freudian ideas that appear to be dramatized, willy-nilly, within those classics. As Goodheart puts it, "[t]he whole panoply of Freudian concepts—the oedipal drama in the family romance, the degraded split sexuality that results in impotence, the return of the repressed, and so on—are confirmed in the imaginative literature of the modern period."[18]

Goodheart acknowledges that such confirmation is not scientific proof that Freud was right. But is it any proof at all? The critic's inference suffers from two crucial errors that I myself committed thirty years ago, when I first concluded that the correspondence I found between Freud's writings and those of the author I was then studying, Nathaniel Hawthorne, constituted a validation of psychoanalytic laws.[19] Those errors cost me a good deal of wasted intellectual effort and eventual embarrassment, but I see that my public penitence has had a negligible dissuasive effect on other literary academics who are bent upon embracing the very same fallacies.

My first mistake was to overlook the difference between psychological description and psychological explanation. Hawthorne's writings indeed dwell on the themes of incest, patricide, and hidden sexual guilt that Freud explains by reference to the universal and perennial Oedipus complex, but this means only that Freud and Hawthorne shared a preoccupation with the same domain of obsessive-compulsive psychological phenomena. A number of historically and biographically delimited explanations, falling well short of the Promethean claims of psychoanalysis, could be consonant with what Hawthorne, the brooding Romantic artist and ironic student of his Puritan forebears' attempts to stifle sin, fictively represented. A comparable stricture applies to Goodheart's observation of the symmetry between Freudian notions and the rendered worlds of Conrad, Lawrence, Proust, and Mann—and doubly so for Lawrence and Mann, whose conscious interest in Freud subtracts much of the uncanniness from their exemplification of his ideas.

My mistake number two was a failure to think skeptically about Freud as a late-Romantic literary intellectual in his own right. Freud's sly rhetoric, along with the hagiography that he posthumously channeled through Ernest Jones,[20] promoted a misperception of him as a mere physician and scientist-drudge who had no time for literature and who was amazed to learn, long after the fact, that his reluctant discoveries about the devious and sex-bespattered human mind had all been anticipated in fiction and drama. That is the bait that Goodheart still finds irresistible in 1995. But the truth is that Freud was a voracious and deeply cultivated reader and that his ideas owed more to Shakespeare, Cervantes, Milton, Goethe, Heine, Schiller, Feuerbach, Schopenhauer, Dostoevsky, and Nietzsche than they did to the turn-of-the-century sexologists whose formulations he also liberally raided.

Great imaginative writers did indeed anticipate Freud, but there was nothing mysterious or confirmatory in the tallying of his ideas with theirs. Quite simply, he plucked their darker intimations about egoism, guilt, and inward drivenness and set them out to dry on the gears and pulleys of an arbitrary explanatory contraption that does not work and never did. The "confirmatory" parallelism between literary and psychoanalytic insight has proved to be the most successful of Freud's many hoaxes—made more delicious, in this instance, by a further punishment of the writers for having gotten there first. Although he professed a humble awe before artistic genius, Freud took satisfaction in reducing it to the demeaning banality of infantile determinism. The literary giants, he taught us after he admitted to having read some of them, were only adult babies after all, spinning out symptomatic fantasies that replayed the same oedipal riddle in every instance—the riddle that he alone had solved in a single flash of titanic self-exploration.

Can we—can I—ever have believed this megalomaniacal claim? Yes, the record cannot be willed away. And even now, when Freud's Napoleonic grandiosity and investigative peremptoriness are more widely acknowledged, literary intellectuals harbor such a thoroughgoing confusion between the mind's complexities and Freud's causal suppositions about those complexities that, from their perspective, defending Freud and honoring mental depth amount to the very same thing. This is why "iconoclasts" like me are perceived by Goodheart and others as "resentful renegade[s]"[21] who, out of malice against a great discoverer of hard truths, would strand us with a soulless positivism.

Yet before long, I predict, it is Goodheart's current position that will look irrational to most academics. The change is being brought about not by iconoclasts but by sober historical researchers who keep discovering that none of the great thinker's claims can be taken at face value. A would-be scientist who turns out to have drawn putative psychological laws from samples of a few cases, or one, or even none will not forever be treated as a sacrosanct founder of discursivity.

Indeed, as I hinted earlier, in order to arrive at a radically altered conception of our indebtedness to Freud we need only attend to already extant studies of his borrowings from other writers. Humanists have not been mistaken in looking to Freud for a sense that the human mind is vastly egoistic and cunning. Their mistake has simply been to accede to his subtle but persistent urging that he be considered apart from all other philosophers of the same stripe. The source studies indicate that his real originality is much narrower than has been thought and that it is confined on the whole to points that are either gratuitous or plainly erroneous.

Consider, for example, what is now known about Freud's relation to Nietzsche, the writer toward whom he most strenuously protested a complete lack of indebtedness. As Ernest Gellner[22] and Didier Anzieu[23] indicated a decade ago, and as Ronald Lehrer has now extensively documented in a new book entitled *Nietzsche's Presence in Freud's Life and Thought*, most of what could be termed "insight" in Freud is already there in Nietzsche. In Anzieu's words, Nietzsche

> had some understanding of the economic point of view, which comprises discharge, and transfer of energy from one drive to another. . . . On several occasions he used the word sublimation (applying it to both the aggressive and the sexual instincts). He described repression, but called it inhibition; he talked of the superego and of guilt feelings, but called them resentment, bad conscience and false morality. Nietzsche also described . . . the turning of drives against oneself, the paternal image, the maternal image, and the renunciation imposed by civilization on the gratification of our instincts.[24]

But this is just the beginning. As Lehrer shows, Nietzsche's preemption of Freud includes the idea that all actions and intellectual choices are egoistic; that we remain unconscious of the conflicts in our motives; that "[g]ood actions are sublimated evil ones";[25] that "[t]he content of our conscience is everything that was during the years of our childhood regularly *demanded* of us without reason by people we honoured and feared";[26] that forgetting is an active step taken to preserve psychic order;[27] that dreams employ symbols to express our primeval selves and our disapproved desires;[28] that comedy results from a sudden release of anxiety; and that laughter entails "being malicious but with a good conscience."[29] (All quoted words in this sentence are Nietzsche's.) Lehrer's book also places beyond reasonable doubt Freud's acquaintance with Nietzsche's ideas at various stages of his own development. Paul Ricoeur and others have taught us to think of Nietzsche and Freud together, along with Marx, as partaking of a common "school of suspicion,"[30] but it has now come to light that one of those parties attended classes in the other's school.

Freud's exalted reputation will surely dwindle as subscribers to the conventional wisdom begin to absorb findings such as Lehrer's. Deprived of the misconception that Freud unearthed the mind's affinity for self-deceit, they will be obliged to realize that his contribution was that of a systematizer, not a discoverer. And then they will find themselves facing the mortifying question that I myself could not shake off during my own little paradigm crisis in the 1970s. What if the Freudian system—the portion of it that was actually new—consists almost entirely of unwarranted reifications and crackpot dogmas?

Look, for instance, at what Freud did with the already familiar insight that our minds operate largely beneath the threshold of consciousness and that our thoughts are shaped in part by bodily needs, self-regard, and ill will toward rivals. In his discourse these dispositions have congealed into "the unconscious," an inner personage that itself allegedly contains homunculi (the ego, the id, the superego) who employ various stratagems in vying with one another for control over our actions. Neither in Freud's case histories nor anywhere else do we find evidence of behavioral manifestations that point unambiguously to the need for such mythic entities. Their usefulness is not empirical but rhetorical. Each posited subset of "the unconscious" permits another strand of contrary motivation to be added to the already tangled explanatory skein, leaving us, if we are sufficiently gullible, so awestruck by the psychoanalytic interpreter's diagnostic acumen that we think we are witnessing elegant and validated feats of deduction instead of being told a self-serving detective story in which the very mystery itself—which of the selves checkmated which others to generate the symptom or dream or error?—is an artifact of question-begging maneuvers.

Even professors of literature, I believe, will someday be constrained to recognize that this surplus of ingenuity, divorced as it is from any means of potential disconfirmation, represents the very opposite of progress in psychological understanding. As Freud remarked to his friend Wilhelm Fliess, his idea of "the splitting of consciousness" signaled a reversion to the intellectual style of the church fathers and their successors, the witch persecutors, who thought of the psyche as a battleground between warring powers and who regarded every action as symptomatic of possession by one contender or the other.[31] But what Freud could not admit was that he had no more basis for such demonizing than Cotton Mather did. As his amused critic Ludwig Wittgenstein brightly put it, to dispense with the Freudian unconscious would be like having to say, truthfully, "We do not know who did that" instead of the more impressive but specious "Mr. Donotknow did that."[32]

Every other distinctive feature of Freudian thought is vulnerable to this same objection. In Freud's discourse, facts are never more than a warm-up act for the featured performance of Mr. Donotknow. A freewheeling inventiveness pervades Freud's assertions about the causes and cure of neuroses, the interpretability of symptoms as somatic compromise formations, the central importance of incestuous and murderous urges and castration fears in the socialization of children, the necessarily envious and amoral character of women, the wish-fulfillment theory of dreaming, the origin of medical compassion in childhood sadism, the emergence of sexual identity from a primary base of bisexuality, the insidious operation of the death instinct, and the indestructibility of all memories, to say nothing of our enervating Lamarckian inheritance of traumatic memory traces from the earliest days of the human race.

This last item belongs, of course, not to the supposedly indispensable core of psychoanalytic theory but to what Freudians somewhat sheepishly designate as Freud's speculative or philosophical side. Gnostic divination, however, was his entire métier. The Oedipus complex was disclosed to him through the same kind of hermetic insight that left him convinced that Shakespeare had not written Shakespeare's plays, that Moses was an Egyptian, and that weaving originated when one of our early ancestors twisted together her pubic hairs in chagrin over her nonpossession of the only sexual organ worthy of the name. And in drawing inferences about the exact infantile causes of symptoms exhibited by free-associating patients in his consulting room, Freud the psychoanalyst was engaging in the same cavalier dismissal of unremarked factors that characterized Freud the anthropologist of prehistory. In both exercises of guesswork his corroborative method was to focus on a single hunch, to allow it to exfoliate into further hunches that were equally remote from observation, and then to take the thematic consilience of all those constructions as strong evidence of their basis in reality.

By thus mistaking convergences within his own fantasy life for deeply lawful regularities in the world outside it, Freud marked himself not just as error prone but as chronically delusional. Such a diagnostic label, however, leaves the most fundamental question about his career farther than ever from a plausible solution. If he was delusional, or even if he was simply wrong in ways that now appear easy to discern, how has he succeeded in imposing his reveries on much of the literate West?

This is one of the central puzzles of modern cultural history, and it calls for a far more complex analysis than I can offer here. A strong beginning has been made in Ernest Gellner's admirable but little-read book of 1985, *The Psychoanalytic Movement*—a work that will soon reach a wider audience in a reprint by Northwestern University Press. Meanwhile, and very much in Gellner's spirit, I would like to propose that Freud has had his way with us chiefly by personifying a uniquely seductive style of hermeneutic license. It is, I believe, his peculiarly Janus-faced relation to the scientific ethos that has made many of us—and academic humanists in particular—willing to overlook his habitual preference for far-fetched explanations.

Of course there are many reasons for being a Freudian, the most common and compelling of which is simply that one has been indoctrinated within the lengthy dependency of a therapeutic "transference." With or without therapy, however, many educated nonscientists have in common an uneasy attitude toward mainstream science, leaving them with a susceptibility to irrational persuasion that Freud exploited with formidable brilliance. On the one hand, few secular skeptics are willing to embrace doctrines that appear flagrantly antiscientific. On the other hand, it is common for them to feel at once discouraged and diminished by science, with its grinding laboriousness, its opaqueness to lay comprehension, its strong purchase on technology, its power to command respect and money, and, above all, its relentless chipping away at the felt scope of free will and the soul. For such a sense of envy and exclusion the treatment of choice may be Freud's intellectual nostrum, a dose of prescientific and even animistic belief that is made palatable by gestures of scientific prudence.

Now, this hypothesis—that psychoanalysis serves as a kind of metaphysical morale booster for the science impaired—may meet with an immediate objection from readers who remember Freud's none too bashful grouping of himself with Copernicus and Darwin as the greatest of all deflators of human pride. His own contribution, he explained, was to show that the ego of the frail evolved creature man is not even master in its own house. But do we really find ourselves humbler when we have acceded to Freud's balkanization of the psyche? True, his system makes the determiner of volition look less like an executive than like a wrangling troika of petulant schemers; but the troika's charge, we recall, has been broadened to cover the generation of symbolically—that is, humanistically!—readable symptoms as well as thoughts and deeds. And we, Freud's recruits, are tacitly exempted, just as he exempted himself, from his fatalism about the necessarily beclouded human mind. We are promised a chance to help our leader track meanings in the virgin terrain of the symptomatic, and in doing so, not incidentally, to set ourselves decisively apart from the uninitiated as exercisers of godless objectivity and stoicism.

Once we have grasped the evangelistic or recruiting character of Freud's discourse, much that appears anomalous and confused within it begins to make rhetorical sense. Most commentators who have noted the mythic element in Freud's thought, for example, have tended to characterize him as torn between poetical and positivistic ways of addressing the dynamics of motivation. On one side stands the witches' cauldron of the unconscious; on the other, a rigorous materialist determinism focused on cathexes, quantities of excitation, and thresholds of discharge. But rigor is completely absent from Freud's psychological writings, which contain no data that have not already been tailored to the argumentative point they were enlisted to support. As for determinism, we ought to distinguish between the assumption that all phenomena bear causes and the pretense that through sheer cogitation one can trace each human deed to its precise motivational origin. The latter idea is a marker not of science but of magical thinking. Where Freud most sternly excludes chance from the realm of the mental, he may sound like he is submitting himself to a trying discipline, but what he actually asserts is his limitless freedom to assign real-world causal significance to his byzantine and arbitrary interpretations. And that is just the giddy freedom that he dangles before us, his readers and potential disciples.

In this light it is telling that Freud's love of science had been aroused in the first place by Goethe's visionary *Naturphilosophie* and that Goethe's Faust, the archetypal trader of his soul for illicit knowledge and power, was the foremost of his several key heroes.[33] Indeed, his chief complaint against his early mentor Josef Breuer was that there was "nothing Faustian in his nature."[34] Psychoanalysis as Freud conceived it was a truly Faustian program for removing obstacles to the omnipotence of his own thoughts. His professed rationalism was little more than a fig leaf for his sympathy with folk superstitions, or what he called "the obscure but indestructible surmises of the common people" as opposed to "the obscurantism of educated opinion."[35] And privately, he was devoted to numerology and telepathy—other arts that leapfrogged purported laws of nature to ratify immemorial beliefs in the uncanny control of matter by mind. In Freud's lurid dreamworld, initially fueled by cocaine as Faust's had been fueled by a narcotic witch's brew,[36] even tuberculosis and cancer could be brushed aside as mere "somatic compliance" with a psychic leverage that fell barely short of levitation.

All in all, Freud was not an empiricist but a Neo-Platonist—that is, a believer that the tangible world is merely a deceptive screen for innumerable interlinked meanings that can be accessed by anointed persons through direct intuition. His rhetorical guile, however, prompted him to cloak his hermetic urge in materialist atomism, enabling him to wage his stirring war against science while wearing the enemy's own uniform. Likewise, his lofty emphasis on ascetic renunciation and tragic fatalism made a perfect foil for the gossipy prurience that lends his case histories much of their literary allure. Any charlatan can offer us interpretive wildness and voyeuristic titillation, but Freud alone knew how to package them as a courageous adherence to the reality principle.

How, you may well ask, does all this relate to the survival of Freud and psychoanalysis in our universities? Obviously, one need not be a Neo-Platonist to be mesmerized by the siren song of Freudianism. An academic can still embrace Freud in the misty way that, for example, Lionel Trilling and his circle did a half-century ago, merely out of a sense that it is the highbrow thing to do. That is roughly where Eugene Goodheart appears to stand today. We can now perceive, however, that Goodheart's curious two-step with respect to validation—removing Freud's propositions from the scientific arena while ratifying some of them as actual psychological laws—bears a distant kinship to Freud's hollow boast of clinical proof as he indulged an unfettered willfulness of assertion. The strategies are opposite but the intended result is the same, a free ride for some of Freud's controversial dogmas about the mind.

The more specific uses to which our colleagues have been putting Freud lie beyond the scope of this paper, but I would not want to leave unmentioned the most fundamental affinity between psychoanalysis and the institutionalized pursuit of the humanities. Both professions place a value on the sheer production of discourse without end. Thanks to its lax and overlapping rules for drawing inferences, its hospitality to several layers of possible exegesis, its readiness to generalize without regard for counterexamples, and its tolerance of self-contradiction, the Freudian hermeneutic leaves an academic interpreter without even a mathematical chance of having nothing to say. All that is required for self-assurance is the "suggestiveness" that Goodheart commends. Not just the single interpretation at hand but also the whole theory behind it can be considered justified if it allows the interpreter to feel that corroboration is popping up wherever he looks—and of course it always does.

Given this advantage among others, I may have been too sanguine here in looking forward to the demise of academic Freudianism. An alternative scenario could be that Freud as an individual will be treated even more rudely while humanists continue to put his hermeneutic to the purposes that suit them. Today, for example, many radical professors welcome psychoanalytically framed challenges to the given social and psychic order, but since the bourgeois Freud and the bourgeois practice of psychotherapy are part of that same order, Freud personally must be taken down a peg. Also, his patriarchal priorities must be inverted. Where he favored the father, correct-thinking academics will now favor the mother. If classic psychoanalysis exaggerates the oedipal, the indicated remedy will now be to exaggerate the pre-oedipal. Where penis envy was, there shall womb envy be; and so forth.

Precisely such an ideologically driven makeover of psychoanalysis has enabled the recovered memory movement, outside the academy, to masquerade as psychological science—and hence to work its mischief not just on therapists and patients but on judges and juries as well. That sobering fact should give pause to anyone who thinks that our apparatus for detecting "suggestiveness" can reliably tell us which parts of the Freudian revelation ought to survive. After all, Freudianism itself continually mutates according to cultural and political suggestion. Academic humanists in particular have not compiled an outstanding record of immunity to such influence, and many among them now cheerfully maintain, as Foucault did, that "knowledge" is just another means of exercising power.

It can be seen, then, that academic disputes over Freud take much of their meaning and importance from a broader struggle over the ethics of intellectual assertion. Insofar as psychoanalysis remains a live issue, I believe that the real choice facing us is not between one point of theory and another but between the rational-empirical ethos and a Faustian pseudoscience that defies it at every turn. The flourishing of academic Freudianism could serve as a kind of fever chart for the sickness of empiricism, which has been weakened by explicit theoretical assault over this past quarter-century. If psychoanalysis fails to survive, it will be because empiricism itself is on the mend at last.

## NOTES

1 See Richard Ofshe & Ethan Watters. *Making Monsters: False Memories, Psychotherapy, and Sexual Hysteria.*
2 Abundant evidence for this claim can be found in a forthcoming special issue of the journal *Psychoanalytic Dialogues*, especially in the lead articles by Adrienne Harris and Jody Messler Davies and in my own commentary on those articles.
3 Frank Cioffi, "Psychoanalysis, Pseudo-Science and Testability"; "Was Freud a Liar?"
4 Malcolm Macmillan, *Freud Evaluated: The Completed Arc.*
5 Max Scharnberg, *The Non-Authentic Nature of Freud's Observations.*
6 Eugene Goodheart, "Freud on Trial," pp. 236–237.
7 Ibid., p. 237.
8 Ibid.
9 Ibid., p. 236.
10 Ibid., p. 237.
11 Ibid.
12 Ibid., p. 238.
13 Ibid., p. 239.
14 Ibid., p. 242.
15 Ibid., p. 241.
16 Ibid., p. 243.
17 Eugene Goodheart, *The Skeptic Disposition in Contemporary Criticism.*
18 Goodheart, "Freud on Trial," p. 239.
19 Frederick Crews, *The Sins of the Fathers: Hawthorne's Psychological Themes.* A recent reprint of this book permitted me the opportunity to add an afterword detailing the respects in which my argument had been historically inhibited by its Freudian premises.
20 Ernest Jones, *The Life and Work of Sigmund Freud.*
21 Goodheart, "Freud on Trial," p. 236.
22 Ernest Gellner, *The Psychoanalytic Movement: Or The Coming of Unreason.*
23 Didier Anzieu, *Freud's Self-Analysis.*
24 Ibid., pp. 88–89.
25 Ronald Lehrer, *Nietzsche's Presence in Freud's Life and Thought: On the Origins of a Psychology of Dynamic Unconscious Mental Functioning*, p. 36.
26 Ibid., p. 37.
27 Ibid., p. 58.
28 Ibid., p. 83.
29 Ibid., p. 40.
30 See Paul Ricoeur, *Freud and Philosophy*, passim.
31 Sigmund Freud, *The Complete Letters of Sigmund Freud to Wilhelm Fliess 1887–1904*, pp. 224–225, 227–228.
32 J. Bouveresse, *Wittgenstein Reads Freud: The Myth of the Unconscious*, pp. 33–34.
33 Another such model was the Antichrist, who must have been rendered attractive to Freud by a bolder rebel than himself, Nietzsche. See P. J. Swales, "Freud, Cocaine, and Sexual Chemistry: The Role of Cocaine in Freud's Conception of the Libido"; and P. C. Vitz, *Sigmund Freud's Christian Unconscious*, pp. 161–165.
34 E. Freud, L. Freud & I. Grubrich-Simitis, *Sigmund Freud: His Life in Pictures and Words*, p. 139.

35 Sigmund Freud, *The Standard Edition of the Complete Psychological Works of Sigmund Freud*, vol. 18, p. 178.

36 Swales, "Freud, Cocaine, and Sexual Chemistry."

## REFERENCES


ANZIEU, DIDIER. *Freud's Self-Analysis.* London: The Hogarth Press, 1986.

BOUVERESSE, J. *Wittgenstein Reads Freud: The Myth of the Unconscious.* Translated by C. Cosman. Princeton, NJ: Princeton University Press, 1995.

CIOFFI, FRANK. "Was Freud a Liar?" *The Listener* 91 (1974):172–174.

———. "Psychoanalysis, Pseudo-science and Testability." In *Popper and the Human Sciences*, edited by G. Currie & A. Musgrave. Dordrecht: Nijhoff, 1985.

CREWS, FREDERICK. *The Sins of the Fathers: Hawthorne's Psychological Themes.* Princeton, NJ: Princeton University Press, 1966.

———. *The Sins of the Fathers: Hawthorne's Psychological Themes.* 2d ed, with a new afterword by the author. Berkeley, CA: University of California Press, 1989.

FREUD, E., L. FREUD & I. GRUBRICH-SIMITIS. *Sigmund Freud: His Life in Pictures and Words.* Translated by C. Trollope. New York, NY: Harcourt Brace Jovanovich, 1978.

FREUD, SIGMUND. *The Standard Edition of the Complete Psychological Works of Sigmund Freud.* 24 vols. Edited and translated by J. Strachey. London: The Hogarth Press, 1953–1974.

———. *The Complete Letters of Sigmund Freud to Wilhelm Fliess 1887–1904.* Edited and translated by J. M. Masson. Cambridge, MA: Harvard University Press, 1985.

GELLNER, ERNEST. *The Psychoanalytic Movement: Or The Coming of Unreason.* London: Paladin, 1985.

GOODHEART, EUGENE. *The Skeptic Disposition in Contemporary Criticism.* Princeton, NJ: Princeton University Press, 1984.

———. "Freud on Trial." *Dissent* 42 (Spring 1995):236–243.

GRÜNBAUM, ADOLF. *The Foundations of Psychoanalysis: A Philosophical Critique.* Berkeley, CA: University of California Press, 1984.

JONES, ERNEST. *The Life and Work of Sigmund Freud.* 3 vols. New York, NY: Basic Books, 1953–1957.

LEHRER, RONALD. *Nietzsche's Presence in Freud's Life and Thought: On the Origins of a Psychology of Dynamic Unconscious Mental Functioning.* Albany, NY: State University of New York Press, 1995.

MACMILLAN, MALCOLM. *Freud Evaluated: The Completed Arc.* Amsterdam: North-Holland, 1991.

OFSHE, RICHARD & ETHAN WATTERS. *Making Monsters: False Memories, Psychotherapy, and Sexual Hysteria.* New York, NY: Scribner's, 1994.

*Psychoanalytic Dialogues* [Special issue on recovered memory of sexual abuse, forthcoming.] 6, 2 (1996).

RICOEUR, PAUL. *Freud and Philosophy.* New Haven, CT: Yale University Press, 1970.

SCHARNBERG, MAX. *The Non-Authentic Nature of Freud's Observations.* 2 vols. Uppsala: Textgruppen i Uppsala, 1993.

SWALES, P. J. "Freud, Cocaine, and Sexual Chemistry: The Role of Cocaine in Freud's Conception of the Libido." Privately published by the author, 1983.

VITZ, P. C. *Sigmund Freud's Christian Unconscious.* New York, NY: Guilford Press, 1988.

# ECOSENTIMENTALISM
## The Summer Dream beneath the Tamarind Tree

GERALD WEISSMANN

*Il y a à parier, que tout idée publique, toute convention reçue, et une sottise, car elle convenu du plus grand nombre.*

—CHAMFORT

(It is safe to bet that every public idea, all conventional wisdom, is foolish because it is acceptable to the mass.)

THE FRENCH EPIGRAPH begins Edgar Allan Poe's review of Longfellow's ballads in Graham's Magazine of March/April, 1842.[1] Poe wasn't very hot on *Hiawatha*, and even less happy with Longfellow and his abolitionist friends. These days *everyone* is hot on *Hiawatha* and the forest primeval, not to speak of the birds, the bees, and the everloving planet. Indeed folks from the new left and old right, from Al Gore to Clint Eastwood, hold in common only one *idée publique*, share only one *convention reçue*, that modern science and Western civilization have ruined the Noble Savage in the State of Nature. In American letters, that myth has had a long, semi-distinguished history reaching its zenith in the work of Henry David Thoreau. It derives, of course, from those archantinomians Petrarch and Rousseau, a point appreciated by James Russell Lowell. Chiding Thoreau for his neglect of the North during the Civil War, Lowell complained in 1865 that "while he studied with respectful attention the minks and woodchucks, his neighbors, he looked with utter contempt on the august drama of destiny of which his country was the scene, and on which the curtain had already risen."[2] It is a charge we might have leveled against Lowell's nature-loving and Chrysler-hating kinsman, Robert Lowell, who sat out the war against Hitler. But James Russell Lowell's chief argument was with sentimentalism per se:

> I look upon a great deal of the modern sentimentalism about Nature as a mark of disease. It is one more symptom of the general liver-complaint. To a man of wholesome constitution the wilderness is well enough for a mood or a vacation, but not for a habit of life. Those who have most loudly advertised their passion

for seclusion and their intimacy with nature, from Petrarch down, have been mostly sentimentalists, unreal men, misanthropes on the spindle side, solacing an uneasy suspicion of themselves by professing contempt for their kind.[3]

The most recent display of ecosentimentalism is in E. L. Doctorow's new novella *The Waterworks*, which hit the paperback best-seller list this week. A freelance writer and critic named Martin Pemberton serves Doctorow as a stand-in for Edgar Allan Poe. The book is set in 1871, twenty-two years after Poe's death of alcoholism and tuberculosis in Baltimore. (One might note that the grape and the tubercle bacillus are natural products as much as yogurt or granola.) Doctorow's slight book—thickened for the trade by its avalanche of ellipses—is a skillful gloss on what Poe's biographer Kenneth Silverman has called "the Longfellow Wars." That contretemps between Longfellow, the well-heeled Harvard professor, and Poe, the impoverished West Pointer, touched on conflicts not only between classes, but also between art and science, reason and feeling, urge and action—between North and South, for that matter.

On October 16, 1845, a disheveled Edgar Allen Poe gave a poetry reading to a Lyceum audience in Boston. He had already launched a series of polemics against the "transcendentalists," "Swedenborgians," "abolitionists," and "Harvard professors"—Longfellow in particular—who dominated the New England literary scene. But that evening, facing the Brahmins, he was more confused than combatative. Thomas Wentworth Higginson writes that Poe, "in a sort of persistent, querulous way, which . . . impressed me at the time as nauseous flattery," went on to throw his audience into sheer stupor. After an extensive apology for not having delivered new material, Poe read in toto his early, befuddled poem "Al Aaraaf." The audience dwindled rapidly, presumably because, like Higginson, they were mystified and perplexed. Not even an impromptu recital of "The Raven" could rescue the evening. The occasion confirmed Lowell's final estimate:

There comes Poe, with his raven, like Barnaby Rudge,
Three-fifth of him genius and two-fifth sheer fudge.[4]

The months after the debacle were followed by the Lyceum Wars. The unsparing Boston reviews of his Lyceum performance were answered by Poe with a torrent of invective, shotgun critique and *ad personem* attacks, mostly published in his own *Broadway Journal.* He went out of his way to accuse the Boston literati of cronyism, muddleheadedness, and plagiarism. Predictably, Longfellow was the chief target. But he also called Margaret Fuller "a detestable old maid," James Russell Lowell "a ranting abolitionist," and Longfellow a plagiarist who appealed only to "those negrophilic old ladies of the North." Poe then committed the unforgivable: he insulted Boston itself—its food, hotels, the architecture of its State House, and the squalor of its duck pond, now invaded by frogs. Longfellow himself paid only passing attention to those attacks, as he had brushed off Poe's 1842 attack on his metrical schemes:

In hexameter sings serenely the Harvard Professor
In pentameter him damns censorious Poe.

Dr. Oliver Wendell Holmes, Parkman Professor of Anatomy and Physiology at the Harvard Medical School—and the man who invented the term "Brahmin" in his 1861 novel *Elsie Venner*—eventually had the last word in "The Professor at the Breakfast Table":

> After a man begins to attack the State House, when he gets bitter about the Frog-pond, you may be sure that there is not much left of him. Poor Edgar Poe died in the hospital soon after he got into this way of talking. Remember poor Edgar! He is dead and gone, but the State House has its cupola fresh-gilded, and the Frog-pond has got a fountain that squirts up a hundred feet into the air.[5]

Poe's quarrel with the Boston literary establishment of reformers, abolitionists, and feminists was another aspect of his internal war between fact and fancy, science and art. As a cadet officer, he had been drawn to many aspects of science and mathematics; as a Virginian, he had squelched them in favor of Arcadian aspirations and poetry. Poe's earliest book of verse (1829) contains this "Sonnet to Science":

> Science! true daughter of Old Time thou art!
> Who alterest all things with thy peering eyes.
> Why preyest thou thus upon the poet's heart,
> Vulture, whose wings are dull realities? . . .
> Hast thou not torn the Naiad from her flood,
> The Elfin from the green grass, and from me
> The summer dream beneath the tamarind tree.[6]

Poe's "Sonnet to Science" is not just a clumsy gloss on Keats's "Lamia," ("Philosophy will clip an Angel's wings") any more than *The Waterworks* is only a gloss on "The Fall of the House of Usher" or "The Premature Burial." "Sonnet to Science" helps us to understand why Poe was so vexed with Longfellow, Lowell, Higginson, Fuller, and the rest. The summer dream beneath the tamarind tree is the hope of a New Age, and the New Age that is dreamt of is the age of Unreason. The politics are not difficult to untangle. In the antebellum period, the North in general, and Boston in particular, was regarded by Southerners as a rampart of roundhead values, concerned with the "dull realities" of science, machines and trade. Poor Poe, on the other hand, fancied himself a son of the cavalier South. A sometime soldier, disinherited but awash in plantation longings, he yearned on Broadway for the summer dream beneath the tamarind tree. These days, the North–South axis, the meridian between Boston and Richmond, between duty and languor, attracts at its southern extreme many of those disenchanted by modern science. Among them are the fans of magic realism, the audiences that love films like *The Piano* and *The Last of the Mohicans*, and every anorexic teenager guzzling Evian water straight from its polypropylene screw-top bottle. Those folk are the ecosentimentalists of our time, deserving the scorn Lowell heaped on Rousseau and friends:

> In the whole school there is a sickly taint. The strongest mark which Rousseau has left upon literature is a sensibility to the picturesque in Nature, not with Nature as a strengthener and consoler, a wholesome tonic for mind ill at ease with itself, but with Nature as kind of feminine echo to the mood, flattering it

> with sympathy rather than correcting it with rebuke or lifting it away from its unmanly depression, as in the wholesomer fellow-feeling of Wordsworth. They seek in her an accessory, and not a reproof. It is less a sympathy with Nature than a sympathy with ourselves as we compel her to reflect us. It is solitude, Nature for her estrangement from man, not for her companionship with him; it is desolation and ruin, Nature as she has triumphed over man, with which this order of mind seeks communion and in which it finds solace. The sentimentalist does not think of what he does so much as of what the world will think of what he does. He translates should into would, looks upon the spheres of duty and beauty as alien to each other, and can never learn how life rounds itself to a noble completeness between these two opposite but mutually sustaining poles of what we long for and what we must.[7]

*The Waterworks* helps us to understand why the aging remnants of the New Left—as opposed to the Old Left of dialectical materialism—are now solidly arrayed in favor of the Noble Savage in a State of Nature as against the Pale Dead Male in an Age of Reason. Doctorow's *The Book of Daniel* was his attempt to justify the Old Left to the New, to tell the kids who burned their draft cards what Paul Robeson sang at Peekskill. In that novel's final lyrical outburst, Doctorow concludes that capitalism and science are enemies of the young and of the spirits of the Earth; only Arcadian Nature can be trusted. *Loon Lake* was another of Doctorow's sermons on that prelapsarian text. From *The Greening of America* to *Loon Lake*, the New Left has learned its lesson well: Mother Earth makes fewer demands than Mother Russia.

Perhaps Doctorow's most overt iteration of the Mother Earth theme in *The Waterworks* deals with our friend the Noble Savage:

> Ever since this day I have dreamt sometimes . . . I, a street rat in my soul, dream even now . . . that if it were possible to lift this littered, paved Manhattan from the earth . . . A season or two of this and the mute, protesting culture buried for so many industrial years under the tenements and factories . . . would rise again . . . of the lean, religious Indians of the bounteous earth, who lived without money or lasting architecture, flat and close to the ground—hunting, trapping, fishing, growing their corn and praying . . . always praying in solemn thanksgiving for their clear and short life in his quiet universe. Such love I have for those savage polytheists of my mind . . . those friends of light and leaf.[8]

This remarkable passage, set in the fictional New York of McIlvaine, is clearly a sentiment prompted by E. L. Doctorow's New York of 1995. It is difficult to square those notions with the facts either of the author's life or of his city. "Street rat in his soul"? Without money or lasting architecture"? According to my *Who's Who in America*, Doctorow is the Glucksman Professor of English and American Letters at New York University, a former editor-in-chief and vice-president of Dial Press, author of at least five best sellers, and recipient of the National Book Critics Circle Award (twice), the National Book Award, the PEN/Faulkner Award, the Edith Wharton Citation for Fiction, and the William Dean Howells medal of the American Academy of Arts and Letters (of which he is a member of the Institute). He has received several honorary degrees, is on the board of directors of the Author's Guild and a member of the Century Association. Nevertheless, in *The Waterworks*—a book advertised as a "haunting tale" by The Book-of-the-Month Club—Doctorow proclaims the perfidy of institutions, claiming that "the institutional mind has only one mental operation: It abhors truth."

Professor Doctorow's ecological reflections on pre-Stuyvesant Indian life in Manhattan also seem disingenuous. "Thanksgiving for their clear and short life"? Who would presume to tell *anyone* that early and violent death from infective disease of one's wife and children is preferable to their survival? It has been estimated that the Native American population had a life expectancy of one generation. And as for those clear and short lives spent "so many industrial years under the tenements and factories," the actual numbers may be worth noting. The death rate in New York City in 1845, that pretenement and preindustrial era of Edgar Allan Poe, was 27.3 per thousand per year, and the average life expectancy was in the 40s. In our day, Doctorow's New York City, the death rate has dropped to 9.1 per thousand per year and life expectancy risen to almost 80. This was no achievement of street rats, but of those who became the sanitarians of New York, men who cleaned our water, built our sewers, and paved our streets, among them Frederick Law Olmsted and George Templeton Strong. Those men sought Nature as accessory, not a reproof—in Lowell's sense. Their monument is Central Park, not a wilderness. Theirs is the old legend of man conquering Nature and disease, to serve the uses of man rather than the fancies of poets.

Ecosentimentalists aside, the old legend may still have life in it. For generations Americans were taught that brave explorers found a New World peopled by savages, to which the white man brought the word of God and the ways of the West. When Columbus first met the Tainos, in 1492, he wrote:

> At daybreak there came to the beach many of these men as I have said, and all of good stature, very handsome people. Their hair is not kinky but loose and coarse like horsehair; and the whole forehead and head is very broad. . . . They bear no arms, nor know thereof; for I showed them swords and they grasped them by the blade and cut themselves through ignorance; they have no iron. . . . I saw some who had marks of wounds on their bodies, and made signs to them to ask what it was, and they showed me how people of other islands which are near came there and wished to capture them, and they defended themselves. And I believed and now believe that people do come here from the mainland to take them as slaves.[9]

Anthropologists are not certain that the gentle Tainos were actually enslaved by their mainland tormentors, the fierce Caribs. But life in pre-Columbian America was no New-Age paradise. The cannibalistic Caribs had moved into the lands of the Mayas and driven the Tainos (Arawaks) to refuge in the outer Caribbean islands. This followed a long and dreadful tradition of Native American imperialism that required no lessons from Europe. The Aztecs displaced the Toltecs, the Incas destroyed the Nascas and Chimus, Toltecs squabbled with Mayas, Seminoles with Choctaws, Algonquins with Iroquois, and so forth: homelands were looted, treasure stolen, tribes decimated. Plunder, pillage, and the plague were here before Columbus.

The specialty of pre-Columbian American religion was human sacrifice, in the course of which cardiectomy without anesthesia (extirpation of the beating heart) was perfected; the chosen few could nibble uncooked ventricles. At one great Aztec celebration 20,000 victims were dispatched to the service of their gods. Even the gentle Tainos indulged in practices that might make a revisionist wince. The male warriors, but not such nonpersons as females, owed their fine sloping foreheads to having these squeezed at infancy between two flat boards, a practice that extended to the Choctaws of our Mississippi. Slavery also anteceded the arrival of Europeans. When Cortez entered Tenochtitlan (Mexico City), he found slave auctions which reminded him of those conducted by the Portuguese who had captured Negroes on the African coast.

"Columbus has been portrayed as some great hero! Anyone with a bit of knowledge can read between the lines. He wasn't such a great guy," argued John Peters (alias Slow Turtle) Supreme Medicine Man of the Mashpee Wampanoag Indians in the Falmouth (MA) *Enterprise* of August 18, 1992. My colleague in the healing arts, who was at the time Commissioner of Indian Affairs for the State of Massachusetts, was pleading for adoption of his program "Relearning Columbus" as part of the cinquecentennial of the Admiral's discovery of America, an occasion that this country celebrated with as much enthusiasm as the Germans showed for V-E day. Said the good doctor about his revisionist program, "It's about nature; it's about reality: the waters and airs and all the things around us. It's about the animals and how we are all interlocked with one another." What a fine sentiment, elegantly put by Dr. Slow Turtle, whose salary derives in part from taxes paid by Raytheon and Genzyme, institutions that may also be said to be "about nature and reality."

Before Columbus, the tribes of both the Old World and New had for generation after generation been savagely interlocked with one another in rituals of attack and revenge. In the Old World, Christian and Moslem, Ethiop and Nubian killed each other on one or another tribal excuse; in the New World, Aztec and Toltec, Carib and Taino bashed skull against war club for one or another aboriginal cause. Those "waters and airs and all the things around us," beloved by medicine man and workout waif alike, brought yellow fever, plague, malaria, and cholera before Columbus set foot on these shores. And as for the interlocking of animals and the Mashpee Indians; the native Americans of Cape Cod had already eliminated most of the deer on the peninsula before the Puritans came. Not until the ecosentimentalists of our own day favored Bambi over venison, not until tick-bearing deer roamed every back yard from Falmouth to New Haven, did we have to undo the mischief of sentiment to battle the spirochete of Lyme disease. At the end of the twentieth century, rabies, tuberculosis, Lyme disease, and Hanta virus—not to speak of AIDS—are not due to industrial pollution but to a misguided choice of sentiment over reason. Nature is not always our friend, as those who dwell by the rivers Ganges and Ebola might tell us. Nature can, of course, serve human needs as—in Lowell's words—"a strengthener and consoler, a wholesome tonic for the mind ill at ease with itself."[10] It ought not remain the preserve of the ecosentimentalist, the new Rousseau who "translates should into would, looks upon the spheres of duty and beauty as alien to each other, and can never learn how life rounds itself to a noble completeness between these two opposite but mutually sustaining poles of what we long for and what we must."

## NOTES

1 Edgar Allan Poe, *The Complete Poems and Stories with Selected Critical Writings*, pp. 189–190.
2 James Russell Lowell, "Thoreau," p. 294.
3 Ibid., p. 298.
4 James Russell Lowell, "Fable for Critics," p. 72.
5 Oliver Wendell Holmes, *Professor at the Breakfast Table*, p. 281.
6 Poe, *Complete Poems and Stories*, p. 1090.
7 James Russell Lowell, "Rousseau," p. 266.
8 E. L., Doctorow, *The Waterworks*, p. 238.
9 S. E. Morison, *Admiral of the Ocean Sea: A Life of Columbus*, pp. 273 ff.
10 Lowell, "Rousseau," p. 266.

## REFERENCES

BROOKS, VAN WYCK. *The Flowering of New England.* New York, NY: E.P. Dutton, 1936.

DOCTOROW, E. L. *The Waterworks.* New York, NY: Random House, 1994.

DUFFY, JOHN. *A History of Public Health in New York 1866–1966.* New York, NY: Russel Sage Foundation, 1974.

HOLMES, OLIVER WENDELL. *Medical Essays.* Vol. 9 of *Collected Works.* Boston, MA: Houghton Mifflin, 1892.

———. *Professor at the Breakfast Table.* Vol. 7 of *Collected Works.* Boston, MA: Houghton Mifflin, 1892.

LOWELL, JAMES RUSSELL. "Fable for Critics." In *Collected Works*, vol. 3. Boston, MA: Houghton Mifflin, 1899.

———. "Rousseau." In *Collected Works*, vol. 2. Boston, MA: Houghton Mifflin, 1899.

———. "Thoreau." In *Collected Works*, vol. 4. Boston, MA: Houghton Mifflin, 1899.

MORISON, S. E. *Admiral of the Ocean Sea: A Life of Columbus.* Boston, MA: Little, Brown, 1942.

MORSE, JR., JOHN T. *Oliver Wendell Holmes: Life and Letters.* Boston, MA: Houghton Mifflin, 1893.

POE, EDGAR ALLEN. *The Complete Poems and Stories with Selected Critical Writings.* New York, NY: Knopf, 1946.

SALE, KIRKPATRICK. *The Conquest of Paradise.* New York, NY: Knopf, 1990.

SILVERMAN, KENNETH. *Edgar A. Poe.* New York, NY: Harper Perennial, 1991.

# RELIGION

THE HEALTH of liberal democracy depends on the general use of reason. Reason must not be the cognitive tool of the few: if the integrity of science and reason are undermined among the majority, then democracy itself is in peril. The merits of this generalization are evident in the records of modern history. Also evident is the ambiguous role of religion, whose discomfort with reason, sometimes open and sometimes concealed, has nevertheless long been evident.

An exchange of views between Paul Kurtz and Eugenie C. Scott points up the difficulties of broadening rational inquiry in a world where religion and cult belief continue to be overwhelmingly popular. Kurtz sees linked threats to democracy in the widespread fascination with paranormal phenomena and the studied obsequiousness toward all religious truth claims, no matter how extreme. As Kurtz shows, this attitude reigns in the media and is unchallenged by those—including liberal religionists and intellectuals—who should know better. He calls for open confrontation and challenge, especially when the issues involve realities about which science has something to say.

Eugenie C. Scott, long a campaigner against educational nonsense in the form of creationism, presents a conflicting view. About the diagnosis of nonsense she has no doubt; she offers a catalog of creationism and the activities of its biblical-literalist troops, with analysis of the social anxieties that encourage it. But Scott believes that confrontation will lead only to political defeat. Instead, she argues, scientists should make common cause with religious moderates—and hence with the majority of the American population—by admitting the limits of science and by emphasizing the complementary, rather than the conflicting, functions of science and religion.

Langdon Gilkey, from the standpoint of a long career in religious studies, reaches yet another conclusion. This is that the diffuse dangers of antireason and antiscience are insignificant in comparison with the immediate danger of Christian fundamentalism in its politically aggressive forms. Theocracy is their goal, he insists, admitted or not; and where there is theocracy, there can be no democracy; neither reason nor science can survive. He too, therefore, urges resistance, but of a far more focused kind than does Kurtz.

Oscar Kenshur, finally, examines the complex relation between rational inquiry and religion in a historical context: the turn of the eighteenth century. He examines the work of the Huguenot philosopher Pierre Bayle, demonstrating the ingenious and convoluted way in which he interwove conflicting traditions: fideism—the rigid insistence on the primacy of faith, ignoring its improbabilities and the urgings of reason—alongside the scientifically informed rationalism that was coming to dominate European thought.

# TWO SOURCES OF UNREASON IN DEMOCRATIC SOCIETY: THE PARANORMAL AND RELIGION

PAUL KURTZ

REPRESENTATIVE DEMOCRACY entrusts to the people the ultimate power to elect the chief officials and to determine the main directions of government. It requires widespread participation at all levels of civic life. Its key policies are based on the freely given consent of a majority of adults exercising the franchise. Democracy relies not upon aristocratic elites, oligarchs, or ruling classes but upon ordinary persons to exercise power. It assumes that the rulers can be turned out of office after a full discussion of the issues. It is receptive to the opinions of dissident minorities in debating alternative courses of action. It thus presupposes a free market of ideas and an open society, in which both orthodox and dissenting views are heard.

Democracy is rooted in a *method* of inquiry. It denies that any one sector of society has a monopoly of truth or virtue. It thus draws on the pooled intelligence of the broader public. Slow and laborious at times, it has nonetheless proven itself to be the most effective method of governing society—at least in comparison with alternative systems. If democracy is to function well, it must assume that there is some degree of competence and common sense and some level of education in the average person. In this view of democracy, the well-informed citizen is the best guarantee against the abuses of power from any sector of social life—political, economic, military, or ecclesiastic. Ideally, this means that we need to cultivate in the ordinary person the arts of intelligence, an appreciation for critical thinking, and some rationality. Where passion or caprice, power or violence predominate, the ability to make reflective judgments is curtailed. John Stuart Mill believed that democracy works best where a society has reached sufficient maturity in its judgment. Here Mill was critically evaluating the underdeveloped colonial areas of the world of the nineteenth century.[1] We reject Mill's limitations and believe that democracy should be permitted to develop in all societies. Nevertheless, this presupposes that great efforts will be made to raise the level of education of the citizens of a democracy. Ever since the Enlightenment, democrats have believed that with an increase in education and literacy, and the application of reason and science, the progressive amelioration of the human condition would eventuate.

Unfortunately, America and other mature democratic societies have seen in recent decades an erosion of this reflective process. A number of technological and social developments have compromised our ability to make rational choices; and this may undermine the viability of democratic societies. Paradoxically, this is occurring at the same time that liberal democracy seems to be everywhere on the ascendent, having vindicated itself over totalitarian and authoritarian methods of rule. Fukuyama maintains that we are at "the end of history," for liberal democratic ideals have prevailed virtually everywhere. This view has been grossly overstated; for there are always new problems to be resolved if democracy is to prosper.[2]

There are many new challenges to reason and the democratic ethos today. Postmodernists reject the ideals of the Enlightenment, and they deny that any form of objective scientific knowledge is possible. Waves of multicultural critics and extreme feminists have sought to undermine the integrity of science and rationality. I wish to concentrate in this paper on two other challenges to rational democracy that are present in contemporary society and are all too rarely identified. The first area of concern is the steady growth of belief in a paranormal and occult universe, and the second concerns the unchallenged persistence of dogmatic religious beliefs and practices.

## THE PARANORMAL

The term "paranormal" refers to phenomena that allegedly cannot be accounted for or explained in terms of normal science and that thus transcend the limits of a naturalistic framework. A precursor of the currently popular paranormal paradigm was no doubt the spiritualist movement of the nineteenth century. The immediate founders of spiritualism were the Fox sisters, two young girls of eight and ten who in 1848 allegedly were able to communicate with discarnate spirits by receiving messages rapped from "the other side." The Fox sisters were heralded as mediums, and within a decade thousands of mediums appeared in America and Europe exhibiting similar "psychical powers." The Fox sisters were tested by scientific bodies, which pointed out that the so-called rappings could have been produced by crackings of the toe and knee bones against wooden floors, without the need for invoking a nonnaturalistic interpretation. The distinguished physicist Michael Faraday attempted to explain table levitation by involuntary psychological and muscular movements rather than occult forces.

The Society for Psychical Research was founded in 1882 in Cambridge, England by Henry Sidgwick and a group of eminent scientists and philosophers, and in the United States in 1885 by William James, founder of the first psychological laboratory at Harvard. Many of these researchers were disturbed by the growth of Darwinism and the challenge that evolution gave to the spiritualist picture of reality. They thought that if physical research were treated rigorously and experimentally, objective truth could be elicited. Thus they dealt with thought transference, apparitions, poltergeists, and other "psychical phenomena." A number of mediums were tested by scientists—including Eusapia Palladino, Leonora Piper, D. D. Home, Margery Crandon; and many of these were caught in trickery, so much so that by the 1920s the entire field had been discredited.

A second wave of paranormalism was stimulated in 1927 by the founding of the first parapsychological laboratory at Duke University by J. B. Rhine. Rhine deferred the question of survival after death and attempted to establish the existence of ESP (clairvoyance, precognition, telepathy) and PK (psychokinesis) by using laboratory methods. The above- or below-chance calls on a run of Zener ESP cards could not be explained by reference to normal sensory perception, he maintained, and could be attributed only to some underlying extrasensory paranormal cause. Rhine explicitly stated that one of the primary motives for the rise of parapsychology was "to refute mechanistic philosophy."[3] He held a dualistic theory of reality in which mind was independent, in some sense, of its material basis. Rhine's interest in this regard was similar to the earlier founders of the Society for Psychical Research, who had hoped to provide a scientific basis for the "spiritual principle of man."

Scientific skepticism about the work of Rhine was widespread. He was criticized by psychologists because of loose protocol, questionable grading techniques, and sensory leakage. The apparent presence of trickery by several subjects and experimenters was also raised. For example, G. S. Soal claimed to demonstrate the presence or precognition and telepathy during the Second World War in London. Fraudulent manipulation of the data by Soal has been clearly demonstrated, so that all of his work has been held suspect. The history of psychical research and parapsychology continually provides similar tales of announced or impending breakthroughs demonstrating the existence of paranormal phenomena and their rejection after laborious examination by skeptics, who show the paucity of the evidential base and especially the inability to replicate the phenomena by other scientists. This has occurred in recent years with regard to dream research, remote viewing, and even the use of Ganzfield data.

Nonetheless, since World War II paranormal claims have proliferated. The mass media continually feeds a receptive public with modern tales of paranormal miracles, and science fiction has opened the imagination to new forms of fantasy and mystery. The term *paranormal* has been extended beyond parapsychology to include space-age claims, such as astrology, UFOlogy, and a wide range of pseudoscientific anomalous claims, including alternative medicines, faith healing, and a return of occult demons and monsters of the deep.

The mass media are in the business of packaging and selling the paranormal as a product. And this is true of book and magazine publishers as well as TV and film producers. If one visits any bookstore in America, one will find that the New Age, inspirational, spiritual, and paranormal shelves far outdistance the science books made available.

Virtually every major newspaper carries an astrology column, and there are dozens of mass-market astrology magazines and hundreds of astrology books that have been published in recent years. In the same period very few books have been published that are critical of astrological claims. Yet astrology has no basis in empirical fact, and all efforts by objective scientists to confirm it have had negative results. Still, large sectors of the American public read their horoscopes daily, and many may not venture forth without doing so. Many Wall Street pundits claim to base their recommendations on astrology, and even AT&T has offered astrological readings to gullible telephone callers. A similar tale can be told for so-called psychic phenomena. There are tens of thousands of psychics offering readings and prognostications, and they are pandered to by the mass media. Literally thousands of books have been written about psychic healing, therapeutic touch, levitation, channeling, fire walking, poltergeists, etc. They claim wondrous powers, with very little scientific dissent. Books by serious publishers critical of this tidal wave are almost nonexistent.

Three recent illustrations of this phenomenon graphically demonstrate the problem. First is the widespread fascination with near-death experience (NDE). There are a great number of books portraying the experience.[4] Raymond A. Moody offers us a collage of various NDE accounts. He tells us that a dying person, overcome by fear and pain, perhaps lying on an operating table or hospital bed, begins to hear loud buzzing and ringing sounds. He feels himself falling rapidly through a tunnel. He finds himself outside of his physical body, which he is able to view as an onlooker. He may even hear the doctor pronounce him dead or attempt resuscitative efforts. As he enters the tunnel he has a panoramic review of his life. Eventually he encounters a bright light and meets the spirits of friends and relatives who welcome him, perhaps even Jesus or Krishna. There is a beam of light that he encounters, and he feels peace and joy. He discovers that he must return to his physical body on earth, though he is reluctant to do so. Suddenly he is back in his body, and his consciousness revives. The experience is so powerful that it may have a transforming effect on his life thereafter.[5]

Such accounts of near-death experience are highly reminiscent of the kinds of experience reported by classic mystics, who claim that they have encountered a being of light. These reports are introspective and subjective, so that most of the evidence is anecdotal. The spiritual significance attributed to the experience is unmistakable. Does this provide evidence that there is a paranormal realm? Is there scientific validity for the claim? Michael Sabom, Melvin Morse, and other researchers maintain that there is, and these views are widely broadcast.[6]

There have been very few skeptical studies published. Psychologist Susan Blackmore points out that there are alternative naturalistic explanations—physiological and psychological—that can be given without postulating the existence of an afterlife.[7] Not everyone who is dying and has been resuscitated has had the same experience. Moreover, elements of the collage can be experienced in other contexts of life. Out-of-body experiences are fairly common. We find them in hypnogogic and hypnopompic sleep. Alpinists who fall from mountains yet land safely, or people in car accidents who survive unscathed, sometimes experience an out-of-body experience and a panoramic review. Ronald K. Siegel reports that he chemically produced similar NDE-type phenomena by administering drugs.[8] Susan Blackmore has summarized a plurality of possible causes: the experience of an hallucination triggered by the distress of the dying process, the deprivation of oxygen, increased levels of carbon dioxide in the blood, the release of endorphins and other natural opiates, seizures in the temporal lobe and lymbic systems, the dissolution of the sense of the self, etc. Skeptics point out that the spirits encountered on "the other side" are colored by the cultural experience of the person having and NDE. The proponents of the survival hypothesis deny this. They point out that even atheists, such as A. J. Ayer, have reported NDEs. They maintain that there is a remarkable consistency in all such reports. Skeptics maintain that if there is a consistency in reports, it is because we share a common physical and psychological structure. The person who has had an ND experience has not died (there is no clear case of brain death), but is undergoing a dying process. The transforming effect that the experience has can be simply explained by the fact that a harrowing brush with death can help us to alter our priorities in life. The dying brain hypothesis more parsimoniously accounts for the altered states of consciousness.

I cannot in this brief paper do justice to the scientific case pro and con. The point I wish to make is that this type of near-death experience illustrates the spiritualist-paranormalist outlook that pervades our culture and the paucity of skeptical critiques available to the public.

A second popular area that illustrates the same point is the reemergence of reincarnation. This classical doctrine presented in Hindu, Buddhist, and ancient Egyptian literature, and indeed even in Greek philosophy, held that the soul was separable from the body, preexisted birth, and will exist as separated from the body after death. Similarly, animals and plants were said to be possessed of souls. Upon death the soul could be transformed from one organism to another and could be released only after a period of time. This was dependent upon *karma*, the behavior of a person in his past lives.

Interestingly, these doctrines, held on a basis of faith, have now been given some kind of pseudoscientific credence by a number of parapsychologists and psychiatrists, and reincarnation is presented by the media as if it has been verified by science. For example, Ian Stevenson, American doctor and former president of the Parapsychological Society, maintains that there is evidence for this claim, especially in India, based upon the memories of young children of their previous lives.[9] I find the evidence he adduces highly questionable. Alternative naturalistic explanations fit the data more parsimoniously. Such memories have been suggested by those in the immediate circle of relatives, friends, and the cultural milieu; predisposed to believe in reincarnation, they interpret the experience of children in terms of it.

Another popular line of research often referred to today is the experience of "regression" that enables a person to recall his past lives. It is interesting to note that most of this evidence is based upon hypnosis. Indeed, there is a growing school of hypnotherapists who use regression as a therapeutic technique. Under an hypnotic state individuals allegedly recall their prior lives. There is an abundant literature of this kind of experience. For example, Dr. Brian Weiss, a Columbia- and Yale-trained psychiatrist practicing at the University of Miami, in his book *Many Lives, Many Masters*,[10] maintains that he has regressed a great number of people and that by regressing them he is able to cure them of present neurotic or pathological conditions. His regressions take people back to the Civil War, the Middle Ages, or earlier. Without going into this alleged evidence in detail, it has been pointed out by skeptics that one does not have to postulate prior existence to account for the bizarre tales that are related. The subjects are rendering unconscious materials, hidden and repressed, and what they report is most likely a product of confabulation and cryptomnesia. Moreover, psychotherapists who believe in past lives tend to induce by suggestion similar beliefs in their subjects.

A third area in which hypnosis is used to support paranormal claimants is UFOlogy. A number of researchers have reported that in hypnotic sessions they are able to elicit tales of abductions of earthlings by extraterrestrials. These creatures from afar allegedly are performing sexual, reproductive, and genetic experiments on their helpless victims. There has been an extensive literature about this. Most noteworthy are the books by David Jacobs, associate professor of history at the University of Pennsylvania,[11] and John Mack, professor of psychiatry at Harvard.[12] Mack, in his book *Abductions*, claims to have investigated 76 abductees, though in the book he focuses on only 13. He calls them "experiencers," for they are undergoing anomalous experiences. His narratives of these experiences are based on hypnotic sessions with them. The phenomenological contours of such cases are that a person who is asleep is suddenly awakened by a bright light that infuses the bedroom, that he or she is paralyzed and unable to move. Visited by small creatures with huge heads and large eyes, he or she is levitated out of the bed, through the walls, and brought aboard a spacecraft, where sexual and genetic experiments are performed.

What are we to make of these reports? Perhaps we are being visited by extraterrestrials from other galaxies, and perhaps thousands or millions of people are being abducted daily, as proponents maintain. Skeptics seek alternative naturalistic explanations of the experiences: again, they point out that the hypnotic session is not a reliable basis for truth claims and that what is being reported is often suggested by a hypnotherapist who believes in the UFO ET hypothesis. Here both cryptomnesia and creative imagination seem to provide a more credible explanation.

Interestingly, John Mack believes that this phenomenon is quasireligious. He states that many abduction experiences are "unequivocably spiritual, which involves some sort of powerful encounter with or immersion in divine light. . . . The alien beings . . . may also be seen as intermediaries, closer than we are to God or to the source of Being."[13] The UFO abduction experience, says Mack, "while unique in many respects, bears many resemblances to the dramatic transformation experiences undergone by shamans, mystics, and ordinary citizens, who have had encounters with the paranormal."[14]

I have touched on only some of the more sensational aspects of the paranormal universe. Nonetheless, if one reads the popular literature, views film and TV productions, and reads popular magazine and newspaper accounts—and, indeed, listens to our students in the colleges and universities—one is impressed by the fact that, by and large, the proparanormal viewpoint is accepted with very little dissent. A serious problem that we face is the fact that the mass media do not give adequate attention to the skeptical and scientific critique. We surely cannot prejudge such paranormal questions—astrology may be true, ESP may exist, the Loch Ness Monster may be lurking underneath, near-death experiences may put us in touch with another reality, the reincarnation of souls may be true, and earthlings perhaps are being kidnapped by extraterrestrials. Surely scientists need an open mind and careful investigation. But some degree of skepticism is essential in the process of inquiry; and we must distinguish between a possible speculative conjecture and objective, reliable, confirmatory evidence. All too often eyewitness testimony is deceptive. Only by careful examination can one find a causal basis for an anomalous phenomenon. Moreover, there is a difference between an open mind and an open sink. The open mind allows for the critical examination of ideas, and it is receptive to new ones; the open sink is willing to accept anything and everything as worthy of examination without any responsible filtering process.

Regrettably, the public does not always appreciate the nature of science, particularly the open yet skeptical character of scientific inquiry. The scientific and academic community needs to convey to editors, publishers, and film and TV producers an appreciation for the methods of scientific inquiry. I am afraid that we have often failed in this regard; for the dispassionate scientific attitude is very rarely or adequately represented or understood. Bizarre claims are uncritically touted as true, often without the least scientific skepticism being heard. It seems to me that anyone concerned about public education should call for balance in the mass media. We believe that producers and publishers have an obligation to provide the public with some appreciation for critical thinking. If the democratic society is to function, it needs an educated public. But how is it possible to develop this when the public is fed mystery and purely speculative nonsense as true, without any inkling of there being scientific explanations for the so-called "unsolved mysteries" or the "X-files." What we are dealing with, I believe, is a religious or quasireligious spiritualist movement, which is constantly fed by a media industry and is growing in spite of evidence against its central beliefs.

## RELIGION

Why is this occurring? There are surely many complex explanations for this sociological phenomenon. One possible cause is the powerful new electronic, computer, and information technologies that we have developed. Many commentators have observed that we may be at a pivotal point in human history; perhaps we are at the beginning of the end of the age of books. The oral tradition gave way to the written word, and the invention of the printing press led to a vast expansion in the publication of books. We seem to be at another historic turning point in the mode of communication, one as dramatic as any of the above; and the effects of this upon the public mind and the education of citizens is incalculable. In my view it is a cause for some concern about the future of democracy. We may ask: Are the cognitive symbols of reading texts being replaced by the visual and auditory arts? Is reflective thought being engulfed by fleeting images and sound bytes? And is this why the defenses against gullibility provided by critical thinking can so easily be broken down?

Another reason for the failure to convey an appreciation for scientific skepticism and critical thinking, I submit, is that there is a vast area of belief and practice that remains largely unexamined. I am here referring to religion. In present-day America it is usually considered to be in bad taste to question the claims of religion, even though many religionists present highly questionable theories about human behavior and reality and may even attempt to censor or suppress alternative naturalistic theories based on scientific inquiry. It is a paradox that America, the most advanced scientific technological society, is also the most religious and that numerous religious sects thrive and flourish—from the orthodox faiths to the bizarre cults of unreason. Interestingly, in countries such as Britain, the Netherlands, and France, church attendance has declined drastically, and the number of people who maintain membership in a church or temple or who profess belief in God is a minority of the population, whereas in America it is a predominant majority. A recent sociological study has shown a steady increase in this phenomenon since the eighteenth century—from Revolutionary days through the Civil War until the twentieth century.[15] Clearly, liberty of thought and conscience and the right to profess and to practice one's religion is not at issue; what is at issue is the reticence to criticize religion in the public square or to subject its basic premises to scrutiny. I am not talking about the criticism of the political role of religion or about attacks on clergy who engage in sexual or other improprieties. Rather, I am talking about the failure to examine critically the basic religious claims themselves: whether the Old Testament idea of the "chosen people" has any validity, whether the alleged revelations of the Ten Commandments by Moses can be sustained historically, whether New Testament claims of the divinity of Jesus hold up under critical examination, whether Mohammed's claims to divine revelation in the Koran are reliable, whether the evidence for the Book of Mormon adduced by Joseph Smith can be corroborated, etc. No doubt religious intolerance and warfare over the millennia by competing sects and doctrines have taught us all a lesson. The Spanish Inquisition, bitter conflicts between Catholics and Protestants, the suppression of the Mormons of the nineteenth century, the Holocaust, warfare between Muslims, Christians, Jews, Sikhs, and Hindus have bloodied human history. Thus the desire to seek a kind of accommodation by mutual tolerance is understandable, even commendable.

Nonetheless, speaking as a secularist and skeptic, I believe this should not preclude others within the community from questioning the claims of Biblical, Koranic, or other absolute faiths, particularly since massive efforts are constantly undertaken by missionaries to recruit members to the fold—whether Jehovah's Witness, Mormons, evangelical Protestant fundamentalists, conservative Roman Catholics, Orthodox Hassidim, or devout Muslims. Simply to accept their admonitions and recommendations without dissent is, in my judgment, irresponsible. This posture is especially questionable given the constant effort by militant religionists to apply their doctrines in the political process, thus seeking to impose their views on others. For example, in the United States the Moral Majority and its successor, the Christian Coalition, are attempting to bridge the historic separation of church and state, and introduce prayer and other religious symbols into the public square; they wish to subvert secular neutrality. Their desire to mandate "creationism" in the public schools is especially threatening to the integrity of science education.

Since this volume is devoted to the "Flight from Science and Reason," I submit that one of the key offenders is the persistence of orthodox religiosity and fundamentalism. A graphic illustration of this is the recent campaign launched by a coalition of religious leaders that seeks to reverse the United States government's fifteen-year-old policy of granting patents for human and animal genes. The petition to the United States Patent and Trade Mark Office was signed by 180 leaders of mainstream Protestant, Roman Catholic, Evangelical Christian, Jewish, Muslim, Buddhist, and Hindu faiths. The proponents maintain the following: "We believe that humans and animals are creations of God, not humans, and as such should not be patented as inventions."[16] "One of the basic principles of our church is that life is a gift from God," said Methodist spokesperson Bishop Kenneth Carter. "The patenting of life forms reduces life to its marketability. Gone is the fundamental principle that life is a gift. . . ."[17] Those who support the patenting of life forms say that new medical technologies require considerable investment of venture capital and that without patents scientific research into new gene therapies (such as that for cystic fibrosis) may not develop. Until now the courts have permitted such patents. Perhaps the patents of the biogenetic industry need to be given public scrutiny and even challenged. This is not the issue. What I question are the reasons given for the opposition. These are all too reminiscent of historic efforts of religionists to suppress scientific research in the past—from Galileo and Darwin to the present—in the name of God.

This raises the broader issue that skeptical, agnostic, or atheist viewpoints in regard to religion really have no effective platform in America. There are literally hundreds and perhaps thousands of proreligious sermons delivered from television and radio pulpits every week. And there are hundreds of books and magazines promoting religious propaganda, with almost no response in kind to the claims made. All too many teachers in the universities and colleges of America, and surely in primary and secondary schools, are reticent about critically examining the existence of God, the doctrine of eternal salvation, the immortality of the soul, the efficacy of prayer. Consequently, large sectors of the American public have no knowledge at all of the skeptical scientific or naturalistic outlook, which is devoid of supernatural overtones.

How explain the fact that in America the majority of Americans still believe in the viability of miracles, the existence of angels, satanic forces, and other supernatural entities, etc? Most Americans are unaware of the fact that there is a rich literature of biblical and scientific criticism going back at least two centuries, in which the best scholarship drawing upon archaeology, linguistics, anthropology, etc., has questioned the very basis of the Bible. One reason for this, I submit, is that there is no existing free market of ideas in the area of religious belief; hence, a selective application of the methods of science to highly specialized and safe domains of inquiry is permitted, but little beyond that.

## EDUCATION

In conclusion, the premise with which I began this paper is that democratic society presupposes thoughtful, educated citizens who can deliberate reflectively and think critically. Yet the mass media continually bombard the public with paranormal pseudoscientific claims as if they were scientifically valid, and religious beliefs and practices are largely immune to sustained rationalist critiques.

Totalitarian societies in the twentieth century have used pseudoscientific ideologies destructively. Hitler's Third Reich was based on spurious theories of Aryan racial superiority and inferiority, and Stalin's Soviet Union drew upon Lysenkian environmentalism and dialectical materialism. The flight from reason can endanger democracy. The failure to distinguish between genuine science and pseudoscience could erode the decision-making process. Similarly, theological dogmas have been appealed to in the past to justify theocratic suppression. Today there are ominous challenges from religious militants who wish to remake America.

Do I have any recommendations to counter these threats? May I suggest two courses of action that I think scientists, scholars, academics, and all those concerned with rational democracy can undertake. These proposals, though modest, are vital.

First, we need to convey to the public an appreciation for the methods of science. This means that we need to encourage the teaching of critical thinking and the methods of scientific inquiry in the schools. Science education is essential if we are to counteract scientific illiteracy, but to be a specialist in a subject matter is not sufficient. One may know how to master his field of expertise—whether mathematics or physics, biology or economics, medical science or engineering—and yet not know how to apply the methods of scientific thinking outside of it. We need to extend the range of rational thought to as many areas of life as possible. The Japanese cult Aum Shinrikyo, which has been accused of gassing Japanese subways and which has several thousand disciples, is an interesting phenomenon. Many of the members of this cult were apparently trained in the sciences and are highly sophisticated in their fields of expertise, yet they were apparently unable to translate what they have learned in their field of competence to other areas and were swept up by the guru and followed him without dissent. The same is true of popular cults, such as Scientology, that are able to attract bright young people who do not apply critical thinking to challenge the cults' claims. The Creationist Institute, which rejects the theory of evolution, has many Ph.D.'s in engineering, chemistry, and other sciences affiliated with its efforts.

We need to convey the idea that the methods of scientific inquiry are not for the esoteric few, but are continuous with common sense. Ideas should be considered as hypotheses, supported by evidence, tested by experimental consequences, and validated by their consistency with other tested hypotheses. Beliefs should not be treated as absolutes, but rather should be open to modification in the light of evidence. It is the tentative, fallible character of science that needs to be appreciated along with an awareness that by using rigorous standards, objective and reliable knowledge can be achieved.

Second, we need to convey an appreciation for the scientific outlook itself and especially the implications of the sciences for our view of the human species and its place within nature. Thus we need some scientists who are generalists able to interpret the sciences and integrate what we know into a coherent framework. Philosophy and the liberal sciences can assist in this endeavor. This understanding can play an important role in evaluating paranormal, spiritualist, and metaphysical theories, especially those that have no basis in empirical fact. The evolutionary view of nature competes with the creationist doctrine today; yet how many scientists are willing to defend evolutionary theories and criticize fundamentalist biblical claims? There is insufficient evidence for the survival of the soul or for a teleological purpose in the universe. Yet many otherwise-thoughtful people still cling to these doctrines, perhaps because they have not read the critiques by scientific skeptics and biblical scholars.

This role of education should be twofold. We need to modify school curricula at all levels of education, from primary and secondary schools to universities and colleges, so that they include programs in critical thinking and an appreciation of the scientific outlook. All too many students today graduate with very little understanding of the sciences or their methods of inquiry. Liberal arts programs are not sufficiently rounded unless they also provide some rational comprehension of the scientific outlook.

But we also need to educate the leaders of the corporate and political world about the importance of this kind of general science education. We especially need to persuade those who control the media of communication of their obligation. This means that we need to impart to publishers, editors, producers, and writers some sense of their responsibility not to distort the truth and their obligation to raise the level of scientific comprehension and critical thinking. Not only is our postmodern industrial and information society dependent upon science and technology, but, even more, our democracy presupposes that our citizens will be well informed, and that they will be capable of exercising reflective judgment.

## NOTES

1 John Stuart Mill, *On Liberty.*
2 Francis Fukuyama, *The End of History and the Last Man.*
3 J. B. Rhine, "Comments on the Science of the Supernatural."
4 The most popular longest-running bestseller in America is Betty J. Eadie's *Embraced by the Light.* In it she reports on an experience she says she underwent twenty years ago when she nearly died and came back. Eadie claims she not only went to "the other side," but met Jesus. None of this is corroborated by independent observers.
5 Raymond A. Moody, *Life After Life* and *Reflections on Life After Life.*
6 M. B. Sabom, *Recollections of Death*; Melvin Morse, *Closer to the Light*; Melvin Morse and Paul Perry, *Parting Visions: Uses and Meanings of the Pre-Death, Psychic, and Spiritual Experiences.* See also Kenneth J. Ring, *Heading Toward Omega: In Search of the Meaning of the Near-Death Experience.*
7 Susan J. Blackmore, *Dying to Live.*
8 Ronald K. Siegel, "The Psychology of Life After Death"; Ronald K. Siegel & A. E. Hirschman, "Hashish Near-Death Experiences."
9 Ian Stevenson, *Twenty Cases Suggestive of Reincarnation.*

10 Brian Weiss, *Many Lives, Many Masters.*
11 David Jacobs, *Secret Life: First-Hand Accounts of UFO Abductions.*
12 John E. Mack, *Abductions: Human Encounters with Aliens.*
13 Ibid., p. 397.
14 Ibid., p. 8.
15 Roger Finke & Rodney Start, *The Churching of America, 1776–1990: Winners and Losers in Our Religious Economy.*
16 Barbara Rosewicz & Michael Waldholz, "Human, Animal Gene Patents Targeted by a Religious Coalition in a Petition."
17 Edmund L. Andrews, "Religious Leaders Prepare To Fight Patents on Genes."

## REFERENCES

ANDREWS, EDMUND L. "Religious Leaders Prepare to Fight Patents on Genes." *New York Times*, May 12, 1995, p. 35, col. 3.

BLACKMORE, SUSAN J. *Dying to Live.* Amherst, NY: Prometheus Books, 1993.

EADIE, BETTY J. *Embraced by the Light.* Placerville, CA: Boulderleaf Press, 1992; New York, NY: Bantam Books, 1994.

FINKE, ROGER & RODNEY START. *The Churching of America, 1776–1990: Winners and Losers in Our Religious Economy.* New Brunswick, NJ: Rutgers University Press, 1992.

FUKUYAMA, FRANCIS. *The End of History and the Last Man.* New York, NY: Free Press, 1992.

JACOBS, DAVID. *Secret Life: First-Hand Accounts of UFO Abductions.* New York, NY: Simon and Schuster, 1992.

MACK, JOHN E. *Abductions: Human Encounters with Aliens.* New York, NY: Charles Scribner, 1994.

MILL, JOHN STUART. *On Liberty.*

MOODY, RAYMOND A. *Life After Life.* Atlanta, GA: Mockingbird Books, 1975.

———. *Reflections on Life After Life.* Atlanta, GA: Mockingbird Books, 1978.

MORSE, MELVIN. *Closer to the Light.* New York, NY: Ballantine Books, 1990.

MORSE, MELVIN & PAUL PERRY. *Parting Visions: Uses and Meanings of the Pre-Death, Psychic, and Spiritual Experiences.* New York, NY: Villard Books, 1984.

RING, KENNETH J. *Heading Toward Omega: In Search of the Meaning of the Near-Death Experience.* New York, NY: Quill, 1984.

RHINE, J. B. "Comments on the Science of the Supernatural." *Science* 123, no. 3184 (1956): 11–14.

ROSEWICZ, BARBARA & MICHAEL WALDHOLZ. "Human, Animal Gene Patents Targeted by a Religious Coalition in a Petition." *Wall Street Journal*, May 15, 1995, p. B2, cols. 3–4.

SABOM, M. B. *Recollections of Death.* London: Corgi, 1982.

SIEGEL, RONALD K. "The Psychology of Life after Death." *American Psychologist* 35 (1980): 911–931.

SIEGEL, RONALD K. & A. E. HIRSCHMAN. "Hashish Near-Death Experiences." *Anabiosis: The Journal of Near-Death Studies* 4 (1984): 69–85.

STEVENSON, IAN. *Twenty Cases Suggestive of Reincarnation.* Charlottesville, VA: University of Virginia Press, 1980.

WEISS, BRIAN. *Many Lives, Many Masters.* New York, NY: Simon and Schuster, 1988.

# CREATIONISM, IDEOLOGY, AND SCIENCE

EUGENIE C. SCOTT

MANY CAUSES AND MOVEMENTS were discussed during the "Flight from Science and Reason" conference that occasioned this volume. Most reject science as a way of knowing, or denigrate logic or reason. Creationism differs in some important ways: supporters are science fans, not detractors, and they believe science is useful and important, something that students should be exposed to. But creation science illustrates extremely well one of the themes of the conference: the evasion or denial of empirically based knowledge when it conflicts with ideology. Opinions and values are more important than facts and reason for creationists, which is shown in their choice of which data to accept or reject. Whereas T. H. Huxley warned of the naturalistic fallacy—assuming that what *is* is what *ought* to be—creation scientists, like extreme Afrocentrists, radical feminists, and several others discussed at the conference, apparently see what *ought* to be as what *is.*

In this paper, I will first discuss who the creationists are, stressing the many varieties of creationism and how they differ in their approaches to both science and theology. I will outline a brief history of the movement, the emergence of creation "science," and also the current "neocreationism" period. I suggest that a characteristic of neocreationism is the rise of more moderate antievolutionists, some of whom are located on secular campuses. Some of these argue that a "Christian perspective" has equal status with a "feminist perspective," or a "Marxist perspective," or some similar approach extant on campuses today, and thus deserves a place in the curriculum.

The presence of so many "isms" on college campuses today invites the question, "What happens when ideologies become 'scholarly perspectives'?" Can there be a "Christian perspective" that is truly scholarly? I discuss reasons why I doubt that supernatural ideology, especially, can be consistently scholarly.

I then discuss the importance to public science literacy of teaching evolution and suggest some ways that university and professional scientists may assist in this important endeavor. To do so will require teachers to distinguish between where science leaves off and where philosophy begins.

## CREATIONISM, BIBLICAL LITERALISM, AND ANTIEVOLUTION

There is not one creationism, but many varieties, ranging from strict Biblical literalist young-earth creationism, through a variety of old-earth creationisms ("gap creation;" "day-age creationism"), to progressive creationism, to continuous creationism, to theistic evolutionism. Specific terms may have slightly different connotations depending on who is using them, but in general, young-earth creationism posits that the universe was created at one time, within the last 10,000 years. Noah's flood is an essential element to both young-earth theology and creation science. It was, supposedly, a historical occurrence, wherein water covered the whole globe. During the year the flood waters receded, all the geological features of the world (such as the Grand Canyon, the Himalayas, etc.) were established. By contrast, old-earth creationists accept modern geology and radiometric dating and their implication of an old planet. One version of "old earth" theory, "gap creationism," allows for a long period of time before the six days of creation described in Genesis,[1] or alternatively for the six days in Genesis to be separated by thousands or hundreds of thousands of years. Alternatively, "day-age creationism" accommodates some of modern geology by claiming that each of the six days in Genesis is actually an immensely long period of time.

In "progressive creationism," God is supposed to have created the original species, but subsequently they have "progressed" by diverging (i.e., evolving) into new forms. The flood is considered a local, not a universal, event. "Continuous creationism" and "theistic evolutionism" are further along the continuum, referring to a Christian perspective that accepts a considerable amount of evolution. In continuous creationism, God plays a very active role in directing evolution from the created kinds. Theistic evolution in the most general sense is the idea that God created, but through the process of evolution. By and large, theistic evolutionism accepts the evidence of science, and fine-tunes the theology if necessary. In this context, the flood of Noah is not a historical event, but a metaphor of the importance of obedience to God, and ultimately of God's love for humankind. Theology varies as to how involved God is in guiding the evolutionary process.

The above continuum is largely organized by the degree of biblical literalism, with theistic evolutionism (the perspective of most mainline Protestants and the Catholic Church) being the least literal. It can also be organized according to how much of modern science is accepted, with the young-earth creationists being the most out of touch. Some theistic evolutionists, especially those of a more deistic inclination (God created the universe and its laws, and left it to operate without further intervention), are scarcely distinguishable from nonreligious evolutionists, which is why the conservative Christian world, with its stress on a personal God, often speaks harshly of theistic evolutionism.

Whether God created is therefore in fact not the main issue in the creation/evolution controversy, since "God created" does not rule out the possibility that God created through the process of evolution. Catholics, mainline Protestant denominations, and Reformed, Conservative, Reconstructionist, and most Orthodox Jews hold to some form of theistic evolution.[2] The term "special creation" has come to refer to the belief that God created according to a literal interpretation of Genesis: the universe was created all at one time, in essentially its present form.

Special creationists, especially at the more conservative end of this continuum, are fundamentally antievolutionist. They believe that evolution is an evil idea that children should be protected from. Their perspective is that if evolution occurred, then God did not create mankind specially, hence mankind is not particularly special to God, which would make the fall of Adam and Eve irrelevant. Without Adam and Eve's sin, the death of Christ is irrelevant—and the death of Christ is the foundational event of Christian theology. Everything in Christianity, in this view, relies on the literal truth of Genesis: six 24-hour days of creation, a flesh-and-blood Adam and Eve, a literal Noah's flood, and so on. If evolution is true, then, salvation itself is in jeopardy, for how can Revelations be true if the rest of the Bible is not? To protect children from the doctrine of evolution is to save their souls. Obviously, this motivates powerfully.

A further motive is to prevent society from going downhill, inasmuch as strict creationists believe that evolutionary theory (because it supposedly denies God) removes the source of morality. As Henry Morris, arguably the most influential creationist of the late twentieth century, puts it,

> Evolution is at the foundation of communism, fascism, Freudianism, social darwinism, behaviorism, Kinseyism, materialism, atheism, and in the religious world, modernism and neo-orthodoxy. Jesus said "A good tree cannot bring forth corrupt fruit." In view of the bitter fruit yielded by the evolutionary system over the past hundred years, a closer look at the nature of the tree itself is well warranted today.[3]

This is not evolution as seen by scientists: the idea that the universe today is different from what it has been in the past and that change through time has occurred. In regard to organic evolution, the evolution of plants and animals, this means that living things share common ancestors in the past from which they are different. Darwin called organic evolution "descent with modification," and it is still a useful phrase. The contrasting perceptions of evolution—those of Morris and of mainline science—illustrate again how ideology shapes interpretation of empirical data.

## CREATION "SCIENCE"

The antievolution movement has had a long history in the United States, dating from the first introduction of Darwin's ideas during the latter part of the nineteenth century. Creation "science" is only a recent manifestation of this antievolutionism. Creation "science" is a movement of largely biblical literalist Christians who seek to get evolution out of the public school curriculum. They differ from other antievolutionists in their attempt to demonstrate the truth of a literal biblical interpretation of Genesis using data and theory from science—not just through theology. The young-earth creationists are the most numerous, but many old-earth creationists use scientific arguments as well. Although attempts to "prove" the literal truth of the Bible have been around for centuries, the most recent version of this approach hails from the mid-1960s, stimulated by the publication of John Whitcomb and Henry Morris's *The Genesis Flood* and the founding of both the Creation Research Society and the Institute for Creation Research. Why the 1960s? The answer is simple: the post-Sputnik science education panic of the late 1950s and 1960s resulted in improved textbooks that returned evolution to the curriculum at levels not seen since before the Scopes trial of 1925.[4] Giving an extra nudge to the process was the Supreme Court case of *Epperson v. Arkansas*, which overthrew antievolution laws such as Tennessee's under which John T. Scopes had been tried.

With evolution no longer liable to be banned, antievolutionists developed a strategy to ameliorate its "evil effects" by teaching biblical Christianity alongside it. Because the First Amendment of the Constitution clearly disallows advocating sectarian religious views in the classroom, "scientific" creationism was developed to be an "alternate scientific" view that could be taught as a secular subject. During the 1970s and early 1980s, creationists campaigned to pass "equal time" laws wherein creation science would be mandated whenever evolution was taught. This approach had to be abandoned in 1987 when the Supreme Court ruled in *Edwards v. Aguillard* that creationism was inherently a religious concept, and that to advocate it as correct, or accurate, would violate the Establishment Clause of the First Amendment. But Justice Brennan's decision left a couple of loopholes that antievolutionists have been exploiting ever since.[5] Subsequently, antievolutionism has evolved into new forms which are characterized by the avoidance of any variant of the "c word." Phrases like "intelligent design theory," and "abrupt appearance theory" are used instead of "creation science," "creationism," and related terms. I call this newest stage of antievolutionism "neocreationism."

## NEOCREATIONISM

The neocreationism tactic continues the "equal time" for creation and evolution of the creation science era, but with some new wrinkles. A popular neocreationist variant of creation science is "intelligent design theory" (IDT), a lineal descendant of William Paley's "argument from design." William Paley's 1802 *Dialogues Concerning Natural History* attempted to prove the existence of God by examining his works. Paley believed in the traditional God of the literal reading of Genesis: the creator who had produced a perfect world in which everything had its purpose. He used the metaphor of a watch to demonstrate how observing the perfection of nature "proved" there was a God, a "Divine Watchmaker." If you found an intricately contrived watch, he argued, it was obvious that such a thing could not have come together spontaneously; the existence of a watch implied a watchmaker who had designed the watch with a purpose in mind. Similarly, as there was order and purpose and design in the world, so naturally there must be an omniscient designer. The existence of God was proven by the presence of order and intricacy.[6]

The vertebrate eye was Paley's classic example of design in nature, and was well known to educated people of the nineteenth century. In response,

Darwin deliberately used the vertebrate eye in *The Origin of Species* to demonstrate how complexity and intricate design could come about by natural selection. Modern day IDT uses the vertebrate eye and similar structural wonders to demonstrate how evolution could not have possibly occurred, ostensibly because complexity of this sort could never have occurred "by chance." In fact, one will find not just the vertebrate eye, but the structure of DNA or cytochrome *c* held up as "too complex" to have evolved "by chance." Also, post-Paley information-theory is woven into the attempt to prove that evolution did not occur. (Or course, no evolutionist argues that complex structures evolve "by chance," but "chance" is synonymous with "evolution" in creationist literature.)

The best-known statement of IDT is a book by Percival Davis and Dean Kenyon, *Of Pandas and People*, written as a supplement for high school biology courses. Instead of advocating the more familiar creationist "teach evolution, but also teach creation science," *Pandas* offers the notion that teachers should "teach evolution, but also intelligent design theory." In content, there is nothing in IDT that has not already been expressed in earlier creation science literature. "Intelligent design theory" is merely a euphemism for creation science. A discussion of the promotion and use of *Pandas* is in Larson.[7]

## EVIDENCE AGAINST EVOLUTION

A second neocreationist approach not only avoids the use of the term "creation science," but eschews euphemism completely, coming right out with the essence of antievolutionism. The argument here is "teach evolution, but also teach the evidence against evolution." The "evidence against evolution" approach, of course, makes little scientific sense: it is rather like requesting that "evidence against heliocentrism" be presented when the solar system is taught. There is no scientific evidence against evolution. "Evidence against evolution" has always been synonymous with creation science, which specializes in finding anomalous tidbits in the scientific literature that appear to "prove" that evolution did not occur.[8]

In my job as director of a nonprofit clearinghouse for information on the creation/evolution controversy, I receive information on the full gamut of attacks on evolution. For example, during the last six months, I have received requests from New Hampshire for information to help keep old-fashioned scientific creationism out of a school district; requests from Louisiana and Texas for information on "intelligent design theory;" and requests from Ohio and California for assistance in combating "arguments against evolution," among other requests. The Epperson and Edwards Supreme Court decisions have only ended the attempts to outlaw the teaching of evolution, and to mandate the teaching of creation science alongside evolution. They have not ended attempts to prevent students from learning evolution, or even attempts (of individual teachers) to teach creation science.

Even though my office still handles numerous calls about creation science, given trends already apparent, I think that in the future the influence of creation science will wane in favor of more sophisticated neocreationist attacks on evolution. Teachers and school boards will face more pressure to teach not creation science, but "intelligent design theory," or "evidence against evolution." I say this primarily because these approaches are less vulnerable to legal challenges. Intelligent design theory and evidence against evolution do not sound as blatantly sectarian as something with the word "creation" in it. They are merely bad science, and the First Amendment protects against the establishment of religion, not against bad science. If creation science is primarily the result of the efforts of young-earth special creationists such as those at the Institute for Creation Research (ICR) and the Bible Science Association, neocreationism is dominated by more moderate creationists, many of whom are conservative Christian antievolutionists operating at the university level. These are generally "old-earth" creationists, who do not quibble about the age of the earth (and who therefore appear more reasonable to the general public), but who nonetheless deny that evolution took place. I believe their influence will increase: although the ICR reaches thousands of individuals at their "back to Genesis" rallies and through their extensive publishing efforts, these neocreationist faculty members are influencing the attitudes of future college graduates. College graduates vote at higher levels than nongraduates, are the next journalists and politicians, and in general are more influential in shaping public opinion. These academically based "Christian scientists" are far less well known than their young-earth counterparts.

## "CHRISTIAN SCIENTISTS" AT THE UNIVERSITY

By "Christian scientists" I do not mean followers of Mary Baker Eddy, but Christians who are also practicing scientists. How many are there? In the general public, the percentage of self-identified Christians is eighty-six percent.[9] Is the percentage among scientists this high? I do not know of any reliable recent surveys on the religious inclinations of scientists, but even if only half the scientist corps identifies itself as Christian, the number of scientists who are believers would be substantial. As is the case with Christian theology, the degree of belief and specific doctrinal acceptance doubtless varies greatly among these academics. My comments will concern the fraction professing conservative Christianity, a term that overlaps with evangelistic Christianity and fundamentalist Christianity, in referring to belief in a personal God and a belief in the inerrancy (and/or literal truth) of the Bible. I also speak primarily of conservative Christians who teach at secular colleges and universities, not religious ones.

Many of these are associated with Christian Leadership Ministries, an affiliate of Campus Crusade for Christ, an organization that targets college and university faculty members. CLM claims to be active on more than 800 United States campuses. Its newsletter, *The Real Issue,* encourages faculty members to stand up for their academic freedom to express religious beliefs. But the expression of religious beliefs can take many forms, from benign to clearly illegal. In the mildest form of advocacy, religious belief might be expressed (as are many other opinions) in time-honored fashion by proclamations taped to a professor's office door. Conservative Christians can argue that if colleagues can put up gay-rights literature on their office doors, they should be able to put up Christian literature without being harassed. On the other hand, in a far less acceptable effort, some professors try to convert students to Christianity in the classroom, in violation of the First Amendment's Establishment clause.

Two court cases concerning classroom proselytizing have arisen. Phillip A. Bishop, professor of exercise physiology at the University of Alabama, was accused of proselytizing students in class, and teaching "intelligent design theory" in an optional class held outside of regular class time. In this forum, he expressed his views on the "evidence of God in human physiology." Students did not have to attend, but as the special class was offered just before the final exam, at least some students felt coerced. The dean of the college instructed Bishop to quit witnessing in class and not to have extraclass meetings in which religious views of the subject matter were discussed.

Bishop sued his institution on free speech and academic freedom grounds, and won at the Federal District Court level.[10] The Appeals Court, however, reversed the decision, declaring that a classroom, during instructional time, was not an open forum, and that the University can reasonably restrict Bishop's speech during that time. The appeals court did not decide against Bishop on the grounds that he was violating the Establishment clause, but on the grounds, narrower in some ways, of the right of the University to set curricula.

In the second case, in 1991, Dilawar Edwards, an education professor at California University of Pennsylvania, used fundamentalist and religious-Right publications (including creation science books) to "balance" established, secular instructional materials in an educational methods course. Students claimed that he did not teach media resources, the topic of the course, but instead harangued them about how secular humanism in public schools violated Christian principles. The dean of the college told him to "cease and desist using doctrinate material of a religious sort" in the classroom. Edwards sued the college, claiming academic freedom to choose whatever materials he wants to use. The case is still pending.

These cases bring up important concerns regarding academic freedom for religiously-based views. It is clear that teachers at the K–12 level have strictly limited academic freedom regarding religion: they may not express religious views to students because of their special position as authority figures in relation to a "captive audience" of younger students. But faculty authority over adult college students is far weaker (though certainly not absent when grades are involved). Usually college students have options allowing them to avoid particular professors and classes. There are also no laws requiring college attendance such as there are at lower educational levels, and thus college students are not "captive" in the sense that kindergarten through twelfth grade students are. What, then, are the limits of academic freedom and religious speech at the university?

The Bishop case concluded that academic freedom is not absolute at the university level when it comes to religious speech: a professor does not have the right to proselytize students. But what about activities intermediate between posting Biblical literature on one's door, on the one hand, and emphatic "witnessing" in class on the other?

Some conservative Christians distinguish between proselytizing and presenting information "from a Christian perspective." As pointed out elsewhere in this volume, many academics openly teach "from the point of view of" some ideology. If it is considered legitimate for a professor to announce in, say, a history class, that he or she will be teaching the subject from a feminist, or Marxist perspective, should not a Christian professor be permitted to make a similar declaration? The point made by conservative Christians like George Marsden[11] is that sauce for the goose is sauce for the gander: if the university does not protest when Marxist, feminist, gay, or Afrocentric perspectives inform the presentation of scholarly material, it should not protest a Christian professor's analogous methods.

## CHRISTIAN SCHOLARSHIP, IDEOLOGY, AND SCIENCE

Christianity is clearly an ideology, but so arguably are Marxism and feminism, and perhaps some other "isms." The argument is made that Marxism and other "isms" are not only ideologies, but legitimate scholarly approaches to knowledge, as well. Can an ideology be scholarship? The question strikes at a fundamental tenet of scholarship and speaks directly to one of the themes of this volume. If knowledge is to be distinguished from opinion, there must be sources of information beyond the individual. Information comes in many guises, and some of it is contradictory. Scholarship requires the weighing and judging of information in order to come to reliable and valid conclusions. Ideology may or may not be based on scholarship, but what makes ideologies troubling is that they also include a component of belief that always has the potential to overshadow the scholarship. There is always the possibility that empirically based knowledge will be supplanted or denied because of ideology.

It is theoretically possible to have a scholarly feminist perspective, a scholarly Marxist perspective, a scholarly environmental science perspective. The sorts of radical feminism, radical multiculturalism, and radical environmentalism criticized in other essays in this volume are not truly scholarly precisely because they elevate ideology over empirical evidence and logic. In fact, some of them revel in their rejection of logic and reason. When the claim is made that black Egyptians sailed to Mesoamerica and taught the Olmec to build step pyramids, clearly the ideology of black superiority is being promoted at the expense of actual empirical evidence.

Now we come to another question. Ideology and scholarship are at best uncomfortable bedfellows. Can a supernatural ideology be scholarship? Can Christian scholarship meet the standards of scholarship in this sense? We are talking now not about posting religious items on one's office door, much less trying to convert students to Christianity, but something called "teaching from a Christian perspective" as a scholarly endeavor. Does a "Christian perspective" fall inherently into that class of views wherein ideology is the ultimate determinant of truth? Most conservative Christians who have complained that views important to them are given short shrift on campus have not been explicit about how their religious views might inform their scholarship, as opposed to their free speech outside of the classroom. George

Marsden suggests one way a Christian perspective might influence a teacher's presentation in social science:

> Such unproven assumptions [that there is no created order] have a greater effect in the humanities and social sciences than in the largely technical disciplines and natural sciences. Since most scholars today will tolerate only a naturalistic understanding of humans, they have no basis for interpreting moral standards except as survival mechanisms—that is, as social constructions suiting the needs of a particular community. The result of this view is, inevitably, moral relativism of one sort or another. Religiously informed scholars, on the other hand, while acknowledging the cultural forces that shape beliefs, are also open to the possibility of permanent, universal standards of right and wrong established by a creator.[12]

Marsden himself seems to imply that the "Christian perspective" may be more relevant in the more subjective sciences. But what of a "Christian" view of natural science? In truth, all true scholarship requires the collection of information and its logical evaluation against agreed-upon standards. Science may, however, be distinct from other scholarship because of the degree of emphasis it places on empirical information and on whether a given explanation "works." Obviously, there is subjectivity in science, and it is no great revelation to state that historical, political, and social factors influence the course of science, and even the conclusions reached. Science is a human endeavor, which means it is cumbersome, subject to human egos, messy, and slow. At any given time, much of it will be in error. But the saving grace of science is that it is open ended, and conclusions are tentative, subject to later revision. Explanations that do not hold up over time, or that are contradicted by other better data, are eventually rejected. (Sometimes we have to emphasize the "eventually!") Science's great power resides in its ability to reject explanations based on logic and empirical evidence, rather than by virtue of mere opinion, authority, or assertion. It is an untidy procedure, but it has given us enormously more knowledge of the natural world than obtained by any other society in the history of our species. Thus, even granting that "scholarship is scholarship," science is and must be held to different standards from other forms of knowledge. It is harder to argue the "strong constructivist"[13] perspective when discussing cell permeability than when discussing, say, the Civil War. One may correctly argue that the "truth" of Reconstruction varies considerably between southern blacks and southern whites, but there is no Afrocentric Krebs cycle that contrasts with a Eurocentric Krebs cycle. Except for those who deny objective reality, the logical-empirical methodology of science does allow the rejection of some ideas and the tentative acceptance of others.

Christianity is clearly an ideology, and it is apparent from many presentations in this volume that when ideology is given precedence over evidence, science suffers. The extensive literature of creation "science" demonstrates the futility of trying to do empirical science after conclusions based on "revealed truth" have already been posited in advance as unassailable. This fact was recognized by the Federal District Judge in the *McLean v. Arkansas* case.

Is there a "Christian perspective" that is not as extreme as the blatant distortions of science promulgated by the Institute for Creation Research? I have not found any examples that are free from the inclination to subvert science in order to defer to ideology. One largely old-earth creationist proposal, for instance, is that there are two different kinds of science: "operations science" and "origins science."[14] A distinction is made between phenomena that occur "with regularity" and those that occur "singularly." It is argued that regularly occurring phenomena can be studied in the fashion of normal science, or "operations science." But one-time phenomena, such as the Big Bang and other evolutionary events, come within what creationists call "origins science."

Of course, there are differences between the study of repeatable events and nonrepeatable ones, but mainstream philosophers of science agree that phenomena of historical sciences like geology, paleontology, and astronomy can be studied scientifically, and even experimentally. Mount St. Helens erupted as a singular event, but this does not exclude a science of volcanoes. Similarly, even if bears and dogs split from a common ancestor only once, we can still evaluate the hypothesis that bears and dogs are closely related against empirical evidence (from fossils, comparative anatomy, biochemistry, etc.). We can also learn about the processes that influence evolution by looking at the evidence for other such splits. There are many ways to study events of this type scientifically.

Creationists often add an additional factor to this bimodal division of the scientific world, one that I believe sheds light on why the division was invented in the first place. Specifically, they allow the explicit intrusion of the supernatural into scientific explanation. Geisler proposes that corresponding to the two kinds of science, there are two kinds of causation: primary causes and secondary causes. Operations science relies properly on secondary causes, but origins science is allowed to invoke primary causes. Thaxton *et al.* refer to "primary cause" more bluntly as the "God hypothesis," and agree that in operations science, "the appeal to God is quite illegitimate, since by definition God's supernatural action would be willed at His pleasure and not in a recurring manner."[15] But when dealing with "origins science," it is not only permissible, but essential to allow recourse to supernatural causation (i.e., miracles).

Few would argue with excluding miracles from "operations science," but proponents of this artificial division do not make a solid case for resorting to miracles in origins science. Arguably, nonrecurrent events may be more difficult and challenging to study than repeated events, but that in itself is insufficient to require resorting to the supernatural.

Science as it is practiced in the late twentieth century has, on the contrary, emphatically eschewed invoking miracles to explain natural phenomena. Creationists recognize that they are outside of the mainstream in their insistence on supernatural causation, but claim that the rejection of the supernatural in modern science is a function of "naturalism" (materialism), a philosophy that defines reality only in terms of material causes.[16] Because evolutionary scientists supposedly are caught up in a metaphysical viewpoint that rejects the possibility of a creator, creationists contend that evolutionists are unable to countenance evidence for supernatural intervention in the history of life.

Actually, modern science has omitted the supernatural for methodological, not philosophical, reasons. It is not that scientists have an axe to grind with respect to theism. Rather, we simply get better explanations by ignoring the possibility of supernatural intervention or causation. Much confusion exists between materialism as a *philosophy,* and the *methodological* materialism that informs all of modern science. It is logically possible to decouple philosophical and methodological materialism, and individual scientists who are believers do it all the time. Gregor Mendel was certainly not a metaphysical naturalist, but he developed his understanding of the rules of heredity using methodological materialism. Alternatively, I personally am a philosophical materialist, not a believer, but when I teach science to students, I leave my irrelevant philosophy out of my course. I stress methodological materialism as a tool to understand the natural world better, not as a foundation for a personal belief.

But creationists would say that Mendel's laws are examples of "operational" science, which elicits the question "what are the topics of origins science?" In my reading, I find that "origins science" is limited to subjects that have theological importance. Conservative Christian theology is concerned with the special creation of the earth, the special creation of life, the special creation of animals and plants, and the special creation of human beings, all by a personal God who had an ultimate purpose in mind. As a result, "origins science" focuses on the Big Bang (the origin of the universe), the origin of life, the origin of the "kinds" of plants and animals, and the origin of humans.

## THE BIG BANG

Creationists recently have split over the significance of the Big Bang, with "old earth" creationists like Hugh Ross of Reasons to Believe Ministries promoting the idea that the Big Bang is evidence for creation, and "young earth" creationists such as Henry Morris and others from the Institute for Creation Research arguing for a more Biblical, literal, special creation view that the entire physical universe came into being in only a few days. Some conservative Christian scientists thus accept modern physics, chemistry, and radiometric dating, but most have difficulty with modern biology, especially descent with modification. The other three "origins" topics all involve living things, which causes considerable discussion among these more moderate creationists.

Assuming that it is an open question whether life might have occurred through natural or supernatural causes, science, because it is limited to understanding the world only by natural processes, can only attempt to explain the origin of life through natural forces. By definition, special creationists do not accept the possibility of a naturalistic evolution of life from nonlife, though theistic evolutionists may. Theologically liberal creationists allow that God might have devised the first replicating molecule, and that descent with modification occurred thereafter.

## ORIGIN OF LIFE

Origin of life research has always been a mainstay of young-earth creationism, and is also of considerable concern to old-earthers. It seems to be the "soft underbelly" of evolutionary biology, as there is not yet a consensus on precisely how life came about. It is a quite active area of research, however, and there are several vigorously competing explanations. Creationists, however, view the origin of life as an intractable mystery that will never be solved because it is "too complex" to explain through natural causes. They appear to assume that because it is not yet explained, it never will be.

The issue is theologically loaded: if life itself can be explained without recourse to supernatural intervention, to some (though not all) it will appear that the existence of God is thereby refuted. The idea that God could have willed life to have come about naturally (under the notion of God as ultimate force in the universe) is tainted with Deism so far as conservative Christians are concerned. It implies that God is a distant prime mover rather than the personal God of their theology. As a result, every possible breakthrough in the scientific analysis of the origins of organic life (the postulate of an RNA world, for example) is attacked as "too improbable." It is difficult to see how a "Christian perspective" on origin of life research could be scholarly, given this manifest unwillingness even to consider—much less evaluate objectively—opposing opinions.

## PLANT AND ANIMAL DIVERSITY

The explanation of the diversity of plants and animals also varies among conservative creationists. The Bible states that "kinds" (Heb.: *baramin*) were created, but "kinds" is a poorly defined term that may denote species, genus, or family. Much creationist literature concerns the definition of a "kind," and how much "variation within the kind" can take place. For the young-earth creationists, this becomes critical because of flood geology. Noah was instructed to take seven pairs of every clean kind of animal and five pairs of every unclean kind, so that even if the Ark was the size of the Queen Mary (as claimed), the number of "kinds" that could have been squeezed onto this ship is rather limited. Thus, if the "cat kind" is viewed as equivalent to the family Felidae, the "seven pairs taken" might have diverged, after the Ark landed, into lions, bobcats, lynxes, pumas, housecats, and other feline species: an example of "variation within a kind." Other conservative Christians might even more broadly allow a "kind" to encompass all carnivora, and for post-Noachian descent with modification to have taken place—in animals, anyway. Most, however—even the most flexible—draw the line at human evolution.

## HUMAN EVOLUTION

The notion of human evolution postulates that humans descended with modification from nonhuman ancestors. Humans and living apes shared a common ancestor. This view contrasts strongly with the idea that God specially created humans in his image. Even the most liberal of the conservative Christians have difficulty accepting human evolution, though many will accept that nonhuman animals evolved. Conservative Christianity is based on an individual, personal relationship with God. How, it asks, could humans have a special place with God if they result from the same processes that brought about the rest of nature? Humans—at the very minimum—have to have had a separate creation. And yet scientists infer from anatomy, biochemistry, behavior, the fossil record, and even embryology that we shared a common ancestor with modern chimps and gorillas.

It is perhaps most obvious in their insistence upon a separate creation for humankind that conservative Christians present a clear example of how Christian ideology affects the interpretation of scientific data. "Origins science" itself focuses on theologically important topics, allowing the intervention of the supernatural on theological, not scientific grounds.

This is illustrated by the fact that there is a direct correlation between the degree of theological conservatism and the amount of scientific evidence for evolution that is accepted. The most theologically conservative accept only physics and chemistry; the less conservative accept some of biology (for example, some evolution of nonhuman forms), and only the most theologically liberal accept human evolution. Ironically, the scientific evidence for the evolution of our species is far better than that for most other mammalian genera. I believe human evolution is not accepted predominantly because of ideological commitments.

There may indeed be a "Christian perspective" that is scholarly, but it will be difficult to apply it to evolutionary studies and still remain scientific. Whether there are other areas within science where a "Christian perspective" can be applied is an empirical question that has not yet been answered. But as with Marxism, feminism, Afrocentrism, environmentalism, or any other ideology, there is a great risk of subordinating evidence to belief. From the evidence so far, a supernatural ideology is even more fraught with this risk.

## TEACHING EVOLUTION: DEFUSING THE RELIGIOUS ISSUE

Currently, the most active and effective antievolutionists are the grassroots, young-earth proponents from the ICR, the Bible-Science Association, and affiliated groups. It is my contention that more moderate forms of antievolution, such as some of those discussed above, will be having a proportionately greater effect in the future partially because of their presence on secular campuses where they are able to influence the next generation of leaders.

A high percentage of citizens reject evolution (currently, polls consistently show forty-seven percent to forty-nine percent of Americans deny that humans evolved from earlier forms).[17] Evolution, however, is a basic component of science, and essential to the understanding of biology and geology. One is not scientifically literate if one does not understand evolution. I would hope that scientists would do what they can to encourage individuals to accept evolution as science.

Evolution—accurately presented—needs to be in precollege as well as college education. There are a number of things that scientists, especially those at colleges and universities, can do to teach it better. Most of these have little to do with the actual scientific content of the subject, but rather with improving the perception of evolution by the public.

I speak before many public audiences and do a fair number of radio call-in programs. The most common response I get when I ask people "what does evolution mean," is "man evolved from monkeys." The second most common answer is, "evolution means that you can't believe in God." The perception that religious faith and acceptance of evolution are incompatible is, in my experience, widespread. One source of this confusion comes from antievolutionists. Leaders of the Institute for Creation Research proclaim that "one can be either an evolutionist or a Christian."

As mentioned before, reliable polls place the number of self-identified Christians in the United States at upwards of eighty-six percent,[18] the highest percentage of believers of any developed country. If scientists give Americans the same choice as the ICR proposes, there will be scant interest in teaching and learning evolution!! This, to say the least, would be highly detrimental to the science literacy of our nation. To encourage people to learn about evolution, it is necessary to allow them to retain their faith.

This follows from the logic of the limitations governing modern science. It is also good strategy, if our goal is to increase the amount of science literacy in our nation.

Not all creationists are extreme in their rejection of evolution. To many moderate creationists, evidence from science demonstrates that the universe is old, the earth is old, plants and animals evolved, and even human beings had earlier ancestors different from them. They merely want to draw the line at the assumption of evolutionary materialism, the philosophy that evolution (and its material causes) are not only sufficient for explaining the presence and form of the modern universe, but proof that there is no supernatural intervention. The fear that teachers are serving up not just science, but materialism in their lessons on evolution inspires much of the opposition to evolution that I have encountered—and not just from young-earthers who are at the fringes of modern science, but from more moderate conservative Christians who are within the modern scientific mainstream in many respects.

In dealing with hundreds of elementary and high-school teachers, I have found that the number of teachers that actually promote philosophical materialism along with evolution is vanishingly small. At the college level, it is more significant, but it is still not general. Vocal proponents of evolutionary materialism such as William Provine at Cornell, Paul Kurtz at the State University of New York, Buffalo, and Daniel Dennet at Tufts vigorously argue that Darwinism makes religion obsolete, and encourage their colleagues to argue likewise. Although I share a similar metaphysical position, I suggest that it is unwise for several reasons to promote this view as "the" scientific one.[19]

First, science is a limited way of knowing, in which practitioners attempt to explain the natural world using natural explanations. By definition, science cannot consider supernatural explanations: if there is an omnipotent deity, there is no way that a scientist can exclude or include it in a research design. This is especially clear in experimental research: an omnipotent deity cannot be "controlled" (as one wag commented, "you can't put God in a test tube, or keep him out of one."). So by definition, if an individual is attempting to explain some aspect of the natural world using science, he or she must act as if there were no supernatural forces operating on it. I think this methodological materialism is well understood by evolutionists. But by excluding the supernatural from our scientific turf, we also are eliminating the possibility of proclaiming, via the epistemology of science, that there is no supernatural. One may come to a philosophical conclusion that there is no God, and even base this philosophical conclusion on one's understanding of science, but it is ultimately a philosophical conclusion, not a scientific one. If science is limited to explaining the natural world using natural causes, and thus cannot admit supernatural explanations, so also is science self-limited in another way: it is unable to reject the possibility of the supernatural.

Scientists, like other teachers, must be aware of the difference between philosophical materialism and methodological materialism and not treat them as conjoined twins. They are logically and practically decoupled. Furthermore, if it is important for Americans to learn about science and evolution, decoupling the two forms of materialism is essential strategy.

To further defuse the religious issue, scientists can be more careful about how they use terms. For example, evolutionists sometimes confuse the evidence we have for considerable contingency during the course of evolution with evidence for a lack of ultimate purpose in the universe. Futuyma writes,

> Perhaps most importantly, if the world and its creatures developed purely by material, physical forces, it could not have been designed and has no purpose or goal. . . . Some shrink from the conclusion that the human species was not designed, has no purpose, and is the product of more material mechanisms—but this seems to be the message of evolution.[20]

G. C. Simpson is regularly quoted with dismay by creationists as saying "Man is the result of a purposeless and natural process that did not have him in mind. He was not planned."[21] A theist might respond that we do not know what God's purpose is or what he planned. It is possible that if there is an omnipotent, omniscient deity, it was part of its plan to bring humans and every other species about precisely in what seems to us the rather zig-zag, contingency-prone fashion that the fossil evidence suggests. Of course, this would be a theological statement, but that, indeed, is the point. Saying that "there is no purpose to life" is not a scientific statement. We are able to explain the world and its creatures using materialist, physical processes, but to claim that this then requires us to conclude that there is no purpose in nature steps beyond science into philosophy. One's students may or may not come to this conclusion on their own; in my opinion, for a nonreligious professor to interject his own philosophy into the classroom in this manner is as offensive as it would be for a fundamentalist professor to pass off his philosophy as science.

Another way scientists can help defuse the religious issue is by being explicit about what we can and cannot say about design. As with "purpose" in nature, "design" is largely a theological position. But many evolutionists, ever mindful of William Paley's "argument from design," stress not the perfection of structure in nature, but the often erratic, cobbled-together-from-what-is-available nature of many structures, such as the panda's thumb or the anglerfish's lure. This is an important point to make, and helps students realize that natural selection does not result in "perfection" of structure. When we look at either the fossil record or the "design" of many structures, it is difficult to see evidence of advance planning. In terms of proximate cause, then, design in nature is not apparent. But final cause is, as we discussed, outside the boundaries of science. Allowing a student of conservative religious views to continue to believe in final design or purpose need not detract from that student's understanding of the evidence for the contingency of proximate cause. Separating the two types of causation may, indeed, keep a student from being "turned off" by evolution, especially if he or she comes into the class with the idea that "evolution means you cannot believe in God."

## CONCLUSION

Creationism offers some interesting contrasts to other topics discussed in this volume. Creationists (antievolutionists) are not opponents of science, per se. They appreciate reason and believe they are being objective, but just like postmodernists, they reject the Enlightenment traditions that brought about modern scientific epistemology. The distinction is that they are premodernists, rather than postmodernists.[22] They illustrate extremely well one of the concerns that supporters of science and reason have in the face of so much antiscience: the replacement of empirical and logical evidence with ideology and dogmatic belief.

The antievolution movement has as its prime motive the fear that religion is under attack by the study of evolution. Great efforts have been made over the last seventy years to "shield" students from evolution's "evil effects." Like a neutral mutation that replaces the wild type if no selective forces are arrayed against it, antievolution will prevail if not opposed by academics at universities (and even more importantly, at the grass-roots level of the school board and the individual school). To oppose antievolution successfully, scientists need to understand what motivates the movement, and also to recognize that the movement is not monolithic. There is great variation among conservative Christians in the degree to which they reject evolution. Those who spurn evolution out of fear that a hegemonic materialist philosophy is being promoted at the expense of their religion are very different from and more "reachable" than those who reject evolution because of fancied scientific evidence against it.

I suggest that scientists can defuse some of the opposition to evolution by first recognizing that the vast majority of Americans are believers, and that most Americans want to retain their faith. It is demonstrable that individuals can retain religious beliefs and still accept evolution as science. Scientists should avoid confusing the methodological materialism of science with metaphysical materialism. Also, scientists should avoid making theological statements (such as those concerning ultimate purpose in life, or final cause) in the context of their scientific discussions.

Antievolutionism is perhaps the most successful form of irrationalism besetting the American public, though it currently is not represented at the university level to the same degree as some of the other "isms" discussed in this volume. It would be nice for American science literacy if its prevalence in the "real world" outside the university were to match its prevalence within. The efforts of university and professional scientists will be critical in this regard.

## NOTES

1 Ronald L. Numbers, *The Creationists.*
2 See Betty McCollister, *Voices for Evolution.*
3 Henry Morris, *The Twilight of Evolution,* p. 24.
4 Numbers, *The Creationists.*
5 Eugenie C. Scott, "The Evolution of Creationism: The Struggle for the Schools."
6 Richard Dawkins, *The Blind Watchmaker.*
7 Erik Larson, "Darwinian Struggle: Instead of Evolution, a Textbook Proposes 'Intelligent Design.'" See Padian, Ruse & Skoog in Liz Rank Hughes, ed., *Reviews of Creationist Books* for reviews.
8 Douglas Futuyma, *Science on Trial.*
9 Ari L. Goldman, "Portrait of Religion in U.S. Holds Dozens of Surprises."
10 *Bishop v. Aranov.*
11 George Marsden, *The Soul of the American University: From Protestant Establishment to Established Nonbelief.*
12 George Marsden, "Religious Professors Are the Last Taboo."
13 Paul Gross & Norman Levitt, *Higher Superstition: The Academic Left and Its Quarrel with Science.*
14 Charles B. Thaxton, Walter L. Bradley & Roger L. Olsen, *The Mystery of Life's Origin: Reassessing Current Theories*; Norman Geisler & Kerby Anderson, *Origin Science: A Proposal for the Creation-Evolution Controversy.*
15 Thaxton *et al.*, *The Mystery of Life's Origin,* p. 203.
16 Phillip E. Johnson, *Evolution as Dogma: The Establishment of Naturalism* (published under the auspices of the Foundation for Thought and Ethics).
17 American Museum of Natural History poll of 1,255 individuals conducted by Louis Harris and Associates, 1994; Eugenie C. Scott, "Anti-evolutionism, Scientific Creationism, and Physical Anthropology;" Christopher Toumey, *God's Own Scientists.*
18 Goldman, "Portrait of Religion."
19 Eugenie C. Scott, "Science and Christianity Are Compatible—with Some Compromises."
20 Douglas Futuyma, *Science on Trial.*
21 George Gaylord Simpson, *The Meaning of Evolution,* p. 345.
22 Raymond A. Eve & Francis B. Harrold, *The Creationist Movement in Modern America.*

## REFERENCES

*Bishop v. Aranov.* 1991 926 F. 2d 1066; 1991 U.S. App. Lexis 4118.

DARWIN, CHARLES. *The Origin of Species.* London: John Murray, 1859.

DAVIS, PERCIVAL & DEAN KENYON. *Of Pandas and People.* 2d edit. Dallas, TX: Haughton Publishing Co., 1993.

DAWKINS, RICHARD. *The Blind Watchmaker.* New York, NY: Norton, 1987.

*Edwards v. Aguillard.* 1987 482 U.S. 578, 55 *U.S. Law Week* 4860, 107 S. Ct. 2573, 96 L., Ed 2nd 510.

*Epperson v. Arkansas.* 1968 393 U.S. 97, 37 *U.S. Law Week* 4017, 89 S. Ct. 266, 21 L., Ed 228.

EVE, RAYMOND A. & FRANCIS B. HARROLD. *The Creationist Movement in Modern America.* Boston, MA: Twayne Publishers, 1991.

FUTUYMA, DOUGLAS. *Science on Trial.* 2d edit. Sunderland, MA: Sinauer, 1995.

GEISLER, NORMAN & KERBY ANDERSON. *Origin Science: A Proposal for the Creation-Evolution Controversy.* Grand Rapids, MI: Baker Book House, 1987.

GOLDMAN, ARI L. "Portrait of Religion in U.S. Holds Dozens of Surprises." *New York Times*, April 10, 1991, p. A1.

GROSS, PAUL R. & NORMAN LEVITT. *Higher Superstition: The Academic Left and Its Quarrel with Science.* Baltimore, MD: Johns Hopkins Press, 1994.

HUGHES, LIZ RANK, ed. *Reviews of Creationist Books.* Berkeley, CA: National Center for Science Education, 1992.

JOHNSON, PHILLIP E. *Evolution as Dogma: The Establishment of Naturalism.* Dallas, TX: Haughton Publishing Co., 1990.

LARSON, ERIK. "Darwinian Struggle: Instead of Evolution, a Textbook Proposes 'Intelligent Design.'" *Wall Street Journal*, November 14, 1994.

MATSUMURA, MOLLEEN, ed. *Voices for Evolution.* Rev. edit. Berkeley, CA: National Center for Science Education, 1995.

*McClean v. Arkansas Board of Education.* 1982 529 F. Supp. 1255, 50 *U.S. Law Week* 2412.

MARSDEN, GEORGE. "Religious Professors Are the Last Taboo." *Wall Street Journal*, December 22, 1993, p. A12.

———. *The Soul of the American University: From Protestant Establishment to Established Nonbelief.* New York, NY: Oxford University Press, 1994.

MORRIS, HENRY R. *The Twilight of Evolution.* Grand Rapids, MI: Baker Book House, 1963.

NUMBERS, RONALD L. *The Creationists.* New York, NY: Alfred A. Knopf, 1992.

PALEY, WILLIAM. *Natural Theology.* 2nd edit. Oxford: J. Vincent, 1828.

SCOTT, EUGENIE C. "Anti-evolutionism, Scientific Creationism, and Physical Anthropology." *Yearbook of Physical Anthropology* 30 (1987) 21–39.

———. "The Evolution of Creationism: The Struggle for the Schools." *Natural History*, July 1994.

———. "Science and Christianity Are Compatible—with Some Compromises." *The Scientist*, January 9, 1995.

SIMPSON, GEORGE GAYLORD. *The Meaning of Evolution.* New Haven, CT: Yale University Press, 1967.

THAXTON, CHARLES B., WALTER L. BRADLEY & ROGER L. OLSEN. *The Mystery of Life's Origin: Reassessing Current Theories.* New York, NY: Philosophical Library, 1984.

TOUMEY, CHRISTOPHER. *God's Own Scientists.* New Brunswick, NJ: Rutgers University Press, 1994.

WHITCOMB, JOHN C. & HENRY R. MORRIS. *The Genesis Flood: The Biblical Record and Its Scientific Implications.* Philadelphia, PA: Presbyterian and Reformed Publishing Co., 1961.

# THE FLIGHT FROM REASON: THE RELIGIOUS RIGHT

LANGDON GILKEY

FLIGHTS FROM REASON take off in our day as frequently as planes from O'Hare. The most serious and dangerous of these retreats is, however, the one represented by the current Religious Right and its drive—now fast, now slow—towards theocracy, the rule of civil society by one religion, by its creed, its code, its leaders. There are two interrelated levels of this present attack on reason. One is the denial of the authority of science on scientific subjects (or of history or social science) in the name of religious truth—for example, creation science. Here revelation replaces scientific inquiry, reshapes scientific conclusions, and overrules scientific education. In effect this is the elimination of science. The second is the denial of rational reflection in other social matters—in the areas of law, economics, politics, morals, and, of course, religion. The essence of religious fundamentalism is its claim to absolute truth. A politically aggressive fundamentalism in turn expands this claim outward from the areas of religion and personal morals (as in historic fundamentalism) to all matters in society generally—of law, economic policy, literature, art, and so on. If reason means and requires open criticism, debate, and reflection, reason here has effectively vanished. Note that the reason that is *here* threatened is more than science. Reason here is "culture creating" reason—legal, political, social, artistic, philosophical, and religious reason. And it is on *this* level of rational reflection that science as an enterprise is itself dependent. When this level is gone, science is gone.

This expansion is the essence of late twentieth-century fundamentalism—as it has been of modern political ideologies. Formerly in our century fundamentalism had been *a*political; it claimed absolute authority in religious and moral matters, but only for its own adherents. It thus has seemed in the American context to be "nice," if a bit out of date, because it was limited by the First Amendment. Now it has moved out; it sees itself as representing Middle America, "normal" America, and, as it now says, it intends to make this a "Christian nation"—in creation science, its social agenda, its identification of conservative politics and economics with real Christianity; its horrendous support of uncritical nationalism, military intransigence, unconditional private property rights; and its as-yet-implicit racism and sexism.

Without the limitation of the First Amendment, fundamentalism naturally drives towards theocracy—illustrated now, as noted, among *us*, but also evident in Islam, in radical groups in Israel and in Hindu India, and most recently in Japanese sects (note the bizarre marriage of science, advanced technology, and absolutist Buddhism in Aum). When politics becomes absolute, then its implicit religious dimension appears; when religion claims absolute validity and seeks control over all of life, there is political theocracy. In either case, reason at best serves a particular creed, code, and style of life; as a critical, constructive, and so creative factor in culture it has succumbed.

This is a much more serious matter than our many current, but merely irritating, forms of cultural antiscience. When the epistemological, metaphysical, and moral assumptions of a liberal culture are eradicated and a theocracy reigns in its place, there is no science. This has happened continually in the twentieth century, sometimes in secular ideological forms, sometimes in religious theocratic forms. The latter are now on the rise across the globe, not least here at home. In each case this union of secular or religious ideology with the political has gobbled up the rest of culture, as in Germany, Japan, Russia, and Iran, always including scientists and even conservative politicians! Modern American fundamentalism does not stand apart from American culture, as it is so apt to claim; on the contrary it has absorbed and united to itself important elements of contemporary culture—technology, nationalism, militarism, property rights, radical individual liberties. It has absolutized these conservative forms of our common life, and then added to them its own moral and religious agenda. The result, if they achieve their aim, will be a conservative Christian politics, a conservative Christian moral code, and in our schools creation science and a literal biblical history. As Duane Gish told me after a debate in 1985, "Now we have the White House, a decade from now the Republican Party."

This union of politics and religion should be no surprise. It has characterized most of human history; and we moderns are not as different from that history as we once supposed. All serious politics, including humanist democratic politics, presupposes particular views of human beings, of authentic community, and of the nature of history's course. On these assumptions depend directly the more immediate principles, policies, and actions of any political community. These grounding assumptions are articulated through the use of certain essential cultural symbols, symbols that skirt on the religious. Thus when politics becomes frantic, it makes these presuppositions absolute and so veers into absolutist religion, as fascism showed. Almost all religion in turn seeks to enact its convictions and norms in all of life, as the many creative Christian and Jewish movements of social reform have shown. If they represent a majority power, however, such unions of politics and religion become theocratic, unless they are carefully limited from within by self-criticism (the "Prophetic principle") and from without by civil patterns of tolerance and the recognition of diversity (e.g., the First Amendment). Most present forms of Judaism and Christianity—most mainline churches and synagogues—respect these limits and have fought more sturdily than many other institutions against such theocratic movements.

Apparently, modern culture—modern scientific, technological, and industrial culture—continuously produces deep anxieties: fears arising from technological developments, fears of economic insecurity, terror at a seeming abyss of relativism, despair at rootlessness and alienation, and, probably most fundamental of all, the common loss of a firm confidence in progress—progress of the individual, of family, of society generally. All this represents a deep threat, the terror—and it *is* a terror—of the loss of assured meaning, an emptiness occasionally felt even by academics. Such anxieties, if heavy enough, breed fanaticism. Our creative engines of modern progress—science, technology and industry—have also generated a host of deep dilemmas and apparently insoluble problems. And it is these dilemmas and problems that *that* side of reason—technical reason, the reason embodied in inquiry—cannot begin to answer or resolve; for it is just this progressive development of knowledge and technique that has created these dilemmas. If, therefore, a scientific culture limits reason to inquiry and technology, this whole essential, grounding basis of culture is left void—and is thus soon enough filled with horrendous demons. The irony is that a scientific culture seems to be breeding the very forms of religion that will, if they prevail, eliminate science—just as the rootless diversity of culture engenders a religion intent on eliminating diversity.

This drive towards theocracy—the growing power and authority of the Religious Right—must be resisted: by science, which slept during the creationist controversy and has ignored the larger problem of reason (and so in the end is in danger of succumbing to it); by mainline religious communities, which up to now have largely been silent, fearful of alienating conservative members. But especially it must be resisted by the managerial and propertied classes, by moderate Republicans and their political leaders—and not least by the Right's elite articulators, the conservative intellectuals. These latter have been happy to welcome this unexpected support from the Religious Right against the liberal forces both fear; and they continue to defend their absolutist colleagues as harmless and eccentric. They are apparently confident they can use rather than be used by these new allies. So were the faculties, entrepreneurs, and established politicians of the Weimar Republic! All we need now to repeat that tragedy is for the Religious Right to join forces with the National Rifle Association and the militias. A half-century ago in America, the liberal left was finally forced to admit that there was, even for them, an "enemy on the left"—namely, orthodox communism. So now the present moderate Right must recognize this danger on its own right flank, lest they too be gobbled up as quickly as will that other intellectual elite, the liberal religionists, humanists, artists, and scientists represented in this volume.

# DOUBT, CERTAINTY, FAITH, AND IDEOLOGY

OSCAR KENSHUR

IN THE CURRENT STRUGGLE over the status of science and reason, even if one does not have a scorecard, one can tell which position a player plays merely by listening to what he or she says about the Enlightenment. The party that stands in defense of science and in opposition to irrational and antirational beliefs presents itself as being on the side of the Enlightenment. It tends to accept the Enlightenment's own characterization of itself as having achieved the benign separation between knowledge and superstition through the techniques of scientific rationalism. Unfounded traditional beliefs have been discarded in favor of the responsible investigation of things that can be observed, tested, and verified.

The party associated with fashionable relativizing theories challenges both the Enlightenment's claim to have found the way to objective knowledge and its claim to be progressive or benign. These theorists speak of the Enlightenment in terms of totalizing theories, foundationalism, and hierarchical or oppressive structures.

These two opposing groups, despite their differences, have certain things in common. In the first instance, both groups are, in varying degrees, heirs of the skeptical tradition. The emergence of scientific rationalism has been understood in terms of a kind of "mitigated skepticism," which employs techniques of skeptical doubt against certain kinds of knowledge claims and leaves the field to empiricist and experimental techniques.[1] The religious skepticism that we associate with the Enlightenment is often linked to skepticism regarding speculative philosophy—both being premised on the rejection of claims about transcendent or otherwise-unobservable reality—and this conjunction may be seen as a characteristic expression of mitigated skepticism. The anti-Enlightenment party, on the other hand, by dint of its refusal to allow Enlightenment science and reason to escape from the scourge of skeptical doubt, has been seen to hearken back to a more thoroughgoing skepticism that challenges all knowledge claims.

A second conspicuous similarity between the pro-Enlightenment and anti-Enlightenment parties is that they tend to characterize the specific epistemological stances that they favor as emancipatory. On the one side, Enlightenment science and reason are seen as freeing us from religious superstition or despotism, and providing us with knowledge and autonomy. On the other side, science and reason are depicted as instruments of domination; and the skeptical undermining of Enlightenment epistemologies is associated, often only implicitly, with the potential emancipation from this domination.

Nor is this difference in the interpretation and valorization of Enlightenment epistemologies merely symptomatic of the competing theoretical positions. Rather, it might be seen as part of their etiologies. Defenders of science and reason today trace their ancestry back to the Enlightenment, while those who see our intellectual and political liberation in terms of the rejection of modernity often see the Enlightenment heritage as precisely that which must be overcome, and trace their own ancestry to the skeptical anti-Enlightenment gestures of Nietzsche, Heidegger, and their more recent acolytes.

Just as there are interesting similarities between the two opposing camps, there is, I think, at least one important respect in which both parties are wrong—namely, in the assumption that any given epistemology is either intrinsically oppressive or intrinsically liberating. I have argued elsewhere that early modern thought was permeated by "intellectual co-optation," a procedure that allowed thinkers to appropriate epistemological positions that at first glance seemed alien to their own, and also to alter the ideological valences of various epistemologies to suit their own purposes.[2] A particularly piquant example of the ideological mutability of epistemologies is that of skepticism itself. Skepticism, which today is often portrayed as an emancipatory antidote to oppressive rationalism, was often deployed as an argument for submission to secular or religious authority.[3] For example, the demonstration that fundamental tenets of Christianity cannot be philosophically defended against the powerful techniques of skepticism could be presented as a pious and humble activity. Fideists—those who held that faith, not reason or science, was the appropriate source of religious belief—could embrace skepticism by urging that the very incredibility of Christian doctrines, the ease with which they can be demolished by rational critiques, pointed up the need to abandon reason and submit ourselves to divine authority.

The very availability of a conservative justification for employing skeptical techniques against religious beliefs also meant, of course, that it would be possible to mock religious beliefs with the utmost sincerity and then insincerely appeal to fideism as a way of deflecting charges of impiety. This possibility, in turn, meant that pious appeals to fideism could arouse deep suspicion among the defenders of orthodoxy, and that one's skeptical critiques of Christian doctrines could be attacked as impious despite the fact that one looked heavenward and claimed that one's skepticism had been operating in the service of faith.

Once we recognize the difficulty of ascertaining whether skepticism was destructive or pious in a given case, we might be tempted to conclude that skepticism is unstable precisely to the extent that it can be tied both to positions that are intrinsically radical and iconoclastic and to positions, such as fideism, that are intrinsically conservative. But such a conclusion stops short of a full understanding of the ideological instability of competing justifications of belief. For fideism itself—despite the fact that it traditionally stipulates that reason and science should be ignored and that one should humbly submit to the authority of God—is also unstable, both epistemologically and ideologically, and no more inherently emancipatory or conservative than skepticism itself. Indeed, a case study that showed the actual complexity of the interplay between fideism and skepticism in the context of early modern epistemology might help to demonstrate the wrongheadedness of contemporary debates about the political implications of competing epistemologies. It is such a case study that I will be presenting here.

II

In his *Dictionnaire historique et critique* (1697), Pierre Bayle masterfully applies skeptical techniques to various philosophical theories as well as to fundamental tenets of Christianity. His project, he assures his readers—in the best fideistic fashion—is not injurious to religion. He wishes merely to show the impotence of reason and to urge reliance on faith.

This fideistic justification did not succeed in disarming Bayle's critics; his powerful demonstration of the impotence of reason looked—both to orthodox Christian critics and to later mockers of Christianity such as Voltaire and Gibbon—for all the world like a demonstration of the impotence of Christianity when exposed to reason.

In response to his various critics, Bayle appended to the second edition of the *Dictionnaire* (1702) four "éclaircissements," or clarifications—the third of which specifically undertakes to answer charges that Bayle's apparent embrace of the rigorous skeptical position called Pyrrhonism had been injurious to religion.[4] Bayle here appeals to very orthodox Christian notions that faith is meritorious precisely to the degree to which it is different from and repugnant to reason. It is a certain and incontestable maxim Bayle tells us at the beginning of the third clarification,

> that Christianity is of a supernatural order, and that it [has to do with] God presenting us with mysteries not so that we will comprehend them, but so that we will believe them with all the humility that is owed to the Infinite Being . . .[5]

Stated thus, Bayle's maxim allows him to treat Christianity not simply as a set of beliefs that include certain mysteries, but as a separate discourse that is wholly beyond rational comprehension, and therefore wholly incommensurable with the language of philosophy. Accordingly, Bayle soon moves from this maxim to an invocation of the authority of St. Paul:

> [Christ] wanted his disciples and the wise men of this world to be so diametrically at odds that they should consider one another to be madmen; he wished that, just as his Gospel should seem madness to the philosophers, their knowledge, in turn, should seem madness to the Christians.[6]

What is most striking about Bayle's embrace of this Pauline conception of the incommensurability of faith and reason is that it valorizes faith on the ground that it requires adherence to beliefs that do not meet our ordinary criteria for knowledge and that therefore require us to abandon those criteria in favor of humble submission to divine authority:

> the moral worth of faith becomes greater to the degree that the revealed truth that is its object surpasses all the powers of our mind. . . . [W]e show ourselves to be more submissive to God and to give him stronger proof of our respect than we would if it were something only moderately difficult to believe.[7]

Faith is meritorious, in contrast to knowledge, because knowledge requires no effort, while faith is a heroic act of submission by virtue of which we believe what is otherwise unbelievable.

Bayle's invocation of St. Paul is only the beginning of a series of appeals to authority. Indeed, the third clarification is replete with citations and examples whose function is ostensibly to establish Bayle's orthodoxy. All this would appear to underscore the traditional and conservative nature of the discourse. But the examples do not always seem in tune with the principle that faith is utterly different from demonstrative knowledge.

Consider first Bayle's discussion of the English physician Sir Thomas Browne. In his *Religio Medici* (1643), Browne offers a vivid reformulation of the ancient fideistic position, complete with the invocation of Tertullian's celebrated paradox, "*Certum est quia impossibile est*" (It is certain because impossible).

In a passage that Bayle quotes immediately after quoting Browne's invocation of Tertullian, Browne congratulates himself for lacking certain knowledge of the articles of Christian belief, and for having to rely on faith:

> Some believe the better for seeing Christ's sepulchre; and when they have seen the Red Sea, doubt not of the miracle. Now contrarily, I bless myself, and am thankful that I lived not in the days of miracles . . . then had my faith been thrust upon me, nor should I enjoy that greater blessing pronounced to all that believe and saw not.[8]

Browne is alluding to Jesus' reproach to doubting Thomas in *John* (chapter 21, verse 29): "Have you believed because you have seen me? Blessed are those who have not seen and yet believe."

Thus framed between the authority of Tertullian and that of Jesus himself, Browne's fideism seems to fit nicely into the tradition that Bayle is setting forth in the third clarification.

But it needs to be observed that Browne's distinction is not between faith and reason, but between faith and direct sensory experience. His assumption seems to be that those who witnessed the historical miracles of the resurrection and the parting of the Red Sea deserve no credit for their beliefs because the beliefs require no effort of faith. Indeed, in the passage from Browne that Bayle goes on to quote (in Latin translation) Browne says, " 'Tis an easy and necessary belief, to credit what our eye and sense hath examined."[9] And in a passage that Bayle does not quote, Browne states, "to credit ordinary and visible objects is not faith but persuasion."[10] Browne thus treats beliefs that arise from normal sensory experience as outside the province of faith and as amounting to nothing more than mere "persuasion." What is even more interesting for our purposes is the possibility of reading Browne's position as implying something other than an unbridgeable gulf between faith and ordinary knowledge. For it might be possible to postulate an epistemological continuum along which the same beliefs may, by degrees, be more or less meritorious to the extent that they are farther from or closer to sensory knowledge.

Browne does not explicitly draw this implication, but it might be said that John Craige, whom Bayle had introduced in tandem with Browne and whom he discusses next, does precisely that. Craige's *Theologiae Christinae Principia Mathematica*, published in 1699, has generally been considered an almost comical oddity in the early history of probability theory.[11] But it begins with the notions about the credibility of marvelous historical events that we have encountered in our discussion of Browne. If we assume that extraordinary or miraculous events counted as knowledge to those who witnessed them, but lose their credibility progressively as time goes on, then it follows that we can say that the miraculous events associated with the origins of Christianity are less credible now than they were shortly after the events occurred, and will be even less credible in the future. What Craige did was try to express this notion in mathematical terms, by computing the degree of probability for each point in time. Once the probability of belief is quantified, then we can predict precisely when the credibility of Christianity will disappear altogether—namely, in the year 3150—and hence that the second coming of Christ will occur before that date.

At first glance, however, Bayle's inclusion of Craige in the third clarification seems to have nothing to do with the specifics of this bizarre mathematical exercise. Rather, it seems based on the fact that Craige had tried to ward off charges of impiety by appealing to the same fideistic conception of Christian belief that Bayle is invoking on his own behalf. Craige, in his preface, deems it necessary to deflect the objection that he is reducing Christian belief to mere probability. Indeed, it is only from this exculpatory preface that Bayle quotes, and at considerable length. Those who are prejudiced against his undertaking, Craige is quoted as saying,

> have not yet examined carefully enough the foundations of the religion they profess; and . . . they do not rightly understand the nature of faith, which is so much praised in Holy Scripture. For faith is nothing other than that persuasion of the mind, derived from an indeterminate probability, by which we believe certain propositions to be true. If the persuasion arises from certainty, then it is not faith that is being produced, but knowledge. Probability generates faith, but destroys knowledge; certainty, on the other hand, generates knowledge and destroys faith.[12]

In appealing to the foundations of Christianity and to Scripture, Craige, like Browne before him, is making a virtue of believing that of which he does not have certain knowledge. But the content of the dichotomy in which the meritoriousness of faith takes its place has been dramatically altered. Faith is no longer being characterized as the submission to beliefs in impossible events and concepts, in things that are repugnant to reason. Indeed, the dichotomy that previously separated absolutely incommensurable sorts of things is giving way to a kind of continuum, in which faith is understood in terms of shifting degrees of probability. In order to understand the significance of this change, we need to glance at aspects of the broader context of probabilistic thinking in the late seventeenth century.

III

We may profitably begin with John Locke, whose discussion of probability and assent in the *Essay Concerning Human Understanding* includes the following striking passage:

> any testimony, the further off it is from the original truth, the less force and proof it has. The being and existence of the thing itself is what I call the original truth. A credible man vouching his knowledge of it is a good proof; but if another, equally credible, do witness it from his report, the testimony is weaker; and a third, that attests the hearsay of a hearsay, is yet less considerable.[13]

Given Craige's general indebtedness to Locke,[14] and the specific similarity between this passage and the premise of Craige's probabilistic enterprise, one might well suppose that when Craige undertakes to compute the waning credibility of the foundational miracles of Christianity, he is giving mathematical expression to the Lockean observation just quoted. Locke, after all, unlike Browne, postulates a lengthy process whereby events lose their credibility step by step, through each new accretion of testimony. But Locke is in no way suggesting that it is meritorious to accept improbable beliefs *because* of their improbability. Far from giving us an opportunity to demonstrate our faith, the increasing weakness of second- and third- and fourth-hand testimony is adduced by Locke as a way of counteracting those who would valorize beliefs that are the products of a long tradition. Indeed the passage just quoted is followed immediately by this one:

> This I thought necessary to take notice of, because I find amongst some men the quite contrary commonly practiced, who look on opinions to gain force by growing older; and what a thousand years since would not, to a rational man contemporary with the first voucher, have appeared at all probable, is now urged as certain beyond all question, only because several have since, from him, said it one after another. Upon this ground propositions, evidently false or doubtful enough in their first beginning, come by an inverted rule of probability, to pass for authentic truths. . . .[15]

In deriding the "inverted rule of probability" that gives venerable status to beliefs whose truth, when viewed rationally, is very improbable, Locke is not addressing the subject of belief in miracles, but is evidently weighing in on the Protestant side in the long polemical battle against Catholic appeals to tradition as their special source of doctrinal and exegetical authority. Whereas Protestants could claim that the necessary truths of religion were directly available in the Bible, tradition was a pillar of Catholic epistemology. Thus Locke's cultural situation allows him to mount a rational attack on traditional beliefs based on a chain of testimony, while impelling him discreetly to ignore the fact that the same sort of critique of "the hearsay of a hearsay" could easily apply to Protestant doctrines regarding biblical miracles.[16]

Craige, on the other hand, it would seem, was willing to take the next logical step. Indeed, he was willing not only to apply to miracles the Lockean dictum about the increasing diminution of credibility as testimony becomes more and more mediated, but even to quantify the specific degrees of improbability. Seen in this light, it might seem as if Craige's use of fideism is simply a way to protect himself against the consequences of his own daring.

But we need to recall that in the self-justification quoted above from Craige's preface, he not only departs from the ancient notion that the merit of faith comes from the utter incompatibility between Christian beliefs and reason, but he also departs from his own Lockean principle that faith in highly attenuated testimony has a very low probability. Faith is characterized as "that persuasion of the mind, derived from an indeterminate probability, by which we believe certain propositions to be true."[17] Both faith and knowledge are persuasions of the mind, but faith is based on probability rather than certainty. To help us see what is going on here, we might recall that Browne's fideism had included the statement that "to credit ordinary and visible objects is not faith but persuasion." For Browne, faith was something beyond persuasion and beyond probability. In his preface, by contrast, Craige seems to be invoking probability, not in order to separate the realm of faith from that of knowledge, but to place the two upon a single continuum.

This continuum links up with a significant seventeenth-century tendency to try to ground knowledge claims on probability, instead of having to aspire to the perhaps unattainable status of certainty. The extreme skeptics had used certainty as the criterion of knowledge, and had assumed that the failure to achieve certainty indicated the failure to achieve knowledge. Fideism, as we have seen, was one reaction to this skeptical tendency. The new interest in probability was another. While fideism is commonly viewed as an extreme reaction to skepticism, the early modern rise of probability is seen as an intermediate position between "fideist dogmatism on the one hand and the most corrosive skepticism on the other," an intermediate position that insisted that probable knowledge was attainable, and that "probable knowledge was indeed knowledge."[18] This intermediate position is associated with what I referred to above as "mitigated skepticism," and hence with scientific epistemology.

What we have found in our discussion of Browne, Craige, and Bayle, however, is nothing like a clear separation between skeptical and fideistic extremes, or between those extremes and an intermediate probabilism. Craige, after all, is appealing to fideistic tradition to justify his probabilism, and Bayle is appealing to Craige to justify his Pyrrhonian skepticism. In the process, the content of the ancient dichotomy in which the meritoriousness of faith was identified in opposition to worldly wisdom has been dramatically altered. Faith is no longer being characterized as the submission to beliefs in impossible events and concepts, in things that are repugnant to reason. Instead faith is characterized in terms of probability rather than impossibility, accepting the truth of testimony to remarkable events that one has not oneself witnessed. Indeed, the dichotomy that separated absolutely distinct sorts of things is giving way to a kind of continuum, in which faith is understood in terms of shifting degrees of probability.

When we think of probability as an "intermediate position" between skepticism and certainty, we tend to think of it simply as providing a pragmatic criterion of truth that allows us to get on with scientific work without being plagued by epistemological doubts. But if we think of probability in the terms enunciated by Locke and Craige, having to do with degrees of credibility of testimonial evidence, then we can begin to see how the probability of scientific truths and the probability of miracles can fall along the same epistemological continuum. For although we are accustomed to thinking of the scientific revolution as rejecting blind faith in ancient beliefs, in favor of the fruits of our own experience and observation, we need to recall that many of the great discoveries of early modern science do not necessarily confirm what we think we have learned from ordinary experience. Rather, the results are often surprising and incredible, and the acceptance of experimental knowledge as true knowledge often depended upon the testimony of a few trustworthy witnesses to marvelous events that occurred in scientific laboratories. Hence the recent scholarly interest in the relationship between scientific marvels and miracles.[19]

According to Steven Shapin, one of the cultural factors that helped to give probable eyewitness testimony its proper epistemological weight were accounts of religious knowledge, since, according to Shapin, "much religious knowledge manifestly had a historical character." Indeed, as he goes on to observe:

> it was a central concern of Christian apologetics to warrant scriptural testimony as reliable and to show that people might as securely give their assent to it as to formally more certain types of knowledge. This meant that it had to be shown that the probable quality of properly testified matters substantially and practically overlapped with the quality of both demonstrable matters and the facts accessible to personal witness.[20]

Shapin seems to be suggesting that the new justificatory protocols of empirical science are borrowing from the older resources of religious epistemology. He presumably has in mind the sort of rationalist theology that is piously echoed by Locke, when he refers to miracles, "which, well attested, do not only find credit themselves, but give it also to other truths which need such confirmation."[21] However, in Craige's fideistic defense of probabilism and in Bayle's fideistic defense of skepticism, at least, I think we are seeing the reverse process: a fideism that is being altered through the appropriation of the role of probable testimony in the new scientific enterprise.

Bayle introduced Browne and Craige, respectively, as a physician and mathematician, and explicitly suggested that their status as men of science enhanced the force of their testimony. Their testimony, he writes, "will carry more weight because their profession does not pass for a school that teaches one to debase reason and to elevate faith."[22] This would seem to imply that the two men's status as scientists does not prevent them from being fideists. But there is a further sense in which Bayle wants to suggest that fideism is being enhanced by the scientific credentials of its adherents. In embracing their fideistic pronouncements without calling attention to the fact they mark a shift from the meritoriousness of believing the impossible to the alleged meritoriousness of believing the probable, Bayle is trying, I think, to quietly merge the new epistemological prestige of scientific probabilism with the moral prestige traditionally attributed to faith as an act of humble submission.

Bayle's conflation of the divergent poles of a fideistic tradition, in sum, allows him tacitly to uphold the moral approbation earned by opting for Christian beliefs in the face of uncertainty, while at the same time treating those beliefs as empirically justified. Thus, to accept the foundational events of Christianity is epistemologically valid because the beliefs, like the belief in scientific marvels, are, in some sense, probable; meanwhile, the acceptance of these very beliefs is morally meritorious because they are "merely" probable.

## IV

I suggested at the outset that my historical case study would have some bearing on contemporary debates regarding epistemology and ideology. Specifically, I indicated that current tendencies to link specific epistemological stances with specific political stances ran up against the historical fact that ideas can be coopted for various purposes. I further suggested that this difficulty was exacerbated by the fact that attempts to anchor one idea in the supposed ideological fixity of another ran up against the capacity of the supposedly stable idea itself to experience drift. I trust that this point has been demonstrated through our examination of fideism, a doctrine that—by dint of its appeal to the virtue of uncritical submission to authority—had at first glance appeared to be a limiting case of epistemological and ideological conservatism. If fideism can be transformed into probabilism and cloak itself in the language of scientific rationalism, then it is hard to imagine that any epistemological principle whatever can retain a fixed ideological valence.

Nor does this amenability to transformation result simply from the fact that the same theory can be put to different political uses. It also results from the fact that the "same theory" can take a different shape each time it is formulated. For theories are subject to interpretation not only by distant commentators, but also by the historical personages who embrace them.

The protean quality of fideism that we find exemplified in Bayle's deployment of it may be also found in his skepticism. I have argued elsewhere that Bayle interprets Pyrrhonism not as a denial of the possibility of certain knowledge, but as an open-ended search for truth—a search, moreover, that seems more hospitable to factual claims than to metaphysical systems.[23] The reason that someone can entertain such a distinction between metaphysical claims and claims about observable facts and nonetheless consider himself to be a rigorous skeptic is that while the evidence on either side of metaphysical questions can be shown to be equally balanced—and hence as requiring the skeptical *epochē*, or suspension of judgment—the evidence on either side of questions regarding the objects of sensory perception is not so balanced. That is to say, although the possibility that our sense perceptions may not be veridical prevents us from being certain about them, it does not follow that the reasons for doubting a given sense impression are as strong as the reasons for believing it to be probably correct.[24] This interpretation of the process of skeptical doubt, according to which skepticism takes a pragmatic and probabilistic turn, has been associated with the tradition of Academic Skepticism rather than that of Pyrrhonism. But it is entirely possible that Bayle combined the two skeptical traditions so as to empiricize and probabilize skepticism in a manner that echoes the probabilistic treatment of fideism that I have examined above.

I am not concerned here with providing the correct interpretation of Bayle's skepticism or with assaying the relative cogency of competing versions of skepticism. I wish merely to show that the instability of epistemological theories—both with respect to their fundamental postulates and with respect to their ideological uses—makes a mockery of global generalizations to the effect that one general sort of theory is, by its very nature, an instrument of domination and that another sort is, by its very nature, a means of emancipation.

But as I have argued in my other essay in this volume and elsewhere,[25] none of this has the effect of undermining the legitimacy of ideological analysis. Indeed, the approach that I advocate gives ideological analysis a much greater scope and much more subtlety. Instead of forcing its practitioners to confirm ad infinitum an assumption about the oppressive or emancipatory tendency of theory A or theory B, ideological analysis should allow us to find surprising variations. Ideological analysis is a humanistic enterprise that, by requiring the investigations of each individual case, requires the practitioner to read texts carefully and to be attentive to nuances and idiosyncracies. Those interested in ideas and their ideological dimensions should not try to ape physics by discounting variations in order to establish general laws. For what we know in the humanities we have learned from a painstaking version of empiricism—one that focuses its energy and its care on the particular.

## NOTES

1 For the term *mitigated skepticism* and the early history of the interplay between skepticism and scientific rationalism, see Richard H. Popkin, *The History of Skepticism from Erasmus to Spinoza.*

2 Oscar Kenshur, *Dilemmas of Enlightenment: Studies in the Rhetoric and Logic of Ideology.*

3 Ibid., pp. 102–111.

4 Supposedly originated by Pyrrho of Elis (c. 360–c. 270 B.C.E.), but codified by Sextus Empiricus (in the second or third century C.E.), Pyrrhonism presented itself as a mode of inquiry that cultivated doubt by discovering equally powerful arguments on either side of every issue, but that avoided all dogmatism, including that which it attributed to the Academic Skeptics—whom it accused of dogmatically claiming that nothing could be known. See Sextus Empiricus, *Outlines of Pyrrhonism.*

5 Pierre Bayle, *Dictionnaire historique et critique*, 2nd edit., vol. 3, p. 3153. Translations from the French are my own.

6 Ibid., p. 3154.

7 Ibid., p. 3156.

8 I cite Brown's English original (p. 9). Bayle, not a reader of English, paraphrases Browne in French and quotes him in Latin translation (ibid., p. 3157).

9 Sir Thomas Browne, *Religio Medici and Other Works*, pp. 9–10.

10 Ibid., p. 9.

11 Ian Hacking, p. 72. Hacking discusses not only the aspect of Craige's probability theory that I describe here, but also an additional aspect that involves a mathematical version of Pascal's famous wager.

12 Bayle, *Dictionnaire historique et critique*, vol. 3, p. 3158. I give Nash's translation of Craige's Latin (pp. 53–54).

13 John Locke, *An Essay Concerning Human Understanding*, IV, xvi, 10.
14 For Locke's influence on Craige's *Theology*, see Richard Nash's commentary in his *John Craige's Mathematical Principles of Christian Theology*, pp. 33–45.
15 Locke, *An Essay Concerning Human Understanding*, IV, xvi, 10.
16 In his actual discussion of the topic a few paragraphs later, Locke evaluates the truth of miracles in terms of their consonance with our conception of the divine nature and purpose, and hence their capacity to "procure belief" in Christianity (IV, xvi, 13). The criterion, thus, is once again a rational one, but Locke does not concern himself with the reliability of testimony regarding the actual occurrence of miraculous events.
17 Nash, *John Craige's Mathematical Principles*, p. 54.
18 Gerd Gigerenzer *et al.*, *The Empire of Chance: How Probability Changed Science and Everyday Life*, p. 5.
19 Peter Dear, "Miracles, Experiments, and the Ordinary Course of Nature"; Lorraine Daston, "Marvelous Facts and Miraculous Evidence in Early Modern Europe"; Steven Shapin, *The Social History of Truth: Civility and Science in Seventeenth-Century England.*
20 Shapin, *The Social History of Truth: Civility and Science in Seventeenth-Century England*, p. 209.
21 Locke, *An Essay Concerning Human Understanding*, IV, xvi, 13.
22 Bayle, *Dictionnaire historique et critique*, vol. 3, p. 3157.
23 Oscar Kenshur, "Pierre Bayle and the Structures of Doubt."
24 Gisela Striker, "Sceptical Strategies."
25 Oscar Kenshur, *Dilemmas of Enlightenment: Studies in the Rhetoric and Logic of Ideology.*

## REFERENCES


BAYLE, PIERRE. *Dictionnaire historique et critique.* 2nd edit., Rotterdam: 1702.

BROWNE, SIR THOMAS. *Religio Medici and Other Works.* Edited by L. C. Martin. Oxford: Oxford University Press, 1964.

DASTON, LORRAINE. "Marvelous Facts and Miraculous Evidence in Early Modern Europe." *Critical Inquiry* 18 (1991): 93–124.

DEAR, PETER. "Miracles, Experiments, and the Ordinary Course of Nature." *Isis* 81 (1990): 663–683.

GIGERENZER, GERD *et al. The Empire of Chance: How Probability Changed Science and Everyday Life.* Cambridge: Cambridge University Press, 1989.

KENSHUR, OSCAR. "Pierre Bayle and the Structures of Doubt." *Eighteenth-Century Studies* 21 (1988): 297–315.

———. *Dilemmas of Enlightenment: Studies in the Rhetoric and Logic of Ideology.* The New Historicism: Studies in Cultural Poetics, no. 26. Berkeley, CA: University of California Press, 1993.

LOCKE, JOHN. *An Essay Concerning Human Understanding.* Edited by John Yolton. New York, NY: Everyman, 1964.

NASH, RICHARD. *John Craige's Mathematical Principles of Christian Theology. Journal of the History of Philosophy* Monograph Series. Carbondale, IL: Southern Illinois University Press, 1991.

POPKIN, RICHARD H. *The History of Skepticism from Erasmus to Spinoza.* (2nd edit. of *The History of Skepticism from Erasmus to Descartes.*) Berkeley, CA: University of California Press, 1979.

SHAPIN, STEVEN. *The Social History of Truth: Civility and Science in Seventeenth-Century England.* Chicago, IL: University of Chicago Press, 1994.

STRIKER, GISELA. "Sceptical Strategies." In *Doubt and Dogmatism: Studies in Hellenistic Epistemology*, edited by Malcolm Schofield *et al.*, pp. 54–83. Oxford: Clarendon Press, 1980.

SEXTUS EMPIRICUS. *Outlines of Pyrrhonism.* In *Sextus Empiricus*, translated by R. G. Bury. Loeb Classical Library. Cambridge, MA: Harvard University Press, 1933.

# EDUCATION

GIVEN THE OMNIPRESENCE of academic concerns in the previous sections, there might well have been a subsection entitled "Education" at the end of each. Here, however, are four contributions that deal directly with serious problems of education, to the extent that difficulties current in our schools and universities are in some measure consequent to a general decline of respect for reason and science, a decline within, as well as without, the halls of learning.

Henry Rosovsky, distinguished administrator and scholar, here subverts his own assertion that years of administrative chores have undermined his capacity to reason. Clearly, they have not, nor have they diminished his sense of humor (which is all the more remarkable). He deals briefly with a sampling of the more common reproaches to higher education: that its "science" is complicit in the devaluation of the humanities; that rationalist hubris deprives education of its spiritual content; that specialization makes a mockery of the ideal of liberal learning. He finds little merit in these claims, which come now from the left as well as the right.

For Gerald Holton there is a startling contradiction between the small part science plays in most curricula and the part it has played in the improvement of human life and hopes. This underemphasis in most programs of general education is matched, moreover, by the reticence of organized science itself in the face of its critics, a diffidence so conspicuous that the most strenuous attacks have until recently been met with equivocation or not met at all. Holton examines a number of cases where arrant antiscientism has imposed itself on what were supposed to have been science education projects. That a defense of any kind has been so slow in coming speaks volumes.

James Trefil sounds a more positive note. He has no illusions about the quality of science education today, or about the state of scientific literacy, in America. But as one of the early participants in the cultural literacy movement, he has argued effectively that the problem is not intractable. As with cultural literacy in general, standards of scientific literacy are definable, even amidst an explosion of information and disinformation. They point the way to new and more effective ways of teaching science to the great mass of nonscientists.

The final contribution is an epilogue of sorts to the prior exchange of views on Afrocentrism in the section entitled *History, Society, Politics.* Bernard Ortiz de Montellano, an anthropologist with a strong background in the physical sciences, examines one of the most widely implemented and most worrisome products of the contemporary romance with "multicultural" education. Far from *multi*cultural, however, the "science" presented in the Portland Baseline Essays is unicultural Afrocentrist propaganda. Remarkably, it now serves as the science base for curriculum building in a number of public school systems, where its errors and distortions of fact are inevitably harmful to its supposed beneficiaries, and counterproductive for the recognition of important achievements of non-Western peoples in coming to grips with nature.

# SCIENCE, REASON, AND EDUCATION

HENRY ROSOVSKY

I ACCEPTED THE INVITATION to contribute to this conference and volume with great hesitation. My understanding of science is underdeveloped; years of administrative chores in universities have undoubtedly undermined my capacity to reason. This conference deserves philosophers, and I cannot claim membership in that exalted calling. Furthermore, I do not—in general—associate myself with the many critics of higher education who have captivated all the media as well as segments of the academy: Bloom, D'Souza, Anderson, etc. Much of what they have to say seems exaggerated and even untrue.

Nevertheless, I want to comment on certain criticisms of higher education—now quite common—that exhibit distrust and sometimes denigration of reason, objectivity, rationality, and science. These attitudes are not entirely identified with one end of the political spectrum. Right and Left as well as liberals and conservatives can be found in the group that shows some distrust of science and reason. That may make the problem more serious.

Let me begin with the common charge that higher education has devalued the humanities: this is what the critics say. In the "good old days," humanists were the most important voices in colleges and universities. They represented the eternal verities of our calling. Now, prestige and money go to science. Worse, prestige and money even go to pseudoscientists—i.e., social scientists who pretend to rigor and in reality are not much better than sellers of snake oil. (To put the matter in historical perspective, the debate of science vs. humanities raged at Cambridge University almost exactly one hundred years ago.)

In its least objectionable and mildest form, we can point to the recent writings of a most distinguished Yale scholar, Jaroslav Pelikan, who wrote:

> The university has not discharged its moral and intellectual responsibility if, in its heroic achievement of attaining the possibility of putting bread on every table, it ignores the fundamental axiom, which may be biblical in its formulation but is universal in its authority, that man does not live by bread alone. The religions of humanity all have their special versions of that axiom, and both in its teaching and research the modern secular university often ignores these at its peril.[1]

Few will wish to quarrel with these sentiments *unless* science and social science are seen primarily as "bread" and little else.

A much more problematic statement has been made by the philosopher Bruce Wilshire:

> The sciences are fractured from the humanities and exercise disproportionate influence in the university. Under the sway of thinking like Descartes's we tend to view ourselves as heads joined somehow with stomachs and groins—men without chests, calculating and inquisitive animals who no longer rejoice simply for being living parts of nature.[2]

Do these quotes illustrate a certain humanistic hubris? Does the study of science and yes, even social science, not impart values as much as the study of other subjects? "Calculating and inquisitive animals" without hearts is not a description that makes science feel appreciated or welcome.

The name of Descartes has already been mentioned, and that is not accidental. Our critics actively dislike the Enlightenment and its consequences. This philosophical movement—one thinks of Bacon, Montaigne, Descartes, Spinoza, and others—is related to the scientific revolution, skepticism (relativism?), and challenges to religious tenets (family values?). It emphasized breaking tradition and making everything subject to rationality. Figures as diverse as Foucault and Vaclav Havel, playwright and Czech President, have all—in one way or another—attacked the consequences of the Enlightenment. Havel sees industrialization and technology as the "soiling of heavens" and suggests that scientists are "playing God."[3] Concerning Foucault, Alexander Nehamas writes:

> For more than twenty years Foucault seemed dedicated to exposing the seamy underside of the Enlightenment, conceding to it no positive accomplishment and refusing any and all visions of a better future. He explicitly denied that reason can transcend time and accident and lead us out of the impasse, since reason itself was an instrument, a part of the program of the Enlightenment.[4]

There are also internal university voices echoing these sentiments. Once again, the thoughts of Bruce Wilshire are relevant.

> The last decades of the eighteenth century—the period of the enlightenment, *so called*, saw a momentous intersection of events: the founding documents of the United States were drafted, as science, secularism and individualism started to gain official status . . . when religion was separated from the official center of public life, it tends to take ethics with it.[5]

Alan Bloom is more ambivalent, though he does consider Russian communism to be a spiritual deformation caused by the Enlightenment. This is not a new attitude since Descartes has been called the father of the French Revolution and the grandfather of the Russian Revolution. Bloom says:

> Older, more traditional orders that do not encourage the free play of reason contain elements reminiscent of the nobler, philosophic interpretation of reason and helped to prevent its degradation. Those elements are connected with the piety that prevails in such orders. They convey a certain reverence for the higher, a respect for the contemplative life, understood as contemplation of God . . . and a cleaving to eternal beings that mitigates absorption in the merely pressing or current.[6]

The message that Bloom intends to convey needs no elaboration: it is clear. Is a monastery preferable to a modern university? He might have answered "yes," thereby questioning rationality as a basic premise of university life.

Are his words so different from the thoughts of the Bratslaver Rebbe (1770–1811):

> The science of Divine Knowledge is the only true science. The more one acquires of the secular sciences, the less room there is in his brain for holy studies. In the brain cells utilized by the secular sciences, gather and unite all passions and evil traits. Since understanding is the soul, he who devotes his understanding solely to holy subjects sanctifies his soul.[7]

Whatever happened to progress?

As we all know, many of our critics consider research universities as examples of most of what is wrong with higher education. Colleges, especially when they can be described as a few Mr. Chipses surrounded by adoring students in a rural setting, are normally treated more kindly. "Small is beautiful," country is better than city, technology equals pollution—is there a pattern?

There is one aspect of these attitudes that poses a particular danger to the flourishing of science—or so it seems to me. Research universities are the fountainhead of specialization and its natural consequence: academic departments. These are seen as evil influences.

What is bothering the opponents of specialization? Two things. First, their conviction that great truths are to be learned from interdisciplinary studies and that the structure of research universities is an obstacle to the discovery of these truths. Second, they fear the influence of "excessive fragmentation" on undergraduate education.

Is specialization not an inevitable aspect of progress and greater productivity? Adam Smith's lesson in *An Inquiry into the Nature and Causes of the Wealth of Nations* (1776) surely applies beyond pin factories. "Excessive fragmentation" is a pejorative term. Obviously excessive is bad. Anything excessive is bad. The challenge is not how to get rid of specialization but how to take advantage of it. *A specialized faculty is not the same thing as a specialized undergraduate curriculum.* It is even possible to use specialized knowledge to further the cause of general education.

Let us also keep in mind the key role of departments in setting professional standards within the university. This is where interdisciplinary units have the smallest comparative advantage. Who can best determine the quality of a physicist, an economist, or a professor of French literature? Members of other fields? Students? Presidents and deans? Of course, some of these individuals will have review responsibilities, and the opinion of outsiders is also important and desirable. However, I do not know of any group of individuals that is better qualified to make the initial selection than members of a disciplinary department. This applies even to scholars with interdisciplinary orientations because that description does not imply "without discipline"; instead it means the use of two or more disciplines. In other words, the judgment of disciplinary specialists remains necessary.

## NOTES

1 Jaroslav Pelikan, *The Idea of the University: A Reexamination*, p. 18.
2 Bruce Wilshire, *The Moral Collapse of the University: Professionalism, Purity, and Alienation*, p. 40.
3 On Havel, see Jan Vladislav, ed., *Vaclav Havel or Living in Truth*, ch. 4. In some ways, President Havel longs for the world before the Industrial Revolution.
4 Alexander Nehamas, "Subject and Abject: The Examined Life of Michel Foucault," p. 32.
5 Wilshire, *Moral Collapse of the University*, p. 47; italics added.
6 Alan Bloom, *The Closing of the American Mind*, p. 251.
7 See Louis I. Newman, *The Hassidic Anthology*, p. 316.

## REFERENCES

BLOOM, ALAN. *The Closing of the American Mind.* New York, NY: Simon and Schuster, 1987.

NEHAMAS, ALEXANDER. "Subject and Abject: The Examined Life of Michel Foucault." *The New Republic*, February 15, 1993, p. 32.

NEWMAN, LEWIS I. *The Hassidic Anthology.* New York, NY. Bloch Publishing Company, 1994.

PELIKAN, JAROSLAV. *The Idea of the University: A Reexamination.* New Haven, CT: Yale University Press, 1992.

VLADISLAV, JAN, ed. *Vaclav Havel or Living in Truth.* London: Faber and Faber, 1987.

WILSHIRE, BRUCE. *The Moral Collapse of the University: Professionalism, Purity, and Alienation.* Albany, NY: State University of New York Press, 1990.

# SCIENTIFIC LITERACY

JAMES TREFIL

THIS PAPER is about the concept of scientific literacy, and there is one point I want to make right from the start. I am *not* talking about the education of scientists, or the idea that everyone ought to be, in some sense, a scientist. Too often, when scientists start to talk about scientific education, they skip the ninety-nine percent of the population that is not going to be trained in a technical career, and zero in on the one percent that are. The ability to calculate the trajectory of an artillery shell or to estimate the heat flow through a block of solid material are not things that I would include under the phrase "scientific literacy." They are skills that a certain group of working scientists will need to have at their disposal, but they are not part of the scientific knowledge that the average citizen needs to function in the world today.

I want to define scientific literacy, then, in a much more general sense. I will say that a person is scientifically literate if he or she has enough of a background in science to deal with the scientific component of issues that confront him or her daily. You have only to pick up a newspaper and look at the headlines to realize that there are very few aspects of the national debate that deal with science alone. Almost every issue that the citizen confronts deals with science in a much broader context. Thus, when we talk about global warming, we are talking about a scientific issue (has global warming begun? is it due to anthropogenic sources?), but it involves much more; it involves social issues (does the developed world have the right to tell developing countries not to cut down their forests?), economic issues (are we really ready to shut down our economy to save the Kirtland's warbler?). Thus, when the average person confronts social issues, he or she needs to know about a good number of subjects—economics, politics, social policy, . . . *and science.*

I will argue in this paper that what the average person needs to know about science is, in fact, very different from what is normally taught both at the high school and college level, and that, therefore, if we wish to have a scientifically literate population, we are not only going to have to teach more science, we are going to have to teach it differently than we do now. In fact, I would argue that we can probably get to the goal we want without increasing the actual time that people spend learning about science, if we would only teach it in a different way.

I want to avoid preaching to the choir, so I will not spend a great deal of time trying to justify why we want a scientifically literate citizenry in this country. Instead, I will use this paper to discuss a series of increasingly focused topics. Scientific literacy is actually a subset of a much more general topic called "cultural literacy." Once I have established this broad framework, I will focus further and discuss a particular realization of a scheme for teaching scientific literacy.

## CULTURAL LITERACY

In any society, at any time, people make assumptions about what other people already know. No communication begins *ab initio*—every communication takes place in a social context. For example, if you pick up the newspaper on any given day, you may find a reference to something that happened at the United States Senate. The author of that article does not go on to explain what the United States Senate is: he does not say, for example, "the United States Senate is one arm of the legislative branch of the United States federal government. . . ." The author simply assumes that you know what the United States Senate is and then proceeds on that assumption.

This is an example of cultural literacy. Cultural literacy can be defined as that body of knowledge that educated people, in a given society, at a given time, assume that other educated possess. If I assume you know something, I am not going to tell you about it. If I do not tell you about it and you do not know, you are out of luck—you are just not going to understand what I am talking about. This is what cultural literacy is all about. In the United States, in 1995, it is possible to define what this body of knowledge we call cultural literacy is. Indeed, those who make a living by communicating with the public—magazine editors, for example—have a very clear idea in their mind of what kind of terms and concepts can and cannot be introduced to their readership without explanation.

Let me show you some examples of cultural literacy in action using cartoons from *The New Yorker* magazine. I will confess that when I get my copy of *The New Yorker*, the first thing I look at is the cartoons. I will also confess that if someone tells me that they look at something else first, I greet that statement with a certain level of skepticism.

One cartoon that I would like to refer to shows a woman and her husband standing in front of a closet. The wife is saying "I sent your favorite sweater out to be cleaned and carbon dated."

Now this is, on the one hand, a rather typical example of the gentle *New Yorker* humor. But think for a minute about what the author of that cartoon assumed that you knew. He assumed that you knew there was such a thing as carbon dating, and that it is used to date very old objects. He did not, however, assume that you knew the half life of carbon 14, or, indeed, that you knew there was such a thing as carbon 14 in the first place. In other words, the kind of knowledge that the cartoonist assumed was very general.

Here is another cartoon example. A young man is on his knees, obviously proposing marriage. He says, "Marry me, Virginia. My genes are excellent and as yet unpatented."

This is what I call a double whammy. The author is assuming that you know there is something called a gene, that the gene has something to do with inheritance and children, and would therefore be something that might be appropriate to discuss in the context of a marriage proposal. The author also assumes that you are aware that there is a continuing set of skirmishes in the court over the legal status of genetic information. What both of these cartoons illustrate is that any kind of communication—even something as trivial as a cartoon—assumes a certain amount of background knowledge in the readership. Without that background, you just cannot understand what is being said.

In fact, there is a large body of psychological research that bolsters the notion that if you have a lot of information about a particular subject, it is easier for you to assimilate new information. I like to picture cultural literacy as a large matrix of knowledge. The idea is that if a particular area of the matrix is well filled in and richly articulated, it is much easier for you to add new things to that part of the matrix—having a lot there makes it easier to fit new things in.

Let me give you one example of some of the psychological research that bears this out. There was a study done a number of years ago at the University of Illinois in which a group of graduate students was selected. Half of these students were Americans, and half were Indian. Each of the students was given a written description of a wedding, and then tested for reading speed and comprehension.

Half the American students, for example, were given descriptions of American weddings, the other half descriptions of Indian weddings. The results of the tests should surprise no one. The American students read the texts on the American weddings quickly and understood them. They did not do so well with the description of the Indian wedding. The exact opposite, of course, happened with the Indian students.

Another story, concerning the genesis of the idea of cultural literacy, illustrates the role of preexisting knowledge. Don Hirsch was doing research based on a hypothesis he had about the importance of what is generally called "good English." His idea was that these rules make communication easier and more efficient. He tested this hypothesis by taking a particular text that was written in good English, giving it to students to read, and testing them on speed and comprehension. He would then compare those results to those of another group who had read texts that had the same information, but was written in bad English. The tests were going along pretty much as expected, with the students being able to read the well-written texts and comprehend them more easily than the poorly written texts. When he went to Sargent Community College in Richmond, Virginia, however, he was absolutely astonished to find that it made no difference to these students how the texts were written—they had trouble understanding both the well-written text and the poorly written one. Concerned with this anomalous data point, Don went back to Richmond and started interviewing the students. The texts the students had been given had something to do with Jefferson Davis, and what Don found was that these students, growing up in the capital of the old Confederacy, had no idea who Jefferson Davis was. (I am sure you do not need to be reminded, but Jefferson Davis was president of the Confederate states.) In other words, what he found was that the students who did not have the background knowledge required to read the text were unable to do so very well. This was the genesis of the idea of cultural literacy.

So the picture that comes out of this theory is that everyone carries around in his or her mind a matrix of information and knowledge, and that matrix plays a very active part in communication. Whether or not you will understand something depends on whether or not you already know something about the subject.

In fact, as an old-line physics professor, I am incapable of giving a fifty minute lecture without doing a demonstration, so I am going to do a lecture-demonstration of cultural literacy. Consider the following text:

> The procedure is actually quite simple. First you arrange the items in different groups. Of course, one pile may be sufficient depending on how much there is to do. If you have to go somewhere else due to lack of facilities, that is the next step. Otherwise you are pretty well set. . . .

Now if you are like most people when you read this text, you get the sense of slipperiness, of sliding around. You have no idea as to what the writer is talking about.

Now let me give you the matrix of information you need to interpret the text. Imagine a title above that text that says "Doing the Laundry." Now, it all makes sense. That little moment of "a-ha!" that you just experienced is the best proof I could give you of the importance of cultural literacy.

I would argue that no one should be allowed to leave our educational system until their matrix of cultural literacy has been filled in. Of course, at the university level we expect people to do much more than possess a body of information. We expect them to be able to deal with abstractions, to communicate effectively, and so on. Nevertheless, cultural literacy represents the ground work, the basis, the floor beneath which no one should be allowed to fall. Now having made this statement, let me now go on to make several comments that I believe will anticipate some of the questions that normally are raised.

First of all, one of the criticisms that is often leveled at this particular approach is that in some way it represents the attempt by a high-class intellectual elite to impose its standards on the rest of society. In an academic climate in which a very high premium is placed on identifying new victims, this is an important point that has to be dealt with. I like to make my point by recalling an incident that was reported in the *Chicago Tribune* a number of years ago. It concerned one of the interminable battles going on in the city council. (I do not remember the details, and I am sure you have no desire to learn them.) At the end of the debate, the newspaper printed the following quotation: "'This was a gathering of Neanderthals, without any idea that Cro-Magnon may have existed.' said Alderman Timothy Evans, Fourth Ward Democrat."

The point of this quotation is that Alderman Evans not only thought that the reporter to whom he was speaking would know what Neanderthals and Cro-Magnon were, but that the people who would be reading it—the people who voted for Alderman Evans—would understand these terms as well. It is a clear judgment on his part about the level of cultural literacy among his constituents.

Now let me be clear about this. I was born in Chicago, I grew up there, and I love the city. It is my favorite place in the world, and I go there as often as I can. But a Chicago Alderman is not a member of a high-class intellectual elite!

In fact, I would argue that in a culture as diverse as that of the United States, the fact that we need to communicate with each other drives us to the point of understanding cultural literacy. If it did not exist, we would have to invent it.

The second issue I would like to bring up is that the matrix of knowledge I have been talking about is not a collection of "just facts." In fact, when you look at realizations of the concept—the kind of thing I will talk about later with reference to scientific literacy—you will find that it is a mixture of terms, concepts, philosophy, history, literary allusions, and so on, and that all of these items are multiply connected by various threads. It is a very dynamic intellectual architecture, about as far from a list as you can possibly get.

Finally, we have to acknowledge that cultural literacy does change, both from one culture to another, and within one culture over time. In fact, we now have some base-line information on this. In the five years between the publication of the first two editions of *The Dictionary of Cultural Literacy*, the "to and fro" of terms in the dictionary changed at the rate of about one percent a year. This is the same rate of change as the words in a dictionary of standard English words itself. Consequently, although it is true that cultural literacy changes over time, it does not do so at a rate that is any less manageable than that of the language itself. We know how to deal with changes in the language, and I submit that we can use the same techniques to deal with the changes in cultural literacy.

## SCIENTIFIC LITERACY

There are three important questions one can ask about scientific literacy: (1) What is it? (2) What is it for? (3) How do we get it?

Scientific literacy consists of a potpourri of facts, concepts, history, philosophy, and ideas, all connected to each other by strands of logic. The scientifically literate person knows some of the basic facts about the way the universe operates, and has some sense as to how scientists came to that knowledge. The scientifically literate person can deal with those aspects of science and technology that come across his or her horizon as easily as he or she deals with aspects of economics, law, or government.

Notice that I have not included in this definition of scientific literacy the ability to *do* science. When I go to hear a concert, I do not expect to be stopped in the vestibule and asked to display virtuosity on the violin before being allowed into the auditorium. I would suggest that no one should be asked to demonstrate that they can do science as the price of entry into scientific literacy.

In fact, when students encounter science in their daily lives, it is never in the context of "doing." For example, here are a series of headlines I took from a randomly chosen issue of the *Washington Post.*

NURTURING JOBS IN A LAND OF OLD GROWTH FORESTS

ASSESSING THE RISK IN CONTRACT'S "COST-BENEFIT" CURB ON (ENVIRONMENTAL) REGULATIONS

TOXIC SOUTH AFRICAN ARMS RAISE CONCERN

Each of these stories involves science and something else. For example, the story on old growth forests was primarily about the economic impact of environmental regulations on lumbering towns in the Pacific northwest, but required some understanding of notions of ecology and of the Endangered Species Act. Similar comments could be made for all the other headlines. In other words, science presents itself to the average person in the context of a problem or issue, and without the kind of academic boundaries that come as second nature to people in universities.

What that means is that when we talk about preparing people to meet the challenge of the future, we cannot talk in terms of university departments. Nature itself is seamless, and university organizations fail to reflect this fact only because they grew up before the underlying unity could be seen. So if science has come that way, why should science education not come that way as well? This, in a sense, is my approach to scientific literacy. It must deal with nature as it exists and must deal with nature as it is presented to the average citizen.

Let me give you an analogy. I think of the universe as being like a large spiderweb. Around the edge of the web are all physical phenomena—supernovae, beetles, radio transmissions, trees. If you look at any part of that web and start asking questions—"what is this?" "how does it work?" "why is it the way that it is?"—you begin to work your way toward the center. If you keep asking questions, eventually you come to the center of the web, where you find a small number of general overriding principles. These principles are either shared among all the sciences or serve as the guiding principles and provide coherence to the large subset of the physical world. The law of conservation of energy, for example, would be found at the center of the web; and all scientists, no matter what they are working on, are familiar with the concept of energy and use it to some degree in their work.

The important point about the ideas in the center is not so much to produce a list, but simply the fact that they exist. They tell us that science is organized in a hierarchical way, with phenomena leading back to what I would call great ideas. You may call them great ideas, laws of nature, or any other name you wish. The point is that there are a relatively small number of them.

And that, in turn, gives us a way of getting at education in scientific literacy. The internal structures of the sciences themselves give us an internal framework on which we can hang educational efforts in the discipline. Furthermore, someone who understands these ideas is very likely to be scientifically literate.

Here is my list of the central ideas of science in (more or less) rank order:

1. Causality/order
2. Energy/entropy
3. Electricity/magnetism
4. The atom
5. Chemical bonds/reactions
6. Natural selection/evolution
7. Cellular/biochemistry
8. Genetic code/molecular medicine
9. Plate tectonics
10. Big bang
11. Stars and galaxies
12. Mendelian genetics
13. Properties of materials
14. Earth cycles
15. Ecosystems
16. Relativity
17. Quantum mechanics

## CONCLUSION AND VIEW TOWARD THE FUTURE

In the end, then, what we want is a citizenry who has a sense of how the world works. I am not so interested in a student knowing the details of the structure of a particular set of minerals as I am in that student having a view of how the earth works. To be perfectly frank, this view could be represented by a little television snippet showing the mantel connection, the motion of plates, and the constant change of the earth's surface. In the same way, I am not so interested in the student being able to name various kinds of organic chemicals living in different cells as I am in the student having a picture of how a living cell works—of materials being brought in across cell boundaries, of information on DNA being transcribed and being turned into proteins, which in turn drive chemical reactions, and so on. What I am trying to avoid are situations like the one I encountered recently in which I was told by a student at the beginning of a class that she was against genetic engineering of foods because she did not want to eat genes. At the end of the course, that student may have decided she still did not like genetically engineered foods, but she would be aware that everything she ate had a lot of genes in it. This, to me, is the goal of education and scientific literacy—to give people enough information so that they can make their own, rational decisions about how they are going to run their lives.

I would like to conclude with what I see as the prospects for enhancing scientific literacy. My views on this issue arise largely from my experience in developing a course to teach scientific literacy and writing a textbook along with my colleague Bob Hazen. As part of this project, I have had the opportunity to visit many universities and address many professional meetings on the subject of scientific literacy.

I have found that there are few universities and colleges in the land that are not thinking about changing the way they teach science to nonscientists. In fact, I know of well over a hundred institutions that have adopted our textbook since it came out a couple of months ago, and I take this as an index of the interest in this general area. Things seem to be moving, and this development should be encouraged.

Let me close by commenting on a somewhat unexpected pattern I see emerging at American universities. I see a tremendous amount of change in places that are not generally regarded as flagship institutions. I would not be surprised if in ten years' time, students coming out of places like East Texas State University and St. Petersburg Junior College are going to be much more scientifically literate than students coming out of Ivy League schools. As a guy who grew up on the wrong side of the tracks in Chicago, I must say that I find this to be a delightful prospect.

# SCIENCE EDUCATION AND THE SENSE OF SELF

GERALD HOLTON

*Is it self-serving to advocate support for science? Perhaps. But if the "self" is the American people and the position of leadership of the U.S. in all fields of science and technology in the 21st century, then I wouldn't worry too much about appearing self-serving.* [Yet] *one thing that has been striking . . . is the perceived stony silence of the science and technology community.*

—NEAL F. LANE

IF HISTORIANS of the future will look back on the history of the United States, they will find two curious paradoxes.

On the one hand, many of the contributions to science and engineering in the United States in the twentieth century will be counted among the major intellectual achievements. In addition, this better understanding of how the world works has led to remarkable increases in industrial productivity in the human life span, in agricultural yields, and so forth. Science and engineering will also be seen as major factors in World War II in beating back enemies sworn to bring down Western civilization itself.

Of course, more needs to be done, and all too often the human race has either failed to avail itself of useful knowledge or, ever since the invention of fire, has found ways to abuse it. I know of no intellectual who would dispute this, or who adheres to the flagrant scientism that now exists mostly in fiction and in the rhetoric of critics. But precisely within the awareness of their limits, the sciences have achieved the moral authority to be considered among the central components of the twentieth-century world view, on a par with the other major motivating cultural forces, such as the arts, and what William Faulkner called the universal truths, those of "love and honor and pity and pride and compassion and sacrifice."

From this perspective, given the essential place of the scientific view, our hypothetical future scholars might reasonably expect that science education in the schools and colleges of our day will have been of the same high quality as the sciences themselves, and that the relation between science and society was rendered in a sound and balanced way by the usual purveyors of such information, our scholarly societies, academies, museums, and so forth.

But it is just at this point that those future historians will be startled to discover how wrong these reasonable expectations are. That is the first paradox. I need not spell out the details. We are all heartsick about the data, such as the fact that all but the most elementary rudiments of science, mathematics, and engineering have been squeezed out of the curriculum for the majority of college students. In some of our most prestigious colleges, a total of only two out of thirty-two one-semester courses are required in science and math—about six percent of the educational experience. And those courses usually have to build on the extremely tenuous foundation, if any, laid previously in high school. And into this void flow negative and bizarre views of science and technology, from courses these students take outside science and from the general culture.

This failure is traceable in large part to negligent acquiescence by scholars, intellectuals, teachers, administrators, and above all, by the scientists themselves. That in turn would not have even been thinkable if it were not for a profound displacement that has occurred in the position of science in our culture, at this so-called end of the modern era. On this point, I agree with the thoughtful analysis by Sir Isaiah Berlin, in his book entitled *The Crooked Timber of Humanity: Chapters in the History of Ideas.* He documents that many of the fundamental notions that the world view of the West developed since the Enlightenment are being challenged in our time, as in various periods in the past, by the rise of a worldwide movement that has the technical name "the Romantic Rebellion." As he says, no one predicted that the current form of this rebellion would be what dominates "the last third of the twentieth century."[1]

The timidity and ineffectiveness, at least until very recently, of most scientists and their scientific societies to counter what amounts to an attack on their very legitimacy are easy to document. I think it is wrong to decry the attacks themselves. They are part of an evolutionary mechanism by which cultures are formed. But the feebleness of the attempts to understand the attacks, and to oppose them—that constitutes the second paradox.

Here I want to illustrate the challenge to the legitimacy of science that has been appearing within the science education movement itself—that is, not in books by individual gurus or on the crazy corner of the Internet, but right in the publications and activities of teams of intelligent, well-educated and, on the whole, well-intentioned persons charged with the business of education. It is exactly such products that are most revealing of the role science is thought to play in American cultural life today.

The first document is a draft of the *National Science Education Standards* for grades K–12. This project started with excellent intentions and under the best possible patronage. The National Research Council of the National Academy of Sciences was commissioned to draft such standards in 1991 by a group including the United States Secretary of Education, the National Science Teachers Association, the National Science Foundation—all this in the wake of the 1989 declaration of the nation's fifty governors, who announced that internationally competitive national standards of excellence for the schools of America were important national goals. As is usual in such projects, a complex set of advisory committees and other outreaches was arranged to assure that the standards will not only improve the study of science in schools, but also will represent the consensus of teachers and other science educators, scientists, and the general public. The list of members of advisory committees, national committees, working groups, and professional organizations involved during the first two years takes eight pages, as given in the first draft, released in November 1992. It includes the names of some of our best scientists.

But as these things are done, they did not write the report, and the result surprised some of them. For example, the version of the report released in November 1992 to selected commentators announced at the outset that the intellectual foundations of school science education and the national science education standards would not be merely the science disciplines, but also the philosophy, history, and sociology of science. In my own role as science educator, I have always tried to infuse science teaching with just such elements, so I was delighted to read this. But at that point in the document, there was a footnote directing the reader to an appendix that would set forth the "contemporary views of the philosophy of science." That appendix—written, according to one staff member, by two philosophers who had been hired for the purpose—is a confused farrago that includes the following flat statements: "Two competing paradigms of science have been the focus of disagreement among historians, philosophers, and sociologists. The older, referred to as logical positivism, is characterized by arguments for the objectivity of scientific observation and the truth of scientific knowledge." Parenthetically, I should insert that at that point I seemed to hear howls of laughter from the graves of the logical positivists about this confused summary of their intentions.

But to continue to the next sentence in the draft: "A more contemporary approach, often called postmodernism, questions the objectivity of observation and the truth of scientific knowledge." And after the admonition that science is "the mental representation constructed by the individual," this section on philosophy closed with the flat statement: "The National Science Education Standards are based on the postmodernist view of the nature of science."

The constructivist point of view had found its way into the educational standards being prepared under such high auspices. Of course, this 1992 draft caused some consternation among members of the national committee when they read it, at least two of whom, I am told, resigned. So the groups involved in the writing were sent back to the drawing board, and 18 months later, in May 1994, there was released for comment a second draft of the National Science Education Standards, from the same agency.

I received a copy and read it with care, because the adoption of national standards can have the biggest effect on the nation's precollege education since the founding of public schools. And in fact the May 1994 document had some promising features—for example, in the defense of evolution in biology, and in the pedagogic use of the history of science, where appropriate. But while the explicit section of 1992 on philosophy was now absent, its disappearance was merely strategic. The same social constructivist view, sometimes subtly and sometimes blatantly, shaped the document from one end to the other. The pages were practically soaked in phrases centering on the word "constructing"—five times on some pages—from the introduction, which announced the "principle" that "in science, students ask questions [and] construct explanations," to the final pages where we find the admonition that "college science courses [too] should engage students in . . . constructing personal meaning."

Nowhere appeared a statement that scientists seek to find regularities in nature, or to discover and explain new phenomena or laws, or to reach shareable and testable insights about the lawfulness and order of the natural world. By these proposed standards, Marie Curie did not discover radium, she "constructed" it. Doing science is considered there on a par with constructing technical artifacts or social policies. From this extremist point of view, intended for the indoctrination of generations of students, the document elaborated some consequences. For example, nowhere is the student encouraged to consult as sources of information other, "external" resources such as books (not a single science book is named), atlases, science museums, planetaria, or science television programs. Nor is the teacher considered to be a resource person who already knows facts and who will provide them at the appropriate moment. The students supposedly must live in a closed, solipsistic universe. In companionship with one another, these youngsters will have to construct whatever knowledge about science they will get. As a result, one of the early examples given showed it might take days or weeks for a class to construct the concept of density.

One way of understanding what happened to the draft standards was that the "inquiry" method of learning and teaching—a current favorite among teachers, particularly in the lower grades—had been hijacked and carried to extremes. The inquiry style itself, while necessarily far from really "doing science," has much to recommend it for precollege schools (as did its now-defunct series of predecessors, such as case studies, "discovery," etc.). It also seems to make young students feel good about science learning. But that does not mean that inquiry must be limited to the mental "construction" by the learner, or that scientific research as such should be presented by that model. Also, the teacher should not fail to challenge students to perform at the highest level each can achieve. I believe the word "excellence" was mentioned only once in passing in the draft. Nowhere did I see an attempt to get teachers actively to discover and nurture any extra or exceptional ability, the kind that might make it more likely to produce a future scientist of special ability. Nor do the standards suggest that teachers should get more adequate training in science than most of them do now.

The philosophy underlying this version of course forbids using such phrases as the progress of science or the evolution of scientific ideas. The main engine of change in science presented there lay largely in changes in commitment. "Changes in commitments . . . forge change commonly referred to as advances in science." Thus scientific advances are only tautological concepts, because anything scientists make a commitment to is by definition an advance. What counts is not the discovery of some new aspect of nature, but the sociological and psychological interaction of the scientists. This view is part of the famous, banalized picture of scientific work as the jump from one paradigm to another as a result of changes of commitment in the social group, rather than advance through the enlargement of the range of applicability of older conceptions, or the simplifications of theories, or the breakdown of old barriers between fields.

Happily, this is not the end of the road for these National Standards for Science Education. The final draft will be quite different, and acceptable to scientists and students, teachers and parents, as it surely must be, if the standards are to be effective. But it is important to see how long it took for scientists to wake up to what was happening, and to go into action. At a meeting that brought together some scientists to look over the 1994 draft I have just discussed, a few of them finally made clear how unacceptable this version was. *Science* magazine (December 9, 1994) went so far as to say that one of the physicists there confessed that he "went ballistic" after reading the version. Others there also objected strongly to the bizarre flaws. With Bruce Alberts, the new president of the National Academy of Sciences, taking particular interest in education and science, a much better outcome could be expected, and a much improved version was put into circulation for final comment and release in December 1995. If there is a flight from reason and science to extremism, of which the early stages of this episode were to a certain extent an illustration, what is required most of all is simply that some scientists and their organizations begin to pay attention to how the process of the delegitimation of science operates, and to defend themselves.

There is a faint analogy here with a lesson from the Gray board hearings in the matter of Robert Oppenheimer. In that case, it was Oppenheimer himself who was so strangely ineffective when his integrity was attacked. He seemed to acquiesce in his own destruction. But it is important to recall how Oppenheimer explained his inaction later. He said: During the hearings, "I had very little sense of self." His friend David Lilienthal, when also brow-beaten during his testimony on behalf of Oppenheimer, saw the rigged hearing procedure for what it was, and commented: "When I saw what they were doing to Oppenheimer, I was ready to throw chairs."[2] Oppenheimer himself, the architect of enormous weapons, was not a man to throw chairs.

I too do not advise throwing furniture. The point I am making by this useful analogy is that the scientific community, and its organizations, will have to learn quickly that one's moral authority depends on having a good "sense of self," and to act in its self-defense when attacked.

---

I could illustrate in detail a similar distortion of the place given to science and education in the culture of the United States today in another quasiofficial document that is intended to have a profound effect on schooling. That document is the recently released *National Standards for United States History* for grades 5–12, subtitled *Exploring the American Experience.* It, too, has distinguished official patronage, in this case principally by the National Endowment for the Humanities and the Department of Education, and has benefited from the labors of "hundreds of classroom teachers of history [and] dozens of talented and active academic historians in the nation," plus many professional and scholarly organizations. And this document, too, has been in the news, attacked and defended for various reasons. But little has been said about the role given to science and technology in the current draft, which is to guide how history is taught to each youngster during eight years of schooling. Again, the intentions were honorable, the labors onerous, but the results provide a window on the perplexed soul of our culture, and reveal the same animus of which I spoke earlier.

For you would expect to find that science, technology, and medicine would play a major role in the account these standards should urge for the study of our history, just as these fields have in fact played a major role in our history itself, especially so during the past one-and-a-half centuries. But the set of interlocking standards and examples in the published history standards refers to such topics only very infrequently, briefly, and anemically.

What is lacking, first of all, is any sense that science and technology in the United States, certainly since World War II, are in most fields culturally and intellectually strongest compared with the rest of the advanced nations. The *existence* in the United States of the development of this world-class science and technology, achieved in this century by the work of millions, is a historic fact of real importance—quite apart from and in addition to the *effects* on society. (A similar but less severe case can be made against the treatment of the arts, to which that document pays attention in a quite slanted and narrowly selective way.)

The most responsible attempt in the history standards document appears on the page before the very end: "Students should be able to demonstrate understanding of the modern American economy by . . . evaluating the importance of scientific and technological change on the workplace and productivity." A few examples of students' achievements intended to meet such standards are given on that page. But apart from appearing very late in the curriculum, the examples are entirely preoccupied with the "effects of" or "impacts of" technologies, rather than establishing first their existence and how they came into our history in the first place. As one observer noted: "Science is [here] implicitly viewed as something which has descended from elsewhere. . . . Technology, too, is treated as an external body which 'impacts' upon society. . . . An entire standard devoted to 'the second industrial revolution' is concerned solely with how it 'changed the nature and conditions of work,' for example in the employment of children." And he adds, "The proposed standards represent the current thought of a cross section of leaders in American primary and secondary education in academic history. Evidently if students are to learn what society has done in the past to nurture science, how science has affected the development of technology, and where scientific thought had stood within American culture, in the future as in the past, they should not look to the typical high school history course."[3]

A glance through the published volume shows how uncomfortably true this critique is. Thus an example of student activity is to "examine the influence of MTV on popular culture," with a special question posed: "How does Madonna symbolize the popular culture created by MTV?" Madonna and MTV appear to be important; but how did TV itself come to be? And what about the telegraph, the telephone, radio, video, the laser, and the transistor, all surely known to students, and whose economic impacts have been vast? Would it not be well to put a little of the story of their American sources into the Standards for United States History, the more so as we are now in the Information Age?

The producers of these history standards have explained that they do not want to make lists of people, since this may be the job of the ultimate producers of curricula based on these standards. Yet one wonders whether it would have handicapped curriculum designers if Samuel Morse, Alexander Graham Bell, Thomas Edison, and others like them had at least been mentioned (not to speak of Benjamin Franklin, the Newton of electricity).

I inquired from some of my colleagues what they would like to see in such a set of national standards for United States history, and I was flooded by suggestions: electric lighting; a whole slew of chemical inventions—and particularly synthetic rubber, without which the United States, cut off from natural rubber in World War II, could hardly have entered the war successfully; something might be said about the finding and refining of petroleum, on which the development of the automobile depended; the invention and application of digital computers; the major immunizing interventions such as the Salk and Sabine contributions to the conquest of polio; before that, the discovery of penicillin by Fleming, and the age of antibiotic therapy that followed; the role of smallpox in history and the elimination of that scourge; similarly the conquest of diseases transmitted by contaminated drinking water. How about the Wright brothers' experiments and designs? Is it possible to have aircraft missing from the history of the United States?

Under these standards, students should be able to "demonstrate understanding of World War II and how the Allies prevailed."[4] Good. But radar is missing, which really helped win the war. And from grade 7 to grade 12, students are invited to examine the "decision to use atomic weapons during World War II." But there is again nothing here about why it was thought necessary to build such weapons in the first place. And that is more than ever an important part of the story, one that must not be neglected.

During this century, an increasing understanding of genetics has resulted in major advances in biology, medicine, and agriculture. One may leave to the *science* standards the description of the Watson-Crick model. But the fact of the *discovery* of the structure of DNA itself in 1953 was a world-historic event. Might it not merit one line of print in the history standards? Last but not least, students should know not only about the history of racism and Social Darwinism in the 19th century, to which a good deal of space is given in these standards, but also of its refutation by anthropological and related sciences in this century. This would be a good place to show that the efforts of scientists have helped to change and expand the concept of human rights in modern times.

All these examples of gaps indicate that they were not accidental. They stem from a particular contemporary vision of what counts in history, and that science and technology are low indeed on that agenda. The predictable reply to such suggestions will be that the document is already too long, and that nothing could be squeezed out to make room. To which I would gladly respond, "Just let me try." In fact, a few scientists have asked the History Standards writers to reconsider the omissions—again in the spirit of correcting a false image of our culture.

---

I will briefly mention a third example where an enterprise with national scope, distinguished patronage, and years of labor resulted in a severely flawed educational presentation of science and technology in our time. This is of course the exhibit "Science in American Life," now showing as a permanent installation at the National Museum of American History of the Smithsonian Institution in Washington (not to be confused with the plans for the Enola Gay exhibition at the National Air and Space Museum). There has been a great deal of public commentary, pro and con. As *Science* magazine put it (5 August 1994), after the commissioning of this exhibit by the American Chemical Society at the cost of $5.3 million, there ensued a "five-year battle between an advisory committee appointed by the ACS and curators at the Smithsonian's Museum of American History. At the heart of the battle was the portrayal of science." As one of the historians associated with the project in the early stages reported, "the chemists came in for a rude shock." "They wanted something like the Du Pont slogan—'better living through chemistry'—[but the head curator] and others wanted pollution and death." A second member of the advisory committee added, "We spent most of our political capital making sure it wasn't a complete exposé of the hazards of science." And yet a third member of the advisory committee wrote: "Several historians assigned to the curatorial team made no secret of their disdain for 'big science' and of everything they believed it represented. Their political ideology opposed industry, and dismissed chemical manufacturers as 'polluters.' Eventually the lead curators . . . wound up creating a largely negative [exhibit]."

This is not the place to rehearse the flaws of that exhibit, which I have now seen. What I do want to tell here is why, in this case, as with the National Standards for Science, there is, after all, hope that the prejudicial and unbalanced representation of science for the education of a national audience is going to be corrected. The main reason for this rethinking is that while all the individual objections from scientists and others at first made no impression on the Smithsonian and its curators, at last a scientific society—specifically the American Physical Society—blew the whistle loudly. Asked to do so by his elected council of the society, the president of the APS, the Nobelist Burton Richter, reported to the Secretary of the Smithsonian that APS members who had visited the exhibit "Science in American Life" had found the portrayal of science trivializing its accomplishments and exaggerating any negative consequences. They were concerned that the presentation was seriously misleading, and inhibited the public's ability to make informed decisions on the future use of science and technology. Acknowledging that examples of misuse do belong in such an exhibit, he stressed that what was missing was balance.

He offered the services of the APS to work with the Smithsonian and with others to develop a more appropriate exhibition, and indeed arranged for the whole Executive Committee of the APS to visit the Smithsonian staff in Washington. On that occasion, the flaws were pointed out, and the possibility of a rethinking seemed achievable.

Three-and-a-half months later, in February 1995, the chairman of the Board of Directors of the American Chemical Society also made the principal concerns of his society known to the Smithsonian. He described the constant frustration, over several years, as the Smithsonian staff consistently ignored the advice given them by an advisory committee jointly put together by the ACS and the Smithsonian. He spoke of arrogance and high-handedness on the part of some of the museum personnel assigned to the project, the demonstration of a strong built-in bias against science, and the tendency to revise and rewrite history. He too offered to provide advice and personal attention to result in a more balanced presentation, one that would repair the damage done to the reputations of both organizations.

As a result of these communications, the Smithsonian has promised to revise at least the most egregious parts of the "Science in American Life" exhibit. When that actually happens, the principle I am advocating will have been demonstrated again: *The moral authority of science, as of any professional field, depends importantly on asserting its sense of self.*

To our Romantic rebels, and to the other purveyors of the equation that science equals pollution and death, the intervention of the scientists in the Standards story, and of the physical and chemical societies in the Smithsonian exhibit, will probably be only additional proofs of their point of view. Theirs is a win-win position: if scientists do not object to an unbalanced presentation, then it will have its intended effect on the populace. And if scientists do object and bring about reconsiderations of the documents and exhibitions, that will be said only to illustrate how scientists wield undue influence.

But the fact is, as I noted, that the scientists and their organizations until recently have usually been too negligent and reluctant to speak out and defend themselves against the dissemination of false images. They seem to begrudge any moment away from the exciting work in their labs. Perhaps they are now beginning to discover the danger of letting others define the ideological, educational, and cultural context of science in today's society.

## NOTES

1 I refer you to the details in other essays in his book, and also to my analysis in the chapter "What Place for Science at the 'End of the Modern Era'?" in my book *Einstein, History, and Other Passions.*

2 Philip M. Stern, *The Oppenheimer Case*, pp. 128–130.

3 "Role of the History of Science and Technology in the Proposed National Standards for Schools," pp. 1–2.

4 "National Standards for United States History," pp. 201–203.

## REFERENCES

BERLIN, SIR ISAIAH. *The Crooked Timber of Humanity: Chapters in the History of Ideas.* New York, NY: Random House, 1992.

HOLTON, GERALD. ***Einstein, History, and Other Passions.*** New York, NY: American Institute of Physics Press, 1995; Reading, MA: Addison Wesley Publ. Co., 1996.

NATIONAL ENDOWMENT FOR THE HUMANITIES and THE DEPARTMENT OF EDUCATION. "National Standards for United States History," 1995.

NATIONAL RESEARCH COUNCIL OF THE NATIONAL ACADEMY OF SCIENCES. "National Science Education Standards." First draft, November 1992.

NATIONAL RESEARCH COUNCIL OF THE NATIONAL ACADEMY OF SCIENCES. "National Science Education Standards." Second draft, May 1994.

"Role of the History of Science and Technology in the Proposed National Standards for Schools." ***History Newsletter of the American Institute of Physics***, Spring 1995, pp. 1–2.

STERN, PHILIP. ***The Oppenheimer Case.*** New York, NY: Harper & Row, 1969.

# AFROCENTRIC PSEUDOSCIENCE
## The Miseducation of African Americans

BERNARD R. ORTIZ DE MONTELLANO

MINORITIES ARE UNDERREPRESENTED in science and engineering. Children in these groups need examples of the role people like them play and have played in science. In 1987, the Portland, Oregon School District published the *African-American Baseline Essays,* a set of six essays providing resource materials and references for teachers on the knowledge and contributions of Africans and African Americans. Unfortunately, this widely distributed attempt will increase scientific illiteracy and impede the recruitment of African-American children into scientific careers.[1] Our discussion will focus on the Science Baseline Essay written by Hunter Havelin Adams,[2] listed as a research scientist at Argonne National Laboratory. Actually, Mr. Adams was an industrial hygiene technician who "does no research on any topic at Argonne," and whose highest degree was a high school diploma.[3]

There are serious problems with this Baseline Essay, but because of the current pressure on school districts to incorporate multicultural material into the classroom and because of the dearth of this kind of material, it has been widely distributed. Hundreds of copies of the Baseline Essays have been sent to school districts across the country. They have been adopted or are being seriously considered by school districts as diverse as Fort Lauderdale, Detroit, Milwaukee, Atlanta, Chicago, Prince George County, Maryland, and Washington, D.C. Scientists and educators need to be aware of the tone of and the errors in the Baseline Essay, and of the harm its widespread adoption may cause to the goal of diminishing scientific illiteracy in this country.[4]

### USEFULNESS OF THE MATERIAL

One of the goals of the Baseline Essay is "mastering the basic concepts of mathematics and science,"[5] but this text cannot be used as is by a teacher in the classroom to teach *science.* Most of the material is presented as "look how great Egypt and Egyptians were," no in-depth explanations of any scientific principles are provided, and no connection is made between the topics covered and specific K–6 classroom activities. The level of presentation and detail of the Science Baseline Essay resembles a social science or a history of science presentation rather than one that would allow grade school teachers (who have very little science training) to explain any of the scientific concepts mentioned to a class. The lesson plans for grades 5–8 available from the Portland School District are pretty straightforward classroom exercises but, in fact, use very few of the ideas put forth in the Baseline Essay. One of the reasons for this is that many of the topics mentioned in the Baseline Essay (glider flight, batteries, quantum mechanics) are not suitable for a grade school curriculum.

The Baseline Essay does not distinguish between the serious academic sources (i.e., referred journals or books from academic presses), intermediate sources (popular science journals), very old sources (which might be obsolete), and unreliable or very questionable sources (newspapers, magazines, vanity press books, or "New Age" publications). They are all treated as equally reliable. The citation style is not helpful either for teachers who want to get more information or for readers who want to verify quotations. Quotations in the text are often cited only by author and title, and a full reference is not included in the bibliography. Even when books are included in the bibliography, page numbers are not given in the footnote.

## CLAIMS THAT THE NATURE OF SCIENCE DEPENDS ON WHO DOES IT

One of the Afrocentric strategies is to claim that Western science is just one of many equally valid approaches to the study of nature, and that therefore, African or Egyptian mythical, magical, or folk knowledge is science. The Baseline Essay cites Wade Nobles's definition of science: "the formal reconstruction or representation of a people's shared set of systematic [sic] and the cumulative ideas, beliefs, and knowledge stemming from their culture."[5] This is not an acceptable definition of science. Science is both a special kind of information and a method. A. N. Strahler defines it thus,

> *Scientific knowledge*: the best picture of the real world that humans can devise, given the present state of our collective investigative capability. By "best" we mean (a) the fullest and most complete *description* of what we observe, (b) the most satisfactory *explanation* of what is observed in terms of relatedness to other phenomena and to basic or universal laws, and (c) description and explanation that carry the greatest probability of being a true picture of the real world . . . scientific knowledge is imperfect and must be continually restudied, modified and corrected . . . *Scientific method*: the method or system by which scientific knowledge is secured. It is designed to minimize the commission of observational errors and mistakes of interpretation.[7]

To this must be added acceptance by the *scientific community* because the attempts at verification or falsification of scientific claims by this community are a crucial part of the scientific method.

The Baseline Essay quotes Ruth Bleier (with no page number), "The dominant categories of cultural experience—*race, gender, religion, and class* will be reflected in the institution of science itself: in its structure, theories, concepts, values, ideologies and practices." (words in italics added by Adams).[8] The Baseline Essay, referring later to this passage, states, "In this light, the common concepts of mathematics, of physical theories such as mass, momentum, and energy; electric charge and magnetic field, the quantum wave function; entropy; distance and time; and even myth, are actually no more than useful organizing strategies our consciousness has developed for ordering the chaos of information it receives from its environment." The clear implication is that concepts such as mass, entropy, momentum, and distance will depend on and vary with the sex and/or race of the person dealing with them. But serious feminist critiques of science do not say this. Feminists criticize science because it is organized and run patriarchally and because this then determines research strategies and interpretations of data, particularly biological data, as in some sociobiological explanations of the role of males and females. No feminist scientist has claimed that the structure of DNA will differ if the analysis is done by a man or by a woman. Similarly the definition of momentum as mass times velocity or of mass as force times acceleration will not differ regardless of the race or culture of the observer. It is also irresponsible to put myth in the same category and level as concepts such as the quantum theory. This postmodern perspectivist model of science has been likened by Gross and Levitt to the Melanesian cargo cults of World War II: "It bears only a vague and superficial resemblance to the real thing, and its internal logic is laughably different. Still, those who built it hope with the aid of their theoretical magical rituals, to gain control over the real thing."[9] My fear is that grade school teachers, the intended audience for the Baseline Essay, will not be sophisticated enough in science to make these distinctions.

## CLAIMS THAT SCIENCE ALSO INVOLVES THE SUPERNATURAL

The Baseline Essay claims that the Egyptians had a concept called *Ma'at* that guided their research and represented the "first set of scientific paradigms." These "scientific paradigms" include "existence via Divine Self-Organization," "acknowledgement of a Supreme Consciousness," and the existence of "transmaterial cause and effect," i.e., supernatural causation. Therefore, "the fundamental scientific paradigms of the ancient Egyptians . . . are antithetical to contemporary ones. . . . Many Western scientists conduct their process of science from a totally different basis, one of which has as its 'main concern,' non-ethical considerations such as cost effectiveness."[10] Apart from the non sequitur of the main concern of Western scientific methodology being "cost effectiveness," which has nothing to do with the basic principles of science, Adams is correct that *Ma'at* in antithetical to Western science. The key question is whether children in public schools are going to be taught that religion (under the guise of "Egyptian science") equals science. This is the same question that was roundly forbidden by both lower courts and the United States Supreme Court in the case of so-called scientific creationism. The essence of the decision of Judge Overton in *McLean vs. Arkansas*[11] was that scientific explanations cannot involve the supernatural.

Apart from the questionable constitutionality of teaching religion (be it Christian or Egyptian) in the public schools, it is a great disservice to give to children such a distorted view of what constitutes science. At some point, these students will enroll in a real science class and will be at a disadvantage because of their quaint and erroneous view of what constitutes science.

One could teach Egyptian religion and ethics in social studies as comparative religion, but that is quite a distance from teaching that Egyptian religion is an essential component of Egyptian "science" and that it is superior or equal to "Western" science. Teaching morality and ethics is compatible with teaching science, i.e., there *are* ethics involved, honesty, truth, respect for others, even ethical contemplation of the consequences of research, but these are independent of the exclusive use of natural laws to explain and understand scientific phenomena. The Second Law of Thermodynamics does not have a supernatural component. Its application in particular cases might have consequences that raise moral and ethical questions, which require discussion, but this is quite different from teaching that supernatural (or transmaterial) causes are acceptable explanations in science.

This whole section of the Baseline Essay is very disturbing. What is its purpose in the curriculum? Do we want to teach *science* as it is usually conceived, or religion? How is a teacher to teach this section to his or her class? To say that African science invokes the supernatural, but that Western science does not, with the implication that "Egyptian science" is equal or superior to Western science will only perpetuate scientific illiteracy. How will students distinguish science from astrology, channeling, crystal healing, telekinesis, psychic surgery, and a myriad of other "New Age" pseudoscience that is floating about?

It is possible to teach that ancient people, Egyptian, Maya, Inca, etc., did scientific observation of the skies (i.e., careful observation, record keeping over the years, mathematical handling of the data, hypothesis testing, etc.). This serves the legitimate and presumed purpose of multicultural education: to show that groups other than European white males engaged in scientific activities and that they have contributed to the advancement of knowledge. What must be done is to point out that, although the activities were scientific, the original motivation for carrying out the activity was religious, because those cultures, just as Europe before the Scientific Revolution, felt that the supernatural influenced everything. It is possible to teach Mayan or Egyptian religious concepts in social science as historical events, while teaching their scientific, astronomical aspects as science stripped of their mystical attributes.

## CLAIMS THAT PARAPSYCHOLOGY AND SUCH ARE SCIENCE

Adams does not distinguish between science and pseudoscience. In 1987, he claimed a scientific basis for astrology, "At birth every living thing has a celestial serial number or frequency power spectrum . . . this is the basis for astrology."[12] The Baseline Essay clearly implies that the zodiac and astrology are scientific ("astropsychological treatises")[13] and later claims that "the Ancient Egyptians were known the world over as the masters of "magic" (psi): precognition, psychokinesics, remote viewing and other undeveloped human capabilities."[14] The Baseline Essay claims that there is a misunderstanding and that

> we first must know the extremely significant distinction between (non-science) "magic" and (science) psychoenergetics. . . . Psychoenergetics (also known in the scientific community as parapsychology and psychotronics) is the multidisciplinary study of the interface and interaction of human consciousness with energy and matter. . . . Psi, as a true scientific discipline, is being seriously investigated at prestigious universities all over the world (e.g., Princeton and Duke). We are concerned here only with psi in Egypt, not "magic" . . . its efficacy depended on a precise sequence of actions, performed at specific times and under controlled environmental conditions, facilitated by the "hekau" (the Egyptian term for professional psi engineers). . . . Today in a similar manner, psi is researched and demonstrated in controlled laboratory and field experiments.[15]

The only evidence for these claims is one of Adams's previous lectures.[16]

The overwhelming majority of scientists consider parapsychology, by whatever name, to be pseudoscience. The problem is that this essay, presented with the imprimatur of the Portland School Board, supposedly written by a scientist, and supposedly reviewed by other scientists, will lead scientifically unsophisticated teachers to believe that magic *is* science. The Baseline Essay compounds the problem by including Aaron T. Curtis, an "electrical engineer and psychoenergeticist," along with George Washington Carver, Benjamin Banneker, and Ernest E. Just in a list of African-American contributors to science and technology.[17]

A further example is the claim that "Egyptians diagnosed and treated trans-material disturbances. To the minds of the Egyptians, in fact to most African people, all elements of life, whether human beings, animal, plants, gods, and even inanimate matter such as stones and stars, are imbued with trans-material primordial energy."[18] The ancient Egyptians call this creative energy "za" (known as "prana" to the Hindus and "chi" to the Chinese,[19] which human beings could manipulate. . . . This "magico-spiritual" aspect of African medicine still baffles Western-trained scholars and practitioners, and as such is dismissed, belittled, and downplayed. However, recent research is uncovering the "hard" evidence of the validity of this aspect of African medicine." According to the Baseline Essay, the laying-on-of-hands procedure called "therapeutic touch" promoted by Doloris Krieger has been shown to work and is a rediscovery of ancient Egyptian practices. Therapeutic touch is very controversial and is far from being "proven to work in clinical trials" or of being accepted in mainstream medicine. All medicine, including both Western and non-Western, consisted primarily of placebo effects until the twentieth century. The effects cited by the Baseline Essay (reduction of blood pressure, reduction of stress, enhancing [as well as harming] the immune system) can be explained by the placebo effect and by psychoneuroimmunology and can be elicited by any number of procedures including touch, therapeutic or not.[20] A literature survey of therapeutic touch[21] concluded, "the current research base supporting continued nursing practice of therapeutic touch is, at best, weak. Well-designed, double-blind studies have thus far shown transient results,[22] no significant results,[23] or are in need of independent replication."[24] The practice of therapeutic touch by nurses will never gain professional credibility without clear, objective evidence to support it. Without this evidence, the nurse practitioners of therapeutic touch will be relegated to the practice of "placebo mumbo jumbo." There is a very long distance between claims that the mind and emotions or people can affect their immune system and lower their resistance to illness, which is a legitimate area of study, psychoneuroimmunology, and claims that rocks have a mind and that Egyptians, or anybody, can manipulate physical reality from afar by psychic powers.

## EXAGGERATED CLAIMS WITH LITTLE EVIDENCE

The Baseline Essay also makes some extraordinary claims about the past and present astronomical knowledge of the Dogon of Mali.

> They knew of the rings of Saturn, and the moons of Jupiter, the spiral structure of the Milky Way, where our star system lies. They claimed that billions of stars spiral in space like the circulation of blood in the human body. . . . Perhaps the most remarkable facet of their knowledge is their knowing intricate details of the Sirius star system, which presently can only be detected with powerful telescopes. The Dogon knew of the white dwarf companion star of Sirius, the brightest star in the sky. They knew its approximate mass ("it is composed of 'sagala,' an extremely heavy, dense metal such that all the earthly beings combined cannot lift it") its orbital period (50 years) and its axial rotation period (one year). Furthermore, they knew of a third star that orbits Sirius and its planet [sic]. The X-ray telescope aboard the Einstein Orbiting Observatory recently confirmed the existence of the third star.[25] The Dogon with no apparent instrument at their disposal appear to have known these facts for at least 500 years.[26]

No evidence is presented to support these claims. The sole citation is a previous paper by Adams.[27] In it, Adams made additional claims, which were endorsed by Van Sertima,[28] that the Dogon knew these things for at least 700 years (not 500 as in the Baseline Essay) and that the ancient Egyptians also possessed this knowledge. The sole source for all of these claims about Dogon astronomical knowledge is the research of two French anthropologists, Marcel Griaule and Germaine Dieterlen, and more directly a book by Temple.[29] Griaule and Dieterlen do not claim that this knowledge was ancient or Egyptian, and Temple argues that this knowledge comes from aquatic inhabitants of a planet in the Sirius system that have visited earth in the past.

Whatever the source of Dogon cosmology it was not based on a visual observation of Sirius B. The visual magnitude of stars is a logarithmic scale—the larger the number the dimmer the star. The usual limit under the best of conditions for naked eye observations is 6.5, although some exceptional individuals with much training can observe stars of 7.8 magnitude. Sirius A, a very bright star, has a magnitude of $-1.4$ while Sirius B has a magnitude of 8.7.[30] It is impossible to see Sirius B with the naked eye.[31] Adams and Van Sertima further claim that the Egyptians used telescopes to see Sirius B. "The Russians have recently discovered a crystal lens, perfectly spherical and of great precision, used in ancient Egypt.[32] It is a short and simple step to place one lens in front of another to make a basic telescope, and chances are that it could have happened many times." Because of its proximity to the bright Sirius A, Sirius B is very hard to see even with a telescope. Under the best conditions at least a 5-inch telescope is required and when it is closest to Sirius A, about half the time, a 100-inch telescope would be needed.[33]

The Baseline Essay claims that, "Egyptians had a theory of species evolution at least 2000 years before Charles Darwin developed his theory of species evolution." The sole evidence for this is a quotation from "The Book of Knowing the Evolutions (the becomings) of Ra (the creator sun god)":

> The words of Neb-er-ter who speaks concerning his coming into existence: "I am he who *evolved* himself under the form of the god Khepra (scarab beetle), that was *evolved* at the "first time." I the *evolver* of evolutions, *evolved* myself from the primordial matter which I had made . . . which has *evolved* multitudes of evolutions at their "first time."[34]

This is an unremarkable cosmological text. Students of comparative religions could produce dozens of similar texts from various religions in which the gods create (evolve?) themselves. For example, in Aztec mythology, Teotihuacan is "the place where gods made themselves." The use of the word *evolve* in its ordinary dictionary meaning of "to develop, to achieve gradually; to undergo change" in no way proves that the user meant the same thing as Charles Darwin's extended theory. The concept of evolution of the species, as proposed by Darwin, involves a whole series of other concepts such as (1) a species being defined as a reproductively isolated group of individuals; (2) that species produce many more offspring than those that survive; (3) that depending on particular environments, certain characteristics will confer advantages on some individuals, and that these individuals will have a reproductive advantage over other individuals; (4) that these characteristics will be inherited by offspring and that over long periods of time this will result in the "evolution, i.e., change over time" of new species. It is ludicrous to claim that the passage cited in the Baseline Essay encompasses or preempts Darwin. As Samuel Gill points out,[35] there are two basic principles that apply to this situation: (1) the burden of proof is on the claimant, and (2) the extraordinary claims require extraordinary proof. The Baseline Essay has clearly failed this test.

Under the heading of "Egyptian Aeronautics," the Baseline Essay argues[36] that a model of a glider was discovered in a tomb near Saquara, and that "this ancient Egyptian glider looks contemporary and bears a strong resemblance to the American Hercules transport aircraft . . . it definitely was not a toy. It was a scale model of a full-sized glider!" Without any further evidence, the Baseline Essay cites the *Guinness Book of World Records,* "[Dietches] has researched evidence of the use of gliders in Ancient Egypt from 2500 B.C.–1500 B.C. He says that the Egyptians used their early planes for travel, expeditions, and recreation!" This is the extent of proof for this extraordinary claim. In fact, the model is that of a bird and the ratio of the length of its wingspan to its body length is approximately 1:1 whereas most gliders have a ratio of 2:1.[37]

There are numerous other extraordinary claims backed up by little or no evidence, such as the ability of ancient Egyptians to predict pregnancy by urinating on barley seeds,[38] that they anticipated many of the philosophical aspects of quantum theory[39] that they knew the particle/wave nature of light,[40] and that "enclosed with the Great Pyramid are the value of pi, the principle of the golden section, the number of days in the tropical year, the relative diameters of the earth at the equator and the poles, and ratiometric distances of the planets from the sun, the approximate mean length of the earth's orbit around the sun, the 26,000–year cycle of the equinoxes, and the acceleration of gravity.[41]

## WHAT CAN BE DONE?

The pressure to teach Afrocentric science differs from that pushing "scientific creationism." Afrocentric science is pushed from the top down, from school boards, superintendents, and math-science departments in urban school districts. It is difficult for teachers to resist, and in many cases, particularly in grade schools, teachers do not have enough scientific knowledge to criticize the material.

One possible strategy is to use the Baseline Essay to teach critical thinking, one of the principal aims of the reform in teaching science, by subjecting its claims to scientific scrutiny. By doing this, students can learn science and the process of inquiry, and as a by-product will expose the fatuity of the Baseline Essay. For example, students could be asked to verify what is required for full size gliders to fly.[42]

The Baseline Essay asserts, with no evidence or citation, that the ancient Egyptians had discovered the law governing the oscillation of the pendulum: "The heartbeat also provided them with a convenient measure of time; from its average rhythm they determined the length of the plumb line that would cause it to oscillate to that rhythm. They discovered the rate of oscillation of a pendulum (plumb bob) varies in inverse ratio to the square of its length, which in this case is 0.69 meters, or about 27 inches."[43] Students could calculate the period of a 69-cm pendulum or very simply make a pendulum with this length and test whether it oscillates at 72 oscillations/minute.

The Baseline Essay claims that the Egyptians could electroplate gold: "Electroplated gold and silver objects have been found in Egypt from roughly the Middle Kingdom (2100 B.C.) to the Greek era of the Ptolemies (350 B.C.)."[44] One could teach a lot of chemistry in debunking this statement. Students should be asked to find out how to dissolve gold and how gold is electroplated, what voltage would be required, and how the Egyptians would have achieved this.[45]

Gerald Holton[46] proposes that antiscience be fought by instituting broad reforms to improve teaching science such as Project 2061. Unfortunately, this is a long-term project, and in our specific case it is insufficient because Project 2061 has no multicultural content. There is a great need to develop multicultural materials in science to fill the demonstrated need that is currently being filled by materials such as the Baseline Essay. These materials, however, must be absolutely accurate historically as well as scientifically.

Holton urges a vigorous opposition to the inclusion of pseudoscience into the schools. As a first step, school systems that claim to be using an "African-centered education," "Afrocentric," or "multicultural" curricula should be urged to divulge the sources they use to teach science. Almost invariably they will include works discussed here.

People who are genuinely concerned with improving science education in the schools and with increasing the number of minority scientists should vigorously oppose the inclusion of material into the curriculum that makes unsupported claims, introduces religion under the guise of science, and claims that the paranormal exists. The critical need to increase the supply of minority scientists requires that they be taught science at its best rather than a parody.

## NOTES

1 There are other Afrocentric works on science that are also severely flawed. One is *Blacks in Science*, edited by Ivan Van Sertima, which is a mixture of worthwhile articles, biographical sketches of African-American scientists, and problematical articles by Hunter Adams, Beatrice Lumpkin, John Pappademos, and Charles Finch as well as an article claiming the existence of gliders in ancient Egypt. Lumpkin's essay has been critiqued by Lowe (*Skeptical Inquirer*), and Pappademos's scholarship has been questioned by Palter ("*Black Athena*, Afrocentricity, and the History of Science"). The dean of Afrocentric studies, Molefi Kete Asante, has written a short, poorly documented imitation of the Portland Baseline Essays ("African–Puerto Rican Centric Curriculum Guide"). The title is misleading because nothing is included about the Spanish or the Taino Indian background of Puerto Ricans in this work.

2 Hunter Havelin Adams, "African and African-American Contributions to Science and Technology," hereafter cited as Baseline Essay.

3 D. Baurac, Letter to Christopher Trey; M. Marriot.

4 B. R. Ortiz de Montellano, "Multicultural Pseudoscience. Spreading Scientific Illiteracy among Minorities—Part I"; "Magic Melanin. Spreading Scientific Illiteracy among Minorities—Part II"; "Afrocentric Creationism"; "Avoiding Egyptocentric Pseudoscience: Colleges Must Help Set Standards for Schools"; "Melanin, Afrocentricity and Pseudoscience"; I. M. Klotz, "Multicultural Perspectives in Science Education: Our Prescription for Failure."

5 Baseline Essay, p. 4.

6 Ibid., p. v.

7 A. N. Strahler, *Understanding Science*, pp. 27–28.

8 P. vi. Adams also deleted various phrases and several paragraphs from the original text (Ruth Bleier, "Introduction," in *Feminist Approaches to Science,* pp. 2–4) with no indication of having done so.

9 Paul R. Gross & Norman Levitt, *Higher Superstition: The Academic Left and Its Quarrels with Science,* p. 41.

10 Baseline Essay, pp. 11–14.

11 *McLean v. Arkansas Board of Education.*

12 Hunter Havelin Adams, Lecture, 1st Melanin Conference, broadcast, "African-American World View."

13 Baseline Essay, pp., 29–30.

14 Ibid., pp. 41–42.

15 Ibid. *HkEy* (hekau) means magician (R. O. Faulkner, *A Concise Dictionary of Middle Egyptian*, p. 179; G. Pinch, *Magic in Ancient Egypt,* pp. 50–53).

16 Hunter Havelin Adams, "Psychoenergetic Aspects of Ancient Egyptian Lifeways."

17 Baseline Essay, pp. 81–82.

18 Ibid., p. 44. This is nothing more than the widespread religious belief of *animism,* well known to anthropologists, but which no one claims to be an objective reality.

19 *Za, prana,* and *chi* are not equivalent and are certainly not coterminous with animism. What we have here is the usual semidigested mixture of beliefs common to New Agers.

20 H. Brody, *Placebos and the Psychology of Medicine,* p. 55; Morton, pp. 46–50.

21 P. E. Clark & M. J. Clark, "Therapeutic Touch: Is There a Scientific Basis for the Practice?"

22 B. Grad, R. Cadoret & G. Paul, "An Unorthodox Method of Treatment of Wound Healing in Mice."
23 G. Randolph, "The Difference in Physiological Response of Female College Students Exposed to Stressful Stimulus, When Simultaneously Treated by either Therapeutic or Casual Touch."
24 B. Grad, "A Telekinetic Effect on Plant Growth."
25 The paper cited as evidence for this (T. Chlebowski, J. P. Halpern & J. E. Steiner, "Discovery of a New X-ray Emitting Dwarf Nova 1E 0643.0-1648") does not claim that the X-ray-emitting dwarf 9′ south of Sirius is a third companion. Sirius is 8.7 light years away from Earth, while this new star is 325 light years away and could hardly "orbit Sirius."
26 Baseline Essay, p. 60.
27 Hunter Havelin Adams, "African Observers of the Universe: The Sirius Question."
28 In I. Van Sertima, ed., *Blacks in Science. Ancient and Modern*, pp. 7–26.
29 M. Griaule & G. Dieterlin, "Un Système Soudanais de Sirius"; *Le renard pâle*; R. G. Temple, *The Sirius Mystery.*
30 C. W. Allen, *Astrophysical Quantities*, p. 235.
31 Bradley Schaefer, personal communication, May 18, 1995.
32 A sphere would be useless as a lens because the focal length would be extremely short, and because the image produced would be greatly distorted by spherical and chromatic aberration. This baseless claim has been endorsed by Sandra Harding, the feminist philosopher of science (*Whose Science? Whose Knowledge?*, pp. 223–224).
33 B. E. Schaefer, "Glare and Celestial Visibility."
34 Baseline Essay, p. 19. Emphasis added.
35 Samuel Gill, "Carrying the War in to the Never-Never Land of Psi."
36 Baseline Essay, pp. 52–54.
37 R. S. Barnaby, *Gliders and Gliding,* p. 132.
38 Baseline Essay, p. 48.
39 Ibid., p. 20.
40 Ibid., p. 26.
41 Ibid., p. 38. The latter is the same old discredited pyramidology that has been around for over a hundred years since the publications of Charles Piazzi Smyth (R. S. Bianchi, "Pyramidiots"; M. Gardner, *Fads and Fallacies in the Name of Science*, pp. 173–185). Gardner showed that the same kind of numerology can derive the speed of light from the dimensions of the Washington Monument, and Jager ("Adventures in Science and Cyclosophy") showed that quantities such as the ratio of the masses between a proton and an electron, the gravity constant, the velocity of light, and the distance between the sun and the earth can be derived from measurements on his bicycle.
42 At that time, in a flat country like Egypt, a catapult would be the only feasible way to launch a glider. Making the reasonable assumption of a 200-kilogram glider flying 10 miles per hour into a 10-mile-per-hour headwind (Barnaby, *Gliders and Gliding*, p. 50; L. B. Barringer, *Flight Without Power*, p. 102), one can calculate the kinetic energy involved. In turn, because of conservation of energy, the equivalent potential energy can be found; and from that the size of the catapult required can also be found. The catapult required to launch this small glider would be many times the size of the largest catapult constructed in antiquity (E. W. Marsden, *Greek and Roman Artillery. Historical Development,* pp. 86–88). Further complicating the problem is that there were no catapults in 2000 B.C.; and that if, in fact, the scheme had been carried out, the force of the launch would almost certainly have stripped the wings from the glider.
43 Baseline Essay, p. 39.
44 Ibid., p. 51
45 Objects can be plated with gold by dipping them into a solution of gold (S. Field and A. D. Weill, *Electro-Plating. A Survey of Modern Practice,* pp. 286–294). Unfortunately, dissolving gold requires aqua regia (a mixture of nitric and hydrochloric acids), and making these acids requires the ability to distill, which was not invented

until after A.D. 1100 (A. J. Ihde, *The Development of Modern Chemistry*, pp. 16–18; C. Singer & E. J. Holmyard, *A History of Technology*, vol. 1, p. 254). The Baseline Essay (p. 51) claims that the Egyptians had iron-copper batteries on the basis of batteries discovered in Iraq (not Egypt). However, electrodeposition from a gold anode to a copper cathode would require at least 1.5 volts. Keyser's extensive study of these batteries flatly points out that electroplating gold by electric current was impossible ("The Purpose of the Parthian Galvanic Cells: A First-Century A.D. Battery Used for Anesthesia"). Keyser's model of a battery using acetic acid (the strongest acid available in antiquity) produced a voltage of 0.49 volts.

46 Gerald Holton, *Science and Anti-Science*, pp. 178–180.

## REFERENCES

ADAMS, HUNTER HAVELIN. "African and African-American Contributions to Science and Technology." [Cited as Baseline Essays.] In *African-American Baseline Essays*. Portland, OR: Multnomah School District, 1990.

———. "African-American World View." Broadcast, WDTR, 90.9 FM, Detroit Public School's Radio, September 25, 1990.

———. "African Observers of the Universe: The Sirius Question." In *Blacks in Science. Ancient and Modern*, edited by I. Van Sertima, pp. 27–46. New Brunswick, NJ: Transaction Books, 1983.

———. Lecture presented at the 1st Melanin Conference, San Francisco, CA, September 16–18, 1987.

———. "Psychoenergetic Aspects of Ancient Egyptian Lifeways." Lecture presented at the Association for Classical African Civilizations, 3rd conference, City College of New York, March 23, 1986.

ALLEN, C. W. *Astrophysical Quantities*. London: The Athlone Press, 1973.

ASANTE, M. K. "African–Puerto Rican Centric Curriculum Guide," vol. 1. Camden, NJ: Camden Public Schools Division of Curriculum and Assessment, n.d.

BARNABY, R. S. *Gliders and Gliding*. New York, NY: Ronald Press, 1930.

BARRINGER, L. B. *Flight without Power*. New York, NY: Pitman, 1940.

BARUAC, D. Letter to Christopher Trey. *New York Times* (May 22, 1991).

BIANCHI, R. S. "Pyramidiots." *Archaeology* (November/December 1991): 84.

BLEIER, RUTH. "Introduction." In *Feminist Approaches to Science*, edited by Ruth Bleier, pp. 1–17. New York, NY: Pergamon Press, 1986.

BRODY, H. *Placebos and the Psychology of Medicine*. Chicago, IL: University of Chicago Press, 1980.

CHLEBOWSKI, T., J. P. HALPERN & J. E. STEINER. "Discovery of a New X-Ray Emitting Dwarf Nova 1E 0643.0-1648." *Astrophysical Journal* 247 (1981): L35–L38.

CLARK, P. E. & M. J. CLARK. "Therapeutic Touch: Is There a Scientific Basis for the Practice?" In *Examining Holistic Medicine*, edited by D. Stalker & C. Glymour, pp. 287–296. Buffalo, NY: Prometheus Books, 1989.

FAULKNER, R. O. *A Concise Dictionary of Middle Egyptian*. Oxford: Griffith Institute, Ashmolean Museum, 1991.

FIELD, S. and A. D. WEILL, *Electro-Plating. A Survey of Modern Practice*. 6th edit. London: Sir Isaac Pitman, 1951.

GARDNER, M. *Fads and Fallacies in the Name of Science*. Rev. edit. New York, NY: Dover Publications, 1957.

GILL, SAMUEL, "Carrying the War into the Never-Never Land of Psi." *Skeptical Inquirer* 15 (1991): 269–273.

GRAD, B. "A Telekinetic Effect on Plant Growth." *International Journal of Parapsychology* 5 (1963): 117–133.

GRAD, B., R. CADORET & G. PAUL. "An Unorthodox Method of Treatment of Wound Healing in Mice." *International Journal of Parapsychology* 3 (1961): 5–24.

GRIAULE, M. & G. DIETERLEN. "Un Système Soudanais de Sirius." *Journal de la Societé des Africanistes* 20 (1950): 273–294.

———. *Le renard pâle*. Paris: Musée de l'Homme, 1965.

GROSS, PAUL R. & NORMAN LEVITT. ***Higher Superstition: The Academic Left and Its Quarrels with Science.*** Baltimore, MD: Johns Hopkins University Press, 1994.

HARDING, SANDRA. ***Whose Science? Whose Knowledge?*** Ithaca, NY: Cornell University Press, 1991.

HOLTON, GERALD. ***Science and Anti-Science.*** Cambridge, MA: Harvard University Press, 1993.

IHDE, A. J. ***The Development of Modern Chemistry.*** New York, NY: Harper and Row, 1964.

JAGER, D. DE. "Adventures in Science and Cyclosophy." ***Skeptical Inquirer*** 16,2 (1992): 167–172.

KEYSER, P. T. "The Purpose of the Parthian Galvanic Cells: A First-Century A.D. Battery Used for Anesthesia." ***Journal of Near Eastern Studies*** 52,2 (1993): 81–98.

KLOTZ, I. M. "Multicultural Perspectives in Science Education: One Prescription for Failure." ***Phi Delta Kappan*** (November 1993): 266–269.

***McLean v. Arkansas Board of Education.*** 529 F. Supp: 1258–1264 (Ed. Ark), 1982. ***Science*** 215 (1982): 934–943.

MARRIOT, M. ***New York Times,*** August 11, 1991, K1, 12.

MARSDEN, E. W. ***Greek and Roman Artillery. Historical Development.*** Oxford: Oxford University Press, 1969.

MORTON, P. ***New Scientist*** (April 9, 1987): 46–50.

ORTIZ DE MONTELLANO, BERNARD R. "Afrocentric Creationism." ***Creation/Evolution*** 29 (1991–1992): 1–8.

———. "Avoiding Egyptocentric Pseudoscience: Colleges Must Help Set Standards for Schools." ***Chronicle of Higher Education***, March 22, 1992, pp. B1–2.

———. "Magic Melanin. Spreading Scientific Illiteracy among Minorities—Part II." ***Skeptical Inquirer*** 16,2 (1991): 162–166.

———. "Melanin, Afrocentricity and Pseudoscience." ***Yearbook of Physical Anthropology*** 36 (1993): 33–58.

———. "Multicultural Pseudoscience. Spreading Scientific Illiteracy among Minorities—Part I." ***Skeptical Inquirer*** 16,1 (1991): 46–50.

PALTER, R. "***Black Athena***, Afrocentrism, and the History of Science." ***History of Science*** 31 (1993): 227–287.

PINCH, G. ***Magic in Ancient Egypt.*** Austin, TX: University of Texas Press, 1994.

RANDOLPH, G. "The Difference in Physiological Response of Female College Students Exposed to Stressful Stimulus, When Simultaneously Treated by either Therapeutic or Casual Touch." Ph.D. diss., New York University, 1980.

ROWE, W. E. "School Daze: A Critical Review of the 'African-American Baseline Essays' for Science and Mathematics." ***Skeptical Inquirer*** 19,4 (1995): 27–32.

SCHAEFER, B. E. "Glare and Celestial Visibility." ***Publication of the Astronomical Society of the Pacific*** 103 (1991): 645–660.

SINGER, C. & E. J. HOLMYARD. ***A History of Technology***, vol. 1. Oxford: Oxford University Press, 1954.

STRAHLER, A. N. ***Understanding Science.*** Buffalo, NY: Prometheus Books, 1992.

TEMPLE, R. G. ***The Sirius Mystery.*** London: Sidwick and Jackson, 1976.

VAN SERTIMA, I., ed. ***Blacks in Science. Ancient and Modern.*** New Brunswick, NJ: Transaction Books, 1983.

# NOTES ON CONTRIBUTORS

JAMES ALCOCK, Professor of Psychology, York University (Toronto), is the author of *Science and Supernature: A Critical Appraisal of Parapsychology.*

GEORGE BORNSTEIN, Professor of English at the University of Michigan, is the author of *Ezra Pound Among the Poets* and *Poetic Remaking: The Art of Browning, Yeats and Pound.*

JEAN BRICMONT is Professor of Theoretical Physics at the University of Louvain (Belgium).

MARIO BUNGE, Professor of Philosophy at McGill University, is the author of numerous books, including *Philosophy of Science and Technology* and *Ethics: The Good and the Right.*

PAUL A. CANTOR, Professor of English at the University of Virginia, is the author of *Shakespeare: Hamlet* and *Creature and Creator: Myth-making and English Romanticism.*

STEPHEN COLE, Professor of Sociology at the University of Queensland (Australia), is the author of *Making Science: Between Nature and Society.*

FREDERICK CREWS, Professor of English (emeritus), University of California at Berkeley, is the author of *The Critics Bear it Away, The Memory Wars: Freud's Legacy in Dispute*, and *The Pooh Perplex.*

RENE DENFELD is the author of *The New Victorians: A Young Woman's Challenge to the Old Feminist Order.*

LOREN FISHMAN, M.D., served as President, New York Society for Physical Medicine and Rehabilitation, and is the author, with Carol Ardman, of *Back Talk.*

ROBIN FOX, Professor of Anthropology, Rutgers University, is the author of *Encounter with Anthropology.*

LANGDON GILKEY, Professor of Theology (emeritus) at the University of Chicago, is the author of *Nature, Reality, and the Sacred: The Nexus of Science and Religion.*

DAVID GOODSTEIN is Professor of Physics and Vice-provost, California Institute of Technology. He is the creator of a celebrated television series of elementary physics lectures.

SHELDON GOLDSTEIN is Professor of Mathematics and Physics at Rutgers University.

BARRY GROSS, who died, tragically, shortly after this conference, was Professor of Philosophy, York College, CUNY, and the author of *Reverse Discrimination.*

PAUL R. GROSS, Professor of Biology at the University of Virginia, is the author (with Norman Levitt) of *Higher Superstition: The Academic Left and Its Quarrels with Science.*

SUSAN HAACK, Professor of Philosophy at the University of Miami, is the author of *Evidence and Inquiry* and *Philosophy of Logics.*

BARBARA S. HELD, Professor of Psychology at Bowdoin College, is the author of *Back to Reality: A Critique of Postmodern Theory in Psychotherapy.*

DUDLEY HERSCHBACH, is Professor of Chemistry at Harvard University, and was awarded the Nobel Prize in Chemistry.

GERALD HOLTON is Professor of Physics and History of Science (emeritus) at Harvard University. His books include *Science and Anti-Science* and *Thematic Origins of Scientific Thought.*

SIMON JACKMAN, teaches Political Science at the University of Chicago.

OSCAR KENSHUR, Professor of Philosophy and Comparative Literature, Indiana University, is the author of *Dilemmas of Enlightenment.*

DANIEL KLEPPNER is Professor of Physics, Massachusetts Institute of Technology.

NORETTA KOERTGE, Professor of Philosophy, Indiana University, is the author (with Daphne Patai) of *Professing Feminism.*

PAUL KURTZ is Professor of Philosophy (emeritus) at the State University of New York, Buffalo, and is Chairman of the Committee for the Scientific Investigation of Claims of the Paranormal (CSICOP). His books include *The New Skepticism: Inquiry and Reliable Knowledge* and *Challenges to the Enlightenment: In Defense of Reason and Science.*

MARY LEFKOWITZ, Professor of Classics, Wellesley College, is the author *Women's Life in Greece and Rome* and *Not Out of Africa.*

NORMAN LEVITT, Professor of Mathematics, Rutgers University, is the author (with Paul R. Gross) of *Higher Superstition: The Academic Left and Its Quarrels with Science.*

S. ROBERT LICHTER is the Director of the Center for Media and Public Affairs and the author of *Keeping the News Media Honest.*

MARTIN W. LEWIS, Professor of Geography at Duke University, is the author of *Green Delusions: An Environmentalist Critique of Radical Environmentalism.*

MEERA NANDA holds a doctorate in biology and is a member of the Science Studies program at Rensselaer Polytechnic Institute.

BERNARD ORTIZ DE MONTELLANO, Professor of Anthropology at Wayne State University, is the author of *Aztec Medicine, Health and Nutrition.*

JANET RADCLIFFE RICHARDS, Professor of Philosophy at The Open University (U.K.), is the author of *The Sceptical Feminist: A Philosophical Enquiry.*

HENRY ROSOVSKY, Professor of Economics (emeritus) at Harvard University, was for many years Dean of Faculty at Harvard and is currently a Fellow of the Harvard Corporation. He is the author of *The University: An Owner's Manual.*

ANN MACY ROTH, an Egyptologist, is Professor of Anthropology at Howard University.

STANLEY ROTHMAN is Professor of Government (emeritus) at Smith College and Director of the Center for the Study of Social and Political Change. He is the author of *The Mass Media in Liberal Democratic Societies.*

MARY BETH RUSKAI is Professor of Mathematics at the University of Massachusetts at Lowell.

WALLACE SAMPSON, M.D., is a Professor of Medicine at Stanford University Medical School.

EUGENIE C. SCOTT, who holds a Ph.D. in physical anthropology, is Director of the National Center for Science Education.

CHRISTINA HOFF SOMMERS is a Professor of Philosophy at Clark University and the author of *Who Stole Feminism?*

JAMES TREFIL is Professor of Physics at George Mason University. His books include *A Scientist in the City* and (with E. D. Hirsch) *The Dictionary of Cultural Literacy.*

GERALD WEISSMANN, M.D., is Professor of Medicine at New York University and the author of *The Doctor with Two Heads and Other Essays* and *Democracy and DNA: American Dreams and Medical Progress.*

# INDEX OF CONTRIBUTORS


Alcock, J. E., 64–78

Bornstein, G., 459–469
Bricmont, J., 131–175
Bunge, M., 96–115

Cantor, P. A., 445–458
Cole, S., 274–287
Crews, F., 470–482

Denfeld, R., 246–255

Fishman, L., 87–95
Fox, R., 327–345

Gilkey, L., 523–525
Goldstein, S., 119–125
Goodstein, D., 31–38
Greenberg, H., ix–xi
Gross, B. R., 79–86
Gross, P. R., 1–7

Haack, S., 57–63, 259–265
Held, B. S., 198–206
Herschbach, D. R., 11–30
Holton, G., 551–560

Jackman, S., 346–368

Kenshur, O., 288–297, 526–536
Kleppner, D., 126–130
Koertge, N., 266–273, 413–419
Kurtz, P., 493–504

Lefkowitz, M., 301–312
Levitt, N., 39–53
Lewis, M. W., 209–230
Lichter, S. R., 231–245

Nanda, M., 420–436

Ortiz de Montellano, B. R., 561–572

Radcliffe Richards, J., 385–412
Rosovsky, H., 539–542
Roth, A. M., 313–326
Rothman, S., 231–245
Ruskai, M. B., 437–441

Sampson, W., 188–197
Scott, E. C., 505–522
Sommers, C. H., 369–381

Trefil, J., 543–550

Weissmann, G., 179–187, 483–489

# SUBJECT INDEX

AAUW, 375, 376, 378
academic freedom, and religious views, 511
ACS, 558, 559
*Acta Eruditorum*, 42
actor-actant theory, 282, 283
acupuncture, 193
  placebo effects of, 195
Adams, Hunter Havelin, 561, 562, 564–566
affirmative action, 414, 415
Africa
  as cultural model of Egypt, 325
  as ignored by Egyptologists, 325
  influence of Egypt on, 323
  as origin of Greek culture, 411
African civilization, 303, 304 (*See also* Egypt, ancient *and* Egyptologists)
  and ancient Greece, 304
  and European civilization, 321
African medicine, 565
African science, 564
*African-American Baseline Studies,* 561
  and increase of scientific illiteracy, 561
  Science Baseline Essay of, 561
African-Americans, 319, 320, 361
  and Afrocentric perspective, 325
  and Egyptian cultural heritage, 321, 323
Afrocentrism, 305, 313, 323–325, 537
  teaching of, 326
  and works on science, 569
Afrocentrists, 299, 307, 320, 505
  Eurocentric approach of, 324
  and Nubia, 317, 318
  and propaganda, 538
AIDS virus, 23
Alexander, 308, 310
Alexandria, 304, 319
allopathy, 188
alternative medicine, 14, 188, 192, 194
  definition of, 190
  terminology of, 190
alternativism, 5
American Association for Public Opinion Research, 353
American Association for the Advancement of Science, 106, 193
American Association of University Women (AAUW), 374 (*See also* AAUW)
American Cancer Society, 192
American Chemical Society (ACS), 558 (*See also* ACS)
American Council for Science and Health, 243
American Enterprise Institute, 107
American Medical Association, 107, 192
American Physical Society, 243, 369, 558
American Revolution, 26
*American Scientist*, 35
ANN, 94, 95
Anthony, Susan B., 438
anthropology, 327, 336, 341
  and epistemological relativism, 336
  humanistic, 334, 336, 339, 343, 345
  and knowledge systems, 423
  as science, 333
  symbolic, 337
antienvironmentalist, 221, 222, 228
antievolutionism, 505, 520
  influence of, 517
  as irrationalism, 520
  and teaching evidence against evolution, 509
antimodernism, 299, 421
antipositivism, 353, 355
antirationalism, 443
antirealism, 2, 199, 257
  and active knower, 202
  and constructivist/constructionist psychotherapy, 198, 200
  and psychotherapy, 201
  and quantum physics, 203
antireason, 1, 491
antiscience, 1, 97, 126, 491, 537, 568, 524
  academic, 97, 110
  and anti-intellectualism, 84
  and background beliefs, 81, 82
  and political paranoia, 84
Aquinas, Saint Thomas, 79
Archimedes, 89, 108
Arendt, Hannah, 354
Aristotle, 47, 79, 81, 83, 303, 304, 306,
  and Egypt, 307
Arnold, Matthew, 47
  and science education, 44
arrow of time, 131, 143, 148, 167, 168
  and irreversibility, 137

subjectivity of, 149
and unstable dynamic systems, 141
artificial neural network (ANN), 93 (*See also* ANN)
Asante, Molefi Kete, 304, 305
asbestos, as carcinogen, 233–235
Association of Women in Psychology, 418
astrology, 106, 403, 495, 496, 498
scientific basis for, 564
authoritarianism, 109
Ayer, A. J., 329

Bacon, Francis, 14, 15, 18, 211, 213, 214, 266, 347, 540
Baruch, Bernard, 26
Barzun, Jacques, 40, 41, 44
Baseline Essay, 562–565, 567, 570 (*See also* *African-American Baseline Studies*, Science Baseline Essay of)
Baudrillard, Jean, 150, 356, 357, 361
Bayesians, 103
Bayle, Pierre, 491, 528, 530, 532, 533, 535
beams, supersonic, 22–24
Becker, Gary, 102, 103
behaviorism, 331, 338
belief
and constructivism, 90
and fear, 90, 91
and feelings, 87
and intuition, 92
religious, abundance of, 68
social transmission of, 66
transcendental, 70
Bell, Alexanber Graham, 557
Bell, J. S., 117, 121–123
and incompleteness of quantum description, 129
Bénard cell, 155, 171
Bénard instability, 154
Benhabib, Seyla, 421, 422, 431
Bentham, Jeremy, 348
Bergson, Henri, 147
Berkeley, George, 18, 43
Berlin, Isaiah, 552
Bernal, Martin, 321–323
Bernoullis, the, 44
bias, 323, 332, 333
as distinguishable from fact, 341
and historians, 301, 302
and history, bias in, 301, 302
metaphor of, 331
political unavoidability of, 321
Bible Science Association, 510, 517
Bible, 501
Big Bang, 52, 514
as evidence of creation, 515
"Big Science," 5
biomedicine, 177
Blake, William, 1, 40, 459, 460, 464
hostility to science of, 15
hostility to cosmology of, 43
Boas, F., 339
Bohm, David, 117, 121
and completion of quantum physics, 123
Bohr, Niels, 117, 119
and atomic model, 19, 20
Boltzmann entropy, 150–52, 156, 158, 170
Boltzmann, Ludwig, 117, 131, 141, 142, 144, 147–149, 153, 155, 157, 158
and irreversibility, 132, 138
Born, Max, 121, 123
and irreversibility, 143
Boyle's Law, 275, 336
Boyle, Robert, 42
and epistemological validity of laboratory science, 290
Breuer, Josef, 479
Browne, Thomas, 529, 533
Brownian motion, 149
Brownowski, Jacob, 12, 15, 27
Buckyball, 23
Buffon, Georges, 42
Buridan's ass, 80, 154
Butler, Samuel, 58

Calvin, John, 80
Campus Crusade for Christ, 510
cancer
and animal studies, 237
and capitalism, 241
controversies regarding, 237, 238
environmentalist survey of, 234, 235
and institutional ratings of, 241
and man-made agents, 231, 234
and scientists versus environmentalists, 241
and the media, 235–237, 239, 240
and the *New York Times*, 240
and *Scientific American*, 240
rates of, 233
and reputations of scientists, 240
risk of, and public information, 242
and scientific and environmentalist communities, 237
and television news, 240
Cancer Control Society, 189
cancer scientists, survey of, 231–234
Caribs, 487, 488
Carroll, Lewis, 44
Cartesian dualism, 215, 217, 225
chaos theory, 50, 104, 153, 162
chaos, 161, 162, 169, 170, 172
and deterministic worldview, 132

and irreversibility, 131, 141
and order, 152
chaotic dynamic systems, 134, 146
and universal determinism, 135
charm theory and social factors, 279
Chaucer, Geoffrey, 41
chemical bonds, 22
Christ, 528, 529, 506
Christian Coalition, 501
Christian Leadership Ministries, 510
Christian perspective
and scholarship, 505, 516, 517
in science, 513
and teaching , 512, 513
Christianity, 2
credibility of, 530
as ideology, 512
as beyond reason, 528
and skepticism, 527
Clement of Alexandria, 306, 309
Cleopatra, 318
Coleridge, Samuel Taylor, 460
Columbus, Christopher, 487, 488
Committee for Freedom of Choice, 189
complementarity, 119, 120, 129
conditioning, 66, 341
classical, 65
enculturation as, 331
operant, 65, 70
superstitious, 70
construct, 198, 199
and confusion arising from the term, 202
construction, 199, 202
cultural, 459
and literature, 465
linguistic, 201
and reality, 202
literary, 467
social, 266
constructionism, 87, 204
cultural, 18
social, 198, 200, 267
and application of scientific results, 270
as antirealist doctrine, 201
constructivism (constructivist view), 198, 200, 257, 275, 276, 296
as antirealist doctrine, 201
and belief, 90
failure of, 280
and knowledge, 89
and Bruno Latour's rejection of, 289
and psychic pressures, 87
research program of, 281
and science as irrational, 87
scientififc, 289
social, 83, 275, 277, 553
and absence of evidence, 285
and sociology of science, 291
and social constructionism, 201
constructivist-relativist, 105, 106
constructivists, 274–279, 293
and reviewing books, 281
cultural, 20
rhetorical trick of, 280
social, 161, 274, 279
Copenhagen interpretation, 117, 120, 129, 148
Copernican revolution, 13
Cortez, Hernando, 488
Craige, John, 530, 533
Creation Research Society, 507
creation science, 510, 523, 524
and denial of empirical knowledge, 505
and school curricula, 507
teaching of, 509
creation, special, 506, 515
and antievolutionists, 5–6
creationism, 403, 491, 501, 505, 506, 508, 515, 518, 520
continuous, 506
and division of science, 514
progressive, 506
scientific, 509, 563, 568
and teaching Biblical Christianity, 508
and supernatural causation, 514
Creationist Institute, 502
Crick, Francis, 12, 280
critical thinking, 499
and the media
in the schools, 502, 503
and science, 74
criticism, contemporary literary
and hegemony of critic, 455
literary texts, 445
cultural determinism, 331
cultural literacy, 537, 544
change in, 547
genesis of idea of, 545
as matrix of knowledge, 545
culture
and rationalist aspiration, 426
as source of rationality, 425
Curie, Madame Pierre, 554

Dallas School of humanism, 344
Darwin, Charles, 153, 155, 156, 501, 507, 508, 567
Darwinism, 518
de Broglie, Louis Victor, 121, 123
de Chardin, Pierre Teilhard, 105
De Man, Paul, and support of Nazis, 461
deconstruction, 188, 194, 334, 339, 342, 443, 452, 461
and the New Historicism, 453

and Bruno Latour's rejection of, 289
and postmodernism, 226
and relativizing knowledge claims, 289
deconstructionists, 210, 303, 330, 453
deism, 516
Delaney clause, 237–239
democracy
as rooted in a method of inquiry, 493
Derrida, J., 335, 342
Descartes, René, 211–214, 289, 420–423, 425, 427, 429, 549,
and determinism, 154, 155, 161
cultural, 331
and fundamental laws, 134
and predictability, 133
and probabilities, 134
and reductionism, 162
determinism
and chaotic dynamic systems, 235
cultural, 331
and Descartes, 154, 155, 161
and Hobbes, 293
and Laplace, 134
Dewey, John, 301,
Diderot, Denis, 43
Dilthey, W., 335, 337
Diodorus of Sicily, 307
discursivity, 472, 475
and M. Foucault, 472
DNA, 22
Doctorow, E. L., 486, 487
Dogon people, astronomical knowledge of, 566
Dostoevsky, Fëdor, 81, 333, 474
double-blind studies, 181, 195
Dryden, John, 42
Durkheim, Emil, 100, 291, 339, 423, 425, 428, 473

ecofeminism, 207, 212, 246
conservatism of, 251, 254
and Plumwood, Val, 214
as religion, 250
technophobic aspect of, 253
and women's intelligence, 249
and women's rights, 250
Ecofeminist Visions Emerging, 248
ecofeminist, 215
ecology, 209, 218
spiritualized, 219
ecophilosophy, 210–212, 217, 221
conservative roots of, 222
ecoradical, 211, 214, 217, 222, 225, 227
and attack on reason, 217
and modernity, 219,
ecoradicalism, 210, 211, 213, 216, 218–220, 222, 224, 228
and conservatism, 228
ecosentimentalism, 483–485, 488
ecosystem, global 210
Eddy, Mary Baker, 510
Edison, Thomas, 557
education (*See also* science education, *and* scientific literacy)
and Afrocentric curricula, 568
and challenge to science, 552
and critical thinking, 503
cultural and moral dimension of, 51
and decline in respect for reason and science, 537
higher, and devaluation of the humanities, 539
improving science teaching in, 569
and "inquiry" method of teaching, 554
liberal arts, 25, 27
multicultural, 537
and science, central ideas of, 549
and scientific literacy, 548
and spiritual content, 537
value of mathematics in, 47
Egypt, ancient, 299, 301, 303–305, 314, 315, 318, 321, 322, 326
aeronautics of, and glider, 567
and blackness of Egyptians, 315, 317
and Egyptian science as religion in the schools, 563
and electroplating technology, 568, 570
greatness of, 319, 561
and Greek culture, 321, 322
and Greek philosophy, 305–307
and influence on Africa, 323
and influence on Europe, 321–323
and mystery system, 309, 310
and particle/wave nature of light, 567
and Plato, 308
and Pythagoras, 308
and quantum theory, 567
religion of, and teaching of social studies, 564
and scientific paradigms, 563
and supernatural causation, 563
and theory of evolution, 567
and Western culture, 323
Egypt, modern, 317, 319
and color, 315, 316
Egyptian Exploration Society, 324
Egyptians (*See* Egypt, ancient)
Egyptologists, 306, 317–319, 323
and Afrocentrism, 313–315, 317, 319–321, 325
as Eurocentric, 318, 324
and Eurocentric motives, 304
and scholarly conspiracy, 324
and goals of Afrocentric Egyptology, 325
Einstein, Albert, 26, 105, 127, 333, 335
and quantum theory, 120, 121

Elementary and Secondary Education Act, 378, 379
Eliot, T. S., 459
Elizabethans, 463
  and relation of cultural practices to historic texts, 464
  stagecraft of, 450
empiricism, 338
  as cultural construct, 459
Enlightenment, 153, 209, 210, 211, 214, 221, 223, 226, 253,, 290, 300, 346, 347, 362, 365, 421, 423, 426, 431, 432, 493, 520, 527, 540
  and the autonomous self, 422
  double vision of, 291
  and dualism, 214, 292
  feminist and postcolonial critics of, 424
  and the Gulags
  and Bruno Latour, 289, 290, 431
  liberal politics of, 222
  and modern science, 425
  and Nazi death camps, 431
  and procedural principles of knowledge, 429
  and progress, 433
  second, 223
  and rejection of science, 1
  and "Romantic Rebellion," 552
  and Russian communism, 540
  and scientific rationalism, 526
  and value of reason, 222
entropy of Clausius, 151, 152
entropy, 150, 152–154, 160, 170–172 (*See also* Boltzmann entropy *and* Gibbs entropy)
  and Boltzmann entropy, 150
  of Clausius, 151, 152
  contextual, 150
  and thermodynamics, 151, 163, 169
environmental philosophy, 209–211, 214, 218, 219
  and cosmology, 220
  and the Enlightenment, 223
environmental science, 220, 221
environmentalism, 223, 228, 231, 243
  and deconstruction, 227
  and ecofeminism, 246
  and the media, 242
  and relationship with science, 209
Epictetus, 80
epistemological positions, mutability of, 527, 534, 535
epistemology, 80, 292, 366, 386, 401, 408–411
  and amenability to transformation, 534
  antirealist, 198
  changes in, and feminism, 394, 396, 397–399, 401, 409, 410
  constructivist, 266
  and co-optation of ideas, 534
  as democratic, 264
  and emancipation, 396, 407, 526, 527
  feminist, 385–387, 391–399, 401, 403, 404, 406, 409–411, 413, 414–417, 433
    aims of, 395
    content of, 400
    impossibility of, 399
    justification of, 398
  and first-order beliefs, 394, 409
  objectivist, 200
  and politics, 395, 404, 409
  postmodern, 200
  realist, 199
  relativistic, 335
  of science, and the supernatural, 518
  as secondary to ethics, 398
  and second-order beliefs, 388
  and social negotiation, 264
  "sociological turn" in, 267
  traditional, 399
  traditional, and feminism, 394–396, 398, 399
ergodicity, 132, 145, 146, 160, 166
  and irreversibility, 147
ESP, 495, 498
ethnocentrism, 336
ethnomethodology, 99, 100, 275
ethology, 331, 339
Euclid, 40, 42
Eurocentrism, 177
Europe, western, 215, 216 (*See also* the West, Western . . .)
Europeans, medieval, 212, 215
evidence, 301, 305, 306
  and evolution, theory of, 153
  historical, 301, 320
  legal, 103
  and revelation, 423
  as social construct, 218
  varying standards of, 262
evolution theory, 153
  and source of morality, 507
evolution, 155, 172, 267, 494
  evidence against, 509
  human, 516
  opposition to, 520
  as science, 517
  teaching of, and sciency literacy, 505, 517
  theistic, 506
  and theological conservatism, 517
  theory of, 153
existentialism, 97, 99, 329

faith
  and knowledge, 529

and probabilism, 532
and probability, 530, 532
and reason, 432
False Memory Syndrome Foundation, 107
falsifiability and credibility, 342
Fascism, 61, 222, 329, 459, 461, 462, 524
Faulkner, William, 551
FDA, 189, 192, 193
feminism, 100, 228, 339, 385, 386, 403, 406, 410, 438
as an applied subject, 392, 399
and changes in belief, 391, 392
derivativeness of, 406
and epistemology, 385–387, 391–399, 404, 406, 409–411, 414–417
aims of, 395
and first-order beliefs, 390
foundations of, 387, 398
and impossibility of feminist theories, 402
and justification of theories, 402
progress in, 390–392
as revolutionary ideology, 246
and radical environmentalism, 225
and second-order beliefs, 390
traditional, 389, 396, 397
and traditional epistemology, 394
and traditional views, 388
and truth, 433
feminist epistemology (*See* epistemology)
feminist theory, radical, 100
feminists, 207, 215, 219, 267, 330, 387, 505
and moral values, 388
early challenges to, 389
of the Third World, 421
and postmodernism, 422
Feyerabend, Paul, 81, 82, 149, 335
and sociology of science, 109
Feynman, Richard, 14, 18
fideism, 491, 532, 534
and skepticism, 527, 528
First Amendment, 523, 524
Establishment clause of, 508, 511
first-order beliefs, 387, 394
changes in, 394, 399
and epistemology, 394, 396, 409, 410
and feminism, 390
absence of feminist reasons for, 400
and standards, 392
Firth, Raymond, 337, 341
Fishman, Loren, 5, 55
Food and Drug Administration (FDA), 181, 270 (*See also* FDA)
Foucault, Michel, 101, 180, 335, 355, 361, 427, 461, 472
and discursivity, 472
and knowledge as power, 480
Frankfurt School, 354, 364
Franklin, Benjamin, 14, 25, 26, 557
fraud, 32
and the biomedical sciences, 31, 37
Caltech rules for, 31, 34
and career pressure, 33
civil, 31, 32
scientific, 31, 36
and funding, 36
motives for, 33
and peer review, and conflict of interest, 36
rarity of, 31, 36
free speech, 349, 360
free will, 293
Freud, Sigmund, 470, 472–474 , 477, 478
and Wilhelm Fliess, 477
hermeneutics of, 480
and D.H. Lawrence, 474
and magical thinking, 479
and Mann, Thomas, 474
and mythical thought, 479
as Neo-Platonist, 479
and Friedrich Nietzsche, 475
philosophical side of, 477
Freudianism, 443, 473
and the universities, 471, 479
Freudians, 328
Friedman, Milton, 103
Fuller, Buckminster, 23
Fuller, Margaret, 484, 485
functionalism, 337, 338, 339–340
and objective standards, 341
fundamentalism, 221, 227, 251, 430, 433, 501, 519
and absolute truth, 523
and contemporary culture, 524
and evidence, 423

Gaia, 211, 213
galaxies, 22
Galileo, 89, 90–92, 211, 347, 501
Gauss, Karl Friedrich, 44
Geertz, Clifford, 337, 338, 425
Gellner, Ernest, 420, 422, 423, 425, 427–432, 478
and Sigmund Freud, 475
and independent reality, 426
and truth as transcendent, 421
gender difference theory, 437–439
gender differences
as attributable to nature, 389
and moral psychology, 370
Gender Equity Package, 378, 379
Genesis, 212
Genetics Society of America, 106
germ theory of disease, 13
Gibbon, Edward, 528

Gibbs entropy, 152
Gibbs, Josiah Willard, 143
Gilligan, Carol, 369–374, 376, 377
Gimbutas, Marija, 212, 250
girls, adolescent
  and self-esteem, 375
    and boy/girl gap, 375
  and "self silencing," 370
  and standardized tests, boy-girl differences in, 377
Gnostics, 307
God, 501, 518, 540
  authority of, 528
  belief in, 500
    and evolution, 517, 520
  and certain knowledge, 424
  as creator, 506, 508, 515
  existence of, 516
  as hypothesis, 514
  and purpose, 519
  and special place of humans, 516
  submission to, 529
Goethe, Johann Wolfgang von, 1, 459, 479
  hostility to cosmology of, 43
Goodheart, Eugene, 471, 473, 475, 480
*Great Instauration, The*, 15, 266
Greece, ancient, 212, 301, 317
  and African civilizations, 305
  and Aristotle, 303
  and invention of history, 310
  and origin of culture from Africa, 411
Greek philosophers, 303, 308
  and the Egyptians, 305–307, 310
Greens, 209, 210, 218–220, 222, 223
Gross, Paul R., 16, 126, 243, 413, 459, 460, 563
Gulags, 431

Haack, Susan, 55, 79, 81, 459
harassment, sexual, 373
Harding, Sandra, 18, 101
Harvard Center for Risk Analysis, 243
Hawthorne, Nathaniel, 474
  and Oedipus complex, 474
Heidegger, Martin 97, 329, 342, 527
  and support of Nazis, 461
Heisenberg uncertainty principle, 83, 335
  and accuracy of knowledge, 129
  and objective knowledge, 128
  and reality, 126
Heisenberg, Werner, 117, 119, 143
hermeneutics, 339
Herodotus, 307, 392
Higginson, Thomas Wentworth, 484, 485
*Higher Superstition: The Academic Left and its Quarrels with Science*, 16, 126, 301
Himmler, Heinrich, 185
historians
  Afrocentrist, 303, 304
  and bias, 301, 302
historicism, 301, 454
  "New," 303
history
  ancient, 303
  bias in, 301, 302
  as cultural projection, 302
  as fiction, 302
  and science, 296
history-without-facts, 303
Hitler, Adolf, 332, 502
Hobbes, Thomas, 42, 347, 348, 351, 357
  and determinism, 293
  and epistemological validity of laboratory science, 290
  and political science, 350
Holmes, Oliver Wendell, 186
homeopathy, 188, 195, 393, 408
Homer, 302
Hooke, Robert, 42
humanism, 337, 338, 339–343, 443
  and hubris, 423, 540
  and Bruno Latour, 288, 289, 295
  and mathematics, 9
  secular, in public schools, 511
  and science, 289
Hume, David, 43, 80, 329, 330, 423, 424, 427
Husserl, Edmund, 97, 329, 342
Huxley, T. H., 505
Huygens, Christian, 42
hybrids, 289, 293, 294
  and the Edinburgh School, 292
hypotheses, falsifiable, 328, 340

ICR, 517, 518
idealism, 210
  transcendental, 98
ideology, 292
  and evidence, 243
  and hypothesis formation, 271
  as scholarship, 512
  and science, 295
  supernatural, 505
IDT, 508, 509
ignorance, 107–109
  political, rationality of, 357, 359
  willful, 108, 109
*Iliad*, 302
imperialism, Native American, 487
industrial revolution, 211, 213, 351, 430, 542
  and Cartesian dualism, 215
  and scientific revolution, 225

Institute for Creation Research (ICR), 507, 510, 514, 515, 518 (*See also* ICR)
intellectual integrity, 57, 59, 60, 61
  and advance of inquiry, 59
  and moral value, 59
intellectuals, 48
  definition of, 45
  and mathematics, 46
intelligent design theory (ID), 508–511 (*See also* IDT)
  as creation science, 509
International Association for Cancer Victims and Friends, 189
intuition, 71, 72, 89
  and belief, 87, 92
  and pattern recognition, 93, 94
  as rational, 94
irrationality
  and feminism, 386
  rescue from, 405
irreversibility, 139, 142, 146, 151–153, 155, 156, 158, 160, 163–165, 169–171
  and arrow of time, 137
  and chaos, 141
  and ergodicity, 147
  explanation of, 138
  and French philosophy, 149
  and macroscopic laws, 140
  misconceptions about, 144
  subjectivity of, 142, 143, 148

James, George G. M., 308–310
James, William, 74, 185, 494
Jaynes, E. T., 134, 144, 145, 150, 152
Jefferson, Thomas, 26, 27
Jensen, Arthur, 106
Jesus, 500, 529
Johnson, Samuel, 18
Jones, Ernest, 474
Judeo-Christian tradition, 212

Kac Ring Model, 144, 146, 151, 156, 160
Kant, Immanuel, 43, 337, 423–425, 427
Keats, John, 460, 485
Kierkegaard, Søren, 80, 329
*King Lear*, 451, 452, 454, 463–465
  conflated text of, 455
  and contemporary productions, 450
  modern texts of, as constructions, 464
  restoration of, 446
  and revisionists, 448
  two texts of, 447
Kipling, Rudyard, 420, 421
Kleppner, David, 4, 117
knowledge, objective, 18, 199
  African, 562
  and the Enlightenment, 526
  and faith, 529
  patriarchal standards of, 404
  phallocentric, 394
  possibility of, 423
  as power, 355
  as relative, 200, 201
  scientific, 416
    definition of, 562
    social construction of, 423
  as social construction, 3, 263, 264, 427
  subjective, 200
  transcultural, 428
  and women, 397
Krebs, Ernst, Jr., 189
Kuhn, Thomas, 18, 50, 331, 335, 340, 423, 325, 472, 473
Kurgans, 251, 254

La Mettrie, Julien, 214
Laetrile, 188, 189, 192, 193
Lagrange, Joseph Louis, 44
Laplace, Pierre Simon de, 44, 131, 132, 133, 136
  and determinism, 134
laser, 20
  light of, 22
Latour, Bruno, 81, 89, 91, 92, 105, 149, 257, 275–280, 282, 283
  and animus against social science, 294
  and anthropology, 292
  and critique of sociology, 219
  and the Enlightment, 289, 290, 431
  and humanists, 288, 289
  and hybrids, 295 (*See also* hybrids)
  and methodology, 293
  and the study of science, 288
Lavoisier, Antoine, 12
learning, experiential, 65, 66, 74
    and magical thinking, 71
    and transcendental belief systems, 68
  intellectual, 66
    and scientific-humanist belief systems, 68
  relation between experiential and intellectual, 66, 67
Lehrer, Ronald, 475, 476
Leibniz, Gottfried Wilhelm von, 42, 45
Lerner, Max, 181, 191, 192
Levi, Primo, 61
Lévi-Strauss, Claude, 337
Levitt, Norman, 5, 9, 16, 126, 243, 413, 459, 460, 563
liberalism
  as consistent with science and relativism, 357
  and democracy, 347
  as methodology, 361

and political science, 348
and relativism, 347, 348
and science, 346–348
*Limits of Science, The*, 14
Liouville's theorem, 152, 170
literacy, scientific (*See* scientific literacy)
literary studies, politicization of, 445
Locke, John, 350, 531, 533
logic, as patriarchal, 417
logical positivism, 329, 336, 553
London School of Economics, 328, 337, 338
Longfellow, Henry Wadsworth, 483, 484
Louis XVI, 27
Lowell, James Russell, 177, 179, 483, 484, 487, 488
Luddites, 80, 84
Luther, Martin, 80
Lysenko, T. D., 331, 332

Magic, 55, 565
Egyptians as masters of, 564
magical thought, 64, 69, 70, 77
and Sigmund Freud, 479
and experiential learning, 71
and parapsychology, 76
magnetic resonance imaging, 21
Malinowski, Bronislaw, 337, 341
Malone, Adrian, 12, 13
Marcuse, Herbert, 101, 354
Markov chains, 167
Marsden, George, 512,513
Marx, Karl, 330, 359
Marxist revolution, and educated elite, 462
Marxists (Marxism), 331, 335, 338, 339, 350, 353
and contemporary literary criticism, 457
materialism, historical, 362
mathematical research, and funding, 39
mathematics
and education, 51
and emergence of technocrats, 43, 44
feminist perspectives in, 438
and humanistic culture, 6, 39, 43, 45
illiteracy in
global consequences of, 48
and inexactitude of thought, 48, 51
and the liberal arts, 40
as prophylactic against deficiencies of thought, 49
resentment and envy of, 50
of the 17th century, 41
and understanding the world, 46, 47
Mather, Cotton, 477
Maxwell, James Clerk, 134, 149
McLuhan, M., 356, 357
Medawar, Peter, 14, 18, 27
medicine (*See also* alternative medicine)
holistic, 189, 190, 192, 196
unconventional, 183
Mendel, Gregor, 515
Mendelian genetics, 329, 330
Merchant, Carolyn, 211, 249, 426
Merton, Robert K., 105, 257, 274–278, 416, 417
Mertonians, 278, 281
and relativism, 282
Mesmer, F., 180, 185
Michelangelo, 445, 446
midwives, 409, 410, 416
Mill, John Stuart, 335, 348–350, 357, 360, 401
and democracy, 493
and women's diferences, 389, 393
Mind Projection Fallacy, 134, 136, 154
miracles, 531, 533, 536
and science, 514
modernity, 209, 210, 215, 217, 220, 221, 223, 225, 227, 283, 291, 420, 432
and ecoradicals, 219
and Bruno Latour, 290
postcolonial critics of, 433
and reactionary politics, 221
and the West, 213
Mohammed, 500
molecular amplifiers, 22
beam, 19, 20, 22
oscillator, 22
structure, 22
Monod, Jacques, 148
Monte Carlo fallacy, 71
Moral Majority, 501
moral standards, as social constructions, 513
Morris, Henry, 507, 508, 515
Morrison, Philip, 25
Morse, Samuel F. B., 557
Moses, 500
Moyers, Bill, 180–185
Mulkay, Michael, 105, 106
Mussolini, Benito, 110, 461

N-K cells, 181, 182
Nabokov, Vladimir, 18
narrative, 199
and psychotherapy, 200, 201
National Academy of Sciences, 180, 544, 552, 555
National Association of Science, 83
National Endowment for the Humanities, 555
National Institutes of Health, 32, 34, 184
National Organization for Women, 246
National Research Council, 552

National Science Education Standards, 4, 552, 555, 558
  first draft of, as based on postmodernism, 553
  second draft of, and social constructivism, 553
National Science Foundation (NSF), 34, 284, 329, 552
  Inspector General of, 32, 34, 35
National Science Teachers Association, 552
National Science Teaching Standards, 544
*National Standards for United States History*
  and slighting of role of science and technology, 556, 557
natural selection, 153, 156, 169
Nature, 486, 487
  design in, 519
  language of, 16
  and objective knowledge, 18
  rules of, 18
  and scientific literacy, 548
  sentimentalism about, as a disease, 483
  and truth, 19
  as unfriendly, 488
Nazis, 108, 185, 251, 262
  death camps of, and the Enlightenment, 431
  and Martin Heidegger, 97
  and racism, 106
near-death experience (NDE), 65, 496, 497, 499
negotiation, social, 264, 267, 275, 278, 279
neocreationism, 508
  and attacks on evolution, 509
  at the university level, 510
New Age, 2, 177, 180, 183, 185, 186, 219, 221, 246, 253, 487, 495, 562
  as age of unreason, 485
  and feminism, 250
  and Gaia principle, 378
  and healing, 181, 182, 184
New Criticism, 451, 453, 459
New Critics, 459, 466
New Historicism, 445
New York Academy of Sciences, 4, 5, 378
Newton, Isaac, 11, 18–20, 26, 27, 42, 44, 101, 211, 212, 214, 275
  and Christianity, 214
  *Principia* of, 18, 27, 42
  *Method of Fluxions* of, 43
Nietzsche, Friedrich, 425, 527
  and preemption of Freud, 476
  and Marx, 476
nihilism, 443
nihilists, postmodern, 12, 299
NMR, 20, 21
  spectroscopy, 20
"Nobel Legacy," the, 11, 12
noble savage, 483, 486
non-Western societies, knowledge systems of, 426
nonreality, construction of, 199, 201
NSF, 35
  Inspector General of, 32, 34, 35
Nubia, 317, 318, 323
nuclear magnetic resonance, 20 (*See also* NMR)
nuclear spin, 20, 21
  structure, 21

Objectivism, 218–220
objectivity, 2, 122, 416, 417
  assault on, 378
  challenges to, 188
  and Sigmund Freud, 478
  and context of justification, 270
  Kantian, 204
  postmodern critique of, 461
  and quantum physics, 123
  and reality, 204
  scientific, 343
Office of Alternative Medicine, 188, 191, 194
Office of Research Integrity of the Public Health Service, 32, 35
Office of Technology Assessment, 190
Office of Unconventional Medical Practices (UMP), 184 (*See also* UMP)
Olmsted, Frederick Law, 487
Oppenheimer, Robert, 555
Oxford Catholics, 337, 338

Paley, William, 519
  and IDT, 508
paradisciplines, 414, 415, 416
paranormal phenomena, 491, 494–496, 498
  media coverage of, 502
Parapsychological Society, 497
parapsychology, 74–76, 106, 495, 497, 565
  and pseudoscience, 564, 565
  and religion, 75
  as a science, 75
Parsons, Talcott, 416, 417
Parteo, Vilfredo, 101, 102
Pascal, Blaise, 72
Pasteur, Louis, 13
patriarchal institutions, 431
patriarchal practices, 421
patriarchal thinking, 413
patriarchy, Western, 422
Paul, Saint, 80, 528, 529
Pauling, Linus, 189, 190, 280
Paulos, John Alan, 47, 48

Pavlov, I. P., 331
peer review, 36, 37
  and referees, 36, 37
    and conflict of interest, 36
Peirce, Charles Sanders, 57, 58, 61
Penrose, R., 140, 141, 146, 153
Pepys, Samuel, 42
*Personal Knowledge*, 18
perspectivism, 473
Petrarch, 483, 484
phallocentricity, 386
phenomenological sociology, 98
phenomenology, 97–99
philosophy, 79, 95
  Greek, 303, 308
    and the Egyptians, 305–307, 310
PHS, 35
physics
  foundations of, 117
  and reality, 117
Piaget, Jean, 68
Piltdown man, 33, 105
placebo effect, 565
  and acupuncture, 193
  and therapeutic touch, 565, 566
plagiarism, 34, 35
Planck, Max, 143
Plato, 79, 212, 304, 426
  and Egypt, 307, 308
Plumwood, Val, 210–212, 214, 215
Poe, Edgar Allan, 483, 484, 487
Poincaré cycles (recurrences), 157, 159, 164, 165
Poincaré recurrence theorem, 144, 149
Poincaré, Jules Henri, 48, 103, 104, 134, 138
Polanyi, Michael, 17, 18
political science
  as liberal science, 350
  and public opinion, 347
  and survey research, 353
political scientist, 304
  and generalizations, 303
politics and meaning, 361
Popper, Karl, 109, 143, 148, 149, 328, 334
  and scientific method, 341
  and the Second Law, 153
positivism, 99, 338, 354, 360
postmodernism, 49, 97 117, 120, 188, 194, 198, 200, 210, 216–219, 221 223–225, 227, 257, 357, 362, 420, 433, 520
  and assault on science, 95
  and cognitive equivalence of all knowledge claims, 423
  and deconstruction, 226
  ecological, 228
  and evidence as construct, 218
  Green version of, 223
  and reason, 55
  and rejection of Enlightenment, 494
  and rejection of objective scientific knowledge, 494
  roots of, 226
  and science, 220
  and science as socially constructed, 1
  and truth of scientific knowledge, 553
poststructuralism, 210, 219, 433, 453, 459
  and personal agency, 467
  and support of murderous regimes, 461
pragmatism, 349
Prigogine, Ilya, 117, 131, 132, 135, 147, 150, 153, 154
  and Boltzmann, 143
  and irreversibility, 138, 141, 144
  and Liouville's theorem, 152
probabilism, fideistic defense of, 533
probabilities, and ignorance, 136
probability, 531
  and degrees of credibility, 533
  and faith, 530
  and knowledge claims, 532
propaganda, 188, 192, 196
  Afrocentrist, 538
  religious, 501
Protestantism, 425
pseudobelief, 58
pseudoinquiry, 57, 58, 61
pseudoscience, 104, 105, 564, 481
  academic, 101, 107, 110
  and Herbert Spencer, 337
pseudoscientists, 97, 378, 539
psychic phenomena, 496
psychoanalysis, 328, 470
  and artistic endeavor, 474
  and empiricism, 481
  as Faustian, 479
  and the humanities, 480
  and "repressed abuse," 471
  and the science impaired, 478
psychokinesis, 495
psychology
  adolescent, 375
  moral, and gender differences, 370
  and Oedipus complex, 471
    and Gnostic divination, 477
psychotherapists, 177
psychotherapy, 178, 198 (*See also* therapy)
  as narrative, 200, 201
Ptolemaic system, 90
Public Health Service, 34 (*See also* PHS)
public opinion
  American, incoherence of, 356

and poll, 352
and social change, 354
and specificity, lack of, 357
study of, 350, 352
and "Old Left" critique, 353
progress in, 353
Purcell, Henry, 21, 22
Pyrrhonism, 528, 532, 534
Pythagoras, 47
Pythagoras, and visit to Egypt, 308

Quantum mechanics, 117, 119, 121, 122, 126, 128, 133, 161–163, 164, 335
and antirealism, 203
and atomic and molecular identity, 129
and common sense, 127
limitations of, 130
and objectivity, 123
quantum philosophy, 117, 119, 120–123
quantum theory, 20, 120, 162 (*See also* quantum mechanics)
Quetelet, Adolphe, 351
and average man, 353, 357
and statistics, 364
Quine, Willard, 335

Rabi, Isidor Isaac, 20, 21, 24, 25, 27
race as social concept, 317, 318
racial categories
and black/white distinction, 316, 317
Eurocentric criteria for, 316
social nature of, 318
racism, "scientific," 106
and Mendelian ratios, 333
radio frequencies, 21
radioastronomy, 22
and radiofrequency spectroscopy, 22
rationalism, 491
and hubris, 537
scientific, 526
rationality
and culture, 423, 425
and gynophobia, 426
and independent reality, 428
and individual autonomy, 427
and science, 55
scientific, 420, 421
feminist critiques of, 425
and transcendence of culture, 426
and universality, 427
substantive, 350, 363
and women, 424, 426
realism, 198, 199, 257
scientific, 299
reality
construction of, 199
creation of, 201
distortion of, 200, 201
experience of, 200
extralinguistic, 204
independent, 199, 200
access to, 205
knowable, 199
as linguistic construction, 202, 205
mental constructions as, 205
objective, 198, 205, 427, 433
and methodology of science, 513
social construction of, 347
subjective, 199
as text, 289
reason
affirmation of, 420
and autonomy, 422
and common sense, 80
as culture, 426
and democracy, 491
and emancipation, 422
and faith, 432
masculine bias of, 420
postcolonial critics of, 426
and postcolonial women, 421
postmodernist critique of, 421
rejection of, 2
scientific, and power, 427
and truth, 424
Western male concept of, 423
and women's autonomy, 431
reasoning
fake, 61
and evidence, 58
and indifference to truth, 59
probabilistic, 71, 73
recovered memory, 470, 480
reductionism, 164, 171, 186, 194
reductivism, 292, 294
Latour's rejection of, 292
and sociologists, 291, 293
Religious Right, and theocracy, 523, 525
reincarnation, 497, 499
relativism, 177, 192, 257, 276, 281, 282, 300, 335, 336, 346, 435, 461
antiscientific, 341
cognitive, falsity of, 434
of constructivists, 284
cultural, 339
epistemological, 275, 299, 340
and liberalism, 347, 348
and John Stuart Mill, 349
moral, 513
and science, 347
relativist-constructivist, 274–276, 282
program of, 283
relativity, 126, 128
and common sense, 127

general, 127
special, 127, 167
religion, 499
beliefs of, 294
persistence of beliefs of, 494
and the political process, 500
and rational inquiry, 491
and scientific skepticism, 500
Renaissance, 303
reporting (media), "balanced," 195, 196
and environmental issues, 242
resonant changes of state, 21, 22
Rhine, Joseph Banks, 75, 495
Roman Catholic church, 216
and evolution, 506
Romantic style, and reaction against rational thought, 43
Romanticism, 460
and the New Critics, 461
and reason, 461
Rome, ancient, 303
Rorty, Richard, 57, 60, 61, 81, 82, 335
Rousseau, Jean Jacques, 373, 483, 485, 488
Russell, Bertrand, 40, 50, 79
Ryle, Gilbert, 79, 88

Sagan, Carl, 12, 25, 106
Salem witch trials, 179
Sartre, Jean-Paul, 81
Schrödinger, Erwin, 120
Sci-Tech-Studies (STS), 6
science (*See also* sociology of science)
accessibility of, 25
Afrocentric, 568
and authority, 81, 82
as biased, 323
and changes in commitment, 554
cognitive content of, and social factors, 277, 279, 280
and common sense, 80, 502
and comprehension, 14
and control, 14
and critical thinking, 74
creativity in, 18
and culture, 226
as egalitarian, 82
and emancipation, 432
epistemology of, and the supernatural, 518
as epistemological system, 333
feminist perspectives in, 438
feminist critiques of, 563
and fraud, 31
and free inquiry, 271
freedom of, 17
goal of, 16
history of, 25
hostility towards, 9
and hubris, 333, 334
ideals of, 329
and independent reality, 426
as irrational, 87, 92
and liberalism, 346–348, 362
and male domination, 81, 82
metaphors for, 15
as language, 16
as path finding, 16, 17
as problem solving, 16, 17
method of, 388, 562
changing standards of, 394
as ethos, 429
misconduct in, 31, 37
and "fabrication, falsification [and] plagiarism" (ffp), 34
organization of, 261
outcome of, and social processes, 278
pathological, 378
as patriarchal, 563
perspectivist model of, 563
as politics, 266
pragmatic success of, and postmodernist critics, 428
postmodern attack on, 360
public perception of, 9
and public policy, 14
and religion, 74, 75, 491
reproducibility of results of, 33
resentment of, 207
and "science criticism," 24
as self-correcting, 31, 33
of society, 290 (*See also* social science, sociology, sociology of science, sociologists of science)
as social enterprise, 74, 106, 257, 280
and the social order, 431
and sociobiological explanations, 563
study of, 288
and technology, 333
confusion between, 333
as texts, 289
and theism, 515
truth of, 332, 334, 335
and deconstruction, 334
misuse of, 333
as relationship between knower and known, 337
and triviality, 329, 332
and ultimate questions, 27
uses of, 329, 333, 334
***Science and Antiscience***, 16
***Science and Human Values***, 15
science education, 14, 24
and Arnold, Matthew, 44
and economics, 25

and general culture, 25
and history, 26
and mathematics, 27
and political science, 26
quality of, 551
"Science in American Life" (exhibit), 4
at Smithsonian Institution, and political ideology, 558
science literacy, and teaching of evolution, 505
science studies
and the Edinburgh school, 292
and the political left, 49
scientific community, 562
scientific culture, 424
scientific illiteracy, 561, 564
scientific literacy, 537, 543
and ability to do science, 547
definition of, 543
and Nature, 548
scientific revolution, 212, 214, 421, 426, 432, 540, 564
and patriarchal thinking, 413
scientists
community of, 17
political views of, 232
Scientology, 502
Scopes, John T., 508
Second Law of Thermodynamics, 144, 147, 153, 154, 163, 165, 169, 564
and entropy, 151, 152
second-order beliefs
absence of feminist reasons for, 400
and feminism, 390
secular humanism, in public schools, 511
self-esteem
and African-American boys and girls, 376
fake crisis in, 378
and gap between boys and girls, 375
and girls, adolescent, 375
sexism, male, 247
Shakespeare, William, 24, 445, 446–448, 451, 455, 463, 464
artistic judgment of, 449
and historicism, 454
as modernist, 456
Sheldrake, Rupert, 211
Shelley, Percy Bysshe, 460
Sidgwick, Henry, 74, 75, 494
skepticism, 296, 526
Academic, 534
and Pierre Bayle, 528
and Christianity, 527
as emancipatory, 527
and fideism, 527, 528, 533
mitigated, 526, 532, 535
Pyrrhonian, 532
Skinner, B. F., 214
Smith, Adam, 541
Smith, Joseph, 500
Smithsonian Institution
and bias against science, 559
and trivializing portrayal of science, 558
Snow, C. P., 24, 45, 128
social sciences, 290, 291
and exact sciences, 49
and Bruno Latour, 295
as pathological, 369
and reductionism, 291
social construction of, 284
and Bruno Latour, 291
and survey research, 354
Society for Psychical Research (SPR), 74, 494, 495 (*See also* SPR)
Society for the Social Study of Science, 274, 281
sociobiology, 332, 340
sociologists of science, 276, 277
and social constructivists, 284
sociology, 292
classical, 294
elitist, 293
sociology of knowledge, 87, 259, 264, 330, 335
sociology of science, 87, 109, 257, 259, 262, 264, 274, 294
constructivist, 282
Edinburgh School of, 262, 292, 296
and evidence, 260
post-Mertonian, 105
and social construction, 284
sociologists of science, 276, 277
and social constructivists, 284
Socrates, 304, 305, 308
SPR, 75
St. Vincent Millay, Edna, 40
Stalin, 332, 502
"standpoint" criticism, 3,4
statistical mechanics, 117, 162, 164, 169
statistics, 351, 353
delevopment of, 351
Steinem, Gloria, 247, 377
Stengers, Isabelle, 131, 135, 147, 150, 154
and irreversibility, 141
and Liouville's theorem, 152
Stern, Otto, 19–22
Stevens, Wallace, 467
Stich, Stephen, 57, 60, 61
structuralism, 350
structuralists, and social anthropology, 328
subjective probability, 103
subjectivism, 98, 105
subjectivity, 120, 201

supernatural causes
  and creationism, 514
  Egyptian science and, 563
  as explanations in science, 564
survey research, criticisms of, 352
Swift, Jonathan, 80, 328

Tacitus, 302
Tainos, 487, 488, 569
Tate, Nahum, 450, 465
technology
  feminist, 107
  and science, 333
Tertullian, 80, 529
theocracy
  and Islam, 524
  and Religious Right, 523, 525
theories
  impossibility of, as feminist, 402
  and knowledge, 3
  patriarchal, 402
  phallocentric, 402
theory ladenness, 427
theory saturation, 427, 428
therapies (medical)
  complementary, 191
  nontraditional, 191
  unconventional, 191
therapy (psychotherapy), 198
  antirealism in, 204
  and constructivism, 203
  constructivist/constructionist, 200
    and antirealism, 201
    logical inconsistency of, 201
  individualizing of, 203
    and antirealism, 204
  narrative paradigm for, 200
  systematization of, and completeness, 204
  systems of, 203
thermodynamics, 165, 170, 171
  and entropy, 151, 163, 169
Thomas, Lewis, 25
Thoreau, Henry David, 483
tobacco, as carcinogen, 233, 234, 237
trajectory, 131, 136, 139, 142, 145, 160, 163, 167
  and probabilities, 135
TRF, 278–280, 283
Trilling, Lionel, 180, 480
truth, 16, 57
  assault on, 378
  as context-dependent, 100
  as epistemologically relative, 336
  and evidence, 58
  and feminists, 433
  indifference to, 59
  and the inquirer, 58, 59
  and instrumental value, 60
  and moral value, 60
  and need for science, 341
  objective, 96
  as purpose of inquiry, 55
  and social context, 335
  and truth, 335

UFOlogy, 495, 498
UMP, 184, 185
Urkowitz, Steven, 449, 450
utilitarianism, 348
utility, subjective, 104

Velikovsky, Immanuel, 105, 106
Verlaine, Paul, 84
Voltaire, 27, 40, 43, 528
von Neumann, John, 121, 123, 150

Walter of Chatillion, 40
Watson, James, 12, 280
wave-particle duality, 123
  and antiscientists, 129
Weber, Max, 330, 335, 336, 423–425, 428, 429, 473
West, the, 213, 221, 224, 225, 421
  and colonialism, 431
  critics of, 79
  and Freud, 478
  intellectual history of, 60
Western civilization, 216, 217, 303, 483
Western culture, 216, 370, 372
Western philosophy, 215, 219, 226
Western rationalist tradition, 426
Western science, 216, 562, 564
Western thought, 213, 214
Whitman, Walt, 460
Wigner, E. P., 121, 122, 150
Wittgenstein, Ludwig, 79, 95, 423, 425
  and Freudian unconscious, 477
women
  and knowledge, 397
  psychological development of, 372
  and rationality, 424, 426
Women's Environment and Development Organization, 247
Women's Foreign Policy Council, 247
women's studies
  and paradisciplines, 414
  place of science and epistemology in, 402
Woolgar, Steve, 275, 27–280, 282
Wordsworth, William, 459, 460, 486

XY fetus, 19

Yeats, W. B., 463, 465–467

Zermelo, E., 138, 144, 149